D0161284

Traffic and Highway Engineering

THIRD EDITION

Nicholas J. Garber
Lester A. Hoel

University of Virginia

BROOKS/COLE

THOMSON LEARNING

Australia • Canada • Mexico • Singapore • Spain • United Kingdom • United States

BROOKS/COLE

THOMSON LEARNING

Publisher: *Bill Stenquist*
Marketing Team: *Christopher Kelly and Mona Weltmer*
Editorial Coordinator: *Valerie Boyajian*
Production Coordinator: *Mary Vezilich*
Production Service: *RPK Editorial Services*
Permissions Editor: *Mary Kay Polsemen*

Cover Design: *Lisa Henry*
Cover Photo: *Kathryn Hollinrake/Masterfile*
Interior Design: *Carmela Pereira*
Print Buyer: *Vena Dyer*
Typesetting: *Bookwrights*
Printing and Binding: *R.R. Donnelley & Sons—Crawfordsville*

Library of Congress Cataloging-in-Publication Data

Garber, Nicholas
    Traffic and highway engineering
  Nicholas J. Garber, Lester A. Hoel—3rd ed.
      p.   cm.
    Includes references and index.
    ISBN 0-534-38743-8
    1. Highway engineering   2. Traffic engineering.   I. Hoel, Lester A. II. Title
TE145.G35   2001
625.7—dc21   2001052615

This book is dedicated to our wives,
Ada and Unni
and to our daughters,
Allison, Elaine, and Valerie
and
Julie, Lisa, and Sonja
With appreciation for the support, help, and encouragement that we received
during the years that were devoted to writing this textbook.

Contents

PART 4 ■ LOCATION, GEOMETRICS, AND DRAINAGE   635

PART 5 ■ MATERIALS AND PAVEMENTS   825

Preface

Traffic and Highway Engineering, Third Edition, is intended primarily for use as an undergraduate text. The book is designed for students in engineering programs where introductory courses in transportation, highway, or traffic engineering are offered. In most cases, these courses are taught in the third or fourth year to students who have little knowledge or understanding of the importance of transportation. Further, they are often unaware of the professional opportunities for engineers in this field.

Since the subject of transportation is a broad one, several approaches can be used to introduce this topic to undergraduates. One approach is to attempt to cover the technical details of all transportation modes—air, highway, pipeline, public, rail, and water—in an overview-type course. This approach ensures comprehensive coverage, but tends to be superficial and lacking in depth. Another approach that can be effective is covering the subject of transportation by generic elements, such as the vehicle, guideway, terminal, cost, demand, supply, human factors, administration, finance, operations, and so forth. This approach poses problems for some students, however, because each of the modes does not always fit neatly within generic categories, and there may be fundamental differences between modes even within each generic topic. A third approach is to emphasize one mode, such as highways, airports, or railroads. There is considerable merit in focusing on one mode. While some material may be useful in other contexts, the material is specific and unambiguous, and the subject matter can be used later in practice.

In this book we have followed this third approach and have elected to emphasize the area of traffic and highway engineering. This focus represents the major area within the civil engineering field that deals with transportation, and undergraduates can relate directly to problems created by motor vehicle travel. We believe that such an approach is appropriate for an introductory transportation course. It provides an opportunity to present material that is not only useful to engineering students who are pursuing careers in transportation engineering, but is also interesting and challenging to those who intend to work in other areas. While written at an introductory level, this book can also serve as a reference for practicing transportation engineers and for students in graduate courses. Our overall objective is to provide a way for students to get into an area, develop a feel for what it is about, and thereby experience some of the challenges of the profession.

The introductory chapters present material that will help students understand the basis for transportation, its importance in society, and the extent to which transportation pervades our daily lives. Later chapters present information about the basic areas in which transportation engineers work: traffic operations and management, planning, design, construction, and maintenance. Thus, the chapters in this book have been categorized into five parts: Part 1, Introduction; Part 2, Traffic Operations; Part 3, Transportation Planning; Part 4, Location, Geometrics, and Drainage; and Part 5, Materials and Pavements.

The topical division of the book organizes the material so that it may be used in two separate courses. For a course in transportation engineering, which is usually offered in the third year and in which the instructor prefers to emphasize traffic and highway aspects, we recommend that material from Parts 1, 2, and 3 (Chapters 1–14) be covered. For a course in highway engineering, in which the emphasis is on highway location, design, materials, and pavements, we recommend that material from Parts 2, 4, and 5 (Chapters 3 and 15–22) be used.

The book is also appropriate for use in a two-semester sequence in transportation engineering in which traffic engineering and planning (Chapters 3–14) would be covered in the first course, and highway design (Chapters 15–22) would be covered in the second course.

The success of our textbook has been a source of great satisfaction because we believe that it has contributed to the better understanding of highway transportation in all its dimensions. We wish to thank our colleagues and their students for selecting this book for use in transportation courses taught in colleges and universities throughout the United States. The third edition builds on this experience and the success of our pedagogic approach, which is to include many examples in each chapter that illustrate basic concepts, a list of references and a comprehensive problem set at the end of each chapter (with complete instructors manual), an organizational structure that subdivides the material into logical and easy-to-understand elements, and a large number of tables and diagrams that augment the text and ensure completeness of material.

Transportation is a fast-moving field, and the third edition reflects many changes that have occurred since the book was first published in 1988. We have added chapters on safety and intersection design, and have expanded and updated each chapter to reflect new methods, procedures, and technology. The number and variety of homework problems have been increased. Chapters have been updated to reflect the Transportation Research Board *Highway Capacity Manual 2000,* and the 2001 Edition of the AASHTO *Policy of Geometric Design of Highways and Streets,* and the 2000 Edition of the *Manual on Uniform Traffic Control Devices.* Recent references in traffic, planning, design, materials, and pavements have been added. We have also added an appendix that compares new geometric design standards and formulas in metric units.

The authors are indebted to many individuals who assisted in reviewing various chapters and drafts of the original manuscript. We especially wish to thank the following for their helpful comments and suggestions: Edward Beimborn, David Boyce, Christian Davis, Michael Demetsky, Richard Gunther, Jerome Hall, Jotin Khisty, Lydia Kostyniak,

Michael Kyte, Winston Lung, Kenneth McGhee, Carl Monismith, Ken O'Connell, Anthony Saka, Robert Smith, Egons Tons, Joseph Wattleworth, Hugh Woo, and Robert Wortman.

In the preparation of later editions, we are indebted to many colleagues who provided helpful comments and suggestions. We wish to especially acknowledge those who reviewed the entire textbook and furnished input regarding the value of the book and advice regarding how it could be made even better. We also thank several of our colleagues and students who read specific chapters and suggested new end-of-chapter problems. Those whom we particularly wish to acknowledge are Maher Alghazzawi, Rakim Benekohal, Stephen Brich, Conrad Dudek, Lily Elefteriadou, Thomas Freeman, Per Garder, Jiwan Gupta, Marvin Hilton, Feng-Bor Lin, Catherine McGhee, John H. Page, Brian Park, Adel Sadek, Gerald Seeley, Ed Sullivan, Joseph Vidunas, F. Andrew Wolfe, Shaw Yu, Yihua Ziong, Rod Turochy, Brian Park, John Miller, and Winston Lung.

And finally, we wish to thank the following reviewers of this third edition: Ron Gallagher, Per Garder, Kathleen Hancock, Richard G. McGinnis, James W. Stoner, James I. Taylor, and Peter T. Weiss.

Nicholas J. Garber
Lester A. Hoel

Introduction

Transportation is essential for a nation's development and growth. In both the public and private sector, opportunities for engineering careers in transportation are exciting and rewarding. Elements are constantly being added to the world's highway, rail, airport, and mass transit systems, and new techniques are being applied for operating and maintaining the systems safely and economically. Many organizations and agencies exist to plan, design, build, operate, and maintain the nation's transportation system.

CHAPTER 1

The Profession of Transportation Engineering

For as long as the human race has existed, transportation has consumed a considerable portion of its time and resources. The primary need for transportation is economic; it has involved personal travel in search of food or work, travel for trade or commerce, travel for exploration, conquest, or personal fulfillment, and travel for the improvement of one's status in life. The movement of people and goods, which is what is meant by transportation, is undertaken to accomplish those basic objectives or tasks that require transfer from one location to another. For example, a farmer must transport produce to market, a doctor must see a patient in the office or in the hospital, and a salesman must visit clients located throughout a territory. Every day, millions of workers leave their homes and travel to factories or offices.

IMPORTANCE OF TRANSPORTATION

Tapping natural resources and markets and maintaining a competitive edge over other regions and nations are closely linked to the quality of the transportation system. The speed, cost, and capacity of available transportation have a significant impact on the economic vitality of an area and the ability to make maximum use of its natural resources. Examination of most developed and industrialized societies indicates that they are noted for their high-quality transportation services. Nations with well-developed maritime systems (such as the British Empire in the 1900s) once ruled vast colonies located around the globe. In more modern times, countries with advanced transportation systems, such as the United States, Canada, Japan, and those in Western Europe, are leaders in industry and commerce. Without the ability to transport manufactured goods, raw materials, and technical know-how, a country is simply unable to maximize the comparative advantage it may have in the form of natural or human resources. A country such as Japan, for example, with little in the way of natural resources, relies heavily on transportation in order to import raw materials and to export manufactured products.

Transportation and Economic Growth

Good transportation, in and of itself, will not assure success in the marketplace; however, the absence of excellent transportation services will contribute to failure. Thus, if a society wishes to develop and grow, it must have a strong internal transportation system as well as excellent linkages to the rest of the world. Transportation is a derived demand, created by the needs and desires of people to move themselves or their goods from one place to another. It is a necessary condition for human interaction and economic survival.

The availability of transportation facilities can strongly influence the growth and development of a region or nation. Good transportation permits the specialization of industry or commerce, reduces costs for raw materials or manufactured goods, and increases competition between regions, resulting in lower costs and greater choice for the consumer. Transportation is also a necessary element of government services such as delivering mail, defending a nation, and retaining control of its territories. Throughout history, transportation systems, such as those that existed in the Roman Empire and those that exist now in the United States, were developed and built to ensure easy mobilization of armies in the event of a national emergency.

Social Costs and Benefits of Transportation

The improvement of a region's economic position by virtue of improved transportation does not come without costs. Building vast transportation systems requires enormous resources of energy, material, and land. In major cities, transportation can consume as much as half of all the land area. An aerial view of any major metropolis will reveal vast acreage used for railroad terminals, airports, parking lots, and freeways. Transportation has other negative effects as well. Travel is not without danger; every mode of transportation brings to mind some major disaster, be it the sinking of the *Titanic*, the disaster of the zeppelin *Hindenburg*, the infrequent but dramatic passenger air crashes, and the highway fatalities that each year claim about 40,000 lives. In addition, transportation creates noise, spoils the natural beauty of an area, changes the environment, pollutes air and water, and consumes energy resources.

Society has indicated a willingness to accept some risks and some change of the natural environment to gain the benefits of efficient transportation systems. A major task for the modern transportation engineer is to balance society's need for fast and efficient transportation with the costs involved so that the most efficient and cost-effective system is created. In carrying out this task, the transportation engineer must work closely with the public and with elected officials and must be aware of modern engineering practices to ensure that the highest-quality transportation systems are built consistent with available funds and accepted social policy.

Society also values many social benefits of transportation. Bringing medical and other services to rural areas and enabling people to socialize who live some distance apart are only a few of the benefits that transportation provides.

Transportation in the United States

The importance of transportation in the United States is illustrated by several statistics. Approximately 16 percent of the U.S. gross domestic product (GDP) is accounted for by expenses related to transportation. In 1999, expenditures on transportation totaled $1.48 trillion, representing a 51 percent increase over outlays in 1990. The percentage of the GDP that is transportation-related has declined from a high of 20.0 percent in 1960, reflecting increased competition created by deregulation of both freight and passenger transportation. Almost 70 percent of petroleum used in the United States is for transportation. Highway transportation, in 1999, consumed 82.7 percent of all the petroleum used by intercity transport modes. United States citizens travel an average of 1 h/day. Over 80 percent of eligible drivers have driver's licenses; they travel an average of 12,000 mi/year. Transportation industries in the United States employ 11 percent of the workforce in jobs as diverse as those of gas station attendants, airline pilots, truck drivers, highway construction workers, barge operators, pipeline welders, railroad train workers, bus drivers, and highway patrol officers.

Figure 1.1 depicts the trends in transport employment between 1960 and 1998. Total employment increased from 8.9 million to 14.3 million, while during the same period the

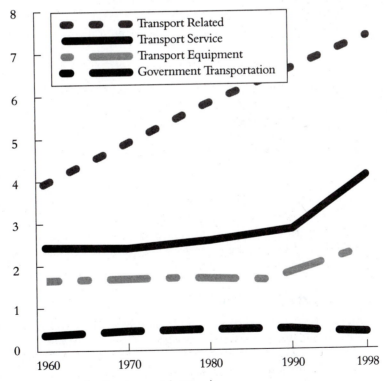

Figure 1.1 Transportation Employment (millions)

SOURCE: *Transportation in America,* 17th ed., Eno Transportation Foundation, Inc., Washington, D.C., 1999. Used with permission.

percentage of transportation employment declined from 13.5 to 11.1 percent of total employment.

OVERVIEW OF U.S. TRANSPORTATION HISTORY

Transportation in the United States could be the subject of a separate book. Transportation history covers a 200-year period and involves the development of many separate modes of transportation. Among the principal topics are travel by foot and horseback, automobile and truck travel, the development of roads and highways, the building of canals and inland waterways, the expansion of the West, the construction of railroads, the use of public transportation such as bus and metro systems in cities, and the development of air transportation, including the aircraft, airports, and air navigation facilities that compose the system.

Population Changes

In its early years, the United States was primarily rural, with a population of about 4 million in the late 1700s. Only about 200,000 (5 percent) lived in cities; the remainder inhabited rural areas and small communities. That pattern remained until the early 1900s. During the twentieth century, the urban population continued to increase such that at present about 75 percent of the U.S. population lives in urban or suburban areas. Large cities have been declining in population; increases have occurred in suburban areas. These changes have a significant impact on the need for highway transportation.

Early Roadbuilding and Planning

During the eighteenth century, travel was by horseback or in animal-drawn vehicles on dirt roads. As the nation expanded westward, roads were built to accommodate the settlers. In 1794, the first toll road, the Lancaster Turnpike, was built to connect the Pennsylvania cities of Lancaster and Philadelphia.

The nineteenth century brought further expansion of U.S. territorial boundaries, and the population increased from 3 million to 76 million. Transportation continued to expand with the nation. In 1808, Secretary of the Treasury Albert Gallatin, who served under President Thomas Jefferson, recommended a national transportation plan. Although his plan was not officially adopted, it did develop a strong case for investing in transportation by the federal government and was the basis for construction projects such as the Erie Canal. The remainder of the nineteenth century saw considerable activity, particularly in canal and railroad building.

The Short-Lived Canal Boom

An era of canal construction began in the 1820s, when the Erie Canal was completed and other inland waterways were constructed. This efficient means of transporting goods was soon replaced by the railroads, which were being developed at the same time. In 1840, the number of miles of canal and railroads were equal (3200 mi), but railroads, which could be constructed almost anywhere in this vast, undeveloped land at a much lower cost, superseded canals as a form of intercity transportation. Thus, after a short-lived period of intense activity, the era of canal construction came to an end.

The Railroad Era

The dominant mode of transportation during the late 1800s was the railroad; railway lines spanned the entire continent (the first transcontinental railroad was completed in 1869). Railroads dominated intercity passenger and freight transportation from the late 1800s to the early 1920s. Railroad transportation enjoyed a resurgence during World War II but has steadily declined since then, owing to the competitiveness of automobile and truck transport. Railroad mileage reached its peak of about 265,000 miles by 1915 but has reduced to about 122,000 statute miles in 1999.

Transportation in Cities

Each decade has seen continual population growth within cities, and with it the demand for improvements in urban transportation systems has increased. City transportation began with horse-drawn carriages on city streets; these later traveled on steel tracks. They were succeeded by cable cars, electric streetcars, underground electrified railroads, and bus transportation. City travel by public transit has been largely replaced by the use of automobiles on urban highways, although new rapid transit systems have been built in San Francisco, Washington, D.C., and Atlanta.

The Automobile and Interstate Highways

The invention and development of the automobile created a revolution in transportation in the United States during the twentieth century. No facet of American life has been untouched by this invention; the automobile, together with the airplane, has changed the way we travel within and between cities. Only four automobiles were produced in the year 1895. By 1901, there were 8000 registered vehicles, and by 1910 over 450,000 cars and trucks. Between 1900 and 1910, 50,000 miles of surfaced roads were constructed, but major highway-building programs did not begin in earnest until the late 1920s. By 1920, more people traveled by private automobile than by rail transportation. By 1930, 23 million passenger cars and 3 million trucks were registered. In 1956, Congress authorized a 42,500-mile interstate highway network. Today, that network is virtually complete.

The Birth of Aviation

Aviation was in its infancy at the beginning of the twentieth century; the Wright brothers' first flight took place in 1903. Both World War I and World War II were catalysts in the development of air transportation. The carrying of mail by air provided a reason for government support of this new industry. Commercial airline passenger service began to grow, and by the mid-1930s, coast-to-coast service was available. After World War II, the expansion of air transportation was phenomenal. The technological breakthroughs that developed during the war, coupled with the training of pilots, created a new industry that replaced both oceangoing steamships and passenger railroads. A summary of the historical highlights of transportation development is shown in Table 1.1.

Table 1.1 Significant Events in Transportation History

1794: First toll road, the Lancaster Turnpike, is completed.

1807: Robert Fulton demonstrates a steamboat on the Hudson River. Within several years, steamboats are operating along the East Coast, on the Great Lakes, and on many major rivers.

1808: Secretary of Treasury Albert Gallatin recommends a federal transportation plan to Congress, but it is not adopted.

1825: Erie Canal is completed.

1830: Operations begin on Baltimore and Ohio Railroad, first railroad constructed for general transportation purposes.

1838: Steamship service on the Atlantic Ocean begins.

1857: First passenger elevator in the United States begins operation, presaging high-density urban development.

1865: First successful petroleum pipeline is laid, between a producing field and a railroad terminal point in western Pennsylvania.

1866: Bicycles are introduced in the United States.

1869: Completion of first transcontinental railroad.

1887: First daily railroad service from coast to coast.

1888: Frank Sprague introduces the first regular electric streetcar service in Richmond, Va.

1903: The Wright brothers fly first airplane 120 ft at Kitty Hawk, N.C.

1914: Panama Canal opens for traffic.

1915–18: Inland waters and U.S. merchant fleet play prominent roles in World War I freight movement.

1916: Interurban electric-rail mileage reaches peak of 15,580 mi.

1919: U.S. Navy and Coast Guard crew crosses the Atlantic in a flying boat.

1927: Charles Lindbergh flies solo from New York to Paris.

1956: Construction of the 42,500-mile Interstate and Defense Highway System begins.

1959: St. Lawrence Seaway is completed, opening nation's fourth seacoast.

1961: Manned spaceflight begins.

1967: U.S. Department of Transportation established.

1969: Men land on moon and return.

1972: San Francisco's Bay Area Rapid Transit System is completed.

1981: Space shuttle *Columbia* orbits and lands safely.

1991: Interstate highway system is essentially complete.

1992: Intelligent transportation systems (ITS) usher in a new era of research and development in transportation.

1995: 161,000-mile National Highway System (NHS) is approved.

1998: Electric vehicles are introduced as an alternative to internal combustion engines.

2000: New millennium ushers in a transportation-information technology revolution.

Transportation Facilities Today

The nation's transportation system is now largely in place, having been built over a period of 200 years. However, much of this system (particularly the physical facilities) is becoming old and deteriorated. This is particularly true of the nation's highway system and railroads. The United States has constructed one of the most extensive transportation systems in the world, and new developments will be focused on keeping the existing system in serviceable condition and on improving the system's capacity through advanced technology. Environmental concerns will also have a strong influence as we attempt to improve air quality. Several initiatives that began in the 1990s will influence the United States' transportation system well into the twenty-first century. These are the the Clean Air Act Amendments of 1990 (CAAA), the Intermodal Surface Transportation Efficiency Act of 1991 (ISTEA), the Strategic Plan for Intelligent Vehicle-Highway Systems of 1992, and the 1998 Transportation Equity Act for the 21st Century (TEA-21).

The surface transportation legislation has provided funding of over $155 billion for 1992 through 1997, and $217 billion for 1998 through 2004. The purpose is to develop a national intermodal transportation system that is economically efficient and environmentally sound, provides the foundation to compete in a global economy, and moves people and goods in an energy-efficient manner. Among the most important features of the legislation are the designation of a national highway system, which comprises approximately 161,000 miles of interstate and principal arterial highways, and funding for intelligent transportation systems (ITS) and prototype high-speed rail systems.

Nonhighway legislation also influences our nation's highway program, particularly the National Environmental Policy Act of 1969 and the CAAA. The CAAA are intended to ensure that in areas experiencing air quality problems, the transportation systems are designed to improve air quality as well as mobility. Thus, emphasis will be placed on reducing auto emissions and devising programs to increase the use of public transit, bicycling, walking, and multioccupant driving.

ISTEA authorized $660 million for the development of ITS, which involves the integration of advanced technologies to alleviate congestion and to improve all aspects of highway safety, productivity, and air quality. The essence of ITS is the use of information technology, including hardware, software, electronics, controls, and communications. In 1992, a strategic plan was formulated to guide the development of these technologies over the next 20 years. Among the areas of application envisioned are the integration of the management of various roadway functions, such as freeway ramp monitoring and signal control. Collection of real-time data will permit the rerouting of drivers around accident sites, thus reducing costs of traffic congestion, and also provide information to travelers regarding road and weather conditions and trip planning. It will also provide collision warning systems and develop new automated highways. The applications will be extended to commercial vehicles, public transportation, and transport needs of rural areas. In each of these cases, information technology will be used for automatic vehicle identification, route information, and fare collection. Many experiments and demonstrations of ITS technology are being undertaken that will result in new ways of managing and operating our nation's transportation system.

EMPLOYMENT IN TRANSPORTATION

Transportation represents one of the broadest opportunities for employment because it involves many disciplines and modes. In the United States in 1999 the number of miles of paved roadway, for intercity travel was 832,000 and there were approximately 122,000 miles of railroads, 394,000 airway miles, 26,000 miles of inland waterways, and 177,000 miles of pipeline. This vast system of infrastructure and vehicles requires a significant workforce to plan, build, operate, and maintain it.

Transportation services are provided within and between cities by rail, air, bus, truck, and private automobile. The automobile is the major mode of travel between cities; the movement of intercity passenger traffic by public carriers is mainly by air and, to a much lesser extent, by bus and rail. Within cities, passenger travel is by personal automobile, bus, and rail rapid transit. Freight movement is largely the province of trucking firms within cities, but truck, rail, water, and pipelines provide intercity freight movement over the continental United States. The professional skills required to plan, build, and operate this extensive transportation system require a variety of disciplines, including engineering, planning, law, and the economic, management, and social sciences.

The physical-distribution aspect of transportation, known as business logistics or physical distribution management, is concerned with the movement and storage of freight between the primary source of raw materials and the location of the finished manufactured product. Business logistics is the process of planning, implementing and controlling the efficient and effective flow and storage of goods, services, and related information from origination to consumption as per customer requirements. An expansion of the logistics concept is called supply chain management, a process that coordinates the product, information, and cash flows to maximize consumption satisfaction and minimize organization costs.

Vehicle design and manufacture is a major industry in the United States and involves the application of mechanical, electrical, and aerospace engineering skills as well as those of technical trained mechanics and workers in other trades.

The service sector provides jobs for vehicle drivers, maintenance people, flight attendants, train conductors, and other necessary support personnel. Other professionals, such as lawyers, economists, social scientists, and ecologists, also work in the transportation fields when their skills are required to draft legislation, to facilitate right-of-way acquisition, or to study and measure the impacts of transportation on the economy, society, and the environment.

Although a transportation system requires many skills and provides a wide variety of job opportunities, we will concentrate on the opportunities for civil engineers. They are responsible primarily for the planning, design, construction, operation, and maintenance of the transportation system within the United States. It is the engineer's responsibility to ensure that the system functions efficiently from an economic point of view and that it meets external requirements concerning energy, air quality, safety, congestion, noise, and land use.

Specialties in Transportation Engineering

Transportation engineers typically are employed by the agency responsible for building and maintaining a transportation system, such as the federal, state, or local government, a railroad, or a transit authority. They also work for consulting firms that carry out the planning and engineering tasks for these organizations. During the past century, transportation engineers have been employed to build the nation's railroads, the interstate highway system, rapid transit systems in major cities, airports, and turnpikes. Each decade has seen a new national need for improved transportation services.

It can be expected that in the twenty-first century, heavy emphasis will be placed on the rehabilitation of the highway system, including its surface and bridges, as well as on devising means to ensure improved safety and utilization of the existing system through traffic control, information technology and systems management. Highway construction will be required, particularly in suburban areas. Building of roads, highways, airports, and transit systems is likely to accelerate in less developed countries, and the transportation engineer will be called on to furnish the services necessary to plan, design, build, and operate highway systems throughout the world. Each of the subspecialties within transportation engineering is described below.

Planning

Transportation planning deals with the selection of projects for design and construction. The transportation planner begins by defining the problem, gathering and analyzing data, and evaluating various alternative solutions. Also involved in the process are forecasts of future traffic; estimates of the impact of the facility on land use, the environment, and the community; and determination of the benefits and costs that will result if the project is built. The transportation planner investigates the physical feasibility of a project and makes comparisons between various alternatives to determine which one will accomplish the task at the lowest cost, consistent with other criteria and constraints.

A transportation planner must be familiar with engineering economics and other means of evaluating alternative systems. The planner also must be knowledgeable in statistics and data-gathering techniques, as well as in computer programming for data analysis and travel forecasting. The transportation planner must be able to communicate with the public and with his or her client and have the ability to write and speak well.

Design

Transportation design involves the specification of all features of the transportation system so that it will function smoothly, efficiently, and in accord with physical laws. The design process results in a set of detailed plans that can be used for estimating the facility costs and for carrying out its construction. For a highway, the design process involves the selection of dimensions for all geometrical features, such as the longitudinal profile, vertical curves and elevations, the highway cross section such as pavement widths, shoulders, right-of-way, drainage ditches, and fencing. It also involves the design of the pavement itself, including the structural requirements for base and subbase courses and the pavement material, such as concrete or asphalt. Also involved in the design are bridges and drainage structures, as well as the provision for traffic control devices, roadside rest areas, and landscaping. The highway designer must be proficient in civil engineering subjects such as soil mechanics, hydraulics, land surveying, pavement design, and structural design. The highway design engineer is concerned primarily with the geometric layout of the road, its cross section, paving materials, roadway thickness, and traffic control devices. Special appurtenances, such as highway bridges and drainage structures, usually are designed by specialists in these areas.

The most important aspect of the highway designer's work is to establish the standards that relate the speed of the vehicle to the geometric characteristics of the road. A balanced design is produced in which all elements of the geometry of the highway—its curve radii, sight distance, superelevation, grade, and vertical curvature—are consistent with a chosen design speed such that if a motorist travels at that speed, he or she can proceed safely and comfortably throughout the entire highway system.

Construction

Transportation construction is closely related to design and involves all aspects of the building process, beginning with clearing of the native soil, preparation of the surface, placement of the pavement material, and preparation of the final roadway for use by traffic. Originally, highways were built with manual labor assisted by horse-drawn equipment for grading and moving materials. Today, modern construction equipment is used for clearing the site, grading the surface, compacting the subbase, transporting materials, and placing the final highway pavement. Advances in construction equipment have made possible the rapid building of large highway sections. Nuclear devices for testing compaction on soil and base courses, laser beams for establishing line and grade, and specialized equipment for handling concrete and bridge work are all innovations in the construction industry. Large, automatically controlled mix plants have been constructed; new techniques for improving durability of structures and the substitutions for scarce materials have been developed.

Traffic Operations and Management

The operation of the nation's highway system is the responsibility of the traffic engineer. Traffic engineering involves the integration of vehicle, driver, and pedestrian characteristics to improve the safety and capacity of streets and highways. All aspects of the transportation system are included after the street or highway has been constructed and opened for operation. Among the elements of concern are traffic accident analyses, parking, loading, design of terminal facilities, traffic signs, markings, signals, speed regulation, and highway lighting. The traffic engineer works to improve traffic flow and safety, using engineering methods and information technology to make decisions that are supported by enforcement and education. Traffic engineers work directly for municipalities, county governments, and private consulting firms.

Maintenance

Highway maintenance involves all the work necessary to ensure that the highway system is kept in proper working order. Maintenance is work such as pavement patching, repair, and other actions necessary to maintain the roadway pavement at a desired level of serviceability. Maintenance also involves record keeping and data management for work activities and project needs, as well as analyses of maintenance activities to determine that they are carried out in the most economical manner. Scheduling of work crews, replacement of worn or damaged signs, and repair of damaged roadway sections are important elements of maintenance management. The work of the civil engineer in the area of maintenance involves redesign of existing highway sections, economic evaluation of maintenance programs, testing of new products, and scheduling of manpower to minimize delay and cost. The maintenance engineer must also maintain an inventory of traffic signs and markings and ensure that they are in good condition.

The Challenge of Transportation Engineering

The transportation engineer is a professional who is concerned with the planning, design, construction, operations, and management of a transportation system (Figure 1.2). As a

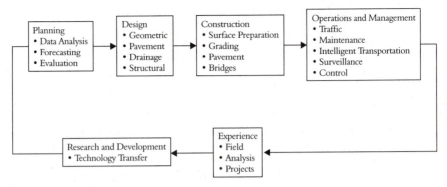

Figure 1.2 The Profession of Transportation Engineering

professional, he or she must make critical decisions about the system that will affect the thousands of people who use it. The work depends on the results of experience and research and is challenging and ever-changing as new needs emerge and new technologies replace those of the past. The challenge of the transportation engineering profession is to assist society in selecting the appropriate transportation system consistent with its economic development, resources, and goals and to construct and manage the system in a safe and efficient manner.

SUMMARY

Transportation is an essential element in the economic development of a society. Without good transportation, a nation or region cannot achieve the maximum use of its natural resources or the maximum productivity of its people. Progress in transportation is not without its costs, both in human lives and in environmental damage, and it is the responsibility of the transportation engineer working with the public to develop high- quality transportation consistent with available funds and social policy and to minimize damage. Transportation is a significant element in our national life, accounting for 16.3 percent of the gross national product and employing over 10 percent of the workforce.

The history of transportation illustrates that the way people move is affected by technology, cost, and demand. The past 200 years have seen the development of several modes of transportation: waterways, railroads, highway, and air. Each mode has been dominant during one period of time; several have been replaced or have lost market share when a new mode emerged that provided a higher level of service at a competitive price.

The career opportunities in transportation that engineering students have are exciting. In the past, transportation engineers planned and built the nation's railroads, highways, mass transit systems, airports, and pipelines. In the coming decades, additional system elements will be required, as will efforts toward maintaining and operating in a safe and economical manner the vast system that is already in place. New systems, such as magnetically levitated high-speed trains or ITS, will also challenge the transportation engineer in the future.

The specialties in transportation engineering are planning, design, construction, traffic management and operations, and maintenance. Planning involves the selection of projects for design and construction; design involves the specification of all features of the transportation project; construction involves all aspects of the building process; traffic management and operations involves studies to improve capacity and safety; and maintenance involves all work necessary to ensure that the highway system is kept in proper working order.

PROBLEMS

1-1   To illustrate the importance of transportation in our national life, identify a transportation-related article that appears in a local or national newspaper. Discuss the issue involved and explain why the item was newsworthy.

1-2   Arrange an interview with a transportation professional in your city or state (that is, someone working for a consulting firm, city, county, state transportation department, transit, or rail agency). Inquire about the job he or she performs, why he or she entered the profession, and what he or she sees as future challenges in his field.

1-3   Keep a diary of all trips you make for a period of three to five days. Record the purpose of each trip, how you traveled, the approximate distance traveled, and the trip time. What conclusions can you draw from the data?

1-4   Identify one significant transportation event that occurred in your city or state. Discuss the significance of this event.

1-5   Describe how transportation influenced the initial settlement and subsequent development of your home city or state.

1-6   Describe your state's transportation infrastructure. Include both passenger and freight transportation.

1-7   What is the total number of miles in your state's highway system? What percent of the highway system is composed of Interstate highways?

1-8   Estimate the number of personal motor vehicles in your city or state. What is the total number of miles driven each year? How much revenue is raised per vehicle for each 1 cent/gallon tax? Assume that the average vehicle achieves 25 mpg.

1-9   How many railroad trains pass through your city each week? What percentage of these are passenger trains?

1-10   Review the classified section of the telephone directory and identify ten different jobs or industries that are related to transportation.

1-11   Estimate the proportion of your monthly budget that is spent on transportation.

1-12   Identify an ITS project or application that is underway in your home state. Describe the project, its purpose, and the way it is operated.

1-13   Most departments of transportation incorporate at least five major transportation engineering subspecialties within their organization. List and briefly indicate at least three tasks falling under each specialty.

1-14   There are many benefits related to our highway system, but there are also many costs, or detrimental effects, that have come into focus in recent years. List four major detrimental effects that are directly related to the construction and use of our highway transportation system.

1-15   Cite four statistics that demonstrate the importance of transportation in the United States.

1-16   A state has a population of 17 million people and an average ownership of 1.5 cars per person, each driven an average of 10,000 mi/year, at 20 mi/gal of gasoline (mpg). Officials estimate that an additional $75 million per year in revenue will be required to improve the state's highway system, and they have proposed an increase in the gasoline tax to meet this need. Determine the required tax in cents per gallon.

1-17   Review Table 1.1 and select a single event which in your opinion is the most significant. Explain the importance of this transportation achievement.

ADDITIONAL READINGS

America's Highways 1776–1976, U.S. Department of Transportation, Federal Highway Administration, Washington, D.C., 1976.

Coyle, John J., Edward J. Bardi, and Robert A. Novack, *Transportation,* 5th edition, South Western College Publishing Co., 2000.

Hoel, Lester A. "Historical Overview of U.S. Passenger Transportation" and Sussman, Joseph M., "Transportation's Rich History and Challenging Future—Moving Goods." Transportation Research Board Circular 461, National Research Council, Washington, D.C., August 1996.

Transportation in America, 18th edition, Eno Transportation Foundation, Washington, D.C., 2000.

CHAPTER 2

Transportation Systems and Organizations

The transportation system in a developed nation is an aggregation of vehicles, guideways, terminal facilities, and control systems that move freight and passengers. These systems are usually operated according to established procedures and schedules in the air, on land, and on water. The set of physical facilities, control systems, and operating procedures referred to as the nation's transportation system is not a system in the sense that each of its components is part of a grand plan or was developed in a conscious manner to meet a set of specified regional or national goals and objectives. Rather, the system has evolved over a period of time and is the result of many independent actions taken by the private and public sectors, which act in their own or in the public's interest.

Each day, decisions are made that affect the way transportation services are used. The decisions of a firm to ship its freight by rail or truck, of an investor to start a new airline, of a consumer to purchase an automobile, of a state or municipal government to build a new highway or airport, of Congress to deny support to a new aircraft, and of a federal transportation agency to approve truck safety standards are just a few examples of how transportation services evolve and a transportation system takes shape.

DEVELOPING A TRANSPORTATION SYSTEM

Over the course of a nation's history, attempts are made to develop a coherent transportation system, usually with little success. A transportation plan for the United States was proposed by Secretary of the Treasury Gallatin in 1808, but this and similar attempts have had little impact on the overall structure of the U.S. transportation system. Planners of the interstate highway system, which is national in scope, failed to recognize or account for the impact of that system on other transportation modes or on urbanization. The creation of the U.S. Department of Transportation (DOT) in 1967 had the beneficial effect of focusing national transportation activities and policies within one cabinet level agency. In turn, many states followed by forming their own transportation departments.

The Interstate Commerce Commission (ICC), created in 1887 to regulate the railroads, was given additional powers in 1940 to regulate water, highway, and rail modes, preserving the inherent advantages of each and promoting safe, economic, and efficient service. The intent of Congress was to develop, coordinate, and preserve a national transportation system; however, the inability to implement vague and often contradictory policy guidelines has not helped to achieve desired results. More recently, regulatory reform has been introduced, and transportation carriers are developing new and innovative ways of providing services. The ICC was abolished in 1996.

Advantages and Complementarity of Modes

The transportation system that evolves in a developed nation may not be as economically efficient as one that is developed in a more analytical fashion, but it is one in which each of the modes usually complements the others in carrying the nation's freight and passengers. A business trip across the country may involve travel by taxi, airplane, and auto; transportation of freight often requires trucks for pickup and delivery and railroads for long-distance hauling.

Each mode has inherent advantages of cost, travel time, convenience, and flexibility that make it "right for the job" under a certain set of circumstances. The automobile is considered to be a reliable, comfortable, flexible, and ubiquitous form of personal transportation for many people. However, when distances are great and time is at a premium, air transportation will be selected, supplemented by the auto for local travel. If cost is important and time is not at a premium, or if an auto is not available, then the intercity bus may be used.

Selecting a mode to haul freight follows a similar approach. Trucks have the advantages of flexibility and the ability to provide door-to-door service. They can carry a variety of parcel sizes and usually can pick up and deliver to meet the customer's schedule. Waterways can ship heavy commodities at low cost, but at slow speeds and only between points on a river or canal. Railroads can haul an immense variety of commodities between any two points, but usually require truck transportation to deliver the goods to a freight terminal or to their final destination. In each instance, a shipper must decide whether the cost and time advantages are such that the goods should be shipped by truck alone or by a combination of truck and rail.

Many industries have been trying to reduce their parts and supplies inventories, preferring to transport them from the factory when needed rather than stockpiling them in a warehouse. This practice has meant shifting transportation modes from rail to truck. Rail shipments usually are made once or twice a week in carload lots, whereas truck deliveries can be made in smaller amounts and on a daily basis, depending on demand. In this instance, lower rail freight rates do not compete with truck flexibility, since the overall result of selecting trucking is a cost reduction for the industry. Recently, a trend toward intermodalism has arisen, which has combined the capabilities of both modes.

Example 2.1   Selecting A Transportation Mode

An individual is planning to take a trip between the downtown area of two cities, A and B, that are 400 miles apart. There are three options available:

Travel by air. This trip will involve driving to the airport near city A, parking, waiting at the terminal, flying to airport B, walking to a taxi stand and taking a taxi to the final destination.

Travel by auto. This trip will involve driving 400 miles through several congested areas, parking in the downtown area and walking to the final destination.

Travel by rail. This will involve taking a cab to the railroad station in city A, direct rail connection to the downtown area in city B, and a short walk to the final destination.

Since this is a business trip, the person making the trip is willing to pay up to $25 for each hour that is saved. (For example, if one mode is two hours faster than another, the traveler is willing to pay $50 more to use the faster mode.) After examining all direct costs involved in making the trip by air, auto or rail (including parking, fuel, fares, tips and taxi charges) the traveler concludes that the trip by air will cost $250 with a total travel time of 5 hours, the trip by auto will cost $200 with a total travel time of 8 hours and the trip by rail will cost $150 with a total travel time of 12 hours.

Which mode is selected based on cost factors alone?

What other factors might be considered by the traveler in making her final selection.

Solution: Since travel time is valued at $25/h, the following costs would be incurred:

Air: 250 + 25(5) = $375

Auto: 200 + 25(8) = $400

Rail 150 + 25(12) = $450

In this instance, the air alternate is cheapest and is the selected mode.

The traveler may have other reasons to select another alternative. Among them are:

Safety. While each of these modes are safe, the traveler may feel "safer" in one mode over the other. For example, she may prefer rail because of concerns regarding air safety issues due to congestion or weather.

Reliability. If it is very important to attend the meeting, the traveler may select the mode that will provide the highest probability of an on-time arrival. If the drive involves travel through work zones and heavily congested areas, rail or air would be preferred.

Convenience. The number of departures and arrivals provided by each mode could be a factor. For example, if the railroad provides only two trains a day, and the airline has six flights/day, the traveler may prefer to go by air.

Interaction of Supply and Demand

The transportation system that exists at any point in time is the product of two factors that act on each other. These are (1) the state of the economy, which produces the demand for transportation, and (2) the extent and quality of the system that is currently in place, which constitutes the supply of transportation facilities and services. In periods of high un-employment or rising fuel costs, the demand for transportation tends to decrease. On the other hand, if a new transportation mode is introduced that is significantly cheaper to use than those modes that already exist, the demand for the new mode will increase, decreasing demand for the existing modes.

These ideas can be illustrated in graphic terms by considering two curves, one describing the demand for transportation at a particular point in time and the other describing how the available transportation service or supply is affected by the volume of traffic that uses that system.

The curve in Figure 2.1 shows how demand in terms of veh/day could vary with cost. The curve is representative of a given state of the economy and of the present population. As is evident, if the transportation cost per mile, C, decreases, then, since more people will use it at a lower cost, the volume, V, will increase. In Figure 2.1, when the traffic vol/day is 6000, the cost is \$0.75/mi. If cost is decreased to \$0.50/mi, the vol/day increases to 8000. In other words, this curve shows us what the demand will be for our product (transportation) under a given set of economic and social conditions.

Demand can occur only if transportation services are available between the desired points. Suppose the demand shown in Figure 2.1 represents the desire to travel between the mainland of Florida and an island that is located off the coast and currently inaccessible, as shown in Figure 2.2.

If a bridge is built, people will use it, but the amount of traffic will depend on cost. The cost to cross the bridge will depend on the bridge toll and the travel time for cars and trucks. If only a few vehicles cross, little time is lost waiting at a tollbooth or in congested traffic. However, as more and more cars and trucks use the bridge, the time required to

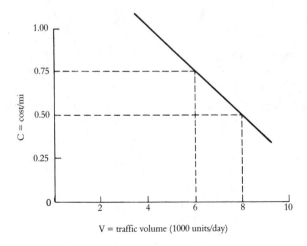

Figure 2.1   Relationship Between Transportation Demand and Cost

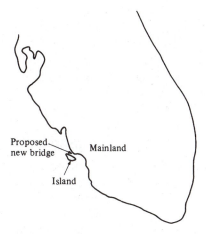

Figure 2.2 Location of a New Bridge Between the Mainland and an Island

cross will increase. Lines will be long at the tollbooth; there might also be traffic congestion at the other end. The curve in Figure 2.3 illustrates how the cost of using the bridge could increase as the volume of traffic increases, assuming that the toll is $0.25/mi. In this figure, if the volume is less than 2000 units/day, there is no delay due to traffic congestion. However, as traffic volumes increase beyond 2000 units/day, delays occur and the travel time increases. Since "time is money," we have converted the increased time to cost/mile. If 4000 units/day use the bridge, the cost is $0.50/mi; at 6000 units/day, the cost is $0.75/mi.

The curve in Figure 2.3 shows how much transportation volume we can supply at various levels of cost to the traveler.

We can now use the two curves to determine what volume (V) can be expected to use the bridge. This value will be found where the demand curve intersects the supply curve,

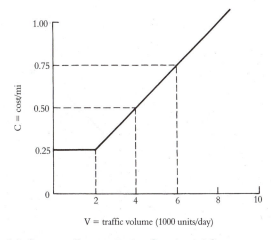

Figure 2.3 Relationship Between Transportation Supply and Cost

because any other value of V will create a shift in demand either upward or downward, until the equilibrium point is reached. If the volume increased beyond the equilibrium point, cost would go up and demand would drop. Likewise, if the volume dropped below equilibrium, cost would go down and demand would increase (see Figure 2.4). Thus in both instances we are approaching equilibrium. In this example, the number of units crossing the bridge would be 6000 units/day. Obviously, we can raise or lower the traffic volume by changing the toll.

Forces That Change the Transportation System

At any point in time, the nation's transportation system is in a state of equilibrium as expressed by the traffic carried (or market share) for each mode and the levels of service provided (expressed as travel attributes such as time, cost, frequency, and comfort). This equilibrium is the result of market forces (state of the economy, competition, costs and prices of service), government actions (regulation, subsidy, promotion), and transportation technology (speed, capacity, range, reliability). As these forces shift over time, the transportation system changes as well, creating a new set of market shares (levels of demand) and a revised transportation system. For this reason, the nation's transportation system is in a constant state of flux, causing short-term changes due to immediate revisions in levels of service (such as raising the tolls on a bridge or increasing the gasoline tax) and long-term changes in lifestyles and land-use patterns (such as moving to the suburbs after a highway is built or converting auto production from large to small cars).

If gasoline prices were to increase significantly, there would be an immediate shift of long-haul freight from truck to rail. In the long run, if petroleum prices remained high, we might see shifts to coal or electricity or to more fuel-efficient trucks and autos.

Government actions also influence transportation equilibrium to a great extent. The federal government's decision to build the national interstate system affected the truck-rail balance in favor of truck transportation. It also encouraged long-distance travel by auto and was a factor in the decline of intercity bus service to small communities.

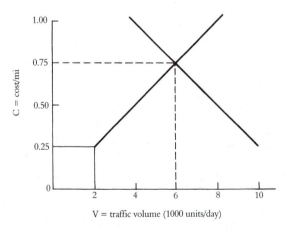

Figure 2.4 Equilibrium Volume for Traffic Crossing a Bridge

Technology has also contributed to substantial shifts in transportation equilibrium. The most dramatic has been the introduction of jet aircraft, which essentially eliminated passenger train travel in the United States and passenger steamship travel between the United States and the rest of the world.

Example 2.2   Computing the Toll to Maximize Revenue Using a Supply-Demand Curve

A toll bridge carries 5000 veh/day. The current toll is 150 cents. When the toll is increased by 25 cents, traffic volume decreases by 500 veh/day. Determine the amount of toll that should be charged such that revenue is maximized. How much additional revenue will be received?

Solution: Let x = the toll increase in cents.

Assuming a linear relation between traffic volume and cost, the expression for V is

$$V = 5000 - x/25 \ (500)$$

The toll is

$$T = 150 + x$$

Revenue is the product of toll and volume:

$$\begin{aligned}
R &= (V)\ (T) \\
&= \{5000 - x/25\ (500)\}\ (150 + x) \\
&= (5000 - 20x)\ (150 + x) = 750{,}000 - 3000\,x + 5000\,x - 20x^2 \\
&= 750{,}000 + 2000\,x - 20\,x^2
\end{aligned}$$

For maximum value of x, compute the first derivative and set equal to zero:

$$\begin{aligned}
dR/dt &= 2000 - 40\,x = 0 \\
x &= 50 \text{ cents}
\end{aligned}$$

The new toll is the current toll plus the toll increase.
Toll for maximum revenue = 150 + 50 = 200 cents or $2.00 (answer).
The additional revenue is

$$(V_{max})(T_{max}) - (V_{current})\ (T_{current})$$
$$\{(5000 - (50/25)(500)\}\{2\} - (5000)(1.50)$$
$$(4000)\ (2) - 7500 = 8000 - 7500 = \$500 \text{ (answer)}$$

MODES OF TRANSPORTATION

The U.S. transportation system today is a highly developed, complex network of modes and facilities that furnishes shippers and travelers with a wide range of choices in terms of services provided. Each mode offers a unique set of service characteristics in terms of travel time, frequency, comfort, reliability, convenience, and safety. The term *level of service* is used to describe the relative values of these attributes. The traveler or shipper must compare the level of service offered with the cost in order to make tradeoffs and mode selection. Furthermore, a shipper or traveler can decide to use a public carrier or to use private (or personal) transportation. For example, a manufacturer can ship goods through a trucking firm or with company trucks; a homeowner who has been relocated can hire a household moving company or rent a truck; and a commuter can elect to ride the bus to work or drive a car. Each of these decisions involves a complex set of factors that require tradeoffs between cost and service.

Freight and Passenger Traffic

The principal modes of intercity freight transportation are highways, railroads, water, and pipeline. Traffic carried by each mode, expressed as ton-miles or passenger-miles, has varied considerably over the 63-year period between 1930 and 1999. These changes are illustrated in Figure 2.5.

Railroads, which accounted for 56 percent of freight traffic in 1950, carried 40.4 percent by 1999. Oil pipelines have increased their share of freight traffic from 12.1 percent in 1950 to 16.8 percent by 1999. Water transportation has remained relatively constant with traffic on rivers and canals at 10.5 percent and Great Lakes traffic 2.6 percent in 1999. Trucking has steadily increased each year, from 16.3 percent in 1950 to 29.4 percent by

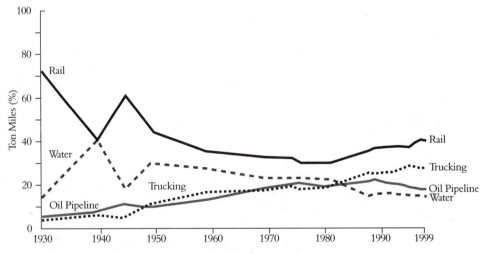

Figure 2.5 Volume of U.S. Domestic Freight Traffic

SOURCE: Data from *Transportation in America*, 17th ed., Eno Transportation Foundation, Inc., Washignton, D.C., 1999. Used with permission.

1999. Air freight is an important carrier for high-value goods, but it is insignificant on a ton-mile basis, representing only about 0.38 percent of total.

The four principal carriers for freight movement (rail, truck, pipeline, and water), account for between about 13 percent and 40 percent of total number of ton-miles of freight. The railroad's share is highest on a ton-mile basis, but it has been reduced significantly due to competition from truck and pipeline. The railroads have lost traffic due to the advances in truck technology and pipeline distribution. Government policies that supported highway and waterway improvements are also a factor. Subsequent to World War II, long-haul trucking was possible because the U.S. highway system developed. As petroleum became more widely used, construction of a network of pipelines for distribution throughout the nation was carried out by the oil industry.

The amount of money spent in the U.S. to move freight has steadily increased in the decades between 1960 and 1998. In 1960 the total freight bill was $47.8 billion whereas in 1999 the amount was $561.8 billion, an almost 12-fold increase. In 1960 railroads' share of revenue was 18.9 percent, and this has declined to 6.4 percent in 1999. In contrast, highway transportation received 67.6 percent of revenues in 1960 which has increased to 81.3 percent in 1999. These figures reflect two factors: the increased dominance of trucking in freight transportation and the higher ton-mile rates that are charged by trucking firms compared with the railroads. It is estimated that although trucks move 29 percent of ton-miles compared with 40 percent by rail, the value of the goods moved by truck comprises about 75 percent of the total value of all goods moved in the United States.

The distribution of passenger transportation is much different from that for freight: one mode—the automobile—accounts for 77.1percent of all domestic intercity passenger-miles traveled in the United States (Figure 2.6). With the exception of the World War II years, when auto use dropped to 63.8 percent of passenger-miles by 1945 and rail

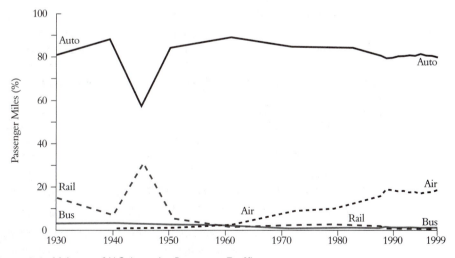

Figure 2.6 Volume of U.S. Intercity Passenger Traffic

SOURCE: Data from *Transportation in America,* 17th ed., Eno Transportation Foundation, Inc., Washington, D.C., 1999. Used with permission.

increased to 27.4 percent between 1958–1999, the automobile, small trucks, and vans have accounted for between 77.1 percent and 90.4 percent of all passenger-miles traveled. The remaining modes, air, bus, and rail, in 1999 shared a market representing about 22.9 percent of the total distributed: 20.3 percent by air, 1.4 percent by intercity bus, 0.6 percent by private air carrier, and 0.6 percent by rail.

Of the four transportation carriers for intercity passenger movement, two—air and auto—are dominant, representing 98 percent of all intercity passenger miles. If the public modes (rail, bus, and air) are considered separately from the auto, a dramatic shift in passenger demand is evident. The passenger railroad, which accounted for 28.9 percent of all revenue traffic in 1960, declined to 2.6 percent in 1999; most of this traffic occurred in the corridor between Washington, D.C., New York, and Boston. Intercity bus transportation also has declined as a percentage of the total; buses now serve a market consisting primarily of passengers who cannot afford to fly or drive. As air fares have decreased with deregulation, bus travel has declined.

The most dramatic increase has occurred in air transportation, which represents 91.3 percent of all intercity passenger-miles traveled using public modes. In cities, buses are the major public transit mode, with the exception of those larger urban areas that have rapid rail systems.

The amount of money spent in the U.S. for passenger transportation has steadily increased in the decades between 1960 and 1999. In 1960 the total passenger bill was $59.7 billion whereas in 1999 the amount was $936.9 billion. Expenditures on private auto transportation were $787.4 billion in 1999 or 84.0 percent of the total. Expenditures for air transportation in 1999 were $95.5 billion or 10.2 percent. The remaining modes, bus, taxi, water and rail, had expenditures in 1999 of $41.7 billion for local transportation and $6 billion for intercity and international transportation.

Public Transportation

Public transportation is a generic term used to describe any and all of the family of transit services available to urban and rural residents. Thus, it is not a single mode but a variety of traditional and innovative services, which should complement each other to provide systemwide mobility. Modes included within the realm of public transportation are

- **Mass transit,** characterized by fixed routes, published schedules, and vehicles, such as buses and light rail or rapid transit, that travel designated routes with specific stops
- **Paratransit,** characterized by more flexible and personalized service than conventional fixed-route, fixed-schedule services, available to the public on demand, by subscription or on a shared-ride basis
- **Ridesharing,** characterized by two or more persons traveling together by prearrangement, such as carpool, vanpool, buspool, or shared-ride taxi

Public transportation is an important element of the total transportation services provided within large and small metropolitan areas. A major advantage of public transportation is that it can provide high-capacity, energy-efficient movement in densely traveled corridors. It also serves medium- and low-density areas by offering an option for auto owners who do not wish to drive, and an essential service to those without access to an automobile—schoolchildren, senior citizens, single-auto families, and others who may be economically or physically disadvantaged.

For most of this century, public transportation was provided by the private sector. However, increases in auto ownership, shifts in living patterns to low-density suburbs, and the relocation of industry and commerce away from the central city, along with changes in lifestyles (which have been occurring since the end of World War II), have resulted in a steady decline in transit ridership. Since the early 1960s, most transit services have been provided by the public sector. Revenues from fares no longer represent their principal source of income, and over a 25- to 30-year period, the proportion of funds for transit provided by federal, state, and local governments has steadily increased.

It is generally believed that highways and motor transport will play a dominant role in providing personal transportation in the beginning decades of the twenty-first century. A conference on long-range trends and requirements for the nation's highway and public transit systems, conducted by the Transportation Research Board, produced a report titled *A Look Ahead: Year 2020*. Among the many observations listed below, it concluded that

> There is no evidence to suggest that the automobile will not be the predominant form of transportation in the year 2020. This hardy little beast is characterized by flexibility for the user. It has a remarkable capacity to survive, changing its color, size and shape to conform to the current environment. Although the vehicle may be technologically unrecognizable as it adapts to the computer age, its essential features will remain.

However, to place this conclusion into perspective, one need only to imagine that similar statements were being made about the predominance of railroads in the year 1920. Few experts in the 1880s could have imagined the dramatic changes in air and automobile transportation that would occur over a 40-year period. There are many unforeseen changes that could alter the balance between public and private transportation. In the TRB *A Look Ahead: 2020* report many of these changes are described. Some could contribute to the further demise of transit, others might cause transit to become stronger, and for the remainder there would be little or no effect. The potential changes that could influence transit usage are categorized in the book *Urban Mass Transportation Planning* as follows:

Bad for Transit

- Growth of suburbs
- Industry and employment moving from the central city
- Increased suburb-to-suburb commuting
- Migration of the population to the south and west
- Loss of population in "frost-belt" cities
- Growth in private vehicle ownership
- Increased diversity in vehicle types such as SUVs, pickup trucks, and RVs
- High cost per mile to construct fixed-rail transit lines
- High labor costs

Good for Transit

- Emphasis by the federal government on air quality
- Higher prices of gasoline

- Depletion of energy resources
- Trends toward higher density living
- Legislation to encourage "livable cities" and "smart growth"
- Location of mega-centers in suburbs
- Need for airport access and circulation within airports
- Increased number of seniors who cannot or choose not to drive

Neutral for Transit

- Increases in telecommuting may require less travel to a work-site
- Internet shopping and e-commerce could reduce shopping trips
- Changes in work schedules to accommodate childcare could increase trip chaining
- Staggering work hours, flex-time, and four-day work weeks reduce peak-hour congestion
- Aging population, most of whom are not transit users, may continue to drive
- Increased popularity in walking and biking could be a substitute for transit riding

Thus the future of public transportation appears to be one of stability and modest growth. In the 1990s and into the year 2000, public support and financing for transit have increased. Political leadership and the citizenry generally agree that transit is essential to the quality of life and as a means to reduce traffic congestion. Furthermore, there remains a significant proportion of the population who must rely primarily on transit because they do not have access to an automobile or are disabled. As with all modes of transportation, change is incremental. If the factors favorable to transit should prevail, then over time transit will continue to improve service, expand its network, and attract an increasing number of riders.

Highway Transportation

Highway transportation is the dominant mode in passenger travel and one of the principal freight modes. The U.S. highway system comprises approximately 3.9 million miles of highways, ranging from high-capacity, multilane freeways to urban streets to unpaved rural roads. Figure 2.7 illustrates the total disbursements for highways, by function, in the United States between 1945 and 1997. Although the total number of roadway miles has not increased greatly, the quality of roadway surfaces and their capacity has improved due to increased investment in maintenance and construction.

The Federal Highway System

The federal-aid system, which includes the interstate and other federal-aid routes, consists of a network of roads totaling approximately 930,000 miles. These roads are classified as rural or urban and as arterials or collectors. Urban roads are located in cities of 25,000 or more, and they serve areas of high-density land development. Urban roads are used primarily for commuting and shopping trips. Rural roads are located outside of cities and serve as links between population centers.

Arterial roads are intended to serve travel between areas and provide improved mobility. Collector roads and minor arterials serve a dual purpose of mobility and access. Local roads, supported by state and local funds, serve primarily as access routes and for short trips. Arterial roads have higher speeds and capacities, whereas local roads, which serve

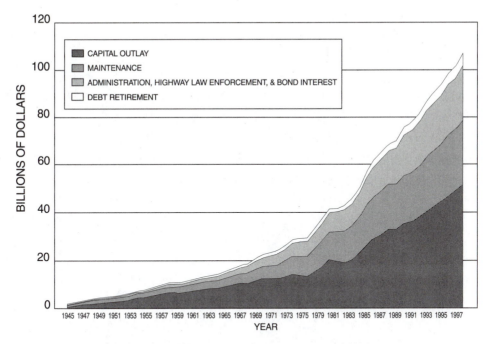

Figure 2.7 Total Disbursements for Highways, by Function 1945–1998

SOURCE: *Highway Statistics,* U.S. Department of Transportation, Federal Highway Administration, Washington, D.C., 1998.

abutting-land uses, and collector roads are slower and not intended for through traffic. A typical highway trip begins on a local road and continues on to a collector and then to a major arterial. This hierarchy of road classification is useful in allocating funds and establishing design standards. Typically, local and collector roads are paid for by property owners and through local taxes, whereas collectors and arterials are paid for jointly by state and federal funds.

The interstate highway system consists of approximately 45,000 miles of limited-access roads. This system, which is essentially complete, as is illustrated in Figure 2.8, connects major metropolitan areas and industrial centers by direct routes, connects the U.S. highway system with major roads in Canada and Mexico, and provides a dependable highway network to serve in national emergencies. Each year, about 2000 route-miles of the national highway system become 20 years old, which is their design life, and a considerable portion of the system must undergo repair, resurfacing, and bridge replacement. Ninety percent of the funds to build the interstate system were provided by the federal government; 10 percent were provided by the states. The completion of the system was originally planned for 1975, but delays resulting from controversies, primarily in the urban sections of the system, caused the completion date to be extended into the 1990s.

In 1995, Congress designated approximately 161,000 miles of routes on the National Highway System (NHS) as required by the Intermodal Surface Transportation Efficiency Act of 1991 (ISTEA). The system represents highways of national significance and consists

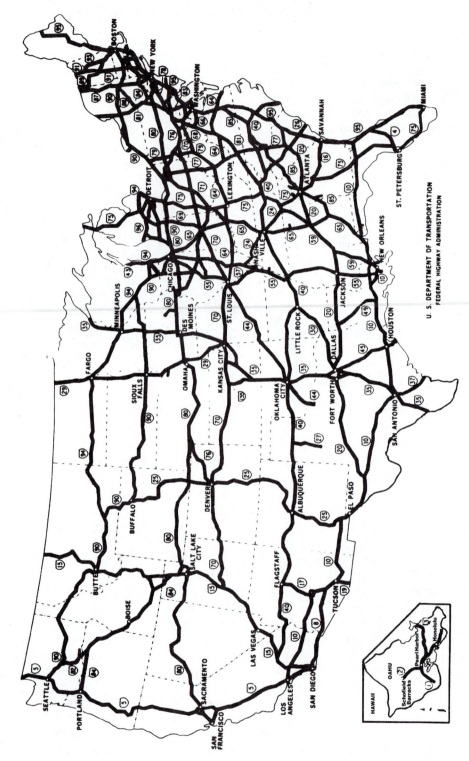

Figure 2.8 Dwight D. Eisenhower System of Interstate and Defense Highways

SOURCE: Federal Highway Administration, Washington, D.C., 1992.

primarily of existing highways. Designated routes are intended to be the major focus of future expenditures on highways by the federal government. The NHS includes the interstate highway system (45,000 mi), the Strategic Highway System (15,700 mi), connectors to the Strategic Highway System (1,900 mi), high priority corridors identified by Congress (4,500 mi), and principal routes (92,000 mi), proposed by the Secretary of the U.S. Department of Transportation in consultation with state and local officials.

Intercity Bus Transportation

Intercity bus transportation services have benefited from the interstate highway system. Buses provide the greatest nationwide coverage of any public carrier by connecting about 15,000 cities and towns with bus routes. Intercity buses carried 27.3 percent of common carrier passengers, but since these trips are relatively short, bus transportation represented only about 6 percent of total passenger-miles by public carriers in 1997. Intercity bus service is provided by private companies; the principal one is Greyhound, which operates at a profit, with little or no support from state or federal governments.

Bus transportation is a highly energy-efficient mode, averaging 300 seat-mi/gal. Buses are also very safe. Their accident rate of 12 fatalities per 100 billion passenger- miles is over 100 times better than that of automobiles.

In spite of its positive characteristics of safety and energy efficiency, bus travel is generally viewed unfavorably by the riding public. Buses are slower and less convenient than other modes and often terminate in downtown stations that are located in the less attractive parts of cities. Other factors, such as lack of through ticketing, comfortable seats, and systemwide information, which the riding public is accustomed to receiving when traveling by air, reinforce the overall negative image of intercity bus transportation.

Truck Transportation

The nation's highway system also carries about 29 percent of all intercity ton-miles of freight, which represents three-fourths of intercity freight revenue. The truck population is very diverse in terms of size, ownership, and use. Most vehicles registered as trucks are less than 10,000 lbs in gross weight, and over half of these are used for personal transportation. Only about 9 percent are heavy trucks used for over-the-road intercity freight. Many states consider recreational vehicles, mobile homes, and vans as trucks. Approximately 20 percent of trucks are used in agriculture; 10 percent are used by utilities and service industries to carry tools and people rather than freight. Personal truck ownership varies by region and is more prevalent in the mountain and western states than in the mid-Atlantic and eastern states.

Trucks are manufactured by the major auto companies as well as by specialty companies. The configurations of trucks are diverse; a truck is usually produced to an owner's specifications. Trucks are classified by gross vehicle weight. The heaviest trucks, weighing over 26,000 lbs, are widely used in intercity freight; lighter ones transport goods and services for shorter distances. Until 1982, truck size and weight were regulated by the states, with little consistency and uniformity. Trucks were required to conform to standards for height, width, length, gross weight, weight per axle, and number of axles, depending on the state. The absence of national standards for size and weight reflected the variety of truck dimensions that existed. An interstate truck shipment that met requirements in one state could be overloaded in another along the same route. A trucker would then either carry a minimal load, use a circuitous route, change shipments at state lines, or travel illegally. With the passage of the Surface Transportation Act of 1982, states were required to permit trucks pulling 28-foot trailers on interstate highways and other principal roads. The law also permits the use of 48-foot semitrailers and 102-inch-wide trucks, carrying up to 80,000 lbs.

The trucking industry has been called a giant composed of midgets because, although it is a major force in the U.S. transportation picture, it is very fragmented and diverse. Regulation of the trucking industry has been limited to those carriers that transport under contract (known as *for hire*), with the exception of carriers operating within a single state or in a specified commercial zone, or those carrying exempt agricultural products. Private carriers, which transport their own company's goods or products, are not economically regulated but must meet federal and state safety requirements. The for-hire portion of the trucking industry accounts for about 45 percent of intercity ton-miles and includes approximately 15,000 separate carriers. With deregulation of various transportation industries, the trucking industry has become more competitive.

The trucking industry has benefited significantly from the improvements in the U.S. highway transportation system. The diversity of equipment and service types is evidence of the flexibility furnished by truck transportation. Continued growth of this industry will depend on how well it integrates with other modes, on the future availability of energy, on limitations on the sizes and weights of trucks, on highway speed limits, and on safety regulations.

TRANSPORTATION ORGANIZATIONS

The operation of the vast network of transportation services in the United States is carried out by a variety of organizations. Each has a special function to perform and serves to create a network of individuals who, working together, furnish the transportation systems and services that presently exist. The following sections will describe some of the organizations and associations involved in transportation. The list is illustrative only and is intended to show the wide range of organizations active in the transportation field. The following seven categories, described briefly in the sections that follow, outline the basic purposes and functions that these organizations serve:

- Private companies that are available for hire to transport people and goods
- Regulatory agencies that monitor the behavior of transportation companies in areas such as pricing of services and safety
- Federal agencies such as the Department of Transportation and the Department of Commerce, which, as part of the executive branch, are responsible for carrying out legislation dealing with transportation at the national level
- State and local agencies and authorities that are responsible for the planning, design, construction, and maintenance of transportation facilities such as roads and airports
- Trade associations, each of which represents the interests of a particular transportation activity, such as railroads or intercity buses, and which serve these groups by furnishing data and information, by representing them at congressional hearings, and by furnishing a means for discussing mutual concerns

- Professional organizations composed of individuals who may be employed by any of the transportation organizations but who have a common professional bond and benefit from meeting with colleagues at national conventions or in specialized committees to share the results of their work, learn about the experience of others, and advance the profession through specialized committee activities
- Organizations of transportation users who wish to influence the legislative process and furnish its members with useful travel information

Other means of exchanging information about transportation include professional and research journals, reports and studies, and university research and training programs.

Private Transportation Companies

Transportation by water, air, rail, highway, or pipeline is furnished either privately or on a for-hire basis. Private transportation, such as automobiles or company-owned trucks, must conform to safety and traffic regulations. For-hire transportation, regulated until recently by the government, is classified as common carriers (available to any user), contract carriers (available by contract to particular market segments), and exempt (for-hire carriers that are exempt from regulation). Examples of private transportation companies are Greyhound, Smith Transfer, United Airlines, and Yellow Cab, to name a few.

Regulatory Agencies

Common carriers have been regulated by the government since the late 1800s, when abuses by the railroads created a climate of distrust toward the railroad "robber barons" who used their monopoly powers to grant preferential treatment to favored customers or charged high rates to those who were without alternative routes or services. The ICC was formed to make certain that the public received dependable service at reasonable rates without discrimination. It was empowered to control the raising and lowering of rates, to require that the carriers had adequate equipment and maintained sufficient routes and schedules, and to certify the entry of a new carrier and control the exit of any certified carrier. Today these concerns are less valid because the shipper has alternatives other than shipping by rail, and the entry and exit restrictions placed on companies no longer tend to favor the status quo or limit innovation and opportunities for new service. For these reasons, the ICC, which had regulated railroads since 1887 and trucking since 1935, was abolished; now private carriers can operate in a more independent economic environment that should result in lower costs and improved services. Similarly, the Civil Aeronautics Board (CAB), which regulated the airline industry starting in 1938, is no longer empowered to certify routes and fares of domestic airline companies; it was phased out in 1985. Other regulatory agencies are the Federal Maritime Commission, which regulates U.S. and foreign vessels operating in international commerce, and the Federal Energy Regulatory Commission, which regulates certain oil and natural gas pipelines.

Federal Agencies

Because transportation pervades our economy, each agency within the executive branch of the federal government is involved in some aspect of transportation. For example, the

Department of State develops policy recommendations concerning international aviation and maritime transportation, and the Department of Defense, through the Army Corps of Engineers, constructs and maintains river and harbor improvements and administers laws protecting navigable waterways. The Department of Transportation is the principal assistant to the president in all matters relevant to federal transportation programs. Within the department are eleven major administrations that deal with programs for highways, aviation, railroads, urban mass transportation, motor carrier safety, highway traffic safety, research, and maritime. Other elements of the Department of Transportation are the Coast Guard, the St. Lawrence Seaway Development Corporation, and the Bureau of Transportation Statistics. The department also deals with special problems such as public and consumer affairs, civil rights, and international affairs.

The U.S. Congress, in which the people are represented by the Senate and the House of Representatives, has jurisdiction over transportation activities through the budget and legislative process. Two committees in the Senate are concerned with transportation: the Commerce, Science, and Transportation Committee and the Committee on Environment and Public Works. The two House committees are Commerce, and Transportation and Infrastructure. The Appropriations Committees in both the Senate and the House have transportation subcommittees. Figure 2.9 illustrates the interrelationship between the U.S. Department of Transportation and other agencies.

State and Local Agencies and Authorities

Each of the 50 states has its own highway or transportation department, which is responsible for planning, building, operating, and maintaining its highway system and for administering funds and programs in other modes such as rail, transit, air, and water. These departments may also be responsible for driver licensing, motor vehicle registration, policing, and inspection. The organization and functions of these departments vary considerably from state to state, but, since highway programs are a direct state responsibility, whereas other modes are either privately owned and operated (such as air and rail) or of greater concern at a local level (such as public transit), highway matters tend to predominate at the state level. In many states, the responsibilities of state transportation departments closely parallel those of the U.S. Department of Transportation.

Local agencies are responsible for carrying out specific transportation functions within a prescribed geographic area. Most larger cities have a regional transportation authority that operates the bus and rapid transit lines, and many communities have a separate traffic department responsible for operating the street system, its signing, traffic timing, and parking controls. Local roads are often the responsibility of county or township agencies, which vary considerably in the quality of engineering staff and equipment. The jurisdiction and administration of the road in a state can create severe economic and management problems if the responsibility for road improvements is fragmented and politicized.

Trade Associations

Americans are joiners; for each occupation or business involved in transportation, there is likely to be an organization that represents its interests. These associations are an attempt to present an industrywide front in matters of common interest. They also promote and

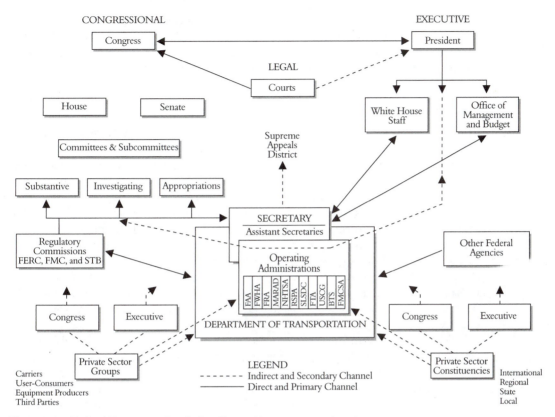

CONGRESSIONAL

EXECUTIVE

LEGAL

Figure 2.9 Federal Transportation Policy Channels

SOURCE: Redrawn from *National Transportation Organizations,* Eno Transportation Foundation, Inc., Washington, D.C., 1998. Used with permission.

Key for Operating Administrations: FAA, Federal Aviation Administration; FHWA, Federal Highway Administration; FRA, Federal Railroad Administration; MARAD, Maritime Administration; NHTSA, National Highway Traffic Safety Administration; RSPA, Research and Special Programs Administration; SLSDC, St. Lawrence Seaway Development Corporation; FTA, Federal Transit Administration; USCG, United States Coast Guard; BTS, Bureau of Transportation Statistics; and FMCSA, Federal Motor Carrier Safety Administration.

develop new procedures and techniques to enhance the marketability of their products and to provide an opportunity for information exchange.

Examples of modally oriented organizations are the Association of American Railroads (AAR), the American Road and Transportation Builders Association (ARTBA), the American Public Transit Association (APTA), and the American Bus Association (ABA).

In addition to carrying out traditional lobbying efforts, trade associations collect and disseminate industry data, perform research and development, provide self-checking for safety matters, publish technical manuals, provide continuing career education, and provide electronic data exchange between carriers and shippers.

Examples of trade associations that improve product performance and marketability are the Asphalt Institute (AI) and the Portland Cement Association (PCA). These groups publish data and technical manuals used by design engineers and maintain technical staffs for special consultations. These and many similar associations attempt to keep abreast of changing conditions and to provide communications links for their members and between the many governmental agencies that use or regulate their products.

Professional Societies

Professional societies are composed of individuals with common professional interests in transportation. Their purposes are to exchange ideas, to develop recommendations for design and operating procedures, and to keep their memberships informed about new developments in transportation practice. Membership in professional organizations is essential for any professional who wishes to stay current in his or her field.

Examples of professional societies are the Womens Transportation Seminar (WTS) and the Institute of Transportation Engineers (ITE), which represent professionals who work in companies or agencies as transportation managers, planners, or engineers. Members of the American Association of State Highway and Transportation Officials (AASHTO) are representatives of state highway and transportation departments and of the Federal Highway Administration. AASHTO produces manuals, specifications standards, and current practices in highway design, which form the basis for practices throughout the country.

The Transportation Research Board (TRB) is a division of the National Research Council and is responsible for encouraging research in transportation and disseminating the results to the professional community. It operates through a technical committee structure composed of knowledgeable practitioners who assist in defining research needs, review and sponsor technical sessions, and conduct workshops and conferences. The Transportation Research Board is supported by the state transportation departments, federal transportation administrations, trade associations, transportation companies, and individual memberships. Its annual meeting in Washington, D.C., attracts over 8000 people each year.

Users of Transport Services

The transportation user has a direct role in the process of effecting change in transportation services and is represented by several transportation groups or associations. The American Automobile Association (AAA) has a wide membership of highway users and serves its members as a travel advisory service, lobbying organization, and insurance agency. The American Railway Passenger Association, the Bicycle Federation of America, and other similar consumer groups have attempted to influence legislation to improve transportation services such as requiring nonsmoking sections in airplanes or building bike lanes. Also, environmental groups have been influential in ensuring that highway planning includes consideration of air quality, land use, wetlands, and noise. In general, however, the average transportation consumer plays a minor role in the transportation decision process and is usually unaware of the way in which the transportation services he or she uses came about or are supported. Quite often the only direct involvement of the consumer is through citizen participation in the selection process for new transportation facilities in the community.

SUMMARY

The transportation system in a developed nation consists of a network of modes that have evolved over many years. The system consists of vehicles, guideways, terminal facilities, and control systems; these operate according to established procedures and schedules in the air, on land, and on water. The system also requires interaction with the user, the operator, and the environment. The systems that are in place reflect the multitude of decisions made by shippers, carriers, government, individual travelers, and affected nonusers concerning the investment in or the use of transportation. The transportation system that has evolved has produced a variety of modes that complement each other. Intercity passenger travel often involves auto and air modes; intercity freight travel involves pipeline, water, rail, and trucking. Urban passenger travel involves auto or public transit; urban freight is primarily by truck.

The nation's transportation system can be considered to be in a state of equilibrium at any given point in time as a result of market forces, government actions, and transportation technology. As these change over time, the transportation system will also be modified. During recent decades, changes in gasoline prices, regulation by government, and new technology have affected the relative importance of each mode. The passenger or shipper thinks of each mode in terms of the level of service provided. Each mode offers a unique set of service characteristics at a given price: travel time, frequency, comfort, convenience, reliability, and safety. The traveler or shipper selects the mode based on how these attributes are valued.

The principal carriers of freight are rail, truck, pipeline, and water. Passenger transportation is by auto, air, rail, and bus. Highway transportation is the dominant mode in passenger travel, accounting for approximately 80 percent of passenger miles. Trucks carry most freight in urban areas and are a principal mode in intercity travel. The United States highway system comprises 3.9 million miles of roadway. The Interstate system, consisting of 42,500 miles of limited-access roads, represents the backbone of the nation's highway network.

A wide range of organizations and agencies provide the resources to plan, design, build, operate, and maintain the nation's transportation system. These include private companies that furnish transportation; regulatory agencies that monitor the safety and service quality provided; federal, state, and local agencies that provide funds to build roads and airports and carry out legislation dealing with transportation at a national level; trade associations that represent the interests of a particular group of transportation providers or suppliers; professional organizations; and transportation user groups.

PROBLEMS

2-1 How would your typical day be changed without availability of your principal mode of transportation? Consider both personal transportation as well as goods and services that you rely on.

2-2 What are the most central problems in your state concerning one of the following: (a) air transportation, (b) railroads, (c) water transportation, (d) highways, or (e) public transportation? (To answer this question, obtain a copy of the governor's plan for transportation in your state or contact a key official in the transportation department.)

2-3   A bridge has been constructed between the mainland and an island. The total cost (excluding tolls) to travel across the bridge is expressed as $C = 50 + 0.5V$, where V is the number of veh/h and C is the cost/vehicle in cents. The demand for travel across the bridge is $V = 2500 - 10C$.

   (a)   Determine the volume of traffic across the bridge.
   (b)   If a toll of 25 cents is added, what is the volume across the bridge?
   (c)   A tollbooth is to be added, thus reducing the travel time to cross the bridge. The new cost function is $C = 50 + 0.2V$. Determine the volume of traffic that would cross the bridge.
   (d)   Determine the toll to yield the highest revenue for demand and supply function in part (a), and the associated demand and revenue.

2-4   A toll bridge carries 10,000 veh/day. The current toll is $3.00/vehicle. Studies have shown that for each increase in toll of 50 cents, the traffic volume will decrease by 1000 veh/day. It is desired to increase the toll to a point where revenue will be maximized.

   (a)   Write the expression for travel demand on the bridge, related to toll increase and current volume.
   (b)   Determine toll charge to maximize revenues.
   (c)   Determine traffic vol/day after toll increase.
   (d)   Determine total revenue increase with new toll.

2-5   Consideration is being given to increasing the toll on a bridge now carrying 4500 veh/day. The current toll is 125¢/veh. It has been found from past experience that the daily traffic volume will decrease by 400 veh/day for each 25¢ increase in toll. Therefore, if x is the increase in toll in cents/veh, the volume equation for veh/day is $V = 4500 - 400\,(x/25)$, and the new toll/veh would be $T = 125 + x$. In order to maximize revenues, what would the new toll charge be per vehicle and what would the traffic vol/day be after the toll increase?

2-6   A large manufacturer uses two factors to decide whether to use truck or rail for movement of its products to market: cost and total travel time. The manufacturer uses a utility formula that rates each mode. The formula is $U = 5C + 10T$, where C is cost ($/ton) and T is time (hours). For a given shipment of goods, a trucking firm can deliver in 16 hours and charges $25/ton, whereas a railroad charges $17/ton and can deliver in 25 hours.

   (a)   Which mode should the shipper select?
   (b)   What other factors should the shipper take into account in making a decision? (Discuss at least two.)

2-7   An individual is planning to take an 800-mile trip between two large cities. Three possibilities exist: air, rail or auto. The person is willing to pay $25 for every hour saved in making the trip. The trip by air costs $600 and travel time is 8 hours, by rail the cost is $450 and travel time is 16 hours, and by auto the cost is $200 and travel time is 20 hours.

   (a)   Which mode is the best choice?
   (b)   What factors other than cost might influence the decision regarding which mode to use?

2-8   Describe the organization and function of your state highway department and/or transportation department.

2-9   List three transportation organizations located in your state. What services do they provide?

2-10   Obtain a copy of a Transportation Research Record (published by the TRB) or a CD of an annual meeting.

    (a) Select one article and write a short summary of its contents.

    (b) Describe the technical area of transportation covered by this article.

2-11   What do the following acronyms mean?

AAA	AAR
AASHTO	AI
APTA	FHWA
ARTBA	PCA
TRB	

2-12   List the seven categories of transportation organization, and give one example of each.

2-13   What are the four principal modes for moving freight? Which of these modes carries the largest share of ton-miles? Which carries the lowest?

2-14   What are the four principal modes for moving people? Which of these modes accounts for the largest share of passenger-miles? Which mode accounts for the lowest?

2-15   **(a)** List four major factors that will determine the future of public transportation in the United States.

    (b) For each of the factors listed indicate if the factor is positive, negative or neutral to the success of transit.

2-16   What are the advantages and disadvantages of using intercity bus transportation?

ADDITIONAL READINGS

A Look Ahead: Year 2020, Special Report 220, Transportation Research Board, National Research Council, Washington, D.C., 1988.

Black, Alan, *Urban Mass Transportation Planning,* McGraw-Hill, Inc., 1995.

Highway Statistics, U.S. Department of Transportation, Federal Highway Administration, Washington, D.C., 1998.

Moving America: New Directions, New Opportunities, U.S. Department of Transportation, Washington, D.C., February 1990.

National Transportation Organizations, Eno Transportation Foundation, Washington, D.C., 1998.

Transportation in America, 18th ed., Eno Transportation Foundation, Washington, D.C., 2000.

Traffic Operations

T he traffic or highway engineer must understand not only the basic characteristics of the driver, the vehicle, and the roadway, but how each interacts with the others. Information obtained through traffic engineering studies serves to identify relevant characteristics and define related problems. Traffic flow is of fundamental importance in developing and designing strategies for intersection control, rural highways, and freeway segments.

CHAPTER 3

Characteristics of the Driver, the Pedestrian, the Vehicle, and the Road

The four main components of the highway mode of transportation are the driver, the pedestrian, the vehicle, and the road. The bicycle is also becoming an important component in the design of urban highways and streets. To provide efficient and safe highway transportation, a knowledge of the characteristics and the limitations of each of these components is essential. It is also important to be aware of the interrelationships that exist among these components in order to determine the effects, if any, that they have on each other. Their characteristics are also of primary importance when traffic engineering measures such as traffic control devices are to be used in the highway mode. Knowing average limitations may not always be adequate; sometimes it may be necessary to obtain information on the full range of limitations. Consider, for example, the wide range of drivers' ages in the United States, which usually begins at 16 and can exceed 80.

It should also be noted that highway statistics provided by the Federal Highway Administration indicate that 65 years and older drivers comprise over 14 percent of the driver population. Sight and hearing vary considerably across age groups, with the ability to hear and see usually decreasing after age 65. In addition, these can vary even among individuals of the same age group.

Similarly, a wide range of vehicles, from compact cars to articulated trucks, have been designed. The maximum acceleration, turning radii, and ability to climb grades differ considerably among the different vehicles. The road therefore must be designed to accommodate a wide range of vehicle characteristics and at the same time to allow use by drivers and pedestrians with a wide range of physical and psychological characteristics.

This chapter discusses the relevant characteristics of the main components of the highway mode and demonstrates their importance and their use in the design and operation of highway facilities.

DRIVER CHARACTERISTICS

One problem that faces traffic and transportation engineers when they consider driver characteristics in the course of design is the varying skills and perceptual abilities of drivers on the highway. This is demonstrated by the wide range of people's abilities to hear, see, evaluate, and react to information. Studies have shown that these abilities may also vary in an individual under different conditions, such as the influence of alcohol, fatigue, and the time of day. Therefore, it is important that criteria used for design purposes be compatible with the capabilities and limitations of most drivers on the highway. The use of an average value, such as mean reaction time, may not be adequate for a large number of drivers. Both the 85th percentile and the 95th percentile have been used to select design criteria; in general, the higher the chosen percentile, the wider the range covered.

The Human Response Process

Actions taken by drivers on a road result from their evaluation of and reaction to information they obtain from certain stimuli that they see or hear. However, evaluation and reaction must be carried out within a very short time, as the information being received along the highways is continually changing. It has been suggested that most of the information received by a driver is visual, implying that the ability to see is of fundamental importance in the driving task. It is therefore important that highway and traffic engineers have some fundamental knowledge of visual perception as well as of hearing perception.

Visual Reception

The principal characteristics of the eye are visual acuity, peripheral vision, color vision, glare vision and recovery, and depth perception.

Visual Acuity.   Visual acuity is the ability to see fine details of an object. Two types of visual acuity are of importance in traffic and highway emergencies: static and dynamic visual acuity. The driver's ability to identify an object when both the object and the driver are stationary depends on his or her static acuity. Factors that affect static acuity include background brightness, contrast, and time. Static acuity increases with an increase in illumination up to a background brightness of about 3 candles (cd)/sq. ft and then remains constant even with an increase in illumination. When other visual factors are held constant at an acceptable level, the optimal time required for identification of an object with no relative movement is between 0.5 and 1.0 sec.

The driver's ability to clearly detect relatively moving objects, not necessarily in his or her direct line of vision, depends on the driver's dynamic visual acuity. Most people have clear vision within a conical angle of 3° to 5° and fairly clear vision within a conical angle of 10° to 12°. Vision beyond this range is usually blurred. This is important when the location of traffic information devices is considered. Drivers will see clearly those devices that are within the 12° cone, but objects outside this cone will be blurred.

Peripheral Vision.   Peripheral vision is the ability of people to see objects beyond the cone of clearest vision. Although objects can be seen within this zone, details and color are not clear. The cone for peripheral vision could be one subtending up to 160°; this value is affected by the speed of the vehicle. Age also influences peripheral vision. For instance, at age 60, a significant change occurs in a person's peripheral vision.

Color Vision. Color vision is the ability to differentiate one color from another, but deficiency in this ability, usually referred to as *color blindness,* is not of great significance in highway driving because other ways of recognizing traffic information devices (e.g., shape) can compensate for it. Combinations of black and white and black and yellow have been shown to be those to which the eye is most sensitive.

Glare Vision and Recovery. There are two types of glare vision: direct and specular. Rowland and others have indicated that *direct glare* occurs when relatively bright light appears in the individual's field of vision and *specular glare* occurs when the image reflected by the relatively bright light appears in the field of vision. Both types of glare result in a decrease of visibility and cause discomfort to the eyes. It is also known that age has a significant effect on the sensitivity to glare, and that at age 40, a significant change occurs in a person's sensitivity to glare.

The time required by a person to recover from the effects of glare after passing the light source is known as *glare recovery.* Studies have shown that this time is about 3 sec when moving from dark to light and can be 6 sec or more when moving from light to dark. Glare vision is of great importance during night driving; it contributes to the problem of serving older people, who see much more poorly at night. This phenomenon should be taken into account in the design and location of street lighting so that glare effects are reduced to a minimum.

Glare effects can be minimized by reducing luminaire brightness and by increasing the background brightness in a driver's field of view. Specific actions taken to achieve this in lighting design include using higher mounting heights, positioning lighting supports farther away from the highway, and restricting the light from the luminaire to obtain minimum interference with the visibility of the driver.

Depth Perception. Depth perception affects the ability of a person to estimate speed and distance. It is particularly important on two-lane highways during passing maneuvers, when head-on accidents may result from a lack of proper judgment of speed and distance.

The ability of the human eye to differentiate between objects is fundamental to this phenomenon. It should be noted, however, that the human eye is not very good at estimating absolute values of speed, distance, size, and acceleration. This is why traffic control devices are standard in size, shape, and color. Standardization not only aids in distance estimation but also helps the color-blind driver to identify signs.

Hearing Perception

The ear receives sound stimuli, which is important to drivers only when warning sounds, usually given out by emergency vehicles, are to be detected. Loss of some hearing ability is not a serious problem, since it can normally be corrected by a hearing aid.

PERCEPTION-REACTION PROCESS

The process through which a driver, cyclist or pedestrian evaluates and reacts to a stimulus can be divided into four subprocesses:

1. *Perception:* the driver sees a control device, warning sign, or object on the road
2. *Identification:* the driver identifies the object or control device and thus understands the stimulus
3. *Emotion:* the driver decides what action to take in response to the stimulus; for example, to step on the brake pedal, to pass, to swerve, or to change lanes
4. *Reaction or volition:* the driver actually executes the action decided on during the emotion subprocess

Time elapses during each of these subprocesses. The time that elapses from the start of perception to the end of reaction is the total time required for perception, identification, emotion, and volition, sometimes referred to as PIEV time or, more commonly, as perception-reaction time.

Perception-reaction time is an important factor in the determination of braking distances, which in turn dictates the minimum sight distance required on a highway and the length of the yellow phase at a signalized intersection. Perception-reaction time varies among individuals and may, in fact, vary for the same person as the occasion changes. These changes in perception-reaction time depend on how complicated the situation is, the existing environmental conditions, age, whether the person is tired or under the influence of drugs and/or alcohol, and whether the stimulus is expected or unexpected.

Triggs and Harris described this phenomenon in detail. They noted that the 85th-percentile times to brake, obtained from several situations, varied from 1.26 sec to over 3 sec. The reaction time selected for design purposes should, however, be large enough to include reaction times for most drivers using the highways. Recommendations made by the American Association of State Highway and Transportation Officials (AASHTO) stipulate 2.5 sec for stopping-sight distances. This encompasses the decision times for about 90 percent of drivers under most highway conditions. Note, however, that a reaction time of 2.5 sec may not be adequate for unexpected conditions or for some very complex conditions, such as those at multiphase at-grade intersections and ramp terminals. For example, when signals are unexpected, reaction times can increase by 35 percent.

Example 3.1   Distance Traveled During Perception-Reaction Time

A driver with a perception-reaction time of 2.5 sec is driving at 65 mi/h when she observes that an accident has blocked the road ahead. Determine the distance the vehicle would move before the driver could activate the brakes. The vehicle will continue to move at 65 mi/h during the perception-reaction time of 2.5 sec.

Solution:

- Convert mi/h to ft/sec:

$$65 \text{ mi/h} = \left(65 \times \frac{5280}{3600}\right) \text{ft/sec} = 65 \times 1.47 = 95.55 \text{ ft/sec}$$

- Find the distance traveled:

$$D = vt$$
$$= 95.55 \times 2.5$$
$$= 238.9 \text{ ft}$$

where v = velocity and t = time.

PEDESTRIAN CHARACTERISTICS

Pedestrian characteristics relevant to traffic and highway engineering practice include those of the driver, discussed in the preceding section. In addition, other pedestrian characteristics may influence the design and location of pedestrian control devices. Such control devices include special pedestrian signals, safety zones and islands at intersections, pedestrian underpasses, elevated walkways, and crosswalks. Apart from visual and hearing characteristics, walking characteristics play a major part in the design of some of these controls. For example, the design of an all-red phase, which permits pedestrians to cross an intersection with heavy traffic, requires a knowledge of the walking speeds of pedestrians. Observations of pedestrian movements have indicated that walking speeds vary between 3.0 and 8.0 ft/sec. Significant differences have also been observed between male and female walking speeds. At intersections, the mean male walking speed has been determined to be 4.93 ft/sec, and for females, 4.63 ft/sec. [The Highway Capacity Manual (HCM) suggests the use of a more conservative value of 4.0 ft/s for design purposes.] When the percentage of elderly pedestrians is higher than 20%, a speed of 3.3 ft/s is recommended.

It has also been suggested that the value of 4 ft/sec may be rather high for 60 years and older pedestrians. This factor should therefore be taken into consideration for the design of intersection pedestrian signals at locations where a high number of older pedestrians is expected. Consideration should also be given to the characteristics of handicapped pedestrians, such as the blind. Studies have shown that accidents involving blind pedestrians can be reduced by installing special signals. The blind pedestrian can turn the signal to a red phase by using a special key, which also rings a bell, indicating to the pedestrian that it is safe to cross. Ramps are also now being provided at intersection curbs to facilitate the crossing of the intersection by the occupant of a wheelchair. Also, consideration should be given to the relatively lower average walking speed of the handicapped pedestrian, which can vary from a low of 1.97 ft/sec to 3.66 ft/sec.

Bicyclists and Bicycles

As noted earlier, bicycles are now an important component of the highway mode, especially for highways located in urban areas. It is therefore essential that highway and traffic engineers have an understanding of the characteristics of bicycles and bicyclists. The basic human factors discussed for the automobile driver also apply to the bicyclist, particularly with respect to perception and reaction. It should, however, be noted that unlike the automobile driver, the bicyclist is not only the driver of the bicycle, but he/she also provides the power to move the bicycle. The bicycle and the bicyclist therefore unite to form a system, thus requiring that both be considered jointly.

Three classes of bicyclists (A, B, and C) have been identified in the *Guide for the Development of Bicycle Facilities* by AASHTO. Experienced or advanced bicyclists are within class A, while less experienced bicyclists are within class B, and children riding on their own or with parents are classified as C. Class A bicyclists typically consider the bicycle as a motor vehicle and can comfortably ride in traffic. Class B bicyclists prefer to ride on neighborhood streets and are more comfortable on designated bicycle facilities, such as bicycle paths. Class C bicyclists use mainly residential streets that provide access to schools, recreational facilities, and stores.

In designing urban roads and streets it is useful to consider the feasibility of incorporating bicycle facilities that will accommodate class B and class C bicyclists.

The bicycle, like the automobile, also has certain characteristics that are unique. For example, based on the results of studies conducted in Florida, Pein suggested that the minimum design speed for bicycles on level terrain is 20 mi/h, but downgrade speeds can be as high as 31 mi/h while upgrade speeds can be as low as 8 mi/h. Pein also suggested that the mean speed of bicycles when crossing an intersection from a stopped position is 8 mi/h and the mean acceleration rate is 3.5 ft/sec^2.

VEHICLE CHARACTERISTICS

Criteria for the geometric design of highways are partly based on the static, kinematic, and dynamic characteristics of vehicles. Static characteristics include the weight and size of the vehicle; kinematic characteristics involve the motion of the vehicle, without considering the forces that cause the motion; dynamic characteristics involve the forces that cause the motion of the vehicle. Since nearly all highways carry both passenger-automobile and truck traffic, it is essential that design criteria take into account the characteristics of different types of vehicles. A thorough knowledge of these characteristics will aid the highway and/or traffic engineer in designing highways and traffic control systems that allow the safe and smooth operation of a moving vehicle, particularly during the basic maneuvers of passing, stopping, and turning. Therefore, designing a highway involves the selection of a *design vehicle,* whose characteristics will encompass those of nearly all vehicles expected to use the highway. The characteristics of the design vehicle are then used to determine criteria for geometric design, intersection design, and sight-distance requirements.

Static Characteristics

The size of the design vehicle for a highway is an important factor in the determination of design standards for several physical components of the highway. These include lane width, shoulder width, length and width of parking bays, and lengths of vertical curves. The axle weights of the vehicles expected on the highway are important when pavement depths and maximum grades are being determined.

For many years, each state prescribed by law the size and weight limits for trucks using its highways, and in some cases local authorities also imposed more severe restrictions on some roads. Table 3.1 shows some features of static characteristics for which limits were prescribed. A range of maximum allowable values is given for each feature.

Table 3.1   Range of State Limits on Vehicle Lengths by Type and Maximum Weight of Vehicle

Type	Allowable Lengths (ft)
Bus	35–60
Single truck	35–60
Trailer, semi/full	35–48
Semitrailer	55–85
Truck trailer	55–85
Tractor semitrailer trailer	55–85
Truck trailer trailer	65–80
Tractor semitrailer, trailer, trailer	60–105

Type	Allowable Weights (lb)
Single-axle	18,000–24,000
Tandem-axle	32,000–40,000
State maximum gross vehicle weight	73,280–164,000
Interstate maximum gross vehicle weight	73,280–164,000

SOURCE: Adapted from *State Maximum Sizes and Weights for Motor Vehicles,* Motor Vehicle Manufacturer's Association of the United States, Detroit, Mich., May 1982.

Since the passage of the Surface Transportation Assistance Act of 1982, the maximum allowable truck sizes and weights on Interstate and other qualifying federal-aid highways are at most

- 80,000 lb gross weight, with axle loads of up to 20,000 lb for single-axles and 34,000 lb for double-axles
- 102 in. width for all trucks
- 48 ft length for semitrailers and trailers
- 28 ft length for each twin trailer

(*Note:* Those states that had higher weight limits before this law was enacted are allowed to retain them for intrastate travel.)

States are no longer allowed to set limits on overall truck length. These provisions will most probably result in wider, longer, and heavier trucks on these highways, which raises the basic safety question of whether there will be an increase in injuries and fatalities resulting from crashes involving larger trucks. Another effect of these provisions is that reconstruction of deteriorating highways to meet these new weight and length dimensions will be extremely expensive.

As stated earlier, the static characteristics of vehicles expected to use the highway are factors that influence the selection of design criteria for the highway. It is therefore necessary that all vehicles be classified so that representative static characteristics for all vehicles within a particular class can be provided for design purposes. AASHTO has selected three general classes of vehicles: passenger cars, trucks, and buses/recreational vehicles. Included in the passenger-car class are compacts and subcompacts, all light vehicles, and all light delivery trucks (vans and pickups). Included in the truck class are single-unit trucks, truck tractor–semitrailer combinations, and trucks or truck tractors with semitrailers in combination with full trailers. Within the class of buses/recreational vehicles are single-unit buses, articulated buses, motor homes, and passenger cars or motor homes pulling trailers or boats. A total of 15 different design vehicles have been selected to represent the different categories of vehicles within all three classes. Table 3.2 shows the physical dimensions for each of these design vehicles.

The *single-unit truck* (SU) design vehicle represents all single-unit trucks and small buses. The *single-unit bus* (BUS) represents intercity and transit buses with a wheelbase of 25 ft and an overall length of 40 ft. The articulated bus (A-BUS) category represents larger-than-conventional buses that have a permanent hinge near the center, which allows better maneuverability. The intermediate semitrailer (WB-40) category represents the majority of medium tractor-semitrailer combinations; the large semitrailer category (WB-50) represents larger tractor-semitrailer combinations in use. The "double-bottom" semitrailer–full trailer (WB-60) category is representative of the larger tractor-semitrailer–full trailer combinations commonly in use. The interstate semitrailer (WB-62) category represents a larger tractor–semitrailer combination permitted by the Surface Transportation Assistance Act (STAA) of 1982; the interstate semitrailer (WB-67) category represents a larger tractor-semitrailer grandfathered on some selected highways by the STAA. The triple semitrailer (WB-96) category represents tractor-semitrailer–full trailer–full trailer combinations (triples); the turnpike double semitrailer (WB-114) category represents the larger tractor-semitrailer–full trailer (turnpike double). Figure 3.1 shows examples of different types of trucks.

Minimum turning radii at low speeds (10 mi/h or less) are dependent mainly on the size of the vehicle. The turning-radii requirements for passenger and WB-60 design vehicles are given in Figures 3.2 and 3.3 respectively. The turning-radii requirements for the other vehicles can be found in AASHTO's *Policy on Geometric Design of Highways and Streets.* These turning paths were selected by conducting a study of the turning paths of scale models of the representative vehicles of each class. It should be emphasized, however, that the minimum turning radii shown in Figures 3.2 and 3.3 are for turns taken at speeds less than 10 mi/h. When turns are made at higher speeds, the lengths of the transition curves are increased, so radii greater than the specified minimum are required. These requirements will be described later.

Kinematic Characteristics

The primary element among kinematic characteristics is the acceleration capability of the vehicle. Acceleration capability is important in several traffic operations, such as passing maneuvers and gap acceptance. Also, the dimensioning of highway features such as freeway ramps and passing lanes is often governed by acceleration rates. Acceleration is also important in determining the forces that cause motion. Therefore, a study of the kinematic characteristics of the vehicle primarily involves a study of how acceleration rates influence the elements of motion, such as velocity and distance. We therefore review in this section the mathematical relationships among acceleration, velocity, distance, and time.

Let us consider a vehicle moving along a straight line from point o to point m, a distance x in a reference plane T. The position vector of the vehicle after time t may be expressed as

$$r_{om} = x\hat{i} \tag{3.1}$$

Table 3.2 Design Vehicle Dimensions

Design Vehicle Type	Symbol	Dimensions (ft)											
		Overall			Overhang		WB_1	WB_2	S	T	WB_3	WB_4	Typical Kingpin to Center of Rear Axle
		Height	Width	Length	Front	Rear							
Passenger car	P	4.25	7	19	3	5	11	—	—	—	—	—	—
Single-unit truck	SU	11–13.5	8.0	30	4	6	20	—	—	—	—	—	—
Buses													
Intercity bus (motor coaches)	BUS-40	12.0	8.5	40	6	6.3[e]	24	3.7	—	—	—	—	—
	BUS-45	12.0	8.5	45	6	8.5[e]	26.5	4.0	—	—	—	—	—
City transit bus	CITY-BUS	10.5	8.5	40	7	8	25	—	—	—	—	—	—
Conventional school bus (65 pass.)	S–BUS 36	10.5	8.0	35.8	2.5	12	21.3	—	—	—	—	—	—
Large school bus (84 pass.)	S–BUS 40	10.5	8.0	40	7	13	20	—	—	—	—	—	—
Articulated bus	A-BUS	11.0	8.5	60	8.6	10	22.0	19,4	6.2[a]	13.2[a]	—	—	—
Combination trucks													
Intermediate semitrailer	WB-40	13.5	8.0	45.5	3	2.5[e]	12.5	27.5	—	—	—	—	27.5
Intermediate semitrailer	WB-50	13.5	8.5	55	3	2[e]	14.6	35.4	—	—	—	—	37.5
Interstate semitrailer	WB-62★	13.5	8.5	68.5	4	2.5[e]	21.6	40.4	—	—	—	—	42.5
Interstate semitrailer	WB-65★★ or WB-67	13.5	8.5	73.5	4	4.5–2.5[e]	21.6	43.4–45.4	—	—	—	—	45.5–47.5
"Double-bottom" semitrailer/trailer	WB-67D	13.5	8.5	73.3	2.33	3	11.0	23.0	3.0[b]	7.0[b]	23.0	—	23.0
Triple semitrailer/trailer	WB-100T	13.5	8.5	104.8	2.33	3	11.0	22.5	3.0[c]	7.0[c]	23.0	23.0	23.0
Turnpike double semitrailer/trailer	WB-109D★	13.5	8.5	114	2.33	2.5	14.3	39.9	2.5[d]	10.0[d]	44.5	—	42.5
Recreation vehicle													
Motor home	MH	12	8	30	4	6	20	—	—	—	—	—	—
Car and camper trailer	P/T	10	8	48.7	3	10	11	—	5	19	—	—	—
Car and boat trailer	P/B	—	8	42	3	8	11	—	5	15	—	—	—
Motor home and boat trailer	MH/B	12	8	53	4	8	20	—	6	15	—	—	—
Farm tractor[f]	TR	10	8–10	16[g]	—	—	10	9	3	6.5	—	—	—

★ = Design vehicle with 48 ft trailer as adopted in 1982 Surface Transportation Assistance Act (STAA).

★★ = Design vehicle with 53 ft trailer as grandfathered in 1982 Surface Transportation Assistance Act (STAA).

a = Combined dimension is 19.4 ft and articulating section is 4 ft wide. b = Combined dimension is typically 10.0 ft. c = Combined dimension is typically 10.0 ft. d = Combined dimension is typically 12.5 ft. e = This is overhang from the back axle of the tandem axle assembly. f = Dimensions are for a 150–200 hp tractor excluding any wagon length. g = To obtain the total length of tractor and one wagon, add 18.5 ft to tractor length. Wagon length is measured from front of drawbar to rear of wagon, and drawbar is 6.5 ft long. WB_1, WB_2, WB_3, and WB_4 are the effective vehicle wheelbases, or distances between axle groups, starting at the front and working towards the back of each unit. S is the distance from the rear effective axle to the hitch point or point of articulation measured back to the center of the next or center of tandem axle assembly. T is the distance from the hitch point or point of articulation measured back to the center of the next or center of tandem axle assembly.

SOURCE: Adapted from *A Policy on Geometric Design of Highways and Streets*, American Association of State Highway and Transportation Officials, Washington, D.C., 2000. Used by permission.

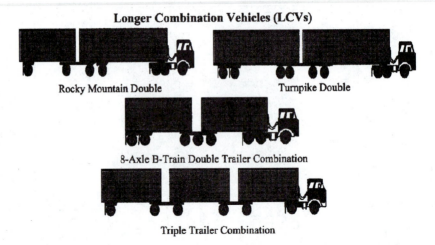

Single Unit Trucks

Conventional Combination Vehicles

5-Axle Tractor Semi-Trailer   6Axle Tractor Semi-Trailer

STAA or "Western" Double

Longer Combination Vehicles (LCVs)

Rocky Mountain Double   Turnpike Double

8-Axle B-Train Double Trailer Combination

Triple Trailer Combination

Figure 3.1   Examples of Different Types of Trucks

where

r_{om} = position vector for m in T

$\hat{\imath}$ = a unit vector parallel to line om

x = distance along the straight line

The velocity and acceleration for m may be simply expressed as

$$u_m = \dot{r}_{om} = \dot{x}_{\hat{\imath}}$$ (3.2)
$$a_m = \ddot{r}_{om} = \ddot{x}_{\hat{\imath}}$$ (3.3)

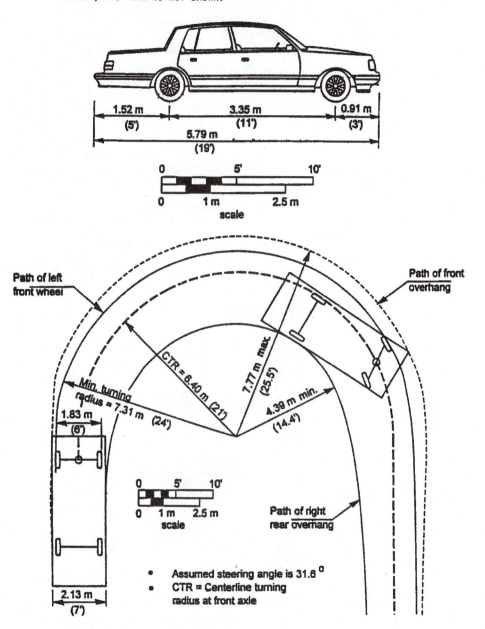

Figure 3.2   Minimum Turning Path for Passenger Design Vehicle

SOURCE: Texas State Department of Highways and Public Transportation, reprinted in *A Policy on Geometric Design of Highways and Streets,* American Association of State Highway and Transportation Officials, Washington, D.C., 2001. Used by permission.

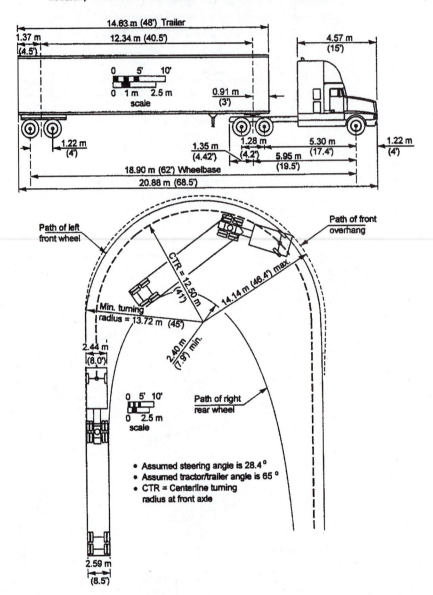

THIS TURNING TEMPLATE SHOWS THE TURNING PATHS OF THE AASHTO DESIGN
VEHICLES. THE PATHS SHOWN ARE FOR THE LEFT FRONT OVERHANG AND THE
OUTSIDE REAR WHEEL. THE LEFT FRONT WHEEL FOLLOWS THE CIRCULAR CURVE,
HOWEVER, ITS PATH IS NOT SHOWN.

- Assumed steering angle is 28.4°
- Assumed tractor/trailer angle is 65°
- CTR = Centerline turning
  radius at front axle

Figure 3.3 Minimum Turning Path for WB-62 Design Vehicle

SOURCE: Texas State Department of Highways and Public Transportation, reprinted in *A Policy on Geometric Design of Highways and Streets,* American Association of State Highway and Transportation Officials, Washington, D.C. Used by permission.

where

u_m = velocity of the vehicle at point m

a_m = acceleration of the vehicle at point m

$$\dot{x} = \frac{dx}{dt}$$

$$\ddot{x} = \frac{d_2 x}{dt^2}$$

Two cases are of interest: (1) acceleration is assumed constant; (2) acceleration is a function of velocity.

Acceleration Assumed Constant

When the acceleration of the vehicle is assumed to be constant,

$$\ddot{x}_i = a$$

$$\frac{d\dot{x}}{dt} = a \tag{3.4}$$

$$\dot{x} = at + C_1 \tag{3.5}$$

$$x = \tfrac{1}{2} at^2 + C_1 t + C_2 \tag{3.6}$$

The constants C_1 and C_2 are determined either by the initial conditions on velocity and position or by using known positions of the vehicle at two different times.

Acceleration as a Function of Velocity

The assumption of constant acceleration has some limitations, because the accelerating capability of a vehicle at any time t is related to the speed of the vehicle at that time (u_t). The lower the speed, the higher the acceleration rate that can be obtained. Figures 3.4a and 3.4b show maximum acceleration rates for passenger cars and tractor-semitrailers at different speeds on level roads. One model that is commonly used in this case is

$$\frac{du_t}{dt} = \alpha - \beta u_t \tag{3.7}$$

where α and β are constants.

In this model, the maximum acceleration rate that can be achieved is theoretically α, which means that α has units of acceleration as its unit. The term βu_t should also have units of acceleration as its unit, which means that β has the inverse of time (for example, $\sec^{-1}$) as its unit.

Integrating Eq. 3.7 gives

$$-\frac{1}{\beta} \ln (\alpha - \beta u_t) = t + C$$

If the velocity is u_o at time zero,

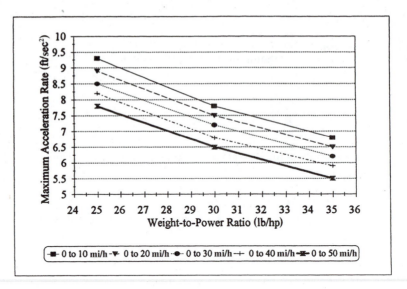

(a) Passenger Cars

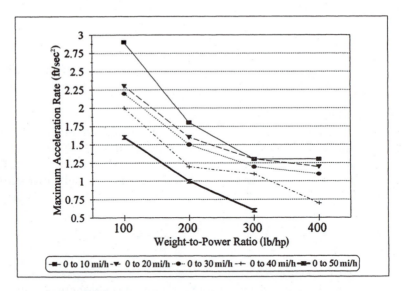

(b) Tractor-Semitrailers

Figure 3.4 Acceleration Capabilities of Passenger Cars and Tractor-Semitrailers on Level Roads

SOURCE: Modified from *Traffic Engineering Handbook,* 5th ed., Institute of Transportation Engineers, Washington, D.C., 1999.

$$C = -\frac{1}{\beta} \ln (\alpha - \beta u_o)$$

and

$$-\frac{1}{\beta} \ln (\alpha - \beta u_t) = t - \frac{1}{\beta} \ln (\alpha - \beta u_o)$$

$$\ln \frac{(\alpha - \beta u_t)}{\alpha - \beta u_o} = -\beta t$$

$$\alpha - \beta u_t = (\alpha - \beta u_o)e^{-\beta t}$$

$$u_t = \frac{\alpha}{\beta} (1 - e^{-\beta t}) + u_o e^{-\beta t} \tag{3.8}$$

The distance $x(t)$ traveled at any time t may be determined by integrating Eq. 3.8:

$$x = \int_0^t u_t dt = \int_0^t \frac{\alpha}{\beta} (1 - e^{-\beta t}) + u_o e^{-\beta t}$$

$$= \left(\frac{\alpha}{\beta}\right)t - \frac{\alpha}{\beta^2} (1 - e^{-\beta t}) + \frac{u_o}{\beta} (1 - e^{-\beta t}) \tag{3.9}$$

Example 3.2   Distance Traveled and Velocity Attained for Variable Acceleration

The acceleration of a vehicle can be represented by the following equation:

$$\frac{du_t}{dt} = 3.3 - 0.04u$$

where u is the vehicle speed in ft/sec. If the vehicle is traveling at 45 mi/h, determine its velocity after 5 sec of acceleration and the distance traveled during that time.

Solution:

- Convert 45 mi/h to ft/sec:
  $45 \times 1.47 = 66.15$ ft/sec
- Use Eq. 3.8 to determine velocity u_t after time t:

$$u = \frac{\alpha}{\beta} (1 - e^{-\beta t}) + u_o e^{-\beta t}$$

$$\alpha = 3.3$$

$$\beta = 0.04$$

$$u_t = \frac{3.3}{0.04} (1 - e^{-(0.04 \times 5)}) + 66.15e^{-(0.04 \times 5)}$$

$$= 82.5(1 - 0.82) + 66.15 \times 0.82$$

$$= 14.85 + 54.24$$

$$= 69.09 \text{ ft/sec}$$

- Convert ft/sec to mi/h:

$$u_t = 69.09/1.47$$
$$= 47.00 \text{ mi/h}$$

- Use Eq. 3.9 to determine distance traveled:

$$x = \left(\frac{\alpha}{\beta}\right)t - \frac{\alpha}{\beta^2}(1 - e^{-\beta t}) + \frac{u_o}{\beta}(1 - e^{-\beta t})$$

$$= \left(\frac{3.3}{0.04}\right)5 - \frac{3.3}{(0.04)^2}(1 - e^{-0.04 \times 5}) + \frac{66.15}{0.04}(1 - e^{-0.04 \times 5})$$

$$= 412.5 - 2062.5(1 - 0.82) + 1653.75(1 - 0.82)$$

$$= 412.5 - 371.25 + 297.68$$

$$= 338.93 \text{ ft}$$

Dynamic Characteristics

Several forces act on a vehicle while it is in motion: air resistance, grade resistance, rolling resistance, and curve resistance. The extents to which these forces affect the operation of the vehicle are discussed in this section.

Air Resistance

A vehicle in motion has to overcome the resistance of the air in front of it as well as the force due to the frictional action of the air around it. The force required to overcome these is known as the *air resistance* and is related to the cross-sectional area of the vehicle in a direction perpendicular to the direction of motion and to the square of the speed of the vehicle. Claffey has shown that this force can be estimated from the formula

$$R_a = 0.5 \frac{(2.15 \, p C_D A u^2)}{g} \tag{3.10}$$

where
    R_a = air resistance force (lb)
    p = density of air (0.0766 lb/ft³) at sea level; less at higher elevations
    C_D = aerodynamic drag coefficient (current average value for passenger cars is 0.4; for trucks, this value ranges from 0.5 to 0.8, but a typical value is 0.5)
    A = frontal cross-sectional area (ft²)
    u = vehicle speed (mi/h)
    g = acceleration of gravity (32.2 ft/sec²)

Grade Resistance

When a vehicle moves up a grade, a component of the weight of the vehicle acts downward, along the plane of the highway. This creates a force acting in a direction opposite that of the motion. This force is the grade resistance. A vehicle traveling up a grade will therefore tend to lose speed unless an accelerating force is applied. The speed achieved at any point along the grade for a given rate of acceleration will depend on the grade. Figure 3.5 shows the relationships between speed achieved and distance traveled on different grades by a typical heavy truck of 200 lb/hp during maximum acceleration. *Note:* grade resistance = weight × grade, in decimal.

Rolling Resistance

There are forces within the vehicle itself that offer resistance to motion. These forces are due mainly to frictional effect on moving parts of the vehicle, but they also include the frictional slip between the pavement surface and the tires. The sum effect of these forces on motion is known as *rolling resistance*. The rolling resistance depends on the speed of the vehicle and the type of pavement. Rolling forces are relatively lower on smooth pavements than on rough pavements.

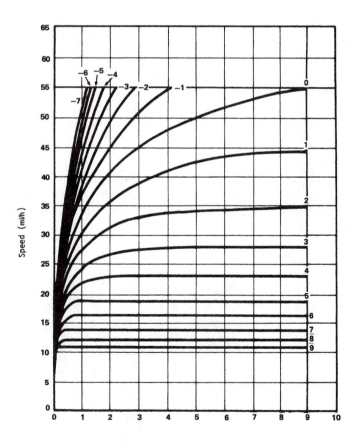

Figure 3.5 Speed Distance Curves for a Typical Heavy Truck of 200 lb/hp for Maximum Acceleration (on percent grades up and down)

SOURCE: *A Policy on Geometric Design of Highways and Streets,* American Association of State Highway and Transportation Officials, Washington, D.C., 2001.

The rolling resistance force for passenger cars on a smooth pavement can be determined from the relation

$$R_r = (C_{rs} + 2.15C_{rv} \, u^2)W \qquad (3.11)$$

where

R_r = rolling resistance force (lb)
C_{rs} = constant (typically 0.012 for passenger cars)
C_{rv} = constant (typically 0.65×10^{-6} sec^2/ft^2 for passenger cars)
u = vehicle speed (mi/h)
W = gross vehicle weight (lb)

For trucks, the rolling resistance can be obtained from

$$R_r = (C_a + 1.47C_b \, u)W \qquad (3.12)$$

where

R_r = rolling resistance force (lb)
C_a = constant (typically 0.02445 for trucks)
C_b = constant (typically 0.00044 sec/ft for trucks)
u = vehicle speed (mi/h)
W = gross vehicle weight

The surface condition of the pavement has a significant effect on the rolling resistance. For example, at a speed of 50 mi/h on a badly broken and patched asphalt surface, the rolling resistance is 51 lb/ton of weight, whereas at the same speed on a loose sand surface, the rolling resistance is 76 lb/ton of weight.

Curve Resistance

When a passenger car is maneuvered to take a curve, external forces act on the front wheels of the vehicle. These forces have components that have a retarding effect on the forward motion of the vehicle. The sum effect of these components constitutes the curve resistance. This resistance depends on the radius of the curve, the gross weight of the vehicle, and the velocity at which the vehicle is moving. It can be determined as

$$R_c = 0.5 \frac{(2.15 \, u^2 \, W)}{gR} \qquad (3.13)$$

where

R_c = curve resistance (lb)
u = vehicle speed (mi/h)
W = gross vehicle weight (lb)
g = acceleration of gravity (32.2 ft/sec^2)
R = radius of curvature (ft)

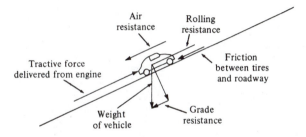

Figure 3.6 Forces Acting on a Moving Vehicle

SOURCE: Redrawn from D. R. Drew, *Traffic Flow Theory and Control,* McGraw-Hill, 1968.

Power Requirements

Power is the rate at which work is done. It is usually expressed in horsepower (a U.S. unit of measure), where 1 horsepower is 550 lb-ft/sec. The performance capability of a vehicle is measured in terms of the horsepower the engine can produce to overcome air, grade, curve, and friction resistance forces and put the vehicle in motion. Figure 3.6 shows how these forces act on the moving vehicle. The power delivered by the engine is

$$P = \frac{1.47\ Ru}{550} \tag{3.14}$$

where

P = horsepower delivered (hp)
R = sum of resistance to motion (lb)
u = speed of vehicle (mi/h)

Example 3.3   Vehicle Horsepower Required to Overcome Resistance Forces

Determine the horsepower produced by a passenger car traveling at a speed of 65 mi/h on a straight road of 5 percent grade with a smooth pavement. Assume the weight of the car is 4000 lb and the cross-sectional area of the car is 40 ft².

Solution: The force produced by the car should be at least equal to the sum of the acting resistance forces:

$$R = (\text{air resistance}) + (\text{rolling resistance}) + (\text{upgrade resistance})$$

Note: There is no curve resistance since the road is straight.

- Use Eq. 3.10 to determine air resistance:

$$R_a = 0.5 \left(\frac{2.15\, p C_D A u^2}{g} \right)$$

$$= 0.5\, \frac{2.15 \times 0.0766 \times 0.4 \times 40 \times 65 \times 65}{32.2}$$

$$= 172.9 \text{ lb}$$

- Use Eq. 3.11 to determine rolling resistance:

$$R_r = (C_{rs} + 2.15 C_{rv}\, u^2)(4000)$$
$$= (0.012 + 2.15 \times 0.65 \times 10^{-6} \times 65 \times 65)(4000)$$
$$= (0.012 + 0.006)(4000)$$
$$= 0.018 \times 4000$$
$$= 72 \text{ lb}$$
$$\text{grade resistance} = 4000 \times \frac{5}{100} = 200 \text{ lb}$$

- Determine total resistance:

$$R = R_a + R_r + \text{grade resistance} = 172.9 + 72 + 200 = 444.9 \text{ lb}$$

- Use Eq. 3.14 to determine horsepower produced:

$$P = \frac{1.47\, R u}{550}$$

$$= \frac{1.47 \times 444.9 \times 65}{550}$$

$$= 77.3 \text{ hp}$$

Braking Distance

The action of the forces (shown in Figure 3.6) on the moving vehicle and the effect of perception-reaction time are used to determine important parameters related to the dynamic characteristics of the vehicles. These include the braking distance of a vehicle and the minimum radius of a circular curve required for a vehicle traveling around a curve with speed u where $u > 10$ mi/h. Also, relationships among elements, such as the acceleration, the coefficient of friction between the tire and the pavement, the distance above ground of the center of gravity of the vehicle, and the track width of the vehicle, could be developed by analyzing the action of these forces.

Braking

Consider a vehicle traveling downhill with an initial velocity of u, in mi/h, as shown in Figure 3.7. Let

W = weight of the vehicle
f = coefficient of friction between the tires and the road pavement
γ = angle between the grade and the horizontal
a = deceleration of the vehicle when the brakes are applied
D_b = horizontal component of distance traveled during braking (that is, from time brakes are applied to time the vehicle comes to rest)

Note that the distance referred to as the braking distance is the horizontal distance and not the inclined distance x. The reason is that measurements of distances in surveying are horizontal and, therefore, distance measurements in highway design are always made with respect to the horizontal plane. Since the braking distance is an input in the design of highway curves, the horizontal component of the distance x is used:

$$\text{Frictional force on the vehicle} = Wf \cos \gamma$$

The force acting on the vehicle due to deceleration is Wa/g, where g is acceleration due to gravity. The component of the weight of the vehicle is $W \sin \gamma$. Substituting into $\Sigma f = ma$, we obtain

$$W \sin \gamma - Wf \cos \gamma = \frac{Wa}{g} \tag{3.15}$$

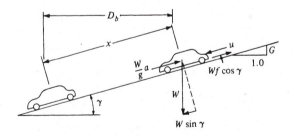

W = weight of vehicle
f = coefficient of friction
g = acceleration of gravity
a = vehicle acceleration
u = speed when brakes applied
D_b = braking distance
γ = angle of incline
G = tan γ (% grade/100)
x = distance traveled by the vehicle along the road during braking

Note: The absolute value of G used as the downward slope is already taken into consideration in developing the equation.

Figure 3.7 Forces Acting on a Vehicle Braking on a Downgrade

The deceleration that brings the vehicle to a stationary position can be found in terms of the initial velocity u as $a = -u^2/2x$ (assuming uniform deceleration), where x is the distance traveled in the plane of the grade during braking. Equation 3.15 can then be written as

$$W \sin \gamma - Wf \cos \gamma = -\frac{Wu^2}{2gx} \tag{3.16}$$

However, $D_b = x \cos \gamma$, and we therefore obtain

$$\frac{Wu^2}{2gD_b} \cos \gamma = Wf \cos \gamma - W \sin \gamma$$

giving

$$\frac{u^2}{2gD_b} = f - \tan \gamma$$

and

$$D_b = \frac{u^2}{2g(f - \tan \gamma)} \tag{3.17}$$

Note, however, that $\tan \gamma$ is the grade G of the incline (that is, percent of grade/100), as shown in Figure 3.6.

Equation 3.17 can therefore be written as

$$D_b = \frac{u^2}{2g(f - G)} \tag{3.18}$$

If g is taken as 32.2 ft/sec^2 and u is expressed in mi/h, Eq. 3.18 becomes

$$D_b = \frac{u^2}{30(f - G)} \tag{3.19}$$

and D_b is given in feet.

A similar equation could be developed for a vehicle traveling uphill, in which case the following equation is obtained:

$$D_b = \frac{u^2}{30(f + G)} \tag{3.20}$$

A general equation for the braking distance can therefore be written as

$$D_b = \frac{u^2}{30(f \pm G)} \tag{3.21}$$

where the plus sign is for vehicles traveling uphill, the minus sign is for vehicles traveling downhill, and G is the absolute value of tan γ.

AASHTO represents the friction coefficient as a/g, and noted that to ensure that the pavement will have and maintain these coefficients, it should be designed to meet the criteria established in the AASHTO *Guidelines for Skid Resistance Pavement Design.* These include guidelines on the selection, quality, and testing of aggregates. AASHTO also recommends that a deceleration rate of 11.2 ft/sec^2 be used, as this is a comfortable deceleration rate for most drivers. This rate is further justified because many studies have shown that when most drivers need to stop in an emergency, the rate of deceleration is greater than 14.8 ft/sec^2. Equation 3.21 then becomes

$$D_b = \frac{u^2}{30\left(\dfrac{a}{g} \pm G\right)} \tag{3.22}$$

Similarly, it can be shown that the horizontal distance traveled in reducing the speed of a vehicle from u_1 to u_2 in mi/h during a braking maneuver is given by

$$D_b = \frac{u_1^2 - u_2^2}{30\left(\dfrac{a}{g} \pm G\right)} \tag{3.23}$$

The distance traveled by a vehicle between the time the driver observes an object in the vehicle's path and the time the vehicle actually comes to rest is longer than the braking distance, since it includes the distance traveled during perception-reaction time. This distance is referred to in this text as the *stopping sight distance S* and is given as

$$S\text{ (ft)} = 1.47ut + \frac{u^2}{30\left(\dfrac{a}{g} \pm G\right)} \tag{3.24}$$

where the first term in Eq. 3.24 is the distance traveled during the perception-reaction time t (sec) and u is the velocity in mi/h at which the vehicle was traveling when the brakes were applied.

Example 3.4   Determining Braking Distance

A student trying to test the braking ability of her car determined that she needed 18.5 ft more to stop her car when driving downhill on a road segment of 5 percent grade than when driving downhill at the same speed along another segment of 3 percent grade. Determine the speed at which the student conducted her test and the braking distance on the 5 percent grade if the student is traveling at the test speed in the uphill direction.

Solution:

- Let x = downhill braking distance on 5 percent grade
- $(x - 18.5)$ ft = downhill braking distance on 3 percent grade
- Use Eq. 3.22:

$$D_b = \frac{u^2}{30\left(\dfrac{a}{g} - G\right)}$$

$$x = \frac{u^2}{30\left(\dfrac{a}{g} - 0.05\right)}$$

$$x - 18.5 = \frac{u^2}{30\left(\dfrac{a}{g} - 0.03\right)}$$

$$18.5 = \frac{u^2}{30\left(\dfrac{a}{g} - 0.05\right)} - \frac{u^2}{30\left(\dfrac{a}{g} - 0.03\right)}$$

- Determine u, the test velocity. Using $a = 11.2$ ft/sec^2 and $g = 32.2$ ft/sec^2 yields

$$\frac{a}{g} = \frac{11.2}{32.2} = 0.35$$

$$\frac{u^2}{30(0.35 - 0.05)} - \frac{u^2}{30(0.35 - 0.03)} = 18.5$$

which gives

$$0.111u^2 - 0.104u^2 = 18.5$$

from which we obtain, $u = 51.4$ mi/h. The test velocity is therefore 51.4 mi/h.

- Determine braking distance on downhill grade (use Eq. 3.20):

$$D_b = \frac{51.4^2}{30(0.35 - 0.05)} = 291.6 \text{ ft}$$

Example 3.5 Exit Ramp Stopping Distance

A motorist traveling at 65 mi/h on an expressway intends to leave the expressway using an exit ramp with a maximum speed of 35 mi/h. At what point on the expressway should the motorist step on her brakes in order to reduce her speed to the maximum allowable on the ramp just before entering the ramp, if this section of the expressway has a downgrade of 3 percent?

Solution: Use Eq. 3.23:

$$D_b = \frac{u_1^2 - u_2^2}{30\left(\dfrac{a}{g} - 0.03\right)}$$

$$a/g = 11.2/32.2 = 0.35$$

$$D_b = \frac{65^2 - 35^2}{30(0.35 - 0.03)} = 312.5 \text{ ft}$$

The brakes should be applied at least 312.5 ft from the ramp.

Example 3.6 Distance Required to Stop for an Obstacle in the Roadway

A motorist traveling at 55 mi/h down a grade of 5 percent on a highway observes an accident ahead of him, involving an overturned truck that is completely blocking the road. If the motorist was able to stop his vehicle 30 ft from the overturned truck, what was his distance from the truck when he first observed the accident? Assume perception-reaction time = 2.5 sec.

Solution:

- Use Equation 3.24 to obtain the stopping distance:

$$\frac{a}{g} = 0.35$$

$$S = 1.47ut + \frac{u^2}{30(0.35 - 0.05)}$$

$$= 1.47 \times 55 \times 2.5 + \frac{55^2}{30 \times 0.30}$$

$$= 202.13 + 336.11$$

$$= 538.2 \text{ ft}$$

- Find the distance of the motorist when he first observed the accident:

$$S + 30 = 568.2 \text{ ft}$$

Estimate of Velocities.   It is sometimes necessary to estimate the speed of a vehicle just before it is involved in an accident. This may be done by using the braking-distance equations if skid marks can be seen on the pavement. The steps taken in making the speed estimate are as follows:

Step 1.   Measure the length of the skid marks for each tire and determine the average. The result is assumed to be the braking distance D_b of the vehicle.

Step 2.   Determine the coefficient of friction f by performing trial runs at the site under similar weather conditions, using vehicles whose tires are in a state similar to that of the tires of the vehicle involved in the accident. This is done by driving the vehicle at a known speed u_k and measuring the distance traveled D_k while braking the vehicle to rest. The coefficient of friction f_k can then be estimated by using Eq. 3.21:

$$f_k = \frac{u_k^2}{30D_k} \mp G$$

Alternatively, a value of 0.35 for a/g can be used for f_k.

Step 3.   Use the value of f_k obtained in step 2 to estimate the unknown velocity u_u just prior to impact; that is, the velocity at which the vehicle was traveling just before it was involved in the accident. This is done by using Eq. 3.23. If it can be assumed that the application of the brakes reduced the velocity u_u to zero, then u_u may be obtained from

$$D_b = \frac{u_u^2}{30(f_k \pm G)}$$

or

$$D_b = \frac{u_u^2}{30\left(\dfrac{u_k^2}{30D_k} \mp G \pm G\right)} = \left(\frac{u_u^2}{u_k^2}\right) D_k \tag{3.25}$$

giving

$$u_u = \left(\frac{D_b}{D_k}\right)^{1/2} u_k \tag{3.26}$$

However, if the vehicle involved in the accident was traveling at speed u_1 when the impact took place and the speed u_1 is known, then using Eq. 3.23, the unknown speed u_u just prior to the impact may be obtained from

$$D_b = \frac{u_u^2 - u_1^2}{30\left(\dfrac{u_k^2}{30D_k} \mp G \pm G\right)} = \frac{u_u^2 - u_1^2}{u_k^2} D_k$$

giving

$$u_u = \left(\frac{D_b}{D_k} u_k^2 + u_1^2 \right)^{1/2} \tag{3.27}$$

Note that the unknown velocity just before the impact, obtained from either Eq. 3.26 or Eq. 3.27, is only an estimate, but it will always be a conservative estimate because it will always be less than the actual speed at which the vehicle was traveling before impact. The reason is that some reduction of speed usually takes place before skidding commences, and in using Eq. 3.26, the assumption that the initial velocity u_u is reduced to zero is always incorrect. The lengths of the measured skid marks do not reflect these factors.

Example 3.7 Estimating the Speed of a Vehicle from Skid Marks

In an attempt to estimate the speed of a vehicle just before it hit a traffic signal pole, a traffic engineer measured the length of the skid marks made by the vehicle and performed trial runs at the site to obtain an estimate of the coefficient of friction. Determine the estimated unknown velocity if the following data were obtained:

Length of skid marks	= 585 ft, 590 ft, 580 ft, 595 ft
Speed of trial run	= 30 mi/h
Distance traveled during trial run	= 300 ft

Examination of the vehicle just after the crash indicated that the speed of impact was 35 mi/h.

Solution:

$$\text{Average length of skid marks} = \frac{585 + 590 + 580 + 595}{4}$$

$$= 587.5 \text{ ft}$$

(This is assumed to be the braking distance D_b.)

Use Eq. 3.27 to determine the unknown velocity:

$$u_u = \left(\frac{D_b}{D_k} u_k^2 + u_1^2 \right)^{1/2}$$

$$= \left(\frac{587.5}{300} 30^2 + 1225 \right)^{1/2}$$

$$= (1762.5 + 1225)^{1/2}$$

$$= 54.66 \text{ mi/h}$$

Minimum Radius of a Circular Curve

When a vehicle is moving around a circular curve, there is an inward radial force acting on the vehicle, usually referred to as the centrifugal force. There is also an outward radial force imagined by the driver as a result of the centripetal acceleration acting toward the center of curvature. In order to balance the effect of the centripetal acceleration, the road is inclined toward the center of the curve. The inclination of the roadway toward the center of the curve is known as *superelevation*. The centripetal acceleration depends on the component of the vehicle's weight along the inclined surface of the road and the side friction between the tires and the roadway. The action of these forces on a vehicle moving around a circular curve is shown in Figure 3.8.

The minimum radius of a circular curve R for a vehicle traveling at u mi/h can be determined by considering the equilibrium of the vehicle with respect to its moving up or down the incline. If α is the angle of inclination of the highway, the component of the weight down the incline is $W \sin \alpha$, and the frictional force also acting down the incline is $Wf \cos \alpha$. The centrifugal force F_c is

$$F_c = \frac{Wa_c}{g}$$

where

a_c = acceleration for curvilinear motion = u^2/R (R = radius of the curve)
W = weight of the vehicle
g = acceleration of gravity

When the vehicle is in equilibrium with respect to the incline (that is, the vehicle moves forward but neither up nor down the incline), we may equate the three relevant forces and obtain

$$\frac{Wu^2}{gR} \cos \alpha = W \sin \alpha + Wf_s \cos \alpha \qquad (3.28)$$

where f_s = coefficient of side friction and $(u^2/g) = R(\tan \alpha + f_s)$, which gives

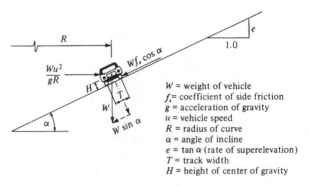

Figure 3.8 Forces Acting on a Vehicle Traveling on a Horizontal Curve Section of a Road

$$R = \frac{u^2}{g(\tan \alpha + f_s)} \tag{3.29}$$

Tan α, the tangent of the angle of inclination of the roadway, is known as the *rate of superelevation e.* Equation 3.29 can therefore be written as

$$R = \frac{u^2}{g(e + f_s)} \tag{3.30}$$

Again, if g is taken as 32.2 ft/sec² and u is measured in mi/h, the minimum radius R is given in feet as

$$R = \frac{u^2}{15(e + f_s)} \tag{3.31}$$

Equation 3.31 shows that to reduce R for a given velocity, either e or f_s or both should be increased.

There are, however, stipulated maximum values that should be used for either e or f_s. Several factors control the maximum value for the rate of superelevation. These include the location of the highway (that is, whether it is in an urban or rural area), weather conditions (such as the occurrence of snow), and the distribution of slow-moving traffic within the traffic stream. For highways located in rural areas with no snow or ice, a maximum superelevation rate of 0.10 is generally used. For highways located in areas with snow and ice, values ranging from 0.08 to 0.10 are used. For expressways in urban areas, a maximum superelevation rate of 0.08 is used. Because of the relatively low speeds on local urban roads, these roads are usually not superelevated.

The values used for side friction f_s generally vary with the design speed and the superelevation. Table 3.3 gives values recommended by AASHTO for use in design.

Table 3.3 Coefficient of Side Friction for Different Design Speeds

Design Speed (mi/h)	Coefficients of Side Friction, f_s★
30	0.16
40	0.15
50	0.14
60	0.12
70	0.10

★Limiting values for superelevation rates of .008 and 0.10.
SOURCE: Adapted from *A Policy on Geometric Design of Highways and Streets,* American Association of State Highway and Transportation Officials, Washington, D.C., 2000. Used by permission.

Example 3.8   Minimum Radius of a Highway Horizontal Curve

An existing horizontal curve on a highway has a radius of 465 ft, which restricts the posted speed limit on this section of the road to only 61.5 percent of the design speed of the highway. If the curve is to be improved so that its posted speed will be the design speed of the highway, determine the minimum radius of the new curve. Assume that the rate of superelevation is 0.08 for both the existing curve and the new curve to be designed.

Solution:

- Use Eq. 3.31 to find the posted speed limit on the existing curve. Since the posted speed limit is not known, assume $f_s = 0.15$:

$$R = \frac{u^2}{15(e + f_s)}$$

$$465 = \frac{u^2}{15(0.08 + 0.15)}$$

$$u = 40.05 \text{ mi/h}$$

- The posted speed limit is 40 mi/h, as speed limits are usually posted at intervals of 5 mi/h
- Check assumed f_s for 40 mi/h = 0.15
- Determine the design speed of the highway:

$$\frac{40}{0.615} = 65.04 \text{ mi/h}$$

- Design speed = 65 mi/h
- Find the radius of the new curve by using Eq. 3.31 with the value of f_s for 65 mi/h from Table 3.3 ($f_s = 0.11$):

$$R = \frac{65^2}{15(0.08 + f_s)}$$

$$= \frac{(65)^2}{15(0.08 + 0.11)} = 1482.45 \text{ ft}$$

ROAD CHARACTERISTICS

The characteristics of the highway discussed in this section are related to stopping and passing because these have a more direct relationship to the characteristics of the driver and the vehicle discussed earlier. This material, together with other characteristics of the highway, will be used in Chapter 16, where geometric design of the highway is discussed.

Sight Distance

Sight distance is the length of the roadway a driver can see ahead at any particular time. The sight distance available at each point of the highway must be such that when a driver is traveling at the highway's design speed, adequate time is given, after an object is observed in the vehicle's path, to make the necessary evasive maneuvers without colliding with the object. The two types of sight distance are (1) stopping sight distance and (2) passing sight distance.

Stopping Sight Distance

The *stopping sight distance* (SSD), for design purposes, is usually taken as the minimum sight distance required for a driver to stop a vehicle after seeing an object in the vehicle's path without hitting that object. This distance is the sum of the distance traveled during perception-reaction time and the distance traveled during braking. The SSD for a vehicle traveling at *u* mi/h is therefore the same as the stopping distance given in Eq. 3.24. The SSD is therefore

$$SSD = 1.47ut + \frac{u^2}{30\left(\dfrac{a}{g} \pm G\right)} \tag{3.32}$$

It is essential that highways be designed such that sight distance along the highway is at least equal to the SSD. Table 3.4 shows SSDs for different design speeds. The SSD requirements dictate the minimum lengths of vertical curves and minimum radii for horizontal curves that should be designed for any given highway. It should be noted that the values given for SSD in Table 3.4 are for horizontal alignment and the grade is zero. On upgrades, the SSDs are shorter; on downgrades, they are longer (see Eq. 3.32).

Decision Sight Distance

The SSDs given in Table 3.4 are usually adequate for ordinary conditions, when the stimulus is expected by the driver. However, when the stimulus is unexpected or when it is necessary for the driver to make unusual maneuvers, longer SSDs are usually required, since the perception-reaction time is much longer. This longer sight distance is the *decision sight distance;* it is defined by AASHTO as the "distance required for a driver to detect an unexpected or otherwise difficult-to-perceive information source or hazard in a roadway environment that may be visually cluttered, recognize the hazard of its threat potential, select an appropriate speed and path, and initiate and complete the required safety maneuvers safely and efficiently.

The decision sight distances depend on the type of maneuver required to avoid the hazard on the road, and also on whether the road is located in a rural or in an urban area. Table 3.5 gives AASHTO's recommended decision sight distance values for different avoidance maneuvers, which can be used for design.

Passing Sight Distance

The *passing sight distance* is the minimum sight distance required on a two-lane, two-way highway that will permit a driver to complete a passing maneuver without colliding with an

Table 3.4 SSDs for Different Design Speeds

		US Customary		
			Stopping Sight Distance	
Design Speed (mi/h)	Brake Reaction Distance (ft)	Braking Distance on Level (ft)	Calculated (ft)	Design
15	55.1	21.6	76.7	80
20	73.5	38.4	111.9	115
25	91.9	60.0	151.9	155
30	110.3	86.4	196.7	200
35	128.6	117.6	246.2	250
40	147.0	153.6	300.6	305
45	165.4	194.4	569.8	360
50	183.8	240.0	423.8	425
55	202.1	290.3	492.4	495
60	220.5	345.5	566.0	570
65	238.9	405.5	644.4	645
70	257.3	470.3	727.6	730
75	275.6	539.9	815.5	820
80	294.0	614.3	908.3	910

Note: brake reaction distance predicated on a time of 2.5s; deceleration rate of 11.2 ft/s^2 used to determine calculated sight distance.
SOURCE: Adapted from *A Policy on the Geometric Design of Highways and Streets,* American Association of State Highway and Transportation Officials, Washington, D.C., 2001. Used by permission.

opposing vehicle and without cutting off the passed vehicle. The passing sight distance will also allow the driver to successfully abort the passing maneuver (that is, return to the right lane behind the vehicle being passed) if he or she so desires. In determining minimum passing sight distances for design purposes, only single passes (that is, a single vehicle passing a single vehicle) are considered. Although it is possible for multiple passing maneuvers (that is, more than one vehicle pass or are passed in one maneuver) to occur, it is not practical for minimum design criteria to be based on them.

In order to determine the minimum passing sight distance, certain assumptions have to be made regarding the movement of the passing vehicle during a passing maneuver:

1. The vehicle being passed (impeder) is traveling at a uniform speed.
2. The speed of the passing vehicle is reduced and it is behind the impeder as the passing section is entered.
3. On arrival at a passing section, some time elapses during which the driver decides whether to undertake the passing maneuver.
4. If the decision is made to pass, the passing vehicle is accelerated during the passing maneuver, and the average passing speed is about 10 mi/h more than the speed of the impeder vehicle.

Table 3.5 Decision Sight Distances for Different Design Speeds and Avoidance Maneuvers

	US Customary				
	Decision Sight Distance (ft)				
Design	*Avoidance Maneuver*				
Speed (mi/h)	*A*	*B*	*C*	*D*	*E*
30	220	490	450	535	620
35	275	590	525	625	720
40	330	690	600	715	825
45	395	800	675	800	930
50	465	910	750	890	1030
55	535	1030	865	980	1135
60	610	1150	990	1125	1280
65	695	1275	1050	1220	1365
70	780	1410	1105	1275	1445
75	875	1545	1180	1365	1545
80	970	1685	1260	1455	1650

Note: brake reaction distance predicted on a time of 2.5 sec; deceleration rate of 11.2 ft/sec^2 used to determine calculated sight distance.
Avoidance Maneuver A: Stop on rural road—$t = 3.0$s
Avoidance Maneuver B: Stop on urban road—$t = 9.1$ s
Avoidance Maneuver C: Speed/path/direction change on rural road—t varies between 10.2 and 11.2 s
Avoidance Maneuver D: Speed/path/direction change on suburban road—t varies between 12.1 and 12.9 s
Avoidance Maneuver E: Speed/path/direction change on urban road—t varies between 14.0 and 14.5 s
SOURCE: Adapted from *A Policy on the Geometric Design of Highways and Streets,* American Association of State Highway and Transportation Officials, Washington, D.C., 2001. Used by permission.

5. A suitable clearance exists between the passing vehicle and any opposing vehicle when the passing vehicle reenters the right lane.

These assumptions have been used by AASHTO to develop a minimum passing sight distance requirement for two-lane, two-way highways.

The minimum passing sight distance is the total of four components, as shown in Figure 3.9:

d_1 = distance traversed during perception-reaction time and during initial acceleration to the point where the passing vehicle just enters the left lane

d_2 = distance traveled during the time the passing vehicle is traveling in the left lane

d_3 = distance between the passing vehicle and the opposing vehicle at the end of the passing maneuver

d_4 = distance moved by the opposing vehicle during two thirds of the time the passing vehicle is in the left lane (usually taken to be $\frac{2}{3} d_2$)

The distance d_1 is obtained from the expression

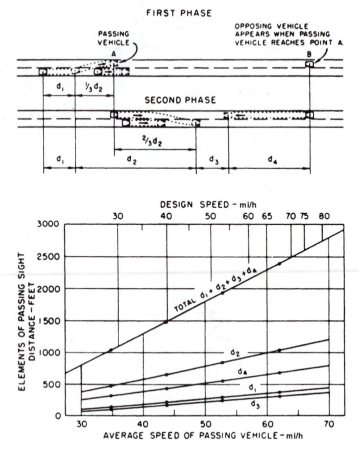

Figure 3.9 Elements of and Total Passing Sight Distance on Two-Lane Highways

SOURCE: *A Policy on Geometric Design of Highways and Streets,* American Association of State Highway and Transportation Officials, Washignton, D.C., 2001. Used by permission.

$$d_1 = 1.47t_1\left(u - m + \frac{at_1}{2}\right) \tag{3.33}$$

where

t_1 = time for initial maneuver (sec)
a = average acceleration rate (mi/h/sec)
u = average speed of passing vehicle (mi/h)
m = difference in speeds of passing and impeder vehicles

The distance d_2 is obtained from

$$d_2 = 1.47\, ut_2$$

where

t_2 = time passing vehicle is traveling in left lane (sec)

u = average speed of passing vehicle (mi/h)

The clearance distance d_3 between the passing vehicle and the opposing vehicle at the completion of the passing maneuver has been found to vary between 100 ft and 300 ft.

Table 3.6 shows these components calculated for different speeds. It should be made clear that values given in Table 3.6 are for design purposes and cannot be used for marking passing and no-passing zones on completed highways. The values used for that purpose are obtained from different assumptions and are much shorter. Table 3.7 shows values recommended for that purpose in the *Manual of Uniform Control Devices*. However, recent studies have shown that these values are inadequate. Table 3.7 also shows minimum sight distance requirements obtained from a study on two-lane, two-way roads in mountainous areas of Virginia.

SUMMARY

The highway or traffic engineer needs to study and understand the fundamental elements that are important in the design of traffic control systems. This chapter has presented concisely the basic characteristics of the driver, the pedestrian, the bicyclist, the vehicle, and the road that should be known and understood by transportation and/or traffic engineers.

Table 3.6 Components of Safe Passing Sight Distance on Two-Lane Highways

	Speed Range in mi/h (Average Passing Speed in mi/h)			
Component	30–40 (34.9)	40–50 (43.8)	50–60 (52.6)	60–70 (62.0)
Initial maneuever:				
a = average acceleration (mi/h/sec)[a]	1.40	1.43	1.47	1.50
t_1 = time (sec)[a]	3.6	4.0	4.3	4.5
d_1 = distance traveled (ft)	145	216	289	366
Occupation of left lane:				
t_2 = time (sec)[a]	9.3	10.0	10.7	11.3
d_2 = distance traveled (ft)	477	643	827	1030
Clearance length:				
d_3 = distance traveled (ft)[a]	100	180	250	300
Opposing vehicle:				
d_4 = distance traveled (ft)	318	429	552	687
Total distance, $d_1 + d_2 + d_3 + d_4$ (ft)	1040	1468	1918	2383

a. For consistent speed relation, observed values are adjusted slightly.
SOURCE: Adapted from *A Policy on Geometric Design of Highways and Streets,* American Association of State Highway and Transportation Officials, Washington, D.C., 2001. Used by permission.

Table 3.7 Suggested Minimum Passing Zone and Passing Sight Distance Requirements for Two-Lane, Two-Way Highways in Mountainous Areas

85th-Percentile Speed (mi/h)	Available Sight Distance (ft)	Minimum Passing Zone		Minimum Passing Sight Distance	
		Suggested (ft)	MUTCD* (ft)	Suggested (ft)	MUTCD* (ft)
30	600–800	490		630	
	800–1000	530		690	
	1000–1200	580	400	750	500
	1200–1400	620		810	
35	600–800	520		700	
	800–1000	560		760	
	1000–1200	610	400	820	550
	1200–1400	650		880	
40	600–800	540		770	
	800–1000	590		830	
	1000–1200	630	400	890	600
	1200–1400	680		950	
45	600–800	570		840	
	800–1000	610		900	
	1000–1200	660	400	960	700
	1200–1400	700		1020	
50	600–800	590		910	
	800–1000	630		970	
	1000–1200	680	400	1030	800
	1200–1400	730		1090	

Manual on Uniform Traffic Control Devices, published by FHWA.
SOURCE: Adapted from N. J. Garber and M. Saito, *Centerline Pavement Markings on Two-Lane Mountainous Highways,* Research Report No. VHTRC 84-R8, Virginia Highway and Transportation Research Council, Charlottesville, Va., March 1983.

The most important characteristic of the driver is the driver response process, which consists of four subprocesses: perception, identification, emotion, and reaction or volition. Each of these subprocesses requires time to complete, the sum of which is known as the perception-reaction time of the driver. The actual distance a vehicle travels before coming to rest is the sum of the distance traveled during the perception-reaction time of the driver and the distance traveled during the actual braking maneuver. Perception-reaction times vary from one person to another, but the recommended value for design is 2.5 seconds. The static, kinematic, and dynamic characteristics of the vehicle are *also* important because they are used to determine minimum radii of horizontal curves for low speeds ($u < 10$ mi/h), the acceleration and deceleration capabilities of the vehicle (through which distance traveled and velocities attained can be determined), and the resistance forces that act on the moving vehicle. The characteristic of the road that has a direct relationship to the characteristics of the driver is the sight distance on the road. Two types of sight distances are considered to be part of the characteristics of the road: the stopping sight distance, which is normally taken as the minimum sight distance required for a driver to stop a vehicle after seeing an object in the vehicle's path without hitting that object, and the passing sight distance, which is the minimum sight distance required on a two-lane, two-way highway that will permit a driver to complete a passing maneuver without colliding with an opposing vehicle and without cutting off the passed vehicle.

Although these characteristics are presented in terms of the highway mode, several of these are also used for other modes. For example, the driver and pedestrian characteristics also apply to other modes in which vehicles are manually driven and some possibility exists for interaction between the vehicle and pedestrians. It should be emphasized again that because of the wide range of capabilities among drivers and pedestrians, the use of average limitations of drivers and pedestrians in developing guidelines for design may result in the design of facilities that will not satisfy a significant percentage of the people using the facility. High-percentile values (such as 85th- or 95th-percentile values) are therefore normally used for design purposes.

PROBLEMS

3-1   Briefly describe the two types of visual acuity.

3-2   **(a)** What color combinations are used for regulatory signs (e.g., speed limit signs) and general warning signs (e.g., advance railroad crossing signs).
(b) Why are these combinations used?

3-3   Determine your average walking speed. Compare your results with that of the suggested walking speed in the *Manual of Uniform Traffic Control Devices*. Is there a difference? Which value is more conservative and why?

3-4   Describe the three types of vehicle characteristics.

3-5   The design speed of a multilane highway is 60 mi/h, what is the minimum stopping sight distance that should be provided on the road if (a) the road is level and (b) the road has a maximum grade of 4 percent? Assume the perception-reaction time = 2.5 sec.

3-6   The acceleration of a vehicle takes the form

$$\frac{du}{dt} = 3.6 - 0.06u$$

where u is the vehicle speed in ft/sec. The vehicle is traveling at 45 ft/sec at time T_o.

(a)   Determine the distance traveled by the vehicle when accelerated to a speed of 55 ft/sec.
(b)   Determine the time at which the vehicle attains the speed of 55 ft/sec.
(c)   Determine the acceleration of the vehicle after 3 sec.

3-7   The driver of a vehicle on a level road determined that she could increase her speed from rest to 50 mi/h in 34.8 seconds and from rest to 65 mi/h in 94.8 seconds. If it can be assumed that the acceleration of the vehicle takes the form

$$\frac{du}{dt} = \alpha - \beta u_t$$

determine the maximum acceleration in each case.

3-8   A 2500-pound passenger vehicle originally traveling on a straight and level road gets onto a section of the road with a horizontal curve of Radius=850 ft. If the vehicle was originally traveling at 55 mi/h, determine (a) the additional horsepower on the curve the vehicle must produce to maintain the original speed, (b) the total resistance force on the vehicle as it traverses the horizontal curve and the total horsepower. Assume that the vehicle is travel-ing at sea level and has a front cross-sectional area of 30 ft^2.

3-9   A horizontal curve is to be designed for a section of a highway having a design speed of 60 mi/h.

(a) If the physical conditions restrict the radius of the curve to 500 ft, what value is required for the superelevation at this curve?

(b) Is this a good design?

3-10   Determine the minimum radius of a horizontal curve required for a highway if the design speed is 70 mi/h and the superelevation rate is 0.08.

3-11   What is the distance required to stop an average passenger car when brakes are applied on a 2 percent downgrade if that vehicle was originally traveling at 40 mi/h?

3-12   The radius of a horizontal curve on an existing highway is 750 ft. The superelevation rate at the curve is 0.08 and the posted speed limit on the road is 65 mi/h. Is this a hazardous loca-tion? If so, why? What action will you recommend to correct the situation?

3-13   A curve of radius 250 ft and $e = 0.08$ is located at a section of an existing rural highway, which restricts the safe speed at this section of the highway to 50 percent of the design speed. This drastic reduction of safe speed resulted in a high accident rate at this section. To reduce the accident rate, a new alignment is to be designed with a horizontal curve. Determine the minimum radius of this curve if the safe speed should be increased to the design speed of the highway. Assume $f_s = 0.17$ for the existing curve, and the new curve is to be designed with $e = 0.08$.

3-14   A temporary diversion has been constructed on a highway of +4 percent gradient due to major repairs that are being undertaken on a bridge. The maximum speed allowed on the diversion is 10 mi/h. Determine the minimum distance from the diversion that a road sign should be located informing drivers of the temporary change on the highway.

Maximum allowable speed on highway = 70 mi/h
Letter height of road sign = 4 in.
Perception-reaction time = 2.5 sec

Assume that a driver can read a road sign within his or her area of vision at a distance of 40 ft for each inch of letter height.

3-15   An elevated expressway goes through an urban area, and crosses a local street as shown in Figure 3.10. The partial cloverleaf exit ramp is on a 2 percent downgrade and all vehicles leaving the expressway must stop at the intersection with the local street. Determine (a) minimum ramp radius and (b) length of the ramp for the following conditions:

Maximum speed on expressway = 60 mi/h
Distance between exit sign and exit ramp = 260 ft
Letter height of road sign = 3 in.
Perception-reaction time = 2.5 sec

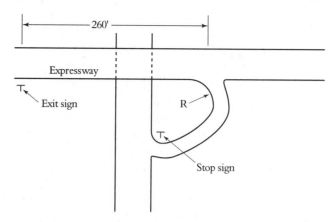

Figure 3.10 Layout of Elevated Expressway Ramp, and Local Street for Problem 3-15

Maximum superelevation $= 0.08$
Expressway grade $= 0\%$

Assume that a driver can read a road sign within his or her area of vision at a distance of 50 ft for each inch of letter height, and the driver sees the stop sign immediately on entering the ramp.

3-16 A section of highway has a superelevation of 0.05 and a curve with a radius of only 300 ft. What speed limit will you recommend at this section of the highway? Assume $f_s = 0.15$.

3-17 Calculate the minimum passing sight distance required for a two-lane rural roadway that has a posted speed limit of 45 mi/h. The local traffic engineer conducted a speed study of the subject road and found the following: average speed of the passing vehicle was 47 mi/h with an average acceleration of 1.43 mi/h/sec, and average speed of impeder vehicles was 40 mi/h.

REFERENCES

A Policy on Geometric Design of Highways and Streets, American Association of State Highway and Transportation Officials, Washington, D.C., 2001.

Claffey, P., *Running Costs of Motor Vehicles as Affected by Road Design and Traffic,* National Cooperative Highway Research Program Report III, Highway Research Board, Washington, D.C., 1971.

Fambro, D.B., et al., *Determination of Stopping Sight Distances,* NCHRP Report 400, Transportation Research Board, National Research Council, Washington, D.C., 1997.

Garber, N.J., and M. Saito, *Centerline Pavement Markings on Two-Lane Mountainous Highways,* Research Report No. VHTRC 84-R8, Virginia Highway and Transportation Research Council, Charlottesville, Va., March 1983.

Garber, N.J. and Subram Nyan, "Feasibility of Developing Congestion Mitigation Strategies that Incorporate Crash Risk. A Case Study: Hampton Roads Area." Virginia Transportation Research Council, 2001.

Guide for the Development of New Bicycle Facilities. American Association of State Highway and Transportation Officials, Washington, D.C., 1999.

Harwood, D.W., *Traffic and Operating Characteristics,* Institute of Transportation Engineers, Washington, D.C., 1992.

Highway Capacity Manual Transportation Research Board, National Research Council, Washington, D.C., 2000.

Highway Statistics 1996, Washington D.C., Federal Highway Administration, 1997.

Manual on Uniform Traffic Control Devices, U.S. Department of Transportation, Federal Highway Administration, Washington, D.C., 2000.

Peripheral Vision Horizon Display (PHVD), proceedings of a conference held at NASA Ames Research Center, Dryden Flight Research, March 15–16, 1983, National Aeronautics and Space Administration, Scientific and Technical Information Branch, 1984.

Rockwell, T.H. et al., Systems Research Group, Department of Industrial and Systems Engineering, Ohio State University, *The Utility of Peripheral Vision to Motor Drivers,* National Highway Traffic Safety Administration, Springfield, Va., 1977.

Rowland, G., E.S. Moretti, and M.L. Patton, *Evaluation of Glare from Following Vehicle's Headlights,* prepared for U.S. Department of Transportation, National Highway Traffic Safety Administration, Washington, D.C., 1981.

Triggs, T., and W.G. Harris, *Reaction Time of Drivers to Road Stimuli,* Human Factors Report HFR-12, Monash University, Clayton, Australia, 1982.

Twin Trailer Trucks: Effects on Highways and Highway Safety, National Research Council, Transportation Research Board, Washington, D.C., 1986.

Visual Characteristics of Navy Drivers, Naval Submarine Medical Research Laboratory, Groton, Conn., 1981.

ADDITIONAL READINGS

Hulbert, S.F., and Burg, A., "Human Factors in Transportation Systems," in *Systems Psychology,* Kenyon B. Greene, ed., McGraw-Hill Book Company, New York, 1970.

Shina, D., *Psychology on the Road: The Human Factor in Traffic Safety,* John Wiley & Sons, New York, 1977.

"Symposium on Geometric Design of Large Trucks," *Transportation Research Record* 1052, Transportation Research Board, National Research Council, Washington, D.C., 1986.

Ziedman, K., and Burger, W., "Effect of Ambient Lighting and Daytime Running Light (DRL) Intensity on Peripheral Detection of DRL," *Transportation Research Record* 1403, Transportation Research Board, National Research Council, Washington, D.C., 1993, 28–35.

CHAPTER 4

Traffic Engineering Studies

The availability of highway transportation has provided several advantages that contribute to a high standard of living. However, several problems related to the highway mode of transportation exist. These problems include highway-related crashes, parking difficulties, congestion, and delay. To reduce the negative impact of highways, it is necessary to adequately collect information that describes the extent of the problems and identifies their locations. Such information is usually collected by organizing and conducting traffic surveys and studies. This chapter introduces the reader to the different traffic engineering studies that are conducted to collect traffic data. Brief descriptions of the methods of collecting and analyzing the data are also included.

Traffic studies may be grouped into three main categories: (1) inventories, (2) administrative studies, and (3) dynamic studies. Inventories provide a list or graphic display of existing information, such as street widths, parking spaces, transit routes, traffic regulations, and so forth. Some inventories—for example, available parking spaces and traffic regulations—change frequently and therefore require periodic updating; others, such as street widths, do not.

Administrative studies use existing engineering records, available in government agencies and departments. This information is used to prepare an inventory of the relevant data. Inventories may be recorded in files but are usually recorded in automated data processing (ADP) systems. Administrative studies include the results of surveys, which may involve field measurements and/or aerial photography.

Dynamic traffic studies involve the collection of data under operational conditions and include studies of speed, traffic volume, travel time and delay, parking, and crashes. Since dynamic studies are carried out by the traffic engineer to evaluate current conditions and develop solutions, they are described in detail in this chapter.

SPOT SPEED STUDIES

Spot speed studies are conducted to estimate the distribution of speeds of vehicles in a stream of traffic at a particular location on a highway. The speed of a vehicle is defined as the rate of movement of the vehicle; it is usually expressed in miles per hour (mi/h) or kilometers per hour (km/h). A spot speed study is carried out by recording the speeds of a sample of vehicles at a specified location. Speed characteristics identified by such a study will be valid only for the traffic and environmental conditions that exist at the time of the study. Speed characteristics determined from a spot speed study may be used to:

- Establish parameters for traffic operation and control, such as speed zones, speed limits (85th percentile speed is commonly used as the speed limit on a road), and passing restrictions.
- Evaluate the effectiveness of traffic control devices, such as variable message signs at work zones.
- Monitor the effect of speed enforcement programs such as the use of drone radar and the use of differential speed limits for passenger cars and trucks.
- Evaluate and or determine the adequacy of highway geometric characteristics such as radii of horizontal curves and lengths of vertical curves.
- Evaluate the effect of speed on highway safety through the analysis of crash data for different speed characteristics.
- Determine speed trends.
- Determine whether complaints about speeding are valid.

Locations for Spot Speed Studies

The locations for spot speed studies depend on the anticipated use of the results. The following locations are generally used for the different applications listed:

1. Locations that represent different traffic conditions on a highway or highways are used for *basic data collection.*
2. Midblocks of urban highways and straight, level sections of rural highways are sites for *speed trend analyses.*
3. Any location may be used for the solution of a *specific traffic engineering problem.*

When spot speed studies are being conducted, it is important that unbiased data be obtained. This requires that drivers be unaware that such a study is being conducted. Equipment used should therefore be concealed from the driver, and observers conducting the study should be inconspicuous. Since the speeds recorded will eventually be subjected to statistical analysis, it is important that a statistically adequate number of vehicle speeds be recorded.

Time of Day and Duration of Spot Speed Studies

The time of day for conducting a speed study depends on the purpose of the study. In general, when the purpose of the study is to establish posted speed limits, to observe speed trends, or to collect basic data, it is recommended that the study be conducted when traffic is free-flowing, usually during off-peak hours. However, when a speed study is conducted in response to citizen complaints, it is useful if the time period selected for the study reflects the nature of the complaints.

The duration of the study should be such that the minimum number of vehicle speeds required for statistical analysis is recorded. Typically, the duration is at least 1 hour and the sample size is at least 30 vehicles.

Sample Size for Spot Speed Studies

The calculated mean (or average) speed is used to represent the true mean value of all vehicle speeds at that location. The accuracy of this assumption depends on the number of vehicles in the sample. The larger the sample size, the greater the probability that the estimated mean is not significantly different from the true mean. It is therefore necessary to select a sample size that will give an estimated mean within acceptable error limits. Statistical procedures are used to determine this minimum sample size. Before discussing these procedures, it is first necessary to define certain significant values that are needed to describe speed characteristics. They are:

1. **Average speed,** which is the arithmetic mean of all observed vehicle speeds (which is the sum of all spot speeds divided by the number of recorded speeds). It is given as

$$\bar{u} = \frac{\sum f_i u_i}{\sum f_i} \tag{4.1}$$

where
$\bar{u}$ = arithmetic mean
f_i = number of observations in each speed group
u_i = midvalue for the ith speed group
N = number of observed values

The formula can also be written as:

$$\bar{u} = \frac{\sum u_i}{N}$$

u_i = speed of the ith vehicle

2. **Median speed,** which is the speed at the middle value in a series of spot speeds that are arranged in ascending order. Fifty percent of the speed values will be greater than the median; 50 percent will be less than the median.
3. **Modal speed,** which is the speed value that occurs most frequently in a sample of spot speeds.
4. The ***i*th-percentile spot speed,** which is the spot speed value below which i percent of the vehicles travel; for example, 85th-percentile spot speed is the speed below which 85 percent of the vehicles travel and above which 15 percent of the vehicles travel.
5. **Pace,** which is the range of speed—usually taken at 10-mi/h intervals—that has the greatest number of observations. For example, if a set of speed data includes speeds

between 30 and 60 mi/h, the speed intervals will be 30 to 40 mi/h, 40 to 50 mi/h, and 50 to 60 mi/h, assuming a range of 10 mi/h. The pace is 40 to 50 mi/h if this range of speed has the highest number of observations.

6. **Standard deviation of speeds,** which is a measure of the spread of the individual speeds. It is estimated as

$$S = \sqrt{\frac{\sum (u_j - \bar{u})^2}{N-1}} \tag{4.2}$$

where

S = standard deviation
$\bar{u}$ = arithmetic mean
u_j = jth observation
N = number of observations

However, speed data are frequently presented in classes where each class consists of a range of speeds. The standard deviation is computed for such cases as

$$S = \sqrt{\frac{\sum (f_i u_i^2) - (\sum f_i u_i)^2 / \sum f_i}{\sum f_i - 1}} \tag{4.3}$$

where

u_i = midvalue of speed class i
f_i = frequency of speed class i

Probability theory is used to determine the sample sizes for traffic engineering studies. Although a detailed discussion of these procedures is beyond the scope of this book, the simplest and most commonly used procedures are presented. Interested readers can find an in-depth treatment of the topic in publications listed in "Additional Readings," at the end of this chapter.

The minimum sample size depends on the precision level desired. The precision level is defined as the degree of confidence that the sampling error of a produced estimate will fall within a desired fixed range. Thus, for a precision level of 90–10, there is a 90 percent probability (confidence level) that the error of an estimate will not be greater than 10 percent of its true value. The confidence level is commonly given in terms of the level of significance (α), where $\alpha = (100 - \text{confidence level})$. The commonly used confidence level for speed counts is 95 percent.

The basic assumption made in determining the minimum sample size for speed studies is that the normal distribution describes the speed distribution over a given section of highway. The normal distribution is given as

$$f(x) = \frac{1}{\sigma\sqrt{2\pi}} e^{-(x-\mu)^2/2\sigma^2} \text{ for} -\infty < x < \infty \tag{4.4}$$

where

μ = true mean of the population

σ = true standard deviation

σ^2 = true variance

The properties of the normal distribution are then used to determine the minimum sample size for an acceptable error d of the estimated speed. The following basic properties are used (see Figure 4.1):

1. The normal distribution is symmetrical about the mean.
2. The total area under the normal distribution curve is equal to 1, or 100 percent.
3. The area under the curve between $\mu + \sigma$ and $\mu - \sigma$ is 0.6827.
4. The area under the curve between $\mu + 1.96\sigma$ and $\mu - 1.96\sigma$ is 0.9500.
5. The area under the curve between $\mu + 2\sigma$ and $\mu - 2\sigma$ is 0.9545.
6. The area under the curve between $\mu + 3\sigma$ and $\mu - 3\sigma$ is 0.9971.
7. The area under the curve between $\mu + \infty$ and $\mu - \infty$ is 1.0000.

The last five properties are used to draw specific conclusions about speed data. For example, if it can be assumed that the true mean of the speeds in a section of highway is 50 mi/h and the true standard deviation is 4.5 mi/h, it can be concluded that 95 percent of all vehicle speeds will be between $(50 - 1.96 \times 4.5) = 41.2$ mi/h and $(50 + 1.96 \times 4.5) = 58.82$ mi/h. Similarly, if a vehicle is selected at random, there is a 95 percent chance that its speed is between 41.2 and 58.8 mi/h. The properties of the normal distribution have been used to develop an equation relating the sample size to the number of standard variations corresponding to a particular confidence level, the limits of tolerable error, and the standard deviation.

The formula is

$$N = \left(\frac{Z\sigma}{d} \right)^2 \tag{4.5}$$

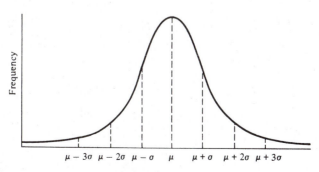

Figure 4.1 Shape of the Normal Distribution

where

N = minimum sample size

Z = number of standard deviations corresponding to the required confidence

= level 1.96 for 95 percent confidence level (Table 4.1)

σ = standard deviation (mi/h)

d = limit of acceptable error in the average speed estimate (mi/h)

The standard deviation can be estimated from previous data, or a small sample size can first be used.

Table 4.1   Constant Corresponding to Level of Confidence

Confidence Level (%)	Constant Z
68.3	1.00
86.6	1.50
90.0	1.64
95.0	1.96
95.5	2.00
98.8	2.50
99.0	2.58
99.7	3.00

Example 4.1   Determining Spot Speed Sample Size

As part of a class project a group of students collected a total of 120 spot speed samples at a location and determined from this data that the standard variation of the speeds was $\pm$ 6 mi/h. If the project required that the confidence level be 95.0 percent and the limit of acceptable error was $\pm$ 1.5 mi/h, determine whether these students satisfied the project requirement.

Solution:

- Use Eq. 4.5 to determine the minimum sample size to satisfy the project requirements.

$$N = \left(\frac{Z\sigma}{d} \right)^2$$

where

Z = 1.96 (from Table 4.1)

σ = $\pm$6

d = 1.5

$$N = \left(\frac{1.96x6}{1.5} \right)^2$$
$$= 61.45$$

Therefore, the minimum number of spot speeds collected to satisfy the project requirement is 62. Since the students collected 120 samples they satisfied the project requirements.

Methods for Conducting Spot Speed Studies

The methods used for conducting spot speed studies can generally be divided into two main categories: manual and automatic. Since the manual method is seldom used, automatic methods will be described.

Several automatic devices that can be used to obtain the instantaneous speeds of vehicles at a location on a highway are now available on the market. These automatic devices can be grouped into three main categories: (1) those that use road detectors, (2) those that use Doppler principle meters (radar type), and (3) those that use the principles of electronics.

Road Detectors

Road detectors can be classified into two general categories: pneumatic road tubes and induction loops. These devices can be used to collect data on speeds at the same time as volume data are being collected. When road detectors are used to measure speed, they should be laid such that the probability of a passing vehicle closing the connection of the meter during a speed measurement is reduced to a minimum. This is achieved by separating the road detectors by a distance of 3 to 15 ft.

The advantage of the detector meters is that human errors are considerably reduced. The disadvantages are that (1) these devices tend to be rather expensive, and (2) when pneumatic tubes are used, they are rather conspicuous and may, therefore, affect driver behavior, resulting in a distortion of the speed distribution.

Pneumatic road tubes are laid across the lane in which data are to be collected. When a moving vehicle passes over the tube, an air impulse is transmitted through the tube to the counter. When used for speed measurements, two tubes are placed across the lane, usually about 6 ft apart. An impulse is recorded when the front wheels of a moving vehicle pass over the first tube; shortly afterward a second impulse is recorded when the front wheels pass over the second tube. The time elapsed between the two impulses and the distance between the tubes are used to compute the speed of the vehicle.

An **inductive loop** is a rectangular wire loop buried under the roadway surface. It usually serves as the detector of a resonant circuit. It operates on the principle that a disturbance in the electrical field is created when a motor vehicle passes across it. This causes a change in potential that is amplified, resulting in an impulse being sent to the counter.

Doppler-Principle Meters

Doppler meters work on the principle that when a signal is transmitted onto a moving vehicle, the change in frequency between the transmitted signal and the reflected signal is proportional to the speed of the moving vehicle. The difference between the frequency of the transmitted signal and that of the reflected signal is measured by the equipment and then converted to speed in mi/h. In setting up the equipment, care must be taken to reduce the angle between the direction of the moving vehicle and the line joining the center of the transmitter and the vehicle. The value of the speed recorded depends on that angle. If the angle is not zero, an error related to the cosine of that angle is introduced, resulting in a lower speed than that which would have been recorded if the angle had been zero. However, this error is not very large, because the cosines of small angles are not much less than 1.

The advantage of this method is that because pneumatic tubes are not used, if the equipment can be located at an inconspicuous position, the influence on driver behavior is considerably reduced.

Enforcement officers in enforcing speed limits often use this type of meter. Figure 4.2a shows the SpeedAce®, manufactured by Monitron. It is a pocket-sized, hand-held laser speed detection system and can be used to measure speeds of individual vehicles at a distance of up to 1312 ft. Figure 4.2b shows the RTMS multilane presence radar manufactured by Electronic Integrated System. The main advantage of this detector over other systems is that it can be mounted on the side of the highway and obtain data on speeds of vehicles in up to eight lanes separately.

Electronic-Principle Detectors

In this method, the presence of vehicles is detected through electronic means, and information on these vehicles is obtained, from which traffic characteristics such as speed, volume, queues, and headways are computed. The great advantage of this method over the use of road detectors is that it is not necessary to physically install loops or any other type of detector on the road. The most promising technology using electronics is video image processing, sometimes referred to as a machine-vision system. This system consists of an electronic camera overlooking a large section of the roadway and a microprocessor. The electronic camera receives the images from the road; the microprocessor determines the vehicle's presence or passage. This information is then used to determine the traffic characteristics in real time. One such system is the **autoscope**.

Figure 4.3a schematically illustrates the configuration of the autoscope, which was developed in the United States. It has a significant advantage over loops in that it can detect traffic in many locations within the camera's field of view. The locations to be monitored are selected by the user through interactive graphics, which normally takes only a few minutes. This flexibility is achieved by placing detector lines along or across the roadway lanes on the monitor showing the traffic. The detector lines are therefore not fixed on the roadway because they are not physically located on the roadway but are placed on the monitor. A detection signal, which is similar to that produced by loops, is generated whenever a vehicle crosses the detector lines, indicating the presence or passage of the vehicle. The autoscope is therefore a wireless detector with a single camera that can replace many loops, thereby providing a wide-area detection system. The device can therefore be installed

(a) The SpeedAce® Meter

(b) The RTMS Meter

Figure 4.2 Examples of Meters Using Doppler Principle

without disrupting traffic operations, as often occurs with loop installation, and the detection configuration can be changed either manually or by using a software routine that provides a function of the traffic conditions. The device is also capable of extracting traffic parameters, such as volume and queue lengths. Figure 4.3b shows a photograph of the autoscope deployed at a study sight.

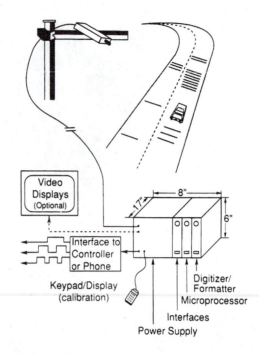

(a) Schematic Illustration of the Autoscope

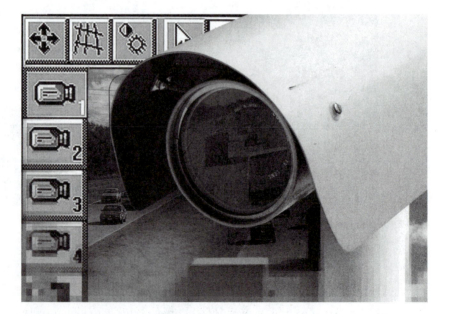

(b) The Autoscope Deployed

Figure 4.3 The Autoscope

Presentation and Analysis of Spot Speed Data

The data collected in spot speed studies are usually taken only from a sample of vehicles using the section of the highway on which the study is conducted, but these data are used to determine the speed characteristics of the whole population of vehicles traveling on the study site. It is therefore necessary to use statistical methods in analyzing these data. Several characteristics are usually determined from the analysis of the data. Some of them can be calculated directly from the data; others can be determined from a graphical representation of the data. Thus, the data must be presented in a form suitable for analysis.

The presentation format most commonly used is the frequency distribution table. The first step in the preparation of a frequency distribution table is the selection of the number of classes—that is, the number of velocity ranges—into which the data are to be fitted. The number of classes chosen is usually between 8 and 20, depending on the data collected. One technique that can be used to determine the number of classes is to determine first the range for a class size of 8 and then that for a class size of 20. Finding the difference between the maximum and minimum speeds in the data and dividing this number by 8, and then by 20, gives the maximum and minimum ranges in each class. A convenient range for each class is then selected and the number of classes determined. Usually the midvalue of each class range is taken as the speed value for that class. The data can also be presented in the form of a frequency histogram, or as a cumulative frequency distribution curve. The frequency histogram is a chart showing the midvalue for each class as the abscissa and the observed frequency for the corresponding class as the ordinate. The frequency distribution curve shows a plot of the cumulative percentage against the upper limit of each corresponding speed class.

Example 4.2   Determining Speed Characteristics from a Set of Speed Data

Table 4.2 shows the data collected on a rural highway in Virginia during a speed study. Develop the frequency histogram and the frequency distribution of the data and determine:

1. the arithmetic mean speed
2. the standard deviation
3. the median speed
4. the pace
5. the mode or modal speed
6. the 85th percentile speed

The speeds range from 34.8 to 65.0 mi/h, giving a speed range of 30.2. For 8 classes, the range per class is 3.75 mi/h; for 20 classes, the range per class is 1.51 mi/h. It is convenient to choose a range of 2 mi/h per class, which will give 16 classes. A frequency distribution table can then be prepared, as shown in Table 4.3, in which the speed classes are listed in column 1 and the midvalues are in column 2. The number of observations for each class is listed in column 3; the cumulative percentages of all observations are listed in column 6.

Table 4.2 Speed Data Obtained on a Rural Highway

Car No.	Speed (mi/h)	Car No.	Speed (mi/h)	Car No.	Speed (mi/h)	Car No.	Speed (mi/h)
1	35.1	23	46.1	45	47.8	67	56.0
2	44.0	24	54.2	46	47.1	68	49.1
3	45.8	25	52.3	47	34.8	69	49.2
4	44.3	26	57.3	48	52.4	70	56.4
5	36.3	27	46.8	49	49.1	71	48.5
6	54.0	28	57.8	50	37.1	72	45.4
7	42.1	29	36.8	51	65.0	73	48.6
8	50.1	30	55.8	52	49.5	74	52.0
9	51.8	31	43.3	53	52.2	75	49.8
10	50.8	32	55.3	54	48.4	76	63.4
11	38.3	33	39.0	55	42.8	77	60.1
12	44.6	34	53.7	56	49.5	78	48.8
13	45.2	35	40.8	57	48.6	79	52.1
14	41.1	36	54.5	58	41.2	80	48.7
15	55.1	37	51.6	59	48.0	81	61.8
16	50.2	38	51.7	60	58.0	82	56.6
17	54.3	39	50.3	61	49.0	83	48.2
18	45.4	40	59.8	62	41.8	84	62.1
19	55.2	41	40.3	63	48.3	85	53.3
20	45.7	42	55.1	64	45.9	86	53.4
21	54.1	43	45.0	65	44.7		
22	54.0	44	48.3	66	49.5		

The frequency histogram for the data shown in Table 4.3 is shown in Figure 4.4. The values in columns 2 and 3 of Table 4.3 are used to draw the frequency histogram, where the abscissa represents the speeds and the ordinate the observed frequency in each class.

The data can also be presented by preparing a frequency distribution curve, as shown in Figure 4.5. In this case, a curve showing percentage of observations against speed is drawn by plotting values from column 5 of Table 4.3 against the corresponding values in column 2. The total area under this curve is 1, or 100 percent.

The cumulative frequency distribution curve, another form of presenting the data, is shown in Figure 4.6. In this case, the cumulative percentages in column 6 of Table 4.3 are plotted against the upper limit of each corresponding speed class. This curve, therefore, gives the percentage of vehicles that are traveling at or below a given speed.

The characteristics of the data can now be given in terms of the formula defined at the beginning of this section.

Table 4.3 Frequency Distribution Table for Set of Speed Data

1	2	3	4	5	6	7
Speed Class (mi/h)	Class Midvalue, u_i	Class Frequency (Number of Observations in Class), f_i	$f_i u_i$	Percentage of Observations in Class	Cumulative Percentage of All Observations	$f(u_i - \bar{u})^2$
34–35.9	35.0	2	70	2.3	2.30	420.5
36–37.9	37.0	3	111	3.5	5.80	468.75
38–39.9	39.0	2	78	2.3	8.10	220.50
40–41.9	41.0	5	205	5.8	13.90	361.25
42–43.9	43.0	3	129	3.5	17.40	126.75
44–45.9	45.0	11	495	12.8	30.20	222.75
46–47.9	47.0	4	188	4.7	34.90	25.00
48–49.9	49.0	18	882	21.0	55.90	9.0
50–51.9	51.0	7	357	8.1	64.0	15.75
52–53.9	53.0	8	424	9.3	73.3	98.00
54–55.9	55.0	11	605	12.8	86.1	332.75
56–57.9	57.0	5	285	5.8	91.9	281.25
58–59.9	59.0	2	118	2.3	94.2	180.50
60–61.9	61.0	2	122	2.3	96.5	264.50
62–63.9	63.0	2	126	2.3	98.8	364.50
64–65.9	65.0	1	65	1.2	100.0	240.25
Totals		86	4260			3632.00

Solution:

- The arithmetic mean speed is computed from Eq. 4.1:

$$\bar{u} = \frac{\sum f_i u_i}{\sum f_i}$$

$$\sum f_i = 86$$

$$\sum f_i u_i = 4260$$

$$\bar{u} = \frac{4260}{86} = 49.5 \text{ mi/h}$$

- The standard deviation is computed from Eq. 4.2:

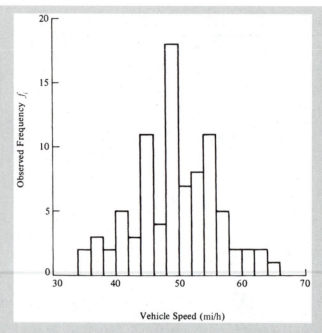

Figure 4.4 Histogram of Observed Vehicles' Speeds

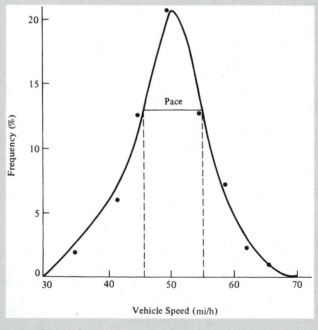

Figure 4.5 Frequency Distribution

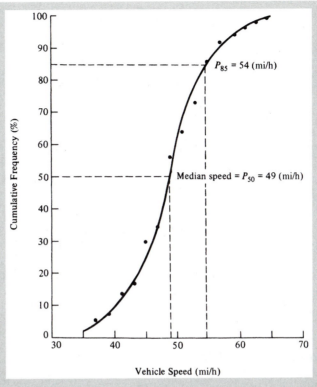

Figure 4.6 Cumulative Distribution

$$S = \sqrt{\frac{\sum f_i(u_i - \bar{u})^2}{N - 1}}$$

$$\sum f_i(u_i - \bar{u})^2 = 3632$$

$$(N - 1) = \sum f_i - 1 = 85$$

$$S^2 = \frac{3632}{85} = 42.73$$

$$S = \pm 6.5 \text{ mi/h}$$

- The median speed is obtained from the cumulative frequency distribution curve (Figure 4.6) as 49 mi/h, the fiftieth-percentile speed.
- The pace is obtained from the frequency distribution curve (Figure 4.5) as 45 to 55 mi/h.
- The mode or modal speed is obtained from the frequency histogram as 49 mi/h (Figure 4.4). It may also be obtained from the frequency distribution curve shown

in Figure 4.5, where the speed corresponding to the highest point on the curve is taken as an estimate of the modal speed.
- 85th-percentile speed is obtained from the cumulative frequency distribution curve as 54 mi/h (Figure 4.6).

Other Forms of Presentation and Analysis of Speed Data

Certain applications of speed study data may require a more complicated presentation and analysis of the speed data. For example, if the speed data are to be used in research on traffic flow theories, it may be necessary for the speed data to be fitted into a suitable theoretical frequency distribution, such as the normal distribution or the Gamma distribution. This is done first by assuming that the data fit a given distribution and then by testing this assumption using one of the methods of hypothesis testing, such as chi-square analysis. If the test suggests that the assumption can be accepted, specific parameters of the distribution can be found using the speed data. The properties of this distribution are then used to describe the speed characteristics, and any form of mathematical computation can be carried out using the distribution. Detailed discussion of hypothesis testing is beyond the scope of this book, but interested readers will find additional information in any book on statistical methods for engineers.

Comparison of Mean Speeds

It is also sometimes necessary to determine whether there is a significant difference between the mean speeds of two spot speed studies. This is done by comparing the absolute difference between the sample mean speeds against the product of the standard deviation of the difference in means and the factor Z for a given confidence level. If the absolute difference between the sample means is greater, it can then be concluded that there is a significant difference in sample means at that specific confidence level.

The standard deviation of the difference in means is given as

$$S_d = \sqrt{\frac{S_1^2}{n_1} + \frac{S_2^2}{n_2}} \tag{4.6}$$

where
$\quad n_1$ = sample size for study 1
$\quad n_2$ = sample size for study 2
$\quad S_d$ = square root of the variance of the difference in means
$\quad S_1^2$ = variance about the mean for study 1
$\quad S_2^2$ = variance about the mean for study 2

If $\bar{u}_1$ = mean speed of study 1, $\bar{u}_2$ = mean speed of study 2, and $|\bar{u}_1 - \bar{u}_2| > ZS_d$, where $|\bar{u}_1 - \bar{u}_2|$ is the absolute value of the difference in means, it can be concluded that the mean speeds are significantly different at the confidence level corresponding to Z. This analysis assumes that $\bar{u}_1$ and $\bar{u}_2$ are estimated means from the same distribution. Since it is usual to use the 95 percent confidence level in traffic engineering studies, the conclusion will, therefore, be based on whether $|\bar{u}_1 - \bar{u}_2|$ is greater than $1.96S_d$.

Example 4.3   Significant Differences in Average Spot Speeds

Speed data were collected at a section of highway during and after utility maintenance work. The speed characteristics are given as $\bar{u}_1$, S_1 and $\bar{u}_2$, S_2, as shown below. Determine whether there was any significant difference between the average speed at the 95 percent confidence level.

$$\bar{u}_1 = 35.5 \text{ mi/h} \qquad \bar{u}_2 = 38.7 \text{ mi/h}$$
$$S_1 = 7.5 \text{ mi/h} \qquad S_2 = 7.4 \text{ mi/h}$$
$$n_1 = 250 \qquad n_2 = 280$$

Solution:

- Use Eq. 4.6:

$$S_d = \sqrt{\frac{S_1^2}{n_1} + \frac{S_2^2}{n_2}}$$

$$= \sqrt{\frac{(7.5)^2}{250} + \frac{(7.4)^2}{280}} = 0.65$$

- Find the difference in means:

$$38.7 - 35.5 = 3.2 \text{ mi/h}$$
$$33.2 > (1.96)(0.65)$$
$$33.2 > 1.3 \text{ mi/h}$$

It can be concluded that the difference in mean speeds is significant at the 95 percent confidence level.

VOLUME STUDIES

Traffic volume studies are conducted to collect data on the number of vehicles and/or pedestrians that pass a point on a highway facility during a specified time period. This time period varies from as little as 15 min to as much as a year, depending on the anticipated use of the data. The data collected may also be put into subclasses which may include directional movement, occupancy rates, vehicle classification, and pedestrian age. Traffic volume studies are usually conducted when certain volume characteristics are needed, some of which follow:

1. **Average Annual Daily Traffic** (AADT) is the average of 24-hr counts collected every day in the year. AADTs are used in several traffic and transportation analyses for
   a. Estimation of highway user revenues
   b. Computation of accident rates in terms of accidents per 100 million vehicle-miles
   c. Establishment of traffic volume trends
   d. Evaluation of the economic feasibility of highway projects
   e. Development of freeway and major arterial street systems
   f. Development of improvement and maintenance programs

2. **Average Daily Traffic** (ADT) is the average of 24-hour counts collected over a number of days greater than 1 but less than a year. ADTs may be used for
   a. Planning of highway activities
   b. Measurement of current demand
   c. Evaluation of existing traffic flow

3. **Peak Hour Volume** (PHV) is the maximum number of vehicles that pass a point on a highway during a period of 60 consecutive minutes. PHVs are used for
   a. Functional classification of highways
   b. Design of the geometric characteristics of a highway, for example, number of lanes, intersection signalization, or channelization
   c. Capacity analysis
   d. Development of programs related to traffic operations, for example, one-way street systems or traffic routing
   e. Development of parking regulations

4. **Vehicle Classification** (VC) records volume with respect to the type of vehicles, for example, passenger cars, two-axle trucks, or three-axle trucks. VC is used in
   a. Design of geometric characteristics, with particular reference to turning-radii requirements, maximum grades, lane widths, and so forth
   b. Capacity analyses, with respect to passenger-car equivalents of trucks
   c. Adjustment of traffic counts obtained by machines
   d. Structural design of highway pavements, bridges, and so forth

5. **Vehicle Miles of Travel** (VMT) is a measure of travel along a section of road. It is the product of the traffic volume (that is, average weekday volume or ADT) and the length of roadway in miles to which the volume is applicable. VMTs are used mainly as a base for allocating resources for maintenance and improvement of highways.

Methods of Conducting Volume Counts

Traffic volume counts are conducted using two basic methods: manual and automatic. A description of each counting method follows.

Manual Method

Manual counting involves one or more persons recording observed vehicles using a counter. Figure 4.7 shows the TMC/48 electronic manual counter, which may be used to conduct manual traffic volume counts at an intersection. With this type of counter, both the turning movements at the intersection and the types of vehicles can be recorded by using more than one counter. For example, truck volumes may be collected by one person using one counter while passenger car volumes are recorded by another person using

Figure 4.7 The TMC/48 Manual Counter

SOURCE: Photograph by Ed Deasy, Virginia Transportation Research Council, Charlottesville, Va. Used with permission.

another counter. Note that in general, the inclusion of pickups and light trucks with four tires in the category of passenger cars does not create any significant deficiencies in the data collected, since the performance characteristics of these vehicles are similar to those of passenger cars. In some instances, however, a more detailed breakdown of commercial vehicles may be required, which would necessitate the collection of data according to number of axles and/or weight. However, the degree of truck classification usually depends on the anticipated use of the data collected.

The TMC/48 electronic manual counter, produced by Time Lapse, Inc., is powered by two separate built-in rechargeable batteries, which may be recharged using a 120V AC 60-Hz wall outlet. Several buttons are provided, each of which can be used to record volume data for different movements and different types of vehicles. The data for each movement are automatically separated into 48 distinct time intervals and stored in a semi-conductor memory. The data may then be manually read out directly. The numbers are shown sequentially on a liquid crystal display on the front of the device. Alternatively, a time-lapse telephone transmission coupler can be used to transmit the data over a tele-phone line. A software package is then used to transfer the data to a computer, where they are processed and printed. Figure 4.7 shows the hookup of the equipment, the coupler, and the modem at the transmittal end.

The main disadvantages of the manual count method are that (1) it is labor-intensive and can therefore be expensive, (2) it is subject to the limitations of human factors, and (3) it cannot be used for long periods of counting.

Automatic Method

Some automatic counters use a counting method that involves the laying of surface detec-tors (such as pneumatic road tubes) or subsurface detectors (such as magnetic or electric contact devices) on the road. These detect the passing vehicle and transmit the information to a recorder, which is connected to the detector at the side of the road. An example of counters using pneumatic road tubes as detectors is the Phoenix Vehicle Traffic Classifier manufactured by Diamond Traffic Products. Figure 4.8 shows an example setup of a surface detector using pneumatic road tubes. An example of counters using magnetic detectors is the Hi-Star NC-90A, manufactured by Nu-Metrics.

Phoenix Vehicle Traffic Classifier.   Figure 4.9a shows the Phoenix Vehicle Classifier. It is capable of obtaining vehicle counts, speeds, and classification data on 1 to 8 lanes with road tubes, and vehicle counts on 1 to 16 lanes when presence inductive loops are used. It has a 68k counter memory, which can be expanded internally in 128K increments to 960K. This expanded memory enables the counter to store individual data for up to 105,000 vehicles. In addition, the memory can be expanded to 16 megabytes by plugging in an optional TAM card as shown in Figure 4.9b.

Hi-Star NC-90A.   The operation of this counter is based on the technology known as vehicle magnetic imaging (VMI), which uses the basic fact that magnetic lines of force can pass through most media. In this case, a magnetic sensor detects distortions in the earth's field caused by a vehicle passing over or next to it. Electrical images are then reproduced that represent the different magnetic masses of the passing vehicles. These are analyzed by the VMI computer to determine vehicle speed, vehicle length, and so on, which are then retrieved using a computer. This counter can simultaneously collect data on volume, speed, length, occupancy, and weather.

Figure 4.10 shows the Hi-Star NC-90A. It is powered by rechargeable nickel-cadmium batteries and can operate in any of three modes: verify, frame, or sequential mode. When it is in the verify mode, data are collected on individual vehicles while the counter is connected by cable to a monitoring computer. This allows for the data on volume and speed to be viewed during the data collection process, but the data are not stored.

Figure 4.8 An Example of a Sensor Setup of a Surface Detector Using Pneumatic Road Tubes

SOURCE: Photograph by Lewis Woodson, Virginia Transportation Research Council, Charlottesville, Va. Used with permission.

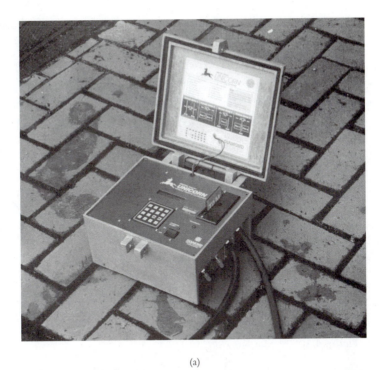

(a)

(b)

Figure 4.9 Phoenix Vehicle Traffic Classifier

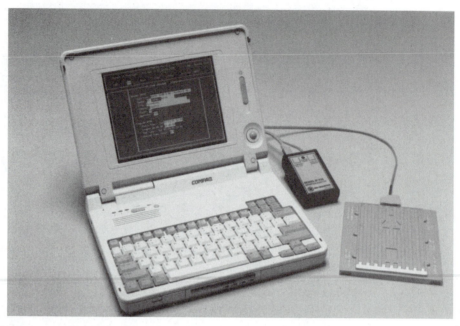

Figure 4.10 Hi-Star NC-90A

SOURCE: Photograph courtesy of Nu-Metrics An International Company, Uniontown, Penna. Used with permission.

In the frame mode, the data are stored in bins of preset intervals, which can be transferred later for analysis. The sequential mode allows in-depth studies that track vehicle movements in seconds of time, and has the capacity of tracking speed and length for about 8000 vehicles per study.

It should also be emphasized that some of the equipment described earlier that are used for collecting speed data could also be used to obtain vehicle counts. For example, the RTMS presence radar detector and the autoscope are capable of obtaining vehicle counts while obtaining speed data.

Types of Volume Counts

Different types of traffic counts are carried out, depending on the anticipated use of the data to be collected. These different types will now be discussed briefly.

Cordon Counts

When information is required on vehicle accumulation within an area, such as the central business district (CBD) of a city, particularly during a specific time, a cordon count is undertaken. The area for which the data are required is cordoned off by an imaginary closed loop; the area enclosed within this loop is defined as the *cordon area*. Figure 4.11 shows such an area, where the CBD of a city is enclosed by the imaginary loop ABCDA. The intersection of each street crossing the cordon line is taken as a count station; volume

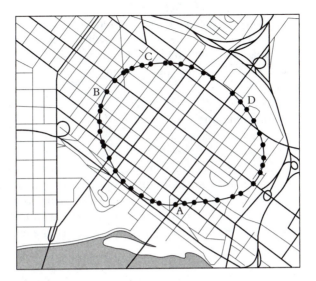

Figure 4.11   Example of Station Locations for a Cordon Count

counts of vehicles and/or persons entering and leaving the cordon area are taken. The information obtained from such a count is useful for planning parking facilities, updating and evaluating traffic operational techniques, and making long-range plans for freeway and arterial street systems.

Screen Line Counts

In screen line counts, the study area is divided into large sections by running imaginary lines, known as screen lines, across it. In some cases, natural and man-made barriers, such as rivers or railway tracks, are used as screen lines. Traffic counts are then taken at each point where a road crosses the screen line. It is usual for the screen lines to be designed or chosen such that they are not crossed more than once by the same street. Collection of data at these screen line stations at regular intervals facilitates the detection of variations in the traffic volume and traffic flow direction due to changes in the land-use pattern of the area.

Intersection Counts

Intersection counts are taken to determine vehicle classifications through movements and turning movements at intersections. These data are used mainly in determining phase lengths and cycle times for signalized intersections, in the design of channelization at intersections, and in the general design of improvements to intersections.

Pedestrian Volume Counts

Volume counts of pedestrians are made at locations such as subway stations, midblocks, and crosswalks. The counts are usually taken at these locations when the evaluation of existing or proposed pedestrian facilities is to be undertaken. Such facilities may include pedestrian overpasses or underpasses.

Pedestrian counts could be made using the TCM/48 manual counter described earlier and shown in Figure 4.7. The locations at which pedestrian counts are taken include intersections, along sidewalks, and mid-block crossings. These counts could be used for crash analysis, capacity analysis, and determining minimum signal timings at signalized intersections.

Periodic Volume Counts

In order to obtain certain traffic volume data, such as AADT, it is necessary to obtain data continuously. However, it is not feasible to collect continuous data on all roads because of the cost involved. To make reasonable estimates of annual traffic volume characteristics on an areawide basis, different types of periodic counts, with count durations ranging from 15 min to continuous, are conducted; the data from these different periodic counts are used to determine values that are then used to estimate annual traffic characteristics. The periodic counts usually conducted are continuous, control, or coverage counts.

Continuous Counts. These counts are taken continuously using mechanical or electronic counters. Stations at which continuous counts are taken are known as permanent count stations. In selecting permanent count stations, the highways within the study area must first be properly classified. Each class should consist of highway links with similar traffic patterns and characteristics. A highway link is defined for traffic count purposes as a homogeneous section that has the same traffic characteristics, such as AADT and daily, weekly, and seasonal variations in traffic volumes, at each point. Broad classification systems for major roads may include freeways, expressways, and major arterials. For minor roads, classifications may include residential, commercial, and industrial streets.

Control Counts. These counts are taken at stations known as control count stations, which are strategically located so that representative samples of traffic volume can be taken on each type of highway or street in an areawide traffic counting program. The data obtained from control counts are used to determine seasonal and monthly variations of traffic characteristics so that expansion factors can be determined. These expansion factors are used to determine year-round average values from short counts.

Control counts can be divided into major and minor control counts. Major control counts are taken monthly, with 24-hr directional counts taken on at least three days during the week (Tuesday, Wednesday, and Thursday) and also on Saturday and Sunday to obtain information on weekend volumes. It is usual to locate at least one major control count station on every major street. The data collected give information regarding hourly, monthly, and seasonal variations of traffic characteristics. Minor control counts are five-day weekday counts taken every other month on minor roads.

Coverage Counts. These counts are used to estimate ADT, using expansion factors developed from control counts. The study area is usually divided into zones that have similar traffic characteristics. At least one coverage count station is located in each zone. A 24-hr nondirectional weekday count is taken at least once every four years at each coverage station. The data indicate changes in areawide traffic characteristics.

Traffic Volume Data Presentation

The data collected from traffic volume counts may be presented in one of several ways, depending on the type of count conducted and the primary use of the data. Descriptions of some of the conventional data presentation techniques follow.

Traffic Flow Maps

These maps show traffic volumes on individual routes. The volume of traffic on each route is represented by the width of a band, which is drawn in proportion to the traffic volume it represents, providing a graphic representation of the different volumes that facilitates easy visualization of the relative volumes of traffic on the different routes. When flows are significantly different in opposite directions on a particular street or highway, it is advisable to provide a separate band for each direction. In order to increase the usefulness of such maps, the numerical value represented by each band is listed near the band. Figure 4.12 shows a typical traffic flow map.

Intersection Summary Sheets

These sheets are graphic representations of the volume and directions of all traffic movements through the intersection. These volumes can be either ADTs or PHVs, depending on the use of the data. Figure 4.13 shows a typical intersection summary sheet, displaying peak-hour traffic through the intersection.

Time-Based Distribution Charts

These charts show the hourly, daily, monthly, or annual variations in traffic volume in an area or on a particular highway. Each volume is usually given as a percentage of the average volume. Figure 4.14 shows typical charts for monthly, daily, and hourly variations.

Summary Tables

These tables give a summary of traffic volume data such as PHV, VC, and ADT in tabular form. Table 4.4 is a typical summary table.

Traffic Volume Characteristics

A continuous count of traffic at a section of a road will show that traffic volume varies from hour to hour, from day to day, and from month to month. However, the regular observation of traffic volumes over the years has identified certain characteristics showing that although traffic volume at a section of a road varies from time to time, this variation is repetitive and rhythmic. These characteristics of traffic volumes are usually taken into consideration when traffic counts are being planned so that volumes collected at a particular time or place can be related to volumes collected at other times and places. A knowledge of these characteristics can also be used to estimate the accuracy of traffic counts.

Monthly variations are shown in Figure 4.14a, where very low volumes are observed during January and February, mainly because of the winter weather, and the peak volume is observed during August, mainly due to vacation traffic. This suggests that traffic

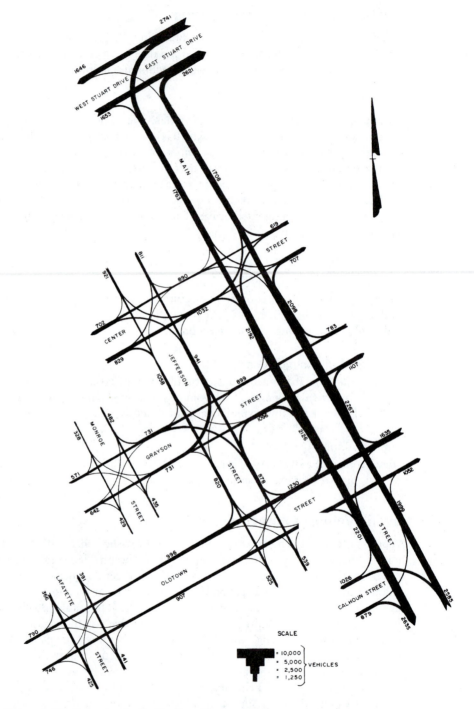

Figure 4.12 Example of a Traffic Flow Map

SOURCE: *Galax Traffic Study Report,* Virginia Department of Transportation, Richmond, Va., 1965.

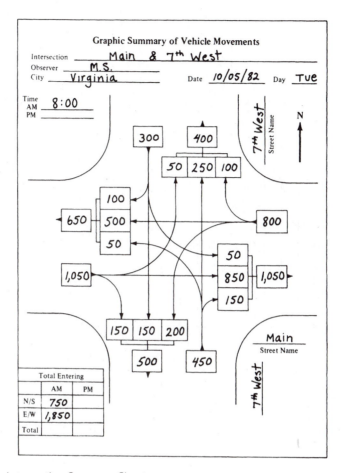

Figure 4.13 Intersection Summary Sheet

volumes taken either during the winter months of January or February or during the summer months of July and August cannot be representative of the average annual traffic. If this information is presented for a number of consecutive years, the repetitive nature of the variation will be observed, since the pattern of the variation will be similar for all years, although the actual volumes may not necessarily be the same.

Daily variations are shown in Figure 4.14b, where it is seen that traffic volumes on Tuesday, Wednesday, and Thursday are similar, but a peak is observed on Friday. This indicates that when short counts are being planned, it is useful to plan for the collection of weekday counts on Tuesday, Wednesday, and Thursday and, when necessary, to plan for the collection of weekend counts separately on Friday and Saturday.

Hourly variations in traffic volume are shown in Figure 4.14c, where the volume for each hour of the day is represented as a percentage of the ADT. It can be seen that there is hardly any traffic between 1 a.m. and 5 a.m., and that peak volumes occur between 8 a.m. and 9 a.m., noon and 1 p.m., and 4 p.m. and 6 p.m. It can be inferred that work trips are

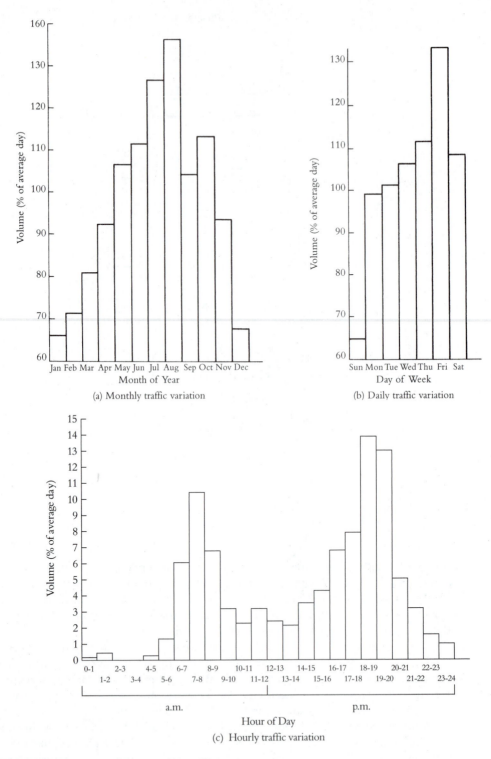

Figure 4.14 Traffic Volumes on an Urban Highway

Table 4.4 Summary of Traffic Volume Data for a Highway Section

PHV	430
ADT	5375
Vehicle Classification	
Passenger cars	70%
Two-axle trucks	20%
Three-axle trucks	8%
Other trucks	2%

primarily responsible for the peaks. If such data are collected on every weekday for one week, the hourly variations will be similar to each other, although the actual volumes may not be the same from day to day.

Sample Size and Adjustment of Periodic Counts

The impracticality of collecting data continuously every day of the year at all counting stations makes it necessary to collect sample data from each class of highway and to estimate annual traffic volumes from periodic counts. This involves the determination of the minimum sample size (number of count stations) for a required level of accuracy and the determination of daily, monthly, and/or seasonal expansion factors for each class of highway.

Determination of Number of Count Stations

The minimum sample size depends on the precision level desired. The commonly used precision level for volume counts is 95-5. When the sample size is less than 30 and the selection of counting stations is random, a distribution known as the student's t distribution may be used to determine the sample size for each class of highway links. The student's t distribution is unbounded, with a mean of zero, and has a variance that depends on the scale parameter, commonly referred to as the *degrees of freedom* (v). The degrees of freedom (v) is a function of the sample size; $v = N - 1$ for the student's t distribution. The variance of the student's t distribution is $v/(v - 2)$, which indicates that as v approaches infinity, the variance approaches 1. The probabilities (confidence levels) for the student's t distribution for different degrees of freedom are given in Appendix A.

Assuming that the sampling locations are randomly selected, the minimum sample number is given as

$$n = \frac{t_{\alpha/2, N-1}^2 (S^2/d^2)}{1 + (1/N)(t_{\alpha/2, N-1}^2)(S^2/d^2)} \tag{4.7}$$

where
    n = minimum number of count locations required
    t = value of the student's t distribution with $(1 - \alpha/2)$ confidence level
      ($N - 1$ degrees of freedom)

N = total number of links (population) from which a sample is to be selected
α = significance level
S = estimate of the spatial standard deviation of the link volumes
d = allowable range of error

To use Eq. 4.7, estimates of the mean and standard deviation of the link volumes are required. These estimates can be obtained by taking volume counts at a few links or by using known values for other, similar highways.

Example 4.4   Minimum Number of Count Stations

To determine a representative value for the ADT on 100 highway links that have similar volume characteristics, it was decided to collect 24-hr volume counts on a sample of these links. Estimates of mean and standard deviation of the link volumes for the type of highways in which these links are located are 32,500 and 5500, respectively. Determine the minimum number of stations at which volume counts should be taken if a 95-5 precision level is required.

Solution:

- Establish the data:

$$\alpha = (100 - 95) = 5 \text{ percent}$$
$$S = 5500$$
$$m = 32,500$$
$$d = 0.1 \times 32,500 = 3250 \text{ (allowable range of error)}$$
$$v = 100 - 1 = 99$$
$$t_{\alpha/2,99} \approx 1.984 \text{ (from Appendix A)}$$

- Use Eq. 4.7 to solve for n:

$$n = \frac{t_{\alpha/2,\,N-1}^2 \,(S^2/d^2)}{1 + (1/N)(t_{\alpha/2,\,N-1}^2)(S^2/d^2)}$$

$$= \frac{(1.984^2 \times 5500^2)/3250^2}{1 + (1/100)(1.984^2 \times 5500^2)/3250^2} = \frac{11.27}{1.11} = 10.1$$

Counts should be taken at a minimum of 11 stations. When sample sizes are greater than 30, the normal distribution is used instead of the student's t distribution.

However, the Federal Highway Administration (FHWA) has suggested that although it is feasible to develop a valid statistical sample for statewide traffic counts independent of the Highway Performance Monitoring System (HPMS) sample design, it is more realistic to use the HPMS sample design. This results in much less effort, because it is available, is clearly defined, and has been implemented. The HPMS sample has been implemented in each state, the District of Columbia, and Puerto Rico; it provides a statistically valid, reliable, and consistent database for analysis within states, between states, and for any aggregation of states up to the national level.

The HPMS sample design is a stratified simple random sample based on AADT, although about 100 data items are collected. The population from which the sample is obtained includes all public highways or roads within a state but excludes local roads. The sampling element is defined as a road section that includes all travel lanes and the volumes in both directions. The data are stratified by (1) type of area (rural, small urban, and individual or collective urbanized areas), and (2) functional class, which in rural areas includes interstate highways, other principal arterials, minor arterials, major collectors, and minor collectors, and which in urban areas includes interstate highways, other freeways or expressways, other principal arterials, minor arterials, and collectors.

Adjustment of Periodic Counts

Expansion factors, used to adjust periodic counts, are determined either from continuous count stations or from control count stations.

Expansion Factors from Continuous Count Stations. Hourly, daily, and monthly expansion factors can be determined using data obtained at continuous count stations.

Hourly expansion factors (HEFs) are determined by the formula

$$HEF = \frac{\text{total volume for 24-hr period}}{\text{volume for particular hour}}$$

These factors are used to expand counts of durations shorter than 24 hr to 24-hr volumes by multiplying the hourly volume for each hour during the count period by the HEF for that hour and finding the mean of these products.

Daily expansion factors (DEFs) are computed as

$$DEF = \frac{\text{average total volume for week}}{\text{average volume for particular day}}$$

These factors are used to determine weekly volumes from counts of 24-hr duration by multiplying the 24-hr volume by the DEF.

Monthly expansion factors (MEFs) are computed as

$$MEF = \frac{AADT}{\text{ADT for particular month}}$$

The AADT for a given year may be obtained from the ADT for a given month by multiplying this volume by the MEF.

Tables 4.5, 4.6, and 4.7 give expansion factors for a particular primary road in Virginia. Such expansion factors should be determined for each class of road in the classification system established for an area.

Table 4.5 Hourly Expansion Factors for a Rural Primary Road

Hour	Volume	HEF	Hour	Volume	HEF
6:00–7:00 a.m.	294	42.00	6:00–7:00 p.m.	743	16.62
7:00–8:00 a.m.	426	29.00	7:00–8:00 p.m.	706	17.49
8:00–9:00 a.m.	560	22.05	8:00–9:00 p.m.	606	20.38
9:00–10:00 a.m.	657	18.80	9:00–10:00 p.m.	489	25.26
10:00–11:00 a.m.	722	17.10	10:00–11:00 p.m.	396	31.19
11:00–12:00 p.m.	667	18.52	11:00–12:00 a.m.	360	34.31
12:00–1:00 p.m.	660	18.71	12:00–1:00 a.m.	241	51.24
1:00–2:00 p.m.	739	16.71	1:00–2:00 a.m.	150	82.33
2:00–3:00 p.m.	832	14.84	2:00–3:00 a.m.	100	123.50
3:00–4:00 p.m.	836	14.77	3:00–4:00 a.m.	90	137.22
4:00–5:00 p.m.	961	12.85	4:00–5:00 a.m.	86	143.60
5:00–6:00 p.m.	892	13.85	5:00–6:00 a.m.	137	90.14

Total daily volume = 12,350.

Table 4.6 Daily Expansion Factors for a Rural Primary Road

Day of Week	Volume	DEF
Sunday	7,895	9.515
Monday	10,714	7.012
Tuesday	9,722	7.727
Wednesday	11,413	6.582
Thursday	10,714	7.012
Friday	13,125	5.724
Saturday	11,539	6.510

Total weekly volume = 75,122.

Table 4.7 Monthly Expansion Factors for a Rural Primary Road

Month	ADT	MEF
January	1350	1.756
February	1200	1.975
March	1450	1.635
April	1600	1.481
May	1700	1.394
June	2500	0.948
July	4100	0.578
August	4550	0.521
September	3750	0.632
October	2500	0.948
November	2000	1.185
December	1750	1.354

Total yearly volume = 28,450.
Mean average daily volume = 2370.

Example 4.5   *Calculating AADT Using Expansion Factors*

A traffic engineer urgently needs to determine the AADT on a rural primary road that has the volume distribution characteristics shown in Tables 4.5, 4.6, and 4.7. She collected the data shown below on a Tuesday during the month of May. Determine the AADT of the road.

7:00–8:00 a.m.	400
8:00–9:00 a.m.	535
9:00–10:00 a.m.	650
10:00–11:00 a.m.	710
11:00–12 noon	650

Solution:

- Estimate the 24-hr volume for Tuesday using the factors given in Table 4.5:

$$\frac{(400 \times 29.0 + 535 \times 22.05 + 650 \times 18.80 + 710 \times 17.10 + 650 \times 18.52)}{5} \approx 11{,}959$$

- Adjust the 24-hr volume for Tuesday to an average volume for the week using the factors given in Table 4.6:

$$\text{Total 7-day volume} = 11{,}959 \times 7.727$$

$$\text{Average 24-hr volume} = \frac{11{,}959 \times 7.727}{7} = 13{,}201$$

- Since the data were collected in May, use the factor shown for May in Table 4.7 to obtain the AADT:

$$\text{AADT} = 13{,}201 \times 1.394 = 18{,}402$$

TRAVEL TIME AND DELAY STUDIES

A travel time study determines the amount of time required to travel from one point to another on a given route. In conducting such a study, information may also be collected on the locations, durations, and causes of delays. When this is done, the study is known as a travel time and delay study. Data obtained from travel time and delay studies give a good indication of the level of service on the study section. These data also aid the traffic engineer in identifying problem locations, which may require special attention in order to improve the overall flow of traffic on the route.

Applications of Travel Time and Delay Data

The data obtained from travel time and delay studies may be used in any one of the following traffic engineering tasks:

- Determination of the efficiency of a route with respect to its ability to carry traffic
- Identification of locations with relatively high delays and the causes for those delays
- Performance of before-and-after studies to evaluate the effectiveness of traffic operation improvements
- Determination of relative efficiency of a route by developing sufficiency ratings or congestion indices
- Determination of travel times on specific links for use in trip assignment models
- Compilation of travel time data that may be used in trend studies to evaluate the changes in efficiency and level of service with time
- Performance of economic studies in the evaluation of traffic operation alternatives that reduce travel time

Definition of Terms Related to Time and Delay Studies

Let us now define certain terms commonly used in travel time and delay studies:

1. **Travel time** is the time taken by a vehicle to traverse a given section of a highway.
2. **Running time** is the time a vehicle is actually in motion while traversing a given section of a highway.
3. **Delay** is the time lost by a vehicle due to causes beyond the control of the driver.
4. **Operational delay** is that part of the delay caused by the impedance of other traffic. This impedance can occur either as side friction, where the stream flow is interfered with by other traffic (for example, parking or unparking vehicles), or as internal friction, where the interference is within the traffic stream (for example, reduction in capacity of the highway).
5. **Stopped-time delay** is that part of the delay during which the vehicle is at rest.
6. **Fixed delay** is that part of the delay caused by control devices such as traffic signals. This delay occurs regardless of the traffic volume or the impedance that may exist.
7. **Travel-time delay** is the difference between the actual travel time and the travel time that will be obtained by assuming that a vehicle traverses the study section at an average speed equal to that for an uncongested traffic flow on the section being studied.

Methods for Conducting Travel Time and Delay Studies

Several methods have been used to conduct travel time and delay studies. These methods can be grouped into two general categories: (1) those using a test vehicle and (2) those not requiring a test vehicle. The particular technique used for any specific study depends on the reason for conducting the study and the available personnel and equipment.

Methods Requiring a Test Vehicle

This category involves three possible techniques: floating-car, average-speed, and moving-vehicle techniques.

Floating-Car Technique. In this method, the test car is driven by an observer along the test section so that the test car "floats" with the traffic. The driver of the test vehicle attempts to pass as many vehicles as those that pass his test vehicle. The time taken to traverse the study section is recorded. This is repeated, and the average time is recorded as the travel time. The minimum number of test runs can be determined using an equation similar to Eq. 4.5, using values of the t distribution rather than the z values. The reason is that the sample size for this type of study is usually less than 30, which makes the t distribution more appropriate. The equation is

$$N = \left(\frac{t_\alpha \times \sigma}{d} \right)^2 \tag{4.8}$$

where

N = sample size (minimum number of test runs)
σ = standard deviation (mi/h)
d = limit of acceptable error in the speed estimate (mi/h)
t_α = value of the student's t distribution with $(1 - \alpha/2)$ confidence level
and $(N - 1)$ degrees of freedom
α = significance level

The limit of acceptable error used depends on the purpose of the study. The following limits are commonly used:

- Before-and-after studies: ± 1.0 to ± 3.0 mi/h
- Traffic operation, economic evaluations, and trend analyses: ± 2.0 to ± 4.0 mi/h
- Highway needs and transportation planning studies: ± 3.0 to ± 5.0 mi/h

Average-Speed Technique.   This technique involves driving the test car along the length of the test section at a speed that, in the opinion of the driver, is the average speed of the traffic stream. The time required to traverse the test section is noted. The test run is repeated for the minimum number of times, determined from Eq. 4.8, and the average time is recorded as the travel time.

In each of these methods, it is first necessary to clearly identify the test section. The way the travel time is usually obtained is that the observer starts a stopwatch at the beginning point of the test section and stops at the end. Additional data may also be obtained by recording the times at which the test vehicle arrives at specific locations which have been identified before the start of the test runs. A second stopwatch also may be used to determine the time that passes each time the vehicle is stopped. The sum of these times for any test run will give the stopped-time delay for that run. Table 4.8 shows an example of a set of data obtained for such a study.

Alternatively, the driver alone can collect the data by using a laptop computer with internal clock and distance functions. The predetermined locations (control points) are first programmed into the computer. At the start of the run, the driver activates the clock and distance functions; then the driver presses the appropriate computer key for each specified location. The data are then recorded automatically. The causes of delay are then recorded by the driver on a tape recorder.

Moving-Vehicle Technique.   In this technique, the observer makes a round trip on a test section like the one shown in Figure 4.15, where it is assumed that the road runs east–west. The observer starts collecting the relevant data at section X-X, drives the car eastward to

Table 4.8 Speed and Delay Information

Street Name: 29 North Date: July 7, 1994 Time: 2:00–3:00 p.m.
Weather: Clear Non-peak

Cross Streets	Distance (ft)	Travel Time (sec)	Segment Speed (mi/h)	Stop Time (sec)	Reason for Stoppage	Speed Limit (mi/h)	Ideal Travel Time (sec)	Segment Delay (sec)	Net Speed (mi/h)
Ivy Road	0	0.0	–	0.0		–	0.0	0.0	–
Massie Road	1584	42.6	25.4	20.1	Signal	40	27.0	15.6	17.2
Arlington Blvd.	1320	27.7	32.5	0.0		40	22.5	5.2	32.5
Wise Street	792	19.7	27.4	8.9	Signal	40	13.5	6.2	18.9
Barracks Road	1320	32.1	28.0	15.4	Signal	40	22.5	9.6	18.9
Angus Road	2244	49.8	30.7	9.2	Signal	40	38.3	11.5	25.9
Hydraulic Road	1584	24.4	44.3	0.0		45	24.0	0.4	44.3
Seminole Court	1584	42.6	25.4	19.5	Signal	45	24.0	18.6	17.4
Greenbrier Drive	1848	41.5	30.4	15.6	Signal	45	28.0	13.5	22.1
Premier Court	1320	37.4	24.1	11.8	Signal	45	20.0	17.4	18.3
Fashion Square I	1584	23.6	45.8	4.9	Signal	45	24.0	–0.4	37.9
Fashion Square II	1056	19.7	36.5	0.0		45	16.0	3.7	36.5
Rio Road	1056	20.2	35.6	14.1	Signal	45	16.0	4.2	21.0
Totals	17292	381.3	30.9	119.5			275.8	105.5	23.5

Note: Segment delay is the difference between observed travel time and calculated ideal travel time.
SOURCE: Study conducted in Charlottesville, Va., by Justin Black and John Ponder.

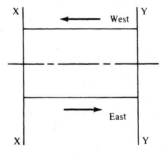

Figure 4.15 Test Site for Moving-Vehicle Method

section Y-Y, and then turns the vehicle around and drives westward to section X-X again. The following data are collected as the test vehicle makes the round trip:

- The time it takes to travel east from X-X to Y-Y (T_e), in minutes
- The time it takes to travel west from Y-Y to X-X (T_w), in minutes
- The number of vehicles traveling west in the opposite lane while the test car is traveling east (N_e)

- The number of vehicles that overtake the test car while it is traveling west from Y-Y to X-X, that is, traveling in the westbound direction (O_w)
- The number of vehicles that the test car passes while it is traveling west from Y-Y to X-X, that is, traveling in the westbound direction (P_w)

The volume (V_w) in the westbound direction can then be obtained from the expression

$$V_w = \frac{(N_e + O_w - P_w)60}{T_e + T_w} \qquad (4.9)$$

where $(N_e + O_w - P_w)$ is the number of vehicles traveling westward that cross the line X-X during the time $(T_e + T_w)$. Note that when the test vehicle starts at X-X, traveling eastward, all vehicles traveling westward should get to X-X before the test vehicle, except those that are passed by the test vehicle when it is traveling westward. Similarly, all vehicles that pass the test vehicle when it is traveling westward will get to X-X before the test vehicle. The test vehicle will also get to X-X before all vehicles it passes while traveling westward. These vehicles have, however, been counted as part of N_e or O_w and should therefore be subtracted from the sum of N_e and O_w to determine the number of westbound vehicles that cross X-X during the time the test vehicle travels from X-X to Y-Y and back to X-X. These considerations lead to Eq. 4.9.

Similarly, the average travel time $\overline{T}_w$ in the westbound direction is obtained from

$$\frac{\overline{T}_w}{60} = \frac{T_w}{60} - \frac{O_w - P_w}{V_w}$$

$$\overline{T}_w = T_w - \frac{60(O_w - P_w)}{V_w} \qquad (4.10)$$

If the test car is traveling at the average speed of all vehicles, it will most likely pass the same number of vehicles as the number of vehicles that overtake it. Since it is probable that the test car will not be traveling at the average speed, the second term of Eq. 4.10 corrects for the difference between the number of vehicles that overtake the test car and the number of vehicles that are overtaken by the test car.

Example 4.6   Volume and Travel Time Using Moving-Vehicle Technique

The data in Table 4.9 were obtained in a travel time study on a section of highway using the moving-vehicle technique. Determine the travel time and volume in each direction at this section of the highway.

Mean time it takes to travel eastward (T_e) = 2.85 min

Mean time it takes to travel westbound (T_w) = 3.07 min

Average number of vehicles traveling westward when test vehicle is traveling eastward (N_e) = 79.50

Table 4.9 Data from Travel Time Study Using the Moving-Vehicle Technique

Run Direction/ Number	Travel Time (min)	No. of Vehicles Traveling in Opposite Direction	No. of Vehicles That Overtook Test Vehicle	No. of Vehicles Overtaken by Test Vehicle
Eastward				
1	2.75	80	1	1
2	2.55	75	2	1
3	2.85	83	0	3
4	3.00	78	0	1
5	3.05	81	1	1
6	2.70	79	3	2
7	2.82	82	1	1
8	3.08	78	0	2
Average	2.85	79.50	1.00	1.50
Westward				
1	2.95	78	2	0
2	3.15	83	1	1
3	3.20	89	1	1
4	2.83	86	1	0
5	3.30	80	2	1
6	3.00	79	1	2
7	3.22	82	2	1
8	2.91	81	0	1
Average	3.07	82.25	1.25	0.875

Average number of vehicles traveling eastward when test vehicle is traveling westward (N_w) = 82.25

Average number of vehicles that overtake test vehicle while it is traveling westward (O_w) = 1.25

Average number of vehicles that overtake test vehicle while it is traveling eastward (O_e) = 1.00

Average number of vehicles the test vehicle passes while traveling westward (P_w) = 0.875

Average number of vehicles the test vehicle passes while traveling eastward (P_e) = 1.5

Solution:

- From Eq. 4.9, find the volume in the westbound direction:

$$V_w = \frac{(N_e + O_w - P_w)\,60}{T_e + T_w}$$

$$= \frac{(79.50 + 1.25 - 0.875)\,60}{2.85 + 3.07} = 809.5 \qquad (\text{or } 810 \text{ veh/h})$$

- Similarly, calculate the volume in the eastbound direction:

$$V_e = \frac{(82.25 + 1.00 - 1.50)\,60}{2.85 + 3.07} = 828.5 \qquad (\text{or } 829 \text{ veh/h})$$

- Find the average travel time in the westbound direction:

$$\overline{T}_w = 3.07 - \frac{(1.25 - 0.875)}{810}\,60 = 3.0 \text{ min}$$

- Find the average travel time in the eastbound direction:

$$\overline{T}_e = 2.85 - \frac{(1.00 - 1.50)}{829}\,60 = 2.9 \text{ min}$$

Methods Not Requiring a Test Vehicle
This category includes the license-plate method and the interview method.

License-Plate Observations.   The license-plate method requires that observers be positioned at the beginning and end of the test section. Observers can also be positioned at other locations if elapsed times to those locations are required. Each observer records the last three or four digits of the license plate of each car that passes, together with the time at which the car passes. The reduction of the data is accomplished in the office by matching the times of arrival at the beginning and end of the test section for each license plate recorded. The difference between these times is the traveling time of each vehicle. The average of these is the average traveling time on the test section. It has been suggested that a sample size of 50 matched license plates will give reasonably accurate results.

Interviews.   The interviewing method is carried out by obtaining information from people who drive on the study site regarding their travel times, their experience of delays, and so forth. This method facilitates the collection of a large amount of data in a relatively short time. However, it requires the cooperation of the people contacted, since the result depends entirely on the information given by them.

PARKING STUDIES
Any vehicle traveling on a highway will at one time or another be parked for either a relatively short time or a much longer time, depending on the reason for parking. The

provision of parking facilities is therefore an essential element of the highway mode of transportation. The need for parking spaces is usually very great in areas where land uses include business, residential, or commercial activities. The growing use of the automobile as a personal feeder service to transit systems ("park-and-ride") has also increased the demand for parking spaces at transit stations. In areas of high density, where space is very expensive, the space provided for automobiles usually has to be divided between that allocated for their movement and that allocated for parking them.

Providing adequate parking space to meet the demand for parking in the CBD may necessitate the provision of parking bays along curbs, which reduces the capacity of the streets and may affect the level of service. This problem usually confronts a city traffic engineer. The solution is not simple, since the allocation of available space will depend on the goals of the community, which the traffic engineer must take into consideration when trying to solve the problem. Parking studies are therefore used to determine the demand for and the supply of parking facilities in an area, the projection of the demand, and the views of various interest groups on how best to solve the problem. Before we discuss the details of parking studies, it is necessary to discuss the different types of parking facilities.

Types of Parking Facilities

Parking facilities can be divided into two main groups: on-street and off-street.

On-Street Parking Facilities

These are also known as curb facilities. Parking bays are provided alongside the curb on one or both sides of the street. These bays can be unrestricted parking facilities if the duration of parking is unlimited and parking is free, or they can be restricted parking facilities if parking is limited to specific times of the day for a maximum duration. Parking at restricted facilities may or may not be free. Restricted facilities may also be provided for specific purposes, such as to provide handicapped parking or as bus stops or loading bays.

Off-Street Parking Facilities

These facilities may be privately or publicly owned; they include surface lots and garages. Self-parking garages require that drivers park their own automobiles; attendant-parking garages maintain personnel to park the automobiles.

Definitions of Parking Terms

Before discussing the different methods for conducting a parking study, it is necessary to define some terms commonly used in parking studies, including *space-hour, parking volume, parking accumulation, parking load, parking duration,* and *parking turnover.*

1. A **space-hour** is a unit of parking that defines the use of a single parking space for a period of 1 hr.
2. **Parking volume** is the total number of vehicles that park in a study area during a specific length of time, usually a day.
3. **Parking accumulation** is the number of parked vehicles in a study area at any specified time. These data can be plotted as a curve of parking accumulation against time, which shows the variation of the parking accumulation during the day.

4. The **parking load** is the area under the accumulation curve between two specific times. It is usually given as the number of space-hours used during the specified period of time.

5. **Parking duration** is the length of time a vehicle is parked at a parking bay. When the parking duration is given as an average, it gives an indication of how frequently a parking space becomes available.

6. **Parking turnover** is the rate of use of a parking space. It is obtained by dividing the parking volume for a specified period by the number of parking spaces.

Methodology of Parking Studies

A comprehensive parking study usually involves (1) inventory of existing parking facilities, (2) collection of data on parking accumulation, parking turnover, and parking duration, (3) identification of parking generators, and (4) collection of information on parking demand. Information on related factors, such as financial, legal, and administrative matters, may also be collected.

Inventory of Existing Parking Facilities

An inventory of existing parking facilities is a detailed listing of the location and all other relevant characteristics of each legal parking facility, private and public, in the study area. The inventory includes both on- and off-street facilities. The relevant characteristics usually listed include the following:

- Type and number of parking spaces at each parking facility
- Times of operation and limit on duration of parking, if any
- Type of ownership (private or public)
- Parking fees, if any, and method of collection
- Restrictions on use (open or closed to the public)
- Other restrictions, if any (such as loading and unloading zones, bus stops, or taxi ranks)
- Probable degree of permanency (can the facility be regarded as permanent or is it just a temporary facility?)

The information obtained from an inventory of parking facilities is useful both to the traffic engineer and to public agencies, such as zoning commissions and planning departments. The inventory should be updated at regular intervals of about four to five years.

Collection of Parking Data

Accumulation. Accumulation data are obtained by checking the amount of parking during regular intervals on different days of the week. The checks are usually carried out on an hourly or 2-hr basis between 6:00 a.m. and 8:00 p.m. The selection of the times depends on the operation times of land-use activities that act as parking generators. For example, if a commercial zone is included, checks should be made during the times when retail shops are open, which may include periods up to 9:30 p.m. on some days. The information obtained is used to determine hourly variations of parking and peak periods of parking demand. (See Figure 4.16.)

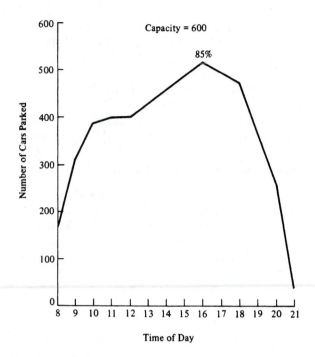

Figure 4.16 Parking Accumulation at a Parking Lot

Turnover and Duration.   Information on turnover and duration is usually obtained by collecting data on a sample of parking spaces in a given block. This is done by recording the license plate of the vehicle parked on each parking space in the sample at the ends of fixed intervals during the study period. The length of the fixed intervals depends on the maximum permissible duration. For example, if the maximum permissible duration of parking at a curb face is 1 hr, a suitable interval is every 20 min. If the permissible duration is 2 hr, checking every 30 min would be appropriate. Turnover is then obtained from the equation

$$T = \frac{\text{number of different vehicles parked}}{\text{number of parking spaces}} \qquad (4.11)$$

Identification of Parking Generators
This phase involves identifying parking generators (for example, shopping centers or transit terminals) and locating these on a map of the study area.

Parking Demand
Information on parking demand is obtained by interviewing drivers at the various parking facilities listed during the inventory. An effort should be made to interview all drivers using the parking facilities on a typical weekday between 8:00 a.m. and 10:00 p.m. Information sought should include (1) trip origin, (2) purpose of trip, and (3) driver's destina-

tion after parking. The interviewer must also note the location of the parking facility, the times of arrival and departure, and the vehicle type.

Parking interviews can also be carried out using the postcard technique, in which stamped postcards bearing the appropriate questions and a return address are handed to drivers or placed under windshield wipers. When this technique is used, usually only about 30 to 50 percent of the cards distributed are returned. It is therefore necessary to record the time and the number of cards distributed at each location, because this information is required to develop expansion factors, which are later used to expand the sample.

Analysis of Parking Data

Analysis of parking data includes summarizing, coding, and interpreting the data so that the relevant information required for decision making can be obtained. The relevant information includes the following:

- Number and duration for vehicles legally parked
- Number and duration for vehicles illegally parked
- Space-hours of demand for parking
- Supply of parking facilities

The analysis required to obtain information on the first two items is straightforward; it usually involves simple arithmetical and statistical calculations. Data obtained from these items are then used to determine parking space-hours.

The space-hours of demand for parking are obtained from the expression

$$D = \sum_{i=1}^{N} (n_i t_i) \tag{4.12}$$

where

D = space vehicle-hours demand for a specific period of time
N = number of classes of parking duration ranges
t_i = midparking duration of the ith class
n_i = number of vehicles parked for the ith duration range

The space-hours of supply are obtained from the expression

$$S = f \sum_{i=1}^{N} (t_i) \tag{4.13}$$

where

S = practical number of space-hours of supply for a specific period of time
N = number of parking spaces available
t_i = total length of time in hours when the ith space can be legally parked on during the specific period
f = efficiency factor

The efficiency factor f is used to correct for time lost in each turnover. It is determined on the basis of the best performance a parking facility is expected to produce. Efficiency factors should therefore be determined for different types of parking facilities—for example, surface lots, curb parking, and garages. Efficiency factors for curb parking, during highest demand, vary from 78 percent to 96 percent; for surface lots and garages, from 75 percent to 92 percent. Average values of f are 90 percent for curb parking, 80 percent for garages, and 85 percent for surface lots.

Example 4.7   Space Requirements for a Parking Garage

The owner of a parking garage located in a CBD has observed that 20 percent of those wishing to park are turned back every day during the open hours of 8 a.m. to 6 p.m. because of lack of parking spaces. An analysis of data collected at the garage indicates that 60 percent of those who park are commuters, with an average parking duration of 9 hr, and the remaining are shoppers, whose average parking duration is 2 hr. If 20 percent of those who cannot park are commuters and the rest are shoppers, and a total of 200 vehicles currently park daily in the garage, determine the number of additional spaces required to meet the excess demand. Assume parking efficiency is 0.80.

Solution:

- Calculate the space-hours of demand using Eq. 4.12:

$$D = \sum_{i=1}^{N}(n_i t_i)$$

Commuters now being served = $0.6 \times 200 \times 9 = 1080$ space-hr
Shoppers now being served = $0.4 \times 200 \times 2 = 160$ space-hr

Total number of vehicles turned away = $\dfrac{200}{0.8} - 200 = 50$

Commuters not being served = $0.2 \times 50 \times 9 = 90$ space-hr
Shoppers not being served = $0.8 \times 50 \times 2 = 80$ space-hr
Total space-hours of demand = $(1080 + 160 + 90 + 80) = 1410$
Total space-hours served = $1080 + 160 = 1240$
Number of space-hours required = $1410 - 1240 = 170$

- Determine the number of parking spaces required from Eq. 4.13:

$$S = f\sum_{i=1}^{N} t_i = 170 \text{ space} - \text{hr}$$

Use the length of time each space can be legally parked on (8 a.m. through 6 p.m. = 10 hr) to determine the number of additional spaces:

$$0.8 \times 10 \times N = 170$$
$$N = 21.25$$

At least 22 additional spaces will be required, since a fraction of a space cannot be used.

SUMMARY

Highway transportation has provided considerable opportunities for people, particularly the freedom to move from place to place at one's will and convenience. The positive aspects of the highway mode, however, go hand in hand with numerous negative aspects, which include traffic congestion, crashes, pollution, and parking difficulties. Traffic and transportation engineers are continually involved in determining ways to reduce these negative effects. The effective reduction of the negative impact of the highway mode of transportation at any location can be achieved only after adequate information is obtained to define the problem and the extent to which the problem has a negative impact on the highway system. This information is obtained by conducting studies to collect and analyze the relevant data. These are generally referred to as traffic engineering studies.

This chapter has presented the basic concepts of the different traffic engineering studies: spot speed studies, volume studies, travel time and delay studies, and parking studies. Spot speed studies are conducted to estimate the distribution of speeds of vehicles in a traffic stream at a particular location on a highway. This is done by recording the speeds of a sample of vehicles at the specified location, from which speed characteristics are derived. These characteristics are the average speed, the median speed, the modal speed, the 85th-percentile speed, the pace, and the standard deviation of the speed. Important factors that should be considered in planning a speed study include the location for the study, the time of day, the duration of the study, and the minimum sample size necessary for the limit of acceptable error. Traffic volume studies entail the collection of data on the number of vehicles and/or pedestrians that pass a point on a highway during a specified time period. The data on vehicular volume can be used to determine the average daily traffic, the average peak-hour volume, vehicle classification, and vehicle-miles of travel. Volume data are usually collected manually or by using electronic or mechanical counters; video imaging can also be used. It should be noted, however, that traffic volume varies from hour to hour and from day to day. It is therefore necessary to use expansion factors to adjust periodic counts to obtain representative 24-hour, weekly, monthly, and annual volumes. A travel time study determines the amount of time required to travel from one point to another on a given route. This information is used to determine the delay, which gives a good indication of the level of service on the study section. The methods used to conduct travel time and delay data can be grouped into two general categories: (1) those that require a test vehicle and (2) those that do not. Parking studies are used to determine the demand for and supply of parking facilities in an area. A comprehensive parking study usually involves (1) inventory of existing parking facilities, (2) collection of data on parking accumulation, parking turnover, and parking duration, (3) identification of parking generators, and (4) collection of information on parking demand.

It should be emphasized here that no attempt has been made to present an in-depth discussion of any of these studies, as such a discussion is beyond the scope of this book. However, enough material has been provided to introduce the reader to the subject so that he or she will be able to understand the more advanced literature on the subject.

PROBLEMS

4-1   Describe the different traffic count programs carried out in your state. What data are collected in each program?

4-2   Speed data collected on an urban roadway yielded a standard deviation in speeds of ±4.8 mi/h.

(a) If an engineer wishes to estimate the average speed on the roadway at a 95 percent confidence level so that the estimate is within ±2 mi/h of the true average, how many spot speeds should be collected?

(b) If the estimate of the average must be within ±1 mi/h, what should the sample size be?

4-3   What are the advantages and disadvantages of machine vision (video image detection) when compared with other forms of detection?

4-4   Define the following terms and cite examples of how they are used.

Average annual daily traffic (AADT)

Average daily traffic (ADT)

Vehicle-miles of travel (VMT)

Peak hour volume (PHV)

4-5   A traffic engineer, wishing to determine a representative value of the ADT on 250 highway links having similar volume characteristics, conducted a preliminary study from which the following estimates were made:

Mean volume = 45,750 veh/day

Standard deviation = 3750 veh/day

Determine the minimum number of stations for which the engineer should obtain 24-hour volume counts for a 95-5 precision level.

4-6   The accompanying data show spot speeds collected at a section of highway located in a residential area. Using the student's t test, determine whether there was a significant difference in the average speeds at the 95 percent confidence level.

Before	After	Before	After
40	23	38	25
35	33	35	21
38	25	30	35
37	36	30	30
33	37	38	33
30	34	39	21
28	23	35	28
35	28	36	23
35	24	34	24
40	31	33	27
33	24	31	20
35	20	36	20
36	21	35	30
36	28	33	32
40	35	39	33

4-7 Using the data furnished in Problem 4-6, draw the histogram frequency distribution and cumulative percentage distribution for each set of data and determine: (a) average speed, (b) 85th-percentile speed, (c) 15th-percentile speed, (d) mode, (e) median, and (f) pace.

4-8 Describe the following types of traffic volume counts and explain when they are used: (a) screen line counts, (b) cordon counts, (c) intersection counts, and (d) control counts.

4-9 How are travel time and delay studies used? Describe one method for collecting travel time and delay data at a section of a highway. Explain how to obtain the following information from the data collected: (a) travel time, (b) operational delay, (c) stopped time delay, (d) fixed delay, and (e) travel time delay.

4-10 Table 4.10 shows data obtained in a travel time study on a section of highway using the moving-vehicle technique. Estimate (a) the travel time and (b) the volume in each direction at this section of the highway.

4-11 An engineer, wishing to determine the travel time and average speed along a section of an urban highway as part of an annual trend analysis on traffic operations, conducted a travel time study using the floating-car technique. He carried out 10 runs and obtained a standard deviation of ± 3 mi/h in the speeds obtained. If a 5 percent significance level is assumed, is the number of test runs adequate?

Table 4.10 Travel Time Data for Problem 4-10

Run Direction/ Number	Travel Time (min)	No. of Vehicles Traveling in Opposite Direction	No. of Vehicles That Overtook Test Vehicle	No. of Vehicles Overtaken by Test Vehicle
Northward				
1	5.25	100	2	2
2	5.08	105	2	1
3	5.30	103	3	1
4	5.15	110	1	0
5	5.00	101	0	0
6	5.51	98	2	2
7	5.38	97	1	1
8	5.41	112	2	3
9	5.12	109	3	1
10	5.31	107	0	0
Southward				
1	4.95	85	1	0
2	4.85	88	0	1
3	5.00	95	0	1
4	4.91	100	2	1
5	4.63	102	1	2
6	5.11	90	1	1
7	4.83	95	2	0
8	4.91	96	3	1
9	4.95	98	1	2
10	4.83	90	0	1

4-12   Briefly describe the tasks you would include in a comprehensive parking study for your college campus, indicating how you would perform each task and the way you would present the data collected.

4-13   Select a parking lot on your campus. For several hours, conduct a study of the lot using the methods described in this chapter. From the data collected, determine the turnover and duration. Draw a parking accumulation curve for the lot.

4-14   Data collected at a parking lot indicate that a total of 300 cars park between 8 a.m. and 6 p.m. Ten percent of these cars are parked for an average of 2 hr, 30 percent for an average of 4 hr, and the remaining cars are parked for an average of 10 hr. Determine the space-hours of demand at the lot.

4-15   If 10 percent of the parking bays are vacant on average (between 8 a.m. and 6 p.m.) at the parking lot of problem 4-14, determine the number of parking bays in the parking lot. Assume an efficiency factor of 0.85.

4-16   The owner of the parking lot of problems 4-14 and 4-15 is planning an expansion of her lot to provide adequate demand for the following 5 years. If she has estimated that parking demand for all categories will increase by 5 percent a year, determine the number of additional parking bays that will be required.

REFERENCES

Highway Performance Monitoring Systems (HPMS) Field Manual, U.S. Department of Transportation, Federal Highway Administration, Washington, D.C., April 1994.

IEEE Transactions on Vehicular Technology, vol. 40, no. 1, February 1991.

Manual of Transportation Engineering Studies, Institute of Transportation Engineers, Prentice Hall, Inc., Englewood Cliffs, New Jersey, 1994.

Traffic Monitoring Guide, U.S. Department of Transportation, Federal Highway Administration, Washington, D.C., October 1992.

ADDITIONAL READINGS

Brown, A.M., and A. Churly, "Parking Surveys and the Use of VISTA in West Sussex," *Traffic Engineering and Control* 23(12):590–594, December 1982.

Hague, P.W., "Photographic Parking Duration Studies," *Traffic Engineering and Control* 21(3):123–126, March 1980.

Higgins, T.J., "Parking Requirements for Transit-oriented Development," *Transportation Research Record* 1404(1993):50–54.

Hoang, L.T., and Poteat, V.P., "Estimating VMT by Using Random Sampling Technique," *Transportation Research Record* 779(1980):6–10.

MacCarley, C.A., Hockaday, S.L.M., Need, D., and Taff, S., "Evaluation of Video Image Processing Systems for Traffic Detection," *Transportation Research Record* 1360(1992):46–49.

Phillips, G., and Blake, P., "Estimating Total Annual Traffic Flow from Short Period Count," *Transportation Planning and Technology* 6(1980):169–174.

Robertson, D.H., Hummer, J.E., and Nelson, D.C., *Manual of Transportation Engineering Studies,* Institute of Transportation Engineers, 1994.

CHAPTER 5

Highway Safety

As the number of motor vehicles and vehicle-miles of travel increases throughout the world, the exposure of the population to traffic crashes also increases. Traffic exposure in the United States was 2625 billion vehicle-miles per year in 1997. Highway safety is a worldwide problem; with over 500 million cars and trucks in use, more than 500,000 people die each year in motor vehicle crashes, and about 15 million are injured. In the United States, motor vehicle crashes are the leading cause of death for people between the ages of 1 to 34 years and rank third as the most significant cause of years of potential life lost, after cardiac disease and cancer. In the United States, between 1966 and 1997, the number of vehicle-miles traveled has increased from about one trillion to 2.6 trillion, whereas fatality rates have declined from 5 per 100 million vehicle-miles to less than 2 per 100 million vehicle-miles. In 1998, there were approximately 40,000 fatalities on the nation's highways, compared with 55,000 in the mid-1970s.

Traffic and highway engineers are continually engaged in working to ensure that the street and highway system is designed and operated such that highway accident rates can be reduced. They also work with law enforcement officials and educators in a team effort to ensure that traffic laws, such as those regarding speed limits and drinking, are enforced, and that motorists are educated about their responsibility to drive defensively and to understand and obey traffic regulations.

States develop, establish, and implement systems for managing highway safety. There are five major safety programs that are addressed by states in developing a safety management program. They are:

1. Coordinating and integrating broad-base safety programs, such as motor carrier–, corridor-, and community-based safety activities, into a comprehensive management approach for highway safety

2. Identifying and investigating hazardous highway safety problems and roadway locations and features, including railroad-highway grade crossings, and establishing countermeasures and priorities to correct the identified hazards or potential hazards

3. Ensuring early consideration of safety in all highway construction programs and projects

4. Identifying safety needs of special user groups (such as older drivers, pedestrians, bicyclists, motorcyclists, commercial motor carriers, and hazardous materials carriers) in the planning, design, construction, and operation of the highway system

5. Routinely maintaining and upgrading safety hardware (including highway–rail crossing warning devices), highway elements, and operational features.

ISSUES INVOLVED IN TRANSPORTATION SAFETY

Crashes or Accidents

"Accident" is the commonly accepted word for an occurrence involving one or more transportation vehicles in a collision that results in property damages, injury, or death. The term "accident" implies a random event that occurs for no apparent reason other than "it just happened." Have you ever been in a situation where something happened that was unintended? Your immediate reaction might have been "sorry, it was just an accident."

In recent years, the National Highway Traffic Safety Administration has suggested replacing the word "accident" with the word "crash" because "crash" implies that the collision could have been prevented or its effect minimized by modifying driver behavior, vehicle design (called "crashworthiness"), roadway geometry, or the traveling environment. The word "crash" is not universally accepted terminology for all transportation modes and is most common in the context of highway and traffic incidents. In this chapter both terms, "crashes" and "accidents," are used because while crashes is the preferred term, in some situations the word "accident" may be more appropriate.

What Causes Transportation Crashes

The occurrence of a transportation crash presents a challenge to safety investigators. In every instance the question arises, "What sequence of events or circumstances contributed to the incident that resulted in injury, loss of lives, or property damage?" In some cases the answer may be a simple one. For example, the cause of a single car crash may be that the driver fell asleep at the wheel, crossed the highway shoulder, and crashed into a tree. In other cases the answer may be complex, involving many factors that, acting together, caused the crash to occur.

Most people know that the *Titanic,* an "unsinkable" ocean liner, went to the bottom of the sea with nearly 1200 passengers and crew. Common belief is that the cause of this tragedy was that the ship struck an iceberg. However, the actual reason is much more complex and involved many factors. These include too few life boats, lack of wireless information regarding ice fields, poor judgment by the captain, an inadequate on-board warning system, overconfidence in the technology of ship construction, and flaws in the rivets that fastened the ship's steel plates.

Based on these illustrations and other similar cases it is possible to construct a general list of the categories of circumstances that could influence the occurrence of transportation crashes. If the factors that have contributed to crash events are identified, it is then possible to modify and improve the transportation system. In the future, with the reduction or elimination of the crash-causing factor, a safer transportation system is likely to result.

Factors Involved in Transportation Crashes

While the causes of crashes are usually complex and involve several factors, they can be considered in four separate categories. These are actions by the driver or operator, mechanical condition of the vehicle, geometric characteristics of the roadway, and the physical or climatic environment in which the vehicle operates. These factors will be reviewed in the following section.

The Driver.   The major contributing causes of many crash situations is the performance of the driver of one or both (in multiple vehicle crashes) of the vehicles involved. Driver error can occur in many ways. These include inattention to the roadway and surrounding traffic, failure to yield the right of way and or traffic laws. These "failures" can occur as a result of unfamiliarity with roadway conditions, traveling at high speeds, drowsiness, drinking, and using a cell phone or other distractions within the vehicle.

The Vehicle.   The mechanical condition of a vehicle can be the cause of transportation crashes. Faulty brakes in heavy trucks have caused crashes. Other reasons are failure of the electrical system, worn tires, and the location of the vehicle's center of gravity. Off-road crashes require early notification to rescue teams, especially in rural areas.

The Roadway.   The condition and quality of the roadway, which includes the pavement, shoulders, intersections, and the traffic control system, can be a factor in a crash. Highways must be designed to provide adequate sight distance at the design speed or motorists will be unable to take remedial action to avoid a crash. Traffic signals must provide adequate decision sight distance when the signal goes from green to red. Railroad grade crossings must be designed to operate safely and thus minimize crashes between highway traffic and rail cars. The superelevation on highway and railroad curves must be carefully laid out with the correct radius and appropriate transition sections to assure that vehicles can negotiate curves safely.

The Environment.   The physical and climatic environment surrounding a transportation vehicle can also be a factor in the occurrence of transportation crashes. The most common is weather. All transportation systems function at their best when the weather is sunny and mild and the skies are clear. Weather on roads can contribute to highway crashes; for example, wet pavement reduces stopping friction and can cause vehicles to hydroplane. Many severe crashes have been caused by fog because vehicles traveling at high speed are unable to see other vehicles ahead that may have stopped or slowed down, creating a multivehicle pile-up. Geography is another environmental cause of transportation crashes. Mountain ranges have been the site of air crashes. Flooded river plains, swollen rivers, and mud slides on the pavement have caused railroad and highway crashes.

This chapter deals with the efforts and the methodology through which highway and traffic engineers evaluate crash data, then redesign and reconstruct the highway system where the potential for high crash rates exists.

THE HIGHWAY SAFETY IMPROVEMENT PROGRAM

In order to evaluate the success or failure of highway improvements, it is necessary to collect data regarding the frequency and severity of crashes at specific locations. The Federal Highway Administration (FHWA) has provided leadership in this area since the 1960s, with legislative actions that have resulted in considerable research in highway safety. In partnership with individual states, FHWA has developed the Highway Safety Improvement Program (HSIP) with their overall objectives of reducing the number and severity of crashes and decreasing the potential for crashes on all highways. The HSIP consists of three components: (1) planning, (2) implementation, and (3) evaluation.

The planning component of the HSIP consists of four processes as shown in Figure 5.1. These are (1) collecting and maintaining data, (2) identifying hazardous locations and elements, (3) conducting engineering studies, and (4) establishing project priorities. Figure 5.1 shows that the information obtained under the planning component serves as input to the two other components, and that results obtained from the evaluation component may also serve as input to the planning component.

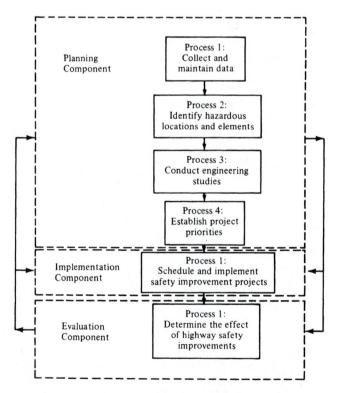

Figure 5.1 Highway Safety Improvement Program at the Process Level

SOURCE: Redrawn from *Highway Safety Engineering Studies Procedural Guide,* U.S. Department of Transportation, Washington, D.C., June 1981.

Collecting and Maintaining Data

Crash data are usually obtained from state and local transportation and police agencies. All relevant information is usually recorded by the police on an accident report form. The type of form used differs from state to state, but a typical completed form will include information on the location, the time of occurrence, roadway and environmental conditions, types and number of vehicles involved, a sketch showing the original paths of the maneuver or maneuvers of the vehicles involved, and the severity (fatal, injury, or property damage only). Figure 5.2 shows the Virginia report form, which is completed by the investigating police officer. Information on minor crashes that do not involve police investigation may be obtained from routine reports given at the police station by the drivers involved, as is required in some states. Sometimes drivers involved in crashes are also required to complete accident report forms, even though the crash is investigated by the police. Figure 5.3 shows the type of form used for this purpose in Virginia.

Storage and Retrieval of Crash Data

Basically, two techniques are used in the storage of crash data. The first technique involves the manual filing of each completed accident report form in the offices of the appropriate police agency. These forms are usually filed either by date, by the name or number of the routes, or by location, which may be identified by intersection and roadway links. Summary tables, which give the number and percentage of each type of crash occurring during a given year at a given location, are also prepared. The location can be a specific spot on the highway or an identifiable length of the highway. This technique is suitable for areas where the total number of crashes is less than 500 per year, although it may be used when the total number is between 500 and 1000 annually. This technique, however, becomes time consuming and inefficient when there are more than 1000 crashes per year.

The second technique involves the use of a computer, where each item of information on the report form is coded and stored in a computer file. This technique is suitable for areas where the total number of crashes per year is greater than 500. With this technique, facilities are provided for storing a large amount of data in a small space. The technique also facilitates flexibility in the choice of methods used for data analysis and permits the study of a large number of crash locations in a short time. There are, however, some disadvantages associated with this technique. These include the high cost of equipment and the requirement of trained computer personnel for the operation of the system. Several agencies are avoiding the necessity of investing large amounts to purchase computer systems by sharing computer facilities. The advent of microcomputers has also made it feasible for relatively small agencies to purchase individual systems.

Several national databanks use computerized systems to store data on national crash statistics. These include the following:

- Highway Performance Monitoring System (HPMS), compiled by the FHWA
- Fatal Analysis Reporting System (FARS), compiled by the National Center for Statistics and Analysis (NCSA) of the National Highway Traffic Safety Administration (NHTSA)

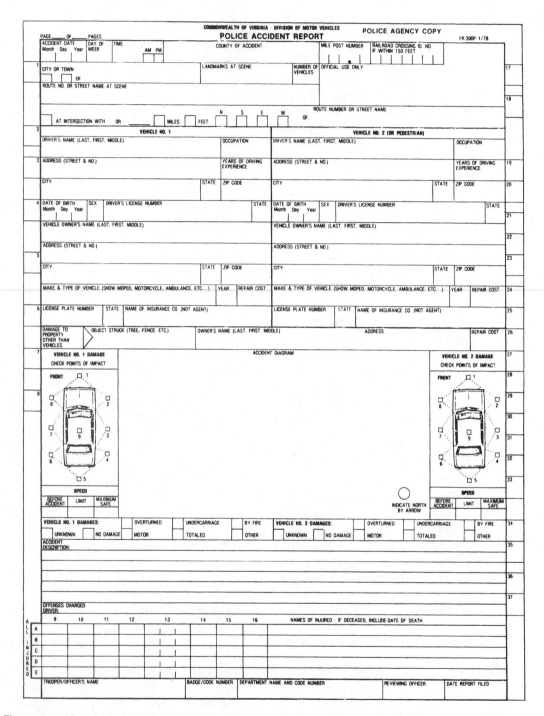

Figure 5.2 Virginia Accident Report Form

SOURCE: Commonwealth of Virginia, Division of Motor Vehicles, Richmond, Va.

COMMONWEALTH OF VIRGINIA
DIVISION OF MOTOR VEHICLES
CITIZEN ACCIDENT REPORT   DMV COPY

ACCIDENT
INFORMATION   (SEE REVERSE SIDE FOR INSTRUCTIONS AND PENALTY FOR NOT FILING)

ACCIDENT DATE MONTH DAY YEAR	DAY OF WEEK	TIME AM PM	INVESTIGATED AT SCENE BY POLICE?	NO. OF VEHICLES	WAS THERE AN INJURY?	WAS THERE A DEATH?	CITY OR COUNTY OF ACCIDENT

ROUTE NO. OR STREET NAME AT SCENE	OR	MILES N E FEET S W	OR	ROUTE NO. OR STREET NAME
	AT INTERSECTION WITH			

VEHICLE INFORMATION

YOUR VEHICLE	OTHER VEHICLE OR PEDESTRIAN INVOLVED
DRIVER'S NAME (LAST, FIRST, MIDDLE)	DRIVER'S NAME (LAST, FIRST, MIDDLE)
ADDRESS (NO. & STREET)	ADDRESS (NO. & STREET)
CITY STATE ZIP CODE	CITY STATE ZIP CODE
DATE OF BIRTH MONTH DAY YEAR SEX DRIVER'S LICENSE NUMBER STATE	DATE OF BIRTH MONTH DAY YEAR SEX DRIVER'S LICENSE NUMBER STATE
VEHICLE OWNER'S NAME (LAST, FIRST, MIDDLE)	VEHICLE OWNER'S NAME (LAST, FIRST, MIDDLE)
ADDRESS (NO. & STREET)	ADDRESS (NO. & STREET)
CITY STATE ZIP CODE	CITY STATE ZIP CODE
DATE OF BIRTH MONTH DAY YEAR SEX OWNER'S DRIVER LICENSE NUMBER STATE	DATE OF BIRTH MONTH DAY YEAR SEX OWNER'S DRIVER LICENSE NUMBER STATE
MAKE & TYPE OF VEHICLE YEAR VEHICLE PARKED?	MAKE & TYPE OF VEHICLE YEAR VEHICLE PARKED?
LICENSE PLATE NUMBER STATE COST TO REPAIR $	LICENSE PLATE NUMBER STATE COST TO REPAIR $
DAMAGE TO PROPERTY OTHER THAN VEHICLES	EST. AMOUNT OF DAMAGES $
WAS VEHICLE INSURED? NAME OF OWNER'S LIABILITY INSURANCE COMPANY (NOT AGENT)	POLICY NUMBER
INSURED'S NAME (LAST, FIRST, MIDDLE)	POLICY PERIOD MONTH DAY YEAR MONTH DAY YEAR

SIGNATURE OF DRIVER	DATE FILLED	IF SIGNED BY PERSON OTHER THAN DRIVER, GIVE REASON

Figure 5.3 Virginia Citizen Accident Report Form

SOURCE: Commonwealth of Virginia, Division of Motor Vehicles, Richmond, Va.

- National Injury Surveillance System (NISS), which compiles data collected at emergency rooms of hospitals
- Bureau of Motor Carrier Series, which contains data on crashes involving passengers and properties of members of the Bureau of Motor Carriers.

Information provided by these databanks may be retrieved by computer techniques for research purposes.

The technique used for retrieving specific crash data depends on the method of storage of the data. When the data are stored manually, the retrieval is also manual. In this case, the file is examined by a trained technician, who then retrieves the appropriate report forms. When the data are stored on computer, retrieval requires only the input of appropriate commands into the computer for any specific data required, and those data are given immediately as output.

Collision Diagrams

These diagrams present pictorial information on individual crashes at a location. Different symbols are used to represent different types of maneuvers, types of crashes, and severity of crashes. The date and time (day or night) at which the crash occurs are also indicated. Figure 5.4 shows a typical collision diagram. One advantage of collision diagrams is that they give information on the location of the crash, which statistical summaries do not give. Collision diagrams may be prepared manually either by retrieving data filed manually or by a computer when data are stored in a computer file.

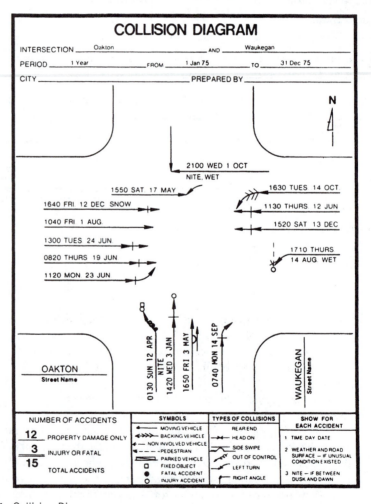

Figure 5.4 Collision Diagram

SOURCE: *Traffic Engineering Handbook,* 4th ed., Institute of Transportation Engineers, Washington, D.C., 1992.

Analysis of Crash Data

The reasons for analyzing traffic data are (1) to identify patterns that may exist, (2) to determine the probable causes with respect to drivers, highways, and vehicles, and (3) to develop countermeasures that will reduce the rate and severity of future crashes. To facilitate the comparison of results obtained from the analysis of crashes at a particular location with those of other locations, one or more crash rates are used. These rates are determined on the basis of exposure data, such as traffic volume and the length of road section being considered. Commonly used rates are rate per million of entering vehicles and rate per 100 million vehicle-miles.

The **rate per million of entering vehicles** (RMEVs) is the number of crashes per million vehicles entering the study location during the study period. It is expressed as

$$RMEV = \frac{A \times 1,000,000}{V} \qquad (5.1)$$

where

$RMEV$ = crash rate per million entering vehicles

A = number of crashes, total or by type occurring in a single year at the location

V = average daily traffic $(ADT) \times 365$

This rate is very often used as a measure of crash rates at intersections.

Example 5.1 Computing Crash Rates at Intersections

The number of all crashes recorded at an intersection in a year was 23, and the average 24-hour volume entering from all approaches was 6500. Determine the crash rate per million entering vehicles (RMEV).

Solution:

$$RMEV = \frac{23 \times 1,000,000}{6500 \times 365} = 9.69 \text{ crashes / million entering vehicles}$$

The **rate per 100 million vehicle miles** (RMVM) is the number of crashes per 100 million vehicle miles of travel. It is obtained from the expression

$$RMVM = \frac{A \times 100,000,000}{VMT} \qquad (5.2)$$

where

A = number of crashes, total or by type at the study location, during a given period

VMT = vehicle miles of travel during the given period

= $ADT \times$ (number of days in study period) $\times$ (length of road)

This rate is very often used as a measure of crash rates on a stretch of highway with similar traffic and geometric characteristics.

Example 5.2   Computing Crash Rates on Roadway Sections

It is observed that 40 traffic crashes occurred on a 17.5-mile-long section of a highway in one year. The ADT on the section was 5000 vehicles.

a.   Determine the rate of total crashes per 100 million vehicle-miles.
b.   Determine the rate of fatal crashes per 100 million vehicle-miles, if 5 percent of the crashes involved fatalities.

Solution: a $RMVM_T = \dfrac{40 \times 100{,}000{,}000}{17.5 \times 5000 \times 365} = 125.24$ crashes/100 million veh/mi

   b $RMVM_F = 125.24 \times 0.05 = 6.26$ crashes/100 million veh/mi

---

Note that any crash rate may be given in terms of the total number of crashes occurring or in terms of a specific type of crash. Therefore, it is important that the basis on which crash rates are determined be clearly stated. Comparisons between two locations can be made only using results obtained from an analysis based on similar exposure data.

Crash Patterns

Two commonly used techniques to determine crash patterns are (1) expected value analysis and (2) cluster analysis. A suitable summary of crash data can also be used to determine patterns.

Expected value analysis is a mathematical method used to identify locations with abnormal crash characteristics. It should be used only to compare sites with similar characteristics (for example, geometrics, volume, traffic control), since the analysis does not consider exposure levels. The analysis is carried out by determining the average number of a specific type of crash occurring at several locations with similar geometric and traffic characteristics. This average, adjusted for a given level of confidence, indicates the "expected" value for the specific type of crash. Locations with values higher than the expected value are considered as overrepresenting that specific type of crash. The expected value can be obtained from the expression

$$EV = \bar{x} \pm ZS \tag{5.3}$$

where

EV = expected range of crash frequency
$\bar{x}$ = average number of crashes per location
S = estimated standard deviation of crash frequencies
Z = the number of standard deviations corresponding to the required confidence level

Example 5.3   Identifying High-Crash-Rate Locations

Data collected for three consecutive years at an intersection study site show that 14 rear-end collisions and 10 left-turn collisions occurred during the 3-year period. Data collected at 10 other intersections with similar geometric and traffic characteristics give the information shown in Table 5.1. Determine whether any type of crash is over-represented at the study site for a 95 percent confidence level $Z = 1.96$).

Solution:

Average number of rear-end collisions at 10 control sites = 7.40

Standard deviation of rear-end collisions at 10 control sites = 1.50

Expected range (95% confidence level) = $7.40 \pm 1.5 \times 1.96 = 10.34$

Number of rear-end collisions at study site = 14

Rear-end collisions are therefore overrepresented at the study site at 95 percent confidence level, since 14 > 10.34.

Average number of left-turn collisions at 10 control sites = 6.90

Standard deviation of left-turn collisions at 10 control sites = 3.07

Expected range (95% confidence level) = $6.90 \pm 3.07 \times 1.96 = 12.92$

Number of left-turn collisions at study site = 10

Left-turn collisions are not overrepresented at the study site, since 10 < 12.92.

Table 5.1   Number of Rear-End and Left-Turn Collisions at Ten Control Stations for Three Consecutive Years

Control Site	Rear-End Collisions	Left-Turn Collisions
1	8	11
2	5	12
3	7	4
4	8	5
5	6	8
6	8	3
7	9	4
8	10	9
9	6	7
10	7	6
Average	7.40	6.90
Standard deviation	1.50	3.07

Cluster analysis involves the identification of a particular characteristic from the crash data obtained at a site. It identifies any abnormal occurrence of a specific crash type in comparison with other types of crashes at the site. For example, if there are two rear-end, one right-angle, and six left-turn collisions at an intersection during a given year, the left-turn collisions could be defined as a *cluster* or grouping, with abnormal occurrence at the site.

However, it is very difficult to assign discrete values that can be used to identify crash patterns. This is because crash frequencies, which are the basis for determining patterns, differ considerably from site to site. It is sometimes useful to use exposure data, such as traffic volumes, to define patterns of crash rates. Care must be taken, however, to use the correct exposure data. For example, if total intersection volume is used to determine left-turn crash rates at different sites, these rates are not directly comparable because the percentages of left-turn vehicles at these sites may be significantly different. Because of these difficulties, it is desirable to use good engineering judgment when this approach is being used.

Methods of Summarizing Crash Data

A summary of crashes can be used to identify safety problems that may exist at a particular site. It can also be used to identify the crash pattern at a site, from which possible causes may be identified, leading to the identification of possible remedial actions (countermeasures).

There are five different ways in which a crash at a site can be summarized:

- Type
- Severity
- Contributing circumstances
- Environmental conditions
- Time periods

Summary by Type.   This method of summarizing crashes involves the identification of the pattern of crashes at a site, based on the specific types of crashes. The types of crashes commonly used are

- Rear-end
- Right-angle
- Left-turn
- Fixed object
- Sideswipes
- Pedestrian-related
- Run off road
- Head-on
- Parked vehicle
- Bicycle-related

Summary by Severity.   This method involves listing each crash occurring at a site under one of three severity classes: fatal (F), personal injury (PI), and property damage (PD). Fatal crashes are those that result in at least one death. Crashes that result in injuries, but no deaths, are classified as personal injury. Crashes that result in neither death nor injuries but involve damage to property are classified as property damage.

This method of summarizing crashes is commonly used to make comparisons at different locations by assigning a weighted scale to each crash based on its severity. Several weighting scales have been used, but a typical one is given as

Fatality = 12
Personal injury = 3
Property damage only = 1

For example, if 1 fatal crash, 3 personal injury crashes, and 5 property damage crashes occurred during a year at a particular site, the severity number of the site is obtained as follows:

$$\text{Severity number} = (12 \times 1) + (3 \times 3) + (1 \times 5) = 26$$

The disadvantage in using the severity number is the large difference between the severity scales for fatal and property damage crashes. This may overemphasize the seriousness of crashes resulting in fatalities over those resulting in property damage. For example, a site with only one fatal crash will be considered much more dangerous than a site with nine property damage crashes. This effect can be reduced by using a lower weighting, for example, 8 for fatal crashes, especially at locations where fatal crashes are very rare in comparison with other crashes.

Summary by Contributing Circumstances. In this method, each crash occurring at a site is listed under one of three contributing factors: (1) human factors, (2) environmental factors, and (3) vehicle-related factors. The necessary information is usually obtained from accident reports.

Summary by Environmental Conditions. This method categorizes crashes based on the environmental conditions that existed at the time of the crashes. Two main categories of environmental conditions are (1) lighting condition, that is, daylight, dusk, dawn, or dark, and (2) roadway surface condition, that is, dry, wet, snowy/icy. This method of summarizing crashes facilitates the identification of possible causes of crashes and safety deficiencies that may exist at a particular location. The expected value method may be used to ascertain whether crash rates under a particular environmental condition are significantly greater at one site than at other similar sites.

Summary by Time Period. This method categorizes all crashes under different time periods to identify whether crash rates are significantly higher during any specific time periods. Three different time periods can be used: (1) hour, (2) day, and (3) month. This method of summarizing data also facilitates the use of the expected value method to identify time periods during which crash occurrences are.

Determiniung Possible Causes of Crashes

Having identified the hazardous locations and the crash pattern, the next stage in the data analysis is to determine possible causes. The types of crashes identified are matched with a list of possible causes from which several probable causes are identified. Table 5.2 shows a list of possible causes for different types of crashes. The environmental conditions existing at the time may also help in identifying possible causes.

Table 5.2 Probable Causes for Different Types of Crashes

Pattern	Probable Cause
Left-turn head-on collisions	• Large volume of left-turns • Restricted sight distance • Too short amber phase • Absence of special left-turning phase • Excessive speed on approaches
Right-angle collisions at signalized intersections	• Restricted sight distance • Excessive speed on approaches • Poor visibility of signal • Inadequate signal timing • Inadequate roadway lighting • Inadequate advance intersection warning signs • Large total intersection volume
Right-angle collisions at unsignalized intersections	• Restricted sight distance • Large total intersection volume • Excessive speed on approaches • Inadequate roadway lighting • Inadequate advance intersection warning signals • Inadequate traffic control devices
Rear-end collisions at unsignalized intersections	• Driver not aware of intersection • Slippery surface • Large number of turning vehicles • Inadequate roadway lighting • Excessive speed on approach • Lack of adequate gaps • Crossing pedestrians
Rear-end collisions at signalized intersections	• Slippery surface • Large number of turning vehicles • Poor visibility of signals • Inadequate signal timing • Unwarranted signals • Inadequate roadway lighting
Pedestrian-vehicle collisions	• Restricted sight distance • Inadequate protection for pedestrians • School crossing area • Inadequate signals • Inadequate phasing signal

SOURCE: Adapted from *Highway Safety Engineering Studies Procedural Guide,* U.S. Department of Transportation, Washington, D.C., June 1981.

Identifying Hazardous Locations and Elements

Hazardous locations are sites where crash frequencies, calculated on the basis of the same exposure data, are higher than the expected value for other similar locations or conditions. Any of the crash rates or summaries described earlier may be used to identify hazardous locations. A common method of analysis involves the determination of crash rates based on the same exposure data for the study site with apparent high rates and several other sites with similar traffic and geometric characteristics. An appropriate statistical test such as the expected value analysis is then performed to determine whether the apparent high crash rate at the study site is statistically significant. If the statistical test shows that the apparent high crash rate is significantly higher, an abnormal rate of crashes at the test location is likely and that site is to be considered a hazardous location.

A technique that is used to identify possible hazardous locations is known as the critical crash rate factor method. Since traffic crashes are random occurrences and can be considered as "rare events," it is not possible to identify hazardous locations simply on the basis of the number of crashes. Rather the critical rate method incorporates the traffic volume to determine if the crash rate at a particular location is significantly higher than the average for the type of facility. Statistics are typically maintained by facility type, which is determined by factors such as traffic volume, traffic control, number of lanes, land-use density, and functional classification (see Chapter 16).

The critical crash rate factor method involves the following expression:

$$CR = AVR + \frac{0.5}{TB} + TF \sqrt{\frac{AVR}{TB}} \tag{5.4}$$

where
$\quad CR$ = critical crash rate, per 100 million vehicle-miles or per million entering vehicles
$\quad AVR$ = average crash rate for the facility type
$\quad TF$ = test factor, standard deviation at a given confidence level (S in Eq. 5.3)
$\quad TB$ = traffic base, veh/mi 100 million vehicle-miles or veh/million entering vehicles

---

Example 5.4   Identifying Hazardous Locations

An urban arterial street segment, 0.2 miles long, has an average annual daily traffic (AADT) of 15,400 vehicles per day. In a three-year period, there have been 8 crashes resulting in death and/or injuries and 15 involving property damage only. The statewide average crash experience for similar types of roadway is 375 per 100 million veh/mi, for a three-year period, of which 120 involved death and/or injury and 255 caused property damage only. In identifying hazardous locations, consider that a single death/injury crash is equivalent to 3 property damage crashes. Use a 95 percent confidence level.

Solution:

Step 1.   Calculate the traffic base, TB.

$$TB = \frac{\text{Years} \times AADT \times \text{segment length} \times 365 \text{ days per year}}{100 \text{ million}}$$

$$TB = \frac{3 \times 15{,}400 \times 0.20 \times 365}{100 \times 10^6}$$

$$TB = 0.0337 \text{ veh/mi}/100 \text{ million veh/mi}$$

Step 2. Calculate the 3 year average crash rate for this type of facility.

$$AVR = 3 \times 120 + 255 = 615 \text{ equivalent crashes/100 million veh/mi}$$

Step 3. Select a test factor based on confidence level. Since a confidence level of 95 percent is specified, the test factor is 1.96.

Step 4. Compute the critical rate.

$$CR = AVR + \frac{0.5}{TB} + TF\sqrt{\frac{AVR}{TB}}$$

$$= 615 + \frac{0.5}{0.0337} + 1.96\sqrt{\frac{615}{0.0337}}$$

$$= 895 \text{ crashes per 100 mvm per 3 years}$$

Step 5. Determine the ratio of actual crash occurrence for the segment with respect to the critical rate.

$$\text{segment crash history} = \frac{3 \times 8 + 15}{0.0337}$$

$$= 1157 \text{ equivalent crashes/100 million veh/mi/3 years}$$

$$\text{crash ratio} = \frac{\text{segment crash history}}{\text{statewide crash history}}$$

$$= \frac{1157}{895} = 1.29$$

Since the ratio exceeds one, a safety problem is likely to exist. Specific crash records for the segment should be reviewed so that appropriate measures can be recommended.

Conducting Engineering Studies

After a particular location has been identified as hazardous, a detailed engineering study is performed to identify the safety problem. Once the safety problem is identified, suitable safety-related countermeasures can be developed.

The first task in this subprocess is an in-depth study of the crash data obtained at the hazardous site. The results of the analysis will indicate the type or types of crashes that predominate or that have abnormal frequency rates. Possible causes can then be identified from Table 5.2. However, the list of possible causes obtained at this stage is preliminary, and personal knowledge of the site, field conditions, and police accident reports should all be used to improve this list.

The next task is to conduct a field review of the study site. This review involves an inspection of the physical condition and an observation of traffic operation at the site. Information obtained from this field review is then used to confirm the existence of physical deficiencies, based on the pattern of crashes, and to refine the list of possible causes.

The refined list is used to determine what data will be required to identify the safety deficiencies at the study site. Table 5.3 gives a partial list of data needs for different possible causes of crashes. A complete list is given in a Department of Transportation publication. After identifying the data needs, existing records will then be reviewed to determine whether the required data are available. Care must be taken to ensure that any existing data are current and are related to the time for which the study is being conducted. In cases where the necessary data are available, it will not be necessary to carry out specific engineering studies. When the appropriate data are not available, the engineering studies identified from Table 5.3 will then be conducted. Some of these studies are described earlier in this chapter.

Table 5.3   Data Needs for Different Possible Causes of Crashes

Possible Causes	Data Needs	Procedures to Be Performed
Left-turn head-on collisions		
Large volume of left turns	• Volume data • Vehicle conflicts • Roadway inventory • Signal timing and phasing • Travel time and delay data	• Volume Study • Traffic Conflict Study • Roadway Inventory Study • Capacity Study • Travel Time and Delay Study
Restricted sight distance	• Roadway inventory • Sight distance characteristics • Speed characteristics	• Roadway Inventory Study • Sight Distance Study • Spot Speed Study
Too short amber phase	• Speed characteristics • Volume data • Roadway inventory • Signal timing and phasing	• Spot Speed Study • Volume Study • Roadway Inventory Study • Capacity Study
Absence of special left-turning phase	• Volume data • Roadway inventory • Signal timing and phasing • Delay data	• Volume Study • Roadway Inventory Study • Capacity Study • Travel Time and Delay Study
Excessive speed on approaches	• Speed characteristics	• Spot Speed Study

Continued

Table 5.3 Data Needs for Different Possible Causes of Crashes (*continued*)

Rear-end collisions at unsignalized intersections		
Driver not aware of intersection	• Roadway inventory • Sight distance characteristics • Speed characteristics	• Roadway Inventory Study • Sight Distance Study • Spot Speed Study
Slippery surface	• Pavement skid resistance characteristics • Conflicts resulting from slippery surface	• Skid Resistance Study • Weather-Related Study • Traffic Conflict Study
Large number of turning vehicles	• Volume data • Roadway inventory • Conflict data	• Volume Study • Roadway Inventory Study • Traffic Conflict Study
Inadequate roadway lighting	• Roadway inventory • Volume data • Data on existing lighting	• Roadway Inventory Study • Volume Study • Highway Lighting Study
Excessive speed on approaches	• Speed characteristics	• Spot Speed Study
Lack of adequate gaps	• Roadway inventory • Volume data • Gap data	• Roadway Inventory Study • Volume Study • Gap Study
Crossing pedestrians	• Pedestrian volumes • Pedestrian/vehicle conflicts • Signal inventory	• Volume Study • Pedestrian Study • Roadway Inventory Study
Rear-end collisions at signalized intersections		
Slippery surface	• Pavement skid resistance characteristics • Conflicts resulting from slippery surface	• Skid Resistance Study • Weather-Related Study • Traffic Conflict Study
Large number of turning vehicles	• Volume data • Roadway inventory • Conflict data • Travel time and delay data	• Volume Study • Roadway Inventory Study • Traffic Conflict Study • Delay Study
Poor visibility of signals	• Roadway inventory • Signal review • Traffic conflicts	• Roadway Inventory Study • Traffic Control Device Study • Traffic Conflict Study

SOURCE: Adapted from *Highway Safety Engineering Studies Procedural Guide,* U.S. Department of Transportation, Washington, D.C., June 1981.

The results of these studies are used to determine traffic characteristics of the study site, from which specific safety deficiencies at the study site are determined. For example, a sight distance study at an intersection may reveal inadequate sight distance at that intersection, which results in an abnormal rate of left-turn head-on collisions. Similarly, a volume study, which includes turning movements at an intersection with no separate left-turn phase, may indicate a high volume of left-turn vehicles, which suggests that a deficiency is the absence of a special left-turn phase.

Having identified the safety deficiencies at the study site, the next task is to develop alternative countermeasures to alleviate the identified safety deficiencies. A partial list of general countermeasures for different types of possible causes is shown in Table 5.4. The

Table 5.4 General Countermeasures for Different Safety Deficiencies

Probable Cause	*General Countermeasure*
Left-turn head-on collisions	
Large volume of left turns	• Create one-way street • Widen road • Provide left-turn signal phases • Prohibit left turns • Reroute left-turn traffic • Channelize intersection • Install stop signs (see MUTCD)★ • Revise signal sequence • Provide turning guidelines (if there is a dual left-turn lane) • Provide traffic signal if warranted by MUTCD★ • Retime signals
Restricted sight distance	• Remove obstacles • Provide adequate channelization • Provide special phase for left-turning traffic • Provide left-turn slots • Install warning signs • Reduce speed limit on approaches
Too short amber phase	• Increase amber phase • Provide all-red phase
Absence of special left-turning phase	• Provide special phase for left-turning traffic
Excessive speed on approaches	• Reduce speed limit on approaches
Rear-end collisions at unsignalized intersections	
Driver not aware of intersection Slippery surface	• Install/improve warning signs • Overlay pavement • Provide adequate drainage • Groove pavement • Reduce speed limit on approaches • Provide "slippery when wet" signs

Continued

Table 5.4 General Countermeasures for Different Safety Deficiencies (*continued*)

Rear-end collisions at unsignalized intersections	
Large numbers of turning vehicles	• Create left- or right-turn lanes • Prohibit turns • Increase curb radii
Inadequate roadway lighting	• Improve roadway lighting
Excessive speed on approach	• Reduce speed limit on approaches
Lack of adequate gaps	• Provide traffic signal if warranted (see MUTCD)★ • Provide stop signs
Crossing pedestrians	• Install/improve signing or marking of pedestrian crosswalks

Rear-end collision at signalized intersections	
Slippery surface	• Overlay pavement • Provide adequate drainage • Groove pavement • Reduce speed limit on approaches • Provide "slippery when wet" signs
Large number of turning vehicles	• Create left- or right-turn lanes • Prohibit turns • Increase curb radii • Provide special phase for left-turning traffic
Poor visibility of signals	• Install/improve advance warning devices • Install overhead signals • Install 12-in. signal lenses (see MUTCD)★ • Install visors • Install back plates • Relocate signals • Add additional signal heads • Remove obstacles • Reduce speed limit on approaches
Inadequate signal timing	• Adjust amber phase • Provide progression through a set of signalized intersections • Add all-red clearance
Unwarranted signals	• Remove signals (see MUTCD)★
Inadequate roadway lighting	• Improve roadway lighting

★*Manual on Uniform Traffic Control Devices,* published by FHWA.
SOURCE: Adapted from *Highway Safety Engineering Studies Procedural Guide,* U.S. Department of Transportation, Washington, D.C., June 1981.

selection of countermeasures should be made carefully by the traffic engineer based on his or her personal knowledge of the effectiveness of each countermeasure considered in reducing the rate at similar sites for the specific types of crashes being considered. Note that countermeasures that are very successful in achieving significant major benefits in one part of the country may not be that successful in another part of the country due to the complexity of the interrelationship that exists among the traffic variables.

Crash Reduction Capabilities of Countermeasures

Crash reduction capabilities are used to estimate the expected reduction that will occur during a given period as a result of implementing a proposed countermeasure. This estimate can be used to carry out an economic evaluation of alternate counter-measures. Reduction capabilities are given in terms of factors that have been developed by several states and agencies. These factors are usually based on the evaluation of data obtained from safety projects and can be obtained from state agencies involved in crash analysis.

In using the crash reduction factor to determine the reduction in crashes due to the implementation of a specific countermeasure, the following equation is used:

$$\text{Crashes prevented} = N \times CR \; \frac{(ADT \text{ after improvement})}{(ADT \text{ before improvement})} \tag{5.5}$$

where

N = expected number of crashes if countermeasure is not implemented and if the traffic volume remains the same

CR = crash reduction factor for a specific countermeasure (some states use the term AR for accident reduction)

ADT = average daily traffic

Example 5.5   Expected Reduction in Crashes Due to a Safety Improvement

The CR factor for a specific type of countermeasure = 30 percent; the ADT before improvement = 7850 (average over 3-year period); and the ADT after improvement = 9000. Over the 3-year before improvement period, the number of specific types of crashes occurring per year = 12, 14, and 13. Determine the expected reduction in number of crashes occurring after the implementation of the countermeasure.

Solution:

$$\text{Average number of crashes / year} = 13$$

$$\text{Crashes prevented} = \frac{13 \times 0.30 \times 9000}{7850}$$

$$= 4.47 \text{ (4 crashes)}$$

This indicates that, if the ADT increases to 9000, the expected number of crashes with the implementation of the countermeasure will be less by 4 every year than the expected number of crashes without the countermeasure. Note that in using Eq. 5.5 the value of N is the average number of crashes during the study period. The value used for the ADT before improvement is also the average of the ADTs throughout the study period.

It is sometimes necessary also to consider multiple countermeasures at a particular site. In such cases, the overall crash reduction factor is obtained from the individual crash reduction factors by using Eq. 5.6, which was proposed by Roy Jorgensen and Associates.

$$CR = CR_1 + (1 - CR_1)CR_2 + (1 - CR_1)(1 - CR_2)CR_3 + \dots$$
$$+ (1 - CR_1)\dots(1 - CR_{m-1})CR_m \quad (5.6)$$

where

CR = overall crash reduction factor for multiple mutually exclusive
    improvements at a single site
CR_i = crash reduction factor for a specific countermeasure i
m = number of countermeasures at the site

In using Eq. 5.6, it is first necessary to list all the individual countermeasures in order of importance. The countermeasure with the highest reduction factor will be listed first, and its reduction factor will be designated CR_1; the countermeasure with the second highest reduction factor will be listed second, with its reduction factor designated CR_2; and so on.

Example 5.6   Crash Reduction Due to Multiple Safety Improvements

At a single location three countermeasures with CRs of 40 percent, 28 percent, and 20 percent are proposed. Determine the overall CR factor.

Solution:

$CR_1 = 0.40$
$CR_2 = 0.28$
$CR_3 = 0.20$

$$CR = 0.4 + (1 - 0.4)0.28 + (1 - 0.4)(1 - 0.28)0.2$$
$$= 0.4 + 0.17 + 0.09$$
$$= 0.66$$

Establishing Project Priorities

Economic Analysis

The purpose of this task is to determine the economic feasibility of each set of countermeasures and to determine the best alternative among feasible mutually exclusive countermeasures. This involves the use of many of the techniques discussed in Chapter 13. Benefits are determined on the basis of expected number of crashes that will be prevented if a specific proposal is implemented, and costs are the capital and continuing costs for constructing and operating the proposed countermeasure.

Table 5.5   National Highway Traffic Safety Administration Accident Unit Costs

PDO	Minor Injury	Moderate Injury	Serious Injury	Severe Injury	Critical Injury	Fatal
$1,481	$6,145	$26,807	$84,189	$158,531	$589,055	$702,281

SOURCE: *The Economic Cost of Motor Vehicle Crashes, 1990,* National Highway Traffic Safety Administration, U.S. Department of Transportation, Washington, D.C., September 1992.

The benefit may be obtained in monetary terms by multiplying the expected number of prevented crashes by an assigned cost for each type of crash severity. Table 5.5 shows costs proposed by the National Highway Traffic Safety Administration (NHTSA). Note, however, that individual states have determined crash costs that are applicable to them. When such information is available, it is advisable to use it.

Implementation and Evaluation

The scheduling and implementation of the proposal selected is the next step. The evaluation component involves the determination of the effect of the highway safety improvement. This includes the collection of data for a period after the implementation of the improvement to determine whether the anticipated benefits are actually accrued. This task is important, since the information obtained will provide valuable data for other similar projects. This is discussed further in Chapter 14, "Transportation Systems Management," since a common reason for implementing a TSM action is to reduce crash rates at a particular location or stretch of highway.

EFFECTIVENESS OF SAFETY DESIGN FEATURES

In 1990, FHWA and NHTSA convened a symposium on effective safety measures, resulting in the identification of 11 short-term countermeasures for emphasis and funding support. These are listed below to illustrate the range of techniques that cities and states are implementing in an ongoing effort to produce a safer highway system. State, local, and federal activities that are underway provide cost-effective information, ideas, and resources which have been used to reduce motor vehicle fatalities.

Pedestrian Safety Improvements
- Initiate a national pedestrian awareness campaign and establish pedestrian safety as a priority area
- Provide engineering improvements to improve pedestrian safety

Driver Behavior and Performance
- Provide improved signing, marking, and delineation
- Improve work zone safety management practices
- Increase speed and other moving violations enforcement through work zones and high-crash locations

Roadway and Roadside Safety
- Work with utility companies to relocate or eliminate utility poles with a history of being hit
- Identify and prioritize a list of high-crash locations for corrective action

Commercial Motor Vehicle Safety
- Reauthorize the Motor Carrier Safety Assistance Program
- Enhance the identification of problem carriers in order to increase selective enforcement
- Increase mobile road enforcement

Corridor Safety Improvement Projects
- Promote and implement Corridor Safety Improvement Projects

The FHWA has published a series of reports that summarize the results of research dealing with safety effectiveness of highway design features. These reports provide useful information about the relationship between crashes and highway geometrics. Among the features to be considered in this chapter are (1) access control, (2) alignment, (3) cross sections, (4) intersections, and (5) pedestrians and bicyclists. Research results that spanned a 30-year period were examined, and in some instances studies dating before 1973 were found to be the most definitive available. Design features are discussed in the following sections based on the information from these FHWA reports.

Access Control

The effects of geometrics on traffic crashes have produced a variety of findings which are not always definitive because often more than one factor may have caused the crash to occur. Furthermore, it is difficult to conduct studies in a controlled environment, and often researchers must rely on data collected by others under a variety of circumstances. Despite these difficulties, research findings over an extended period have confirmed a strong relationship between access control and safety.

Access control is defined as some combination of at-grade intersections, business and private driveways, and median cross overs. For any given highway, access control can range from full control, such as an interstate highway, to no access control, common on most urban highways. The reason why access control improves safety is because there are fewer unexpected events caused by vehicles entering and leaving the traffic stream at slower speeds, resulting in less interference with through traffic.

The effect of control of access is illustrated in Table 5.6, which shows that the total crash rate per million vehicle-miles is almost three times as great on roads in urban areas with no access control than on fully controlled highways. This finding underscores the safety value of the Interstate system compared with other parallel roads where access is either partial or nonexisting.

Similarly, the increase in roadside development, which creates an increased number of at-grade intersections and businesses with direct access to the highway, will also significantly increase crashes. Table 5.7 shows how crash rates increase on a two-lane rural highway when the number of access points increases. For example, when the number of intersections per mile increases from 2 to 20, the crash rate per 100 million vehicle-miles increases by more than 600 percent. Figure 5.5 depicts a similar finding for crash

Table 5.6   Effect of Access Control on Crash Rates

| | Crash Rates per Million Vehicle-Miles | | | |
| | Urban | | Rural | |
Access Control	Total	Fatal	Total	Fatal
Full	1.86	0.02	1.51	0.03
Partial	4.96	0.05	2.11	0.06
None	5.26	0.04	3.32	0.09

SOURCE: *Safety Effectiveness of Highway Design Features,* U.S. Department of Transportation, Federal Highway Administration, Washington, D.C., 1992, Volume I.

Table 5.7   Effect of Access Points on Crash Rates on Two-Lane Rural Highways

Intersections per Mile	Businesses per Mile	Crash Rate★
0.2	1	126
2.0	10	270
20.0	100	1718

★Crashes/100 million vehicle-miles.
SOURCE: *Safety Effectiveness of Highway Design Features,* Volume I, U.S. Department of Transportation, Federal Highway Administration, Washington, D.C., 1992.

rates on highways related to the number of businesses per mile with direct access to the highway.

There are several mechanisms for reducing crashes due to access, all of which require the elimination of access points from through traffic. Examples include (1) the removal of the access point by closing median openings, (2) frontage road access for business driveways, (3) special turning lanes to separate through vehicles from those vehicles using the access point, and (4) proper signing and pavement markings to warn motorists of changing conditions along the roadway.

Alignment

The geometric design of highways, discussed in Chapter 16, involves three elements: (1) vertical alignment, (2) horizontal alignment, and (3) cross section. Design speed is the determining factor in the selection of the alignment needed for the motorist to have sufficient sight distance to safely stop or reduce speed as required by changing traffic and environmental conditions. A safe design ensures that traffic can flow at a uniform speed while traveling on a roadway that changes in a horizontal or vertical direction.

The design of the vertical alignment, which includes tangent grades and sag or crest vertical curves, is influenced by consideration of terrain, cost, and safety. Generally, crash rates for downgrades are higher than for upgrades. One study reported that only 34.6 percent of crashes occurred on level grade, whereas 65.4 percent occurred on a grade or at the location where grades change.

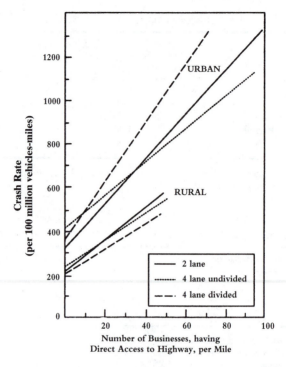

Figure 5.5 Effect of Business Access on Crash Rates on Noninterstate Highways

SOURCE: *Safety Effectiveness of Highway Design Features,* Volume I, U.S. Department of Transportation, Federal Highway Administration, Washington, D.C., November 1992.

The design of the horizontal alignment, which consists of level tangents connected by circular curves, is influenced by design speed and superelevation of the curve itself. Crash rates for horizontal curves are higher than on tangent sections, with rates ranging between 1.5 and 4 times greater than on straight sections. Several factors appear to influence the safety performance of horizontal curves, including (1) traffic volume and mix, (2) geometric features of the curve, (3) cross section, (4) roadside hazards, (5) stopping sight distance, (6) vertical alignment superimposed on horizontal alignment, (7) distance between curves and between curves and the nearest intersection or bridge, (8) pavement friction, and (9) traffic control devices.

The improvement of horizontal curve design involves three steps. First, problem sites must be identified based on crash history and roadway conditions. Second, improvements should be evaluated and implemented. Third, before-and-after studies of crash performance should be conducted to assess the effectiveness of the changes.

Improvements to safety of horizontal curves include (1) reconstructing the curve to make it less sharp, (2) widening lanes and shoulders on curves, (3) adding spiral transitions to curves, (4) increasing the amount of superelevation (up to allowable maximums of 0.08 and 0.10 in urban and rural areas, respectively), (5) increasing the clear roadside recovery distance by relocating utility poles and trees, (6) improving vertical and horizontal alignment by avoiding sharp left-hand curves and sharp downgrades, (7) assuring adequate pavement surface drainage on long radius curves and locations where cross-drainage is longer than one lane in width, and (8) providing increased surface skid resistance on downgrade curve sites. (See Chapter 16 for additional material about horizontal and vertical curves.)

Cross Sections

One of the most important roadway features affecting safety is the highway cross section. As illustrated in Figure 5.6, a rural two-lane highway cross section includes travel lanes, shoulders, sideslopes, clear zones, and ditches. The road may be constructed on an embankment (fill) section or depressed below the natural grade (cut). Cross-section elements, including through and passing lanes, medians, and left-turn lanes, may be added when a two-lane road is inadequate, possibly improving both traffic operations and safety. Safety improvements in the highway cross section are usually focused on two-lane roads, with the exception of clear zone treatments and median design for multilane highways.

In general, wider lanes and/or shoulders will result in fewer crashes. A 1987 study by the FHWA measured the effects of lane width, shoulder width, and shoulder type on highway crash experience, based on data for approximately 5000 miles of two-lane highway. Table 5.8 lists the percentage reduction in crash types related to lane widening.

Related crashes include run-off-road, head-on, and sideswipe occurrences. Not all crash types are "related" to geometric roadway elements. For example, if a lane is widened by 2 feet, from 9 feet to 11 feet, a 23 percent reduction in related crashes can be expected. Table 5.9 provides similar results for shoulders. For example, if an unpaved shoulder is widened by 6 feet, from 2 feet to 8 feet, and the shoulder is paved, then a 40 percent reduction in related crash types can be expected, assuming that other features such as clear zone and sideslopes are unaltered. If both pavement and shoulder width improvements are made simultaneously, the percentage reductions are not additive. Rather, the

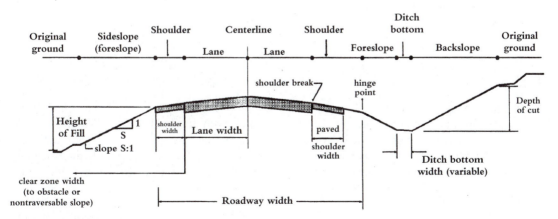

Figure 5.6 Cross Section Elements for Rural Two-Lane Highway

SOURCE: *Safety Effectiveness of Highway Design Features,* Volume III, U.S. Department of Transportation, Federal Highway Administration, Washington, D.C., November 1992.

Table 5.8 Effect of Lane Widening for Related Crash Types on Two-Lane Rural Roads

Amount of Lane Widening (ft)	Crash Reduction (percent)
1	12
2	23
3	32
4	40

SOURCE: *Safety Effectiveness of Highway Design Features,* Volume II, U.S. Department of Transportation, Federal Highway Administration, Washington, D.C., 1992.

Table 5.9 Effect of Shoulder Widening for Related Crash Types on Two-Lane Rural Roads

Shoulder Widening per Side (ft)	Crash Reduction (percent)	
	Paved	Unpaved
2	16	13
4	29	25
6	40	35
8	49	43

SOURCE: *Safety Effectiveness of Highway Design Features,* Volume III, U.S. Department of Transportation, Federal Highway Administration, Washington, D.C., 1992.

contribution of each is computed assuming that the other has taken effect. Crash factors for various combinations of pavement and shoulder width are shown in Table 5.10, and factors that convert total number of crashes to number of related crashes (RC) are shown in Table 5.11.

Example 5.7   Crash Reduction from Lane and Shoulder Widening

Records indicate that there have been a total of 53 crashes per year over a three-year period along a two-lane rural roadway section with 10-foot lanes and 2-foot unpaved shoulders. The highway is located in a mountainous area where average daily traffic is 4000 veh/d. Determine the crash reduction that can be expected (a) if only the lanes are widened to 12 feet, (b) if only the shoulders are paved and widened to 6 feet, and (c) if both measures are implemented together.

Solution: Compute the number of related crashes prevented (CP) using factors in Table 5.11.

$$RC = 53 \times 0.61 = 32 \text{ related crashes/yr.}$$

Table 5.10 Effect of Lane and Shoulder Widening for Related Crashes on Two-Lane Rural Roads

Amount of Lane Widening (in feet)	Existing Shoulder Condition (before period)		Percent Related Crashes Reduced — Shoulder Condition (after period)						
	Shoulder width	Surface type	2-ft Shoulder P	U	4-ft Shoulder P	U	6-ft Shoulder P	U	8-ft Shoulder P	U
3	0	N/A	43	41	52	49	59	56	65	62
	2	Paved	32	—	43	—	52	—	59	—
	2	Unpaved	34	33	44	41	53	49	60	56
	4	Paved	—	—	32	—	43	—	52	—
	4	Unpaved	—	—	36	32	46	41	54	49
	6	Paved	—	—	—	—	32	—	43	—
	6	Unpaved	—	—	—	—	37	32	47	41
	8	Paved	—	—	—	—	—	—	32	—
	8	Unpaved	—	—	—	—	—	—	39	32
2	0	N/A	35	33	45	42	53	50	61	56
	2	Paved	23	—	35	—	45	—	53	—
	2	Unpaved	25	23	37	33	46	42	55	50
	4	Paved	—	—	23	—	35	—	45	—
	4	Unpaved	—	—	27	23	38	33	48	42
	6	Paved	—	—	—	—	23	—	35	—
	6	Unpaved	—	—	—	—	29	23	40	33
	8	Paved	—	—	—	—	—	—	23	—
	8	Unpaved	—	—	—	—	—	—	31	23
1	0	N/A	26	24	37	34	47	43	55	50
	2	Paved	12	—	26	—	37	—	47	—
	2	Unpaved	14	12	28	24	39	34	48	43
	4	Paved	—	—	12	—	26	—	37	—
	4	Unpaved	—	—	17	12	30	24	41	34
	6	Paved	—	—	—	—	12	—	26	—
	6	Unpaved	—	—	—	—	19	12	31	24
	8	Paved	—	—	—	—	—	—	12	—
	8	Unpaved	—	—	—	—	—	—	21	12

SOURCE: *Safety Effectiveness of Highway Design Features,* Volume III, U.S. Department of Transportation, Federal Highway Administration, Washington, D.C., 1992.

a. Crash prevented (CP) due to lane widening alone. From Table 5.8, the reduction is 23 percent.

$$CP = 32 \times 0.23 = 7 \text{ crashes prevented/year}$$

b. Crash prevented due to shoulder widening and paving alone. From Table 5.9, the reduction is 29 percent.

Table 5.11   Ratio of Cross Section Related Crashes to Total Crashes on Two-Lane Rural Roads

| | *Terrain Adjustment Factors* | | |
ADT (vpd)	*Flat*	*Rolling*	*Mountainous*
500	.58	.66	.77
1,000	.51	.63	.75
2,000	.45	.57	.72
4,000	.38	.48	.61
7,000	.33	.40	.50
10,000	.30	.33	.40

Note: Related crashes include run-off-road, head-on, and opposite-direction and same-direction sideswipe.
SOURCE: *Safety Effectiveness of Highway Design Features,* Volume III, U.S. Department of Transportation, Federal Highway Administration, Washington, D.C., 1992.

$$CP = 32 \times 0.29 = 9 \text{ crashes prevented/yr}$$

c.   Crash reduction from both lane and shoulder widening. From Table 5.10, the reduction is 46 percent.

$$CR = 32 \times 0.46 = 15 \text{ crashes prevented/yr}$$

The physical condition along the roadside is also a factor that affects the safety of two-lane highways, since crashes can occur as a result of a vehicle running off the road. A motorist is less likely to experience injury or death under these circumstances if the area adjacent to the pavement is clear of obstructions and has a relatively flat sideslope.

The distance available for a motorist to recover and either stop or return safely to the paved surface, referred to as the "roadside recovery distance" (also called the "clear zone" distance), is a factor in crash reduction. Roadside recovery distance is measured from the edge of pavement to the nearest rigid obstacle, steep slope, nontraversable ditch, cliff, or body of water. Recovery distances are determined by averaging the clear zone distances measured at 3 to 5 locations for each mile. Table 5.12 shows the percent reduction in related crashes as a function of recovery distance. For example, if roadside recovery is increased by 8 feet, from 7 to 15 feet, a 21 percent reduction in related crash types can be expected.

Among the means to increase the roadside recovery distance are (1) relocating utility poles, (2) removing trees, (3) flattening sideslopes to a maximum 4:1 ratio, and (4) removing other obstacles such as bridge abutments, fences, mailboxes, and guardrails. When highway signs or obstacles such as mailboxes cannot be relocated, they should be mounted so as to break away when struck by a moving vehicle, thus minimizing crash severity.

When two-lane roads become more heavily traveled, particularly in suburban, commercial, and recreational areas, crash rates tend to increase. The reasons were dis-

Table 5.12 Effect of Roadside Recovery Distance for Related Crashes

Amount of Increased Roadside Recovery Distance (ft)	Reduction in Related Crash Types (percent)
5	13
8	21
10	25
12	29
15	35
20	44

SOURCE: *Safety Effectiveness of Highway Design Features,* Volume III, U.S. Department of Transportation, Federal Highway Administration, Washington, D.C., 1992.

cussed earlier and include lack of passing opportunities, increased numbers of access points, mixed traffic (including trucks and cars), and local and through destinations. Without the opportunity to create a multilane facility, there are several alternative operational and safety treatments possible. These are (1) passing lanes, (2) short four-lane sections, (3) use of paved shoulders, (4) turnout lanes for slower moving traffic, especially on up-grades, and (5) two-way left-turn lanes. These options, illustrated in Figure 5.7, were evaluated for 138 treated sites, and crash data were compared with standard two-lane sections. The results, shown in Table 5.13 for total and for fatal and injury crashes, are valid for high-volume conditions. For the option "use of paved shoulders," no effect was noted, whereas for the other options, reduction in fatal and injury crashes range from 30 to 85 percent. Since these results are site specific, the operational treatment selected may not always be appropriate, and specific site studies are required.

Intersections

Intersections represent the site of most urban motor vehicle crashes in the United States. The number of crashes at intersections has increased by 14 percent over a twenty-year period. This result is not surprising, since intersections are the confluence of many vehicle

Table 5.13 Effect of Auxiliary Lanes on Crash Reduction on High-Volume, Two-Lane Highways

Multilane Design Alternative	Type of Area	Percent Reduction in Crashes	
		Total	F + I only
Passing lanes	Rural	25	30
Short four-lane section	Rural	35	40
Turnout lanes	Rural	30	40
Two-way, left-turn lane	Suburban	35	35
Two-way, left-turn lane	Rural	70–85	70–85
Shoulder use section	Rural	no known significant effect	

SOURCE: *Safety Effectiveness of Highway Design Features,* Volume III, U.S. Department of Transportation, Federal Highway Administration, Washington, D.C., 1992.

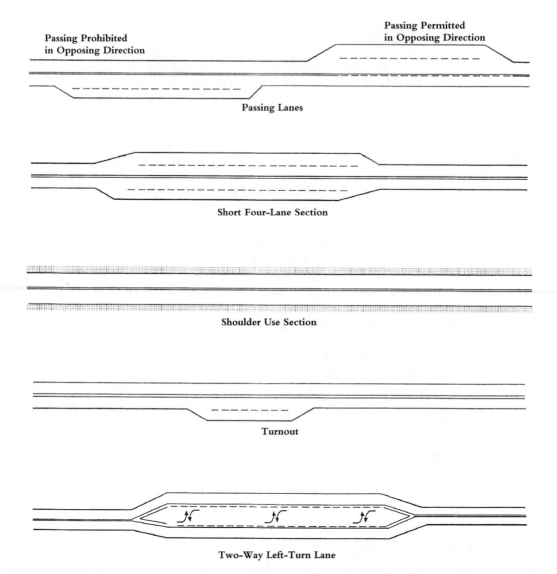

Figure 5.7 Operational and Safety Improvements for Two-Lane Highways

SOURCE: *Safety Effectiveness of Highway Design Features,* Volume III, U.S. Department of Transportation, Federal Highway Administration, Washington, D.C., November 1992.

and pedestrian paths that may conflict with each other. An encouraging trend, however, is the reduction in severity of intersection crashes, such that fatal crashes have reduced by 11 percent over the same twenty-year period, to 28 percent of the total.

The reduction in fatalities is the result of improvements in intersection design, use of passenger restraints, separation of vehicles from pedestrians, enhanced visibility, and improvements in traffic control devices.

In urban areas with high traffic volumes, intersections must accommodate a high volume of turning movements that traverse a large surface area. In this situation, channelization is an effective means to improve safety. Further discussion of channelization is found in Chapter 7. Right-turn-only lanes have become recognized as a simple means to separate through traffic from slow turning traffic. For left-turning traffic, however, several options are available, including left-turn storage lanes and raised dividers to guide traffic through the intersection area. Types of lane dividers used for intersection channelization are raised reflectors, painted lines, barriers, and medians.

Crash rates are also affected by the sight distance available to motorists as they approach an intersection. Stopping sight distance is affected by horizontal and vertical alignment. Vertical curve lengths and horizontal curve radii should be selected to conform with design speeds; when this is not feasible, advisory speed limit signs should be posted. The ability to see traffic that approaches from a cross street is dependent on obtaining a clear diagonal line of sight. When blocked by foliage, buildings, or other obstructions, the sight line may be insufficient to permit a vehicle from stopping in time to avoid colliding with side street traffic. Figure 5.8 illustrates how sight distance is improved when trees are removed near an intersection, and Table 5.14 indicates the expected reduction in number of crashes per year as a function of ADT and increased sight radius. The following example illustrates in general terms the effect of sight distance on safety. However, in the design of an intersection, in order to account for adequate sight distance, the approach velocity must be taken into account, as explained in Chapter 7.

Example 5.8   Calculating the Crash Reduction Due to Increased Sight Radius

A motorist is 50 feet from an intersection and sees a vehicle approaching from the right when it is 20 feet from the intersection. After removal of the foliage that has been blocking the sight line, it is now possible to see the same vehicle when it is 75 feet from the intersection. Average daily traffic volumes on the main roadway are 12,000 vehicles per day. Prior to removal of the obstructing foliage, the average number of crashes per year was 8.6. Determine the expected number of crashes per year after the foliage has been removed, based on the research data provided in Table 5.14.

Solution: From Table 5.14, the crash reduction (CR) is 2.26 crashes per year. The average number of crashes per year = 8.60 − 2.26 = 6.34 crashes/yr.

Table 5.14   Crash Reduction, Per Year, Due to Increased Intersection Sight Distance

ADR[2]	Increased Sight Radius[1]		
	20–49 ft	*50–99 ft*	*>100 ft*
<5000	0.18	0.20	0.30
5000–10000	1.00	1.3	1.40
10000–15000	0.87	2.26	3.46
>15000	5.25	7.41	11.26

[1]At 50 feet from intersection, increasing obstruction on approaching leg from initial <20 feet from intersection.
[2]Average Daily Traffic.
SOURCE: *Safety Effectiveness of Highway Design Features,* Volume V, U.S. Department of Transportation, Federal Highway Administration, Washington, D.C., 1992.

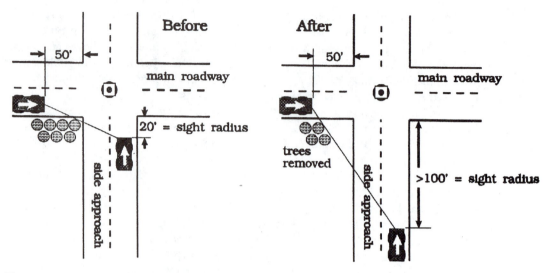

Figure 5.8 Increased Sight Radius by Removal of Obstacles

SOURCE: *Safety Effectiveness of Highway Design Features,* Volume V, U.S. Department of Transportation, Federal Highway Administration, Washington, D.C., November 1992.

Pedestrian Facilities

The safety of pedestrians is of great concern to traffic and highway engineers. In 1989, there were 119,000 pedestrians involved in traffic crashes, of which 6552 were killed. Further, there are about 900 bicyclists killed in collisions with motor vehicles, and over half a million persons treated in hospital emergency rooms for bicycle-related injuries. Efforts to reduce pedestrian and bicycle crashes involve education, enforcement, and engineering measures, as is the case for motor vehicle crashes. In addition, characteristics of pedestrian crashes indicate that factors related to occurrence include age, sex, alcohol use, time of day, urban or rural area type, and intersection or midblock crossing location.

For example, it is known that fatality rates increase sharply for pedestrians over 70 years of age, and that the highest crash rates occur for males 5–9 years old. Alcohol was a factor in 37–44 percent of fatal pedestrian crashes reported between 1980 and 1989. Peak crash periods occur in the afternoon and evening hours, and over 85 percent of all nonfatal crashes occur in urban areas. Approximately 65 percent of all pedestrian crashes occur at locations other than intersections, and many of these involve younger children who dart out into the street. The various types of pedestrian crashes and percentage occurrence are listed in Table 5.15. Note that dart-out at locations account for over one-third of the 14 crash types listed. The most common types of pedestrian crashes are illustrated in Figure 5.9.

The principal geometric design elements that are used to improve pedestrian safety are (1) sidewalks, (2) overpasses or tunnels, (3) raised islands, (4) auto-free shopping

Table 5.15 Pedestrian Crash Types and Frequency

DART-OUT (FIRST HALF) (24%)
    Midblock (not at intersection)
    Pedestrian sudden appearance and short time exposure (driver does not have time to react)
    Pedestrian crossed less than halfway.

DART-OUT (SECOND HALF) (10%)
    Same as above, except pedestrian gets at least halfway across before being struck.

MIDBLOCK DASH (8%)
    Midblock (not at intersection)
    Pedestrian is running but *not* sudden appearance or short time exposure as above.

INTERSECTION DASH (13%)
    Intersection
    Same as dart-out (short time exposure or running) except it occurs at an intersection.

VEHICLE TURN-MERGE WITH ATTENTION CONFLICT (4%)
    Vehicle is turning or merging into traffic.
    Driver is attending to traffic in one direction and hits pedestrian from a different direction.

TURNING VEHICLE (5%)
    Vehicle is turning or merging into traffic.
    Driver attention is *not* documented.
    Pedestrian is *not* running.

MULTIPLE THREAT (3%)
    Pedestrian is hit as he steps into the next traffic lane by a vehicle moving in the same direction
        as vehicle(s) that stopped for the pedestrian, because driver's vision of pedestrian obstructed
        by the stopped vehicle.

BUS STOP–RELATED (2%)
    Pedestrian steps out from in front of bus at a bus stop and is struck by vehicle moving in
        same direction as bus while passing bus.

VENDOR–ICE CREAM TRUCK (2%)
    Pedestrian struck while going to or from a vendor in a vehicle on the street.

DISABLED VEHICLE–RELATED (1%)
    Pedestrian struck while working on or next to a disabled vehicle.

RESULT OF VEHICLE-VEHICLE CRASH (3%)
    Pedestrian hit by vehicle(s) as a result of a vehicle-vehicle collision.

TRAPPED (1%)
    Pedestrian hit when traffic light turned red (for pedestrian) and vehicles started moving.

WALKING ALONG ROADWAY (1%)
    Pedestrian struck while walking along the edge of the highway or on the shoulder.

OTHER (23%)
    Unusual circumstances, not countermeasure corrective.

SOURCE: Adapted from *Safety Effectiveness of Highway Design Features,* Volume VI, U.S. Department of Transportation, Federal Highway Administration, Washington, D.C., 1992.

Figure 5.9 Common Types of Pedestrian Crashes

SOURCE: *Safety Effectiveness of Highway Design Features,* Volume VI, U.S. Department of Transportation, Federal Highway Administration, Washington, D.C., November 1992.

streets, (5) neighborhood traffic control to limit speeding and through traffic, (6) curb cuts that assist wheelchair users and pedestrians with baby carriages, and (7) shoulders that are paved and widened. Other traffic control measures that may assist pedestrians include crosswalks, traffic signs and signals, parking regulations, and lighting.

Sidewalks and pedestrian paths can significantly improve safety in areas where the volumes of automobile and pedestrian traffic are high. The sidewalk provides a safe and separated lane intended for the exclusive use of pedestrians. However, they should not be used by higher-speed nonmotorized vehicles, such as bicycles. Guidelines for the minimum width and location of sidewalks are illustrated in Figure 5.10 based on classification of roadway and residential density. For example, for residential areas with 1 to 4 dwelling units per acre, sidewalks are preferred on both sides of the local street. In commercial and industrial areas, sidewalks are required on both sides of the street.

Sidewalks are often carried on grade-separated structures, such as overpasses or subways, when crossings involve freeways or expressways that carry high-speed and high-volume traffic. They are most effective when pedestrian demand is high, for example, as a connector between a residential area and destinations such as schools, hospitals, and shopping areas. Pedestrians will use these facilities if they are convenient and do not require circuity of travel, but they will select the alternative unsafe path if they are required to walk significantly farther on the overcrossing or through the tunnel.

Traffic safety in residential neighborhoods is a major concern, especially in suburban areas where through traffic uses residential streets as a shortcut, thus bypassing congested arterials and expressways. Typically citizens protest when they perceive that their neighborhoods are becoming more dangerous for children and others who are required to walk along the same roadways as moving traffic. Several geometric designs have been developed in the United States and in Europe to create more friendly pedestrian environments. Several options, some illustrated in Figure 5.11, are as follows:

1. Create cul-de-sacs by closing streets at an intersection or at midblock
2. Reduce the roadway width at the intersection, or provide on-street parking at midblock (a narrower roadway tends to reduce speeds and improve pedestrian crossing)
3. Limit street access to one-way traffic, and narrow the intersections to improve pedestrian crossings; often four-way stop signs are used to further reduce traffic speeds
4. Install diagonal barriers at the intersection to divert traffic and thus discourage through traffic with increased travel time and distance
5. Use mechanisms such as speed humps, photo radar, electronic speed reminders, direct police enforcement, and traffic circles to reduce speeds and eliminate through traffic; these types of traffic control measures for residential areas are used in some U.S. cities, but more extensive experience with these techniques is found in many cities in Europe
6. Use other roadway and geometric features that can assist both pedestrians and bicyclists, such as curb ramps, widened and paved shoulders along rural two-lane highways, stripes to signify separate bicycle lanes, and widened highway lanes in urban areas

Proposed Minimum Sidewalk Widths

Central Business Districts - Conduct level of service analysis according to method in 1985 Highway Capacity Manual.

Commercial/industrial areas outside a central business district - Minimum
5 ft wide with 2 ft planting strip or 6 ft wide with no planting strip.

Residential areas outside a central business district:

Arterial and collector streets - Minimum 5 ft with minimum 2 ft planting strip.

Local Streets:

• Multi-family dwellings and single-family dwellings with densities greater than four dwelling units per acre - Minimum 5 ft with minimum 2 ft planting strip.

• Densities up to four dwelling units per acre - Minimum 4 ft with minimum 2 ft planting strip.

Land-Use/Roadway Functional Classification/Dwelling Unit	New Urban and Suburban Streets	Existing Urban and Suburban Streets
Commercial & Industrial (All Streets)	Both sides.	Both sides. Every effort should be made to add sidewalks where they do not exist and complete missing links.
Residential (Major Arterials)	Both sides.	
Residential (Collectors)	Both sides.	Multi-family - both sides. Single-family dwellings - prefer both sides required at least one side.
Residential (Local Streets) More than 4 Units Per Acre	Both sides.	Prefer both sides, required at least one side.
Residential (Local Streets) 1 to 4 Units Per Acre	Prefer both sides; required at least one side.	One side preferred, at least 4 ft shoulder on both sides required.
Residential (Local Streets) Less than 1 Unit Per Acre	One side preferred, shoulder both sides required.	At least 4 ft shoulder on both sides required.

Figure 5.10 Guidelines for Sidewalk Installation

SOURCE: *Safety Effectiveness of Highway Design Features,* Volume VI, U.S. Department of Transportation, Federal Highway Administration, Washington, D.C., November 1992.

Semi-Diverters   **Chokers/Narrowing**

Diagonal Diverters   **Cul-de-Sac/Street Closures**

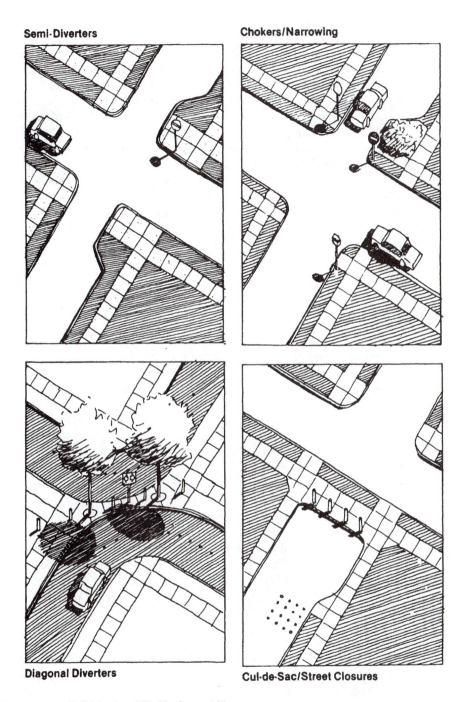

Figure 5.11 Neighborhood Traffic Control Measures

SOURCE: *Safety Effectiveness of Highway Design Features,* Volume VI, U.S. Department of Transportation, Federal Highway Administration, Washington, D.C., November 1992.

SUMMARY

Safety is an important consideration in the design and operation of streets and highways. Traffic and highway engineers, working with law enforcement officials, are constantly seeking better ways to ensure safety for the motorist and pedestrians. A Highway Safety Improvement Process is in place which involves planning, implementation, and evaluation. The planning process requires that the engineer collect and maintain traffic safety data, identify hazardous locations, conduct engineering studies, and establish project priorities. The quality of geometric design will also influence the safety of streets and highways. Roadways should be designed to provide adequate sight distance, to separate through traffic from local traffic, to avoid speed variations, and to ensure that the driver is aware of changes occurring on the roadway and has adequate time and distance to change speed or direction without becoming involved in a collision with another vehicle, a fixed object, or a pedestrian. In this chapter we have described a variety of actions that can be taken to improve the highway itself. However, safety on the highway is also dependent on a well-educated population who drive courteously and defensively, and requires enforcement of rules of the road.

PROBLEMS

5-1   Describe the type of information provided on a collision diagram.

5-2   A local jurisdiction has determined that for a given set of geometric conditions, a maximum rate of 8 crashes/million entering vehicles can be tolerated. At an intersection of 2 roadways with ADTs of 10,000 and 7500, how many crashes can occur before corrective action must be sought?

5-3   Studies were conducted at two sites on rural roads with similar characteristics. The first site was 5.1 miles in length with an ADT of 6500. Over the year-long study period, 28 crashes occurred on this portion of roadway, 5 of them resulting in fatalities. The second site was a 10-mile section with an ADT of 5000. There were 32 crashes in this section with 4 fatalities. Determine the appropriate crash rates for both locations, and discuss the implications.

5-4   Describe the categories used to summarize crash data. Give an example of how each category would be used to evaluate a given location.

5-5   A review of crash records shows that a signalized intersection is a hazardous location because of an abnormally high number of rear-end collisions. What are the possible causes of these crashes, and what data should be collected to determine the actual causes?

5-6   The crash rate on a heavily traveled two-lane rural highway is abnormally high. The corridor is 14 miles long with an ADT of 34,000. An investigation has determined that head-on collisions are most common, with a $RMVM_T$ of 4.5, and are caused by vehicles attempting to pass. Determine an appropriate countermeasure and calculate the estimated yearly reduction in total crashes.

5-7   A state transportation agency has established crash reduction factors shown in the table. An intersection has been identified as having an abnormally high left-turn crash rate (17, 20, and 18 crashes in the last three years), attributed to excessive speeds and the absence of an exclusive left-turn phase. The intersection for the last three years is 8400, and the ADT for after implementation is 9500. Determine (a) the appropriate countermeasures and (b) expected reduction in crashes if the countermeasures are implemented.

Countermeasure	Reduction Factor
Retime signals	0.10
Provide left-turn signal phase	0.30
Reduce speed limit on approaches	0.35
Provide turning guide lines	0.05
Prohibit left turns	0.75

5-8   The geometrics along a portion of roadway have been improved to provide wider travel lanes and paved shoulders. While these improvements have reduced the crash rate, the local traffic engineer is still concerned that too many crashes are occurring, particularly those involving collisions with fixed objects along the edge of pavement. Thirty such crashes have occurred in the past 12 months. What countermeasure should the engineer undertake to reduce the number of crashes by at least five per year?

5-9   A 10-mile section of a rural two-lane road with 9-foot lanes and 2-foot paved shoulders has an crash rate of 156/100 million vehicle-miles traveled. The current ADT is 20,000, which, due to a nearby commercial development, is anticipated to increase to 37,700. Estimate the additional lane and shoulder widths required to keep the average number of crashes per year at the current level.

5-10   Residents of a local neighborhood have been complaining to city officials that vehicles are using their side streets as shortcuts to avoid rush hour traffic. Discuss the options available to the city transportation officials to address the residents' concerns.

5-11   Estimate the yearly reduction in total and fatal crashes resulting from the upgrade of an 18.4-mile corridor with partially controlled access to one with full control. The road is in an urban area and has an ADT of 62,000.

5-12   An engineer has proposed four countermeasures to be implemented to reduce the high crash rate at an intersection. CR factors for these countermeasures are 0.25, 0.30, 0.17, and 0.28. The number of crashes occurring at the intersection during the past three years were 28, 30, and 31, and the AADTs during those years were 8450, 9150, and 9850, respectively. Determine the expected reductions in number of crashes during the first three years after the implementation of the countermeasures if the AADT during the first year of implementation is 10,850 and the estimated traffic growth rate is 4 percent per annum after implementation.

5-13   Survey your local college campus. What pedestrian facilities are provided? How might pedestrian safety be improved?

REFERENCES

Batchelder, J.H. et al., *Simplified Procedures for Evaluating Low-Cost TSM Projects, Users' Manual,* NCHRP Report 263, Transportation Research Board, National Research Council, Washington, D.C., 1983.

Highway Safety Engineering Procedural Guide.

Highway Safety Engineering Studies Procedural Guide, U.S. Department of Transportation, Federal Highway Administration, Washington, D.C., November 1981.

Safety Effectiveness of Highway Design Features, U.S. Department of Transportation, Federal Highway Administration, Washington, D.C., November 1992. Volume I, *Access Control;* Volume II, *Alignment;* Volume III, *Cross-sections;* Volume IV, *Interchanges;* Volume V, *Intersections;* Volume VI, *Pedestrians and Bicyclists.*

Safety Management System, U.S. Department of Transportation, Federal Highway Administration, FHWA-HI-95-012, February 1995.

Status Report: Effective Highway Accident Countermeasures, U.S. Department of Transportation, Federal Highway Administration, Washington, D.C., June 1993.

Symposium on Effective Accident Countermeasures.

Symposium on Effective Highway Accident Countermeasures, U.S. Department of Transportation, Federal Highway Administration and National Highway Traffic Safety Administration, Washington, D.C., August 1990.

ADDITIONAL READINGS

Evans, L., *Public Safety and the Driver,* Van Nostrand Reinhold, New York, 1991.

Forkenbrock, D.J., Foster, N.S.J., and Pogue, T.F., *Safety and Highway Investment,* Public Policy Center, University of Iowa, Iowa City, Iowa, June 1994.

Roadside Safety Features and Other General Design Issues, Record No. 1599, Transportation Research Board, National Research Council, National Academy Press, Washington, D.C., 1997.

Roadside Safety Issues Revisited, Circular 453, Transportation Research Board, National Research Council, National Academy Press, Washington, D.C., February 1996.

Strategic Highway Safety Plan, American Association of State Highway and Transportation Officials, 1998.

Strategies for Improving Roadside Safety, NCHRP Research Results Digest 220, Transportation Research Board, National Research Council, National Academy Press, Washington, D.C., November 1997.

The Traffic Safety Toolbox: A Primer on Traffic Safety, Institute of Traffic Engineers, Washington, D.C., 1998.

CHAPTER 6

Fundamental Principles of Traffic Flow

Traffic flow theory involves the development of mathematical relationships among the primary elements of a traffic stream: flow, density, and speed. These relationships help the traffic engineer in planning, designing, and evaluating the effectiveness of implementing traffic engineering measures on a highway system. Traffic flow theory is used in design to determine adequate lane lengths for storing left-turn vehicles on separate left-turn lanes, the average delay at intersections and freeway ramp merging areas, and changes in the level of freeway performance due to the installation of improved vehicular control devices on ramps. Another important application of traffic flow theory is simulation, where mathematical algorithms are used to study the complex interrelationships that exist among the elements of a traffic stream or network and to estimate the effect of changes in traffic flow on factors such as accidents, travel time, air pollution, and gasoline consumption.

Methods ranging from physical to empirical have been used in studies related to the description and quantification of traffic flow. This chapter, however, will introduce only those aspects of traffic flow theory that can be used in the planning, design, and operation of highway systems.

TRAFFIC FLOW ELEMENTS

Let us first define the elements of traffic flow before discussing the relationships among them. Before we do that, though, we will describe the time-space diagram, which serves as a useful device for defining the elements of traffic flow.

The time-space diagram is a graph that describes the relationship between the location of vehicles in a traffic stream and the time as the vehicles progress along the highway. Figure 6.1 shows a time-space diagram for six vehicles, with distance plotted on the vertical axis and time on the horizontal axis. At time zero, vehicles 1, 2, 3, and 4 are at respective

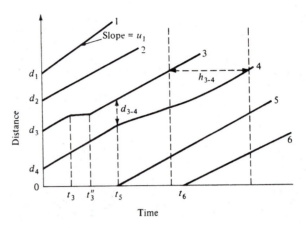

Figure 6.1 Time-Space Diagram

distances d_1, d_2, d_3, and d_4 from a reference point, whereas vehicles 5 and 6 cross the reference point later at times t_5 and t_6, respectively.

The primary elements of traffic flow are flow, density, and speed. Another element, associated with density, is the gap or headway between two vehicles in a traffic stream. The definitions of these elements follow.

Flow (*q*) is the equivalent hourly rate at which vehicles pass a point on a highway during a
time period less than 1 h. It can be determined by

$$q = \frac{n \times 3600}{T} \text{ veh/h} \qquad (6.1)$$

where
n = the number of vehicles passing a point in the roadway in T secs
q = the equivalent hourly flow.

Density (*k*), sometimes referred to as *concentration,* is the number of vehicles traveling over a
unit length of highway at an instant in time. The unit length is usually 1 mi, thereby
making vehicles per mile (veh/mi) the unit of density.

Speed (*u*) is the distance traveled by a vehicle during a unit of time. It can be expressed
in miles per hour (mi/h), kilometers per hour (km/h), or feet per second (ft/sec).
The speed of a vehicle at any time t is the slope of the time-space diagram for that
vehicle at time t. Vehicles 1 and 2 in Figure 6.1, for example, are moving at constant speeds because the slopes of the associated graphs are constant. Vehicle 3 moves
at a constant speed between time zero and time t_3, then stops for the period t_3 to t_3''
(the slope of graph equals zero), and then accelerates and eventually moves at a constant speed. There are two types of mean speeds: time mean speed and space mean
speed.

Time mean speed $(\bar{u}_t)$ is the arithmetic mean of the speeds of vehicles passing a point on a
highway during an interval of time. The time mean speed is found by

$$\bar{u}_t = \frac{1}{n} \sum_{i=1}^{n} u_i \tag{6.2}$$

where

n = number of vehicles passing a point on the highway

u_i = speed of the ith vehicle (ft/sec)

Space mean speed $(\bar{u}_s)$ is the harmonic mean of the speeds of vehicles passing a point on a highway during an interval of time. It is obtained by dividing the total distance traveled by two or more vehicles on a section of highway by the total time required by these vehicles to travel that distance. This is the speed that is involved in flow-density relationships. The space mean speed is found by

$$\bar{u}_s = \frac{n}{\displaystyle\sum_{i=1}^{n}(1/u_i)}$$

$$= \frac{nL}{\displaystyle\sum_{i=1}^{n}t_i} \tag{6.3}$$

where

$\bar{u}_s$ = space mean speed (ft/sec)

n = number of vehicles

t_i = the time it takes the ith vehicle to travel across a section of highway (sec)

u_i = speed of the ith vehicle (ft/sec)

L = length of section of highway (ft)

The time mean speed is always higher than the space mean speed. The difference between these speeds tends to decrease as the absolute values of speeds increase. It has been shown from field data that the relationship between time mean speed and space mean speed can be given as

$$\bar{u}_t = \bar{u}_s + \frac{\sigma^2}{\bar{u}_s} \tag{6.4}$$

Equation 6.5, shows a more direct relationship developed by Garber and Sankar, using data collected at several sites on freeways. Figure 6.2 also shows a plot of time mean speeds against space mean speeds using the same data.

$$\bar{u}_t = 0.966\bar{u}_s + 3.541 \tag{6.5}$$

where

$\bar{u}_t$ = time mean speed, km/h

$\bar{u}_s$ = space mean speed, km/h

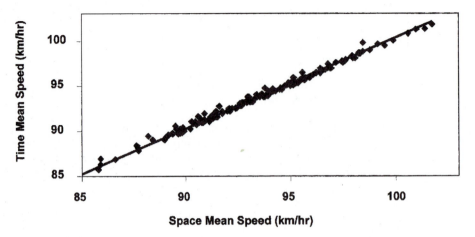

Figure 6.2 Space Mean Speed vs. Time Mean Speed

Time headway (*h*) is the difference between the time the front of a vehicle arrives at a point on the highway and the time the front of the next vehicle arrives at that same point. Time headway is usually expressed in seconds. For example, in the time-space diagram (Figure 6.1), the time headway between vehicles 3 and 4 at d_1 is h_{3-4}.

Space headway (*d*) is the distance between the front of a vehicle and the front of the following vehicle. It is usually expressed in feet. The space headway between vehicles 3 and 4 at time t_5 is d_{3-4} (see Figure 6.1).

Example 6.1   Determining Flow, Density, Time Mean Speed, and Space Mean Speed

Figure 6.3 shows vehicles traveling at constant speeds on a two-lane highway between sections *X* and *Y* with their positions and speeds obtained at an instant of time by photography. An observer located at point *X* observes the four vehicles passing point *X* during a period of *T* sec. The velocities of the vehicles are measured as 45, 45, 40, and 30 mi/h, respectively. Calculate the flow, the density, the time mean speed, and the space mean speed.

Solution: The flow is calculated by

$$q = \frac{n \times 3600}{T}$$

$$= \frac{4 \times 3600}{T} = \frac{14,400}{T} \text{ veh/h} \qquad (6.5)$$

With *L* equal to the distance between *X* and *Y* (ft), density is obtained by

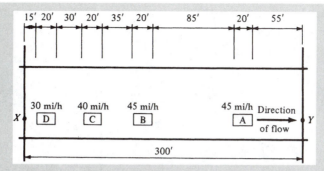

Figure 6.3 Locations and Speeds of Four Vehicles on a Two-Lane Highway at an Instant of Time

$$k = \frac{n}{L}$$

$$= \frac{4}{300} \times 5280 = 70.4 \text{ veh/mi}$$

The time mean speed is found by

$$u_t = \frac{1}{n} \sum_{i=1}^{n} u_i$$

$$= \frac{30+40+45+45}{4} = 40 \text{ mi/h}$$

The space mean speed is found by

$$\bar{u}_s = \frac{n}{\displaystyle\sum_{i=1}^{n} (1/u_i)}$$

$$= \frac{Ln}{\displaystyle\sum_{i=1}^{n} t_i}$$

$$= \frac{300n}{\displaystyle\sum_{i=1}^{n} t_i}$$

where t_i is the time it takes the ith vehicle to travel from X to Y at speed u_i, and L (ft) is the distance between X and Y.

$$t_i = \frac{L}{1.47u_i} \text{ sec}$$

$$t_A = \frac{300}{1.47 \times 45} = 4.54 \text{ sec}$$

$$t_B = \frac{300}{1.47 \times 45} = 4.54 \text{ sec}$$

$$t_C = \frac{300}{1.47 \times 40} = 5.10 \text{ sec}$$

$$t_D = \frac{300}{1.47 \times 30} = 6.80 \text{ sec}$$

$$\bar{u}_s = \frac{4 \times 300}{4.54 + 4.54 + 5.10 + 6.80} = 57 \text{ ft/sec}$$

$$= 39.0 \text{ mi/h}$$

FLOW–DENSITY RELATIONSHIPS

The general equation relating flow, density, and space mean speed is given as

$$\text{flow} = \text{density} \times \text{space mean speed}$$
$$q = k\bar{u}_s \tag{6.6}$$

Each of the variables in Eq. 6.6 also depends on several other factors, including the characteristics of the roadway, the characteristics of the vehicle, the characteristics of the driver, and environmental factors such as the weather.

Other relationships that exist among the traffic flow variables are given below.

$$\text{space mean speed} = (\text{flow}) \times (\text{space headway})$$
$$\bar{u}_s = q\bar{d} \tag{6.7}$$

where

$$\bar{d} = (1/k) = \text{average space headway} \tag{6.8}$$

$$\text{density} = (\text{flow}) \times (\text{travel time for unit distance})$$
$$k = q\bar{t} \tag{6.9}$$

where $\bar{t}$ is the average time for unit distance.

$$\text{average space headway} = (\text{space mean speed}) \times (\text{average time headway}) \tag{6.10}$$
$$\bar{d} = \bar{u}_s\bar{h}$$

$$\text{average time headway} = (\text{average travel time for unit distance})$$
$$\times (\text{average space headway}) \tag{6.11}$$
$$\bar{h} = \bar{t}\bar{d}$$

Fundamental Diagram of Traffic Flow

The relationship between the density (veh/mi) and the corresponding flow of traffic on a highway generally is referred to as the fundamental diagram of traffic flow. The following theory has been postulated with respect to the shape of the curve depicting this relationship.

1. When the density on the highway is zero, the flow is also zero because there are no vehicles on the highway.
2. As the density increases, the flow also increases.
3. However, when the density reaches its maximum, generally referred to as the *jam density* (k_j), the flow must be zero because vehicles will tend to line up end to end.
4. It follows that as density increases from zero, the flow will also initially increase from zero to a maximum value. Further continuous increase in density will then result in continuous reduction of the flow, which will eventually be zero when the density is equal to the jam density. The shape of the curve therefore takes the form in Figure 6.4a.

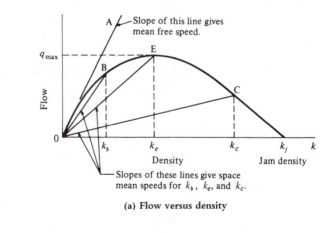

(a) Flow versus density

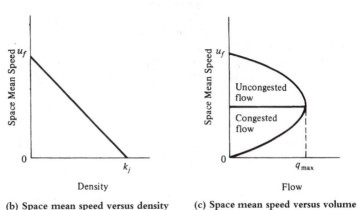

(b) Space mean speed versus density

(c) Space mean speed versus volume

Figure 6.4 Fundamental Diagrams of Traffic Flow

Data have been collected that tend to confirm the argument postulated above, but there is some controversy regarding the exact shape of the curve. A similar argument can be postulated for the general relationship between the space mean speed and the flow. When the flow is very low, there is little interaction between individual vehicles. Drivers are therefore free to travel at the maximum possible speed. The absolute maximum speed is obtained as the flow tends to zero, and it is known as the *mean free speed* (u_f). The magnitude of the mean free speed depends on the physical characteristics of the highway. Continuous increase in flow will result in a continuous decrease in speed. A point will be reached, however, when further addition of vehicles will result in the reduction of the actual number of vehicles that pass a point on the highway (that is, reduction of flow). This results in congestion, and eventually both the speed and the flow become zero. Figure 6.4c shows this general relationship. Figure 6.4b shows the direct relationship between speed and density.

From Eq. 6.6, we know that space mean speed is flow divided by density, which makes the slopes of lines 0B, 0C, and 0E in Figure 6.4a represent the space mean speeds at densities k_b, k_c, and k_e, respectively. The slope of line 0A is the speed as the density tends to zero and little interaction exists between vehicles. The slope of this line is therefore the mean free speed (u_f); it is the maximum speed that can be attained on the highway. The slope of line 0E is the space mean speed for maximum flow. This maximum flow is the capacity of the highway. Thus it can be seen that it is desirable for highways to operate at densities not greater than that required for maximum flow.

Mathematical Relationships Describing Traffic Flow

Mathematical relationships describing traffic flow can be classified into two general categories—macroscopic and microscopic—depending on the approach used in the development of these relationships. The macroscopic approach considers flow density relationships, whereas the microscopic approach considers spacings between vehicles and speeds of individual vehicles.

Macroscopic Approach

The macroscopic approach considers traffic streams and develops algorithms that relate the flow to the density and space mean speeds. The two most commonly used macroscopic models are the Greenshields and Greenberg models.

Greenshields Model.   Greenshields carried out one of the earliest recorded works, in which he studied the relationship between speed and density. He hypothesized that a linear relationship existed between speed and density, which he expressed as

$$\bar{u}_s = u_f - \frac{u_f}{k_j} k \tag{6.12}$$

Corresponding relationships for flow and density and for flow and speed can be developed. Since $q = \bar{u}_s k$, substituting $q/\bar{u}_s$ for k in Eq. 6.12 gives

$$\bar{u}_s^2 = u_f \bar{u}_s - \frac{u_f}{k_j} q \tag{6.13}$$

Also substituting q/k for $\bar{u}_s$ in Eq. 6.12 gives

$$q = u_f k - \frac{u_f}{k_j} k^2 \qquad (6.14)$$

Equations 6.13 and 6.14 indicate that if a linear relationship in the form of Eq. 6.12 is assumed for speed and density, then parabolic relationships are obtained between flow and density and between flow and speed. The shape of the curve shown in Figure 6.4a will therefore be a parabola. Also, Eqs. 6.13 and 6.14 can be used to determine the corresponding speed and the corresponding density for maximum flow.

Consider Eq. 6.13.

$$\bar{u}_s^2 = u_f \bar{u}_s - \frac{u_f}{k_j} q$$

Differentiating q with respect to $\bar{u}_s$, we obtain

$$2\bar{u}_s = u_f - \frac{u_f}{k_j} \frac{dq}{du_s}$$

that is,

$$\frac{dq}{d\bar{u}_s} = u_f \frac{k_j}{u_f} - 2\bar{u}_s \frac{k_j}{u_f} = k_j - 2\bar{u}_s \frac{k_j}{u_f}$$

For maximum flow,

$$\frac{dq}{d\bar{u}_s} = 0 \qquad k_j = 2\bar{u}_s \frac{k_j}{u_f} \qquad u_o = \frac{u_f}{2} \qquad (6.15)$$

Thus, the space mean speed u_o at which the volume is maximum is equal to half the free mean speed.

Consider Eq. 6.14.

$$q = u_f k - \frac{u_f}{k_j} k^2$$

Differentiating q with respect to k, we obtain

$$\frac{dq}{dk} = u_f - 2k \frac{u_f}{k_j}$$

For maximum flow,

$$\frac{dq}{dk} = 0$$

$$u_f = 2k \frac{u_f}{k_j} \tag{6.16}$$

$$\frac{k_j}{2} = k_o$$

Thus, at the maximum flow, the density k_o is half the jam density. The maximum flow for the Greenshields relationship can therefore be obtained from Eqs. 6.6, 6.15, and 6.16, as shown in Eq. 6.17.

$$q_{max} = \frac{k_j u_f}{4} \tag{6.17}$$

Greenberg Model. Several researchers have used the analogy of fluid flow to develop macroscopic relationships for traffic flow. One of the major contributions using the fluid-flow analogy was developed by Greenberg in the form

$$\bar{u}_s = c \ln \frac{k_j}{k} \tag{6.18}$$

Multiplying each side of Eq. 6.18 by k, we obtain

$$\bar{u}_s k = q = ck \ln \frac{k_j}{k} \tag{6.19}$$

Differentiating q with respect to k, we obtain

$$\frac{dq}{dk} = c \ln \frac{k_j}{k} - c$$

For maximum flow,

$$\frac{dq}{dk} = 0$$

giving

$$\ln \frac{k_j}{k} = 1$$

Substituting 1 for $\ln (k_j/k_o)$ in Eq. 6.18 gives

$$u_o = c$$

Thus, the value of c is the speed at maximum flow.

Model Application

Use of these macroscopic models depends on whether they satisfy the boundary criteria of the fundamental diagram of traffic flow at the region that describes the traffic conditions. For example, the Greenshields model satisfies the boundary conditions when the density k is approaching zero as well as when the density is approaching the jam density k_j. The Greenshields model therefore can be used for light or dense traffic. The Greenberg model, on the other hand, satisfies the boundary conditions when the density is approaching the jam density, but it does not satisfy the boundary conditions when k is approaching zero. The Greenberg model is therefore useful only for dense traffic conditions.

Calibration of Macroscopic Traffic Flow Models. The traffic models discussed thus far can be used to determine specific characteristics such as the speed and density at which maximum flow occurs and the jam density of a facility. This usually involves collecting appropriate data on the particular facility of interest and fitting the data points obtained to a suitable model. The most common method of approach is *regression analysis*. This is done by minimizing the squares of the differences between the observed and expected values of a dependent variable. When the dependent variable is linearly related to the independent variable, the process is known as *linear regression analysis,* and when the relationship is with two or more independent variables, the process is known as *multiple linear regression analysis.*

If a dependent variable y and an independent variable x are related by an estimated regression function, then

$$y = a + bx \tag{6.20}$$

The constants a and b could be determined from Eqs. 6.21 and 6.22. (For development of these equations, see Appendix B.)

$$a = \frac{1}{n}\sum_{i=1}^{n} y_i - \frac{b}{n}\sum_{i=1}^{n} x_i = \bar{y} - b\bar{x} \tag{6.21}$$

and

$$b = \frac{\sum_{i=1}^{n} x_i y_i - \frac{1}{n}\left(\sum_{i=1}^{n} x_i\right)\left(\sum_{i=1}^{n} y_i\right)}{\sum_{i=1}^{n} x_i^2 - \frac{1}{n}\left(\sum_{i=1}^{n} x_i\right)^2} \tag{6.22}$$

where
    n = number of sets of observations
    x_i = ith observation for x
    y_i = ith observation for y

A measure commonly used to determine the suitability of an estimated regression function is the coefficient of determination (or square of the estimated correlation coefficient) R^2, which is given by

$$R^2 = \frac{\sum_{i=1}^{n}(Y_i - \bar{y})^2}{\sum_{i=1}^{n}(y_i - \bar{y})^2} \qquad (6.23)$$

where Y_i is the value of the dependent variable as computed from the regression equations. The closer R^2 is to 1, the better the regression fits.

Example 6.2   Fitting Speed and Density Data to the Greenshields Model

Let us now use the data shown in Table 6.1 (columns 1 and 2) to demonstrate the use of the method of regression analysis in fitting speed and density data to the macroscopic models discussed earlier.

Solution: Let us first consider the Greenshields expression

$$\bar{u}_s = u_f - \frac{u_f}{k_j} k$$

- Comparing this expression with our estimated regression function, Eq. 6.20, we see that the speed $\bar{u}_s$ in the Greenshields expression is represented by y in the estimated regression function, the mean free speed u_f is represented by a, and the value of the mean free speed u_f divided by the jam density k_j is represented by $-b$. We therefore obtain

$$\sum y_i = 404.8 \qquad \sum x_i = 892 \qquad \bar{y} = 28.91$$
$$\sum x_i y_i = 20619.8 \qquad \sum x_i^2 = 66,628 \qquad \bar{x} = 63.71$$

- Using Eqs. 6.21 and 6.22, we obtain

$$a = 28.91 - 63.71b$$

$$b = \frac{20,619.8 - \dfrac{(892)(404.8)}{14}}{66,628 - \dfrac{(892)^2}{14}} = -0.53$$

Table 6.1   Speed and Density Observations at a Rural Road

(a) Computations for Example 6.2

Speed, u_s (mi/h) y_i	Density, k (veh/mi) x_i	$x_i y_i$	x_i^2
53.2	20	1064.0	400
48.1	27	1298.7	729
44.8	35	1568.0	1,225
40.1	44	1764.4	1,936
37.3	52	1939.6	2,704
35.2	58	2041.6	3,364
34.1	60	2046.0	3,600
27.2	64	1740.8	4,096
20.4	70	1428.0	4,900
17.5	75	1312.5	5,625
14.6	82	1197.2	6,724
13.1	90	1179.0	8,100
11.2	100	1120.0	10,000
8.0	115	920.0	13,225
$\Sigma = 404.8$	$\Sigma = 892$	$\Sigma = 20{,}619.8$	$\Sigma = 66{,}628.0$
$\bar{y} = 28.91$	$\bar{x} = 63.71$		

(b) Computations for Example 6.3

Speed, u_s (mi/h) y_i	Density, k (veh/mi)	$\ln k_i$ x_i	$x_i y_i$	x_i^2
53.2	20	2.995732	159.3730	8.974412
48.1	27	3.295837	158.5298	10.86254
44.8	35	3.555348	159.2796	12.64050
40.1	44	3.784190	151.746	14.32009
37.3	52	3.951244	147.3814	15.61233
35.2	58	4.060443	142.9276	16.48720
34.1	60	4.094344	139.6171	16.76365
27.2	64	4.158883	113.1216	17.29631
20.4	70	4.248495	86.66929	18.04971
17.5	75	4.317488	75.55605	18.64071
14.6	82	4.406719	64.33811	19.41917
13.1	90	4.499810	58.94750	20.24828
11.2	100	4.605170	51.57791	21.20759
8.0	115	4.744932	37.95946	22.51438
$\Sigma = 404.8001$		$\Sigma = 56.71864$	$\Sigma = 1547.024$	$\Sigma = 233.0369$
$\bar{y} = 28.91$		$\bar{x} = 4.05$		

or

$$a = 28.91 - 63.71(-0.53) = 62.68$$

Since $a = 62.68$ and $b = -0.53$, then $u_f = 62.68$ mi/h, $u_f/k_j = 0.53$, and so $k_j = 118$ veh/mi, and $\bar{u}_s = 62.68 - 0.53k$.

- Using Eq. 6.23 to determine the value of R^2, we obtain $R^2 = 0.95$.
- Using the above estimated values for u_f and k_j, we can determine the maximum flow from Eq. 6.17 as

$$q_{max} = \frac{k_j u_f}{4} = \frac{118 \times 62.68}{4}$$
$$= 1849 \text{ veh/h}$$

- Using Eq. 6.15 we also obtain the velocity at which flow is maximum, that is, $(62.68/2) = 31.3$ mi/h, and Eq. 6.16, the density at which flow is maximum, or $(118/2) = 59$ veh/h.

Example 6.3 Fitting Speed and Density Data to the Greenberg Model

The data in Table 6.1b can also be fitted into the Greenberg model shown in Eq. 6.18:

$$\bar{u}_s = c \ln \frac{k_j}{k}$$

which can be written as

$$\bar{u}_s = c \ln k_j - c \ln k \tag{6.24}$$

Solution:

- Comparing Eq. 6.24 and the estimated regression function Eq. 6.20, we see that $\bar{u}_s$ in the Greenberg expression is represented by y in the estimated regression function, $c \ln k_j$ is represented by a, c is represented by $-b$, and $\ln k$ is represented by x. Table 6.1b shows values for x_i, $x_i y_i$, and x_i^2. (Note that these values are computed to a higher degree of accuracy since they involve logarithmic values.) We therefore obtain

$$\sum y_i = 404.8 \qquad \sum x_i = 56.72 \qquad \bar{y} = 28.91$$
$$\sum x_i y_i = 1547.02 \qquad \sum x_i^2 = 233.04 \qquad \bar{x} = 4.05$$

- Using Eqs. 6.21 and 6.22, we obtain

$$a = 28.91 - 4.05b$$

$$b = \frac{1547.02 - \dfrac{(56.72)(404.8)}{14}}{233.04 - \dfrac{(56.72)^2}{14}} = -28.68$$

or

$$a = 28.91 - 4.05(-28.68) = 145.06$$

Since $a = 145.06$ and $b = -28.68$, the speed for maximum flow is $c = 28.68$ mi/h. Finally, since

$$c \ln k_j = 145.06$$
$$\ln k_j = 5.06$$
$$k_j = 157 \text{ veh/h}$$

then

$$\bar{u}_s = 28.68 \ln \frac{157}{k}$$

- Obtaining k_o, the density for maximum flow from Eq. 6.19, we then use Eq. 6.18 to determine the value of the maximum flow.

$$\ln k_j = 1 + \ln k_o$$
$$\ln 158 = 1 + \ln k_o$$
$$5.06 = 1 + \ln k_o$$
$$58.0 = k_o$$
$$q_{max} = 58.0 \times 28.68 \text{ veh/h}$$
$$q_{max} = 1663 \text{ veh/h}$$

The R^2 based on the Greenberg expression is 0.9, which indicates that the Greenshields expression is a better fit for the data in Table 6.1. Figure 6.5 shows plots of speed versus density for the two estimated regression functions obtained and also for the actual data points. Figure 6.6 shows similar plots for the volume against speed.

Microscopic Approach

The microscopic approach, which is sometimes referred to as the car-following theory or the follow-the-leader theory, considers spacings between and speeds of individual vehicles. Consider two consecutive vehicles, A and B, on a single lane of a highway, as shown in

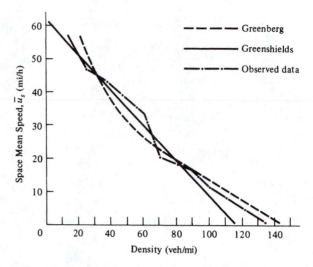

Figure 6.5 Speed Versus Density

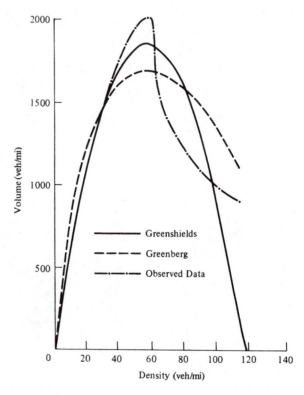

Figure 6.6 Volume Versus Density

Figure 6.7. If the leading vehicle is considered to be the *n*th vehicle and the following vehicle is considered the (*n* + 1)th vehicle, then the distances of these vehicles from a fixed section at any time *t* can be taken as x_n and x_{n+1}, respectively.

If the driver of vehicle B maintains an additional separation distance *P* above the separation distance at rest *S* such that *P* is proportional to the speed of vehicle B, then

$$P = \rho \dot{x}_{n+1} \tag{6.25}$$

where

ρ = factor of proportionality with units of time
$\dot{x}_{n+1}$ = speed of the (*n* + 1)th vehicle

We can write

$$x_n - x_{n+1} = \rho \dot{x}_{n+1} + S \tag{6.26}$$

where *S* is the distance between front bumpers of vehicles at rest.

Differentiating Eq. 6.26 gives

$$\ddot{x}_{n+1} = \frac{1}{\rho} \left[\dot{x}_n - \dot{x}_{n+1} \right] \tag{6.27}$$

Equation 6.27 is the basic equation of microscopic models, and it describes the stimulus response of the models. Researchers have shown that a time lag exists for a driver to respond to any stimulus that is induced by the vehicle just ahead, and Eq. 6.27 can therefore be written as

$$\ddot{x}_{n+1}(t + T) = \lambda \left[\dot{x}_n(t) - \dot{x}_{n+1}(t) \right] \tag{6.28}$$

where

T = time lag of response to the stimulus
λ = $(1/\rho)$ (sometimes called the sensitivity)

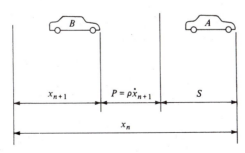

Figure 6.7   Basic Assumptions in Follow-the-Leader Theory

A general expression for λ is given in the form

$$\lambda = a \frac{\dot{x}_{n+1}^{m}(t+T)}{[x_n(t) - x_{n+1}(t)]^{\ell}} \tag{6.29}$$

The general expression for the microscopic models can then be written as

$$\ddot{x}_{n+1}(t+T) = a \frac{\dot{x}_{n+1}^{m}(t+T)}{[x_n(t) - x_{n+1}(t)]^{\ell}} [\dot{x}_n(t) - \dot{x}_{n+1}(t)] \tag{6.30}$$

where a, ℓ, and m are constants.

The microscopic model (Eq. 6.30) can be used to determine the velocity, flow, and density of a traffic stream when the traffic stream is moving in a steady state. The direct analytical solution of either Eq. 6.28 or Eq. 6.30 is not easy. It can be shown, however, that the macroscopic models discussed earlier can all be obtained from Eq. 6.30.

For example, if $m = 0$ and $\ell = 1$, the acceleration of the $(n + 1)$th vehicle is given as

$$\ddot{x}_{n+1}(t+T) = a \frac{\dot{x}_n(t) - \dot{x}_{n+1}(t)}{x_n(t) - x_{n+1}(t)}$$

Integrating the above expression, we find that the velocity of the $(n + 1)$th vehicle is

$$\dot{x}_{n+1}(t+T) = a \ln [x_n(t) - x_{n+1}(t+1)] + C$$

Since we are considering the steady-state condition,

$$\dot{x}_n(t+T) = \dot{x}_n(t) = u$$
$$u = a \ln [x_n - x_{n+1}] + C$$

Also,

$$x_n - x_{n+1} = \text{average space headway} = \frac{1}{k}$$

$$u = a \ln \left(\frac{1}{k}\right) + C$$

Using the boundary condition,

$$u = 0 \ \text{when} \ k = k_j$$

$$0 = a \ln \left(\frac{1}{k_j}\right) + C$$

$$C = -a \ln \left(\frac{1}{k_j}\right)$$

Substituting for C in the equation for u, we obtain

$$u = a \ \ln \left(\frac{1}{k} \right) - a \ \ln \left(\frac{1}{k_j} \right)$$

$$= a \ \ln \left(\frac{k_j}{k} \right)$$

which is the Greenberg model given in Eq. 6.18. Similarly, if m is allowed to be 0 and $\ell = 2$, we obtain the Greenshields model.

SHOCK WAVES IN TRAFFIC STREAMS

The fundamental diagram of traffic flow for two adjacent sections of a highway with different capacities (maximum flows) is shown in Figure 6.8. This figure describes the phenomenon of backups and queuing on a highway due to a sudden reduction of the capacity of the highway (known as a *bottleneck condition*). The sudden reduction in capacity could be due to accidents, reduction in the number of lanes, restricted bridge sizes, work zones, a signal turning red, and so forth, creating a situation where the capacity on the highway suddenly changes from C_1 to a lower value of C_2, with a corresponding change in optimum density from k_o^a to a value of k_o^b.

When such a condition exists and the normal flow and density on the highway are relatively large, the speeds of the vehicles will have to be reduced while passing the bottleneck. The point at which the speed reduction takes place can be approximately noted by the

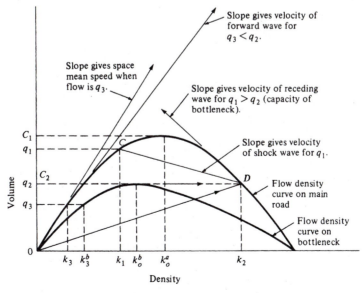

Figure 6.8 Kinematic and Shock Wave Measurements Related to Flow-Density Curve

turning on of the brake lights of the vehicles. An observer will see that this point moves upstream as traffic continues to approach the vicinity of the bottleneck, indicating an upstream movement of the point at which flow and density change. This phenomenon is usually referred to as a *shock wave* in the traffic stream.

Let us consider two different densities of traffic, k_1 and k_2, along a straight highway as shown in Figure 6.9, where $k_1 > k_2$. Let us also assume that these densities are separated by the line w, representing the shock wave moving at a speed u_w. If the line w moves in the direction of the arrow (that is, in the direction of the traffic flow), u_w is positive.

With u_1 equal to the space mean speed of vehicles in the area with density k_1 (section P), the speed of the vehicle in this area relative to the line w is

$$u_{r_1} = (u_1 - u_w)$$

The number of vehicles crossing line w from area P during a time period t is

$$N_1 = u_{r_1} k_1 t$$

Similarly, the speed of vehicles in the area with density k_2 (section Q) relative to line w is

$$u_{r_2} = (u_2 - u_w)$$

and the number of vehicles crossing line w during a time period t is

$$N_2 = u_{r_2} k_2 t$$

Since the net change is zero—that is, $N_1 = N_2$ and $(u_1 - u_w)k_1 = (u_2 - u_w)k_2$—we have

$$u_2 k_2 - u_1 k_1 = u_w(k_2 - k_1) \tag{6.31}$$

If the flow rates in sections P and Q are q_1 and q_2, respectively, then

$$q_1 = k_1 u_1 \qquad q_2 = k_2 u_2$$

Substituting q_1 and q_2 for $k_1 u_1$ and $k_2 u_2$ in Eq. 6.31 gives

$$q_2 - q_1 = u_w(k_2 - k_1)$$

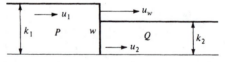

Figure 6.9 Movement of Shock Wave Due to Change in Densities

That is,

$$u_w = \frac{q_2 - q_1}{k_2 - k_1} \qquad (6.32)$$

which is also the slope of the line CD shown in Figure 6.8. This indicates that the velocity of the shock wave created by a sudden change of density from k_1 to k_2 on a traffic stream is the slope of the chord joining the points associated with k_1 and k_2 on the volume density curve for that traffic stream.

Example 6.4   Length of Queue Due to a Moving Shock Wave

The volume at a section of a two-lane highway is 1500 veh/h in each direction, and the density is about 25 veh/mi. A large dump truck loaded with soil from an adjacent construction site joins the traffic stream and travels at a speed of 10 mi/h for a length of 2.5 mi along the upgrade before turning off onto a dump site. Due to the relatively high flow in the opposite direction, it is impossible for any car to pass the truck. Vehicles just behind the truck therefore have to travel at the speed of the truck, which results in the formation of a platoon having a density of 100 veh/mi and a flow of 1000 veh/h. Determine how many vehicles will be in the platoon by the time the truck leaves the highway.

Solution:

- Use Eq. 6.32 to obtain the wave velocity.

$$
\begin{aligned}
u_w &= \frac{q_2 - q_1}{k_2 - k_1} \\
&= \frac{1000 - 1500}{100 - 25} \\
&= -6.7 \text{ mi/h}
\end{aligned}
$$

- Knowing that the truck is traveling at 10 mi/h and that the shock wave is moving backward relative to the road at 6.7 mi/h, determine the growth rate of the platoon.

$$10 - (-6.7) = 16.7 \text{ mi/h}$$

- Calculate the time spent by the truck on the highway—2.5/10 = 0.25—to determine the length of the platoon by the time the truck leaves the highway.

$$0.25 \times 16.7 = 4.2 \text{ mi}$$

- Use the density of 100 veh/mi to calculate the number of vehicles in the platoon.

$$100 \times 4.2 = 420 \text{ vehicles}$$

Special Cases of Shock Wave Propagation

The shock wave phenomenon can also be explained by considering a continuous change of flow and density in the traffic stream. If the change in flow and the change in density are very small, we can write

$$(q_2 - q_1) = \Delta q \qquad (k_2 - k_1) = \Delta k$$

The wave velocity can then be written as

$$u_w = \frac{\Delta q}{\Delta k} = \frac{dq}{dk} \qquad (6.33)$$

Since $q = k\bar{u}_s$, substituting $k\bar{u}_s$ for q in Eq. 6.33 gives

$$u_w = \frac{d(k\bar{u}_s)}{dk} \qquad (6.34)$$

$$= \bar{u}_s + k \frac{d\bar{u}_s}{dk}$$

When such a continuous change of volume occurs in a vehicular flow, a phenomenon similar to that of fluid flow exists, in which the waves created in the traffic stream transport the continuous changes of flow and density. The speed of these waves is dq/dk and is given by Eq. 6.34.

We have already seen that as density increases, the space mean speed decreases (see Eq. 6.6), giving a negative value for $d\bar{u}_s/dk$. This shows that at any point on the fundamental diagram, the speed of the wave is theoretically less than the space mean speed of the traffic stream. Thus, the wave moves in the opposite direction relative to that of the traffic stream. The actual direction and speed of the wave will depend on the point at which we are on the curve (that is, the flow and density on the highway), and the resultant effect on the traffic downstream will depend on the capacity of the restricted area (bottleneck). When both the flow and the density of the traffic stream are very low, that is, approaching zero, the flow is much lower than the capacity of the restricted area and there is very little interaction between the vehicles. The differential of $\bar{u}_s$ with respect to k ($d\bar{u}_s/dk$) then tends to zero, and the wave velocity approximately equals the space mean speed. The wave therefore moves forward with respect to the road, and no backups result. As the flow of the traffic stream increases to a value much higher than zero but still lower than the capacity of the restricted area (say, q_3 in Figure 6.8), the wave velocity is still less than the space mean speed of the traffic stream, and the wave moves forward relative to the road. This results in a reduction in speed and an increase in the density from k_3 to k_3^b as vehicles enter the bottleneck but no backups occur. When the volume on the highway is equal to the capacity of the restricted area (C_2 in Figure 6.8), the speed of the wave is zero and the wave does not move. This results in a much slower speed and a greater increase in the density to k_o^b as the vehicles enter the restricted area. Again, delay occurs but there are no backups. However, when the flow on the highway is greater than the capacity of the restricted area, not only is the speed of the wave less than the space mean speed of the vehicle stream, but it moves backward relative to the road. As vehicles enter the restricted area, a complex queuing condition arises, resulting in an immediate increase in the density from k_1 to k_2 in the upstream section of the road and a considerable decrease in speed. The movement of the wave toward the upstream section of the traffic stream creates a shock wave in the traffic stream, eventually resulting in backups, which gradually moves upstream of the traffic stream.

The expressions developed for the speed of the shock wave, Eqs. 6.32 and 6.34, can be applied to any of the specific models described earlier. For example, the Greenshields model can be written as

$$\bar{u}_{si} = u_f\left(1 - \frac{k_i}{k_j}\right) \qquad \bar{u}_{si} = u_f(1 - \eta_i) \qquad (6.35)$$

where $\eta_i = (k_i/k_j)$ (normalized density).

If the Greenshields model fits the flow density relationship for a particular traffic stream, Eq. 6.32 can be used to determine the speed of a shock wave as

$$
\begin{aligned}
u_w &= \frac{\left[k_2 u_f\left(1 - \frac{k_2}{k_j}\right)\right] - \left[k_1 u_f\left(1 - \frac{k_1}{k_j}\right)\right]}{k_2 - k_1} \\
&= \frac{k_2 u_f(1 - \eta_2) - k_1 u_f(1 - \eta_1)}{k_2 - k_1} \\
&= \frac{u_f(k_2 - k_1) - k_2 u_f \eta_2 + k_1 u_f \eta_1}{k_2 - k_1} \\
&= \frac{u_f(k_2 - k_1) - \frac{u_f}{k_j}(k_2^2 - k_1^2)}{k_2 - k_1} \\
&= \frac{u_f(k_2 - k_1) - \frac{u_f}{k_j}(k_2 - k_1)(k_2 + k_1)}{k_2 - k_1} \\
&= u_f[1 - (\eta_1 + \eta_2)]
\end{aligned}
$$

The speed of a shock wave for the Greenshields model is therefore given as

$$u_w = u_f[1 - (\eta_1 + \eta_2)] \qquad (6.36)$$

Density Nearly Equal

When there is only a small difference between k_1 and k_2 (that is, $\eta_1 \approx \eta_2$),

$$u_w = u_f[1 - (\eta_1 + \eta_2)] \qquad \text{(neglecting the small change in } \eta_1\text{)}$$
$$= u_f[1 - 2\eta_1]$$

Stopping Waves

Equation 6.36 can also be used to determine the velocity of the shock wave due to the change from green to red of a signal at an intersection approach if the Greenshields model is applicable. During the green phase, the normalized density is η_1. When the traffic signal changes to red, the traffic at the stop line of the approach comes to a halt, which results in a density equal to the jam density. The value of η_2 is then equal to 1.

The speed of the shock wave, which in this case is a stopping wave, can be obtained by

$$u_w = u_f[1 - (\eta_1 + 1)] = -u_f\eta_1 \qquad (6.37)$$

Equation 6.37 indicates that in this case the shock wave travels upstream of the traffic with a velocity of $u_f\eta_1$. If the length of the red phase is t sec, then the length of the line of cars upstream of the stopline at the end of the red interval is $u_f\eta_1 t$.

Starting Waves

At the instant when the signal again changes from red to green, η_1 equals 1. Vehicles will then move forward at a speed of $\bar{u}_{s2}$, resulting in a density of η_2. The speed of the shock wave, which in this case is a starting wave, is obtained by

$$u_w = u_f[1 - (1 + \eta_2)] = -u_f\eta_2 \qquad (6.38)$$

Equation 6.35, $\bar{u}_{s2} = u_f(1 - \eta_2)$, gives

$$\eta_2 = 1 - \frac{\bar{u}_{s2}}{u_f}$$

The velocity of the shock wave is then obtained as

$$u_w = -u_f + \bar{u}_{s2}$$

Since the starting velocity $\bar{u}_{s2}$ just after the signal changes to green is usually small, the velocity of the starting shock wave approximately equals $-u_f$.

Example 6.5   Length of Queue Due to a Stopping Shock Wave

Studies have shown that the traffic flow on a single-lane approach to a signalized inter-section can be described by the Greenshields model. If the jam density on the approach is 130 veh/mi, determine the velocity of the stopping wave when the approach signal changes to red if the density on the approach is 45 veh/mi and the space mean speed is 40 mi/h. At the end of the red interval, what length of the approach upstream from the stop line will vehicles be affected if the red interval is 35 sec?

Solution:

- Use the Greenshields model.

$$\overline{u}_s = u_f - \frac{u_f}{k_j} k$$

$$40 = u_f - \frac{u_f}{130} 45$$

$$5200 = 130 u_f - 45 u_f$$

$$u_f = 61.2 \text{ mi/h}$$

- Use Eq. 6.37 for a stopping wave.

$$u_w = -u_f \eta_1$$

$$= -61.2 \times \frac{45}{130}$$

$$= -21.2 \text{ mi/h}$$

Since u_w is negative, the wave moves upstream.

- Determine the approach length that will be affected in 35 sec.

$$21.2 \times 1.47 \times 35 = 1090.7 \text{ ft}$$

GAP AND GAP ACCEPTANCE

Thus far we have been considering the theory of traffic flow as it relates to the flow of vehicles in a single stream. Another important aspect of traffic flow is the interaction of vehicles as they join, leave, or cross a traffic stream. Examples of these include ramp vehicles merging onto an expressway stream, freeway vehicles leaving the freeway onto frontage roads, and the changing of lanes by vehicles on a multilane highway. The most important factor a driver considers in making any one of these maneuvers is the availability of a gap between two vehicles that, in the driver's judgment, is adequate for him or her to complete the maneuver. The evaluation of available gaps and the decision to carry out a specific maneuver within a particular gap are inherent in the concept of gap acceptance.

Following are the important measures that involve the concept of gap acceptance.

1. **Merging** is the process by which a vehicle in one traffic stream joins another traffic stream moving in the same direction, such as a ramp vehicle joining a freeway stream.
2. **Diverging** is the process by which a vehicle in a traffic stream leaves that traffic stream, such as a vehicle leaving the outside lane of an expressway.
3. **Weaving** is the process by which a vehicle first merges into a stream of traffic, obliquely crosses that stream, and then merges into a second stream moving in the same direction; for example, the maneuver required for a ramp vehicle to join the far side stream of flow on an expressway.
4. **Gap** is the headway in a major stream, which is evaluated by a vehicle driver in a minor stream who wishes to merge into the major stream. It is expressed either in units of time (time gap) or in units of distance (space gap).
5. **Time lag** is the difference between the time a vehicle that merges into a main traffic stream reaches a point on the highway in the area of merge and the time a vehicle in the main stream reaches the same point.
6. **Space lag** is the difference, at an instant of time, between the distance a merging vehicle is away from a reference point in the area of merge and the distance a vehicle in the main stream is away from the same point.

Figure 6.10 depicts the time-distance relationships for a vehicle at a stop sign waiting to merge and for vehicles on the near lane of the main traffic stream.

A driver who intends to merge must first evaluate the gaps that become available to determine which gap (if any) is large enough to accept the vehicle, in his or her opinion. In accepting that gap, the driver feels that he or she will be able to complete the merging maneuver and safely join the main stream within the length of the gap. This phenomenon generally is referred to as *gap acceptance.* It is of importance when engineers are considering the delay of vehicles on minor roads wishing to join a major-road traffic stream at unsignalized intersections, and also the delay of ramp vehicles wishing to join expressways.

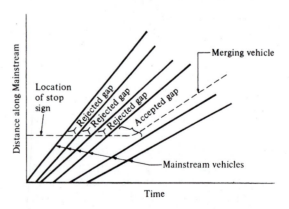

Figure 6.10 Time-Space Diagrams for Vehicles in the Vicinity of a Stop Sign

It can also be used in timing the release of vehicles at an on-ramp of an expressway, such that the probability of the released vehicle finding an acceptable gap in arriving at the freeway shoulder lane is maximum. To use the phenomenon of gap acceptance in evaluating delays, waiting times, queue lengths, and so forth at unsignalized intersections and at on-ramps, the average minimum gap length that will be accepted by drivers should be determined first. Several definitions have been given to this "critical" value. Greenshields referred to it as the "acceptable average minimum time gap" and defined it as the gap accepted by 50 percent of the drivers. The concept of "critical gap" was used by Raff, who defined it as the gap for which the number of accepted gaps shorter than it is equal to the number of rejected gaps longer than it. The data in Table 6.2 are used to demonstrate the determination of the critical gap using Raff's definition. Either a graphical or an algebraic method can be used. In using the graphical method, two cumulative distribution curves are drawn as shown in Figure 6.11. One of them relates gap lengths t with the number of accepted gaps less than t, and the other relates t with the number of rejected gaps greater than t. The intersection of these two curves gives the value of t for the critical gap.

In using the algebraic method, it is necessary first to identify the gap lengths between which the critical gap lies. This is done by comparing the change in number of accepted

Table 6.2   Computation of Critical Gap (t_c)

(a) Gaps Accepted and Rejected		
1 *Length of Gap* *(t sec)*	*2* *Number of Accepted Gaps* *(less than t sec)*	*3* *Number of Rejected Gaps* *(greater than t sec)*
0.0	0	116
1.0	2	103
2.0	12	66
3.0	$m =$ 32	$r = 38$
4.0	$n =$ 57	$p = 19$
5.0	84	6
6.0	116	0

(b) Difference in Gaps Accepted and Rejected			
1 *Consecutive* *Gap Lengths* *(t sec)*	*2* *Change in* *Number of* *Accepted Gaps* *(less than t sec)*	*3* *Change in* *Number of* *Gaps (greater* *than t sec)*	*4* *Difference* *Between Columns* *2 and 3*
0.0–1.0	2	13	11
1.0–2.0	10	37	27
2.0–3.0	20	28	8
3.0–4.0	25	19	6
4.0–5.0	27	13	14
5.0–6.0	32	6	26

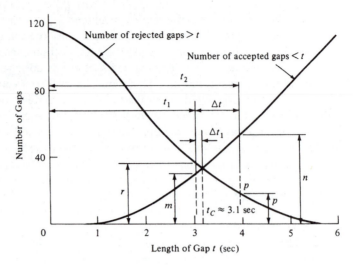

Figure 6.11   Cumulative Distribution Curves for Accepted and Rejected Gaps

gaps less than t sec (column 2 of Table 6.2b) for two consecutive gap lengths, with the change in number of rejected gaps greater than t sec (column 3 of Table 6.2b) for the same two consecutive gap lengths. The critical gap length lies between the two consecutive gap lengths, where the difference between the two changes is minimal. Table 6.2b shows the computation and indicates that the critical gap for this case lies between 3 and 4 sec.

In the example of Figure 6.10, with Δt equal to the time increment used for gap analysis, the critical gap lies

$$\text{between } t_1 \text{ and } t_2 = t_1 + \Delta t$$

where
m = number of accepted gaps less than t_1
r = number of rejected gaps greater than t_1
n = number of accepted gaps less than t_2
p = number of rejected gaps greater than t_2

Assuming that the curves are linear between t_1 and t_2, the point of intersection of these two lines represents the critical gap. From Figure 6.11, the critical gap expression can be written as

$$t_c = t_1 + \Delta t_1$$

Using the properties of similar triangles,

$$\frac{\Delta t_1}{r - m} = \frac{\Delta t - \Delta t_1}{n - p}$$

$$\Delta t_1 = \frac{\Delta t(r - m)}{(n - p) + (r - m)}$$

and we obtain

$$t_c = t_1 + \frac{\Delta t(r - m)}{(n - p) + (r - m)} \tag{6.39}$$

For the data given in Table 6.2, we thus have

$$t_c = 3 + \frac{1(38 - 32)}{(57 - 19) + (38 - 32)} = 3 + \frac{6}{38 + 6}$$

$$\approx 3.14 \text{ sec}$$

STOCHASTIC APPROACH TO GAP AND GAP ACCEPTANCE PROBLEMS

The use of gap acceptance to determine the delay of vehicles in minor streams wishing to merge onto major streams requires a knowledge of the frequency of arrivals of gaps that are at least equal to the critical gap. This in turn depends on the distribution of arrivals of main stream vehicles at the area of merge. It is generally accepted that for light to medium traffic flow on a highway, the arrival of vehicles is randomly distributed. It is therefore important that the probabilistic approach to the subject be discussed. It is usually assumed that for light-to-medium traffic the distribution is Poisson, although assumptions of gamma and exponential distributions have also been made.

Assuming that the distribution of main stream arrival is Poisson, then the probability of x arrivals in any interval of time t sec can be obtained from the expression

$$P(x) = \frac{\mu^x e^{-\mu}}{x!} \qquad \text{for } x = 0, 1, 2 \dots, \infty \tag{6.40}$$

where

$P(x)$ = the probability of x vehicles arriving in time t sec
μ = average number of vehicles arriving in time t

If V represents the total number of vehicles arriving in time T sec, then the average number of vehicles arriving per second is

$$\lambda = \frac{V}{T} \qquad \mu = \lambda t$$

We can therefore write Eq. 6.40 as

$$P(x) = \frac{(\lambda t)^x e^{-\lambda t}}{x!} \tag{6.41}$$

Now consider a vehicle at an unsignalized intersection or at a ramp waiting to merge into the main stream flow, arrivals of which can be described by Eq. 6.41. The minor stream vehicle will merge only if there is a gap of t sec equal to or greater than its critical gap. This will occur when no vehicles arrive during a period t sec long. The probability of this is the probability of zero cars arriving (that is, when x in Eq. 6.41 is zero). Substituting zero for x in Eq. 6.41 will therefore give a probability of a gap $(h \ge t)$ occurring. Thus,

$$P(0) = P(h \ge t) = e^{-\lambda t} \qquad \text{for } t \ge 0 \tag{6.42}$$

$$P(h < t) = 1 - e^{-\lambda t} \qquad \text{for } t \ge 0 \tag{6.43}$$

Since

$$P(h < t) + P(h \ge t) = 1$$

it can be seen that t can take all values from 0 to ∞, which therefore makes Eqs. 6.42 and 6.43 continuous functions. The probability function described by Eq. 6.42 is known as the *exponential distribution*.

Equation 6.42 can be used to determine the expected number of acceptable gaps that will occur at an unsignalized intersection or at the merging area of an expressway on-ramp during a period T, if the Poisson distribution is assumed for the main stream flow and the volume V is also known. Let us assume that T is equal to 1 hr, and that V is the volume in veh/h on the main stream flow. Since $(V-1)$ gaps occur between V successive vehicles in a stream of vehicles, then the expected number of gaps greater or equal to t is given as

$$\text{Freq. } (h \ge t) = (V-1)e^{-\lambda t} \tag{6.44}$$

and the expected number of gaps less than t is given as

$$\text{Freq. } (h < t) = (V-1)(1 - e^{-\lambda t}) \tag{6.45}$$

Example 6.6 Number of Acceptable Gaps for Vehicles on an Expressway Ramp

The peak hour volume on an expressway at the vicinity of the merging area of an on-ramp was determined to be 1800 veh/h. If it is assumed that the arrival of expressway vehicles can be described by a Poisson distribution, and the critical gap for merging vehicles is 3.5 sec, determine the expected number of acceptable gaps for ramp vehicles that will occur on the expressway during the peak hour.

Solution:

- List the data.

$V = 1800$
$T = 3600$ sec
$\lambda = (1800/3600) = 0.5$ veh/sec

- Calculate the expected number of acceptable gaps in 1 hr using Eq. 6.44.

$$(h \geq t) = (1800 - 1)e^{(-0.5 \times 3.5)} = 1799e^{-1.75} = 312$$

The expected number of occurrences of different gaps t for the above example have been calculated and are shown in Table 6.3.

Table 6.3 Number of Different Lengths of Gaps Occurring During a Period of 1 hr for $V = 1800$ veh/h and an Assumed Distribution of Poisson for Arrivals

Gap (t sec)	Probability		No. of Gaps	
	$P(h \geq t)$	$P(h < t)$	$h \geq t$	$h \leq t$
0	1.0000	0.0000	1799	0
0.5	0.7788	0.2212	1401	398
1.0	0.6065	0.3935	1091	708
1.5	0.4724	0.5276	849	950
2.0	0.3679	0.6321	661	1138
2.5	0.2865	0.7135	515	1284
3.0	0.2231	0.7769	401	1398
3.5	0.1738	0.8262	312	1487
4.0	0.1353	0.8647	243	1556
4.5	0.1054	0.8946	189	1610
5.0	0.0821	0.9179	147	1652

The basic assumption made in the above analysis is that the arrival of main stream vehicles can be described by a Poisson distribution. This assumption is reasonable for light-to-medium traffic but may not be acceptable for conditions of heavy traffic. Analyses of the occurrence of different gap sizes when traffic volume is heavy have shown that the main discrepancies occur at gaps of short lengths (that is, less than 1 sec). The reason for this is that although theoretically there are definite probabilities for the occurrence of gaps between 0 and 1 sec, in reality these gaps very rarely occur, since a driver will tend to keep a safe distance between his or her vehicle and the vehicle immediately in front. One alternative used to deal with this situation is to restrict the range of headways by introducing a minimum gap. Equations 6.42 and 6.43 can then be written as

$$P(h \geq t) = e^{-\lambda(t-\tau)} \qquad \text{for } t \geq 0 \qquad (6.46)$$

$$P(h < t) = 1 - e^{-\lambda(t-\tau)} \qquad \text{for } t \geq 0 \qquad (6.47)$$

where τ is the minimum headway introduced.

Example 6.7   Number of Acceptable Gaps with a Restrictive Range, for Vehicles on an Expressway Ramp

Repeat Example 6.6 using a minimum gap in the expressway traffic stream of 1.0 sec and the data below.

$V = 1800$
$T = 3600$
$\lambda = (1800/3600) = 0.5$ veh/sec
$t = 3.5$ sec

Solution:

- Calculate the expected number of acceptable gaps in 1 hr.

$$(h \geq t) = (1800 - 1)e^{-0.5(3.5-1.0)} = 1799e^{-0.5 \times 2.5}$$
$$= 515$$

INTRODUCTION TO QUEUING THEORY

One of the greatest concerns of traffic engineers is the serious congestion that exists on urban highways, especially during peak hours. This congestion results in the formation of queues on expressway on-ramps and off-ramps, at signalized and unsignalized intersections, and on arterials, where moving queues may occur. An understanding of the processes that lead to the occurrence of queues and the subsequent delays on highways is essential for the proper analysis of the effects of queuing. The theory of queuing therefore concerns the use of mathematical algorithms to describe the processes that result in the formation of queues, so that a detailed analysis of the effects of queues can be undertaken. These mathematical algorithms can be used to determine the probability that an arrival will be delayed, the expected waiting time for all arrivals, the expected waiting time of an arrival that waits, and so forth.

Several models have been developed that can be applied to traffic situations such as the merging of ramp traffic to freeway traffic, interactions at pedestrian crossings, and sudden reduction of capacity on freeways. This section will give only the elementary queuing theory relationships for a specific type of queue, that is, the single-channel queue. The theoretical development of these relationships is not included here. Interested readers are referred to the "Additional Readings" section at the end of this chapter for a more detailed treatment of the topic.

A queue is formed when arrivals wait for a service or an opportunity, such as the arrival of an accepted gap in a main traffic stream, the collection of tolls at a tollbooth or of parking fees at a parking garage, and so forth. The service can be provided in a single channel or in several channels. Proper analysis of the effects of such a queue can be carried out only if the queue is fully specified. This requires that the following characteristics of the queue be given: (1) the characteristic distribution of arrivals, such as uniform, Poisson, and so on; (2) the method of service, such as first come–first served, random, and priority; (3) the characteristic of the queue length, that is, whether it is finite or infinite; (4) the distribution of service times; and (5) the channel layout, that is, whether there are single or multiple channels and, in the case of multiple channels, whether they are in series or parallel. Several methods for the classification of queues based on the above characteristics have been used, some of which are discussed below.

> *Arrival Distribution.* The arrivals can be described as either a deterministic distribution or a random distribution. Light-to-medium traffic is usually described by a Poisson distribution, and this is generally used in queuing theories related to traffic flow.
>
> *Service Method.* Queues can also be classified by the method used in servicing the arrivals. These include first come–first served, where units are served in order of their arrivals, and last in–first served, where the service is reversed to the order of arrival. The service method can also be based on priority, where arrivals are directed to specific queues of appropriate priority levels—for example, giving priority to buses. Queues are then serviced in order of their priority level.
>
> *Characteristics of the Queue Length.* The maximum length of the queue, that is, the maximum number of units in the queue, is specified, in which case the queue is a finite or truncated queue, or else there may be no restriction on the length of the queue. Finite queues are sometimes necessary when the waiting area is limited.
>
> *Service Distribution.* This distribution is also usually considered as random, and the Poisson and negative exponential distributions have been used.
>
> *Number of Channels.* The number of channels usually corresponds to the number of waiting lines and is therefore used to classify queues, for example, as a single- channel or multichannel queue.
>
> *Oversaturated and Undersaturated Queues.* Oversaturated queues are those in which the arrival rate is greater than the service rate, and undersaturated queues are those in which the arrival rate is less than the service rate. The length of an undersaturated queue may vary but will reach a steady state with the arrival of units. The length of an oversaturated queue, however, will never reach a steady state but will continue to increase with the arrival of units.

Single-Channel, Undersaturated, Infinite Queues

Figure 6.12 is a schematic of a single-channel queue in which the rate of arrival is q veh/h and the service rate is Q veh/h. For an undersaturated queue, $Q > q$, assuming that both the

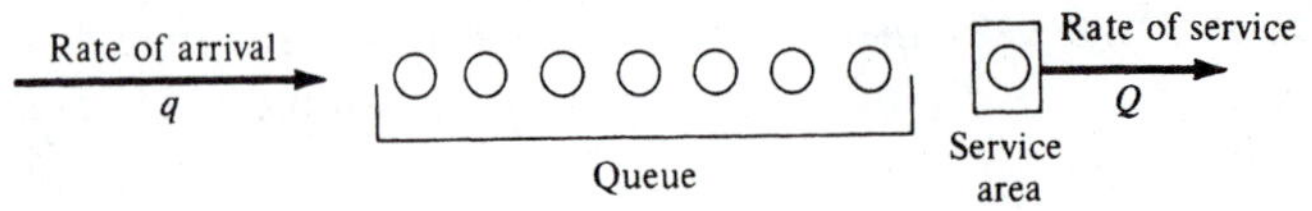

Figure 6.12 A Single-Channel Queue

rate of arrivals and the rate of service are random, the following relationships can be developed:

1. Probability of n units in the system, $P(n)$

$$P(n) = \left(\frac{q}{Q}\right)^n \left(1 - \frac{q}{Q}\right) \tag{6.48}$$

   where n is the number of units in the system, including the unit being serviced.

2. The expected number of units in the system, $E(n)$

$$E(n) = \frac{q}{Q - q} \tag{6.49}$$

3. The expected number of units waiting to be served (that is, the mean queue length) in the system, $E(m)$

$$E(m) = \frac{q^2}{Q(Q - q)} \tag{6.50}$$

   Note that $E(m)$ is not exactly equal to $E(n) - 1$, the reason being that there is a definite probability of zero units being in the system, $P(0)$.

4. Average waiting time in the queue, $E(w)$

$$E(w) = \frac{q}{Q(Q - q)} \tag{6.51}$$

5. Average waiting time of an arrival, including queue and service, $E(v)$

$$E(v) = \frac{1}{Q - q} \tag{6.52}$$

6. Probability of spending time t or less in the system

$$P(v \leq t) = 1 - e^{-\left(1 - \frac{q}{Q}\right)qt} \tag{6.53}$$

7. Probability of waiting for time t or less in the queue

$$P(w \leq t) = 1 - \frac{q}{Q} e^{-\left(1 - \frac{q}{Q}\right)qt} \tag{6.54}$$

8. Probability of more than N vehicles being in the system, that is, $P(n > N)$

$$P(n > N) = \left(\frac{q}{Q}\right)^{N+1} \tag{6.55}$$

Equation 6.49 can be used to produce a graph of the relationship between the expected number of units in the system, $E(n)$, and the ratio of the rate of arrival to the rate of service, $\rho = q/Q$. Figure 6.13 is such a representation for different values of ρ. It should be noted that as this ratio tends to 1 (that is, approaching saturation), the expected number of vehicles in the system tends to infinity. This shows that q/Q, which is usually referred to as the *traffic intensity*, is an important factor in the queuing process. The figure also indicates that queuing is of no significance when ρ is less than 0.5, but at values of 0.75 and above, the average queue lengths tend to increase rapidly. Figure 6.14 also is a graph of the probability of n units being in the system versus q/Q. Equation 6.49 can also be used to produce this graph.

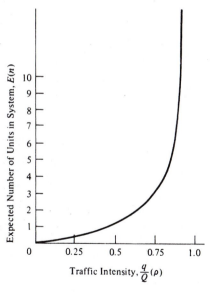

Figure 6.13   Expected Number of Vehicles in the System $E(n)$ vs. Traffic Intensity (ρ)

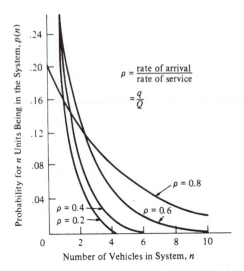

Figure 6.14 Probability of *n* Vehicles Being in the System for Different Traffic Intensities (ρ)

Example 6.8 Application of the Single-Channel, Undersaturated, Infinite Queue Theory to a Tollbooth Operation

On a given day, 425 vehicles per hour arrive at a tollbooth located at the end of an off-ramp of a rural expressway. If the vehicles can be serviced by only a single channel at the service rate of 625 veh/h, determine (a) the percentage of time the operator of the tollbooth will be free, (b) the average number of vehicles in the system, and (c) the average waiting time for the vehicles that wait. (Assume Poisson arrival and negative exponential service time.)

Solution:

a. $q = 425$ and $Q = 625$. For the operator to be free, the number of vehicles in the system must be zero. From Eq. 6.48,

$$P(0) = 1 - \frac{q}{Q} = 1 - \frac{425}{625}$$

$$= 0.32$$

The operator will be free 32 percent of the time.

b. From Eq. 6.49,

$$E(n) = \frac{425}{625 - 425}$$

$$\approx 2$$

c. From Eq. 6.52,

$$E(v) = \frac{1}{625 - 425} = 0.005 \text{ hr}$$
$$= 18.0 \text{ sec}$$

Single-Channel, Undersaturated, Finite Queues

In the case of a finite queue, the maximum number of units in the system is specified. Let this number be N. Let the rate of arrival be q and the service rate be Q. If it is also assumed that both the rate of arrival and the rate of service are random, the following relationships can be developed for the finite queue.

1. Probability of n units in the system

$$P(n) = \frac{1 + \rho}{1 - \rho^{N+1}} \rho^n \tag{6.56}$$

where $\rho = q/Q$.

2. The expected number of units in the system

$$E(n) = \frac{\rho}{1 - \rho} \frac{1 - (N + 1)\rho^N + N\rho^{N+1}}{1 - \rho^{N+1}} \tag{6.57}$$

Example 6.9 Application of the Single-Channel, Undersaturated, Finite Queue Theory to an Expressway Ramp

The number of vehicles that can enter the on-ramp of an expressway is controlled by a metering system, which allows a maximum of 10 vehicles to be on the ramp at any one time. If the vehicles can enter the expressway at a rate of 500 veh/h and the rate of arrival of vehicles at the on-ramp is 400 veh/h during the peak hour, determine (a) the probability of 5 cars being on the on-ramp, (b) the percent of time the ramp is full, and (c) the expected number of vehicles on the ramp during the peak hour.

Solution:

a. Probability of 5 cars being on the on-ramp: $q = 400$, $Q = 500$, and $\rho = (400/500)$ = 0.8. From Eq. 6.56,

$$P(5) = \frac{(1 - 0.8)}{1 - (0.8)^{11}} (0.8)^5$$
$$= 0.072$$

   b.   From Eq. 6.56, the probability of 10 cars being on the ramp is

$$P(10) = \frac{1-0.8}{1-(0.8)^{11}} (0.8)^{10}$$

$$= 0.023$$

That is, the ramp is full only 2.3 percent of the time.
   c.   The expected number of vehicles on the ramp is obtained from Eq. 6.57.

$$E(n) = \frac{0.8}{1-0.8} \frac{1-(11)(0.8)^{10}+10(0.8)^{11}}{1-(0.8)^{11}} = 2.97$$

The expected number of vehicles on the ramp is 3.

SUMMARY

One of the most important current functions of a traffic engineer is to implement traffic control measures that will facilitate the efficient use of existing highway facilities, since extensive highway construction is no longer taking place at the rate it once was. Efficient use of any highway system entails the flow of the maximum volume of traffic without causing excessive delay to the traffic and inconvenience to the motorist. It is therefore essential that the traffic engineer understand the basic characteristics of the elements of a traffic stream, since these characteristics play an important role in the success or failure of any traffic engineering action to achieve an efficient use of the existing highway system.

This chapter has furnished the fundamental theories that are used to determine the effect of these characteristics. The definitions of the different elements have been presented, together with mathematical relationships of these elements. These relationships are given in the form of macroscopic models, which consider the traffic stream as a whole, and microscopic models, which deal with individual vehicles in the traffic stream. Using the appropriate model for a traffic flow will facilitate the computation of any change in one or more elements due to a change in another element. An introduction to queuing theory is also presented to provide the reader with simple equations that can be used to determine delay and queue lengths in simple traffic queuing systems.

PROBLEMS

6-1   Observers stationed at two sections XX and YY, 500 ft apart on a highway, record the arrival times of four vehicles as shown in the accompanying table. If the total time of observation at XX was 15 sec, determine (a) the time mean speed, (b) the space mean speed, and (c) the flow at section XX.

Time of Arrival

Vehicle	Section XX	Section YY
A	T_0	$T_0 + 7.58$ sec
B	$T_0 + 3$ sec	$T_0 + 9.18$ sec
C	$T_0 + 6$ sec	$T_0 + 12.36$ sec
D	$T_0 + 12$ sec	$T_0 + 21.74$ sec

6-2   Data obtained from aerial photography showed six vehicles on a 600-ft-long section of road. Traffic data collected at the same time indicated an average time headway of 4 sec. Determine (a) the density on the highway, (b) the flow on the road, and (c) the space mean speed.

6-3   The data shown below were obtained by time-lapse photography on a highway. Use regression analysis to fit these data to the Greenshields model and determine (a) the mean free speed, (b) the jam density, (c) the capacity, and (d) the speed at maximum flow.

Speed (mi/h)	Density (veh/m)
14.2	85
24.1	70
30.3	55
40.1	41
50.6	20
55.0	15

6-4   Two sets of students are collecting traffic data at two sections, *xx* and *yy*, of a highway 1500 ft apart. Observations at *xx* show that five vehicles passed that section at intervals of 3, 4, 3, and 5 sec, respectively. If the speeds of the vehicles were 50, 45, 40, 35, and 30 mi/h, respectively, draw a schematic showing the locations of the vehicles 20 sec after the first vehicle passed section *xx*. Also determine (a) the time mean speed, (b) the space mean speed, and (c) the density on the highway.

6-5   Researchers have used analogies between the flow of fluids and the movement of vehicular traffic to develop mathematical algorithms describing the relationship among traffic flow elements. Discuss in one or two paragraphs the main deficiencies in this approach.

6-6   Assuming that the expression

$$\bar{u}_s = u_f e^{-k/k_j}$$

can be used to describe the speed-density relationship of a highway, determine the capacity of the highway from the data below using regression analysis.

k (veh/mi)	$\bar{u}_s$ (mi/h)
43	38.4
50	33.8
8	53.2
31	42.3

Under what flow conditions is the above model valid?

6-7   Results of traffic flow studies on a highway indicate that the flow-density relationship can be described by the expression

$$q = u_f k - \frac{u_f}{k_j} k^2$$

If speed and density observations give the data shown below, develop an appropriate expression for speed versus density for this highway, and determine the density at which the maximum volume will occur as well as the value of the maximum volume. Also plot speed versus density and volume versus speed for both the expression developed and the data shown. Comment on the differences between the two sets of curves.

Speed (mi/h)	Density (veh/mi)
50	18
45	25
40	41
34	58
22	71
13	88
12	99

6-8   Studies have shown that the traffic flow on a two-lane road adjacent to a school can be described by the Greenshields model. A length of 0.5 mi adjacent to a school is described as a school zone (see Figure 6.15). The school zone operates for only 20 min. Data collected at the site when the school zone is in operation are given below.

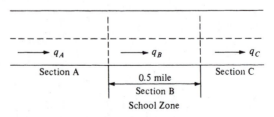

Figure 6.15 Layout of School Zone for Problem 6-8

Determine the speed of the shock waves AB and BC. Also determine both the length of the queue after the 20 min period and the time it will take the queue to dissipate after the 20 min period.

$$q_A \text{ (one direction)} = 1750 \text{ veh/h} \qquad \bar{u}_a = 35 \text{ mi/h}$$
$$q_B \text{ (one direction)} = 1543 \text{ veh/h} \qquad \bar{u}_b = 18 \text{ mi/h}$$
$$q_C \text{ (one direction)} = 1714 \text{ veh/h} \qquad \bar{u}_c = 30 \text{ mi/h}$$

6-9   A developer wants to provide access to a new building from a driveway placed 1000 ft upstream of a busy intersection. He is concerned that queues developing during the red phase of the signal at the intersection will block access. If the speed on the approach averages 35 mi/h, the density is 50 veh/mi, and the red phase is 20 sec, determine if the driveway will be affected. Assume that the traffic flow has a jam density of 110 veh/mi and can be described by the Greenshields model.

6-10   Briefly discuss the phenomenon of gap acceptance with respect to merging and weaving maneuvers in traffic streams.

6-11   The table below gives data on accepted and rejected gaps of vehicles on the minor road of an unsignalized intersection. If the arrival of major road vehicles can be described by the Poisson distribution, and the peak hour volume is 1100 veh/h, determine the expected number of accepted gaps that will be available for minor road vehicles during the peak hour.

Gap (*t*) (s)	Number of Rejected Gaps > *t*	Number of Accepted Gaps < *t*
1.5	92	3
2.5	52	18
3.5	30	35
4.5	10	62
5.5	2	100

6-12   Using appropriate diagrams, describe the resultant effect of a sudden reduction of the capacity (bottleneck) on a highway both upstream and downstream of the bottleneck.

6-13   The capacity of a highway is suddenly reduced to 60 percent of its normal capacity due to closure of certain lanes in a work zone. If the Greenshields model describes the relationship between speed and density on the highway, the jam density of the highway is 112 veh/mi, and the mean free speed is 64.5 mi/h, determine by what percentage the space mean speed at the vicinity of the work zone will be reduced if the flow upstream is 80 percent of the capacity of the highway.

6-14   The arrival times of vehicles at the ticket gate of a sports stadium may be assumed to be Poisson with a mean of 30 veh/h. It takes an average of 1.5 min for the necessary tickets to be bought for occupants of each car.

(a) What is the expected length of queue at the ticket gate, not including the vehicle being served?

(b) What is the probability that there are no more than 5 cars at the gate, including the vehicle being served?

(c) What will be the average waiting time of a vehicle?

6-15   An expressway off-ramp consisting of a single lane leads directly to a tollbooth. The rate of arrival of vehicles at the expressway can be considered to be Poisson with a mean of 50 veh/h, and the rate of service to vehicles can be assumed to be exponentially distributed with a mean of 1 min.

(a) What is the average number of vehicles waiting to be served at the booth (that is, the number of vehicles in queue, not including the vehicle being served)?

(b) What is the length of the ramp required to provide storage for all exiting vehicles 85 percent of the time? Assume the average length of a vehicle is 20 ft and that there is an average space of 5 ft between consecutive vehicles waiting to be served.

(c) What is the average waiting time a driver waits before being served at the tollbooth (that is, the average waiting time in the queue)?

ADDITIONAL READINGS

Derzko, N.A., A.J. Ugge, and E.R. Case, "Evaluation of Dynamic Freeway Flow Model by Using Flow Data," *Transportation Research Record* 905 (1983): 52–60.

Hurdle, V.F., and P.K. Datta, "Speeds and Flows on an Urban Freeway: Some Measurements and a Hypothesis," *Transportation Research Record* 905 (1983): 127–137.

Kikuchi, S. and Chakroborty, P., "Car Following Model Based on Fuzzy Inference System," *Transportation Research Record* 1365 (1992): 82–91.

Intersection Design

An intersection is an area, shared by two or more roads, whose main function is to provide for the change of route directions. Intersections vary in complexity from a simple intersection, which has only two roads crossing at a right angle to each other, to a more complex intersection, at which three or more roads cross within the same area. Drivers therefore have to make a decision at an intersection concerning which of the alternative routes they wish to take. This effort, which is not required at non-intersection areas of the highway, is part of the reason why intersections tend to have a high potential for crashes. The overall traffic flow on any highway depends to a great extent on the performance of the intersections, since intersections usually operate at a lower capacity than through sections of the road.

Intersections are classified into three general categories: grade-separated without ramps, grade-separated with ramps (commonly known as interchanges), and at-grade. Grade-separated intersections usually consist of structures that provide for traffic to cross at different levels (vertical distances) without interruption. The potential for crashes at grade-separated intersections is reduced because many potential conflicts between intersecting streams of traffic are eliminated. At-grade intersections do not provide for the flow of traffic at different levels, and therefore there exist conflicts between intersecting streams of traffic. Figure 7.1 shows different types of grade-separated intersections, and Figures 7.2 and 7.3 show different types of at-grade intersections.

Interchange design is beyond the scope of this book; this chapter presents the basic principles of the design of at-grade intersections.

TYPES OF AT-GRADE INTERSECTIONS

The basic types of at-grade intersections are T or three-leg intersections, which consist of three approaches; four-leg or cross intersections, which consist of four approaches; and multileg intersections, which consist of five or more approaches.

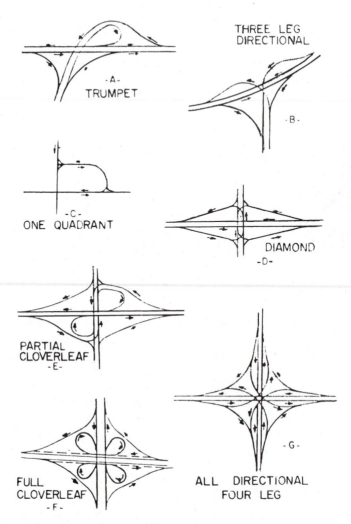

Figure 7.1 Examples of Grade Separated Intersections

SOURCE: *A Policy of Geometric Design of Highways and Streets,* American Association of State Highway and Transportation Officials, Washington, D.C., 2001, p. 854. Used with permission.

T Intersections

Figure 7.4 shows different types of T intersections, ranging from the simplest shown in Figure 7.4a to a channelized one with divisional islands and turning roadways, shown in Figure 7.4d. Channelization involves the provision of facilities such as pavement markings and traffic islands to regulate and direct conflicting traffic streams into specific travel paths. The intersection shown in Figure 7.4a is suitable for minor or local roads and may be used when minor roads intersect important highways with an intersection angle less than 30 degrees from the normal. This type of intersection is also suitable for use in rural

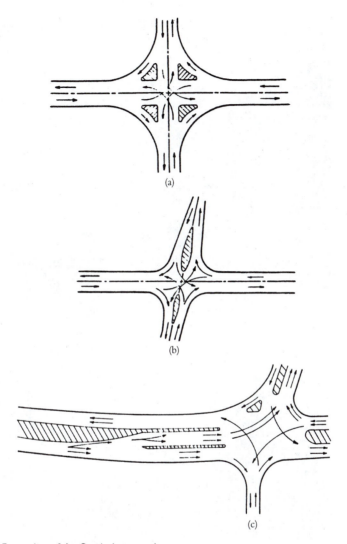

(a)

(b)

(c)

Figure 7.2 Examples of At-Grade Intersections

SOURCE: *A Policy of Geometric Design of Highways and Streets,* American Association of State Highway and Transportation Officials, Washington, D.C., 2001, pp. 680 and 682. Used with permission.

two-lane highways that carry light traffic. At locations with higher speeds and turning volumes, which increase the potential of rear-end collisions between through vehicles and turning vehicles, usually an additional area of surfacing or flaring is provided, as shown in Figure 7.4b. In this case, the flare is provided to separate right-turning vehicles from the through vehicles approaching from the east. In cases where left-turn volume from the through road onto the minor road is sufficiently high but does not require a separate left-turn lane, an auxiliary lane may be provided, as shown in Figure 7.4c. This provides

(a)

(b)

Figure 7.3 Examples of At-Grade Intersections in Urban Areas

(c)

Figure 7.3 Examples of At-Grade Intersections in Urban Areas (*continued*)

SOURCE: Photographs by Lewis Woodson, Virginia Transportation Research Council, Charlottesville, Va. Used with permission.

the space needed for through vehicles to maneuver around the left-turning vehicles, which have to slow down before making their turns. Figure 7.4d shows a channelized T intersection, in which the two-lane through road has been converted into a divided highway through the intersection. The channelized T intersection also provides both a left-turn storage lane for left-turning vehicles from the through road to the minor road and a right-turn lane on the west approach. This type of intersection is suitable for locations where volumes are high, such as high left-turn volumes from the through road and high right-turn volumes onto the minor road. An intersection of this type will probably be signalized.

Four-Leg Intersections

Figure 7.5 shows varying levels of channelization at a four-leg intersection. The unchannelized intersection shown in Figure 7.5a is used mainly at locations where minor or local roads cross, although it can also be used where a minor road crosses a major highway. In these cases the turning volumes are usually low and the roads intersect at an angle that is not greater than 30 degrees from the normal. When turning movements are frequent, right-turning roadways, such as those in Figure 7.5b, can be provided. This type of design is also common in suburban areas where pedestrians are present. The layout shown in Figure 7.5c is suitable for a two-lane highway that is not a minor crossroad and that carries

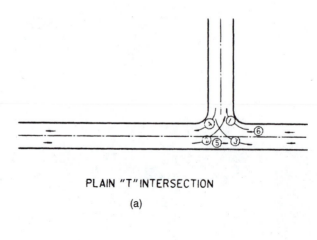

PLAIN "T" INTERSECTION

(a)

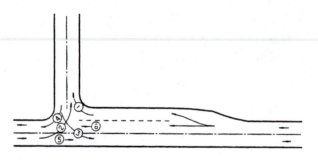

"T" INTERSECTION

(With Right Turn Lane)

(b)

Figure 7.4 Examples of T Intersections

moderate volumes at high speeds or operates near capacity. Figure 7.5d shows a suitable design for four four-lane approaches carrying high through and turning volumes. This type of intersection is usually signalized.

Multileg Intersections

Multileg intersections have five or more approaches, as shown in Figure 7.6. Whenever possible, this type of intersection should be avoided. In order to remove some of the conflicting movements from the major intersection and thereby increase safety and operation, one or more of the legs are realigned. In Figure 7.6a, the diagonal leg of the intersection is realigned to intersect the upper road at a location some distance away from the main intersection. This results in the formation of an additional T intersection, but with the multileg intersection now converted to a four-leg intersection. There are two important factors to consider when realigning roads in this way: the diagonal road should

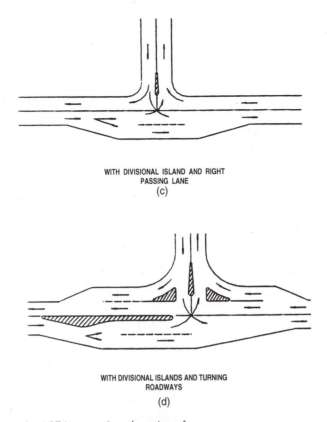

WITH DIVISIONAL ISLAND AND RIGHT
PASSING LANE
(c)

WITH DIVISIONAL ISLANDS AND TURNING
ROADWAYS
(d)

Figure 7.4   Examples of T Intersections (*continued*)

SOURCE: *A Policy of Geometric Design of Highways and Streets,* American Association of State Highway and Transportation Officials, Washington, D.C., 2001. Used with permission.

be realigned to the minor road, and the distance between the intersections should be such that they can operate independently. A similar realignment of a six-leg intersection is shown in Figure 7.6b, resulting in two four-leg intersections. In this case, it is also necessary for the realignment to be made to the minor road. For example, if the road in the right-to-left direction is the major road, it may be better to realign each diagonal road to the road in the top-to-bottom direction, thereby forming two additional T intersections and resulting in a total of three intersections. Again, the distances between these intersections should be great enough to allow for the independent operation of each intersection.

Traffic Circles

A traffic circle is a circular intersection that provides a circular traffic pattern with significant reduction in the crossing conflict points. The Federal Highway Administration publication, *Roundabouts: An Informational Guide,* describes three types of traffic circles: rotaries, neighborhood traffic circles, and roundabouts.

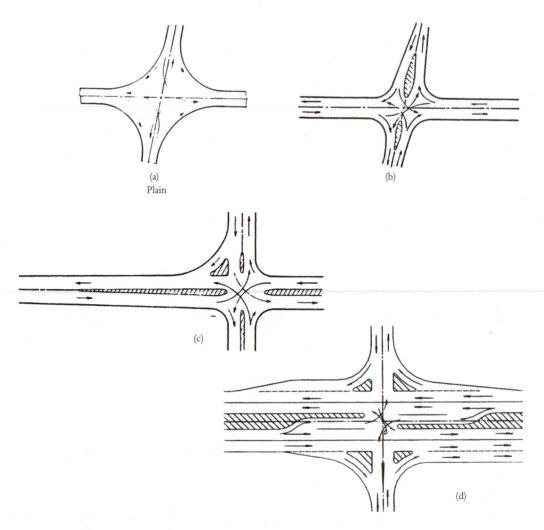

Figure 7.5 Different Levels of Channelization of Four-Leg Intersections

SOURCE: *A Policy of Geometric Design of Highways and Streets,* American Association of State Highway and Transportation Officials, Washington, D.C., 2001. Used with permission.

Rotaries have large diameters that are usually greater than 300 ft, thereby allowing speeds exceeding 30 mi/h, with a minimum horizontal deflection of the path of the through traffic.

Neighborhood traffic circles have diameters that are much smaller than rotaries and therefore allow much lower speeds. Consequently, they are used mainly at the intersections of local streets, as a means of traffic calming and/or as an aesthetic device. As a rule, they consist of pavement markings and do not usually employ raised islands. Traffic circles may

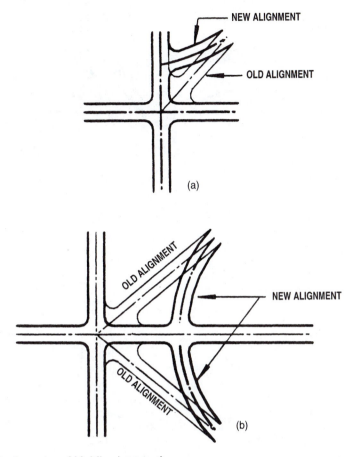

Figure 7.6 Examples of Multileg Intersections

SOURCE: *A Policy of Geometric Design of Highways and Streets,* American Association of State Highway and Transportation Officials, Washington, D.C., 2001. Used with permission.

use stop control or no control at the approaches and may or may not allow pedestrian access to the central circle. Parking may also be allowed within the circulatory roadway.

Roundabouts have specific defining characteristics that separate them from other circular intersections. These include

- Yield control at each approach
- Separation of conflicting traffic movements by pavement markings or raised islands
- Geometric characteristics of the central island that typically allow travel speeds of less than 30 mi/h
- Parking usually not allowed within the circulating roadway

Figure 7.7a shows the key features of a roundabout, and Figure 7.7b shows the key dimensions. Roundabouts can be further categorized into six classes based on the size and the environment in which they are located. These are

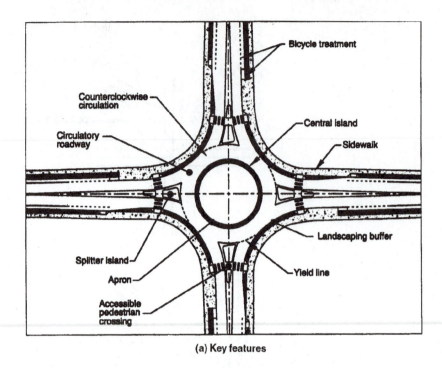

(a) Key features

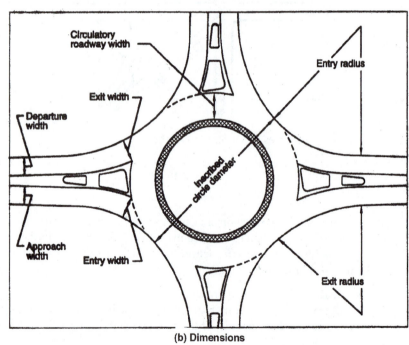

(b) Dimensions

Figure 7.7 Key Features and Dimensions of the Roundabout

SOURCE: *Roundabouts: An Information Guide.* U.S. Dept. of Transportation, Federal Highway Administration Publication No. FHWA-RD-00-067. Washington, D.C., 2000.

- Mini-roundabouts
- Urban compact roundabouts
- Urban single-lane roundabouts
- Urban double-lane roundabouts
- Rural single-lane roundabouts
- Rural double-lane roundabouts

The characteristics of each of these categories are shown in Table 7.1.

DESIGN PRINCIPLES FOR AT-GRADE INTERSECTIONS

The fundamental objective in the design of at-grade intersections is to minimize the severity of potential conflicts among different streams of traffic and between pedestrians and turning vehicles. At the same time, it is necessary to provide for the smooth flow of traffic across the intersection. The design should therefore incorporate the operating characteristics of both the vehicles and pedestrians using the intersection. For example, the corner radius of an intersection pavement or surfacing should not be less than either the turning radius of the design vehicle or the radius required for design velocity of the turning roadway under consideration. The design also should ensure adequate pavement widths of turning roadways and approach sight distances. This suggests that at-grade intersections should not be located at or just beyond sharp crest vertical curves or at sharp horizontal curves.

Table 7.1 Characteristics of Roundabout Categories

Design Element	Mini-Roundabout	Urban Compact	Urban Single-Lane	Urban Double-Lane	Rural Single-Lane	Rural Double-Lane
Recommended maximum entry design speed	25 km/h (15 mi/h)	25 km/h (15 mi/h)	35 km/h (20 mi/h)	40 km/h (25 mi/h)	40 km/h (25 mi/h)	50 km/h (30 mi/h)
Maximum number of entering lanes per approach	1	1	1	2	1	2
Typical inscribed circle diameter[1]	13 to 25 m (45 ft to 80 ft)	25 to 30 m (80 to 100 ft)	30 to 40 m (100 to 130 ft)	45 to 55 m (150 to 180 ft)	35 to 40 m (115 to 130 ft)	55 to 60 m (180 to 200 ft)
Splitter island treatment	Raised if possible, crosswalk cut if raised	Raised, with crosswalk cut	Raised, with crosswalk cut	Raised, with crosswalk cut	Raised and extended, with crosswalk cut	Raised and extended, with crosswalk cut
Typical daily service volumes on 4-leg roundabout (veh/day)	10,000	15,000	20,000	Refer to the source	20,000	Refer to the source

[1] Assumes 90-degree entries and no more than four legs.

SOURCE: *Roundabouts: An Informational Guide.* U.S. Department of Transportation, Federal Highway Administration, Publication No. FHWA-RD-00-067, Washington, D.C., 2000.

The design of an at-grade intersection involves the design of the alignment, the design of a suitable channeling system for the traffic pattern, the determination of the minimum required widths of turning roadways when traffic is expected to make turns at speeds higher than 15 mi/h, and the assurance that the sight distances are adequate for the type of control at the intersection. The methodology presented later in this chapter for determining minimum sight distances should be used to ensure that the minimum required sight distance is available on each approach. The sight distance at an approach of an at-grade intersection can be improved by flattening cut slopes and by lengthening vertical and horizontal curves. Preferably, approaches of the intersection should intersect at angles which are not greater than 30 degrees from the normal.

Alignment of At-Grade Intersections

The best alignment for an at-grade intersection is when the intersecting roads meet at right or nearly right angles. This alignment is superior to acute-angle alignments because much less road area is required for turning at the intersection, there is a lower exposure time for vehicles crossing the main traffic flow, and visibility limitations, particularly for trucks, are not as serious as those at acute-angle intersections. Figure 7.8 shows alternative methods for realigning roads intersecting at acute angles to obtain a nearly right-angle intersection. The dashed lines in this figure represent the original minor road as it intersected the major road at an acute angle. The solid lines that connect both ends of the dashed lines represent the realignment of the minor road across the major road. The methods illustrated in Figures 7.8a and 7.8b have been used successfully, but care must be taken to ensure that the realignment provides for a safe operating speed, which, to avoid hazardous situations, should not be much less than the speeds on the approaches.

The methods illustrated in Figures 7.8c and 7.8d involve the creation of a staggered intersection, in that a single curve is placed at each crossroad leg. This requires a vehicle on the minor road crossing the intersection to turn first onto the major highway and then back onto the minor highway. The realignment illustrated in Figure 7.8d is preferable because the minor-road vehicle crossing the intersection is required to make a right turn rather than a left turn from the major road to reenter the minor road. Therefore, the method illustrated in Figure 7.8c should be used only when traffic on the minor road is light and when most of this traffic is turning onto and continuing on the major road rather than crossing the intersection. A major consideration in roadway realignment at intersections is that every effort should be made to avoid creating short-radii horizontal curves, since such curves result in the encroachment of drivers on sections of the opposite lanes.

Profile of At-Grade Intersections

In designing the profile (vertical alignment) at the intersection, a combination of grade lines should be provided to facilitate the driver's control of the vehicle. For example, wherever possible, large changes in grade should be avoided; preferably, grades should not be greater than 3 percent. The stopping and accelerating distances for passenger cars on grades of 3 percent or less are not much different from those of cars on flat grades; however, significant differences start to occur at grades higher than 3 percent. When it is unavoidable to use grades of 3 percent or more, design factors such as stopping distances and accelerating

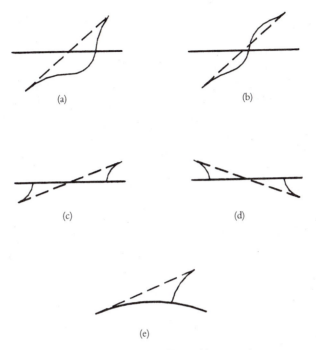

(a)

(b)

(c)

(d)

(e)

Figure 7.8 Alternative Methods of Realigning Skewed Intersections

SOURCE: *A Policy of Geometric Design of Highways and Streets,* American Association of State Highway and Transportation Officials, Washington, D.C., 2001. Used with permission.

distances should be adjusted so that conditions equivalent to those on level ground exist. In any case, it is not advisable to use grades higher than 6 percent at intersections.

When it is necessary to adjust the grade lines of the approaches at an intersection, it is preferable that the grade line of the major highway be continued across the intersection and that of the minor road be altered to obtain the desired result. However, any adjustment to a grade line of an approach should be made at a suitable distance from the intersection in order to provide a smooth junction and proper drainage. It should always be remembered that the combination of alignment and grades at an intersection should produce traffic lanes that are clearly seen by motorists at all times, without the sudden appearance of potential hazards. Also, motorists should be able to easily understand the path they should take for any desired direction.

Curves at At-Grade Intersections

The angle of turn, the turning speed, the design vehicle, and traffic volume are the main factors governing the design of curves at at-grade intersections. When the turning speed at an intersection is assumed to be 15 mi/h or less, the curves for the pavement edges are designed to conform to at least the minimum turning path of the design vehicle. When the turning speed is expected to be greater than 15 mi/h, the design speed is also considered.

The three types of design commonly used when turning speeds are 15 mi/h or less are the simple curve (an arc of a circular curve), the simple curve with taper, and the 3-centered compound curve (three simple curves joined together and turning in the same direction). (Simple curves are discussed in Chapter 16.) Figure 7.9 shows the minimum designs necessary for a passenger car making a 90-degree right turn. Figure 7.9a shows the minimum design using a simple curve. The radius of the inner edge pavement, shown as a solid line, should not be less than 25 ft, since this is the sharpest simple curvature that provides adequate space for the path of the vehicle's inner wheels to clear the pavement's edge. This design will provide for a clearance of 0.75 ft near the end of the arc. Increasing the radius of the inner pavement edge to 30 ft, as shown by the dotted line, will provide clearances of 1.2 ft at the end of the curve and 5.4 ft at the middle of the curve. The design shown in Figure 7.9b is a simple curve with tapers of 1:10 at each end and an offset of 2.5 ft. In this case it is feasible to use the lower radius of 20 ft. The layout of a 3-centered compound curve is shown in Figure 7.9c. This type of curve is composed of three circular curves of radii of 100, 20, and 100 ft, with the center of the middle curve located at a distance of 22.5 ft, including the 2.5-ft offset, from the tangent edges. This design is preferable to the simple curve because it provides for a smoother transition and because the resulting edge of the pavement fits the design vehicle path more closely. In fact, in comparison with the 30-ft radius simple curve, this design results in little additional pavement. The simple curve with taper shown in Figure 7.9b closely approximates the 3-centered curve in the field. Similar designs for single-unit (SU) trucks are shown in Figure 7.10, where the minimum radii are 50 ft for the simple curve, 40 ft for the simple curve with taper, and 120 ft, 40 ft, and 120 ft for the 3-centered curve.

The minimum design for passenger cars shown in Figure 7.9 is used only at locations where the absolute minimum turns will occur, such as the intersections of local roads with major highways where only occasional turns are made, and at intersections of two minor highways carrying low volumes. It is recommended when conditions permit that the minimum design for the SU truck be used. The minimum design layouts for larger design vehicles turning at 90 degrees are given in *A Policy of Geometric Design of Highways and Streets*. Minimum edge-of-pavement designs for different angles of turn and design vehicles are given in Table 7.2 for simple curves and simple curves with taper, and in Table 7.3 for symmetric and asymmetric 3-centered curves. Table 7.2 indicates that it is not feasible to have simple curves for large trucks such as WB-50 and WB-62 when the angle of turn is 45 degrees or greater, and for WB-40 when the angle of turn is 75 degrees or greater.

When the turning speed at an intersection is greater than 15 mi/h, the expected turning speed is used to determine the minimum radius required using the procedure presented in Chapter 3.

Channelization of At-Grade Intersections

AASHTO defines *channelization* as the separation of conflicting traffic movements into definite paths of travel by traffic islands or pavement markings to facilitate the safe and orderly movements of both vehicles and pedestrians. A *traffic island* is a defined area between traffic lanes that is used to regulate the movement of vehicles or to serve as a pedestrian refuge. Vehicular traffic is excluded from the island area. A properly channelized intersection will result in increased capacity, enhanced safety, and increased driver

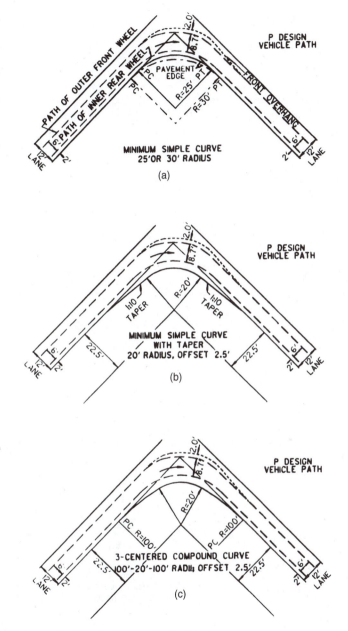

Figure 7.9 Minimum Designs for Passenger Vehicles

SOURCE: *A Policy of Geometric Design of Highways and Streets,* American Association of State Highway and Transportation Officials, Washington, D.C., 2001, p. 695. Used with permission.

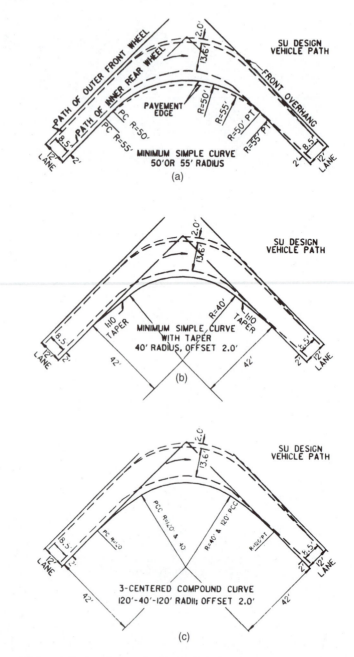

Figure 7.10 Minimum Design Curves for Single Unit Trucks

SOURCE: *A Policy of Geometric Design of Highways and Streets,* American Association of State Highway and Transportation Officials, Washington, D.C., 2001. Used with permission.

Table 7.2 Minimum Edge of Pavement Designs for Turns at Intersections: Simple Curves and Simple Curves with Taper

Angle of Turn (degrees)	Design Vehicle	Simple Curve Radius	Simple Curve Radius with Taper		
			Radius (ft)	Offset (ft)	Taper (ft:ft)
30	P	60	—	—	—
	SU	100	—	—	—
	WB-40	150	—	—	—
	WB-50	200	—	—	—
	WB-62	360	220	3.0	15:1
45	P	50	—	—	—
	SU	75	—	—	—
	WB-40	120	—	—	—
	WB-50	—	120	2.0	15:1
	WB-62	—	140	4.0	15:1
60	P	40	—	—	—
	SU	60	—	—	—
	WB-40	90	—	—	—
	WB-50	—	95	3.0	15:1
	WB-62	—	140	4.0	15:1
75	P	35	25	2.0	10:1
	SU	55	45	2.0	10:1
	WB-40	—	60	2.0	15:1
	WB-50	—	65	3.0	15:1
	WB-62	—	140	4.0	20:1
90	P	30	20	2.5	10:1
	SU	50	40	2.0	10:1
	WB-40	—	45	4.0	10:1
	WB-50	—	60	4.0	15:1
	WB-62	—	120	4.0	30:1
105	P	—	20	2.5	8:1
	SU	—	35	3.0	10:1
	WB-40	—	40	4.0	10:1
	WB-50	—	55	4.0	15:1
	WB-62	—	115	3.0	30:1
120	P	—	20	2.0	10:1
	SU	—	30	3.0	10:1
	WB-40	—	35	5.0	8:1
	WB-50	—	45	4.0	15:1
	WB-62	—	100	5.0	25:1
135	P	—	20	1.5	15:1
	SU	—	30	4.0	8:1

Continued

Table 7.2 Minimum Edge of Pavement Designs for Turns at Intersections: Simple Curves and Simple Curves with Taper (*continued*)

Angle of Turn (degrees)	*Design Vehicle*	*Simple Curve Radius*	*Simple Curve Radius with Taper*		
			Radius (ft)	*Offset (ft)*	*Taper (ft:ft)*
135	WB-40	—	30	8.0	6:1
	WB-50	—	40	6.0	10:1
	WB-62	—	80	5.0	20:1
150	P	—	18	2.0	10:1
	SU	—	30	4.0	8:1
	WB-40	—	30	6.0	8:1
	WB-50	—	35	7.0	6:1
	WB-62	—	60	10.0	10:1
180	P	—	15	0.5	20:1
	SU	—	30	1.5	10:1
	WB-40	—	20	9.5	5:1
	WB-50	—	25	9.5	5:1
	WB-62	—	55	10.0	15:1

SOURCE: *A Policy of Geometric Design of Highways and Streets,* The American Association of State Highway and Transportation Officials, Washington, D.C., 2001. Used by permission.

confidence. On the other hand, an intersection that is not properly channelized may have the opposite effect. Care should always be taken to avoid overchannelization, since this frequently creates confusion for the motorist and may even result in a lower operating level than that for an intersection without any channelization. When islands are used for channelization, they should be designed and located at the intersection without creating undue hazard to vehicles; at the same time, they should be commanding enough to prevent motorists from driving over them.

Channelization at an intersection is normally used to achieve one or more of the following objectives:

1. Direct the paths of vehicles so that not more than two paths cross at any one point
2. Control the merging, diverging, or crossing angle of vehicles
3. Decrease vehicle wander and the area of conflict among vehicles by reducing the amount of paved area
4. Provide a clear indication of the proper path for different movements
5. Give priority to the predominant movements
6. Provide pedestrian refuge
7. Provide separate storage lanes for turning vehicles, thereby creating space away from the path of through vehicles for turning vehicles to wait
8. Provide space for traffic control devices so that they can be readily seen
9. Control prohibited turns

Table 7.3 Minimum Edge of Pavement Designs for Turns at Intersections: 3-Centered Curves

Angle of Turn (degrees)	Design Vehicle	3-Centered Compound Curve Radii (ft)	Symmetric Offset (ft)	3-Centered Compound Curve Radii (ft)	Asymmetric Offset (ft)
30	P	—	—	—	—
	SU	—	—	—	—
	WB-40	—	—	—	—
	WB-50	—	—	—	—
	WB-62	—	—	—	—
45	P	—	—	—	—
	SU	—	—	—	—
	WB-40	—	—	—	—
	WB-50	200-100-200	3.0	—	—
	WB-62	460-240-460	2.0	120-140-500	3.0-8.5
60	P	—	—	—	—
	SU	—	—	—	—
	WB-40	—	—	—	—
	WB-50	200- 75-200	5.5	200- 75-275	2.0-6.0
	WB-62	400-100-400	15.0	110-100-220	10.0-12.0
75	P	100- 75-200	2.0	—	—
	SU	100- 45-120	2.0	—	—
	WB-40	120- 45-120	5.0	120- 45-200	2.0-6.5
	WB-50	150- 50-150	6.0	150- 50-225	2.0-10.0
	WB-62	440- 75-440	15.0	140-100-540	5.0-12.0
90	P	100- 20-100	2.5	—	—
	SU	120- 40-120	2.0	—	—
	WB-40	120- 40-120	5.0	120- 40-200	2.0-6.0
	WB-50	180- 60-180	6.0	120- 40-200	2.0-10.0
	WB-62	400- 70-400	10.0	160- 70-360	6.0-10.0
105	P	100- 20-100	2.5	—	—
	SU	100- 35-100	3.0	—	—
	WB-40	100- 35-100	5.0	100- 55-200	2.0-8.0
	WB-50	180- 45-180	8.0	150- 40-210	2.0-10.0
	WB-62	520- 50-520	15.0	360- 75-600	4.0-10.5
120	P	100- 20-100	2.0	—	—
	SU	100- 30-100	3.0	—	—
	WB-40	120- 30-120	6.0	100- 30-180	2.0-9.0
	WB-50	180- 40-180	8.5	150- 35-220	2.0-12.0
	WB-62	520- 70-520	10.0	80- 55-520	24.0-17.0

Continued

Table 7.3 Minimum Edge of Pavement Designs for Turns at Intersections: 3-Centered Curves (*continued*)

Angle of Turn (degrees)	Design Vehicle	3-Centered Compound		3-Centered Compound	
		Curve Radii (ft)	*Symmetric Offset (ft)*	*Curve Radii (ft)*	*Asymmetric Offset (ft)*
135	P	100- 20-100	1.5	—	—
	SU	100- 30-100	4.0	—	—
	WB-40	120- 30-120	6.5	100- 25-180	3.0-13.0
	WB-50	160- 35-160	9.0	130- 30-185	3.0-14.0
	WB-62	600- 60-600	12.0	100- 60-640	14.0-7.0
150	P	75- 85-75	2.0	—	—
	SU	100- 30-100	4.0	—	—
	WB-40	100- 30-100	6.0	90- 25-160	1.0-12.0
	WB-50	160- 35-160	7.0	120- 30-180	3.0-14.0
	WB-62	480- 55-480	15.0	140- 60-560	8.0-10.0
180	P	50- 15-50	0.5	—	—
	SU	100- 30-100	1.5	—	—
	WB-40	100- 20-100	9.5	85- 20-150	6.0-13.0
	WB-50	130- 25-130	9.5	100- 25-180	6.0-13.0
	WB-62	800- 45-800	20.0	100- 55-900	15.0-15.0

SOURCE: *A Policy of Geometric Design of Highways and Streets,* The American Association of State Highway and Transportation Officials, Washington, D.C., 2001. Used by permission.

10. Separate different traffic movements at signalized intersections with multiple-phase signals
11. Restrict the speeds of vehicles

The factors that influence the design of a channelized intersection are availability of right of way, terrain, type of design vehicle, expected vehicular and pedestrian volumes, cross sections of crossing roads, approach speeds, bus stop requirements, and the location and type of the traffic control device. For example, factors such as right of way, terrain, bus stop requirements, and vehicular and pedestrian volumes influence the extent to which channelization can be undertaken at a given location, while factors such as type of design vehicle and approach speeds influence the design of the edge of pavement.

The design of a channelized intersection should also always be governed by the following principles:

1. Motorists should not be required to make more than one decision at a time.
2. Sharp reverse curves and turning paths greater than 90 degrees should be avoided.
3. Merging and weaving areas should be as long as possible, but other areas of conflict between vehicles should be reduced to a minimum.
4. Crossing traffic streams that do not weave or merge should intersect at 90 degrees, although a range of 60–120 degrees is acceptable.

5. The intersecting angle of merging streams should be such that adequate sight distance is provided.
6. Refuge areas for turning vehicles should not interfere with the movement of through vehicles.
7. Prohibited turns should be blocked wherever possible.
8. Decisions on the location of essential traffic control devices should be a component of the design process.

General Characteristics of Traffic Islands

The definition given for traffic islands in the previous section clearly indicates that they are not all of one physical type. These islands can be formed by using raised curbs, pavement markings, or the pavement edges as shown in Figure 7.11.

Curbed Traffic Islands. A curbed island usually is formed by the construction of a concrete curb that delineates the area of the island, as shown in Figure 7.11a. Curbs are generally classified as mountable or barrier. *Mountable curbs* are constructed with their faces inclined at an angle of 45 degrees or less so that vehicles may mount them without difficulty if necessary. The faces of *barrier curbs* are usually vertical. Specific designs for the different types of curbs are presented in Chapter 16. It should be noted, however, that because of glare, curbed islands may be difficult to see at night, which makes it necessary that intersections with curbed islands have fixed-source lighting. Curbed islands are used mainly in urban highways, where approach speed is not excessively high and pedestrian volume is relatively high.

Traffic Islands Formed by Pavement Markings. This type of island is sometimes referred to as a *flushed island* because it is flush with the pavement, as shown in Figure 7.11b. Flushed islands are formed by pavement markings that delineate the area of the island. Markers include paint, thermoplastic striping, and raised retroreflective markers. Flushed islands are preferred over curbed islands at intersections where approach speeds are relatively high, pedestrian traffic is low, and signals or sign mountings are not located on the island.

Islands Formed by Pavement Edges. These islands are usually unpaved and are used mainly at rural intersections where there is space for large intersection curves.

Functions of Traffic Islands

Traffic islands can also be classified into three categories based on their functions: channelized, divisional, and refuge. *Channelized islands* are used mainly to control and direct traffic. *Divisional islands* are used mainly to divide opposing or same-directional traffic streams. *Refuge islands* are used primarily to provide refuge for pedestrians. In most cases, however, traffic islands perform two or more of these functions rather than a single function, although each island may have a primary function.

Channelized Islands. The objective of channelized islands is to eliminate confusion to motorists at intersections with different traffic movements by guiding them into the correct lane for their intended movement. This is achieved by converting excess space at the intersection into islands in a manner that leaves very little to the discretion of the motorist. A channelized island may take one of many shapes, depending on its specific

(a) Curbed island at an intersection

(b) Island formed by pavement markings (flushed island)

Figure 7.11 Examples of Traffic Islands

SOURCE: Photographs by Lewis Woodson, Virginia Transportation Research Council, Charlottesville, Va. Used with permission.

purpose. For example, a triangularly shaped channelized island is often used to separate right-turning traffic from through traffic (see Figure 7.12a), whereas a curved central island is frequently used to guide turning vehicles (see Figure 7.12b). In any case, the outlines of a channelized island should be nearly parallel to the lines of traffic it is channeling. Where the island is used to separate turning traffic from through traffic, the radii of the curved sections must be equal to or greater than the minimum radius required for the expected turning speed.

The number of islands used for channelization at an intersection should be kept to the practical minimum, since the presence of several islands may cause confusion to the motorist. For example, the use of a set of islands to delineate several one-way lanes may cause unfamiliar drivers to enter the intersection in the wrong lane.

Divisional Islands.   These are frequently used at intersections of undivided highways to alert drivers that they are approaching an intersection and to control traffic at the intersection. They can also be used effectively to control left turns at skewed intersections. Examples of divisional islands are shown in Figure 7.13. When it is necessary to widen a road at an intersection so that a divisional island can be included, every effort should be made to ensure that the path a driver is expected to take is made quite clear. The alignment

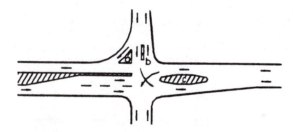

(a) Triangle-shaped, channelized island on north approach of the intersection

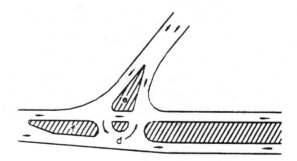

(b) Curve central island

Figure 7.12 Examples of Channelized Islands

SOURCE: *A Policy of Geometric Design of Highways and Streets,* American Association of State Highway and Transportation Officials, Washington, D.C., 2001. Used with permission.

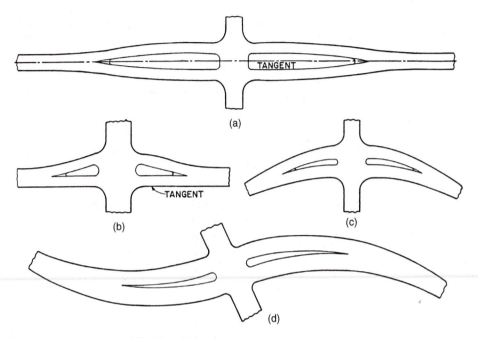

Figure 7.13 Examples of Divisional Islands

SOURCE: *A Policy of Geometric Design of Highways and Streets,* American Association of State Highway and Transportation Officials, Washington, D.C., 2001. Used with permission.

should also be designed so that the driver can traverse the intersection easily without any excessive steering.

It is sometimes necessary to use reverse curves (two simple curves with opposite curvatures, forming a compound curve) when divisional islands are introduced, particularly when the location is at a tangent. At locations where speeds tend to be high, particularly in rural areas, it is recommended that the reversal in curvature be no greater than 1 degree. Sharper curves can be used when speeds are relatively low, but a maximum of 2 degrees is recommended.

Refuge Islands. Refuge islands, sometimes referred to as *pedestrian islands,* are used mainly at urban intersections to serve as refuge areas for wheelchairs and pedestrians crossing wide intersections. They may also be used for loading and unloading transit passengers. Figure 7.14 shows examples of islands that provide refuge as well as function as channelized islands.

Minimum Sizes of Islands

It is essential that islands be large enough to command the necessary attention by drivers. In order to achieve this, AASHTO recommends that curbed islands have a minimum area of approximately 50 sq ft for urban intersections and 75 sq ft for rural intersections,

(a)

(b)

Figure 7.14 Examples of Refuge Islands at Wide Intersections

SOURCE: Photographs by Lewis Woodson, Virginia Transportation Research Council, Charlottesville, Va. Used with permission.

although 100 sq ft is preferable for both. The minimum side lengths recommended are 12 ft (but preferably 15 ft) for triangular islands after the rounding of corners, 20 to 25 ft for elongated or divisional islands, and 100 ft (but preferably several hundred feet) for curbed divisional islands that are located at isolated intersections on high-speed highways. It is not advisable to introduce curbed divisional islands at isolated intersections on high-speed roads, since this may create a hazardous situation unless the island is made visible enough to attract the attention of the driver.

Islands having side lengths near the minimum are considered to be small islands, whereas those with side lengths of 100 ft or greater are considered to be large. Those with side lengths less than those for large islands but greater than the minimum are considered to be intermediate islands.

In general, the width of elongated islands should not be less than 4 ft, although this dimension can be reduced to an absolute minimum of 2 ft in special cases when space is limited. In cases where signs are located on the island, the width of the sign must be considered in selecting the width of the island to ensure that the sign does not extend beyond the limits of the island.

Location and Treatment of Approach Ends of Curbed Islands

The location of a curbed island at an intersection is dictated by the edge of the through traffic lanes and the turning roadways. Figures 7.15 and 7.16 show the locations of curbed islands at intersections without and with shoulders, respectively. Figures 7.15 and 7.16, respectively, illustrate the condition where the curbed island edge is located on one approach by providing an offset to the through traffic lane, and where it is located outside a shoulder that is carried through the intersection. The offset from the through traffic lane should be 2 to 3 ft, depending on factors such as the type of edge treatment, island contrast, length of taper or auxiliary pavement preceding the curbed island, and traffic speed. This offset is required for the sides of curbed islands adjacent to the through traffic lanes for both barrier and mountable curbs, and for the sides of barrier curbs adjacent to turning roadways. However, it is not necessary to offset the side of a mountable curb adjacent to a turning roadway except as needed to provide additional protection for the island. In cases where there are no shoulders (Figure 7.15), an offset of 2 to 3 ft should be maintained when there are no curbs on the approach pavement. When there is a mountable curb on the approach pavement, a similar curb can be used on the curbed island at the edge of the through lane. However, this requires that the length of the curbed island be adequate to obtain a gradual taper from the nose offset. AASHTO recommends that the offset of the approach nose of a curbed island from the travel lane should normally be about 2 ft greater than that of the side of the island from the travel lane. However, for median curbed islands an offset of at least 2 ft but preferably 4 ft of the approach nose from the normal median edge of pavement is recommended. It should be emphasized that in order to prevent the perception of lateral constraint by drivers, the required offset from the edge of through pavement lanes should always be used when barrier curbed islands are introduced. When uncurbed large or intermediate islands are used, the required offsets may be eliminated, although it is still preferable to provide the offsets.

At intersections with approach shoulders but without deceleration or turn lanes, the offset of curbed islands from the through travel lane should be equal to the width of

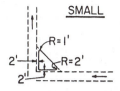

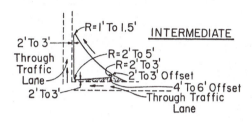

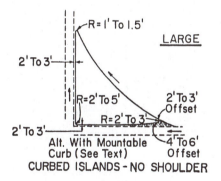

Figure 7.15 Layouts of Curbed Islands Without Shoulders

SOURCE: *A Policy of Geometric Design of Highways and Streets,* American Association of State Highway and Transportation Officials, Washington, D.C., 2001. Used with permission.

the shoulder. When a deceleration lane precedes the curbed island, or when a gradually widened auxiliary pavement exists and speeds are within the intermediate-to-high range, it is desirable to increase the offset of the nose by an additional 2 to 4 ft.

The end treatments for curbed islands are also shown in Figures 7.15 and 7.16. These figures show that the approach noses are rounded using curves of radii 2 to 3 ft, while the merging ends are rounded with curves of radii 1 to 1.5 ft; the other corners are rounded with curves of radii of 2 to 5 ft.

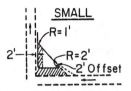

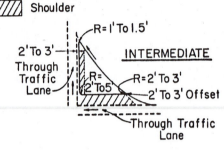

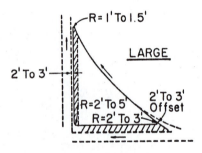

CURBED ISLANDS WITH SHOULDERS

Figure 7.16 Layout of Curbed Islands with Shoulders

SOURCE: *A Policy of Geometric Design of Highways and Streets,* American Association of State Highway and Transportation Officials, Washington, D.C., 2001. Used with permission.

Minimum Pavement Widths of Turning Roadways at At-Grade Intersections

In cases where vehicle speeds are expected to be greater than 15 mi/h, such as at channelized intersections and where ramps intersect with local roads, it is necessary to increase the pavement widths of the turning roadways. Three classifications of pavement widths are used:

- Case I: one-lane, one-way operation with no provision for passing a stalled vehicle
- Case II: one-lane, one-way operation with provision for passing a stalled vehicle
- Case III: two-lane operation, either one-way or two-way

Case I is used mainly at relatively short connecting roads with moderate turning volumes. Case II, which provides for the passing of a stalled vehicle, is commonly used at locations where ramps intersect with local roads, and at channelized intersections. Case III is used at one-way, high-volume locations that require two-lanes, or at two-way locations. The pavement width for each case depends on the radius of the turning roadway and the characteristics of the design vehicle. Figure 7.17 shows the basis for deriving the appropriate pavement width for any design vehicle. The pavement width depends on the widths of the front and rear overhangs of the design vehicle, the total clearance per vehicle, an extra width allowance due to difficulty of driving on curves, and the track width of the vehicle as it moves around the curve. Values for the front overhang F_A for different vehicle types can be obtained directly from Figure 7.18b or computed from the equation given in that figure. The width of the rear overhang F_B is usually taken as 0.5 ft for passenger cars, since the width of the body of a typical passenger car is 1 ft greater than the out-to-out width of the rear wheels. For truck vehicles the rear overhang is 0 ft, since the width of truck bodies is usually the same as the out-to-out width of the rear wheels.

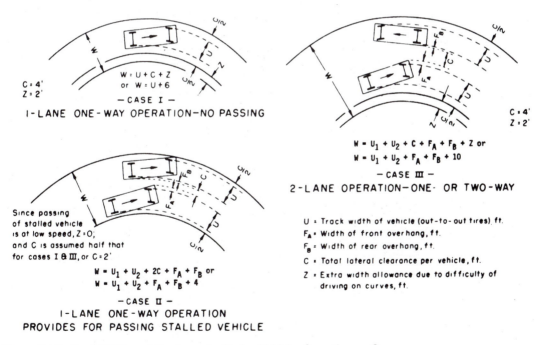

Figure 7.17 Track Width and Overhang for Design Vehicles Operating on Curves

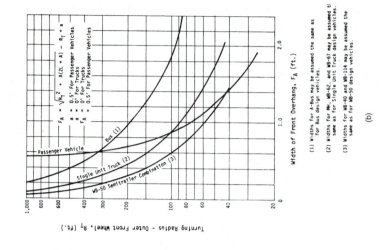

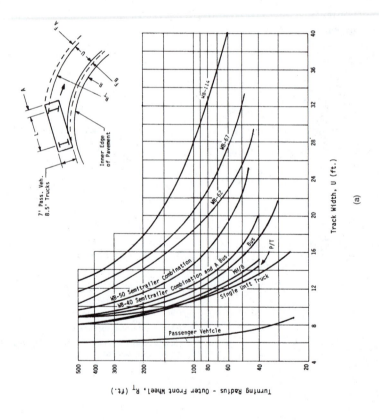

Figure 7.18 Pavement Widths of Curves at Intersections—Basis of Derivation

SOURCE: *A Policy of Geometric Design of Highways and Streets*, American Association of State Highway and Transportation Officials, Washington, D.C., 2001. Used with permission.

The pavement width provides for some clearance both between the edge of the pavement and the nearest wheel path and between the sides of the vehicles passing or meeting. The clearance between the edge of the pavement and the nearest wheel path is usually taken as 2 to 3 ft, and the clearance between the sides of vehicles is taken as 4 ft. The total clearance C per lane of traffic is therefore 4 ft, or 2 ft on either side of the vehicles.

An extra width allowance Z is provided to compensate for the difficulty of maneuvering on a curve and for variation in driver operation. It is obtained from the empirical expression

$$Z = \frac{v}{\sqrt{R}} \tag{7.1}$$

where

Z = extra width allowance to compensate for the difficulty in maneuvering (ft)
v = design speed (mi/h)
R = radius of curve (ft)

Because of the relationship between speed and velocity at intersections, it has been found that Z is usually about 2 ft for radii between 50 and 500 ft.

The track width U for passenger cars and single-unit trucks is given as

$$U = u + R_T - \sqrt{(R_T^{\,2} - L^2)} \tag{7.2}$$

where

u = normal track width, which is 6 ft for passenger cars and 8.5 ft for single-unit trucks

R_T = turning radius of the outer front wheel (ft)
L = wheelbase, which is 11 ft for passenger cars and 20 ft for single-unit trucks

Equation 7.2 cannot be used to directly compute the offtracking for semitrailer combinations, since these vehicles usually have two or more wheelbase lengths, with each involved in some offtracking on curves. Therefore, scale models were used to obtain data for semitrailer combinations. Figure 7.18a shows values for the track width U for different design vehicles derived from these data.

These different width elements are summed to determine the total pavement width W for the roadway at the intersection. Table 7.4 gives derived pavement widths for different design vehicles and pavement edge radii. In practice, however, the pavement width is seldom designed for a single vehicle type, since the highway should be capable of accommodating different types of vehicles, especially when provision is made for passing. The design of the pavement width is therefore based on three types of traffic conditions, listed below, each concerning a specific mix of vehicle types:

- *Traffic Condition A:* Passenger vehicles are predominant, but this traffic condition also provides for the operation of an occasional large truck.
- *Traffic Condition B:* Proportion of SU vehicles warrants this vehicle type to be the design vehicle, but it allows for about 5–10 percent of the total traffic to be semitrailer vehicles.

Table 7.4   Derived Pavement Widths for Turning Roadways for Different Design Vehicles

Radius on Inner Edge of Pavement R (ft)	Case I, One Lane, One-Way Operation, No Provision for Passing a Stalled Vehicle										
	P	PT	MH/B	SU	WB-40	WB-50	BUS	A-BUS	WB-62	WB-67	WB-114
50	13	18	19	18	23	26	20	23	26	29	37
75	13	17	17	17	19	22	18	19	25	26	33
100	13	16	16	16	18	21	18	18	23	26	31
150	12	16	16	16	17	19	16	17	21	24	25
200	12	15	15	16	16	17	16	16	19	21	23
300	12	14	14	15	16	17	16	16	18	19	20
400	12	14	14	15	16	16	15	16	17	18	19
500	12	14	14	15	15	16	15	15	17	18	19
Tangent	12	14	14	15	15	15	15	15	15	15	15

	Case II, One-Lane, One-Way Operation with Provision for Passing a Stalled Vehicle by Another of the Same Type										
50	20	29	30	29	36	44	31	37	45	49	61
75	19	27	28	27	31	36	29	32	42	45	56
100	19	25	25	25	29	34	28	30	38	42	50
150	18	24	24	24	27	29	25	28	34	37	41
200	18	22	22	23	25	27	25	26	31	35	37
300	18	20	20	22	24	25	24	25	28	31	32
400	17	20	20	22	23	24	22	23	22	29	30
500	17	20	20	22	23	24	22	23	25	27	30
Tangent	17	20	20	21	21	21	21	21	21	21	21

	Case III, Two-Lane Operation, Either One- or Two-Way (Same Type Vehicle in Both Lanes)										
50	26	35	36	35	42	50	37	43	51	55	67
75	25	33	34	33	37	42	35	38	48	51	62
100	25	31	31	31	35	40	34	36	44	48	56
150	24	30	30	30	33	35	31	34	40	43	47
200	24	28	28	29	31	33	31	32	37	41	43
300	24	26	26	28	30	31	30	31	34	37	38
400	23	26	26	28	29	30	28	29	33	35	36
500	23	26	26	28	29	30	28	29	31	33	36
Tangent	23	26	26	27	27	27	27	27	27	27	27

SOURCE: *A Policy of Geometric Design of Highways and Streets,* The American Association of State Highway and Transportation Officials, Washington, D.C., 2001. Used by permission.

- *Traffic Condition C:* Proportion of semitrailer (WB-40 or WB-50) vehicles in the traffic stream warrants one of these vehicle types to be the design vehicle.

Table 7.5 gives pavement design widths for the different operational cases and traffic conditions. The pavement widths given in Table 7.5 are further modified with respect to the treatment at the edge of the pavement. For example, when the pavement is designed for small trucks but there are space and stability outside the pavement and no barrier preventing its use, a large truck can occasionally pass another. The pavement width can therefore be a little narrower than the tabulated value. On the other hand, the existence of a barrier curb along the edge of the pavement creates a sense of restriction to the driver, and so the

Table 7.5 Design Widths of Pavements for Turning Roadways

	Pavement Width (ft)								
	Case I One-Lane, One-Way Operation—No Provision for Passing a Stalled Vehicle			Case II One-Lane, One-Way Operation—With Provision for Passing a Stalled Vehicle			Case III Two-Lane Operation—Either One-Way or Two-Way		
Radius on Inner Edge of Pavement	Design Traffic and Condition								
R (ft)	A	B	C	A	B	C	A	B	C
50	18	18	23	23	25	29	31	36	42
75	16	17	19	21	23	27	29	33	37
100	15	16	18	20	22	25	28	31	35
150	14	16	17	19	21	24	27	30	33
200	13	16	16	19	21	23	27	29	31
300	13	15	16	18	20	22	26	28	30
400	13	15	16	18	20	22	26	28	29
500	12	15	15	18	20	22	26	28	29
Tangent	12	15	15	17	19	21	25	27	27
	Width Modification Regarding Edge of Pavement Treatment								
No stabilized shoulder	None			None			None		
Mountable curb	None			None			None		
Barrier curb: one side two side	Add 1 ft Add 2 ft			None Add 1 ft			Add 1 ft Add 2 ft		
Stabilized shoulder, one or both sides	None			Deduct shoulder width; minimum pavement width as under case			Deduct 2 ft where shoulder is 4 ft or wider		

Note: Traffic Condition A = predominately P vehicles, but some consideration for SU trucks.
    Traffic Condition B = sufficient SU vehicles to govern design, but some consideration for semitrailer vehicles.
    Traffic Condition C = sufficient bus and combination-types of vehicles to govern design.
SOURCE: *A Policy of Geometric Design of Highways and Streets,* The American Association of State Highway and Transportation Officials, Washington, D.C., 2001. Used by permission.

occasional truck has no additional space to maneuver. This requires additional space than the widths shown in the table. Consideration of these factors results in the modifications given at the bottom of Table 7.5.

The values given in Table 7.5 are based on the assumption that passing is rather infrequent in Case II, and full offtracking does not necessarily occur for both the stalled and passing vehicles, since the stalled vehicle can be placed closer to the inner edge of the pavement and thus provide additional clearance for the passing vehicle. Therefore, smaller vehicles in combination are used in deriving the values for Case II rather than for Case III. The design vehicles in combination that will provide the full clear width C are given in Table 7.6a, where the vehicle designations indicate the types of vehicles that can pass each other. For example, the designation "P-SU" indicates that the design width will allow a passenger car to pass a stalled single-unit truck, or vice versa. Although it is feasible for larger vehicles to pass each other on turning roadways with the widths shown in the table for Cases II and III, this will occur at lower speeds and requires more caution and skill by drivers. The larger vehicles that can be operated under these conditions are shown in Table 7.6b. This will result in a reduced clearance between vehicles, that will vary from about one-half the value of C for sharper curves to nearly the full value of C for flatter curves.

When larger design vehicles such as WB-62 or WB-114 are expected to be using the turning roadway, the width of the pavement should not be less than the minimum width required for these vehicles for the Case 1 classification in Table 7.4. In such cases, values obtained from Table 7.5 must be compared with those given in Table 7.4 for Case I, and the higher values should be used.

Table 7.6 Design Vehicles in Combination

(a) Design Vehicles in Combination That Provide the Full Clear Width (C)

| Case | *Design Traffic Condition* | | |
	A	*B*	*C*
I	P	SU	WB-40
II	P-P	P-SU	SU-SU
III	P-SU	SU-SU	WB-40-WB-40

(b) Larger Vehicles That Can Be Operated

| Case | *Design Traffic Condition* | | |
	A	*B*	*C*
I	WB-40	WB-40	WB-50
II	P-SU	P-WB-40	SU-WB-40
III	SU-WB-40	WB-40-WB-40	WB-50-WB-50

SOURCE: *A Policy of Geometric Design of Highways and Streets,* The American Association of State Highway and Transportation Officials, Washington, D.C., 2001. Used by permission.

Example 7.1   Determining the Width of a Turning Roadway at an Intersection

A ramp from an urban expressway with a design speed of 30 mi/h connects with a local road, forming a T intersection. An additional lane is provided on the local road to allow vehicles on the ramp to turn right onto the local road without stopping. The turning roadway has a mountable curb and will provide for a one-lane, one-way operation with provision for passing a stalled vehicle. Determine the width of the turning roadway if the predominant vehicles on the ramp are single-unit trucks, but give some consideration to semitrailer vehicles. Use 0.08 for the superelevation.

Solution:

- Determine the minimum radius of the curve.
  Because the speed is greater than 15 mi/h, use Eq. 3.31 of Chapter 3 and a value of 0.16 for f_s (from Table 3.3):

$$R = \frac{u^2}{15(e + f_s)}$$

$$= \frac{30^2}{15(0.08 + 0.16)}$$

$$= 250 \text{ ft}$$

- Determine the type of operation and traffic condition.
  The operational requirements, one-lane, one-way with provision for passing a stalled vehicle, require Case II operation.
  The traffic condition, single-unit trucks but with some consideration given to semitrailer vehicles, requires Traffic Condition B.
- Determine the turning roadway width.
  Use Table 7.5, with $R = 250$ ft, Traffic Condition B, and Case II.

$$\text{pavement width} = 20.5 \text{ ft (interpolating)}$$

Since the turning roadway has a mountable curb, no modification to the width is required.

Sight Distance at Intersections

The high crash potential at an intersection can be reduced by providing sight distances that allow drivers to have an unobstructed view of the entire intersection at a distance great enough to permit control of the vehicle. At signalized intersections, the unobstructed view may be limited to the area where the signals are located, but for unsignalized intersections, it is necessary to provide an adequate view of the crossroads or intersecting highways to reduce the potential of collision with crossing vehicles. The sight distance required depends on the type of control at the intersection (see Chapter 8).

At-grade intersections either have no control or are controlled by one of the following methods: yield control, stop control, or signal control.

Sight Distance Requirements for No-Control Intersections but Allowing Vehicles to Adjust Speed—Case I

In this situation the intersection is not controlled by a yield sign, a stop sign, or a traffic signal, but sufficient sight distance is provided for the operator of a vehicle approaching the intersection to see a crossing vehicle and if necessary to adjust the vehicle's speed so as to avoid a collision. This distance must include the distance traveled by the vehicle both during the driver's perception reaction time and during brake actuation or the acceleration to regulate speed. At intersections, 2.0 sec is usually acceptable for perception reaction time, and an additional 1.0 sec is added for the driver to actuate braking or to accelerate to regulate speed. The distance traveled during this period of 3.0 sec is the limiting distance from which a driver approaching the intersection should first observe an approaching vehicle on the crossroad. Table 7.7 gives average distances for different approach speeds, as suggested by AASHTO. Figure 7.19 shows a schematic of the sight triangle required for the location of an obstruction that will allow for the provision of the minimum distances d_a and d_b. These minimum distances depend on the approaching speed as shown in Table 7.7. For example, if a road with a speed limit of 40 mi/h intersects with a road with a speed limit of 25 mi/h, the distances d_a and d_b are 180 ft and 110 ft, respectively. It should be emphasized that the distances shown in Table 7.7 only allow time for drivers to adjust the vehicle's speed but not to stop. It is preferable to design uncontrolled intersections such that the driver of each vehicle sees the intersection and the traffic on the crossroad in sufficient time for stopping the vehicle before reaching the intersection. When this is done, the safe stopping distances presented in Chapter 3 for the given design speed should be used for d_a and d_b. It can be seen from Figure 7.19 that triangles ABC and ADE are similar, which gives:

Table 7.7 Minimum Distances for Sight Triangle: No Control by Allowing Vehicles to Adjust Speed

Speed (mi/h)	Distance (ft)
10	45
15	70
20	90
25	110
30	130
35	155
40	180
50	220
60	260
70	310

SOURCE: *A Policy of Geometric Design of Highways and Streets,* The American Association of State Highway and Transportation Officials, Washington, D.C., 2001. Used by permission.

$$\frac{CB}{AB} = \frac{ED}{AD} \tag{7.3}$$

$$\frac{d_b}{d_a} = \frac{a}{d_a - b} \tag{7.4}$$

If any three of the variables d_a, d_b, a, and b are known, the fourth can be determined using Eq. 7.4.

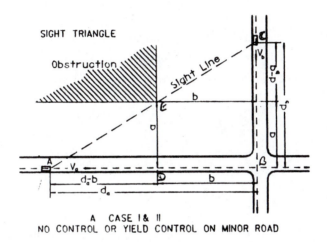

A CASE I & II
NO CONTROL OR YIELD CONTROL ON MINOR ROAD

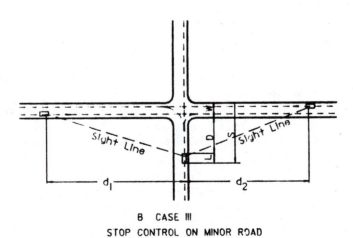

B CASE III
STOP CONTROL ON MINOR ROAD

Figure 7.19 Minimum Sight Triangle at an Intersection

SOURCE: *A Policy of Geometric Design of Highways and Streets,* American Association of State Highway and Transportation Officials, Washington, D.C., 2001. Used with permission.

Example 7.2   Computing Speed Limit on a Local Road

A tall building is located 45 ft from the center line of the right lane of a local road (*b* in Figure 7.19) and 65 ft from the center line of the right lane of an intersecting road (*a* in Figure 7.19). If the maximum speed limit on the intersecting road is 35 mi/h, what should the speed limit on the local road be such that the minimum sight distance is provided to allow the drivers of approaching vehicles to avoid imminent collision by adjusting their speeds?

Solution:

- Determine the distance on the local road at which the driver first sees traffic on the intersecting road.

  Speed limit on intersecting road = 35 mi/h

  Distance required on intersecting road (d_a) = 155 ft (from Table 7.7)

- Calculate the distance available on local road by using Eq. 7.4.

$$d_b = a \frac{d_a}{d_a - b}$$
$$= 65 \frac{155}{155 - 45}$$
$$= 91.6 \text{ ft}$$

- Determine the maximum speed allowable on the local road.

  The maximum speed allowable on local road is 20 mi/h (from Table 7.7).

Sight Distance Requirement for Yield-Control Intersections on Minor Roads—Case II

In this situation the minor road is controlled by a yield sign. Vehicles on the minor road are therefore required to yield to vehicles on the major road, which often requires the vehicle on the minor road to slow down or to stop prior to reaching the intersecting roadway. The sight distance provided on the minor road must therefore be sufficient for the driver to see a vehicle approaching from either the left or right of the major road, and to be able to stop the vehicle before reaching the intersecting roadway, as shown in Figure 7.19. The solution is similar to that for the no-control condition except that in this case the minimum stopping sight distances given for the appropriate speeds in Chapter 3 are always used for d_a and d_b in Eq. 7.4. It should be noted that the grades of the approaches should be taken into consideration when determining the minimum stopping sight distances, as discussed in Chapter 3.

Sight Distance Requirement for Stop-Control Intersections on Minor Roads—Case III

When vehicles are required to stop at an intersection, the drivers of such vehicles should be provided sufficient sight distance to allow for a safe departure from the stopped position for the three basic maneuvers that occur at an average intersection. These maneuvers are:

1. Crossing the intersection, thereby clearing traffic approaching from both sides of the intersection (see Figure 7.20a)
2. Turning left onto the crossroad, which requires clearing the traffic approaching from the left and then joining the traffic stream on the crossroad with vehicles approaching from the right (see Figure 7.20b)
3. Turning right onto the crossroad by joining the traffic on the crossroad with vehicles approaching from the left (see Figure 7.20c)

Crossing Maneuver—Case IIIA.   The sight distance required for this maneuver depends on the time the vehicle will take to cross the intersection and on the distance an approaching vehicle on the crossroad (usually the major highway) will travel during that time. Equation 7.5 can be used to determine this distance.

$$d = 1.47v(J+t_a) \tag{7.5}$$

where

d = sight distance along the major highway from the intersection (ft)
v = design speed on the major highway (mi/h) (note that 1.47 converts mi/h to ft/sec)
J = perception-reaction time plus the time required by the driver to actuate the clutch or an automatic shift (2.0 sec is assumed as recommended by AASHTO)
t_a = time taken to accelerate and clear the intersection

The time t_a depends on the type of vehicle and a distance S, which is given as

$$S = D + W + L \tag{7.6}$$

where

D = distance of the front of the stopped vehicle to the near pavement edge of the crossroad (ft), usually assumed to be 10 ft
W = width of the intersection pavement (ft)
L = overall length of the stopped vehicle (ft)

With the value of S known, the value of t_a is obtained from Figure 7.21.

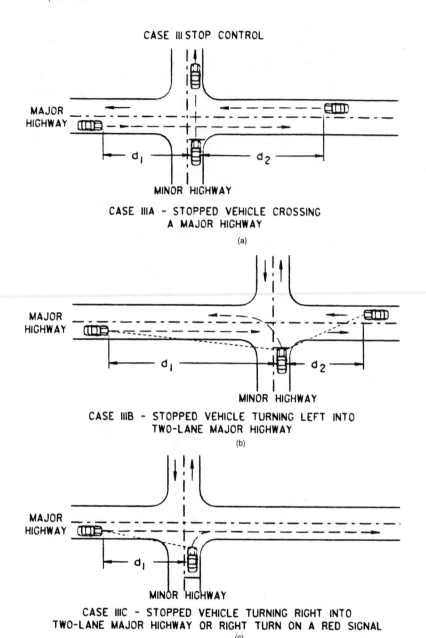

Figure 7.20 Maneuvers at an Unsignalized Intersection—Sight Distance Requirements

SOURCE: *A Policy of Geometric Design of Highways and Streets,* American Association of State Highway and Transportation Officials, Washington, D.C., 2001. Used with permission.

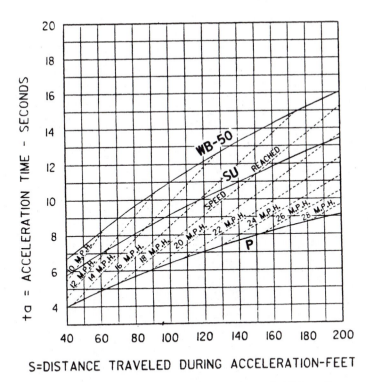

S=DISTANCE TRAVELED DURING ACCELERATION-FEET

Figure 7.21 Acceleration Times (t$_a$) Required to Travel a Distance (S) from Stop

SOURCE: *A Policy of Geometric Design of Highways and Streets,* American Association of State Highway and Transportation Officials, Washington, D.C., 2001. Used with permission.

Figure 7.22 shows the general layout of an intersection of a minor road and a major road with a stop control for the minor road. The figure also indicates the layout of the variables D, W, and L.

When the median of a divided highway is equal to or wider than the length of the design crossing vehicle, the crossing maneuver can be accomplished in two steps. First, the vehicle crosses the first set of lanes and stops in the protected area of the median. Second, the vehicle crosses the remainder of the road when a suitable gap arrives. In cases where the median width is less than the length of the design turning vehicle, the vehicle crosses in one step, thus requiring the inclusion of the median width as part of W.

When it is necessary to determine the available sight distance at an intersection, the measurement should be taken from an eye height of 3.5 ft (for passenger cars) to the top of an object 4.25 ft high from the pavement. When the value computed for the sight distance d along the major highway is less than the stopping sight distance given in Chapter 3 for design speed, the stopping sight distance should be used as the required sight distance.

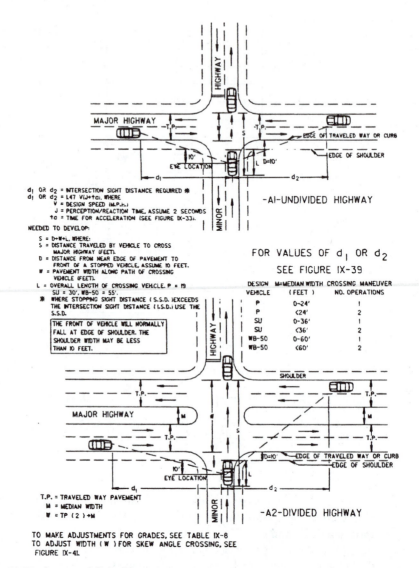

Figure 7.22 relevant text:

MAJOR HIGHWAY

HIGHWAY

EDGE OF TRAVELED WAY OR CURB

EDGE OF SHOULDER

10'
EYE LOCATION

L D=10'

-A1-UNDIVIDED HIGHWAY

MINOR

d_1 OR d_2 = INTERSECTION SIGHT DISTANCE REQUIRED ✱
d_1 OR d_2 = 1.47 V(J+t$_a$), WHERE
  V = DESIGN SPEED (M.P.H.)
  J = PERCEPTION/REACTION TIME, ASSUME 2 SECONDS
  t$_a$ = TIME FOR ACCELERATION (SEE FIGURE IX-33).

NEEDED TO DEVELOP:

  S = D+W+L, WHERE:
  S = DISTANCE TRAVELED BY VEHICLE TO CROSS
    MAJOR HIGHWAY (FEET).
  D = DISTANCE FROM NEAR EDGE OF PAVEMENT TO
    FRONT OF A STOPPED VEHICLE, ASSUME 10 FEET.
  W = PAVEMENT WIDTH ALONG PATH OF CROSSING
    VEHICLE (FEET).
  L = OVERALL LENGTH OF CROSSING VEHICLE. P = 19
    SU = 30', WB-50 = 55'.
  ✱ WHERE STOPPING SIGHT DISTANCE (S.S.D.)EXCEEDS
    THE INTERSECTION SIGHT DISTANCE (I.S.D.) USE THE
    S.S.D.

THE FRONT OF VEHICLE WILL NORMALLY
FALL AT EDGE OF SHOULDER. THE
SHOULDER WIDTH MAY BE LESS
THAN 10 FEET.

FOR VALUES OF d_1 OR d_2
SEE FIGURE IX-39

DESIGN VEHICLE	M=MEDIAN WIDTH (FEET)	CROSSING MANEUVER NO. OPERATIONS
P	0-24'	1
P	<24'	2
SU	0-36'	1
SU	<36'	2
WB-50	0-60'	1
WB-50	<60'	2

SHOULDER

MAJOR HIGHWAY

M

HIGHWAY

S

M

T.P.

T.P.

T.P.

D=10' EDGE OF TRAVELED WAY OR CURB
EDGE OF SHOULDER

10'
EYE LOCATION

L

MINOR

-A2-DIVIDED HIGHWAY

T.P. = TRAVELED WAY PAVEMENT
M = MEDIAN WIDTH
W = TP (2) +M

TO MAKE ADJUSTMENTS FOR GRADES, SEE TABLE IX-8
TO ADJUST WIDTH (W) FOR SKEW ANGLE CROSSING, SEE
FIGURE IX-41.

Figure 7.22 Layout of Sight Distance Requirements for Crossing at an Intersection—Case IIIA

SOURCE: *A Policy of Geometric Design of Highways and Streets,* American Association of State Highway and Transportation Officials, Washington, D.C., 2001. Used with permission.

Example 7.3 Determining the Minimum Sight Distance for Crossing at an Intersection with Stop Control

An urban two-lane minor road crosses a four-lane divided highway with a speed limit of 45 mi/h. If the minor road is controlled by a stop sign, determine the sight distance from the intersection that is required along the major road such that the driver of a stopped vehicle on the minor road can safely cross the intersection. The following conditions exist at the intersection.

Major road lane width = 11 ft
Median width = 11 ft
Design vehicle on minor road is single-unit truck, length = 30 ft

Solution:

- Use Eq. 7.6 to determine S_1, the distance that the crossing vehicle must travel to clear the first part of the major road.

$$S_1 = D + W + L$$
$$= 10 + (2 \times 11) + 30$$
$$= 62 \text{ ft}$$

- Determine time (t_a) required to accelerate and traverse the distance S_1 by a single-unit truck.

From Figure 7.21, for a single-unit truck $S = 62$ ft

$$t_a = 7.1 \text{ sec}$$

- Determine d_1 using Eq. 7.5.

$$d_1 = 1.47v(J + t_a)$$
$$= 1.47(45)(2 + 7.1)$$
$$= 602 \text{ ft}$$

Similarly, the sight distance d_2 on the right of the minor-road driver is

$$d_2 = 1.47v(J + t_a)$$

However, since the median width is only 11 ft, which is too short for the design vehicle, a vehicle must cross in a single step.

- Calculate S_2.

In this case, S_2 is the total of (i) the distance D, which is 10 ft; (ii) the total width of all lanes, which in this case is 4×11 ft; (iii) the median width, which in this case is 11 ft; and (iv) the length of the design crossing vehicle, which in this case is 30 ft.

$$S_2 = D + W + L$$
$$= 10 + [(4 \times 11) + 11] + 30$$
$$= 95 \text{ ft}$$

- From Figure 7.21, determine t_a for the distance S_2 as 9.0.
- Calculate the sight distance d_2 on the right of the minor-road driver.

$$d_2 = 1.47v (J + t_a)$$
$$= 1.47(45)(2 + 9.0)$$
$$= 727.7 \text{ ft}$$

Turning Left into Major Highway—Case IIIB.   In this case, the vehicle on the minor road enters the cross road from a stopped position, first by crossing the near lane or lanes of the major road without interfering with vehicles from the left, and then by turning left onto the major road without significantly interfering with the traffic coming from the right. This requires adequate sight distances from the left and right (d_1 and d_2, respectively), as shown in Figure 7.20b.

Equation 7.5 may be used to compute the sight distance d_1 from the left, with t_a computed for the distance S as determined from Eq. 7.7.

$$S = D + L + W_1 \tag{7.7}$$

where W_1 is the distance traveled to clear traffic coming from the left approach. (For a two-lane major road, W_1 is equal to 1.5 times the lane width.) In computing the sight distance d_2 from the right, the following two assumptions are made: (1) The speed of the vehicle from the right on the major road will be reduced to 85 percent of the design speed of the major road, and (2) a minimum time gap of 2 sec exists between the turning vehicle and the vehicle coming from the right when both vehicles are traveling at 85 percent of the design speed of the major road.

In the case of a two-lane minor road crossing a two-lane major road, Figure 7.23 is used to obtain the sight distance from the right—designated as S_d in the figure—by going through the following steps:

1. From Figure 7.24, determine the distance P traversed by the turning vehicle to attain a speed of 85 percent of the major-road design speed.
2. Determine the time t_a required to travel the distance obtained in step 1. This is done by using either Figure 7.21 or Table 7.8, or by making appropriate assumptions for vehicle acceleration rates and turning paths.
3. Determine the sum of t_a and J, that is, $(t_a + 2)$ sec.
4. Determine the distance Q in ft traveled by the vehicle on the major road coming from the right, by assuming it is traveling at an average speed of 85 percent of the design speed of the major road. That is,

$$Q = 1.47(t_a + 2)(0.85 \times \text{design speed of major road}) \tag{7.8}$$

5. Compute the distance z traveled by turning vehicle from minor road in making the turn. In other words, z is the length of arc traveled by the turning vehicle. For two-lane highways (see Figure 7.23)

$$z_{\text{two-lane}} = \pi \left(\frac{1}{4}\right)(2R) = \pi \frac{R}{2} \tag{7.9}$$

6. Compute the distance h shown in Figure 7.23, that is

$$h = P - z + (\text{radius of turning arc}) - L - VG \tag{7.10}$$

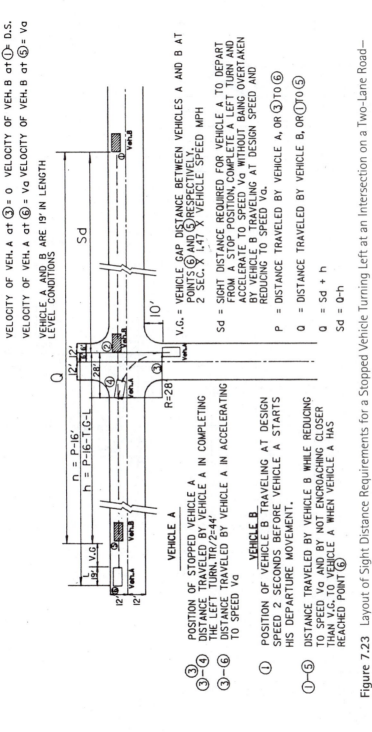

Figure 7.23 Layout of Sight Distance Requirements for a Stopped Vehicle Turning Left at an Intersection on a Two-Lane Road—Case IIIB

SOURCE: *A Policy of Geometric Design of Highways and Streets*, American Association of State Highway and Transportation Officials, Washington, D.C., 2001. Used with permission.

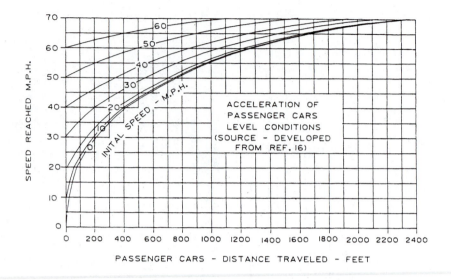

PASSENGER CARS - DISTANCE TRAVELED - FEET

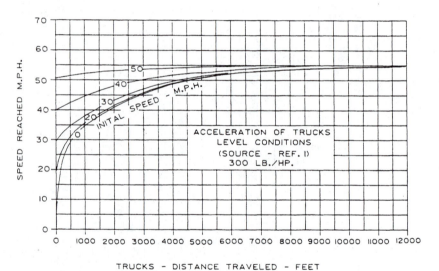

TRUCKS - DISTANCE TRAVELED - FEET

Figure 7.24 Speed-Distance Relationships for Passenger Cars and Trucks

SOURCE: *A Policy of Geometric Design of Highways and Streets,* American Association of State Highway and Transportation Officials, Washington, D.C., 2001. Used with permission.

where

P = distance traveled by turning vehicle to obtain a speed of 85 percent of design speed of the major road (obtained in step 1)

z = length of arc traveled by vehicle from minor road while making the turn

R = radius of turning arc

L = length of vehicle from minor road

Table 7.8 Acceleration Rates from Passenger Cars

Speed (mi/h)	Distance (ft)	T_a (sec)
15	50	4.5
20	90	6.1
25	140	7.3
30	215	9.4
35	305	11.3
40	420	13.5
45	570	15.9
50	760	18.6
55	1000	21.7
60	1315	25.4
65	1735	30.0
70	2320	35.9

SOURCE: *A Policy of Geometric Design of Highways and Streets,* The American Association of State Highway and Transportation Officials, Washington, D.C., 2001. Used by permission.

VG = vehicle gap distance between turning vehicle and vehicle coming from right on major road, when both vehicles are traveling at 85 percent of the major-road design speed (in other words, assuming that a time gap of two seconds exists, $VG = 2 \times 0.85 \times 1.47 \times$ major road design speed)

For a turning radius of 28 ft, as shown in Figure 7.23,

$$h = P - 44 + 28 - VG - L$$
$$= P - 16 - VG - L$$

7. Determine the sight distance S_d using Eq. 7.11.

$$S_d = Q - h \tag{7.11}$$

When a two-lane minor road crosses a divided multilane highway as shown in Figure 7.25, the same procedure described above is used, taking into consideration that the turning vehicle traverses both the length of the turning arc and an additional distance s_1 while making the turn. This additional distance is determined as

$$s_1 = (N - 1)w + M \tag{7.12}$$

where
   N = number of lanes on the left approach of the major road
   w = the width of each lane of the major road (ft)
   M = width of the median (ft)

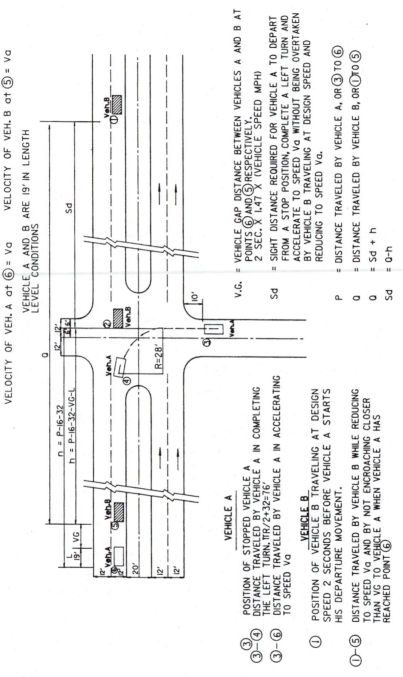

Figure 7.25 Layout of Sight Distance Requirements for a Stopped Vehicle Turning Left at an Intersection on a Multi-Lane Highway—Case IIIB

SOURCE: *A Policy of Geometric Design of Highways and Streets*, American Association of State Highway and Transportation Officials, Washington, D.C., 2001. Used with permission.

This gives

$$z_{\text{multilane}} = \pi\frac{R}{2} + (N - 1)w + M \tag{7.13}$$

For the case shown in Figure 7.25,

$$s_1 = (2 - 1)12 + 20$$
$$= 32 \text{ ft}$$

The equation for h in this case is therefore

$$h = P - 16 - 32 - VG - L \tag{7.14}$$
$$= P - 48 - VG - L$$

Example 7.4   Determining Sight Distances Along the Major Road for Minor-Road Traffic Turning Left onto the Major Road

A two-lane minor road intersects with a four-lane divided highway with a design speed of 55 mi/h. If the design vehicle on the minor road is a passenger car and it is controlled by a stop sign, determine the sight distance on the right along the major road that is required by a driver turning left from the minor road onto the major road without significantly interfering with the traffic on the major road. The following apply:

lane width on major road = 12 ft

median width = 12 ft

radius of turning arc = 28 ft

length of passenger car = 19 ft

Solution: The problem is solved by going through the steps discussed in the previous section.

- Determine the distance P traversed by the turning vehicle to attain a speed of 85 percent of the major-road design speed.

$$85\% \text{ of major-road design speed} = 0.85 \times 55$$
$$= 46.75 \text{ mi/h}$$

From Figure 7.24a, and using the curve for initial speed equal to zero,

$$P = 630 \text{ ft}$$

- Determine the time t_a required for a passenger car to travel the distance P.

$$t_a = 16.85 \text{ sec (interpolating from Table 7.7)}$$

- Determine the sum of t_a and J.

$$t_a + J = 16.85 + 2$$
$$= 18.85 \text{ sec}$$

- Determine the distance Q traveled by the vehicle on the major road during the time determined in step 3, using Eq. 7.8, assuming the average speed is 85 percent of the design speed.

$$Q = 1.47(t_a + J)(0.85 \times \text{design speed of major road})$$
$$= 1.47 \times 18.85 \times 0.85 \times 55$$
$$= 1295.4 \text{ ft}$$

- Compute the distance z traveled by the turning vehicle from the minor road in making the turn.
  Since the major road is a multilane highway, we use Eq. 7.13:

$$z = \pi\frac{R}{2} + (N - 1)w + M$$
$$= \pi\frac{28}{2} + (2 - 1)12 + 12$$
$$= 68 \text{ ft}$$

- Compute the distance h shown in Figure 7.23 by using Eq. 7.10.

$$h = P - z + (\text{radius of turning arc}) - L - VG$$
$$= 630 - 68 + 28 - 19 - (2 \times 1.47 \times 0.85 \times 55)$$
$$= 433.6 \text{ ft}$$

- Determine sight distance S_d using Eq. 7.11.

$$S_d = Q - h$$
$$= 1295.4 - 433.6$$
$$= 861.8 \text{ ft}$$

Turning Right into a Major Highway—Case IIIC. In this case, an adequate sight distance should be provided to allow the minor-road vehicle to complete its right turn and to accelerate to 85 percent of the design speed of the major road before it is overtaken by a vehicle approaching from the left on the major road. The same procedure described for Case IIIB is used (see Figure 7.26) for a two-lane minor road crossing a two-lane major road. In this case the turning radius is taken as 25 ft, and thus h is given as

$$h = P - 14.3 - VG - L \tag{7.15}$$

Sight Distance at Intersections with Signal Control

Although the unobstructed view at signalized intersections may be limited to the area of control, it is recommended that the minimum sight distances based on Case III conditions be made available at these intersections. These minimum sight distances are necessary to avoid the hazardous situations resulting from unanticipated conflicts at signalized intersections, including signal failure, vehicles running the red light, use of the flashing red/yellow mode, and right turns on red. In particular, when right turn on red is permitted at signalized intersections, the sight distance requirements for turning right onto a major road (Case IIIC) should be satisfied. The basic principle of signalized intersections is to provide sight distances that will enable the driver to see the signals early enough to take the necessary action indicated by the signals.

Effect of Skew on Intersection Sight Distance

As noted earlier, it is desirable for roads to intersect at or nearly at 90 degrees. When it is not feasible to obtain a normal or near-normal intersection, it is necessary to consider the way in which the sight distances are computed. In using the sight triangle for Cases I and II, the distances d_a and d_b should be measured along the roads, and the distances a and b should be measured parallel to the road. It is also important to ensure that the area within the sight triangle formed by the legs d_a and d_b is not obstructed from the driver's view. In cases where the angle of intersection makes it difficult for the driver to see the whole of the sight triangle, particularly at acute angle quadrants, it is desirable to consider both the Case II and Case III conditions and to select the larger of the two sight distances used, even when traffic volumes are low.

Effect of Grade on Sight Distances at Intersections

It is desirable for the grades of the approaches at an intersection to be 3 percent or less. When grades are 3 percent or less, the effect of the grade on the stopping distance is not significant and may be ignored. However, the grade of the minor road affects the time required for a vehicle to cross the major highway in the Case III condition. This effect is taken into consideration by using an adjusted value of t_a, which is the product of the appropriate factor obtained from Table 7.9 and the value of t_a for the level condition. This adjusted value of t_a is used in Eq. 7.5 to determine d.

DESIGN OF RAILROAD GRADE CROSSINGS

Railroad crossings are similar to four-leg intersections and can also be either at-grade or grade-separated. When a railroad crosses a highway at grade, it is essential that the highway approaches be designed to provide the safest condition so that the likelihood of a collision between a vehicle crossing the rail tracks and an oncoming train is reduced to the absolute

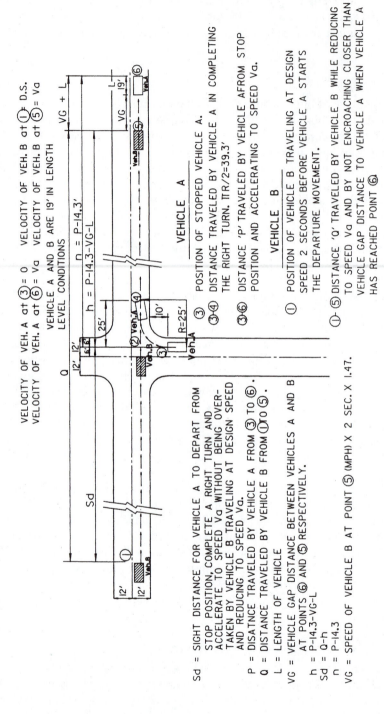

VELOCITY OF VEH. A at ③ = 0   VELOCITY OF VEH. B at ① = D.S.
VELOCITY OF VEH. A at ⑥ = Va   VELOCITY OF VEH. B at ⑤ = Va

VEHICLE A AND B ARE 19' IN LENGTH
LEVEL CONDITIONS

VEHICLE A

③   POSITION OF STOPPED VEHICLE A.

③-④   DISTANCE TRAVELED BY VEHICLE A IN COMPLETING
        THE RIGHT TURN. πR/2=39.3'

③-⑥   DISTANCE 'P' TRAVELED BY VEHICLE A FROM STOP
        POSITION AND ACCELERATING TO SPEED Va.

VEHICLE B

①   POSITION OF VEHICLE B TRAVELING AT DESIGN
        SPEED 2 SECONDS BEFORE VEHICLE A STARTS
        THE DEPARTURE MOVEMENT.

①-⑤   DISTANCE 'Q' TRAVELED BY VEHICLE B WHILE REDUCING
        TO SPEED Va AND BY NOT ENCROACHING CLOSER THAN
        VEHICLE GAP DISTANCE TO VEHICLE A WHEN VEHICLE A
        HAS REACHED POINT ⑥

Sd = SIGHT DISTANCE FOR VEHICLE A TO DEPART FROM
     STOP POSITION, COMPLETE A RIGHT TURN AND
     ACCELERATE TO SPEED Va WITHOUT BEING OVER-
     TAKEN BY VEHICLE B TRAVELING AT DESIGN SPEED
     AND REDUCING TO SPEED Va.
P = DISATNCE TRAVELED BY VEHICLE A FROM ③ TO ⑥.
Q = DISTANCE TRAVELED BY VEHICLE B FROM ① TO ⑤.
L = LENGTH OF VEHICLE
VG = VEHICLE GAP DISTANCE BETWEEN VEHICLES A AND B
     AT POINTS ⑥ AND ⑤ RESPECTIVELY.
h = P-14.3-VG-L
Sd = Q-h
n = P-14.3
VG = SPEED OF VEHICLE B AT POINT ⑤ (MPH) X 2 SEC. X 1.47.

Figure 7.26 Layout of Sight Distance Requirements for a Stopped Vehicle Turning Right at an Intersection on a Two-Lane
Highway—Case IIIC

SOURCE: *A Policy of Geometric Design of Highways and Streets*, American Association of State Highway and Transportation Officials, Washington, D.C., 2001.
Used with permission.

Table 7.9 Ratio of Accelerating Time on Grade to Accelerating Time on Level Section

Design Vehicle	Crossroad Grade (percent)				
	−4	−2	0	+2	+4
P	0.7	0.9	1.0	1.1	1.3
SU	0.8	0.9	1.0	1.1	1.3
WB-50	0.8	0.9	1.0	1.2	1.7

SOURCE: *A Policy of Geometric Design of Highways and Streets,* The American Association of State Highway and Transportation Officials, Washington, D.C., 2001. Used by permission.

minimum. This involves both the proper selection of the traffic control systems and the appropriate design of the horizontal and vertical alignments of the highway approaches.

Selection of Traffic Control Devices

The selection of the appropriate traffic control systems used at the railroad crossing should be made at the same time as the design of the vertical and horizontal alignments. These include both passive and active warning devices. Passive devices include signs, pavement markings, and grade crossing illumination, which warn an approaching driver of the crossing location. This type of warning device allows the driver of an approaching vehicle to determine whether a danger exists because of an approaching train; in fact, in this case the decision on whether to stop or to proceed belongs entirely to the driver. Several factors influence the designer's decision to use passive or active control devices. Passive warning devices are used mainly at low-volume, low-speed roadways with adequate sight distance that cross minor spurs or tracks that are seldom used by trains. Active warning devices include flashing light signals and automatic gates that give the driver a positive indication of an approaching train at the intersection. When it is necessary to use an active control device at a railroad grade crossing, the basic device used is the flashing light signal. An automatic gate is usually added after an evaluation of the conditions at the location. The factors considered in this evaluation include the highway type; the volumes of vehicular, railroad, and pedestrian traffic; the expected speeds of both the vehicular and the railroad traffic; accident history at the site; and the geometric characteristics of the approaches, including available sight distances. For example, when multiple main-line tracks exist, or when the moderate volumes of railroad and vehicular traffic exist with high speeds, frequently the automatic gate is used. AASHTO, however, suggests that the potential of eliminating grade crossings with only passive control devices should be given serious consideration. This can be achieved by closing crossings that are infrequently used and installing active controls at others.

Design of the Horizontal Alignment

Guidelines similar to those for the design of intersections formed by two highways are used in the design of a railroad grade crossing. It is desirable that the highway intersects the rail tracks at a right angle, and that other intersections or driveways be located at distances far enough away from the crossing to ensure that traffic operation at the crossing is not affected by vehicular movements at these other intersections and driveways. In order to enhance the sight distance, railroad grade crossings should not be placed at curves.

Design of the Vertical Alignment

The basic requirements for the vertical alignment at a railway grade crossing are the provision of suitable grades and adequate sight distances. Grades at the approaches should be as flat as possible to enhance the view across the crossing. When vertical curves are used, their lengths should be adequate to ensure that the driver on the highway clearly sees the crossing. In order to eliminate the possibility of low-clearance vehicles being trapped on the tracks, the surface of the crossing roadway should be at the same level as the top of the tracks for a distance of 2 ft from the outside of the rails. Also, at a point on the roadway 30 ft from the nearest track, the elevation of the road should not be higher than 0.25 ft or lower than 0.5 ft than the elevation of the tracks, unless this cannot be achieved because of the superelevation of the track.

Sight Distance Requirements at Railroad Crossings

The provision of adequate sight distances at the crossing is very important, especially at crossings where train-activated warning devices are not installed. The two principal actions the driver of a vehicle on the road can take when approaching a railroad crossing are:

1. The driver, having seen the train, brings the vehicle to a stop at the stop line on the road.
2. The driver, having seen the train, continues safely across the crossing before the train arrives.

The minimum sight distances required to satisfy these conditions should be provided at the crossing, as illustrated in Figure 7.27.

Driver Stops at the Stop Line
The total distance required for the driver to stop a vehicle at the stop line consists of the following:

- The distance traveled during the driver's perception/reaction time, or $1.47v_v t$, where v_v is the velocity of the approaching vehicle in mi/h, and t is the perception reaction time, usually taken as 2.5 sec.
- The actual distance traveled during braking, or $v_v^2/(30\frac{a}{g})$ (assuming a flat grade), as discussed in Chapter 3.

This gives

$$\text{total distance traveled during braking} = 1.47v_v t + \frac{v_v^2}{30\frac{a}{g}}$$

Therefore, the distance d_H the driver's eye should be away from the track to safely complete a stopping maneuver is the sum of the total distance traveled during the braking

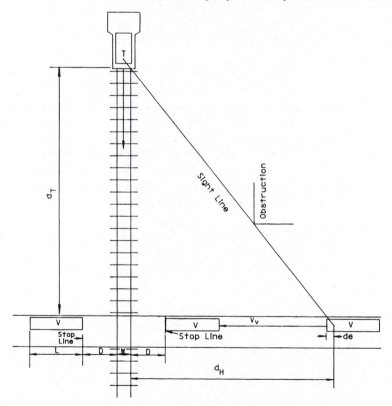

$$d_H = 1.47\ V_V\ t + \frac{V_V^2}{30f} + D + de$$

$$dT = \frac{V_T}{V_V}(1.47\ V_V t + \frac{V_V^2}{30f} + 2D + L + W)$$

d_H = Sight Distance Along Highway

d_T = Sight Distance Along Railroad Tracks

V_V = Velocity of Vehicle

t = Perception/Reaction Time (assumed 2.5 Sec.)

f = Coefficient Of Friction   (See Table III-I)

D = Distance From Stop Line To Near Rail (Assumed 15 ft.)

W = Distance Between Outer Rails   (Single Track W=5ft.)

L = Length Of Vehicle   (Assumed 65ft.)

V_T = Velocity Of Train

de = Distance From Driver To Front Of Vehicle (Assumed 10ft.)

Adjustments Must Be Made For Skew Crossings.
Assumed Flat Highway Grades Adjacent To And At Crossings.

Figure 7.27 Layout of Sight Distance Requirements for a Moving Vehicle to Cross or Stop at a Railroad Crossing Safely—Case I

SOURCE: *A Policy of Geometric Design of Highways and Streets,* American Association of State Highway and Transportation Officials, Washington, D.C., 2001. Used with permission.

maneuver, the distance between the stop line and the nearest rail, and the distance between the driver's eyes and the front of the vehicle.

$$d_H = 1.47v_v t + \frac{v_v^2}{30\left(\dfrac{a}{g}\right)} + D + d_e \qquad (7.16)$$

where

d_H = sight distance leg along the highway that allows a vehicle approaching the crossing at speed v_v to stop at the stop line at a distance D from the nearest rail

D = distance from the stop line to the nearest rail (assumed to be 15 ft)

d_e = distance from the driver's eye to the front of the vehicle at the stop line (assumed to be 10 ft)

v_v = velocity of vehicle mi/h

Driver Continues Safely Across Crossing

In this case the sight distance leg d_T along the railroad depends on both the time it takes the vehicle on the highway to cross the tracks from the time the driver first sees the train and the speed at which the train is approaching the crossing. The time at which the driver first sees the train must, however, provide a distance that will allow the driver either to cross the tracks safely or to stop the vehicle safely without encroachment of the crossing area. This distance must therefore be at least the sum of the distance required to stop the vehicle at the stop line and that between the two stop lines. It can be seen from Figure 7.27 that this total distance is

$$1.47v_v t + \frac{v_v^2}{30\left(\dfrac{a}{g}\right)} + 2D + L + W$$

where

L = length of the vehicle, assumed as 65 ft

W = the distance between the nearest and farthest rails, i.e., 5 ft

The time taken for the vehicle to traverse this total distance is obtained by dividing this distance by the speed of the vehicle (v_v). The distance traveled by the train during this time is then obtained by multiplying the time by the speed of the train. This gives the sight distance leg d_T along the railroad track as

$$d_T = \frac{v_T}{v_v}\left(1.47v_v t + \frac{v_v^2}{30\left(\dfrac{a}{g}\right)} + 2D + L + W\right) \qquad (7.17)$$

where

v_T = velocity of train, mi/h

The sight distance legs d_H along the highway and d_T along the track and the sight line connecting the driver's eye and the train form the sides of the sight triangle, which should be kept clear of all obstructions to the driver's view. Table 7.10 gives computed values (Case I) for these sight distance legs for different approach speeds of the train and vehicle.

Driver Departs from Stopped Position and Safely Clears Crossing

In this case an adequate sight distance should be provided along the track to allow enough time for a driver to accelerate the vehicle and safely cross the tracks before the arrival of a train that comes into view just as this maneuver begins. It is assumed that the driver will go through the crossing in first gear. The time required to go through the crossing, t_s, is the sum of (a) the time it takes to accelerate the vehicle to its maximum speed in the first gear, or v_g/a; and (b) the time it takes to travel the distance from the point at which the vehicle attains maximum speed in first gear to the position where the rear of the vehicle is at a distance D from the farthest tracks (i.e., the time it takes to travel the distance from the stopped position to the position where the rear of the vehicle is at D from the farthest tracks minus the distance traveled when the vehicle is being accelerated to its maximum speed in the first gear) and (c) the perception-reaction time J.

$$t_s = \frac{v_g}{a_1} + (L + 2D + W - d_a)\frac{1}{v_g} + J \qquad (7.18)$$

Table 7.10 Required Design Distances for Combination of Highway and Train Vehicle Speeds; 65-ft Truck Crossing a Single Set of Tracks at 90 Degrees

Train Speed (mi/h)	Case II Departure from Stop	Case I Moving Vehicle						
		Vehicle Speed (mi/h)						
	0	10	20	30	40	50	60	70
	Distance Along Railroad from Crossing, d_T (ft)							
10	240	145	103	99	103	112	122	134
20	480	290	207	197	207	224	245	269
30	719	435	310	296	310	337	367	403
40	959	580	413	394	413	449	489	537
50	1,200	725	517	493	517	561	611	671
60	1,439	870	620	591	620	673	734	806
70	1,679	1,015	723	690	723	786	856	940
80	1,918	1,160	827	789	827	898	978	1,074
90	2,158	1,305	930	887	930	1,010	1,101	1,209
	Distance Along Highway from Crossing, d_H (ft)							
		69	132	221	338	486	659	865

SOURCE: *A Policy of Geometric Design of Highways and Streets,* The American Association of State Highway and Transportation Officials, Washington, D.C., 2001. Used by permission.

where

v_g = maximum velocity of vehicle in first gear, assumed to be 8.8 ft/sec

a_1 = acceleration of vehicle in first gear, assumed to be 1.47 ft/sec^2

L = length of vehicle, assumed to be 65 ft

D = distance of stop line to nearest rail, assumed to be 15 ft

J = perception-reaction time, assumed to be 2.0 sec

W = distance between outer rails (for a single track this is equal to 5 ft)

d_a = distance vehicle travels while accelerating to maximum speed in the first gear, or $v_g^2/2a_1$

The distance d_T traveled by the train along the track during this time is obtained by multiplying this time by the speed of the train.

$$d_T = 1.47v_T \left[\frac{v_g}{a_1} + (L + 2D + W - d_a)\frac{1}{v_g} + J \right] \qquad (7.19)$$

The sight triangle formed by the lengths $2D$, d_T, and the sight line between the eyes of the driver of the stopped vehicle and the approaching train, as shown in Figure 7.28, should therefore be kept clear of any obstruction to the vehicle driver's view. Values for d_T for the stopped condition (Case II) are also given in Table 7.10.

Equations 7.16 through 7.19 have been developed with the assumptions that the approach grades are level and that the road crosses the tracks at 90 degrees. When one or both assumptions do not apply, the necessary corrections should be made (as discussed earlier) for highway–highway at-grade intersections. Also, when it is not feasible to provide the required sight distances for the expected speed on the highway, the appropriate speed for which the sight distances are available should be determined, and speed control signs and devices should be used to advise drivers of the maximum speed allowable at the approach.

SUMMARY

The basic principles involved in the design of at-grade intersections are presented in this chapter. The fundamental objective in the design of at-grade intersections is to minimize the severity of the potential conflicts among different streams of traffic and between pedestrians and turning movements. This design process involves the selection of an appropriate layout of the intersection and the appropriate design of the horizontal and vertical alignments. The selection of the appropriate layout depends on the turning volumes, the available space, and whether the intersection is in an urban or a rural area. An important factor in designing the layout is whether the intersection should be channelized. When traffic volumes warrant channelization, care should be taken not to overchannelize, since this may confuse drivers. Channelization usually involves the use of islands, which may or may not be raised. Raised islands may be either of the mountable or barrier type. It is advisable not to use raised barriers at intersections on high-speed roads. Care should also be taken to ensure that the minimum sizes for the type of island selected be provided.

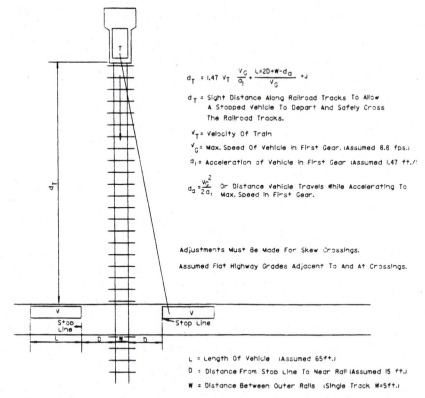

$$d_T = 1.47 \ v_T \left[\frac{v_G}{a_1} + \frac{L = 2D + W - d_a}{v_G} \right] + J$$

d_T = Sight Distance Along Railroad Tracks To Allow A Stopped Vehicle To Depart And Safely Cross The Railroad Tracks.

v_T = Velocity Of Train

v_G = Max. Speed Of Vehicle In First Gear. (Assumed 8.8 fps.)

a_1 = Acceleration of Vehicle In First Gear (Assumed 1.47 ft./

$d_a = \dfrac{v_G^2}{2 a_1}$ Or Distance Vehicle Travels While Accelerating To Max. Speed In First Gear.

Adjustments Must Be Made For Skew Crossings.

Assumed Flat Highway Grades Adjacent To And At Crossings.

L = Length Of Vehicle   (Assumed 65ft.)

D = Distance From Stop Line To Near Rail (Assumed 15 ft.)

W = Distance Between Outer Rails   (Single Track W=5ft.)

Figure 7.28 Layout of Sight Distance Requirements for Departure of a Vehicle at a Stopped Position to Cross a Single Railroad Track

SOURCE: *A Policy of Geometric Design of Highways and Streets,* American Association of State Highway and Transportation Officials, Washington, D.C., 2001. Used with permission.

In designing the horizontal alignments, sharp horizontal curves should be avoided, and every effort should be made to ensure that the intersecting roads cross at 90 degrees or as near to 90 degrees as possible.

The most important factor in the design of the vertical alignment is to provide sight distances that allow drivers to have an unobstructed view of the whole intersection. The minimum sight distance required at any approach of an intersection depends on the design vehicle on the approach, the expected velocity on the approach, and the type of control at the approach.

An at-grade railroad crossing is similar to an at-grade highway-highway intersection in that similar principles are used in their design. For example, in the design of the vertical alignment of a railroad crossing, suitable grades and sight distances must be provided at the approaches. In the design of the railroad crossing, however, additional factors come into play. For example, at a railroad crossing the elevation of the surface of the crossing roadway should be the same as the top of the tracks for a distance of 2 ft from the outside rails, and at a point on the roadway 30 ft from the nearest track, the elevation of the roadway should not be more than 0.25 ft higher or 0.5 ft lower than that of the track.

In general, all intersections are points of conflicts between vehicles and pedestrians, and the basic objective of their design should be to eliminate these conflict points as much as possible.

PROBLEMS

7-1   Briefly describe the different principles involved in the design of at-grade intersections.

7-2   Describe the different functional classes of at-grade intersections. Also give an example of an appropriate location for the use of each type.

7-3   Figure 7.29a illustrates a three-leg intersection of State Route 150 and State Route 30. Both roads carry relatively low traffic, with most of the traffic oriented along State Route 150. The layout of the intersection, coupled with the high-speed traffic on State Route 150, have made this intersection a hazardous location. Drivers on State Route 30 tend to violate the stop sign at the intersection because of the mild turn onto westbound State Route 150, and they also experience difficulty in seeing the high-speed vehicles approaching from the left on State Route 150. Design a new layout for the intersection to eliminate these difficulties for the volumes shown in Figure 7.29b. Design vehicle is a passenger car.

7-4   Figure 7.30a shows the staggered unsignalized intersection of Patton Avenue and Goree Street. The distance between the T intersections is about 160 ft. The general layout and striping of the lanes at this intersection result in confusion to drivers and create multiple conflicts. Design an improved layout for the intersection for the traffic volumes shown in Figure 7.30b. Design vehicle is a passenger car.

7-5   A four-leg intersection with no traffic control is formed by two 2-lane roads, with the speed limits on the minor and major roads being 25 and 35 mi/h, respectively. If the roads cross at 90 degrees and a building is to be located at a distance of 45 ft from the center line of the nearest lane on the minor road, determine the minimum distance at which the building should be located from the center line of the outside lane of the major road so that adequate sight distances are provided.

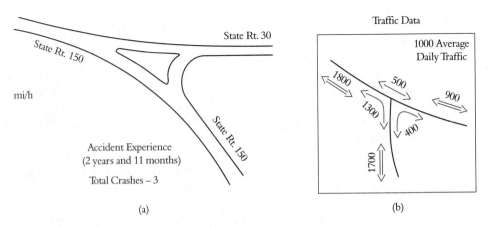

Figure 7.29 Intersection Layout for Problem 7-3

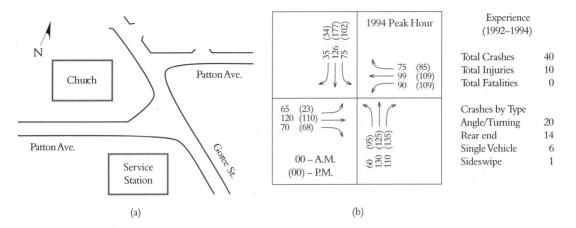

Figure 7.30 Intersection Layout for Problem 7-4

7-6 A two-lane minor road intersects a two-lane major road at 90 degrees, forming a four-leg intersection with traffic on the minor road controlled by a yield sign. A building is located 125 ft from the center line of the outside lane of the major road and 35 ft from the center line of the nearest lane of the minor road. Determine the maximum speed that can be allowed on the minor road if the speed limit on the major road is 45 mi/h.

7-7 If in Problem 7-6 the speed limit on the minor road were 40 mi/h, determine the minimum distance that the building can be located from the center line of the outside lane of the major street.

7-8 A developer has requested permission to build a large retail store at a location adjacent to the intersection of an undivided four-lane major road and a two-lane minor road. Traffic on the minor road is controlled by a stop sign. The speed limits are 35 and 50 mi/h on the minor and major roads, respectively. The building is to be located at a distance of 65 ft from the near lane of one of the approaches of the minor road. Determine where the building should be located relative to the center line of the outside lane of the major road in order to provide adequate sight distance for a driver on the minor road to proceed safely across the intersection after stopping at the stop line. Design vehicle is a single-unit truck. Lanes on the major road are 12 ft wide.

7-9 A minor road intersects a multilane divided highway at 90 degrees, forming a T intersection. The speed limits on the major and minor roads are 55 and 35 mi/h, respectively. Determine the sight distance required for a vehicle on the minor road to depart from a stopped position, complete a right turn, and accelerate to a speed of 85 percent of the design speed of the major road without forcing a vehicle on the major road that is traveling at 85 percent of the design speed to reduce its speed to that of the turning vehicle.

7-10 For the information given in Problem 7-9, determine the sight distance required by the minor-road vehicle to safely complete a left turn onto the major road.

7-11 A ramp from an expressway with a design speed of 35 mi/h connects with a local road, forming a T intersection. An additional lane is provided on the local road to allow vehicles from the ramp to turn right onto the local road without stopping. The turning roadway has stabilized shoulders on both sides and will provide for a one-lane, one-way operation with no provision for passing a stalled vehicle. Determine the width of the turning roadway, if the design vehicle is a single-unit truck. Use 0.08 for superelevation.

7-12   Determine the required width of the turning roadway in Problem 7-11 for a two-lane operation with barrier curbs on both sides.

7-13   Briefly discuss what factors are considered in the design of an at-grade railway crossing in addition to those for an at-grade crossing formed by two highways.

7-14   A two-lane road crosses an at-grade railroad track at 90 degrees. If the design speed of the two-lane road is 45 mi/h and the velocity of the train when crossing the highway is 80 mi/h, determine the sight distance leg along the railroad tracks to permit a vehicle traveling at the design speed to safely cross the tracks when a train is observed at a distance equal to the sight distance leg.

7-15   A stop sign controls all vehicles on the highway at a railroad crossing. Determine the minimum distance a building should be placed from the center line of the tracks to allow a stopped vehicle to safely clear the intersection. Assume that the building is located 40 ft from the center line of the near lane. The velocity of trains approaching the crossing is 85 mi/h.

REFERENCES

A Policy of Geometric Design of Highways and Streets, American Association of State Highway and Transportation Officials, Washington, D.C., 2001.

Manual on Uniform Traffic Control Devices, Federal Highway Administration, Washington, D.C., 2000.

Roundabouts: An Informational Guide, U.S. Department of Transportation, Federal Highway Administration, Publication No. FHWA-RD-00-067, Washington, D.C., 2000.

Taggart, R.C. et al., *Evaluating Grade-Separated Rail and Highway Crossing Alternatives,* NCHRP Report #288, Washington, D.C., 1987.

"Traffic Control Devices and Rail-Highway Grade Crossings," National Research Council, Transportation Research Board, *Transportation Research Record* 1114, Washington, D.C., 1987.

CHAPTER 8

Intersection Control

An intersection is an area shared by two or more roads. Its main function is to allow the change of route directions. While Chapter 7 describes the different types of intersections, this chapter gives a brief description as well, to facilitate a clear understanding of the elements of intersection control. A simple intersection consists of two intersecting roads; a complex intersection serves several intersecting roads within the same area. The intersection is therefore an area of decision for all drivers; each must select one of the available choices to proceed. This requires an additional effort by the driver that is not necessary in nonintersection areas of a highway. The flow of traffic on any street or highway is greatly affected by the flow of traffic through the intersection points on that street or highway because the intersection usually performs at a level below that of any other section of the road.

Intersections can be classified as grade-separated without ramps, grade-separated with ramps (commonly known as interchanges), or at-grade. Interchanges consist of structures that provide for the cross-flow of traffic at different levels without interruption, thus reducing delay, particularly when volumes are high. Figure 8.1 shows some types of at-grade intersections; Figure 8.2 shows some types of interchanges. Several types of traffic control systems are used to reduce traffic delays and crashes on at-grade intersections and to increase the capacity of highways and streets. However, the appropriate regulations must be enforced if these systems are to be effective. This chapter describes the different methods of controlling traffic on at-grade intersections and freeway ramps.

GENERAL CONCEPTS OF TRAFFIC CONTROL

The purpose of traffic control is to assign the right of way to drivers, and thus to facilitate highway safety by ensuring the orderly and predictable movement of all traffic on

(a)

(b)

Figure 8.1 Types of Intersections at Grade

(c)

Figure 8.1 Types of Intersections at Grade (*continued*)

SOURCE: Virginia Department of Transportation. Used with permission.

highways. Control may be achieved by using traffic signals, signs, or markings that regulate, guide, warn, and/or channel traffic. The more complex the maneuvering area, the greater the need for a properly designed traffic control system. Many intersections are complex maneuvering areas and therefore require properly designed traffic control systems. Guidelines for determining whether a particular control type is suitable for a given intersection are provided in the *Manual on Uniform Traffic Control Devices* (MUTCD).

To be effective, a traffic control device must

- Fulfill a need
- Command attention
- Convey a clear simple meaning
- Command the respect of road users
- Give adequate time for proper response

To ensure that a traffic control device possesses these five properties, the MUTCD recommends that engineers consider the following five factors:

1. **Design.** The device should be designed with a combination of size, color, and shape that will convey a message and command the respect and attention of the driver.

(a)

Figure 8.2 Types of Interchanges

2. **Placement.** The device should be located so that it is within the cone of vision of the viewer and the driver has adequate response time when driving at normal speed.
3. **Operation.** The device should be used in a manner that ensures the fulfillment of traffic requirements in a consistent and uniform way.
4. **Maintenance.** The device must be regularly maintained to ensure that legibility is sustained.
5. **Uniformity.** To facilitate the recognition and understanding of these devices by drivers, similar devices should be used at locations with similar traffic and geometric characteristics.

(b)

Figure 8.2 Types of Interchanges (*continued*)

In addition to these considerations, it is essential that engineers avoid using control devices that conflict with one another at the same location. It is imperative that control devices aid each other in transmitting the required message to the driver.

CONFLICT POINTS AT INTERSECTIONS

Conflicts occur when traffic streams moving in different directions interfere with each other. The three types of conflicts are merging, diverging, and crossing. Figure 8.3 shows the different conflict points that exist at a four-approach unsignalized intersection. There are 32 conflict points in this case. The number of possible conflict points at any intersection depends on the number of approaches, the turning movements, and the type of traffic control at the intersection.

The primary objective in the design of a traffic control system at an intersection is to reduce the number of significant conflict points. In designing such a system, it is first necessary to undertake an analysis of the turning movements at the intersection, which will indicate the significant types of conflicts. Factors that influence the significance of a conflict include the type of conflict, the number of vehicles in each of the conflicting streams, and the speeds of the vehicles in these streams. Crossing conflicts, however, tend

(c)

Figure 8.2 Types of Interchanges (*continued*)

SOURCE: Virginia Department of Transportation. Used with permission.

to have the most severe effect on traffic flow and should be reduced to a minimum whenever possible.

TYPES OF INTERSECTION CONTROL

Several methods of controlling conflicting streams of vehicles at intersections are in use. The choice of one of these methods depends on the type of intersection and the volume of traffic in each of the conflicting streams. Guidelines for determining whether a particular control type is suitable for a given intersection have been developed and are given in the MUTCD. These guidelines are presented in the form of warrants, which have to be com-

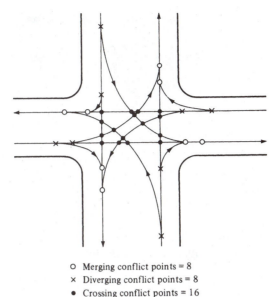

○ Merging conflict points = 8
× Diverging conflict points = 8
● Crossing conflict points = 16

Figure 8.3 Conflict Points at a Four-Approach Unsignalized Intersection

pared with the traffic and geometric characteristics at the intersection being considered. The different types of intersection control are described below.

Yield Signs

All drivers on approaches with yield signs are required to slow down and yield the right of way to all conflicting vehicles at the intersection. Stopping at yield signs is not mandatory, but drivers are required to stop when necessary to avoid interfering with a traffic stream that has the right of way. Yield signs are therefore usually placed on minor-road approaches, where it is necessary to yield the right of way to the major-road traffic. Figure 8.4 shows the regulated shape and dimensions for a yield sign. The most significant factor in the warrant for yield signs is the approach speed on the minor road. This sign is warranted at intersections where there is a separate or channelized right-turn lane without an adequate acceleration lane.

Stop Signs

A stop sign is used where an approaching vehicle is required to stop before entering the intersection. Figure 8.4 also shows the regulated shape and dimensions of a stop sign. Stop signs should be used only when they are warranted, since the use of these signs results in considerable inconvenience to motorists. Stop signs should not be used at signalized intersections or on through roadways of expressways. The warrants for stop signs suggest that a stop sign may be used on a minor road when it intersects a major road, at an unsignalized

Stop Sign Sizes

Conventional Roads 30" × 30"
Expressways   36" × 36"
Minimum   24" × 24"

Yield Sign Sizes

Conventional Roads 36" × 36" × 36"
Expressways   48" × 48" × 48"
Freeways   48" × 48" × 48"
Minimum   30" × 30" × 30"

R1–1   R1–2

Figure 8.4 Stop Sign and Yield Sign

intersection, and where a combination of high speed, restricted view, and serious crashes indicates the necessity for such a control.

Multiway Stop Signs

Multiway stop signs require that all vehicles approaching the intersection stop before entering it. They are used as a safety measure at some intersections and are normally used when the traffic volumes on all the approaches are approximately equal. When traffic volumes are high, however, the use of signals is recommended.

The warrants for this control specify that the total intersection approach volume should not be less than 500 veh/h for 8 hr of an average day, nor should the combined volume of vehicles and pedestrians from the minor approach be less than 200 units per hour for the same 8 hr. The average delay of the vehicles on the minor street also should not be less than 30 sec per vehicle during the maximum hour. The minimum requirement for vehicular volume can be reduced by 30 percent if the 85th-percentile approach speed on the major approach is greater than 40 mi/h.

Example 8.1   Evaluating the Need for a Multiway Stop Sign at an Intersection

A minor road carrying 150 veh/h for 8 hr of an average day crosses a major road carrying 320 veh/h for the same 8 hr, forming a four-leg intersection. Determine whether a multiway stop sign is justified at this location if the following conditions exist:

1. The pedestrian volume from the minor street for the same 8 hr as the traffic volumes is 70 ped/hr.
2. The average delay to minor-street vehicular traffic during the maximum hours is 37 sec/veh.
3. The 85th-percentile approach speed of the major road is 43 mi/h.

Solution:

- Determine whether traffic volume satisfies the warrant. Total vehicular volume entering the intersection from all approaches is 150 + 320 = 470 veh/h. Although this is less than 500, the 85th-percentile speed on the major approach is over 40 mi/h.
- Total minimum entering volume is 0.70 × 500 = 350.

The vehicular volume is satisfied. The average delay to minor street traffic = 37 sec per veh, which is greater than 30 sec. Thus, the delay condition is satisfied.

Combined vehicular and pedestrian traffic from the minor approach is 150 + 70 = 220. This is greater than 200. Thus, the combined vehicle and pedestrian requirement is satisfied. Therefore, all conditions are satisfied and a multiway stop sign is justified.

Intersection Channelization

Intersection channelization is used mainly to separate turn lanes from through lanes. A channelized intersection consists of solid white lines or raised barriers, which guide traffic within a lane so that vehicles can safely negotiate a complex intersection. When raised islands are used, they can also provide a refuge for pedestrians.

Channelization design criteria have been developed by many states individually to provide guidelines for the cost-effective design of channelized intersections. A detailed description of the techniques that have proven effective for both simple and complicated intersections is given in Chapter 7 and in the *Intersection Channelization Design Guide*. Guidelines for the use of channels at intersections include

- Laying out islands or channel lines to allow a natural, convenient flow of traffic
- Avoiding confusion by using a few well-located islands
- Providing adequate radii of curves and widths of lanes for the prevailing type of vehicle

Traffic Signals

One of the most effective ways of controlling traffic at an intersection is the use of traffic signals. Traffic signals can be used to eliminate many conflicts because different traffic streams can be assigned the use of the intersection at different times. Since this results in a delay to vehicles in all streams, it is important that traffic signals be used only when necessary. The most important factor that determines the need for traffic signals at a particular intersection is the intersection's approach traffic volume, although other factors such as pedestrian volume and crash experience may also play a significant role. The *Manual on Traffic Signal Design* gives the fundamental concepts and standard practices used in the design of traffic signals. In addition, the MUTCD describes in detail 11 warrants, at least one of which should be satisfied for an intersection to be signalized. However, these warrants should be considered only as a guide; professional judgment based on experience also should be used to decide whether or not an intersection should be signalized. The factors considered in the warrants are

- Eight-hour vehicular volume
  - minimum vehicular volume
  - interruption of continuous traffic
  - combination of warrants
- Minimum pedestrian volume
- School crossing

- Coordinated signal system
- Crash experience
- Roadway network
- Four-hour vehicular volume
- Peak hour
  - Peak-hour delay
  - Peak-hour volume

A brief discussion of each of these is given below. Interested readers are referred to the MUTCD for details.

Minimum Vehicle Volume

This warrant is applied when the principal factor for considering signalization is the intersection traffic volume. The warrant is satisfied when traffic volumes on major streets and on higher-volume minor street approaches for each of any 8 hr of an average day are at least equal to the volumes specified in Table 8.1. An "average" day is a weekday whose traffic volumes are normally and repeatedly observed at the location.

Combination of Warrants

This warrant, in exceptional cases, justifies the installation of signals when none of the above warrants is satisfied but when the first two warrants are satisfied to the extent of 80 percent of the stipulated volumes, i.e., volumes given in Tables 8.1 and 8.2.

Interruptions of Continuous Traffic

This warrant should be considered when traffic on a minor street suffers excessive delay due to the heavy volume of traffic on a major street. Heavy major-street traffic may also make it hazardous for minor-street traffic to enter or cross the major street. The warrant is satisfied when the traffic volume on the major street and on the higher-volume minor street approach for each of any 8 hr of an average day is at least equal to the volumes specified in Table 8.2.

Table 8.1 Volume Requirements for Minimum Vehicular Volumes Warrant

Number of Lanes for Moving Traffic on Each Approach		*Vehicles per Hour on Major Street (total of both approaches)*	*Vehicles per Hour on Higher-Volume Minor-Street Approach (one direction only)*
Major Street	*Minor Street*		
1	1	500	150
2 or more	1	600	150
2 or more	2 or more	600	200
1	2 or more	500	200

SOURCE: Adapted from the *Manual on Uniform Traffic Control Devices,* U.S. Department of Transportation, Federal Highway Administration, Washington, D.C., 2000.
Note: If posted, statutory or 85th percentile speed on the major street is greater than 40 mi/h or if intersection lies in an area with population of less than 10,000, 70 percent of the volumes shown may be used.

Table 8.2 Minimum Vehicular Volumes for Interruption of Continuous Traffic Warrant

Number of Lanes for Moving Traffic on Each Approach		Vehicles per Hour on Major Street (total of both approaches)	Vehicles per Hour on Higher-Volume Minor-Street Approach (one direction only)
Major Street	Minor Street		
1	1	750	75
2 or more	1	900	75
2 or more	2 or more	900	100
1	2 or more	750	100

SOURCE: Adapted from *Manual on Uniform Traffic Control Devices,* U.S. Department of Transportation, Federal Highway Administration, Washington, D.C., 2000.
Note: If posted, statutory or 85th percentile speed on the major street is greater than 40 mi/h or the intersection lies in an area with population of less than 10,000, 70 percent of the volume shown may be used.

Minimum Pedestrian Volume

This warrant is satisfied when the pedestrian volume crossing the major street on an average day is at least 100 for each of any 4 hr or 190 during any 1 hr and there are fewer than 60 gaps per hr that are acceptable by pedestrians for crossing. In addition, the nearest traffic signal along the major street should be at least 300 ft away from the intersection. When this warrant is used, the signal should be of the traffic-actuated type, with pushbuttons for pedestrian crossing.

School Crossing

When an analysis of gap data at an established school zone shows that the frequency of occurrence of gaps and the lengths of gaps are inadequate for safe crossing of the street by schoolchildren, this warrant is applied. It stipulates that if during the period when schoolchildren are using the crossing the number of acceptable gaps is less than the number of minutes in that period and there are at least 20 students during the highest crossing hour, the use of traffic signals is warranted. The signal in this case should be pedestrian-actuated, and all obstructions to view, such as parked vehicles, should be prohibited for at least 100 ft before and 20 ft after the crosswalk.

Coordinated Signal System

This warrant may justify the installation of traffic lights at an intersection where lights would not otherwise have been installed. It justifies the installation of traffic lights when such an installation will help maintain a proper grouping of vehicles and effectively regulate group speed. This warrant is not applicable when the resultant spacing of the traffic signal will be less than 300 ft.

Crash Experience

This warrant justifies signalization of an intersection when crash frequency has not been reduced by adequate trial of less restrictive measures. It stipulates that the use of traffic signals is warranted at an intersection if 5 or more injury or reportable property-damage-only crashes have occurred within a 12-month period and that signal control is a suitable countermeasure for these crashes. In addition, the traffic and pedestrian volumes should not be less than 80 percent of the requirements specified in the minimum vehicle volume warrant, the interruption of continuous traffic warrant, or the minimum pedestrian volume warrant.

Roadway Network

This warrant justifies the installation of signals at some intersections when such an installation will help to encourage concentration and organization of traffic networks. The warrant can be applied at intersections of two or more major roads when (1) the total existing or immediately projected entering volume is at least 1000 during the peak hour of a typical weekday, and the 5-year projected traffic volumes, based on an engineering study, satisfy the requirements of the following warrants: maximum vehicular volume, interruption of traffic, combination of warrants, 4-hr volume, and peak-hour volume, or (2) when the total existing or projected entering volume is at least 1000 vehicles for each of any 5 hr of a Saturday and/or a Sunday.

Four-Hour Volume

This warrant is based on the comparison of standard graphs given in the MUTCD, one of which is shown in Figure 8.5, with a plot on the same graph of the total volume in veh/h on the approach with the higher volume in the minor street (vertical scale) against the total volume in veh/h on both approaches of the major street (horizontal scale) at the intersection. When the plot for each of any 4 hr of an average day falls above the standard graph, this warrant is satisfied. Standard graphs are given for different types of lane configurations at the intersection. A different set of standard plots is also given for intersections where the

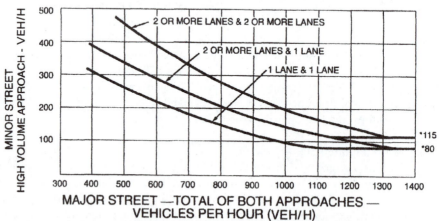

Figure 8.5 Graphs for Four Hour Volume Warrant

SOURCE: Redrawn from *Manual on Uniform Traffic Control Devices,* U.S. Department of Transportation, Federal Highway Administration, Washington, D.C., 2000.

85th-percentile speed of the major-street traffic is higher than 40 mi/h or where the intersection is located in a built-up area of an isolated community whose population is less than 10,000.

Peak-Hour Delay

This warrant is used to justify the installation of traffic signals at intersections where traffic conditions during 1 hr of the day result in undue delay to traffic on the minor street. The warrant is satisfied when the delay during any four consecutive 15-minute periods on one of the minor-street approaches (one direction only) controlled by a stop sign is equal to or greater than specified levels and the same minor-street approach (one direction only) volume and the total intersection entering volume are equal to or greater than the specified levels.

The specified delay is 4 vehicle-hours for a one-lane approach and 5 vehicle-hours for a two-lane approach. The specified volumes are 100 veh/h for one moving lane of traffic, 150 veh/h for two moving lanes of traffic, and 800 veh/h entering volume for intersections with four or more approaches or 650 veh/h for intersections with three approaches.

Peak-Hour Volume

This warrant is also used to justify the installation of traffic signals at intersections where traffic conditions during 1 hr of the day result in undue delay to traffic on the minor street. It is based on the comparison of the standard graphs given in the MUTCD (see Figure 8.6) with a plot of the peak-hour volume on the approach with the higher volume on the minor street against the corresponding peak-hour volume on the major street (both directions) of the intersection. When this plot for an average day lies above the appropriate standard curve, this warrant is satisfied.

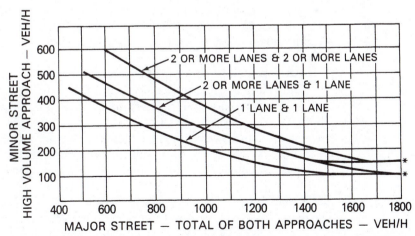

*NOTE: 150 VEH/H APPLIES AS THE LOWER THRESHOLD VOLUME FOR A MINOR STREET
        APPROACH WITH TWO OR MORE LANES AND 100 VEH/H APPLIES AS THE LOWER
        THRESHOLD VOLUME FOR A MINOR STREET APPROACHING WITH ONE LANE.

Figure 8.6 Peak Hour Volume Warrant

SOURCE: Redrawn from *Manual on Uniform Traffic Control Devices,* U.S. Department of Trasportation, Federal Highway Administration, Washington, D.C., 2000.

Example 8.2 Determining Whether the Conditions at an Intersection Warrant Installing Traffic Signals

A two-lane minor street crosses a four-lane major street. If the traffic conditions are as given below, determine whether installing a traffic signal at this intersection is warranted.

1. The traffic volumes for each of 8 hr of an average day (both directions on major street) total 400 veh/h. For the higher-volume minor-street approach (one direction only), the total is 100.
2. The 85th-percentile speed of major-street traffic is 43 mi/h.
3. The pedestrian volume crossing the major street during an average day is 200 ped/hr during peak pedestrian periods (2 hr in the morning and 2 hr in the afternoon).
4. The number of gaps per hour in the traffic stream for pedestrians to cross during peak pedestrian periods is 52.
5. The nearest traffic signal is located 450 ft from this location.

Solution:

Since vehicular and pedestrian volumes are given, the minimum vehicle volumes and minimum pedestrian volumes warrants are checked to determine whether the conditions of any are satisfied.

- Check minimum vehicle volumes warrant:
  Minimum volumes requirements for this intersection (two lanes for moving traffic on major street and one lane on minor street) are 600 veh/h on the major street (total of both approaches) and 150 veh/h on the minor street (one direction only). See Table 8.1. Volumes at the intersection are less than the minimum volumes: that is; 400 < 600 and 100 < 150. However, because the 85th-percentile speed is 43 mi/h > 40 mi/h, the minimum volumes required can be reduced to 70 percent of the requirements.
    Therefore, minimum required volumes are $600 \times 0.7 = 420$ and $150 \times 0.7 = 105$. The volumes at the intersection are still less than the adjusted required volumes. The minimum vehicular volumes warrant is therefore not satisfied.
- Check minimum pedestrian volume:
  The minimum pedestrian volumes required are 100 or more for each of any 4 hr or 190 or more during any 1 hr. With 200 pedestrians per hour during the peak pedestrian volumes, the minimum requirement for pedestrian volumes is satisfied.
- Check minimum number of acceptable gaps:
  The minimum acceptable gaps for pedestrian crossing are 60 gaps per hour. The number of gaps available is 52. The number of gaps for pedestrian crossing is insufficient, and the acceptable-gaps condition is satisfied.
- Check location of nearest traffic signal:
  For the location of the nearest traffic signal along the major street, the minimum distance required is 300 ft. The nearest traffic signal to the intersection is

450 ft away; 450 > 300. Thus, the location of nearest traffic signal condition is satisfied.

A traffic signal is therefore justified under the minimum pedestrian volume warrant. The signal should therefore be of the traffic-actuated type, with pushbuttons for pedestrians who are crossing.

SIGNAL TIMING FOR DIFFERENT COLOR INDICATIONS

The warrants described earlier will help the engineer only in deciding whether a traffic signal should be used at an intersection. The efficient operation of the signal also requires proper timing of the different color indications, which is obtained by implementing the necessary design. Before presenting the different methods of signal timing design, however, it is first necessary to define a number of terms commonly used in the design of signal times.

A *controller* is a device in a traffic signal installation that changes the colors indicated by the signal lamps according to a fixed or variable plan. It assigns the right-of-way to different approaches at appropriate times.

The *cycle (cycle length)* is the time in seconds required for one complete color sequence of signal indication. Figure 8.7 is a schematic of a cycle. In Figure 8.7, for example, the cycle length is the time that elapses from the start of the green indication to the end of the red indication.

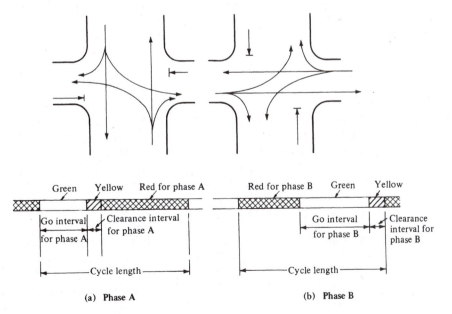

Figure 8.7 Two-Phase Signal System

A *phase (signal phase)* is that part of a cycle allocated to a stream of traffic, or a combination of two or more streams of traffic, having the right-of-way simultaneously during one or more intervals. See Figure 8.7.

An *interval* is any part of the cycle length during which signal indications do not change.

The *offset* is the time lapse in seconds or the percentage of the cycle length between the beginning of a green phase at the intersection and the beginning of a corresponding green phase at the next intersection. It is the time base of the system controller.

The *change and clearance interval* is the total length of time in seconds of the yellow and all-red signal indications. This time is provided for vehicles to clear the intersection after the green interval, before conflicting movements are released.

An *all-red interval* is the display time of a red indication for all approaches. It is sometimes used as a phase exclusively for pedestrian crossing or to allow vehicles and pedestrians to clear very large intersections before opposing approaches are given the green indication.

A *split-phase* is the part of a phase that is set apart from the primary movement, thus forming a special phase that relates to the parent phase.

The *peak-hour factor (PHF)* is a measure of the variability of demand during the peak hour. It is the ratio of the volume during the peak hour to the maximum rate of flow during a given time period within the peak hour. For intersections, the time period used is 15 min, and the PHF is given as

$$\text{PHF} = \frac{\text{volume during peak hour}}{4 \times \text{volume during peak 15 min within peak hour}} \tag{8.1}$$

The PHF may be used in signal timing design to compensate for the possibility that peak arrival rates for short periods during the peak hour may be much higher than the average for the full hour. Design hourly volume (DHV) can then be obtained as

$$\text{DHV} = \frac{\text{peak - hour volume}}{\text{PHF}} \tag{8.2}$$

Not all factors that affect PHF have been identified, but it is generally known that the PHF is a function of the traffic generators being served by the highway, the distances between these generators and the highway, and the population of the metropolitan area in which the highway is located.

Lane Group. A lane group consists of one or more lanes on an intersection approach and having the same green phase. Lane groups for each approach are established using the following guidelines:

1. Separate lane groups should be established for exclusive left-turn lane(s), unless the approach also contains a shared left-turn and through lane. In such a case, consideration should also be given to the distribution of traffic volume between the movements. These same guidelines apply for exclusive right-turn lanes.
2. When exclusive left-turn lane(s) and/or exclusive right-turn lane(s) are provided on an approach, all other lanes are generally established as a single lane group.

3. When an approach with more than one lane also has a shared left-turn lane, the operation of the shared left-turn lane should be evaluated as shown below to determine whether it is effectively operating as an exclusive left-turn lane because of the high volume of left-turn vehicles on it.

Figure 8.8 shows typical lane groups used for analysis. Note that when two or more lanes have been established as a single lane group for analysis, all subsequent computations must consider these lanes a single entity.

The *critical lane group* is the lane group that requires the longest green time in a phase. This lane group, therefore, determines the green time that is allocated to that phase.

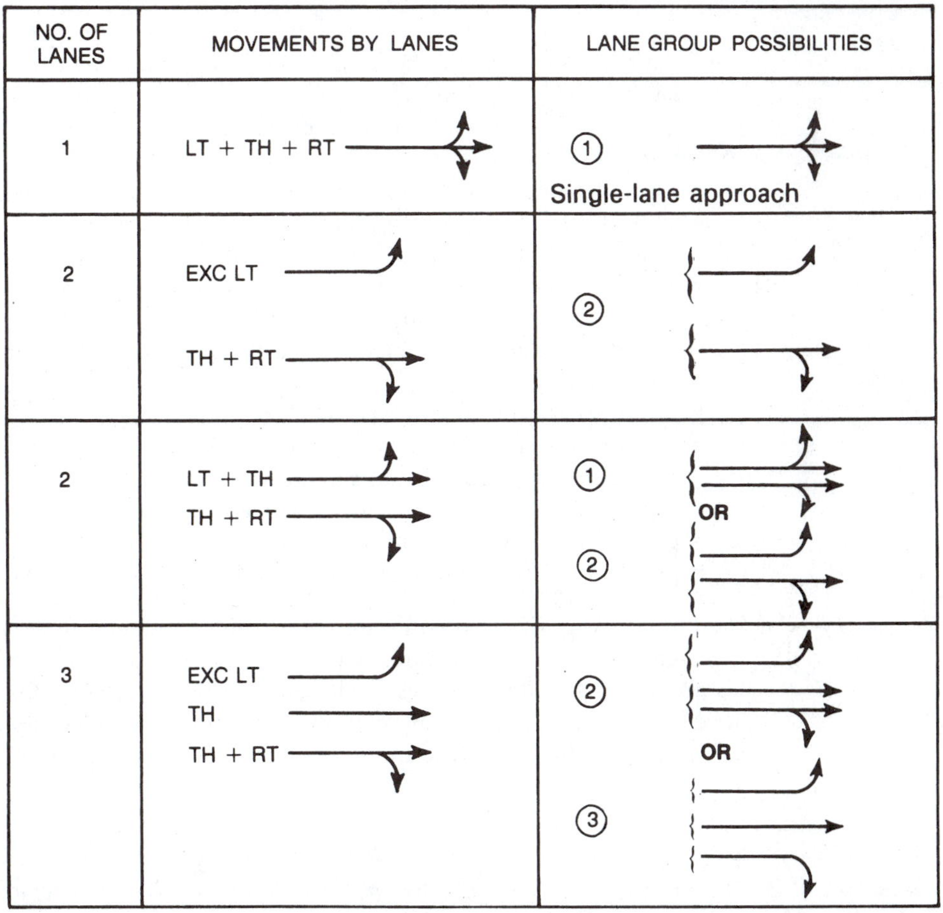

Figure 8.8 Typical Lane Groups for Analysis

SOURCE: *Highway Capacity Manual,* Special Report 209. Copyright 2000 by the Transportation Research Board, National Research Council, Washington, D.C. Used with permission.

The *saturation flow rate* is the flow rate in veh/h that the lane group can carry if it has the green indication continuously, (i.e, if $g/C = 1$). The saturation flow rate(s) depends on an ideal saturation flow (s_o), which is usually taken as 1900 veh/h of green time per lane. The ideal saturation flow is then adjusted for the prevailing conditions to obtain the saturation flow for the lane group being considered. The adjustment is made by introducing factors that correct for the number of lanes, lane width, the percent of heavy vehicles in the traffic, approach grade, parking activity, local buses stopping within the intersection, area type, lane utilization factor, and right and left turns. An equation given in the *Highway Capacity Manual* (HCM) and shown as Eq. 8.3 below can be used to compute the saturation flow rate.

$$s = s_o N f_w f_{HV} f_g f_p f_{bb} f_a f_{LU} f_{LT} f_{RT} f_{Lpb} f_{Rpb} \tag{8.3}$$

where

s = saturation flow rate for the subject lane group, expressed as a total for all lanes in the lane group, veh/h

s_o = base saturation flow rate per lane, pc/h/ln

N = number of lanes in the group

f_w = adjustment factor for lane width

f_{HV} = adjustment factor for heavy vehicles in the traffic stream

f_g = adjustment factor for approach grade

f_p = adjustment factor for the existence of a parking lane and parking activity adjacent to the lane group

f_{bb} = adjustment factor for the blocking effect of local buses that stop within the intersection area

f_a = adjustment factor for area type

f_{LU} = adjustment factor for lane utilization

f_{LT} = adjustment factor for left turns in the lane group

f_{RT} = adjustment factor for right turns in the lane group

f_{Lpb} = pedestrian adjustment factor for left-turn movements

f_{Rpb} = pedestrian/bicycle adjustment factor for right-turn movements

The HCM also gives a procedure for determining saturation flow rate using field measurements. It is recommended that when field measurements can be obtained, that procedure should be used, as results tend to be more accurate than values computed from the equation. An in-depth discussion of how the factors in Eq. 8.3 are obtained and details of the field measurement procedure are given in Chapter 10.

Example 8.3   Determining the Peak-Hour Factor and the Design Hourly Volume at an Intersection

The table below shows 15-minute volume counts during the peak hour on an approach of an intersection. Determine the PHF and the design hourly volume of the approach.

Time	Volume
6:00–6:15 p.m.	375
6:15–6:30 p.m.	380
6:30–6:45 p.m.	412
6:45–7:00 p.m.	390

Total volume during peak hour = (375 + 380 + 412 + 390) = 1557

Volume during peak 15 min = 412

Solution:

- Obtain peak-hour factor from Eq. 8.1:

$$PHF = \frac{\text{volume during peak hour}}{4 \times \text{volume during peak 15 min within peak hour}}$$

$$= \frac{1557}{4 \times 412}$$

$$= 0.945$$

- Obtain design hourly volume from Eq. 8.2:

$$DHV = \frac{\text{peak - hour volume}}{PHF}$$

$$= \frac{1557}{0.945}$$

$$= 1648$$

Objectives of Signal Timing

The main objectives of signal timing at an intersection are to reduce the average delay of all vehicles and the probability of crashes. These objectives are achieved by minimizing the possible conflict points when assigning the right of way to different traffic streams at different times. The objective of reducing delay, however, sometimes conflicts with that of crash reduction. This is because the number of distinct phases should be kept to a minimum to reduce average delay, whereas many more distinct phases may be required to separate all traffic streams from each other. When this situation exists, it is essential that engineering judgment be used to determine a compromise solution. In general, however, it is usual to adapt a two-phase system whenever possible, using the shortest practical cycle length that is consistent with the demand. At a complex intersection, though, it may be necessary to use a multiphase (three or more phases) system to achieve the main design objectives.

Signal Timing at Isolated Intersections

An isolated intersection is one in which the signal time is not coordinated with that of any other intersection and therefore operates independently. The cycle length for an intersection of this type should be short, preferably between 35 and 60 sec, although it may be necessary to use longer cycles when approach volumes are very high. However, cycle lengths should be kept below 120 sec, since very long cycle lengths will result in excessive delay. Several methods have been developed for determining the optimal cycle length at an intersection and, in most cases, the yellow interval is considered as a component of the green time. Before discussing two of these methods, we will discuss the basis for selecting the yellow interval at an intersection.

Yellow Interval

The main purpose of the yellow indication after the green is to alert motorists to the fact that the green light is about to change to red and to allow vehicles already in the intersection to cross it. A bad choice of yellow interval may lead to the creation of a *dilemma zone,* an area close to an intersection in which a vehicle can neither stop safely before the intersection nor clear the intersection without speeding before the red signal comes on. The required yellow interval is the time period that guarantees that an approaching vehicle can either stop safely or proceed through the intersection without speeding.

Figure 8.9 is a schematic of a dilemma zone. For the dilemma zone to be eliminated, the distance X_o should be equal to the distance X_c. Let τ_{min} be the yellow interval (sec) and let the distance traveled during the change interval without accelerating be $u_o(\tau_{min})$, with u_o = speed limit on approach (ft/sec). If the vehicle just clears the intersection, then

$$X_c = u_o(\tau_{min}) - (W + L)$$

where X_c is the distance within which a vehicle traveling at the speed limit (u_o) during the yellow interval τ_{min} cannot stop before encroaching on the intersection. Vehicles within

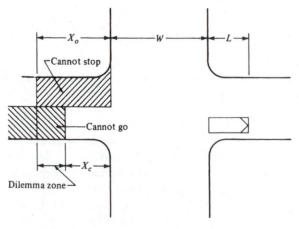

Figure 8.9 Schematic of a Dilemma Zone at an Intersection

this distance at the start of the yellow interval will therefore have to go through the intersection.

W = width of intersection (ft)
L = length of vehicle (ft)

For vehicles to be able to stop, however,

$$X_o = u_o \delta + \frac{u_o^2}{2a}$$

where
X_o = the minimum distance from the intersection for which a vehicle traveling at the speed limit u_o during the clearance interval Y_o cannot go through the intersection without accelerating; any vehicle at this distance or at a distance greater than this has to stop
δ = perception-reaction time (sec)
a = constant rate of braking deceleration (ft/sec^2)

For the dilemma zones to be eliminated, X_o must be equal to X_c. Accordingly,

$$u_o \tau_{min} - (W + L) = u_o \delta + \frac{u_o^2}{2a} \tag{8.4}$$

$$\tau_{min} = \delta + \frac{W + L}{u_o} + \frac{u_o}{2a}$$

If the effect of grade is added, Eq. 8.3 becomes

$$\tau_{min} = \delta + \frac{W + L}{u_o} + \frac{u_o}{2(a + Gg)} \tag{8.5}$$

where G is the grade of the approach, and g is the acceleration due to gravity (32.2 ft/sec^2).

Safety considerations, however, normally preclude yellow intervals of less than 3 sec, and to encourage motorists' respect for the yellow interval, it is usually not made longer than 5 sec. When longer yellow intervals are required as computed from Eq. 8.4 or Eq. 8.5, an all-red phase can be inserted to follow the yellow indication. The change interval, yellow plus all-red, must be at least the value computed from Eq. 8.5.

Example 8.4 Determining the Minimum Yellow Interval at an Intersection

Determine the minimum yellow interval at an intersection whose width is 40 ft if the maximum allowable speed on the approach roads is 30 mi/h. Assume average length of vehicle is 20 ft.

Solution: We must first decide on a deceleration rate. AAIHTO recommends a deceleration rate of 11.2 ft/sec^2. Assuming this value for a and taking δ as 1.0 sec, we obtain

$$\tau_{min} = 1.0 + \frac{40 + 20}{30 \times 1.47} + \frac{30 \times 1.47}{2 \times 11.2}$$

$$= 4.3 \text{ sec}$$

In this case, a yellow period of 4.5 sec will be needed.

Cycle Lengths of Fixed (Pretimed) Signals

The signals at isolated intersections can be pretimed (fixed), semiactuated, or fully actuated. Pretimed signals assign the right of way to different traffic streams in accordance with a preset timing program. Each signal has a preset cycle length that remains fixed for a specific period of the day or for the entire day. Several design methods have been developed to determine the optimum cycle length, two of which—the Webster and the HCM methods—are presented here.

Webster Method. Webster has shown that for a wide range of practical conditions, minimum intersection delay is obtained when the cycle length is obtained by the equation

$$C_o = \frac{1.5L + 5}{1 - \sum_{i=1}^{\phi} Y_i} \qquad (8.6)$$

where
C_o = optimum cycle length (sec)
L = total lost time per cycle (sec)
Y_i = maximum value of the ratios of approach flows to saturation flows for all lane groups using phase i, i.e., q_{ij}/Sj
ϕ = number of phases
q_{ij} = flow on lane groups having the right of way during phase i
S_j = saturation flow on lane group j

Total Lost Time. Figure 8.10 shows a graph of rate of discharge of vehicles at various times during a green phase of a signal cycle at an intersection. Initially, some time is lost before the vehicles start moving, and then the rate of discharge increases to a maximum. This maximum rate of discharge is the saturation flow. If there are sufficient vehicles in the queue to use the available green time, the maximum rate of discharge will be sustained until the yellow phase occurs. The rate of discharge will then fall to zero when the yellow signal changes to red. The number of vehicles that go through the intersection is represented by the area under the curve. Dividing the number of vehicles that go through the intersection by the saturation flow will give the effective green time, which is less than the

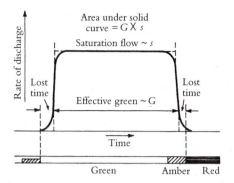

Figure 8.10   Discharge of Vehicles at Various Times During a Green Phase

sum of the green and yellow times. This difference is considered lost time, since it is not used by any other phase for the discharge of vehicles; it can be expressed as

$$\ell_i = G_{ai} + \tau_i - G_{ei} \tag{8.7}$$

where

ℓ_i = lost time for phase i
G_{ai} = actual green time for phase i (not including yellow time)
τ_i = yellow time for phase i
G_{ei} = effective green time for phase i

Total lost time is given as

$$L = \sum_{i=1}^{\phi} \ell_i + R \tag{8.8}$$

where R is the total all-red time during the cycle.

Allocation of Green Times.   In general, the total effective green time available per cycle is given by

$$G_{te} = C - L = C - \left(\sum_{i=1}^{\phi} \ell_i + R \right)$$

where

C = actual cycle length used (usually obtained by rounding off C_o to the nearest 5 sec)
G_{te} = total effective green time per cycle

To obtain minimum overall delay, the total effective green time should be distributed among the different phases in proportion to their Y values to obtain the effective green time for each phase.

$$G_{ei} = \frac{Y_i}{Y_1 + Y_2 + \cdots Y_\phi} G_{te} \qquad (8.9)$$

and the actual green time for each phase is obtained as

$$
\begin{aligned}
G_{a1} &= G_{e1} + \ell_1 - \tau_1 \\
G_{a2} &= G_{e2} + \ell_2 - \tau_2 \\
G_{ai} &= G_{ei} + \ell_i - \tau_i \\
G_{a\phi} &= G_{e\phi} + \ell_\phi - \tau_\phi
\end{aligned}
\qquad (8.10)
$$

Minimum Green Time. At an intersection where a significant number of pedestrians cross, it is necessary to provide a minimum green time that will allow the pedestrians to safely cross the intersection. The length of this minimum green time may be higher than that needed for vehicular traffic to go through the intersection. The green time allocated to the traffic moving in the north-south direction should, therefore, not be less than the green time required for pedestrians to cross the east-west approaches at the intersection. Similarly, the green time allocated to the traffic moving in the east-west direction cannot be less than that required for pedestrians to cross the north-south approaches. The minimum green time can be determined by using the HCM expressions given in the Eqs. 8.11 and 8.12.

$$G_p = 3.2 + \frac{L}{S_p} + \left[2.7\frac{N_{ped}}{W_E}\right] \qquad \text{for } W_E > 10 \text{ ft} \qquad (8.11)$$

$$G_p = 3.2 + \frac{L}{S_p} + (0.27N_{ped}) \qquad \text{for } W_E \leq 10 \text{ ft} \qquad (8.12)$$

where
   G_p = minimum green time (sec)
   L = crosswalk length (ft)
   S_p = average speed of pedestrians, usually taken as 4 ft/sec (assumed to represent 15th percentile pedestrian walking speed)
   3.2 = pedestrian start up time
   W_E = effective crosswalk width
   N_{ped} = number of pedestrians crossing during an interval

Example 8.5   Signal Timing Using the Webster Method

Figure 8.11a shows peak-hour volumes for a major intersection on an expressway. Using the Webster method, determine a suitable signal timing for the intersection using the four-phase system shown below. Use a yellow interval of 3 sec and the saturation flow given.

Phase	Lane Group	Saturation Flow
A	①	1615
	②	3700
B	①	3700
	②	1615
C	①	1615
	②	3700
D	①	1615
	②	3700

Note: The influence of heavy vehicles and turning movements and all other factors that affect the saturation flow have already been considered.

Solution: Determine equivalent hourly flows, by dividing the peak hour volumes by the PHF (e.g., for left turn lane group of phase A, equivalent hourly flow = 222/0.95 = 234). See Figure 8.11b for all equivalent hourly flows.

Phase (ϕ)	Critical Lane Volume (veh/h)
A	488
B	338
C	115
D	519
	$\sum$ 1450

- Compute the total lost time using Eq. 8.7. Since there is not an all-red phase—that is, $R = 0$—and there are four phases,

$$L = \sum \ell_i = 4 \times 3.5 = 14 \text{ sec (assuming lost time per phase is 3.5 sec)}$$

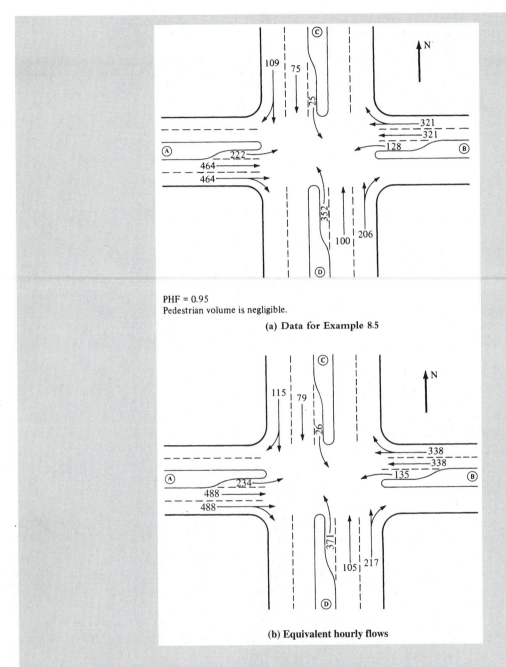

PHF = 0.95
Pedestrian volume is negligible.

(a) Data for Example 8.5

(b) Equivalent hourly flows

Figure 8.11   Signal Timing Using the Webster Method

- Determine Y_i and $\sum Y_i$:

Lane Group	Phase A (EB)		Phase B (WB)		Phase C (SB)		Phase D (NB)	
	1	2	1	2	1	2	1	2
q_{ij}	234	976	676	135	26	194	371	322
S_j	1615	3700	3700	1615	1615	3700	1615	3700
Q_{ij}/S_j	0.145	0.264	0.183	0.084	0.016	0.052	0.230	0.087
Y_i	0.264		0.183		0.052		0.230	

$\sum Y_I = (0.264 + 0.183 + 0.052 + 0.230) = 0.729$

- Determine the optimum cycle length using Eq. 8.6:

$$C_o = \frac{1.5\,L + 5}{1 - \displaystyle\sum_{i=1}^{\phi} Y_i}$$

$$= \frac{(1.5 \times 14) + 5}{1 - 0.729}$$

$$= 95.9 \text{ sec}$$

Use 100, sec as cycle length are usually multiples of 5 or 10 sec.
- Find the total effective green time:

$$G_{te} = C - L$$

$$= (100 - 14)$$

$$= 86 \text{ sec}$$

- Effective time for phase i is obtained from Eq. 8.9:

$$G_{ei} = \frac{Y_i}{Y_1 + Y_2 + \cdots + Y_n}\,G_{te}$$

$$= \frac{Y_i}{0.25 + 0.17 + 0.06 + 0.26}\,86$$

$$= \frac{Y_i}{0.729}\,86$$

Yellow time $\tau = 3.0$ sec; the actual green time G_{ai} for each phase is obtained from Eq. 8.10 as

$$G_{ai} = G_{ei} + \ell_i - 3.0$$

- Actual green time for Phase A $(G_{aA}) = \dfrac{0.264}{0.729} \times 86 + 3.5 - 3.0$

  ≈ 32 sec

- Actual green time for Phase B $(G_{aB}) = \dfrac{0.183}{0.729} \times 86 + 3.5 - 3.0$

  ≈ 22 sec

- Actual green time for Phase C $(G_{aC}) = \dfrac{0.052}{0.729} \times 86 + 3.5 - 3.0$

  ≈ 7 sec

- Actual green time for Phase D $(G_{aD}) = \dfrac{0.23}{0.729} \times 86 + 3.5 - 3.0$

  ≈ 27 sec

The Highway Capacity Method. This method is used to determine the cycle length based on the capacity (the maximum flow based on the available effective green time) of a lane group. Since the saturation flow rate is the maximum flow rate on an approach or lane group when 100 percent effective green time is available, the capacity of an approach or lane group depends on the percentage of the cycle length that is given to that approach or lane group.

The capacity of an approach or lane group is given as

$$c_i = s_i(g_i/C) \tag{8.13}$$

where

$\quad c_i$ = capacity of lane group i (veh/h)

$\quad s_i$ = saturation flow rate for lane group or approach i (veh/h of green, or veh/h/g)

(g_i/C) = green ratio for lane group or approach i

$\quad g_i$ = effective green for lane group i or approach i

$\quad C$ = cycle length

The ratio of flow to capacity (v/c) is usually referred to as the *degree of saturation* and can be expressed as

$$(v/c)_i = X_i = \dfrac{v_i}{s_i(g_i/C)} \tag{8.14}$$

where

$\quad X_i$ = (v/c) ratio for lane group or approach i

$\quad v_i$ = actual flow rate for lane group or approach i (veh/h)

$\quad s_i$ = saturation flow for lane group or approach i (veh/h)

$\quad g_i$ = effective green time for lane group i or approach i (sec)

It can be seen that when the flow rate equals capacity, X_i equals 1.00; when flow rate equals zero, X_i equals zero.

When the overall intersection is to be evaluated with respect to its geometry and the total cycle time, the concept of critical volume-to-capacity ratio (X_c) is used. The critical (v/c) ratio is usually obtained for the overall intersection but considers only the critical lane groups or approaches, which are those lane groups or approaches that have the maximum flow ratio (v/s) for each signal phase. For example, in a two-phase signalized intersection, if the north approach has a higher (v/s) ratio than the south approach, more time will be required for vehicles on the north approach to go through the intersection during the north-south green phase, and the phase length will be based on the green time requirements for the north approach. The north approach will therefore be the critical approach for the north-south phase. The critical v/c ratio for the whole intersection is given as

$$X_c = \sum_i (v/s)_{ci} \frac{C}{C-L} \tag{8.15}$$

where

X_c = critical v/c ratio for the intersection
$\sum_i(v/s)_{ci}$ = summation of the ratios of actual flows to saturation flow for all critical lanes, groups, or approaches
C = cycle length (sec)
L = total lost time per cycle computed as the sum of the lost time, (t_l), for each critical signal phase, $L = \sum_i t_l$

Equation 8.15 can be used to estimate the signal timing for the intersection if this is unknown and a critical (v/c) ratio is specified for the intersection. Alternatively, this equation can be used to obtain a broader indicator of the overall sufficiency of the intersection by substituting the maximum permitted cycle length for the jurisdiction and determining the resultant critical (v/c) ratio for the intersection. When the critical (v/c) ratio is less than 1.00, the cycle length provided is adequate for all critical movements to go through the intersection if the green time is proportionately distributed to the different phases. That is, for the assumed phase sequence, all movements in the intersection will be provided with adequate green times if the total green time is proportionately divided among all phases. If the total green time is not properly allocated to the different phases, it is possible to have a critical (v/c) ratio of less than 1.00, but with one or more individual oversaturated movements within a cycle.

Example 8.6   Determining Cycle Lengths from v/c Criteria

A four-phase signal system is to be designed for a major intersection in an urban area. The flow ratios are:

Phase A $(v/s)_A = 0.25$
Phase B $(v/s)_B = 0.25$
Phase C $(v/s)_C = 0.20$
Phase D $(v/s)_D = 0.15$

If the total lost time (L) is 14 secs/cycle, determine:

1. The shortest cycle length that will avoid oversaturation.
2. The cycle length if the desired critical v/c ratio (X_c) is 0.95.
3. The critical v/c ratio (X_c) if a cycle length of 90 secs is used.

Solution:

- The shortest cycle length that will avoid oversaturation is the cycle length corresponding to the critical (v/c) ratio $X_c = 1$. Determine C from Eq. 8.15 for $X_c = 1$.

$$X_C = \sum \left(\frac{v}{s}\right)_{ci} \frac{C}{C - L}$$

$$1 = (0.25 + 0.25 + 0.20 + 0.15)\frac{C}{C-14}$$

$$1 = 0.85\frac{C}{C - 14}$$

$$C - 14 = 0.85C$$

$$0.15C = 14$$

$$C = 93.3 \text{ sec, say 95 sec}$$

- Determine C if the desired critical (v/c) ratio X_c is 0.95.
  Use Eq. 8.15 for $X_c = 0.95$

$$0.95 = (0.25 + 0.25 + 0.20 + 0.15)\frac{C}{C - 14}$$

$$0.95 = 0.85\frac{C}{C - 14}$$

$$0.95C - 13.3 = 0.85C$$

$$0.95C - 0.85C = 13.3$$

$$0.1C = 13.3$$

$$C = 133 \text{ sec, say 135 sec}$$

- Determine X_c for a cycle length of 90 sec.
  Use Eq. 8.15.

$$X_c = \sum \left(\frac{v}{s}\right)_{ci} \frac{C}{C - L}$$

$$= 0.85\frac{90}{90 - 14}$$

$$= \frac{0.85 \times 90}{76}$$

$$= 1.01 \text{ (This will result in oversaturation)}$$

Example 8.7   Determining Cycle Length using the HCM method and Pedestrian Criteria

Repeat problem 8-5 using the HCM method with the following additional information:

The desired critical v/c ration is 0.85
Number of pedestrians crossing the east approach = 35 per interval in each direction
Number of pedestrians crossing the west approach = 25 per interval in each direction
Number of pedestrians crossing the north approach = 30 per interval in each direction
Number of pedestrians crossing the south approach = 30 per interval in each direction
Effective crosswalk width for each crosswalk = 9 ft
Crosswalk length in E-W direction = 40 ft
Crosswalk length in the N-S direction = 40 ft
Speed limit on each approach = 30 mi/h

Solution:

- Determine $\sum(v/s)$ for the critical lane groups.

  Phase A (EB), $(v/s)_c = 0.264$
  Phase B (WB), $(v/s)_c = 0.183$
  Phase C (SB), $(v/s)_c = 0.052$
  Phase D (NB), $(v/s)_c = 0.230$
  $\sum(v/s)_{ci} = 0.729$

- Determine cycle length for $X_c = 0.85$
  Use Eq. 8.15

$$X_c = 0.85 = \sum\left(\frac{v}{s}\right)\frac{C}{C - L}$$
$$= 0.729(C/(C - 14))$$
$$0.85(C - 14) = 0.729C$$
$$0.85C - 0.729C = 0.85 \times 14$$
$$C = 98.3 \text{ sec}$$
$$\text{Say } C = 100 \text{ sec}$$

- Find the minimum yellow interval by using Eq. 8.5 for the N-S and E-W movement.

$$\tau_{min} = 1.0 + \frac{40 + 20}{30 \times 1.47} + \frac{30 \times 1.47}{2 \times 0.27 \times 11.2}$$
$$= 1.0 + 1.36 + 1.97$$
$$= 4.33 \text{ sec}$$
$$\text{Say 5 sec}$$

- Determine actual green for each phase
  Allow a yellow interval of 5 sec for each phase
  In this case the total effective green time $(G_{te}) = C\text{-}L\text{-}R$
  In this case $R = 0$

$$G_{te} = 100 - 4 \times 3.5 = 86 \text{ sec}$$

Actual green time for (EB) Phase A $(G_{ta}) = (0.264/.729) \times 86 + 3.5 - 5 \approx 30 \text{ sec}$
Actual green time for (WB) Phase B $(G_{tb}) = (0.183/.729) \times 86 + 3.5 - 5 \approx 20 \text{ sec}$
Actual green time for (SB) Phase C $(G_{tc}) = (0.052/.729) \times 86 + 3.5 - 5 \approx 5 \text{ sec}$
Actual green time for (NB) Phase C $(G_{tD}) = (0.23/.729) \times 86 + 3.5 - 5 \approx 25 \text{ sec}$

- Check for minimum green times required for pedestrian crossing.
  Since $W_E < 10$ ft, use Eq. 8.12.

$$G_p = 3.2 + \frac{L}{S_p} + \left(0.27 N_{ped}\right)$$

Minimum time required for east approach $= 3.2 + 40/4 + 0.27 \times 35 = 22.65 \text{ sec}$
Minimum time required for west approach $= 3.2 + 40/4 + 0.27 \times 25 = 19.95 \text{ sec}$
Minimum time required for south approach $= 3.2 + 40/4 + 0.27 \times 30 = 21.30 \text{ sec}$
Minimum time required for north approach $= 3.2 + 40/4 + 0.27 \times 30 = 21.30 \text{ sec}$

Because of the phasing system used, the total time available to cross each approach is:

East-West approaches $= (5 + 25) \text{ sec} = 30 \text{ sec}$
North South approaches $= (30 + 20) \text{ sec} = 50 \text{ sec.}$

The minimum time requirements are therefore satisfied.

Determination of Left-Turn Treatment

Left turn vehicles at signalized intersections can proceed under one of three signal conditions: Permitted, Protected, and Protected/Permissive turning movements.

Permitted turning movements are those made within gaps of an opposing traffic stream or through a conflicting pedestrian flow. For example, when a right turn is made while pedestrians are crossing a conflicting crosswalk, the right turn is a permitted turning movement. Similarly, when a left turn is made between two consecutive vehicles of the opposing traffic stream, the left turn is a permitted turn. The suitability of permitted turns at a given intersection depends on the geometric characteristics of the intersection, the turning volume, and the opposing volume.

Protected turns are those turns protected from any conflicts with vehicles in an opposing stream or pedestrians on a conflicting crosswalk. A permitted turn takes more time than a similar protected turn and will use more of the available green time. *Protected/Permissive* is a combination of the protected and permissive conditions, in which vehicles are first allowed to make left turns under the protected condition and then allowed to make left turns under the permissive condition.

The determination of the specific treatment at a location depends on the transportation jurisdiction as guidelines vary from one jurisdiction to another. The HCM, however, suggests the following guidelines for providing protected left turn treatments:

- A protected left-turn phase should be provided when two or more left turn lanes are on the approach.
- A protected left-turn phase should be provided when the left turn unadjusted volume is higher than 240 veh/h.
- A protected left-turn phase should be provided when the cross product of the unadjusted left turn and the opposing main line volume exceeds the values given in Table 8.3. The opposing volume usually includes the right-turn volume, unless the analyst is confident that the geometry of the intersection is such that the left-turn vehicles can safely ignore the opposing right-turn vehicles.
- A protected left-turn phase should be provided if the left-turn equivalent factor is 3.5 or higher. The left-turn equivalent factor is used to convert left-turning vehicles to equivalent straight-through vehicles, because left-turning vehicles generally require a longer green time than straight-through vehicles. The determination of left-turn equivalent factors is discussed in Chapter 10.

When a protected left-turn phase is provided, an exclusive left-turn lane must also be provided. The length of this storage lane should be adequate for the turning volume so that the safety or capacity of the approach is not affected negatively. The length of the left-turn lane can be determined from Figure 8.12 and Table 8.4.

Phase Plans

The phase plan at a signalized intersection indicates the different phases used and the sequential order in which they are implemented. It is essential that an appropriate phase plan be used at an intersection as this facilitates the optimum use of the effective green time provided. The simplest phase plan is the two-phase plan shown in Figure 8.13. The higher the number of phases, the higher is the total lost time in a cycle. It is therefore

Table 8.3 Minimum Cross-Product Values for Recommending Left-Turn Protection

Number of Through Lanes	*Minimum Cross-Product*
1	50,000
2	90,000
3	110,000

SOURCE: *Highway Capacity Manual,* Special Report 209, Transportation Research Board, National Research Council, Washington, D.C., 2000.

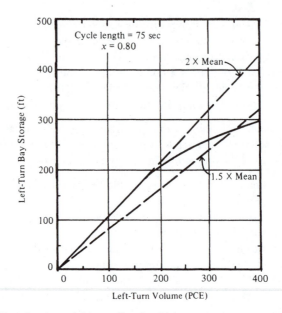

Figure 8.12 Left-Turn Bay Length Versus Turning Volume

Note: The values obtained from this figure are for a cycle length of 75 sec and a *v/c* ratio of 0.80. For other values, the length obtained from this figure is multiplied by a correction factor obtained from Table 8.4.

SOURCE: C.J. Messer, "Guidelines for Signalized Left-Turn Treatments," *Implementation Package,* FHWA-IP-81-4. Federal Highway Administration, Washington, D.C., 1981, Figure 2.

Table 8.4 Left-Turn Bay Length Adjustment Factors

v/c Ratio, X	Cycle Length, C (sec)				
	60	70	80	90	100
0.50	0.70	0.76	0.84	0.89	0.94
0.55	0.71	0.77	0.85	0.90	0.95
0.60	0.73	0.79	0.87	0.92	0.97
0.65	0.75	0.81	0.89	0.94	1.00
0.70	0.77	0.84	0.92	0.98	1.03
0.75	0.82	0.88	0.98	1.03	1.09
0.80	0.88	0.95	1.05	1.11	1.17
0.85	0.99	1.06	1.18	1.24	1.31
0.90	1.17	1.26	1.40	1.48	1.56
0.95	1.61	1.74	1.92	2.03	2.14

SOURCE: C.J. Messer, "Guidelines for Signalized Left-Turn Treatments," *Inplementation Package FHWA-IP-81-4,* Federal Highway Administration, Washington, D.C., 1981, Table 1.

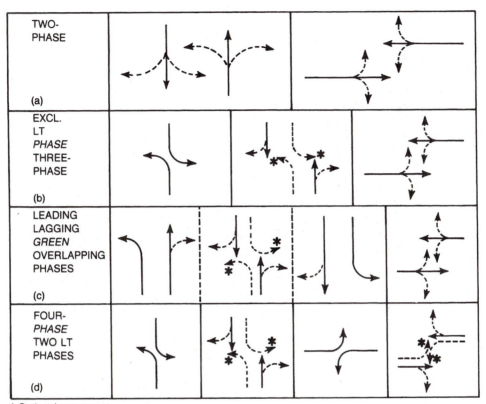

* Optional movement

Figure 8.13 Phase Plans for Pretimed and Actuated Signals

SOURCE: *Highway Capacity Manual,* Special Report 209, Third Edition, copyright 2000 by the Transportation Research Board, National Research Council, Washington, D.C. Used with permission.

recommended that the two-phase plan be used unless the traffic conditions at the intersection dictate otherwise. Figure 8.13 also shows examples of other common phase plans. The dashed lines indicate permitted turning movements, and the solid lines indicate protected movements.

Delay at Pretimed Signalized Intersections

One of the main objectives of installing a signal system at an intersection is to reduce the average delay of vehicles at the intersection. Delay is therefore an important measure of effectiveness to use in the evaluation of a signalized intersection. Delay at a signalized intersection can be estimated by using an expression developed by Webster and given in Eq. 8.16. It gives the average delay experienced per vehicle on the jth approach during the ith phase, assuming a uniform arrival of vehicles at the intersection.

$$d_j = \left(CA + \frac{B}{V_j} \right) \frac{100 - P}{100} \tag{8.16}$$

where

d_j = average delay per vehicle on jth approach during ith phase

$$A = \frac{(1-\lambda_i)^2}{2(1-\lambda_i x_j)} \quad \text{(see Table 8.5)}$$

$$B = \frac{x_j^2}{2(1-x_j)} \quad \text{(see Table 8.6)}$$

C = cycle length (sec)

V_j = actual volume on jth approach (vehicles/lane/sec)

λ_i = proportion of cycle length that is effectively green (that is, G_{ei}/C, where G_{ei} is effective green time for phase i)

x_j = degree of saturation for the jth approach = $V_j/\lambda_i s$;

s_j = saturation flow for the jth approach (vehicles/lane/sec)

P = percentage correction, ranging from 5 percent to 15 percent for normal conditions (see Table 8.7)

The total hourly delay for any approach can be estimated by determining the d_j for each lane on that approach, multiplying each d_j by the corresponding lane volumes, and then summing these. The total intersection hourly delay can then be determined by summing the total delay for each approach. Values for A, B, and P have been calculated and are shown in Tables 8.5, 8.6, and 8.7.

It should be noted that the *Highway Capacity Manual* gives an equation for average control delay that consists of three terms. The first term of that equation is based on the first term of the Webster delay model given in Eq. 8.16. It accounts for the delay that occurs when the arrival in the subject lane is uniformly distributed over time. That equation and its similarity to Eq. 8.16 are discussed in Chapter 10.

Example 8.8   Computing Delay Using the Webster Delay Model

Using the solution obtained for Example 8.5 and Eq. 8.16, determine the total hourly delay of vehicles in the left lane of the eastbound approach (Phase A).

$$d_j = \left(CA + \frac{B}{V_j} \right) \frac{100 - P}{100}$$

Solution:

- Use the following values for the left lane:

Table 8.5   *A* Values for Use in Webster Delay Model

x_j	λ_i												
	0.1	*0.2*	*0.3*	*0.35*	*0.40*	*0.45*	*0.50*	*0.55*	*0.60*	*0.65*	*0.70*	*0.80*	*0.90*
0.1	0.409	0.327	0.253	0.219	0.188	0.158	0.132	0.107	0.085	0.066	0.048	0.022	0.005
0.2	0.413	0.333	0.261	0.227	0.196	0.166	0.139	0.114	0.091	0.070	0.052	0.024	0.006
0.3	0.418	0.340	0.269	0.236	0.205	0.175	0.147	0.121	0.098	0.076	0.057	0.026	0.007
0.4	0.422	0.348	0.278	0.246	0.214	0.184	0.156	0.130	0.105	0.083	0.063	0.029	0.008
0.5	0.426	0.356	0.288	0.256	0.225	0.195	0.167	0.140	0.114	0.091	0.069	0.033	0.009
0.55	0.429	0.360	0.293	0.262	0.231	0.201	0.172	0.145	0.119	0.095	0.073	0.036	0.010
0.60	0.431	0.364	0.299	0.267	0.237	0.207	0.179	0.151	0.125	0.100	0.078	0.038	0.011
0.65	0.433	0.368	0.304	0.273	0.243	0.214	0.185	0.158	0.131	0.106	0.083	0.042	0.012
0.70	0.435	0.372	0.310	0.280	0.250	0.221	0.192	0.165	0.138	0.112	0.088	0.045	0.014
0.75	0.438	0.376	0.316	0.286	0.257	0.228	0.200	0.172	0.145	0.120	0.095	0.050	0.015
0.80	0.440	0.381	0.322	0.293	0.265	0.236	0.208	0.181	0.154	0.128	0.102	0.056	0.018
0.85	0.443	0.386	0.329	0.301	0.273	0.245	0.217	0.190	0.163	0.137	0.111	0.063	0.021
0.90	0.445	0.390	0.336	0.308	0.281	0.254	0.227	0.200	0.174	0.148	0.122	0.071	0.026
0.92	0.446	0.392	0.338	0.312	0.285	0.258	0.231	0.205	0.179	0.152	0.126	0.076	0.029
0.94	0.447	0.394	0.341	0.315	0.288	0.262	0.236	0.210	0.183	0.157	0.132	0.081	0.032
0.96	0.448	0.396	0.344	0.318	0.292	0.266	0.240	0.215	0.189	0.163	0.137	0.086	0.037
0.98	0.449	0.398	0.347	0.322	0.296	0.271	0.245	0.220	0.194	0.169	0.143	0.093	0.042

Note: Values of *A* calculated from

$$A = \frac{(1 - \lambda_i)^2}{2(1 - \lambda_i x_j)}$$

SOURCE: Adapted from F. V. Webster and B. M. Cobbe, *Traffic Signals,* Road Research Technical Paper No. 39, Her Majesty's Stationery Office, London, 1958.

Table 8.6   *B* Values for Use in Webster Method

x_i	*0.00*	*0.01*	*0.02*	*0.03*	*0.04*	*0.05*	*0.06*	*0.07*	*0.08*	*0.09*
0.1	0.006	0.007	0.008	0.010	0.011	0.013	0.015	0.017	0.020	0.022
0.2	0.025	0.028	0.031	0.034	0.038	0.042	0.046	0.050	0.054	0.059
0.3	0.064	0.070	0.075	0.081	0.088	0.094	0.101	0.109	0.116	0.125
0.4	0.133	0.142	0.152	0.162	0.173	0.184	0.196	0.208	0.222	0.235
0.5	0.250	0.265	0.282	0.299	0.317	0.336	0.356	0.378	0.400	0.425
0.6	0.450	0.477	0.506	0.536	0.569	0.604	0.641	0.680	0.723	0.768
0.7	0.817	0.869	0.926	0.987	1.05	1.13	1.20	1.29	1.38	1.49
0.8	1.60	1.73	1.87	2.03	2.21	2.41	2.64	2.91	3.23	3.60
0.9	4.05	4.60	5.28	6.18	7.36	9.03	11.5	15.7	24.0	49.0

Note: Values of *B* calculated from

$$B = \frac{x_j^2}{2(1 - x_j)}$$

SOURCE: Adapted from F. V. Webster and B. M. Cobbe, *Traffic Signals,* Road Research Technical Paper No. 39, Her Majesty's Stationery Office, London, 1958.

Table 8.7 *P* Values for Use in Webster Model

x_j	λ_i	M_i 2.5	5	10	20	40	x_i	λ_i	M_i 2.5	5	10	20	40
0.3	0.2	2	2	1	1	0	0.8	0.2	18	17	13	10	7
	0.4	2	1	1	0	0		0.4	16	15	13	10	9
	0.6	0	0	0	0	0		0.6	15	15	14	12	9
	0.8	0	0	0	0	0		0.8	14	15	17	17	15
0.4	0.2	6	4	3	2	1	0.9	0.2	13	14	13	11	8
	0.4	3	2	2	1	1		0.4	12	13	13	11	9
	0.6	2	2	1	1	0		0.6	12	13	14	14	12
	0.8	2	1	1	1	1		0.8	13	13	16	17	17
0.5	0.2	10	7	5	3	2	0.95	0.2	8	9	9	9	8
	0.4	6	5	4	2	1		0.4	7	9	9	10	9
	0.6	6	4	3	2	2		0.6	7	9	10	11	10
	0.8	3	4	3	3	2		0.8	7	9	10	12	13
0.6	0.2	14	11	8	5	3	0.975	0.2	8	9	10	9	8
	0.4	11	9	7	4	3		0.4	8	9	10	10	9
	0.6	9	8	6	5	3		0.6	8	9	11	12	11
	0.8	7	8	8	7	5		0.8	8	10	12	13	14
0.7	0.2	18	14	11	7	5							
	0.4	15	13	10	7	5							
	0.6	13	12	10	8	6							
	0.8	11	12	13	12	10							

Note: M_j is the average actual flow per lane per cycle for the *j*th approach.

$$M_j = V_j C$$

SOURCE: Adapted from F. V. Webster and B. M. Cobbe, *Traffic Signals*, Road Research Technical Paper No. 39, Her Majesty's Stationery Office, London, 1958.

$$G_{e_A} = \frac{0.264}{0.729} \times 86 = 31.14 \text{ sec}$$

$$V_A = 234 \text{ veh/h} = 0.065 \text{ veh/sec} = 0.093 \text{ pc/sec}$$

$$s = 1615 \text{ veh/h} = 0.449 \text{ veh/h}$$

$$\lambda_A = \frac{31.14}{100} = 0.311$$

$$x_A = \frac{0.065}{0.311 \times 0.449} = 0.465$$

- Determine *A* and *B* either by using Tables 8.5 and 8.6, respectively, or by computing them from the appropriate equations. Using the appropriate equations,

$$A = \frac{(1 - \lambda_i)^2}{2(1 - \lambda_i x_i)} = \frac{(1 - 0.311)^2}{2[1 - (0.311)(0.465)]} = 0.278$$

$$B = \frac{x_1^2}{2(1 - x_j)} = \frac{(0.465)^2}{2(1 - 0.465)} = 0.202$$

- Find M in order to compute P:

$$M_j = V_j C = 0.065 \times 100 = 6.5$$

From Table 8.7, we obtain P 5 by interpolating and rounding.
- Use Eq. 8.17:

$$d_j = \left(CA + \frac{B}{v_j} \right) \frac{100 - P}{100}$$

$$d_1 = \left(100 \times 0.278 + \frac{0.202}{0.065} \right) \frac{100 - 5}{100}$$

$$= 29.36 \text{ sec/vehicle}$$

- Calculate the total hourly delay in the left lane:

$d_1 = d_{T1} = d_1 \times$ hourly volume in the left lane
$= 29.36 \times 222$ sec (note that hourly volume is given in Figure 8.11)
$= 1.81$ hr

Cycle Lengths of Actuated Traffic Signals

A major disadvantage of fixed or pretimed signals is that they cannot adjust themselves to handle fluctuating volumes. When the fluctuation of traffic volumes warrants it, a vehicle-actuated signal is used. These signals are capable of adjusting themselves. When such a signal is used, vehicles arriving at the intersection are registered by detectors, which transmit this information to a controller. The controller then adjusts the phase lengths to meet the requirements of the prevailing traffic condition.

The following terms are associated with actuated signals:

The *demand* is a request for the right of way by a traffic stream through the controller.
The *initial portion* is the first portion of the green phase that an actuated controller has timed out, for vehicles waiting between the detector and the stop line during the red phase.
The *minimum period* is the shortest time that should be provided for a green interval during any traffic phase.

The *extendible portion* is the portion of the green phase that follows the initial portion, to allow for more vehicles arriving between the detector and the stop line during the green phase to go through the intersection.

The *extension limit* is the maximum additional time that can be given to the extendible portion of a phase after actuation on another phase.

The *unit extension* is the minimum time by which a green phase could be increased during the extendable portion after an actuation on that phase. However, the total extendable portion should not exceed the extension limit.

Semiactuated Signals.   Actuated signals can be either semiactuated or fully actuated. A semiactuated signal uses detectors only in the minor stream flow. Such a system can be installed even when the minor-stream volume does not satisfy the volume requirements for signalization. The operation of the semiactuated signal is based on the ability of the controllers to vary the lengths of the different phases to meet the demand on the minor approach. The signals are set as follows:

1. The green signal on the major approach is preset for a minimum period, but it will stay on until the signal is actuated by a minor-stream vehicle.
2. If the green signal on the major approach has been on for a period equal to or greater than the preset minimum, the signal will change to red in response to the actuation of the minor-street vehicle.
3. The green signal on the minor stream will then come on for at least a period equal to the preset minimum for this stream. This minimum is given an extendible green for each vehicle arriving, up to a preset extension limit.
4. The signal on the minor stream then changes to red, and that on the major stream changes to green.

Note that when the volume is high on the minor stream, the signal acts as a pretimed one.

The operation of a semiactuated signal requires certain times to be set for both the minor and major streams. For the minor streams, times should be set for the initial portion, unit extension, maximum green (sum of initial portion and extension limit), and change intervals. For the major streams, times should be set for the minimum green and change intervals. When pedestrian actuators are installed, it is also necessary to set a time for pedestrian clearance. Several factors should be taken into consideration when setting these times. The major factor, however, is that a semiactuated signal works as a pretimed signal during peak periods. It is therefore important that the time set for the maximum green in the minor stream be adequate to meet the demand during the peak period. Similarly, the time set for the minimum green on the major approach should be adequate to provide for the movement through the intersection of the expected number of vehicles waiting between the stop line and the detector whenever the signals change to green during the peak period. However, these settings should not be so large that the resulting cycle length becomes undesirable. In general, the procedures described below can be used to obtain some indication of the required lengths of the different set times.

- **Unit Extension.**   The unit extension depends on the average speed of the approaching vehicles and the distance between the detectors and the stop line. The unit extension should be at least the time it takes a vehicle moving at the average speed to travel from the location of the detectors to the stop line. Therefore,

$$\text{unit extension} = \frac{X}{1.47u} \ (\text{sec}) \qquad (8.17)$$

where

u = average speed (mi/h)

X = distance between detectors and stop line (ft)

This time will allow a vehicle detected at the end of the initial portion to arrive at the stop line just as the signal is changing to yellow and to clear the intersection during the change interval.

However, if the desire is to provide a unit extension time that will also allow the vehicle to clear the intersection, then

$$\text{unit extension time} = \frac{X + W + L}{1.47u} \ (\text{sec}) \qquad (8.18)$$

where

W = width of the cross street (ft)

L = length of the vehicle (ft)

- **Initial Portion.**   This time should be adequate to allow vehicles waiting between the stop line and the detector during the red phase to clear the intersection. This time depends on the number of vehicles waiting, the average headway, and the starting delay. The time for the initial portion can be obtained as

$$\text{initial portion} = (hn + K_1)$$

where

h = average headway (sec)

n = number of vehicles waiting between the detectors and the stop line

K_1 = starting delay (sec)

Suitable values for h and K_1 are 2 sec and 3.5 sec, respectively.

- **Minimum Green.**   This is the sum of the initial portion and the unit extension.

Note that alternative extensions for computing those green times are also given in the HCM.

Fully Actuated Signals.   Fully actuated signals are suitable for intersections at which large fluctuations of traffic volumes occur on all approaches during the day. Maximum and minimum green times are set for each approach. The basic operation of the fully actuated signal can be described using an intersection with four approaches and a two-phase signal system. Let phase A be assigned to the north–south direction and phase B to the east–west direction. If phase A is given the right of way, the green signal will stay on until the minimum green time ends and an approaching vehicle in the east–west direction actuates one of the detectors for phase B. A demand for the right of way in the east–west direction is then registered. If there is no traffic in the north–south direction, the red signal will come on for phase A and the right of way will be given to vehicles in the east–west direction; that is, the green indicator will come on for phase B. This right of way will be held by phase B until at least the minimum green time expires. At the expiration of the minimum green time, the right of way will be given to phase A—that is, the north–south direction—only if during the period of the minimum green a demand is registered by an approaching vehicle in this direction. If no demand is registered in the north–south direction and vehicles continue to arrive in the east–west direction, the right of way will continue to be given to phase B until the maximum green is reached. At this time the right of way is given to phase A, and so on. A procedure similar to that described for semiactuated signals may be used to determine the lengths of the different set times.

Signal Timing of Arterial Routes

In urban areas where two or more intersections are adjacent to each other, the signals should be timed so that when a queue of vehicles is released by receiving the right of way at an intersection, these vehicles will also have the right of way at the adjacent intersections. This coordination will reduce the delay experienced by vehicles on the arterial. To obtain this coordination, all intersections in the system must have the same cycle length. In rare instances, however, some intersections in the system may have cycle lengths equal to half or twice the common cycle length. It is usual for a common cycle length to be set, with an offset (time lapse between the start of the green phases of two adjacent intersections on an arterial) that is suitable for the main street. Traffic conditions at a given intersection are used to determine the appropriate phases of green, red, and yellow times for that intersection. The methods used to achieve the required coordination are the simultaneous system, the alternate system, and the progressive system. The speed of progression is important in determining the cycle length for each of these methods.

The speed of progression is the speed at which a platoon of vehicles released at an intersection will proceed along the arterial. It is usually taken as the mean operating speed of vehicles on the arterial for the specific time of day being considered. This speed is represented by the ratio of the distance between the traffic signals and the corresponding travel time.

Simultaneous System

In a simultaneous system, all signals along a given arterial have the same cycle length and have the green phase showing at the same time. When given the right of way, all vehicles move at the same time along the arterial and stop at the nearest signalized intersection when the right of way is given to the side streets. A simple approximate mathematical relationship for this system is

$$u = \frac{X}{1.47C} \tag{8.19}$$

where
X = average spacing for signals (ft)
u = progression speed (mi/h)
C = cycle length (sec)

Alternate System

With the alternate system, intersections on the arterial are formed into groups of one or more adjacent intersections. The signals are then set such that successive groups of signals are given the right of way alternately. This system is known as the single-alternate when the groups are made up of individual signals—that is, when each signal alternates with those immediately adjacent to it. It is known as the double-alternate when the groups are made up of two adjacent signals, and so on. The speed of progression in an alternate system is given as

$$u = \frac{nX}{1.47C} \tag{8.20}$$

where
  X = average spacing for signals (ft)
  n = 2 for the simple-alternate system
  n = 4 for the double-alternate system
  n = 6 for the triple-alternate system
  u = progression speed (mi/h)
  C = cycle length (sec)

An alternate system is most effective when the intersections are at equal distances from each other.

Example 8.9   Choosing an Appropriate Alternate System for an Urban Arterial

The traffic signals on an urban arterial are to be coordinated to facilitate the flow of traffic. The intersections are spaced at approximately 500-ft intervals, with at least one intersection being part of another coordinated system, whose cycle length is 60 sec. Determine whether a single-, double-, or triple-alternate system is preferable for this arterial if the mean velocity on the arterial is 35 mi/h.

Solution: Since there is one intersection in another coordinated system, it is necessary to use the cycle length at that intersection—that is, $C = 60$ sec. We also want to use Eq. 8.20.

- Find the mean speed for the single-alternate system:

$$u = \frac{nX}{1.47C}$$

$$u = \frac{2 \times 500}{1.47 \times 60} = 11.34 \text{ mi/h}$$

- Find the mean speed for the double-alternate system:

$$u = \frac{4 \times 500}{1.47 \times 60} = 22.68 \text{ mi/h}$$

- Find the mean speed for the triple-alternate system:

$$u = \frac{6 \times 500}{1.47 \times 60} = 34.01 \text{ mi/h}$$

Since the mean speed is currently 35 mi/h, the triple-alternate system is preferable, because this requires a progressive speed that is approximately equal to the existing mean speed.

Progressive System

The progressive system provides for a continuous flow of traffic through all intersections under the system when traffic moves at the speed of progression. The same cycle length is used for all intersections, but the green indication for each succeeding intersection is offset by a given time from that of the preceding intersection, depending on the distance from the preceding intersection and the speed of progression for that section of the street. When the offset and cycle length are fixed, the system is known as the limited or simple progressive system; when the offset and cycle length can be changed to meet the demands of fluctuating traffic at different times of the day, it is known as the flexible progressive system.

Design of Progressive Signal System.   The design of a progressive signal system involves the selection of the best cycle length, using the criterion that the speed of progression is approximately equal to the mean operating speed of the vehicles on the arterial street. This selection is accomplished by a trial-and-error procedure. Equation 8.19 can be used to obtain a suitable cycle length by using the mean operating speed of the arterial for u and the measured distance between intersections as X. In addition, the required cycle length for each intersection should be computed and compared with that obtained from Eq. 8.19. If this cycle length is approximately equal to those obtained for the majority of the intersections, it can be selected on a trial basis. However, it is the usual practice to use cycle lengths that have been established for intersecting or adjacent systems as guides for selecting a suitable cycle length. The actual design of the progressive system is normally conducted by using one of several available computer software packages. The first step in any method used is the collection of adequate data on traffic volumes, intersection spacings, speed limits, on-street parking, operating speed, and street geometrics.

Computer programs have been developed to cope with several problems associated with progressive signal timing, such as large variations in distances between intersections, differences in speeds between traffic streams in opposite directions, variable speed patterns in different sections of the system, and the requirement of a high level of computational effort. The use of computers to reduce the computational effort and increase analysis flexibility has made the design of signalized arterial systems less taxing.

With the advent of microcomputers, several programs have been developed that can be used to design an arterial signalized system. One such program, Arterial Analysis Package/Microcomputer (AAP/M), was developed by the University of Florida Transportation Research Center for Tampa's Division of Traffic Engineering. This program can be used as a computerized tool to design and evaluate signal timing for arterial traffic control systems. Three existing traffic signal timing programs are combined in this single package: SOAP, which deals with individual intersections; PASSER II, which is used to optimize arterial progression; and TRANSYT 7F, which optimizes the stops and delay performance of a coordinated traffic control system. Inputs for AAP/M include the common cycle length, the number of intersections, and the distances between successive intersections. Output includes the offset for each intersection as a percentage of the cycle length.

Other programs include an updated version of PASSER II (PASSER II-90), PASSER IV-94, and INTEGRATION. PASSER II-90 can be used to analyze signal operations with "Permitted," "Protected" and "Permitted/Protected" or "Protected/ Permitted" left-turn signal treatments. PASSER IV-94 can be used for timing signals in networks based on progression bandwidth optimization. It calculates green splits using the Webster method and optimizes cycle lengths and offsets. INTEGRATION is capable of modeling microscopic platoon dispersion using a macroscopic process.

FREEWAY RAMPS

Ramps are usually part of grade-separated intersections, where they serve as interconnecting roadways for traffic streams at different levels. They are also sometimes constructed between two parallel highways to allow vehicles to change from one highway to the other. Freeway ramps can be divided into two groups: entrance ramps, which allow the merging of vehicles into the freeway stream, and exit ramps, which allow vehicles to leave the freeway stream. When it becomes necessary to control the number of vehicles entering or leaving a freeway at a particular location, access to the entrance or exit ramp is controlled in one of several ways.

Freeway Entrance Ramp Control

The control of entrance ramps is essential to the efficient operation of freeways, particularly when volumes are high. The main objective in controlling entrance ramps is to regulate the number of vehicles entering the freeway so that the volume is kept lower than the capacity of the freeway. This will ensure that freeway traffic moves at a speed approximately equal to the optimum speed (which will result in maximum flow rates). The control of entrance ramps also allows a better level of service on the freeway and safer overall operation of both the freeway and the ramp. On the other hand, entrance-ramp control may result in long queues on the ramps, formed by vehicles waiting to join the freeway traffic stream, or to the diversion of traffic to local roads, which may result in serious congestion on those roads. It is therefore essential that the control of freeway entrance ramps be undertaken only when certain conditions are satisfied. The MUTCD provides general guidelines for the successful application of ramp control. The guidelines are mainly qualitative, because there are too many variables that affect the flow of traffic on freeways. It is therefore extremely difficult to develop numerical volume warrants that will be applicable to the wide variety of conditions found in practice.

Installation of entrance-ramp control signals may be justified when it will result in a reduction of the total expected delay to traffic in the freeway corridor, including freeway ramps and local streets, and at least one of the following conditions exists:

1. There is recurring congestion on the freeway due to traffic demand in excess of the capacity; or there is recurring congestion or a severe accident hazard at the freeway entrance because of an inadequate ramp merging area.
2. The signals are needed to accomplish transportation system management objectives identified locally for freeway traffic flow.
3. The signals are needed to reduce (predictable) sporadic congestion on isolated sections of freeway caused by short-period peak traffic loads from special events or from severe peak loads of recreational traffic.

Methods for Controlling Freeway Entrance Ramps

The common methods used in controlling freeway entrance ramps are

- Closure
- Simple metering
- Traffic response metering
- Integrated systems control

Closure. Closure entails the physical closure of the ramp by using "Do Not Enter" signs or by placing barriers at the entrance to the ramp. This is the simplest form of ramp control, but unfortunately it is the most restrictive. It should therefore be used only when absolutely necessary. Factors that suggest the use of closure include inadequacy of the storage area on the ramp for entering vehicles, improper design of the merging area connecting ramp traffic and freeway traffic, and volume of traffic on the freeway being at or approaching capacity. Experience has shown that the use of signs is relatively ineffective when compared to barriers. Barriers can be either manually placed or automated. Manually placed barriers are labor-intensive and therefore are not very efficient for closure at specific times of the day—for example, peak hours over a long period of time. Automatic barriers, however, provide the flexibility of opening and closing of the ramp in response to traffic conditions.

Simple Metering. This form of control consists of setting up a pretimed signal with extremely short cycles at the ramp entrance. The time settings are usually made for different times of the day and/or days of the week. Simple metering can be used to reduce the flow of traffic on the ramp from about 1200 veh/h, which is the normal capacity of a properly designed ramp, to about 250 veh/h. Figure 8.14 shows the layout of a typical simple metering system for an entrance ramp, including some optional features that can be added to the basic system. The basic system consists of a traffic signal, with a two-section (green–red) or three-section (green–yellow–red) indicator located on the ramp, a warning sign, which informs motorists that the ramp is being metered, and a controller, which is actuated by a time clock. The detectors shown are optional, but when used will enhance the efficient operation of the system. For example, the check-in detector allows the signal to change to green only when a vehicle is waiting, which means that the signal will stay red

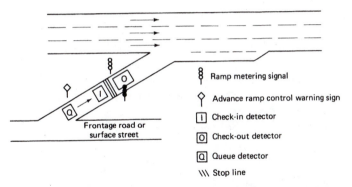

Figure 8.14 Layout of Pretimed Entrance Ramp Metering System

SOURCE: *Traffic Control Systems Handbook,* Federal Highway Administration, U.S. Department of Transportation. Washington, D.C., 1985.

until a vehicle is detected by the check-in detector and the minimum red time has elapsed. The check-out detector is useful when a single-entry system is desired. The green interval is terminated immediately when a vehicle is detected by the check-out detector.

The calculation of the metering rate depends on the primary objective of the control. When this objective is to reduce congestion on the highway, the difference between the upstream volume and the downstream capacity (maximum flow that can occur) is used as the metering rate. This is demonstrated in Figure 8.15, where the metering rate is 400 veh/h. It must be remembered, however, that the guidelines given by the MUTCD must be taken into consideration. For example, if the storage space on the ramp is not adequate to accommodate this volume, the signal should be pretimed for a metering rate that can be accommodated on the ramp. When the objective is to enhance safety in the merging area of the ramp and the freeway, the metering rate will be such that only one vehicle at a time is within the merging area. This will allow an individual vehicle to merge into the freeway traffic stream before the next vehicle reaches the merging area. The metering rate in this case will depend on the average time it takes a stopped vehicle to merge, which in turn depends on the ramp geometry and the probability of an acceptable gap occurring in the freeway stream. If it is estimated that it takes 9 sec to merge on the average, then the metering rate is 3600/9—that is, 400 veh/h.

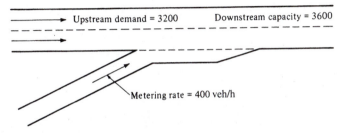

Figure 8.15 Relations Between Metering Rate, Upstream Demand, and Downstream Capacity

One of two methods of metering can be used in this system, depending on the magnitude of the metering rate. When the rate is below 900 veh/h, single-entry is used; platoon-entry is used for rates higher than 900 veh/h. Single-entry allows only one vehicle to merge into the freeway stream during the green interval; platoon-entry allows the release of two or more vehicles per cycle. Platoon-entry can be either parallel (two vehicles abreast of each other on two parallel lanes are released) or tandem (vehicles are released one behind the other). Care should be taken in designing the green interval for tandem platooning; it must be long enough to allow all the vehicles in the platoon to pass the signal.

Traffic Response Metering. This control system is based on the same principles as the simple metering system, but the traffic response system uses actual current information on traffic conditions to determine the metering rates, whereas historical data on traffic volumes are used to determine metering rates in the simple metering system. The traffic response system therefore has the advantage of being capable of responding to short-term changes in traffic conditions. Figure 8.16 shows the basic requirements for a traffic-responsive metering system, with some optional features that can be added to the system. The basic requirements include a traffic signal, detectors, a ramp control sign, and a controller that can monitor the variation of traffic conditions. The optional features include the queue detector, which when continuously actuated indicates that the vehicles queued on the ramp may interfere with traffic on the local road; a merging detector, which indicates whether a merging vehicle is still in the merging area; and a check-out detector, which indicates whether a vehicle uses the green interval to proceed to the merging area.

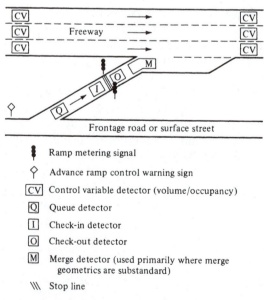

Figure 8.16 Layout of Traffic Responsive Ramp Metering System

SOURCE: *Traffic Control Systems Handbook,* Federal Highway Administration, U.S. Department of Transportation. Washington, D.C., 1985.

Two methods are used to determine the metering rate: (1) the demand capacity control and (2) the occupancy control.

Demand Capacity Control.   In this method, the actual upstream volume is measured at regular short intervals and is then compared with the downstream capacity, which may be preset or calculated using downstream traffic conditions. To ascertain whether the freeway is operating under congested or free-flow conditions, occupancy measurements are also made from at least one upstream detector.

Occupancy Control.   This method uses a predetermined relationship between occupancy rate and lane volume, developed from data previously collected at the freeway adjacent to the ramp being considered. An example of such a relationship is shown in Figure 8.17. This relationship also gives the occupancy rate at capacity. A metering rate is selected for the subsequent control period (usually 1 min), based on the occupancy rate measured during the current period either upstream or downstream of the ramp. The metering rate selected is the difference between the previously determined capacity of the freeway and the current volume. The single-entry metering system is normally used, except when metering rates are higher than 900 veh/h.

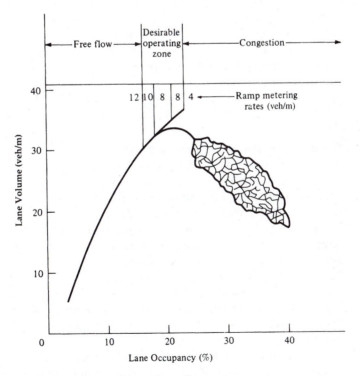

Figure 8.17 Plot of Lane Volume Versus Lane Occupancy

SOURCE: Redrawn from J. M. McDermott, et. al., *Chicago Area Expressway Surveillance and Control,* Final Report, State of Illinois, Department of Transportation, March 1979.

Integrated System Control. The philosophy used in this system is that it is more effective to integrate the control on several ramps rather than dealing with individual ramps independently. The metering rates are therefore based on consideration of the demand and the capacity of the entire stretch of the freeway being considered, rather than on the traffic considerations at individual ramps. A detailed description of the design of an integrated control system is beyond the scope of this book, but interested readers may consult the *Traffic Engineering Handbook* for further discussion of the subject. This system is usually applied to a group of ramps that are installed either with the traffic response or with pretimed metering devices. When the traffic response system is installed, linear programming is used to determine a set of integrated metering rates for each ramp, based on the expected range of capacity and demand conditions. The proper metering rate at each intersection is then selected from the appropriate set of metering rates, based on current traffic conditions on the freeway.

Freeway Exit Ramp Control

Control of exit ramps can be used to reduce the flow of traffic from the freeway to congested local streets, to reduce weaving when the distance between an entrance ramp and an exit ramp is short, and to reduce the volume of traffic on the freeway beyond the point of lane-drop by encouraging more traffic to leave the freeway before the lane-drop. Care should be taken in using the control of exit ramps because this can result in queuing of vehicles onto the freeway and increasing the risk of rear-end collisions. Control of exit ramps can be achieved either by using a metering system or by closing the ramp. Closure of exit ramps usually results in an increase of travel time for some motorists, a situation not likely to be appreciated by them.

SUMMARY

Control of traffic at highway intersections is of fundamental importance to traffic engineers seeking ways to achieve efficient operation of any highway system. Several methods of controlling traffic at intersections are presented in this chapter. These include yield signs, stop signs, multiway stop signs, channelization, and traffic signals. When traffic conditions warrant signalization, it may be an effective means of traffic control at intersections, since the number of conflict points at an intersection could be significantly reduced by signalization. Note, however, that an attempt to significantly reduce the number of conflict points will increase the number of phases required, which may result in increased delay. The two methods—Webster and HCM—commonly used to determine the cycle and phase lengths of signal systems are discussed, together with the methods for determining the important parameters for coordinated signals. Any of these methods could be used to obtain reasonable cycle lengths for different traffic conditions at intersections.

Ramp control can be used to limit the number of vehicles entering or leaving an expressway at an off- or on-ramp. However, care should be taken in using ramp control because this may result in longer travel times or in congestion on the local streets.

Although mathematical algorithms are presented for computing the important parameters required for signalization and ramp control, traffic engineers must always be aware that a good design is based both on correct mathematical computation and on engineering judgment.

PROBLEMS

8-1   Using an appropriate diagram, identify all the possible conflict points at an unsignalized T intersection.

8-2   A two-phase signal system is installed at the intersection described in Problem 8-1, with channelized left-turn lanes and shared through and right-turn lanes. Using a suitable diagram, determine the possible conflict points. Indicate the phasing system used.

8-3   Under what conditions would you recommend the use of each of the following intersection control devices at urban intersections: (a) yield sign, (b) stop sign, and (c) multiway stop sign?

8-4   Both crash rates and traffic volumes at an unsignalized urban intersection have steadily increased during the past few years. Briefly describe the types of data you will collect and how you will use those data to justify the installation of signal lights at the intersection.

8-5   For the geometric and traffic characteristics shown below, determine a suitable signal phasing system and phase lengths for the intersection using the Webster method. Show a detailed layout of the phasing system and the intersection geometry used.

	North Approach	South Approach	East Approach	West Approach
Approach Width	56 ft	56 ft	68 ft	68 ft
Peak-hour approach volume				
Left turn	133	73	168	134
Through movement	420	.373	563	516
Right turn	140	.135	169	178
PHF	0.95	0.95	0.95	0.95
Conflicting pedestrian volume	900 ped/hr	1200 ped/hr	1200 ped/hr	900 ped/hr

Assume the following saturation flows:

Through lanes 1600 veh/lane/hr
Through and right lanes 1400 veh/lane/hr
Left lanes 1000 veh/lane/hr
Left and through lanes 1200 veh/lane/hr
Left, through and right lanes 1100 veh/lane/hr

8-6   Repeat Problem 8-5 using saturation flow rates that are 10 percent higher. What effect does this have on the cycle length?

8-7   Repeat Problem 8-5 using pedestrian flows that are 20 percent higher. What effect does this have on the cycle length and the different phases?

8-8   Using the Webster delay model, determine the total hourly delay for each of the solutions of Problems 8-5, 8-6, and 8-7.

8-9   Repeat problem 8-5 using the HCM method and a critical v/c of 0.9.

8-10   Repeat problem 8-7 using HCM method and a critical v/c of 0.9.

8-11   Using the results for problems 8-5 and 8-9, compare the two different approaches used for computing the cycle length.

8-12   Briefly describe the different ways the signal lights at the intersection of an arterial route could be coordinated, stating under what conditions you would use each of them.

8-13   Briefly discuss the different methods by which freeway entrance ramps can be controlled. Clearly indicate the advantages and/or disadvantages of each method, and give the conditions under which each of them can be used.

REFERENCES

Manual on Uniform Traffic Control Devices, U.S. Department of Transportation, Federal Highway Administration, Washington, D.C., 1988.

Intersection Channelization Design Guide, National Cooperative Highway Research Program Report 279, National Research Council, Transportation Research Board, Washington, D.C., November 1985.

Highway Capacity Manual, Special Report 209, 3rd ed., Transportation Research Board, National Research Council, Washington, D.C., 2000.

Webster, F.V., and B.M. Cobbe, *Traffic Signals,* Road Research Technical Paper No. 39, Her Majesty's Stationery Office, London, 1958.

Webster, F.V., and B.M. Cobbe, *Traffic Signals,* Road Research Laboratory, Ministry of Transport Research Technical Paper 56, Her Majesty's Stationery Office, London, 1966.

Courage, K.G., C.E. Wallace, and D.P. Reaves, *Arterial Analysis Package, AAP Users Manual,* Transportation Research Center, University of Florida, Gainesville, Fla., 1986.

Messer, C.J., *Guidelines for signalized left-turn treatments,* Implementation Package, Federal Highway Administration, Washington D.C., 1981.

Chang, Edmond C.P., and Carroll J. Messer, "Arterial Signal Timing Optimization Using PASSER II-90," Texas Transportation Institute, Texas A & M University System, College Station, Tex., June 1991.

Chaudhavy, Nadeem A., and Carroll J. Messer, "PASSER IV-94 Version 1.0, User Reference Manual," Texas Transportation Institute, Texas A & M University System, College Station, Tex., December 1993.

Bacon, Pinton W., Jr., and David J. Lowell, "Use of INTEGRATION Model to Study HOV Facilities," *Transportation Research Record,* No. 1446, Transportation Research Board, Washington, D.C., 1994.

Traffic Engineering Handbook, 5th edition, James L. Pline, ed., Institute of Transportation Engineering, Washington, D.C., 1999.

CHAPTER 9

Capacity and Level of Service of Two-Lane Highways, Multilane Highways, and Freeway Segments

In Chapter 6 we used the fundamental diagram of traffic flow to show the relationship between flow and density. It was shown that traffic flows reasonably well when the flow rate is less than at capacity, but excessive delay and congestion can occur when the flow rate is at or near capacity. This phenomenon is a primary consideration in the planning and design of highway facilities because a main objective is to design or plan facilities that will operate at flow rates below their optimum rates. This objective can be achieved only if a good estimate of the optimum flow of a facility can be made. Capacity analysis therefore involves the quantitative evaluation of the capability of a road section to carry traffic. It uses a set of procedures to determine the maximum flow of traffic that a given section of highway will carry under prevailing roadway traffic and control conditions.

It was also shown in Chapter 6 that the maximum speed that can be achieved on a uniform section of highway is the mean free speed, which depends solely on the physical characteristics of the highway. This speed can be achieved only when traffic demand volume tends to zero and there is no interaction between any two vehicles on the highway segment. A driver is then free to drive at his or her desired speed up to the mean free speed. Under these conditions, motorists have a very high perception of the quality of flow on the highway. As the demand volume increases, vehicle interaction and density increases, resulting in the gradual lowering of the speed that can be safely achieved by the drivers. As the interaction among vehicles increases, the drivers are increasingly influenced by the action of other drivers, and individual drivers find it more difficult to achieve their desired speeds. Motorists therefore perceive a lowering of the quality of flow as density on the highway increases. In other words, for a given capacity, the level of operating performance—that is, the quality of flow—changes with the traffic density on the highway. The level of operating performance is indicated by the concept of level of service, which uses

329

qualitative measures that characterize both operational conditions within a traffic stream and motorists' and passengers' perception of them. This chapter presents procedures for determining the level of service on two-lane highways, multilane highways, and basic freeway segments.

TWO-LANE HIGHWAYS

The procedures developed for two-lane highways provide for evaluating level of service and capacity at two levels of analysis: (1) operational and (2) planning applications. Two classes of two-lane highways are analyzed. They are defined according to their function in the following manner:

- **Class I.** Two-lane highways that function as primary arterials, daily commuter routes, and links to other arterial highways. Motorists' expectations are that travel will be at relatively high speeds.
- **Class II.** Two-lane highways where the expectation of motorists is that travel speeds will be lower than for Class I roads. These highways may serve as access to Class I two-lane highways; they may serve as scenic byways or may be used by motorists for sightseeing. They may also be located in rugged terrain. Average trip lengths on Class II highways are shorter than on Class I highways.

At an *operational level* of analysis, level of service is determined based on existing or future traffic conditions and specific roadway characteristics. The HCM procedure is designed to analyze two-lane highways for (1) two-way traffic, (2) for a specific direction, or (3) for a directional segment with a passing lane. If the terrain is mountainous, or if the segment length to be analyzed is greater than 0.6 mi and the grade is at least 3 percent, two-lane highways are analyzed as specific upgrades or downgrades.

At a *planning level* of analysis, operational procedures are followed but with estimates, HCM default values, and/or local default values. Annual average traffic (AADT) values are used to estimate directional design hour volume (DDHV).

There are two measures used to describe the service quality of a two-lane highway. These are (1) percent time following another vehicle and (2) average travel speed.

1. **Percent time spent following another vehicle (PTSF)** is the average percentage of time that vehicles are traveling behind slower vehicles. When the time between consecutive vehicles (called the "headway") is less than three seconds, the trailing vehicle is considered to be following the lead vehicle. PTSF is a measure of the quality of service provided by the highway.
2. **Average travel speed (ATS)** is the space mean speed of vehicles in the traffic stream. Space mean speed is the segment length divided by average time for all vehicles to traverse the segment in both directions during a designated interval. ATS is a measure of the degree in which the highway serves its function of providing efficient mobility.

Figure 9.1 depicts the relationship between flow rate, ATS, and PTSF for two-way segments. Figure 9.2 depicts the relationship for directional segments. The relationships shown in these figures are for base conditions, defined as the absence of restrictive

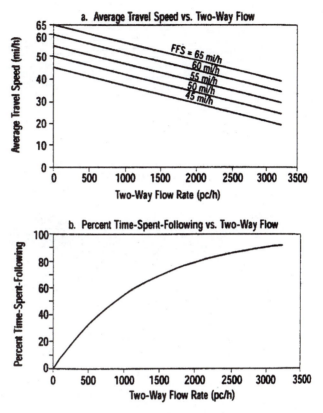

Figure 9.1 Speed-Flow and Percent Time-Spent-Following Flow Relationships for Two-Way Segments with Base Conditions

SOURCE: *Highway Capacity Manual,* Special Report 209, Fourth Edition, Transportation Research Board, National Research Council, Washington, D.C., 2000. Used with permission.

geometric, traffic, or environmental factors. Base conditions exist for the following characteristics:

- Level terrain
- Lane widths of 12 ft or greater
- Clear shoulders 6 ft wide or greater
- Passing permitted with absence of no-passing zones
- No impediments to through traffic due to traffic control or turning vehicles
- Passenger cars only in the traffic stream
- Equal volume in both directions (for analysis of two-way flow)

Capacity of a two-lane highway is 1700 passenger cars per hour (pc/h) for each direction of travel and is nearly independent of the directional distribution of traffic. For extended segments, the capacity of a two-lane highway will not exceed a combined total of 3200 pc/h. Short sections of two-lane highway, such as a tunnel or bridge, may reach a capacity of 3200–3400 pc/h.

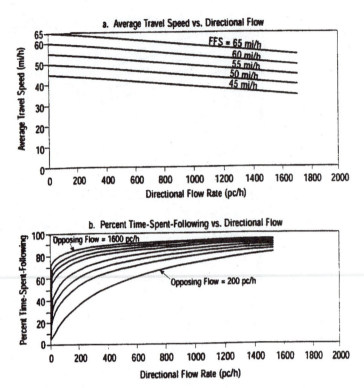

Figure 9.2 Speed-Flow and Percent Time-Spent-Following Flow Relationships for Directional Segments with Base Conditions

SOURCE: *Highway Capacity Manual,* Special Report 209, Fourth Edition, Transportation Research Board, National Research Council, Washington, D.C., 2000. Used with permission.

Level of Service (LOS) expresses the performance of a highway at traffic volumes less than capacity. LOS for Class I highways is based on two measures: PTSF and ATS. LOS for Class II highways is based on a single measure: PTSF. Level of service criteria are applied to travel during the peak 15 minutes of travel and on highway segments of significant length. Level of service designations are from *A* (highest) to *F* (lowest). Definitions of LOS and appropriate ranges for PTSF and ATS values are as follows:

Level of Service A: This is the highest quality of service than can be achieved. Motorists are able to travel at their desired speed. The need for passing other vehicles is well below the capacity for passing, and few (if any) platoons of three or more cars are observed. **Class I** highway average travel speed (ATS) is 55 mi/h or greater, and travel delays (PTSF) occur no more than 35 percent of the time. **Class II** highway maximum delay (PTSF) is 40 percent of the time. Maximum service flow rate (two-way) under base conditions is 490 pc/h.

Level of Service B: At this level of service, if vehicles are to maintain desired speeds the demand for passing other vehicles increases significantly. At the lower level of LOS *B* range the passing demand and passing capacity are approximately equal. **Class I** highway average travel speeds (ATS) are 50–55 mi/h. Travel delays (PTSF) occur between 35 and

50 percent of the time. **Class II** highway maximum delay (PTSF) is 40 to 55 percent of the time. Maximum service flow rate (two-way) under base conditions is 780 pc/h.

Level of Service C: Further increases in flow beyond the LOS *B* range results in a noticeable increase in the formation of platoons and an increase in platoon size. Passing opportunities are severely decreased. **Class I** highway average travel speeds (ATS) are 45–50 mi/h, and travel delays (PTSF) occur between 50 and 65 percent of the time. **Class II** highway maximum delay (PTSF) is 55 to 70 percent of the time. Maximum service flow rate (two-way) under base conditions is 1190 pc/h.

Level of Service D: Flow is unstable and passing maneuvers are difficult, if not impossible, to complete. Since the number of passing opportunities is approaching zero as passing desires increase, each lane operates essentially independently of the opposing lane. It is not uncommon that platoons will form that are five to ten consecutive vehicles in length. **Class I** highway average travel speeds (ATS) are 40–45 mi/h, and travel delays (PTSF) occur between 65 and 80 percent of the time. **Class II** highway maximum delay (PTSF) is 70 to 85 percent of the time. Maximum service flow rate (two-way) under base conditions is 1830 pc/h.

Level of Service E: Passing has now become virtually impossible. Platoons are longer and more frequent as slower vehicles are encountered more often. Operating conditions are unstable and are difficult to predict. **Class I** highway average travel speeds (ATS) are 40 mi/h or less and travel delays (PTSF) occur more than 80 percent of the time. **Class II** highway maximum delay (PTSF) is greater than 85 percent of the time. Maximum service flow rate (two-way) under base conditions is 3200 pc/h, a value seldom encountered on rural highways due to lack of demand.

Level of Service F: Traffic is congested with demand exceeding capacity. Volumes are lower than capacity and speeds are variable.

Table 9.1 summarizes the ranges in values of PTSF and ATS for each level of service category for Class I two-lane roads. Table 9.2 summarizes the ranges in values of PTSF for each level of service category for Class II two-lane roads. For Class I highways, two criteria apply: (1) percent time spent following (PTSF) and (2) average travel speed (ATS). For Class II highways, a single criterion is used: percent time spent following (PTSF).

Example 9.1   Determining the Level of Service of Two-Lane Roads If PTSF and ATS Are Known

The following values of PTSF and ATS have been determined based on the analysis of four roadway segments. (Methods for performing the analysis are described later in this chapter.) Determine the LOS if the roadway segments are: (a) Class I and (b) Class II

Segment	PTSF (%)	ATS (mi/h)
1	36	54
2	54	47
3	72	42
4	90	30

For convenience in problem solving, all tables are located in an appendix at the end of the chapter.

Solution:

(a) Use the values in Table 9.1 (see page 379) to determine Class I LOS. The results are:

Segment 1: LOS B
Segment 2: LOS C
Segment 3: LOS D
Segment 4: LOS E

Note: If values of PTSF and ATS do not correspond to the same LOS, the lower value is used.

(b) Use the values in Table 9.2 (see page 379) to determine Class II LOS. The results are:

Segment 1: LOS A
Segment 2: LOS B
Segment 3: LOS D
Segment 4: LOS E

Note: LOS values for Class II highways are often higher than for Class I highways because average travel speed (ATS) is not considered.

Procedures for Evaluating Levels of Service

The procedures for evaluating the LOS of a two-lane highway is carried out separately for the following cases:

- Two-way segments that are located in level or rolling terrain.
- Specific upgrades or downgrades defined as two-lane highways located in mountainous terrain or with grades that exceed 3 percent in segments exceeding lengths of 0.6 mi.
- Directional segments for which the LOS is determined for traffic in a single direction. Any segment can be analyzed as a directional segment, but the procedure is primarily used to analyze steep grades and also passing lanes for relatively short uniform segments.

Two-Way Segments

The analysis of two-lane roads for two-way segments is usually performed on extended lengths when the segment length is at least 2.0 mi and the segment is located in level or rolling terrain. Definitions of level and rolling terrain are as follows:

- **Level terrain** segments contain flat grades of 2 percent of less. Heavy vehicles are able to maintain the same speed as passenger cars throughout the segment.
- **Rolling terrain** segments contain short or medium length grades of 4 percent or less. Heavy truck speeds are lower than passenger cars but are not at crawl speed. If the grade exceeds 4 percent the two-way segment procedure cannot be used but must be analyzed using the specific grade procedure for directional segments.

Calculating the Value of PTSF for Two-Way Segments. The percent time spent following (PTSF) for a two-way segment is computed using Eq. 9.1.

$$PTSF = BPTSF + f_{d/np} \tag{9.1}$$

BPTSF = the base percent time spent following for both directions and is computed using Eq. 9.2

$$BPTSF = 100\left[1 - e^{-0.000879\,v_p}\right] \tag{9.2}$$

$f_{d/np}$ = adjustment in PTSF to account for the combined effect of (1) percent of directional distribution of traffic and (2) percent of passing zones. Table 9.3 provides appropriate values for various two-way flow rates.

v_p = passenger-car equivalent flow rate for the peak 15-min period and is computed using Eq. 9.3.

$$v_p = \frac{V}{(PHF)(f_G)(f_{HV})} \tag{9.3}$$

V = demand volume for the entire peak hour, veh/h

PHF = peak hour factor, $V/(4)$ (peak 15-min volume)

f_G = grade adjustment factor for level or rolling terrain (Table 9.4)

f_{HV} = adjustment factor to account for heavy vehicles in the traffic stream and is computed using Eq. 9.4

$$f_{HV} = \frac{1}{1 + P_T\,(E_T - 1) + P_R\,(E_R - 1)} \tag{9.4}$$

P_T and P_R = the decimal portion of trucks (and buses) and RVs in the traffic stream. For example if there are 22 percent trucks in the traffic stream, then $P_T = 0.22$

E_T and E_R = the passenger car equivalent for trucks and RVs respectively. Values are provided in Table 9.5.

Since the values of E_T and E_R are a functions of two-way flow rates in pc/h, an iterative process is required in which a trial value of v_p is based on the PHF only. Then a new value of v_p is computed using appropriate values of E_T and E_R. If the second value of v_p is within the range used to determine truck and RV equivalents, the computed value is correct. If not, a second iteration is required using the next higher range of flow rate.

Example 9.2   Computing the Value of Percent Time Spent Following (PTSF) for a Two-Way, Two-Lane Highway

Determine the value of PTSF for a 6-mi two-lane highway in rolling terrain. Traffic data are as follows. (Similar problems are solved using a tabular format in HCM 2000.)

Volume = 1600 veh/h (two-way)
Percent trucks = 14
Percent RVs = 4
Peak hour factor = 0.95
Percent directional split = 50-50
Percent no-passing zones = 50

Solution:

Step 1. Compute peak 15-min hourly passenger car equivalent v_p.

Trial value for v_p is V/PHF = 1600/0.95 = 1684 pc/h
Determine f_G = 1.00 (Table 9.4)
Determine E_T = 1.00 and E_R = 1.00 (Table 9.5)

$$f_{HV} = \frac{1}{1+P_T(E_T-1)+P_R(E_R-1)} = \frac{1}{1+(0.14)(1.0-1.0)+(0.04)(1.0-1.0)} = 1.00$$

$$v_p = \frac{V}{(PHF)(f_G)(f_{HV})} = \frac{1600}{(0.95)(1.00)(1.00)} = 1684 \text{ pc/h}$$

Note: since 1684 < 3200 this section is operating below capacity.

Step 2. Compute base percent time spent following (BPTSF).

$$BPTSF = 100\left[1-e^{-0.000879v_p}\right] = 100\left[1-e^{-0.000879(1684)}\right] = 77.2\%$$

Step 3. Compute Percent Time Spent Following (PTSF)

$$PTSF = BPTSF + f_{d/np}$$
$$f_{d/np} = 4.8. \text{ (By interpolation from Table 9.3)}$$
$$PTSF = 77.2 + 4.8 = 82.0\% \text{ (answer)}$$

Calculating the Value of ATS for Two-Way Segments. The average travel speed (ATS) for a two-way segment is completed using Eq. 9.5

$$ATS = FFS - 0.00776v_p - f_{np} \qquad (9.5)$$

ATS = average travel speed for both directions of travel combined (mi/h)
FFS = free flow speed, the mean speed at low flow when volumes are < 200 pc/h
f_{np} = adjustment for the percentage of no-passing zones (Table 9.6)
v_p = passenger-car equivalent flow rate for the peak 15-min period

(Eq. 9.3 is used to compute v_p with values of f_G from Table 9.7 and E_T and E_R from Table 9.8.)

The determination of free-flow speed can be completed in three ways:

- Field measurements at volumes < 200 pc/h
- Field measurements at volumes > 200 pc/h using the following correction:

$$FFS = S_{FM} + 0.00776 \frac{V_f}{f_{HV}}$$

S_{FM} = mean speed of traffic measured in the field, mi/h
V_f = observed flow rate, veh/h for the period when speed data were obtained
f_{HV} = heavy vehicle adjustment factor (Eq. 9.4)

- Indirect estimation, when field data are unavailable, is computed using Eq. 9.6.

$$FFS = BFFS - f_{LS} - f_A \tag{9.6}$$

FFS = estimated free-flow speed, mi/h
$BFFS$ = base free-flow speed, mi/h
f_{LS} = adjustment for lane and shoulder width (Table 9.9)
f_A = adjustment for number of access points per mi (Table 9.10)

The base free-flow speed (BFFS) depends upon local conditions regarding the desired speeds of drivers. The transportation engineer estimates BFFS based on her knowledge of the area and the speeds on similar facilities. The range of BFFS 45–65 mi/h. Posted speed limits or design speeds may serve as surrogates for BFFS.

Example 9.3 Two-Directional Highway

Use the data provided in Example 9.2 to estimate the average travel speed (ATS). Assume that the base free-flow speed (BFFS) is the posted speed of 60 mi/h. The section length is 6 mi, lane width is 11 ft, shoulder width is 4 ft, and there are 20 access points per mi.

Solution:

Step 1. Compute the free flow speed under the given conditions using Eq. 9.6.

$FFS = BFFS - f_{LS} - f_A$
$f_{LS} = 1.7$ (Table 9.9)
$f_A = 5.0$ (Table 9.10)
$FFS = 60 - (1.7) - (5.0) = 53.3$ mi/h

Step 2. Compute average travel speed.

$ATS = FFS - 0.00776 v_p - f_{np}$
$FFS = 53.3$ mi/h

Calculate v_p as follows:

$$v_p = \frac{V}{(PHF)(f_G)(f_{HV})} = \frac{1600}{(0.95)(0.99)(0.931)} = 1827 \text{ pc/h}$$

To determine the value of f_{hv}:

$f_G = 0.99$ (Table 9.7 since $v > 1200$, rolling terrain)
$E_T = 1.5$; $E_R = 1.1$ (Table 9.8 since $v > 1200$, rolling terrain)

$$f_{HV} = \frac{1}{1+P_T(E_T-1)+P_R(E_R-1)} = \frac{1}{1+(0.14)(1.5-1.0)+(0.04)(1.1-1.0)} = 0.931$$

$f_{np} = 0.8$ (Table 9.6 since $v_p = 1827$ and percent no passing zones = 50)

Use Eq. 9.5 to compute ATS

$$ATS = 53.3 - 0.00776\,(1827) - 0.8 = 53.3 - 14.2 - 0.8 = 38.3 \text{ mi/h (answer)}$$

Calculating Other Performance Measures for Two-Way, Two-Lane Highways. Additional measures that can be computed are as follows:

- Volume-to-capacity ratio, v/c
- Total number of veh/mi during the peak 15-min period, VMT_{15}
- Total number of veh/mi during the peak hour, VMT_{60}
- Total travel time during the peak 15-min period, TT_{15}

The formulas are as follows:
Volume-to-capacity ratios computed using Eq. 9.7.

$$\frac{v}{c} = \frac{v_p}{c} \tag{9.7}$$

v_p/c = volume-to-capacity ratio
  c = two-way segment capacity (3200 for a two-directional segment, 1700 for a directional segment)
  v_p = passenger car equivalent flow rate for peak 15-min period, pc/h

Total number of vehicle-miles during the peak 15-min periods computed using Eq. 9.8.

$$VMT_{15} = 0.25\left(\frac{V}{PHF}\right)L_t \tag{9.8}$$

VMT_{15} = total travel on the analysis segment during the peak 15-min, veh/mi
  L_t = total length of the analysis segment, mi
  V = hourly volume, veh/h

Total number of vehicle-miles during the peak hours computed using Eq. 9.9.

$$VMT_{60} = V(L_t) \qquad (9.9)$$

VMT_{60} = total travel on the analysis segment during the peak hour, veh/mi

Total travel time during the peak 15-min period is computed using Eq. 9.10.

$$TT_{15} = \frac{VMT_{15}}{ATS} \qquad (9.10)$$

TT_{15} = total travel time for all vehicles on the analyzed segment during the peak 15-min, veh/h.

Example 9.4   Level of Service and Performance Measures for Two-Lane, Two-Directional Highways

Use the data and results in Examples 9.2 and 9.3 to determine the following:

- Level of Service if the segment is a Class I or a Class II highway
- Volume-to-capacity ratio, v/c
- Total number of veh/mi during the peak 15-min period, VMT_{15}
- Total number of veh/mi during the peak hour, VMT_{60}
- Total travel time during the peak 15-min period, TT_{15}

Solution:

- Level of Service if the segment is a Class I or a Class II highway
  From Example 9.2 and 9.3:

  $PTSF = 82.0\%$
  $ATS = 38.3$ mi/h
  Class I LOS = E (Table 9.1)
  Class II LOS = D (Table 9.2)

- Volume-to-capacity ratio, v/c
  Eq. 9.7:

  $$v = \frac{v_p}{c} = \frac{1827}{3200} = 0.57$$

- Total number of vehicle-miles during the peak 15-min period, VMT_{15}
  Eq. 9.8:

  $$VMT_{15} = 0.25\left(\frac{V}{PHF}\right)L_t = 0.25\left(\frac{1600}{0.95}\right)(6) = 2526 \text{ veh/mi}$$

- Total number of vehicle-miles during the peak hour, VMT_{60}
  Eq. 9.9:

$$VMT_{60} = V L_T = 1600 \, (6) = 9600 \text{ veh/mi}$$

- Total travel time during the peak 15-min period, TT_{15}
  Eq. 9.10:

$$TT_{15} = \frac{VMT_{15}}{ATS} = \frac{2526}{38.3} = 66 \text{ veh/mi}$$

Directional Segments

Three categories of directional segments are considered. They are:

- **Extended segments** located in level or rolling terrain with a length of at least 2 mi
- **Specific upgrades or downgrades** located in mountainous terrain or with grades of at least 3 percent for segment lengths of at least 0.6 mi long
- **A passing lane** added within a section in level or rolling terrain or as a truck climbing lane

Calculating the Value of PTSF for Directional Segments in Level or Rolling Terrain. The percent time spent following (PTSF) for a directional segment is computed by using Eq. 9.11.

$$PTSF_d = BPTSF_d + f_{np} \tag{9.11}$$

BPTSF is computed by using Eq. 9.12.

$$BPTSF_d = 100\left(1 - e^{av_d^{\,b}}\right) \tag{9.12}$$

$PTSF_d$ = percent time spent following in the direction analyzed
$BPTSF_d$ = base percent time spent following in the direction analyzed (Eq. 9.12)
f_{np} = adjustment for percentage of no-passing zones in the analysis direction (Table 9.11)
v_d = passenger-car equivalent flow rate for the peak 15 minutes in the analysis direction pc/h
a, b = coefficients based on opposing flow rate, v_o (Table 9.12)

Example 9.5 Computing the Value of Percent Time Spent Following (PTSF) for the Peak Direction on a Two-Lane Highway

During the peak hour on a Class I two-lane highway in rolling terrain, volumes northbound are 1200 veh/h and volumes southbound are 400 veh/h. The PHF is 0.95, and there are 14 percent trucks/buses and 4 percent RVs. Lane widths are 11 ft and shoulder widths are 4 ft. The roadway section is 5 mi in length and there are 20 access points per mile. There are 50 percent no-passing zones and the base free-flow speed is 60 mi/h. Determine the percent time spent following in the peak direction of travel.

Solution:

Step 1. Compute Peak 15-min Hourly Passenger-Car Equivalent in the Peak Direction, v_d and in the Opposite Direction v_o.

Trial value for v_d is v_d/PHF = 1200/0.95 = 1263 veh/h
Determine f_G = 1.00 (Table 9.4)
Determine E_T = 1.00 and E_R = 1.00 (Table 9.5)

Compute f_{HV}

$$f_{HV} = \frac{1}{1+P_T(E_T-1)+P_R(E_R-1)} = \frac{1}{1+(0.14)(1.0-1.0)+(0.04)(1.0-1.0)} = 1.00$$

Compute v_d

$$v_d = \frac{V}{(PHF)(f_G)(f_{HV})} = \frac{1200}{(0.95)(1.00)(1.00)} = 1263 \text{ pc/h}$$

Trial value for v_o is v_o/PHF = 400/0.95 = 421 veh/h

Determine f_G = 0.94 (Table 9.4)
Determine E_T = 1.5 and E_R = 1.0 (Table 9.5)

Compute f_{HV}

$$f_{HV} = \frac{1}{1+P_T(E_T-1)+P_R(E_R-1)} = \frac{1}{1+(0.14)(1.5-1.0)+(0.04)(1.0-1.0)} = 0.935$$

Compute v_o

$$v_o = \frac{V}{(PHF)(f_G)(f_{HV})} = \frac{400}{(0.95)(0.94)(0.935)} = 479 \text{ pc/h}$$

Step 2.   Compute Base Percent Time Spent Following (BPTSF)

$$BPTSF_d = 100\left(1 - e^{av_d^b}\right)$$ (9.12)

Determine the values a and b from Table 9.12 by interpolation

$a = -\{0.057 + (0.043)(79/200)\} = -0.074$
$b = 0.479 - (0.066)(79/200) = -0.453$

$$BPTSF_d = 100\left(1 - e^{(-0.074)(1263)^{0.453}}\right) = 84.7 \text{ percent}$$

Step 3.   Compute Percent Time Spent Following (PTSF)

$$PTSF_d = BPTSF_d + f_{np}$$

Use Table 9.11 to determine f_{np}

f_{np} v_o	50% no-passing & FFS 60 mi/h
400	$(12.1 + 14.8)/2 = 13.45$
600	$(7.5 + 9.6)/2 = 8.55$
479	$(13.45 - (79/200)(4.90) = 11.5$

$$PTSF_d = BPTSF_d + f_{np} = 84.7 + 11.5 = 96.2 \text{ percent (answer)}$$

Calculating the Value of ATS for Directional Segments in Level or Rolling Terrain.
The average travel speed (ATS) for a two-way segment is computed by using Eq. 9.13.

$$ATS_d = FFS_d - 0.00776(v_d + v_o) - f_{np}$$ (9.13)

ATS_d = average travel speed in the analysis direction of travel (mi/h)
f_{np} = adjustment for the percentage of no-passing zones in the analysis direction (Table 9.13)
FFS_d = free-flow speed in the analysis direction

Example 9.6   Computing the Value of Average Travel Time (ATS) for the Peak Direction on a Two-Lane Highway

Use the data provided in Example 9.5 to estimate the average travel speed (ATS).

Solution:

Step 1.   Compute the Free-Flow Speed Under the Given Conditions.

$FFS = BFFS - f_{LS} - f_A$
$f_{LS} = 1.7$ (Table 9.9)
$f_A = 5.0$ (Table 9.10)
$FFS = 60 - (1.7) - (5.0) = 53.3$ mi/h

Step 2.   Compute the Average Travel Speed using Eq. 9.13:

$$ATS_d = FFS_d - 0.00776(v_d + v_o) - f_{np}$$

Compute v_d:

$f_G = 0.99$ (Table 9.7 since $v > 600$, rolling terrain)
$E_T = 1.5; E_R = 1.1$ (Table 9.8 since $v > 600$, rolling terrain)

$$f_{HV} = \frac{1}{1 + P_T(E_T - 1) + P_R(E_R - 1)} = \frac{1}{1 + (0.14)(1.5 - 1.0) + (0.04)(1.1 - 1.0)} = 0.931$$

$$v_d = \frac{V}{(PHF)(f_G)(f_{HV})} = \frac{1200}{(0.95)(0.99)(0.931)} = 1370 \text{ pc/h}$$

Compute v_o:

$f_G = 0.93$ (Table 9.7 since $v > 300–600$, rolling terrain)
$E_T = 1.9; E_R = 1.1$ (Table 9.8 since $v > 300–600$, rolling terrain)

$$f_{HV} = \frac{1}{1 + P_T(E_T - 1) + P_R(E_R - 1)} = \frac{1}{1 + (0.14)(1.9 - 1.0) + 0.04(1.1 - 1.0)} = 0.884$$

$$v_o = \frac{V}{(PHF)(f_G)(f_{HV})} = \frac{400}{(0.95)(0.93)(0.884)} = 512 \text{ pc/h}$$

Note that both v_d and v_o are less than 1700, the capacity of a one-way segment.

$f_{np} = 1.6$ (Table 9.13, by interpolation, since $v_o = 512$ pc/h, FFS = 53.3 mi/h and percent no-passing zones = 50)

$ATS_d = 53.3 - 0.00776 (1370 + 512) - 1.6 = 53.3 - 14.6 - 1.6 = 37.1$ mi/h (answer)

Example 9.7   Level of Service and Performance Measures for Two-Lane Directional Highways

Use the data and results in Examples 9.5 and 9.6 to determine the following:

- Level of Service if the segment is a Class I or Class II highway
- Volume-to-capacity ratio, v/c
- Total number of vehicle-mi during the peak 15-min period, VMT_{15}
- Total number of vehicle-km during the peak hour, VMT_{60}
- Total travel time during the peak 15-min period, TT_{15}

Solution:

- Level of Service if the segment is a Class I highway
  From Examples 9.5 and 9.6:

$PTSF = 96.2\%$
$ATS = 37.1$ mi/h
$LOS = E$ (Table 9.1)

- Level of Service if the segment is a Class II highway

$PTSF = 96.2\%$
$LOS = E$ (Table 9.2)

- Volume-to-capacity ratio, v/c Eq. 9.7:

$$v = \frac{v_p}{c} = \frac{1370}{1700} = 0.81$$

- Total number of vehicle-mi during the peak 15-min period, VMT_{15}, Eq. 9.8:

$$VMT_{15} = 0.25\left(\frac{V}{PHF}\right)L_t = 0.25\left(\frac{1200}{0.95}\right)(6) = 1895 \text{ veh/mi}$$

- Total number of vehicle-mi during the peak hour, VMT_{60}, Eq. 9.9:

$$VMT_{60} = V_d L_t = 1200(6) = 7200 \text{ veh/mi}$$

- Total travel time during the peak 15-min period, TT_{15}, Eq. 9.10:

$$TT_{15} = \frac{VMT_{15}}{ATS_d} = \frac{1895}{37.1} = 51.1 \text{ veh/h}$$

Calculating the Value of PTSF and ATS for Directional Segments on Specific Up-grades.   Any grade of 3 percent or more and at least 0.6 mi in length must be analyzed as a specific upgrade. Lengths of 0.25 mi or more and upgrades of 3 percent or more may be analyzed. Segments in mountainous terrain are analyzed as specific upgrades. When grades vary within the section, a composite grade is computed as the total change in elevation divided by the total length expressed as a percentage.

The procedure described in the preceding section for computing PTSF and ATS of directional segments is followed for specific upgrades and downgrades. However, the effect of heavy vehicles, as described by the grade adjustment factor, f_G, and the heavy vehicle factor, f_{HV}, used in Eq. 9.4 are determined based on the average segment grade and the segment length.

To Calculate PTSF:

1. Determine f_G using Table 9.14
2. Determine f_{HV} using Table 9.15.
3. Compute f_{HV} using Eq. 9.4.

To Calculate ATS:

1. Determine f_G using Table 9.16.
2. Determine f_{HV}: using Table 9.17 and Table 9.18.
3. Compute f_{HV} using Eq. 9.4.

$$f_{HV} = \frac{1}{1+P_T(E_T-1)+P_R(E_R-1)} \tag{9.4}$$

4. Determine v_d using Eq. 9.3.

$$v_d = \frac{V_d}{(PHF)(f_G)(f_{HV})} \tag{9.3}$$

v_d = passenger car equivalent flow rate for the peak 15-min period in the direction analyzed, pc/h

V_d = demand volume for the full peak hour in the direction analyzed, veh/h

5. Determine v_o using Eq. 9.3.

$$v_o = \frac{V_o}{(PHF)(f_G)(f_{HV})}$$

v_o = passenger car equivalent flow rate for the peak 15-min period in the opposing direction of travel, pc/h

V_o = demand volume for the full peak hour in the opposing direction of travel, veh/h

When computing v_o, the values used for f_{HV} and *PHF* should apply to the opposing direction of travel.

6. Compute *PTSF* and *ATS* following procedures used for directional analysis as described for level and rolling terrain.

Calculating the Value of PTSF and ATS for Directional Segments on Specific Downgrades. Any downgrade of 3 percent or more and at least 0.6 mi in length is analyzed as a specific downgrade as are all downgrade segments in mountainous terrain. The opposing direction of travel to a specific upgrade should be analyzed as a specific downgrade. For most downgrades, $f_G = 1.0$ and f_{HV} is determined from Table 9.5 for level terrain. For specific downgrades that are long and steep, such that heavy vehicles must travel at crawl speeds to avoid losing control of the vehicle, the value of h_{HV} is computed by using Eq. 9.14.

$$f_{HV} = \frac{1}{1 + P_{TC}P_T(E_{TC}-1) + (1-P_{TC})P_T(E_T-1) + P_R(E_R-1)} \tag{9.14}$$

P_{TC} = decimal proportion of trucks in the traffic stream that travel at crawl speeds on the analysis segment. In the absence of other information, the percentage of tractor-trailer combinations is used in this calculation.

E_{TC} = passenger-car equivalent for trucks in the traffic stream that travel at crawl speeds on the analysis segment. Table 9.19.

Example 9.8   Computing Volumes on Directional Segments for Specific Upgrades and Downgrades

Repeat Examples 9.5 and 9.6 if the grade in the peak direction is + 4.75 percent. Determine the values of v_d and v_o that are needed to compute PTSF and ATS. The data provided are reproduced below.

Class I two-lane highway.

Volumes northbound (peak direction) are 1200 veh/h.

Volumes southbound are 400 veh/h.

PHF is 0.95.

14 percent trucks/busses, of which 15 percent are semi-trailers and 4 percent RVs.

Lane widths are 11 ft.

Shoulder widths are 4 ft.

Roadway section is 5 mi in length.

20 access points per mile.

50 percent no-passing zones.

Base free-flow speed is 60 mi/h.

The difference between free-flow speed and crawl speed is 25 mi/h.

Determine the Value of v_d and v_o for *PTSF*.

v_d for a specific upgrade

Determine f_G:
Refer to Table 9.14. $f_G = 1.00$

Determine f_{HV}:
Refer to Table 9.15 $E_T = 1.8$ $E_R = 1.0$

Compute f_{HV} using Eq. 9.4:

$$f_{HV} = \frac{1}{1 + P_T(E_T - 1) + P_R(E_R - 1)} = \frac{1}{1 + (0.14)(1.8 - 1) + (0.04)(1 - 1)} = 0.899$$

Compute v_d using Eq. 9.3:

$$v_d = \frac{V_d}{(PHF)(f_G)(f_{HV})} = \frac{1200}{(0.95)(1.00)(0.899)} = 1405 \text{ pc/h}$$

v_o for a specific downgrade

Determine f_G:
On downgrades, $f_G = 1.00$

Determine f_{HV}:
Refer to Table 9.19. $E_{TC} = 5.7$
Refer to Table 9.5. $E_T = 1.1$ $E_R = 1.0$

Compute f_{HV} using Eq. 9.14:

$$f_{HV} = \frac{1}{1 + P_{TC}P_T(E_{TC} - 1) + (1 - P_{TC})P_T(E_T - 1) + P_R(E_R - 1)}$$

$$f_{HV} = \frac{1}{1 + (0.15)(0.14)(5.7 - 1) + (1 - 0.15)(0.14)(1.1 - 1) + (0.04)(1 - 1)} = 0.900$$

Compute v_o using Eq. 9.3:

$$v_o = \frac{V_o}{(PHF)(f_G)(f_{HV})} = \frac{400}{(0.95)(1.00)(0.900)} = 468 \text{ pc/h}$$

A similar procedure is followed in computing the value of v_d and v_o for *ATS*. The procedures to determine *PTSF* and *ATS* are as described in Examples 9.5 and 9.6.

Calculating Percent Time Spent Following (PTSF) for Directional Segments When a Passing Lane Has Been Added Within an Analysis Section in Level or Rolling Terrain.   A passing lane is added to a two-lane highway to provide the motorist with additional opportunities to overtake slower vehicles. By adding a lane in one direction of travel, the percentage of time spent following can be reduced in the widened section and in a portion of the section that follows. The net effect of a passing lane is to reduce the overall percent time spent following for the segment being analyzed. The greatest effect would result if the passing lane were as long as the entire length of the segment. The effect decreases as the length of the passing lane is reduced.

Figure 9.3 depicts a plan view of a typical passing lane. The length includes tapers on both ends. Passing lanes are provided in a variety of formats. They may be exclusively for a single direction of traffic, or opposing traffic may be permitted its use. Passing lanes may be provided intermittently or at fixed intervals for each direction of travel. They may also be provided for both directions of travel at the same location resulting in a short section of four-lane undivided highway.

Figure 9.4 illustrates conceptually how a passing lane through the analysis segment influences the PTSF. Four regions are involved in which there are changes to the value of PTSF. They are

- **Region I.** The PTSF in the immediate region preceding the passing lane (upstream of the passing lane) of length L_u, will be $PTSF_d$, the value computed as described earlier for a directional segment. The length of Region I is determined by deciding where the planned or actual passing lane should be placed relative to the beginning point of the analysis segment.
- **Region II.** The passing lane of length L_{pl} is the constructed or planned length including tapers. This section will experience a sudden reduction in PTSF as shown in Figure 9.4. The extent of the reduction, f_{pl}, in PTSF ranges between 0.58 and 0.62, depending on the directional flow rate v_d as shown in Table 9.20. The optimal length of Region II is determined from Table 9.21, and ranges from 0.50 to 2.0 mi depending on the directional flow rate.
- **Region III.** The region immediately downstream of the passing lane of length L_{de} benefits from the effect of the passing lane as the PTSF gradually increases to its original value. This region is considered to be within the effective length of the passing

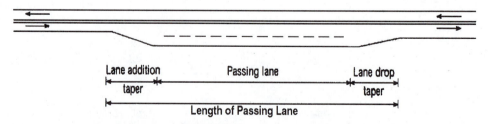

Figure 9.3   Plan View of a Typical Passing Lane

SOURCE: *Highway Capacity Manual,* Special Report 209, Fourth Edition, Transportation Research Board, National Research Council, Washington, D.C., 2000. Used with permission.

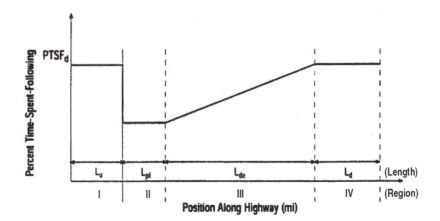

Figure 9.4 Effect of a Passing Lane on Percent Time-Spent-Following as Represented in the Operational Analysis Methodology

SOURCE: *Highway Capacity Manual,* Special Report 209, Fourth Edition, Transportation Research Board, National Research Council, Washington, D.C., 2000. Used with permission.

lane, and it is assumed that the value of PTSF increases linearly throughout its length. The length of Region III, L_{de}, is determined from Table 9.22 as a function of the directional flow rate.

- **Region IV.** This is the remaining downstream region within the segment, and its length is L_d. This section is beyond the effective length of the passing lane, and the value $PTSF_d$ is equal to that of the upstream section of length L_u. The length of this region is simply the difference between the total segment length, L_t, and the sum of the lengths of regions I, II, and III. Thus $L_d = L_t - (L_u + L_{pl} + L_{de})$

The average value of **percent time spent following** for the entire analysis segment with a passing lane in place is $PTSF_{pl}$. The value is determined as the weighted average of the PTSF values in each region weighted by the region length, and computed by using Eq. 9.15.

$$PTSF_{pl} = \frac{1}{L_t}\left[\begin{array}{l} PTSF_d(L_u) + PTSF_d(f_{pl})(L_{pl}) + \left\{ (PTSF_d)(f_{pl}) + (PTSF_d) \right\} \dfrac{L_{de}}{2} + \\ (PTSF_d)(L_d) \end{array} \right]$$

$PTSF_{pl}$ = percent time spent following for the entire segment including the passing lane

$PTSF$ = percent time spent following for the entire segment without the passing lane from Eq. 9.11

f_{pl} = factor for the effect of a passing lane on percent time spent following from Table 9.20

$$PTSF_{pl} = \frac{PTSF_d \left[L_u + (f_{pl})(L_{pl}) + \left(\frac{f_{pl}+1}{2} \right)(L_{de}) + L_d \right]}{L_t} \qquad (9.15)$$

If the full downstream length is not reached because a traffic signal or a town interrupts the analysis section, then L is replaced by a shorter length, L'_{de}. In this instance, Eq. 9.15 is modified as follows:

$$PTSF_{pl} = \frac{PTSF_d \left[L_u + (f_{pl})(L_{pl}) + \left(\frac{1-f_{pl}}{2} \right)\left(\frac{(L'_{de})^2}{L_{de}} \right) \right]}{L_t} \qquad (9.16)$$

L'_{de} = actual distance from end of passing lane to end of analysis segment in miles
L'_{de} must be less than or equal to the value of L_{de} (obtained from Table 9.22).

Calculating Average Travel Speed (ATS) for Directional Segments when a Passing Lane Has Been Added Within an Analysis Section in Level or Rolling Terrain. Figure 9.5 illustrates the changes in the value of average travel speed (ATS) that occur within each of the four regions when a passing lane has been added. The values are:

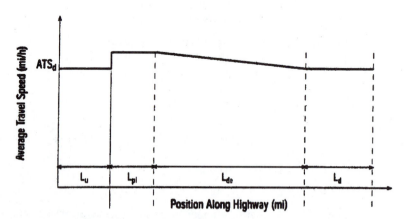

Figure 9.5 Effect of a Passing Lane on Average Travel Speed as Represented in the Operational Analysis Methodology

SOURCE: *Highway Capacity Manual,* Special Report 209, Fourth Edition, Transportation Research Board, National Research Council, Washington, D.C., 2000. Used with permission.

- **Region I.** Average travel speed in the upstream region is ATS_d for length L_u.
- **Region II.** Average travel speed in the passing lane increases by a factor f_{pl} provided in Table 9.20. Values range from 1.08 to 1.11 depending on the directional flow rate v_d.
- **Region III.** Average travel speed within the downstream length decreases linearly from the value in Region II to ATS_d. $L_{de} = 1.7$ mi as shown in Table 9.22.
- **Region IV.** Average travel speed in the downstream region is ATS_d for length L_d. To compute the average speed for the analysis section, it is first necessary to compute the individual travel time within regions t_I, t_{II}, t_{III}, and t_{IV}. The average travel speed is the total analysis length L_t divided by the sum of the four travel times. The derivation below produced Eq. 9.17, the value of ATS_{pl}, which is the effect on the average travel speed in the analysis section as the result of adding a passing lane:

$$ATS_{pl} = \frac{L_t}{t_I + t_{II} + t_{III} + t_{IV}} = \frac{L_t}{\dfrac{L_u}{ATS_d} + \dfrac{L_{pl}}{(f_{pl})(ATS_d)} + \dfrac{(L_{de})}{\dfrac{f_{pl}(ATS_d)+(ATS_d)}{2}} + \dfrac{L_d}{ATS_d}}$$

$$ATS_{pl} = \frac{(ATS_d)(L_t)}{L_u + \dfrac{L_{pl}}{f_{pl}} + \dfrac{2L_{de}}{f_{pl}+1} + L_d} \qquad (9.17)$$

Example 9.9 Computing Level of Service for Two-Lane Directional Highways with a Passing Lane

Use the data and results for Examples 9.5 and 9.6 to determine the level of service if a passing lane is added. The length of the passing lane is 1 mi and it is designed to begin 1 mi downstream from the beginning of the 6-mi analysis section.

Solution:

From Examples 9.5 and 9.6:
$PTSF = 96.2$ percent
$ATS = 37.1$ mi/h

Step 1. Determine the Length of Each Region
Region I L_u = 1 mi
Region II L_{pl} = 1 mi
Region III
    For $PTSF$ L_{de} = 3.6 mi (Table 9.22) f_{pl} = 0.62 (Table 9.20)
    For ATS L_{de} = 1.7 mi (Table 9.22) f_{pl} = 1.11 (Table 9.20)
Region IV
    For $PTSF$ L_d = 6 − 1 − 1 − 3.6 = 0.4 mi
    For ATS L_d = 6 − 2 − 2 − 1.7 = 2.3 mi

Step 2. Compute $PTSF_{pl}$
Use Eq. 9.15

$$PTSF_{pl} = \frac{PTSF_d\left[L_u + (f_{pl})(L_{pl}) + \left(\frac{f_{pl}+1}{2}\right)(L_{de}) + L_d\right]}{L_t} =$$

$$\frac{96.2\left[1 + (0.62)(1) + \left(\frac{0.62+1}{2}\right)(3.6) + 0.4\right]}{10} = 79.4 \text{ percent}$$

Step 3.   Compute ATS_{pl}
Use Eq. 9.17

$$ATS_{pl} = \frac{(ATS_d)(L_t)}{L_u + \dfrac{L_{pl}}{f_{pl}} + \dfrac{2L_{de}}{f_{pl}+1} + L_d} = \frac{(37.1)(6)}{1 + \dfrac{1}{1.11} + \dfrac{1(1.7)}{1.11+1} + 2.3} = 38.3 \text{ mi/h}$$

Step 4.   Determine Level of Service
Use Table 9.1: LOS = E with a passing lane
Note: While the passing lane improved the PTSF and ATS values, the level of service
is unchanged.

Calculating Average Travel Speed (ATS) for Directional Segments when a Climbing Lane Has Been Added Within an Analysis Section on an Upgrade.   Climbing lanes are provided on long upgrade sections to provide an opportunity for faster moving vehicles to pass heavy trucks whose speed has been reduced. Climbing lanes are warranted on two-lane highways when one of the following conditions is met:

- Directional flow rate on the upgrade > 200 veh/h
- Directional flow rate for trucks on the upgrade > 20 veh/h, and when any of the following conditions apply:
  Speed reduction for trucks is 10 mi/h for a typical heavy truck, LOS is E or F on the grade, or the LOS on the upgrade is two or more levels of service values higher than the approach grade.

Operational analysis of climbing lanes follows the same procedural steps as described earlier for computing PTSF and ATS in level and rolling terrain. When applying the directional procedure to the direction without a passing lane, Tables 9.14 and 9.15 should be used to determine the grade adjustment factor, f_G, and the heavy vehicle factor, f_{HV}. If the climbing lane is not long or steep enough to be analyzed as a specific upgrade, it is analyzed as a passing lane. Values for the adjustment factors f_{pl} for percent time spent following (PTSF) and average travel speed (ATS) are shown in Table 9.23. Furthermore, L_u and L_d are zero because climbing lanes are analyzed as part of a specific upgrade. L_{de} is usually zero unless the climbing lane ends before the top of the grade. If the lane does end before the top of grade, L_{de} is a lower value than shown in Table 9.22.

Computing the Combined Value of PTSF and ATS for Contiguous Directional Analysis Segments.   A directional two-lane highway facility is composed of a series of contiguous directional segments, each analyzed separately as described previously. A combined value for PTSF and ATS that can be used to determine the overall LOS is computed by using Eq. 9.18 and 9.19.

$$PTSF_c = \frac{\sum_{i=1}^{n} (TT_i)(PTSF_i)}{\sum_{i=1}^{n} (TT_i)} \tag{9.18}$$

$$ATS_c = \frac{\sum_{i=1}^{n} (VMT_i)}{\sum_{i=1}^{n} (TT_i)} \tag{9.19}$$

n = the number of analysis segments
TT_i = total travel time for all vehicles on analysis segment i during the peak 15-min period (veh/h). (Equation 9.10)
$PTSF_i$ = percent time spent following for segment i
$PTSF_c$ = percent time spent following for all segments combined
ATS_c = average travel time for all segments combined
VMT_i = total travel on analysis segment i during the peak 15-min period (veh/km). (Equation 9.8)

MULTILANE HIGHWAYS

The procedures developed are used to analyze the capacity and Level of Service (LOS) for multilane highways. The results can be used in the planning and design phase to determine lane requirements necessary to achieve a given LOS and to consider the impacts of traffic and design features in rural and suburban environments. Multilane highways differ from two-lane highways by virtue of the number of lanes and from freeways by virtue of the degree of access. They span the range between freeway-like conditions of limited access to urban street conditions with frequent traffic-controlled intersections. Illustrations of the variety of multilane highway configurations are provided in Figure 9.6. Multilane highways may exhibit some of the following characteristics.

- Posted speed limits are usually between 40 and 55 mi/h.
- They may be undivided or include medians.
- They are located in suburban areas or in high-volume rural corridors.
- They may include a two-way left-turn median lane (TWLTL).
- Traffic volumes range from 15,000 to 40,000/day.
- Volumes are up to 100,000/day with grade separations and no cross-median access.

(a) Divided multilane highway in a rural environment.

(b) Divided multilane highway in a suburban environment.

(c) Undivided multilane highway in a rural environment.

(d) Undivided multilane highway in a suburban environment.

Figure 9.6 Typical Multilane Highways

SOURCE: *Highway Capacity Manual,* Special Report 209, Fourth Edition, Transportation Research Board, National Research Council, Washington, D.C., 2000. Used with permission.

- Traffic signals at major crossing points are possible.
- There is partial control of access.

Level of Service (LOS) for Multilane Highways

Any two of the following three performance characteristics can describe the level of service (LOS) for a multilane highway:

- V_p/c: Flow rate (PL/h/ln)
- S: Average passenger car speed, mi/h
- D: Density (number of cars per mi), pc/mi/lane

The relationship between the three performance characteristics can be computed using Eq. 9.20

$$D = \frac{v_p}{S} \qquad (9.20)$$

Table 9.24 lists the level-of-service criteria for multilane highways in terms of maximum density, average speed, and maximum volume-to-capacity ratio. Figure 9.7 illustrates the level-of-service regimes. The definition of each level of service, A through F, is as follows:

- **Level of Service A.** Travel conditions are completely free flow. The only constraint on the operation of vehicles lies in the geometric features of the roadway and individual driver preferences. Maneuverability within the traffic stream is good, and minor disruptions to traffic are easily absorbed without an effect on travel speed.
- **Level of Service B.** Travel conditions are at free flow. The presence of other vehicles is noticed but is not a constraint on the operation of vehicles as are the geometric features of the roadway and individual driver preferences. Minor disruptions are easily absorbed, although localized reductions in LOS are noted.
- **Level of Service C.** Traffic density begins to influence operations. The ability to maneuver within the traffic stream is affected by other vehicles. Travel speeds show some reduction when free-flow speeds exceed 50 mi/h. Minor disruptions may be expected to cause serious local deterioration in service, and queues may begin to form.
- **Level of Service D.** The ability to maneuver is severely restricted due to congestion. Travel speeds are reduced as volumes increase. Minor disruptions may be expected to cause serious local deterioration in service, and queues may begin to form.

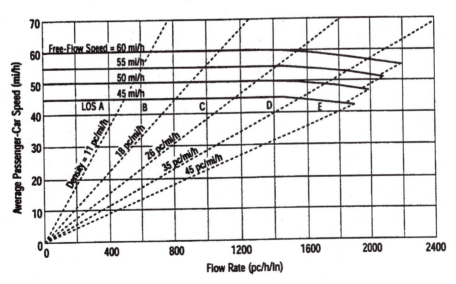

Figure 9.7 Speed-Flow Curves with Level-of-Service Criteria for Multilane Highways

SOURCE: *Highway Capacity Manual,* Special Report 209, Fourth Edition, Transportation Research Board, National Research Council, Washington, D.C., 2000. Used with permission.

- **Level of Service E.** Operations are unstable and at or near capacity. Densities vary depending on the free-flow speed. Vehicles operate at the minimum spacing for which uniform flow can be maintained. Disruptions cannot be easily dissipated and usually result in the formation of queues and the deterioration of service.
- **Level of Service F.** A forced breakdown of flow occurs at the point where the number of vehicles that arrive at a point exceed the number of vehicles discharged or when forecast demand exceeds capacity. Queues form at the breakdown point, while at sections downstream they may appear to be at capacity. Operations are highly unstable.

Calculating the Flow Rate for a Multilane Highway. The flow rate in pc/h/ln for a multilane highway is computed using Eq. 9.21.

$$v_p = \frac{V}{(PHF)(N)(f_p)(f_{HV})} \tag{9.21}$$

where

v_p = 15-minute passenger-car equivalent flow rate, pc/h/ln
V = hourly peak vehicle volume, veh/h in one direction
N = number of travel lanes in one direction (2 or 3)
f_p = driver population factor. Range: 0.85–1.00 Use 1.00 for commuter traffic.
If there is significant recreational or weekend traffic, the value is reduced.
f_{HV} = heavy vehicle adjustment factor (Eq. 9.4)

$$f_{HV} = \frac{1}{1 + P_T(E_T - 1) + P_R(E_R - 1)} \tag{9.4}$$

The computation of f_{HV} requires that the passenger car equivalents for trucks, E_T, and for recreational vehicles, E_R, be determined as a function of grade and length of analysis section. There are two situations that must be considered:

- Extended General Segments in which the terrain is level, rolling or mountainous. The values of E_T and E_R are obtained from Table 9.25.
- Specific Grades. These are defined as segments that are:
    > 1.0 mi and < 3 percent, or
    > 0.5 mi and > 3 percent
    Upgrades: The values of E_T and E_R are obtained from Tables 9.26 and 9.27.
    Downgrades: The value of E_T is obtained from Table 9.28. RVs are treated as if they were on level terrain.

To determine the grade of the highway profile for the analysis section, three situations are possible, each involving a different computational technique:

- Highway profile is a constant grade, computed using Eq. 9.22.
    $$G = (100)\frac{H}{L} \tag{9.22}$$
    G = grade in percent
    H = difference in elevation between the beginning and end of the section, ft.
    L = horizontal distance between the beginning and end of the analysis section, ft.

- Highway profile consists of two or more segments of different grades. The value of the grade for an analysis segment is determined for two possible conditions.
  (1) Segment length is < 4000 ft or grade is < 4 percent.
      Use Eq. 9.22 to compute an average grade.
  (2) Segment length is > 4000 ft and/or grade > 4 percent. Use truck performance curves as shown in Figure 9.13. An explanation of the use of performance curves to compute a composite grade is provided in Example 9.13.

Calculating the Average Passenger Car Speed, (S), Density (D) and Level of Service (LOS) for a Multilane Highway.   Average passenger car speed is depicted in Figure 9.7 and is one of the three variables that can be used to determine the level of service. As shown in the figure, the value of (S) is a constant equal to the free-flow speed (FFS) up to a flow rate of 1400 pc/h/ln. The following steps are used to calculate average passenger car speed:

Step 1.   Compute the Value of Free-Flow Speed. Use Eq. 9.23 to estimate FFS:

$$FFS = BFFS - f_{LW} - f_{LC} - f_M - f_A$$

FFS = estimated free-flow speed, mi/h
$BFFS$ = base free-flow speed, mi/h. In the absence of field data a default value of
    60 mi/h is used for rural/suburban miltilane highways
f_{LW} = adjustment for lane width (Table 9.29)
f_{LC} = adjustment for lateral clearance (Table 9.30)
f_M = adjustment for median type (Table 9.31)
f_A = adjustment for access point density (Table 9.32)

Step 2.   Compute the value of flow rate, v, pc/h/ln. Use Eq. 9.20.
Step 3.   Construct a Speed-Flow Curve of the same shape as shown in Figure 9.17. Interpolate between two curves that span the value of *FFS* obtained in Step 1.
Step 4.   Determine the value of average passenger car speed, *S*.
    Read up from the value of flow rate v, obtained in Step 2, to the intersection of the FFS curve. (Note that if v is 1400 pc/h/ln or less, the value of APCS is the same as that of FFS.)
Step 5.   Compute the Density, pc/mi/in. Use Eq. 9.20.
Step 6.   Determine the Value of Level of Service (LOS). Use the value of density, computed in Step 4 and enter Table 9.24.

Example 9.10   Determining the LOS of a Multilane Highway Segment of Uniform Grade

A 3200 ft segment of 3.25 mi four lane undivided multilane highway in a suburban area is at a 1.5 percent grade. The highway is in level terrain, and lane widths are 11 ft. The measured free-flow speed is 46.0 mi/h. The peak-hour volume is 1900 veh/h, PHF is 0.90, and there are 13 percent trucks and 2 percent RVs.

Determine the LOS, speed, and density, for upgrade and downgrade.

Solution:

Compute v_p using Eq. 9.21
Input data:

$$V = 1900 \text{ veh/h}$$
$$PHF = 0.90$$
$$N = 2$$
$$f_p = 1.00$$

$f_{HV} = 0.935$ computed from Eq. 9.4

$E_T = 1.5, E_R = 1.2$ (Table 9.25) since 1.5 percent grade is considered level terrain

$P_T = 0.13, P_R = 0.02$

$$f_{HV} = \frac{1}{1 + P_T(E_T - 1) + P_R(E_R - 1)} = \frac{1}{1 + 0.13(1.5 - 1) + 0.02(1.2 - 1)} = 0.935$$

$$v_p = \frac{V}{(PHF)(N)(f_p)(f_{HV})} = \frac{1900}{(0.90)(2)(1.00)(0.935)} = 1129 \text{ pc/h/ln}$$

$S = FFS = 46$ mi/h (since $v_p < 1400$)

Compute density from Eq. 9.20

$$D = \frac{v_p}{S} = \frac{1129}{46} = 24.5 \text{ pc/mi/ln}$$

LOS C (Table 9.24)

For the upgrade direction:
Compute v_p using Eq. 9.20
Input data:

$$V = 1900$$
$$PHF = 0.9$$
$$N = 2$$
$$f_p = 1.00$$

$f_{HV} = 0.905$ computed from Eq. 9.4

$E_T = 1.5$, (Table 9.25) $E_R = 3.0$ (Table 9.26)

$P_T = 0.13, P_R = 0.02$

$$f_{HV} = \frac{1}{1 + P_T(E_T - 1) + P_R(E_R - 1)} = \frac{1}{1 + 0.13(1.5 - 1) + 0.02(3.0 - 1)} = 0.905$$

$$v_p = \frac{V}{(PHF)(N)(f_p)(f_{HV})} = \frac{1900}{(0.90)(2)(1.00)(0.905)} = 1166 \text{ pc/h/ln}$$

FFS = 46 mi/h (since $v < 1400$)

Compute density from Eq. 9.20

$$D = \frac{v_p}{S} = \frac{1166}{46} = 25.3 \text{ pc/mi/ln}$$

LOS C (Table 9.24)

BASIC FREEWAY SECTIONS

A freeway is a divided highway with full access control and two or more lanes in each direction for the exclusive use of moving traffic. Signalized or stop-controlled, at-grade intersections or direct access to adjacent land use are not permitted in order to insure the uninterrupted flow of vehicles. Opposing traffic is separated by a raised barrier, an at-grade median, or a raised traffic island. A freeway is composed of three elements: basic freeway sections, weaving areas, and ramp junctions.

Basic freeway sections are segments of the freeway that are outside of the influence area of ramps or weaving areas. Merging or diverging occurs where on- or off-ramps join the basic freeway section. Weaving occurs when vehicles cross each others' path while traveling on freeway lanes. Figure 9.8 illustrates a basic freeway section.

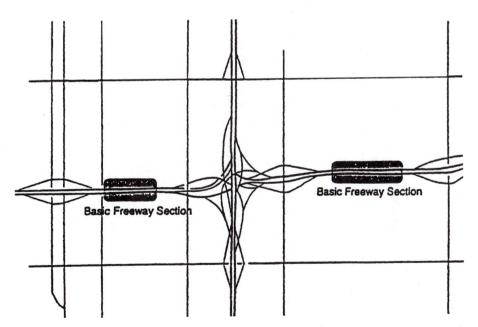

Figure 9.8   Example of Basic Freeway Section

SOURCE: *Highway Capacity Manual,* Special Report 209, Fourth Edition, Transportation Research Board, National Research Council, Washington, D.C., 2000. Used with permission.

The exact point at which a basic freeway section begins or ends—that is, where the influence of weaving areas and ramp junctions has dissipated—depends on local conditions, particularly the level of service operating at the time. If traffic flow is light, the influence may be negligible, whereas under congested conditions, queues may be extensive.

The speed-flow–density relationship existing on a basic freeway section illustrated in Figure 9.9 depends on the prevailing traffic and roadway conditions. Base free-flow conditions include the following freeway characteristics:

- Lanes are 12 ft wide.
- Lateral clearance between the edge of a right lane and an obstacle is 6 ft or greater.
- There are no trucks, buses, or RVs in the traffic stream.
- Urban freeways are five lanes in each direction.
- Interchanges are spaced at least 2 mi apart.
- Grades do not exceed 2 percent.
- Drivers are familiar with the freeway.

FREEWAY CAPACITY

The *capacity* of a freeway is the maximum sustained 15-min rate of flow, expressed in passenger cars per hour per lane (pc/h/ln), which can be accommodated by a uniform freeway segment under prevailing traffic and roadway conditions in one direction. The roadway conditions are the geometric characteristics of the freeway segment under study, which include number and width of lanes, right shoulder lateral clearance, interchange spacing, and grade. The traffic conditions are flow characteristics including the percentage compo-

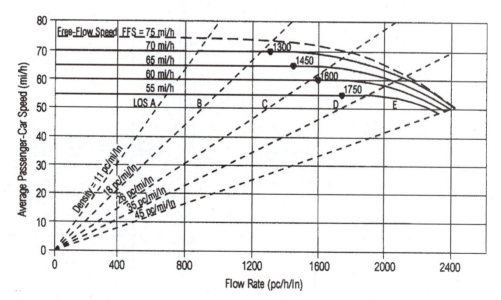

Figure 9.9 Speed-Flow Curves and Level-of-Service for Basic Freeway Segments

SOURCE: *Highway Capacity Manual,* Special Report 209, Fourth Edition, Transportation Research Board, National Research Council, Washington, D.C., 2000. Used with permission.

sition of vehicle types and the extent to which drivers are familiar with the freeway segment. Conditions of free-flow speed occur when flow rates are low to moderate (less than 1300 pc/h/ln at 70 mi/h). As flow rates increase beyond 1300, the mean speed of passenger cars in the traffic stream decreases.

Level of Service for Freeway Sections

Level of service (LOS) qualitatively measures both the operating conditions within a traffic system and how these conditions are perceived by drivers and passengers. It is related to the physical characteristics of the highway and the different operating characteristics that can occur when the highway carries different traffic volumes. Although speed-flow-density relationships are the principal factor affecting the level of service of a highway segment under ideal conditions, factors such as lane width, lateral obstruction, traffic composition, grade, speed, and driver population also affect the maximum flow on a given highway segment. The effects of each of these factors on flow are briefly discussed.

- **Lane Width.** Traffic flow tends to be restricted when lane widths are narrower than 12 ft. This is because vehicles have to travel closer together in the lateral direction, and motorists tend to compensate for this by reducing their travel speed.
- **Lateral Clearance.** When roadside or median objects are located too close to the edge of the pavement, motorists in lanes adjacent to the objects tend to shy away from them, resulting in reduced lateral distances between vehicles, a result similar to lane reduction. Drivers compensate by reducing speed. The effect of lateral clearance is more pronounced for the right shoulder than for the median. Figure 9.10 illustrates how vehicles shy away from both roadside and median barriers.
- **Vehicle Equivalents.** The presence of vehicles other than passenger cars such as trucks, buses, and recreational vehicles in a traffic stream reduces the maximum flow on the highway because of their size, operating characteristics, and interaction with other vehicles. Because freeway capacity is measured in terms of pc/h/ln, the number of heavy vehicles in the traffic stream must be converted into an equivalent number of passenger cars. Figure 9.11 illustrates the effect of trucks and other heavy vehicles on freeway traffic.
- **Grade.** The effect of a grade depends on both the length and the slope of the grade. Traffic operations are significantly affected when grades of 3 percent or greater are longer than $1/4$ mi and when grades are less than 3 percent and longer than $1/2$ mi. The effect of heavy vehicles on such grades is much greater than that of passenger vehicles.
- **Speed.** *Space mean speed,* defined in Chapter 6, is used in level-of-service analysis.
- **Driver Population.** Under ideal conditions, the driver population consists primarily of commuters. However, it is known that other driver populations do not exhibit the same characteristics. For example, recreational traffic capacities can be as much as 20 percent lower than for commuter traffic.
- **Interchange Spacing.** Short freeway sections, as found in urban areas, operate at lower free-flow speeds than longer ones, where interchanges are less frequent. The ideal spacing is 2 mi or more (0.5 interchanges per mi). The minimum average interchange spacing is 0.5 mi (2 interchanges per mi).

(a) Note how vehicles shy away from both roadside and median barriers, driving as close to the lane marking as possible. The existence of narrow lanes compounds the problem, making it difficult for two vehicles to travel alongside each other.

(b) In this case, vehicles shy away from the roadside barrier. This causes the placement of vehicles in each lane to be skewed toward the median. This is also an indication that the median barrier illustrated here does not present an obstruction to drivers.

Figure 9.10   Effect of Lane Width and Lateral Clearance on Traffic Flow

SOURCE: *Highway Capacity Manual,* Special Report 209, Fourth Edition, Transportation Research Board, National Research Council, Washington, D.C., 2000. Used with permission.

(a) Note the formation of large gaps in front of slow-moving trucks climbing the upgrade.

(b) Even on relatively level terrain, the appearance of large gaps in front of trucks or other heavy vehicles is unavoidable.

Figure 9.11   Effect of Trucks and Other Heavy Vehicles on Traffic Flow

SOURCE: *Highway Capacity Manual,* Special Report 209, Fourth Edition, Transportation Research Board, National Research Council, Washington, D.C., 2000. Used with permission.

As with multilane highways, any two of the following three performance characteristics can describe the level of service (LOS) for a basic freeway section:

v_p: Flow rate in passenger, pc/h/ln
S: Average passenger car speed, mi/h
D: Density (number of cars per mi), pc/mi/ln

The relationship between the three performance characteristics is as noted in Eq. 9.20.

$$D = \frac{v_p}{S} \tag{9.20}$$

Table 9.33 lists the level-of-service criteria for basic freeway sections in terms of free-flow speed and density. Figure 9.11 illustrates the level-of-service (LOS) regimes as a function of flow rate and density, and Figure 9.12 illustrates each level of service. The definition of each level of service A through F is as follows:

Level of Service A

Free-flow operations in which vehicles are completely unimpeded in their ability to maneuver. Under these conditions, motorists experience a high level of physical and psychological comfort, and the effects of incidents or point breakdowns are easily absorbed.

Level of Service B

Traffic is moving under reasonably free-flow conditions, and free-flow speeds are sustained. The ability to maneuver within the traffic stream is only slightly restricted. A high level of physical and psychological comfort is provided and the effects of minor incidents and point breakdowns are easily absorbed.

Level of Service C

Speeds are at or near the free-flow speed, but freedom to maneuver is noticeably restricted. Lane changes require more care and vigilance by the driver. When minor incidents occur, local deterioration in service will be substantial. Queues may be expected to form behind any significant blockage.

Level of Service D

Speeds can begin to decline slightly and density increases more quickly with increasing flows. Freedom to maneuver is more noticeably limited, and drivers experience reduced physical and psychological comfort. Vehicle spacings average 165 ft (8 car lengths) and maximum density is 32 pc/mi/ln. Because there is so little space to absorb disruptions, minor incidents can be expected to create queuing.

Level of Service E

Operations are volatile because there are virtually no useable gaps. Maneuvers such as lane changes or merging of traffic from entrance ramps will result in a disturbance of the traffic stream. Minor incidents result in immediate and extensive queuing. Capacity is reached at its highest density value of 45 pc/mi/ln.

LOS A

LOS B

LOS C

LOS D

LOS E

LOS F

Figure 9.12 Levels of Service for Freeways

SOURCE: *Highway Capacity Manual,* Special Report 209, Fourth Edition, Transportation Research Board, National Research Council, Washington, D.C., 2000. Used with permission.

Level of Service F

Operation is under breakdown conditions in vehicular flow. These conditions prevail in queues behind freeway sections experiencing temporary or long-term reductions in capacity. The flow conditions are such that the number of vehicles that can pass a point is less than the number of vehicles arriving upstream of the point, or at merging or weaving areas where the number of vehicles arriving is greater than the number discharged. Breakdown occurs when the ratio of forecasted demand to capacity exceeds 1.00.

Calculating the Flow Rate for a Basic Freeway Section.   The formula for the flow rate in passenger cars per hour per lane for a basic freeway section is

$$v_p = \frac{V}{(PHF)(N)(f_p)(f_{HV})} \tag{9.21}$$

where

v_p = 15 min passenger-car equivalent flow rate, pc/h/ln
V = hourly peak vehicle volume, veh/h in one direction
PHF = peak hour factor
N = number of travel lanes in one direction
f_p = driver population factor. Range: 0.85–1.00 Use 1.00 for commuter traffic. If there is significant recreational or weekend traffic, the value is reduced.
f_{HV} = heavy vehicle adjustment factor (Eq. 9.4)

$$f_{HV} = \frac{1}{1 + P_T(E_T - 1) + P_R(E_R - 1)}$$

The flow rate obtained from Eq. 9.21 will be achieved only if good pavement and weather conditions exist and there are no incidents on the freeway segments. If these conditions do not exist, the actual flow that will be achieved may be less.

The heavy-vehicle volumes in a traffic stream must be converted to an equivalent flow rate expressed in passenger cars per hour per lane. First, the passenger car equivalent (PCE) of each truck/bus (E_T) or recreational vehicle (E_R) is determined for the prevailing traffic and roadway conditions from Table 9.25. These numbers represent the number of passenger cars that would use up the same space on the highway as one truck/bus or recreational vehicle under the prevailing conditions. Trucks and buses are treated identically because it has been determined that their traffic flow characteristics are similar. Second, E_T and E_R are used with the proportions of each type of vehicle, P_T and P_R, to compute the adjustment factor, f_{HV}, using Eq. 9.4.

The extent to which the presence of a truck affects the traffic stream also depends on the grade of the segment being considered. PCEs for trucks can be determined for three grade conditions: extended general freeway segments, specific upgrades, and specific downgrades.

Extended General Freeway Segments.   These occur when a single grade is not too long or too steep to have significant impact on capacity. Instead, upgrades, downgrades, and level sections are all considered to be extended general freeway segments. Grades of at least 3 percent and less than $^1/_4$ mi, *or* grades less than 3 percent and less than $^1/_2$ mi, are included. The PCEs for these conditions are given in Table 9.25. PCE values are affected by the type of terrain, which is classified as level, rolling, or mountainous.

Freeway segments are considered to be on *level terrain* if the combination of grades and horizontal alignment permits heavy vehicles to maintain the same speed as passenger cars. Grades are generally short and no greater than 2 percent.

Freeway segments are considered to be on *rolling terrain* if the combination of grades and horizontal alignment causes heavy vehicles to reduce their speeds to values substantially below those of passenger cars but not to travel at crawl speeds for any significant length of time.

Freeway segments are considered to be in *mountainous terrain* if the combination of grades and horizontal alignment causes heavy vehicles to operate at crawl speeds for a significant distance or at frequent intervals. Crawl speed is the maximum sustained speed that trucks can maintain on an extended upgrade of a given percent.

PCEs for Specific Upgrades.   Any freeway grade of 3 percent or more and longer than $^1/_4$ mi, or a grade of less than 3 percent and longer than $^1/_2$ mi, should be considered as a separate segment. Specific grades are analyzed individually for downgrade and upgrade conditions. The segment is considered either as a grade of constant percentage or as a series of grades.

The variety of trucks and recreational vehicles with varying characteristics results in a wide range of performance capabilities on specific grades. The truck population on freeways has an average weight-to-horsepower ratio of 125 to 150 lb/hp, and this range is used in determining PCEs for trucks and buses on specific upgrades. Recreational vehicles vary considerably in both type and characteristics and range from cars pulling trailers to large self-contained mobile homes. Further, unlike trucks, they are not driven by professionals, and the skill levels of recreational vehicle drivers vary greatly. Typical weight-to-horsepower ratios for recreational vehicles range between 30–60 lb/hp.

Tables 9.26 and 9.27 give PCE values for trucks/buses (E_T) and for recreational vehicles, (E_R) respectively, traveling on specific upgrades for different grades and different percentages of heavy vehicles within the traffic stream. For example, the effect on other traffic of a truck or bus traveling up a grade is magnified with increasing segment length and grade because the vehicle slows down. The PCEs selected should be associated with the point on the freeway where the effect is greatest, which is usually at the end of the grade, although a ramp junction at midgrade could be a critical point. It has been suggested that the grade length be obtained from a profile of the road, including the tangent portion, plus one-fourth of the length of the vertical curves at the beginning and the end of the grade. Where two consecutive upgrades are connected by a vertical curve, half the vertical curve length is added to each portion of the grade. However, this guideline may not apply for some specific conditions. For example, to determine the effect of an on-ramp at an upgrade section of a freeway, the length used is that up to the ramp junction.

PCEs for Specific Downgrades.   If the downgrade is not so severe as to cause trucks to shift into low gear, they may be treated as if they were on level segments. Where grades are severe and require that trucks downshift, the effect on car equivalents (E_T) is greater, as shown in Table 9.28. For recreational vehicles, a downgrade is treated as if it were level.

Composite Grades. When a segment of freeway consists of two or more consecutive upgrades with different slopes, the PCE of heavy vehicles is determined by using one of two techniques. One technique determines the average grade of the segment by finding the total rise in elevation and dividing it by the total horizontal distance. This average grade is then used with Tables 9.26 and 9.27. The average grade technique is valid for conditions where grades in all subsections are less than 4 percent or the total length of the composite grade is less than 4000 ft.

Example 9.11 Computation of PCE for Consecutive Upgrades Using Average Grades

A segment of freeway consists of two consecutive upgrades of 3 percent, 2000 ft long and 2 percent, 1500 ft long. Determine the PCE of trucks/buses and recreational vehicles on this composite upgrade if 6 percent of the vehicles are trucks and buses and 10 percent are recreational vehicles.

Solution: The average grade technique can be used since subsection grades are less than 4 percent and the total length is less than 4000 ft.

$$\text{total rise} = (0.03 \times 2000) + (0.02 \times 1500) = 90 \text{ ft}$$

$$\text{average grade} = 90/(2000 + 1500) = 0.026, \text{ or } 2.6\%$$

$$\text{total length} = 3500/5280 = 0.66 \text{ mi}$$

Enter Table 9.26, with 2–3 percent grade and length $^1/_2$–$^3/_4$ mi, and obtain E_T for trucks/buses of 1.5.
Enter Table 9.27, with 2–3 percent grade and length greater than $^1/_2$ mi, and obtain $E_R = 1.5$.

The second technique for determining the PCE of heavy vehicles on consecutive upgrades is to estimate the value of equivalent continuous grade G_E that would result in the final speed for trucks that is the same as that which would result from the actual series of consecutive grades of the same length. This technique should be used if any single portion of the consecutive grade exceeds 4 percent or if the total length of grade (measured horizontally) exceeds 4000 ft.

Truck acceleration/deceleration curves are used based on a vehicle with an average weight-to-horsepower ratio of 200 lb/hp, which represents a somewhat heavier vehicle than the "typical" truck of 125 to 150 lb/hp used to determine PCE values. This is a conservative approach that accounts for the greater influence heavier vehicles have over light vehicles on operation. Figure 9.13 is a performance curve for a standard 200 lb/hp truck, depicting the relationship between speed (mi/h) and length of grade for upgrades ranging

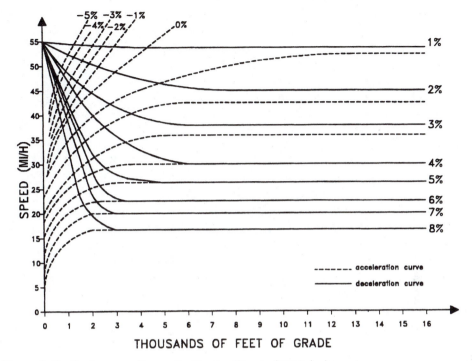

Figure 9.13   Performance Curves for Standard Trucks (200 lb/hp)

SOURCE: *Highway Capacity Manual,* Special Report 209, Fourth Edition, Transportation Research Board, National Research Council, Washington, D.C., 2000. Used with permission.

from 1 to 8 percent and downgrades of 0 to 5 percent for an entry speed of 55 mi/h. For example, at a 5 percent grade, a truck will slow to a crawl speed of 27 mi/h in a distance of 5000 ft.

Figure 9.13 can be used to determine the single constant grade G_E for a series of consecutive grades of length L_{CG}. The procedure, illustrated in the following example, requires that the speed V_{EG} of a truck at the end of the consecutive grades be determined using Figure 9.11. Then, knowing the length of the composite grade, L_{CG}, and the truck speed at the end of the grade, V_{EG}, an equivalent grade G_E is determined, again using Figure 9.13.

Example 9.12   Computation of PCE for a Consecutive Upgrade, Using Performance Curves

A consecutive upgrade consists of two sections, the first of 2 percent grade and 5000 ft long, and the second of 6 percent grade and 5000 ft long. Trucks comprise 10 percent of traffic, and recreational vehicles comprise 6 percent. Determine (a) the equivalent grade and (b) the PCEs. Entry speed is 55 mi/h.

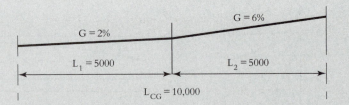

Solution:

(a) Since the lengths of both grades exceed 4000 ft, the average grade technique does not apply. Instead, the performance curve for standard trucks must be used, as illustrated in Figure 9.14.

    1. Find the point on the 2 percent curve that intersects with length = 5000 ft. In this case (point 1), that point signifies that the speed of a truck at the end of the first grade (point 2) is 47 mi/h.

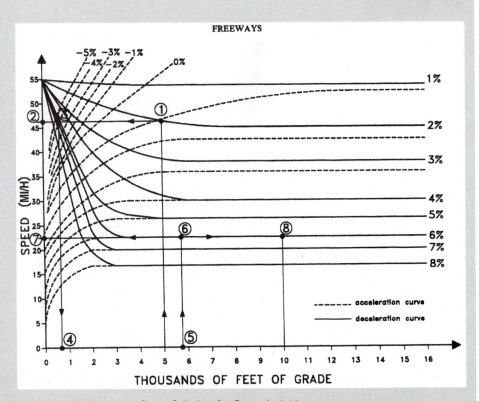

Figure 9.14 Performance Curve Solution for Example 9.12

SOURCE: *Highway Capacity Manual,* Special Report 209, Fourth Edition, Transportation Research Board, National Research Council, Washington, D.C., 2000. Used with permission.

2. Since 47 mi/h is also the speed of the truck at the beginning of the 6 percent grade, draw a line horizontally, at 47 mi/h, that intersects the 6 percent grade curve (point 3). The reference distance is located on the horizontal axis (point 4).
3. Find the point on the horizontal axis that has a value 5000 ft greater than that at point 4. This point (point 5) represents the distance traveled by a truck on a 6 percent grade whose initial speed is 47 mi/h. The final speed for this truck is found to be 23 mi/h (point 7), which is also the crawl speed, since the 6 percent curve is horizontal at that point.
4. Finally, determine the intersection of the lines $L_{CG} = 10,000$ ft and $V_{EG} = 23$ mi/h (point 8). Since this point lies on the 6 percent grade curve, the equivalent grade, G_E, is 6 percent.

(b) PCEs for typical trucks on specific freeway upgrades:

$$L_{CG} = 10,000/5280 = 1.89 \text{ mi}$$
$$G_E = 6 \text{ percent}$$
$$E_T = 3.5 \text{ (from Table 9.26)}$$
$$E_R = 4.5 \text{ (from Table 9.27)}$$

Note that if the average grade technique had been (incorrectly) used, G_A would have been calculated as $(100 + 300)/10,000$, or 4 percent, resulting in values for E_T of 3.0 and E_R of 2.0, both lower than the correct values of 3.5 and 4.5.

Calculating the Average Passenger Car Speed (S), Density (D), and Level of Service (LOS) for a Basic Freeway Section. Average passenger car speed is depicted in Figure 9.9 and is one of the three variables that can be used to determine the level of service. As shown in the figure, the value of APCS is a constant equal to the free-flow speed (FFS) up to a value of 1300 pc/h/ln at a FFS of 70 mi/h. In order to conduct a capacity analysis, compute v_p and obtain the average passenger car speed (S) from Figure 9.9. Compute the density using Eq. 9.20. The LOS is found in Table 9.33. Four types of problems are solved by capacity analysis:

- Type I. Given the highway volume, number of lanes, and free-flow speed, determine level of service.
- Type II. Given highway volume, level of service, and free-flow speed, determine the average speed.
- Type III. Given the level of service and free-flow speed, determine the hourly flow rate.
- Type IV. Given the highway volume, free-flow speed, and the desired level of service, determine the number of freeway lanes required.

To solve these problems, it is necessary to determine the free-flow speed under prevailing physical and geometric conditions. Free-flow speed is the mean speed of passenger cars measured when the equivalent hourly flow rate is no greater than 1300 pc/h/ln. The free-flow speed can be measured in the field with a speed study, using procedures described in Chapter 4. If speed-study data are not available, the free-flow speed can be determined on the basis of specific characteristics of the freeway section including lane width, number of lanes, right shoulder lateral clearance, and interchange density. Equation 9.22 is used to compute the free-flow speed. We have

$$FFS = BFFS - f_{LW} - f_{LC} - f_N - f_{ID} \qquad (9.22)$$

where
$\quad FFS$ = free-flow speed, mi/h
$\quad BFFS$ = base free-flow speed, mi/h. 70 (urban) or 75 (rural)
$\quad f_{LW}$ = adjustment for lane width (Table 9.34)
$\quad f_{LC}$ = adjustment for right-shoulder lateral clearance (Table 9.35)
$\quad f_N$ = adjustment for number of lanes (Table 9.36)
$\quad f_{ID}$ = adjustment for interchange density (Table 9.37)

Adjustment for Lane Width
When the average lane width of the freeway (or multilane highway) is less than the ideal value of 12 ft, the free-flow speed of 70 or 75 mi/h will be reduced. Table 9.34 provides the values of speed reduction in mi/h for lane widths of 10 and 11 ft.

Adjustment for Right Shoulder Lateral Clearance
The ideal clearance conditions are 6 ft on the right (shoulder) side of the freeway and 2 ft on the left (median) side. When right-shoulder clearances have been reduced, free-flow speed will diminish by the values shown in Table 9.35. For example, if the right-shoulder clearance is 4 ft on a two lane freeway section in one direction, the free-flow speed is reduced by 1.2 mi/h.

Adjustment for Number of Lanes
If the number of freeway lanes in one direction is less than five, the free-flow speed is reduced on urban and suburban highways, as shown in Table 9.36. Rural freeways are usually two lanes in one direction, and the speed reduction value is zero.

Adjustment for Interchange Density
The ideal interchange spacing is 2 mi or greater or 0.5 interchanges per mile. When the number of interchanges per mile increases, there is a corresponding decrease in free-flow speed, as shown in Table 9.37. Interchanges are defined as having one or more on-ramps. Interchange spacing for the basic freeway section is determined by the number of interchanges within a distance of 3 mi upstream and 3 mi downstream of the section, and dividing this number by 6.

Example 9.13 Using Adjustment Factors to Calculate Free-Flow Speed

Determine the free-flow speed of an urban freeway section, if the $BFFS = 70$ mi/h. The data are as follows:

> Number of lanes = 3 (one direction)
> Lane width = 11 ft
> Lateral clearance = 3 ft (right side)
> Interchange density = 1.5 per mile

Solution:

Step 1.   Determine the correction factors using the appropriate table.

$BFFS$ = base free-flow speed, 70 mi/h
f_{LW} = adjustment for lane width, 1.9 mi/h (Table 9.34)
f_{LC} = adjustment for right shoulder lateral clearance, 1.2 mi/h (Table 9.35)
f_N = adjustment for number of lanes, 3.0 mi/h (Table 9.36)
f_{ID} = adjustment for interchange density, 5.0 mi/h (Table 9.37)

Step 2.   Substitute adjustment values into Eq. 9.22.

$$FFS = BFFS - f_{LW} - f_{LC} - f_N - f_{ID}$$
$$= 70 - 1.9 - 1.2 - 3.0 - 5.0 = 58.9 \text{ mi/h (Answer)}$$

Example 9.14   Level of Service of a Freeway Section When the Free-Flow Speed and Hourly Flow Rate Are Known

Determine the level of service for the freeway section described in Example 9.13 if the flow rate is 725 pc/h/ln during the peak 15-min period.

Solution: Use the relationship between speed, flow, and density (Eq. 9.20) to calculate the density in vehicles per mile.
Since $FFS = 70$ mph, $v_p < 1300$, $BFFS = S$

$$D = \frac{v_p}{S} = \frac{725}{58.9} = 12.31 \text{ pc/m /ln}$$

Use Figure 9.7 or Table 9.33 to determine LOS.
Since 12.31 lies between 11 and 18, the LOS is B.

Example 9.15   Computation of Speed, Density, and Level of Service for an Extended Section of a Freeway

Determine the LOS on a regular weekday on a 0.40-mi section of a six-lane freeway with a grade of 2 percent, using the following data:

$$\text{Hourly volume, } V = 3000 \text{ veh/h}$$
$$\text{PHF} = 0.85$$

Traffic Composition:
$$\text{Trucks} = 12 \text{ percent}$$
$$\text{RVs} = 2 \text{ percent}$$
$$\text{Lane width} = 11 \text{ ft}$$
$$\text{Terrain} = \text{level}$$
$$\text{Base free-flow speed} = 70 \text{ mi/h}$$
$$\text{Shoulder width} = 6 \text{ ft}$$
$$\text{Interchange spacing} = 1 \text{ mile}$$
$$\text{Driver population adjustment factor } f_p = 1.0$$

Solution:

Step 1.   Compute heavy vehicle factor using Eq. 9.4.

Determine the correction factors for extended general freeway segments (section is less than 3 percent grade and less than $1/2$ mi long):

PCEs:
  $E_T = 1.5$ (from Table 9.25)
  $E_R = 1.2$ (from Table 9.25)

$$
\begin{aligned}
f_{HV} &= \frac{1}{1 + P_T(E_T - 1) + P_R(E_R - 1)} \\
&= \frac{1}{1 + 0.12(1.5 - 1) + 0.02(1.2 - 1)} \\
&= \frac{1}{1 + 0.06 + 0.004} \\
&= 0.94
\end{aligned}
$$

(Note: Since there are five times as many trucks and buses as RVs, this problem could have been solved assuming 14 percent are trucks, with the same result for f_{HV}.)

Step 2.   Use Eq. 9.21 to determine the peak 15-min passenger-car equivalent flow rate.

$$
\begin{aligned}
v_p &= \frac{V}{\text{PHF} \times N \times f_{HV} \times f_p} \\
&= \frac{3000}{0.85 \times 3 \times 0.94 \times 1.00} \\
&= 1251 \text{ pc/h/ln}
\end{aligned}
$$

Step 3.   Compute the free-flow speed and density (Eq. 9.22).

$$FFS = BFFS - f_{LW} - f_{LC} - f_N - f_{ID}$$

(obtain factors from Tables 9.34–9.37)

= 70 − 1.9 − 0 − 3.0 − 2.5
= 62.6 mi/h

$$D = \frac{v}{S} = \frac{1251}{62.6} = 19.98 \text{ pc/m/ln}$$

Step 4.   Use Figure 9.7 or Table 9.33 to determine level of service.

LOS = C (Since 20 lies between 18 and 26 pc/m/ln)

SUMMARY

Traffic engineers frequently are engaged in evaluating the performance of different facilities of the highway system. These facilities include freeway sections, intersections, ramps, and so forth. The level of service is usually taken as a good indication of how well the particular component is operating. It is a qualitative measure of motorists' perceptions of the operational conditions existing on the facility. A primary objective in traffic engineering is to provide highway facilities that operate at levels of service acceptable to the users of those facilities. Regular evaluation of the level of service at the facilities will help the engineers to determine whether acceptable conditions exist and to identify those locations where improvements may be necessary. Levels of different operating conditions are assigned to different levels of service, ranging from level of service A to level of service F for each facility. The different levels of operating conditions are also related to the volume of traffic that can be accommodated by the specific component. This amount of traffic is also related to the capacity of the facility.

This chapter has described the procedures for determining the level of service on freeway sections as presented in the *Highway Capacity Manual 2000*. The procedures are based on the results of several major studies, from which several empirical expressions have been developed to determine the maximum flow rate at a specific facility under prevailing conditions. The chapter has focused on freeways, multilane highways, and two-lane highways. In the next chapter, we discuss capacity and level of service for signalized intersections.

PROBLEMS

9-1   Define the elements of a Class I and a Class II two-lane highway.

9-2   What are the two measures used to describe service quality for a two-lane highway? Which of the measures is used to determine LOS for Class I and Class II highways?

9-3   Describe the traffic characteristics associated with the six levels of service for two-lane highways.

9-4   The following values of PTSF and ATS have been determined for three separate two-lane segments. Determine the LOS if the segments are (a) Class I, and (b) Class II.

Segment	PTSF %	ATS (mi/h)
10	25	52
11	46	39
12	67	39

9-5 Determine the PTSF for a 4.5 mi two-lane highway segment in level terrain. Traffic volumes (two-way) are 1100 veh/h. Trucks: 10 percent, RVs: 7 percent, PHF: 0.97, directional split: 60/40, no passing zones: 40 percent.

9-6 Use the data provided in Problem 9-5 to estimate the ATS. Base free flow speed: 55 mi/h, lane width: 11 ft, shoulder width: 3 ft, access points/mi: 15.

9-7 Use the results of Problems 9-5 and 9-6, to compute: LOS, v/c, veh-mi in peak 15 min and peak hour, and total travel time in peak 15 min.

9-8 Use the data provided in text Examples 9.5 and 9.6 to determine PTSF and ATS in the peak direction if northbound volume is 1000 veh/h and southbound volume is 600 veh/h.

9-9 Use the data and results obtained in Problem 9-8 to determine the level of service of a two-lane section if a passing lane 1.5 mi long is added. The passing lane begins 0.75 mi from the starting point of the analysis segment.

9-10 Briefly describe the traffic characteristics associated with each of the levels of service for basic freeway sections.

9-11 Describe the factors that affect the level of service of a freeway section and the impact each has on flow.

9-12 A freeway is being designed to carry a heavy volume of 5000 veh/h on a regular weekday in a rolling terrain. If the PHF is 0.9 and the traffic consists of 90 percent passenger cars and 10 percent trucks, determine the number of 12-ft lanes required in each direction if the highway is to operate at level of service C. The free-flow speed is 70 mi/h, there is no lateral obstruction, and interchanges are 3 mi apart.

9-13 A section of a four-lane (two lanes in each direction) freeway that is 2 mi long and has a sustained grade of 4 percent is to be improved to carry on a regular weekday a heavy volume of 3000 veh/h, consisting of 85 percent passenger cars, 10 percent trucks, 2 percent buses, and 3 percent recreational vehicles. The PHF is 0.95. Determine the additional number of 12-ft lanes required in each direction if the road is to operate at level of service B. The free-flow speed is 70 mi/h, there is a lateral obstruction 5 ft from the pavement on the right side of the road, and interchange spacing is 1 mi.

9-14 A particular roadway segment has a 6000-ft section of 3 percent upgrade, followed by a 5000-ft section of 5 percent upgrade. The traffic on the road includes 8 percent trucks and 4 percent recreational vehicles. Determine the PCEs.

ADDITIONAL READING

Highway Capacity Manual, Special Report 209, Fourth Edition, Transportation Research Board, National Research Council, Washington, D.C., 2000.

CHAPTER 9 APPENDIX: TABLES

All tables in this appendix are reprinted from *Highway Capacity Manual,* Special Report 209, Fourth Edition, Transportation Research Board, National Research Council, Washington, D.C., 2000. Used with permission.

Table 9.1 Level-of-Service Criteria for Two-Lane Highways in Class I

LOS	Percent Time-Spent-Following	Average Travel Speed (mi/h)
A	≤ 35	> 55
B	> 35–50	> 50–55
C	> 50–65	> 45–50
D	>65–80	> 40–45
E	> 80	≤ 40

NOTE: LOS F applies whenever the flow rate exceeds the segment capacity.

Table 9.2 Level-of-Service Criteria for Two-Lane Highways in Class II

LOS	Percent Time-Spent-Following
A	≤ 40
B	> 40–55
C	> 55–70
D	> 70–85
E	> 85

NOTE: LOS F applies whenever the flow rate exceeds the segment capacity.

Table 9.3 Adjustment ($f_{d/np}$) for Combined Effect of Directional Distribution of Traffic and Percentage of No-Passing Zones on Percent Time-Spent-Following on Two-Way Segments

Two-Way Flow Rate, v_p (pc/h)	Increase in Percent Time-Spent-Following (%)					
	No-Passing Zones (%)					
	0	20	40	60	80	100
Directional Split = 50/50						
≤ 200	0.0	10.1	17.2	20.2	21.0	21.8
400	0.0	12.4	19.0	22.7	23.8	24.8
600	0.0	11.2	16.0	18.7	19.7	20.5
800	0.0	9.0	12.3	14.1	14.5	15.4
1400	0.0	3.6	5.5	6.7	7.3	7.9
2000	0.0	1.8	2.9	3.7	4.1	4.4
2600	0.0	1.1	1.6	2.0	2.3	2.4
3200	0.0	0.7	0.9	1.1	1.2	1.4
Directional Split = 60/40						
≤ 200	1.6	11.8	17.2	22.5	23.1	23.7
400	0.5	11.7	16.2	20.7	21.5	22.2
600	0.0	11.5	15.2	18.9	19.8	20.7
800	0.0	7.6	10.3	13.0	13.7	14.4
1400	0.0	3.7	5.4	7.1	7.6	8.1
2000	0.0	2.3	3.4	3.6	4.0	4.3
≥ 2600	0.0	0.9	1.4	1.9	2.1	2.2
Directional Split = 70/30						
≤ 200	2.8	13.4	19.1	24.8	25.2	25.5
400	1.1	12.5	17.3	22.0	22.6	23.2
600	0.0	11.6	15.4	19.1	20.0	20.9
800	0.0	7.7	10.5	13.3	14.0	14.6
1400	0.0	3.8	5.6	7.4	7.9	8.3
≥ 2000	0.0	1.4	4.9	3.5	3.9	4.2
Directional Split = 80/20						
≤ 200	5.1	17.5	24.3	31.0	31.3	31.6
400	2.5	15.8	21.5	27.1	27.6	28.0
600	0.0	14.0	18.6	23.2	23.9	24.5
800	0.0	9.3	12.7	16.0	16.5	17.0
1400	0.0	4.6	6.7	8.7	9.1	9.5
≥ 2000	0.0	2.4	3.4	4.5	4.7	4.9
Directional Split = 90/10						
≤ 200	5.6	21.6	29.4	37.2	37.4	37.6
400	2.4	19.0	25.6	32.2	32.5	32.8
600	0.0	16.3	21.8	27.2	27.6	28.0
800	0.0	10.9	14.8	18.6	19.0	19.4
≥ 1400	0.0	5.5	7.8	10.0	10.4	10.7

Table 9.4   Grade Adjustment Factor (f_G) to Determine Percent Time-Spent-Following on Two-Way and Directional Segments

Range of Two-Way Flow Rates (pc/h)	Range of Directional Flow Rates (pc/h)	Type of Terrain	
		Level	Rolling
0–600	0–300	1.00	0.77
> 600–1200	> 300–600	1.00	0.94
> 1200	> 600	1.00	1.00

Table 9.5   Passenger-Car Equivalents for Trucks (E_T) and RVs (E_R) to Determine Percent Time-Spent-Following on Two-Way and Directional Segments

Vehicle Type	Range of Two-Way Flow Rates (pc/h)	Range of Directional Flow Rates (pc/h)	Type of Terrain	
			Level	Rolling
Trucks, E_T	0–600	0–300	1.1	1.8
	> 600–1,200	> 300–600	1.1	1.5
	>1,200	> 600	1.0	1.0
RVs, E_R	0–600	0–300	1.0	1.0
	> 600–1,200	> 300–600	1.0	1.0
	>1,200	> 600	1.0	1.0

Table 9.6 Adjustment (f_{np}) for Effect of No-Passing Zones on Average Travel Speed on Two-Way Segments

Two-Way Demand Flow Rate, v_p (pc/h)	Reduction in Average Travel Speed (mi/h) No-Passing Zones (%)					
	0	20	40	60	80	100
0	0.0	0.0	0.0	0.0	0.0	0.0
200	0.0	0.6	1.4	2.4	2.6	3.5
400	0.0	1.7	2.7	3.5	3.9	4.5
600	0.0	1.6	2.4	3.0	3.4	3.9
800	0.0	1.4	1.9	2.4	2.7	3.0
1000	0.0	1.1	1.6	2.0	2.2	2.6
1200	0.0	0.8	1.2	1.6	1.9	2.1
1400	0.0	0.6	0.9	1.2	1.4	1.7
1600	0.0	0.6	0.8	1.1	1.3	1.5
1800	0.0	0.5	0.7	1.0	1.1	1.3
2000	0.0	0.5	0.6	0.9	1.0	1.1
2200	0.0	0.5	0.6	0.9	0.9	1.1
2400	0.0	0.5	0.6	0.8	0.9	1.1
2600	0.0	0.5	0.6	0.8	0.9	1.0
2800	0.0	0.5	0.6	0.7	0.8	0.9
3000	0.0	0.5	0.6	0.7	0.7	0.8
3200	0.0	0.5	0.6	0.6	0.6	0.7

Table 9.7 Grade Adjustment Factor (f_G) to Determine Speeds on Two-Way and Directional Segments

Range of Two-Way Flow Rates (pc/h)	Range of Directional Flow Rates (pc/h)	Type of Terrain	
		Level	Rolling
0–600	0–300	1.00	0.71
> 600–1200	> 300–600	1.00	0.93
> 1200	> 600	1.00	0.99

Table 9.8 Passenger-Car Equivalents for Trucks (E_T) and RVs (E_R) to Determine Speeds on Two-Way and Directional Segments

Vehicle Type	Range of Two-Way Flow Rates (pc/h)	Range of Directional Flow Rates (pc/h)	Type of Terrain	
			Level	Rolling
Trucks, E_T	0–600	0–300	1.7	2.5
	> 600–1,200	> 300–600	1.2	1.9
	> 1,200	> 600	1.1	1.5
RVs, E_R	0–600	0–300	1.0	1.1
	> 600–1,200	> 300–600	1.0	1.1
	> 1,200	> 600	1.0	1.1

Table 9.9 Adjustment (f_{LS}) for Lane Width and Shoulder Width

Lane Width (ft)	Reduction in FFS (mi/h)			
	Shoulder Width (ft)			
	≥ 0 < 2	≥ 2 < 4	≥ 4 < 6	≥ 6
9 < 10	6.4	4.8	3.5	2.2
≥ 10 < 11	5.3	3.7	2.4	1.1
≥ 11 < 12	4.7	3.0	1.7	0.4
≥ 12	4.2	2.6	1.3	0.0

Table 9.10 Adjustment (f_A) for Access-Point Density

Access Points per mi	Reduction in FFS (mi/h)
0	0.0
10	2.5
20	5.0
30	7.5
40	10.0

Table 9.11   Adjustment (f_{np}) to Percent Time-Spent-Following for Percentage of No-Passing Zones in Directional Segments

Opposing Demand Flow Rate, v_o (pc/h)	No-Passing Zones (%)				
	≤ 20	40	60	80	100
FFS = 65 mi/h					
≤ 100	10.1	17.2	20.2	21.0	21.8
200	12.4	19.0	22.7	23.8	24.8
400	9.0	12.3	14.1	14.4	15.4
600	5.3	7.7	9.2	9.7	10.4
800	3.0	4.6	5.7	6.2	6.7
1000	1.8	2.9	3.7	4.1	4.4
1200	1.3	2.0	2.6	2.9	3.1
1400	0.9	1.4	1.7	1.9	2.1
≥ 1600	0.7	0.9	1.1	1.2	1.4
FFS = 60 mi/h					
≤ 100	8.4	14.9	20.9	22.8	26.6
200	11.5	18.2	24.1	26.2	29.7
400	8.6	12.1	14.8	15.9	18.1
600	5.1	7.5	9.6	10.6	12.1
800	2.8	4.5	5.9	6.7	7.7
1000	1.6	2.8	3.7	4.3	4.9
1200	1.2	1.9	2.6	3.0	3.4
1400	0.8	1.3	1.7	2.0	2.3
≥ 1600	0.6	0.9	1.1	1.2	1.5
FFS = 55 mi/h					
≤ 100	6.7	12.7	21.7	24.5	31.3
200	10.5	17.5	25.4	28.6	34.7
400	8.3	11.8	15.5	17.5	20.7
600	4.9	7.3	10.0	11.5	13.9
800	2.7	4.3	6.1	7.2	8.8
1000	1.5	2.7	3.8	4.5	5.4
1200	1.0	1.8	2.6	3.1	3.8
1400	0.7	1.2	1.7	2.0	2.4
≥ 1600	0.6	0.9	1.2	1.3	1.5

Continued

Table 9.11   Adjustment (f_{np}) to Percent Time-Spent-Following for Percentage of No-Passing Zones in Directional Segments (*continued*)

Opposing Demand Flow Rate, v_o (pc/h)	No-Passing Zones (%)				
	≤ 20	40	60	80	100
FFS = 50 mi/h					
≤ 100	5.0	10.4	22.4	26.3	36.1
200	9.6	16.7	26.8	31.0	39.6
400	7.9	11.6	16.2	19.0	23.4
600	4.7	7.1	10.4	12.4	15.6
800	2.5	4.2	6.3	7.7	9.8
1000	1.3	2.6	3.8	4.7	5.9
1200	0.9	1.7	2.6	3.2	4.1
1400	0.6	1.1	1.7	2.1	2.6
≥ 1600	0.5	0.9	1.2	1.3	1.6
FFS = 45 mi/h					
≤ 100	3.7	8.5	23.2	28.2	41.6
200	8.7	16.0	28.2	33.6	45.2
400	7.5	11.4	16.9	20.7	26.4
600	4.5	6.9	10.8	13.4	17.6
800	2.3	4.1	6.5	8.2	11.0
1000	1.2	2.5	3.8	4.9	6.4
1200	0.8	1.6	2.6	3.3	4.5
1400	0.5	1.0	1.7	2.2	2.8
≥ 1600	0.4	0.9	1.2	1.3	1.7

Table 9.12   Values of Coefficients (*a, b*) Used in Estimating Percent Time-Spent-Following for Directional Segments

Opposing Demand Flow Rate, v_o (pc/h)	a	b
≤ 200	−0.013	0.668
400	−0.057	0.479
600	−0.100	0.413
800	−0.173	0.349
1000	−0.320	0.276
1200	−0.430	0.242
1400	−0.522	0.225
≥ 1600	−0.665	0.119

Table 9.13 Adjustment (f_{np}) to Average Travel Speed for Percentage of No-Passing Zones in Directional Segments

Opposing Demand Flow Rate, v_o (pc/h)	No-Passing Zones (%)				
	≤ 20	40	60	80	100
FFS = 65 mi/h					
≤ 100	1.1	2.2	2.8	3.0	3.1
200	2.2	3.3	3.9	4.0	4.2
400	1.6	2.3	2.7	2.8	2.9
600	1.4	1.5	1.7	1.9	2.0
800	0.7	1.0	1.2	1.4	1.5
1000	0.6	0.8	1.1	1.1	1.2
1200	0.6	0.8	0.9	1.0	1.1
1400	0.6	0.7	0.9	0.9	0.9
≥ 1600	0.6	0.7	0.7	0.7	0.8
FFS = 60 mi/h					
≤ 100	0.7	1.7	2.5	2.8	2.9
200	1.9	2.9	3.7	4.0	4.2
400	1.4	2.0	2.5	2.7	2.9
600	1.1	1.3	1.6	1.9	2.0
800	0.6	0.9	1.1	1.3	1.4
1000	0.6	0.7	0.9	1.1	1.2
1200	0.5	0.7	0.9	0.9	1.1
1400	0.5	0.6	0.8	0.8	0.9
≥ 1600	0.5	0.6	0.7	0.7	0.7
FFS = 55 mi/h					
≤ 100	0.5	1.2	2.2	2.6	2.7
200	1.5	2.4	3.5	3.9	4.1
400	1.3	1.9	2.4	2.7	2.8
600	0.9	1.1	1.6	1.8	1.9
800	0.5	0.7	1.1	1.2	1.4
1000	0.5	0.6	0.8	0.9	1.1
1200	0.5	0.6	0.7	0.9	1.0
1400	0.5	0.6	0.7	0.7	0.9
≥ 1600	0.5	0.5	0.6	0.6	0.7

Continued

Table 9.13 Adjustment (f_{np}) to Average Travel Speed for Percentage of No-Passing Zones in Directional Segments (*continued*)

Opposing Demand Flow Rate, v_o (pc/h)	No-Passing Zones (%)				
	≤ 20	40	60	80	100
FFS = 50 mi/h					
≤ 100	0.2	0.7	1.9	2.4	2.5
200	1.2	2.0	3.3	3.9	4.0
400	1.1	1.6	2.2	2.6	2.7
600	0.6	0.9	1.4	1.7	1.9
800	0.4	0.6	0.9	1.2	1.3
1000	0.4	0.4	0.7	0.9	1.1
1200	0.4	0.4	0.7	0.8	1.0
1400	0.4	0.4	0.6	0.7	0.8
≥ 1600	0.4	0.4	0.5	0.5	0.6
FFS = 45 mi/h					
≤ 100	0.1	0.4	1.7	2.2	2.4
200	0.9	1.6	3.1	3.8	4.0
400	0.9	0.5	2.0	2.5	2.7
600	0.4	0.3	1.3	1.7	1.8
800	0.3	0.3	0.8	1.1	1.2
1000	0.3	0.3	0.6	0.8	1.1
1200	0.3	0.3	0.6	0.7	1.0
1400	0.3	0.3	0.6	0.6	0.7
≥ 1600	0.3	0.3	0.4	0.4	0.6

Table 9.14   Grade Adjustment Factor (f_G) for Estimating Percent Time-Spent-Following on
Specific Upgrades

| | | Grade Adjustment Factor, f_G | | |
| | | Range of Directional Flow Rates, v_d (pc/h) | | |
Grade (%)	Length of Grade (mi)	0–300	> 300–600	> 600
≥ 3.0 < 3.5	0.25	1.00	0.92	0.92
	0.50	1.00	0.93	0.93
	0.75	1.00	0.93	0.93
	1.00	1.00	0.93	0.93
	1.50	1.00	0.94	0.94
	2.00	1.00	0.95	0.95
	3.00	1.00	0.97	0.96
	≥ 4.00	1.00	1.00	0.97
≥ 3.5 < 4.5	0.25	1.00	0.94	0.92
	0.50	1.00	0.97	0.96
	0.75	1.00	0.97	0.96
	1.00	1.00	0.97	0.97
	1.50	1.00	0.97	0.97
	2.00	1.00	0.98	0.98
	3.00	1.00	1.00	1.00
	≥ 4.00	1.00	1.00	1.00
≥ 4.5 < 5.5	0.25	1.00	1.00	0.97
	0.50	1.00	1.00	1.00
	0.75	1.00	1.00	1.00
	1.00	1.00	1.00	1.00
	1.50	1.00	1.00	1.00
	2.00	1.00	1.00	1.00
	3.00	1.00	1.00	1.00
	≥ 4.00	1.00	1.00	1.00
≥ 5.5 < 6.5	0.25	1.00	1.00	1.00
	0.50	1.00	1.00	1.00
	0.75	1.00	1.00	1.00
	1.00	1.00	1.00	1.00
	1.50	1.00	1.00	1.00
	2.00	1.00	1.00	1.00
	3.00	1.00	1.00	1.00
	≥ 4.00	1.00	1.00	1.00
≥ 6.5	0.25	1.00	1.00	1.00
	0.50	1.00	1.00	1.00
	0.75	1.00	1.00	1.00
	1.00	1.00	1.00	1.00
	1.50	1.00	1.00	1.00
	2.00	1.00	1.00	1.00
	3.00	1.00	1.00	1.00
	≥ 4.00	1.00	1.00	1.00

Table 9.15 Passenger-Car Equivalents for Trucks (E_T) and RVs (E_R) for Estimating Percent Time-Spent-Following on Specific Upgrades

| Grade (%) | Length of Grade (mi) | Passenger-Car Equivalent for Trucks, E_T | | | RVs, E_R |
| | | Range of Directional Flow Rates, v_d (pc/h) | | | |
		0–300	> 300–600	> 600	
≥ 3.0 < 3.5	0.25	1.0	1.0	1.0	1.0
	0.50	1.0	1.0	1.0	1.0
	0.75	1.0	1.0	1.0	1.0
	1.00	1.0	1.0	1.0	1.0
	1.50	1.0	1.0	1.0	1.0
	2.00	1.0	1.0	1.0	1.0
	3.00	1.4	1.0	1.0	1.0
	≥ 4.00	1.5	1.0	1.0	1.0
≥ 3.5 < 4.5	0.25	1.0	1.0	1.0	1.0
	0.50	1.0	1.0	1.0	1.0
	0.75	1.0	1.0	1.0	1.0
	1.00	1.0	1.0	1.0	1.0
	1.50	1.1	1.0	1.0	1.0
	2.00	1.4	1.0	1.0	1.0
	3.00	1.7	1.1	1.2	1.0
	≥ 4.00	2.0	1.5	1.4	1.0
≥ 4.5 < 5.5	0.25	1.0	1.0	1.0	1.0
	0.50	1.0	1.0	1.0	1.0
	0.75	1.0	1.0	1.0	1.0
	1.00	1.0	1.0	1.0	1.0
	1.50	1.1	1.2	1.2	1.0
	2.00	1.6	1.3	1.5	1.0
	3.00	2.3	1.9	1.7	1.0
	≥ 4.00	3.3	2.1	1.8	1.0
≥ 5.5 < 6.5	0.25	1.0	1.0	1.0	1.0
	0.50	1.0	1.0	1.0	1.0
	0.75	1.0	1.0	1.0	1.0
	1.00	1.0	1.2	1.2	1.0
	1.50	1.5	1.6	1.6	1.0
	2.00	1.9	1.9	1.8	1.0
	3.00	3.3	2.5	2.0	1.0
	≥ 4.00	4.3	3.1	2.0	1.0
≥ 6.5	0.25	1.0	1.0	1.0	1.0
	0.50	1.0	1.0	1.0	1.0
	0.75	1.0	1.0	1.3	1.0
	1.00	1.3	1.4	1.6	1.0
	1.50	2.1	2.0	2.0	1.0
	2.00	2.8	2.5	2.1	1.0
	3.00	4.0	3.1	2.2	1.0
	≥ 4.00	4.8	3.5	2.3	1.0

Table 9.16 Grade Adjustment Factor (f_G) for Estimating Average Travel Speed on Specific Upgrades

| | | Grade Adjustment Factor, f_G | | |
| | | Range of Directional Flow Rates, v_d (pc/h) | | |
Grade (%)	Length of Grade (mi)	0–300	> 300–600	> 600
≥ 3.0 < 3.5	0.25	0.81	1.00	1.00
	0.50	0.79	1.00	1.00
	0.75	0.77	1.00	1.00
	1.00	0.76	1.00	1.00
	1.50	0.75	0.99	1.00
	2.00	0.75	0.97	1.00
	3.00	0.75	0.95	0.97
	≥ 4.00	0.75	0.94	0.95
≥ 3.5 < 4.5	0.25	0.79	1.00	1.00
	0.50	0.76	1.00	1.00
	0.75	0.72	1.00	1.00
	1.00	0.69	0.93	1.00
	1.50	0.68	0.92	1.00
	2.00	0.66	0.91	1.00
	3.00	0.65	0.91	0.96
	≥ 4.00	0.65	0.90	0.96
≥ 4.5 < 5.5	0.25	0.75	1.00	1.00
	0.50	0.65	0.93	1.00
	0.75	0.60	0.89	1.00
	1.00	0.59	0.89	1.00
	1.50	0.57	0.86	0.99
	2.00	0.56	0.85	0.98
	3.00	0.56	0.84	0.97
	≥ 4.00	0.55	0.82	0.93
≥ 5.5 < 6.5	0.25	0.63	0.91	1.00
	0.50	0.57	0.85	0.99
	0.75	0.52	0.83	0.97
	1.00	0.51	0.79	0.97
	1.50	0.49	0.78	0.95
	2.00	0.48	0.78	0.94
	3.00	0.46	0.76	0.93
	≥ 4.00	0.45	0.76	0.93
≥ 6.5	0.25	0.59	0.86	0.98
	0.50	0.48	0.76	0.94
	0.75	0.44	0.74	0.91
	1.00	0.41	0.70	0.91
	1.50	0.40	0.67	0.91
	2.00	0.39	0.67	0.89
	3.00	0.39	0.66	0.88
	≥ 4.00	0.38	0.66	0.87

Table 9.17   Passenger-Car Equivalents for Trucks (E_T) for Estimating Average Travel Speed on Specific Upgrades

| | | Passenger-Car Equivalent for Trucks, E_T | | |
| | | Range of Directional Flow Rates, v_d (pc/h) | | |
Grade (%)	Length of Grade (mi)	0–300	> 300–600	> 600
≥ 3.0 < 3.5	0.25	2.5	1.9	1.5
	0.50	3.5	2.8	2.3
	0.75	4.5	3.9	2.9
	1.00	5.1	4.6	3.5
	1.50	6.1	5.5	4.1
	2.00	7.1	5.9	4.7
	3.00	8.2	6.7	5.3
	≥ 4.00	9.1	7.5	5.7
≥ 3.5 < 4.5	0.25	3.6	2.4	1.9
	0.50	5.4	4.6	3.4
	0.75	6.4	6.6	4.6
	1.00	7.7	6.9	5.9
	1.50	9.4	8.3	7.1
	2.00	10.2	9.6	8.1
	3.00	11.3	11.0	8.9
	≥ 4.00	12.3	11.9	9.7
≥ 4.5 < 5.5	0.25	4.2	3.7	2.6
	0.50	6.0	6.0	5.1
	0.75	7.5	7.5	7.5
	1.00	9.2	9.0	8.9
	1.50	10.6	10.5	10.3
	2.00	11.8	11.7	11.3
	3.00	13.7	13.5	12.4
	≥ 4.00	15.3	15.0	12.5
≥ 5.5 < 6.5	0.25	4.7	4.1	3.5
	0.50	7.2	7.2	7.2
	0.75	9.1	9.1	9.1
	1.00	10.3	10.3	10.2
	1.50	11.9	11.8	11.7
	2.00	12.8	12.7	12.6
	3.00	14.4	14.3	14.2
	≥ 4.00	15.4	15.2	15.0
≥ 6.5	0.25	5.1	4.8	4.6
	0.50	7.8	7.8	7.8
	0.75	9.8	9.8	9.8
	1.00	10.4	10.4	10.3
	1.50	12.0	11.9	11.8
	2.00	12.9	12.8	12.7
	3.00	14.5	14.4	14.3
	≥ 4.00	15.4	15.3	15.2

Table 9.18   Passenger-Car Equivalents for RVs (E_R) for Estimating Average Travel Speed on Specific Upgrades

| | | Passenger-Car Equivalent for RVs, E_R | | |
| | | Range of Directional Flow Rates, v_d (pc/h) | | |
Grade (%)	Length of Grade (mi)	0–300	> 300–600	> 600
≥ 3.0 < 3.5	0.25	1.1	1.0	1.0
	0.50	1.2	1.0	1.0
	0.75	1.2	1.0	1.0
	1.00	1.3	1.0	1.0
	1.50	1.4	1.0	1.0
	2.00	1.4	1.0	1.0
	3.00	1.5	1.0	1.0
	≥ 4.00	1.5	1.0	1.0
≥ 3.5 < 4.5	0.25	1.3	1.0	1.0
	0.50	1.3	1.0	1.0
	0.75	1.3	1.0	1.0
	1.00	1.4	1.0	1.0
	1.50	1.4	1.0	1.0
	2.00	1.4	1.0	1.0
	3.00	1.4	1.0	1.0
	≥ 4.00	1.5	1.0	1.0
≥ 4.5 < 5.5	0.25	1.5	1.0	1.0
	0.50	1.5	1.0	1.0
	0.75	1.5	1.0	1.0
	1.00	1.5	1.0	1.0
	1.50	1.5	1.0	1.0
	2.00	1.5	1.0	1.0
	3.00	1.6	1.0	1.0
	≥ 4.00	1.6	1.0	1.0
≥ 5.5 < 6.5	0.25	1.5	1.0	1.0
	0.50	1.5	1.0	1.0
	0.75	1.5	1.0	1.0
	1.00	1.6	1.0	1.0
	1.50	1.6	1.0	1.0
	2.00	1.6	1.0	1.0
	3.00	1.6	1.2	1.0
	≥ 4.00	1.6	1.5	1.2
≥ 6.5	0.25	1.6	1.0	1.0
	0.50	1.6	1.0	1.0
	0.75	1.6	1.0	1.0
	1.00	1.6	1.0	1.0
	1.50	1.6	1.0	1.0
	2.00	1.6	1.0	1.0
	3.00	1.6	1.3	1.3
	≥ 4.00	1.6	1.5	1.4

Table 9.19 Passenger-Car Equivalents (E_{TC}) for Estimating the Effect on Average Travel Speed of Trucks That Operate at Crawl Speeds on Long Steep Downgrades

Difference Between FFS and Truck Crawl Speeds (mi/h)	Passenger-Car Equivalent for Trucks at Crawl Sppeds, E_{TC}		
	Range of Directional Flow Rates, v_d (pc/h)		
	0–300	> 300–600	> 600
≤ 15	4.4	2.8	1.4
25	14.3	9.6	5.7
≥ 40	34.1	23.1	13.0

Table 9.20 Factors (f_{pl}) for Estimating the Average Travel Speed and Percent Time-Spent-Following Within a Passing Lane

Directional Flow Rate (pc/h)	Average Travel Speed	Percent Time-Spent-Following
0–300	1.08	0.58
> 300–600	1.10	0.61
> 600	1.11	0.62

Table 9.21 Optimal Lengths (L_{pl}) of Passing Lanes

Directional Flow Rate (pc/h)	Optimal Passing Lane Length (mi)
100	≤ 0.50
200	> 0.50–0.75
400	> 0.75–1.00
≥ 700	> 1.00–2.00

Table 9.22 Downstream Length (L_{de}) of Roadway Affected by Passing Lanes on Directional Segments in Level and Rolling Terrain

Directional Flow Rate (pc/h)	Downstream Length of Roadway Affected L_{de} (mi)	
	Percent Time-Spent-Following	Average Travel Speed
≤ 200	13.0	1.7
400	8.1	1.7
700	5.7	1.7
≥ 1000	3.6	1.7

Table 9.23 Factors (f_{pl}) for Estimating the Average Travel Speed and Percent Time-Spent-Following Within a Climbing Lane

Directional Flow Rate (pc/h)	Average Travel Speed	Percent Time-Spent-Following
0–300	1.02	0.20
> 300–600	1.07	0.21
> 600	1.14	0.23

Table 9.24 Level-of-Service Criteria for Multilane Highways

Free-Flow Speed	Criteria	LOS				
		A	B	C	D	E
60 mi/h	Maximum density (pc/mi/ln)	11	18	26	35	40
	Average speed (mi/h)	60.0	60.0	59.4	56.7	55.0
	Maximum volume-to-capacity ratio (v/c)	0.30	0.49	0.70	0.90	1.00
	Maximum service flow rate (pc/h/ln)	660	1080	1550	1980	2200
55 mi/h	Maximum density (pc/mi/ln)	11	18	26	35	41
	Average speed (mi/h)	55.0	55.0	54.9	52.9	51.2
	Maximum v/c	0.29	0.47	0.68	0.88	1.00
	Maximum service flow rate (pc/h/ln)	600	990	1430	1850	2100
50 mi/h	Maximum density (pc/mi/ln)	11	18	26	35	43
	Average speed (mi/h)	50.0	50.0	50.0	48.9	47.5
	Maximum v/c	0.28	0.45	0.65	0.86	1.00
	Maximum service flow rate (pc/h/ln)	550	900	1300	1710	2000
45 mi/h	Maximum density (pc/mi/ln)	11	18	26	35	45
	Average speed (mi/h)	45.0	45.0	45.0	44.4	42.2
	Maximum v/c	0.26	0.43	0.62	0.82	1.00
	Maximum service flow rate (pc/h/ln)	480	810	1170	1550	1900

Table 9.25 Passenger Car Equivalents for Trucks and Buses (E_T) and RVs (E_R) on General Highway Segments: Multilane Highways and Basic Freeway Sections

Factor	Type of Terrain		
	Level	Rolling	Mountainous
E_T (trucks and buses)	1.5	2.5	4.5
E_R (RVs)	1.2	2.0	4.0

Table 9.26   Passenger Car Equivalents for Trucks and Buses (E_T) on Upgrades, Multilane Highways, and Basic Freeway Sections

Upgrade (%)	Length (mi)	E_T Percentage of Trucks and Buses								
		2	4	5	6	8	10	15	20	25
< 2	All	1.5	1.5	1.5	1.5	1.5	1.5	1.5	1.5	1.5
≥ 2–3	0.00–0.25	1.5	1.5	1.5	1.5	1.5	1.5	1.5	1.5	1.5
	> 0.25–0.50	1.5	1.5	1.5	1.5	1.5	1.5	1.5	1.5	1.5
	> 0.50–0.75	1.5	1.5	1.5	1.5	1.5	1.5	1.5	1.5	1.5
	> 0.75–1.00	2.0	2.0	2.0	2.0	1.5	1.5	1.5	1.5	1.5
	> 1.00–1.50	2.5	2.5	2.5	2.5	2.0	2.0	2.0	2.0	2.0
	> 1.50	3.0	3.0	2.5	2.5	2.0	2.0	2.0	2.0	2.0
> 3–4	0.00–0.25	1.5	1.5	1.5	1.5	1.5	1.5	1.5	1.5	1.5
	> 0.25–0.50	2.0	2.0	2.0	2.0	2.0	2.0	1.5	1.5	1.5
	> 0.50–0.75	2.5	2.5	2.0	2.0	2.0	2.0	2.0	2.0	2.0
	> 0.75–1.00	3.0	3.0	2.5	2.5	2.5	2.5	2.0	2.0	2.0
	> 1.00–1.50	3.5	3.5	3.0	3.0	3.0	3.0	2.5	2.5	2.5
	> 1.50	4.0	3.5	3.0	3.0	3.0	3.0	2.5	2.5	2.5
> 4–5	0.00–0.25	1.5	1.5	1.5	1.5	1.5	1.5	1.5	1.5	1.5
	> 0.25–0.50	3.0	2.5	2.5	2.5	2.0	2.0	2.0	2.0	2.0
	> 0.50–0.75	3.5	3.0	3.0	3.0	2.5	2.5	2.5	2.5	2.5
	> 0.75–1.00	4.0	3.5	3.5	3.5	3.0	3.0	3.0	3.0	3.0
	> 1.00	5.0	4.0	4.0	4.0	3.5	3.5	3.0	3.0	3.0
> 5–6	0.00–0.25	2.0	2.0	1.5	1.5	1.5	1.5	1.5	1.5	1.5
	> 0.25–0.30	4.0	3.0	2.5	2.5	2.0	2.0	2.0	2.0	2.0
	> 0.30–0.50	4.5	4.0	3.5	3.0	2.5	2.5	2.5	2.5	2.5
	> 0.50–0.75	5.0	4.5	4.0	3.5	3.0	3.0	3.0	3.0	3.0
	> 0.75–1.00	5.5	5.0	4.5	4.0	3.0	3.0	3.0	3.0	3.0
	> 1.00	6.0	5.0	5.0	4.5	3.5	3.5	3.5	3.5	3.5
> 6	0.00–0.25	4.0	3.0	2.5	2.5	2.5	2.5	2.0	2.0	2.0
	> 0.25–0.30	4.5	4.0	3.5	3.5	3.5	3.0	2.5	2.5	2.5
	> 0.30–0.50	5.0	4.5	4.0	4.0	3.5	3.0	2.5	2.5	2.5
	> 0.50–0.75	5.5	5.0	4.5	4.5	4.0	3.5	3.0	3.0	3.0
	> 0.75–1.00	6.0	5.5	5.0	5.0	4.5	4.0	3.5	3.5	3.5
	> 1.00	7.0	6.0	5.5	5.5	5.0	4.5	4.0	4.0	4.0

Table 9.27 Passenger Car Equivalents for RVs (E_R) on Uniform Upgrades, Multilane Highways, and Basic Freeway Segments

Grade (%)	Length (mi)	E_R Percentage of RVs								
		2	4	5	6	8	10	15	20	25
≤ 2	All	1.2	1.2	1.2	1.2	1.2	1.2	1.2	1.2	1.2
> 2–3	0.00–0.50	1.2	1.2	1.2	1.2	1.2	1.2	1.2	1.2	1.2
	> 0.50	3.0	1.5	1.5	1.5	1.5	1.5	1.2	1.2	1.2
> 3–4	0.00–0.25	1.2	1.2	1.2	1.2	1.2	1.2	1.2	1.2	1.2
	> 0.25–0.50	2.5	2.5	2.0	2.0	2.0	2.0	1.5	1.5	1.5
	> 0.50	3.0	2.5	2.5	2.5	2.0	2.0	2.0	1.5	1.5
> 4–5	0.00–0.25	2.5	2.0	2.0	2.0	1.5	1.5	1.5	1.5	1.5
	> 0.25–0.50	4.0	3.0	3.0	3.0	2.5	2.5	2.0	2.0	2.0
	> 0.50	4.5	3.5	3.0	3.0	3.0	2.5	2.5	2.0	2.0
> 4–5	0.00–0.25	4.0	3.0	2.5	2.5	2.5	2.0	2.0	2.0	1.5
	> 0.25–0.50	6.0	4.0	4.0	3.5	3.0	3.0	2.5	2.5	2.0
	> 0.50	6.0	4.5	4.0	4.5	3.5	3.0	3.0	2.5	2.0

Table 9.28 Passenger Car Equivalents for Trucks (E_T) on Downgrades, Multilane Highways, and Basic Freeway Segments

Downgrade (%)	Length (mi)	E_T Percentage of Trucks			
		5	10	15	20
< 4	All	1.5	1.5	1.5	1.5
4–5	≤ 4	1.5	1.5	1.5	1.5
4–5	> 4	2.0	2.0	2.0	1.5
> 5–6	≤ 4	1.5	1.5	1.5	1.5
> 5–6	> 4	5.5	4.0	4.0	3.0
> 6	≤ 4	1.5	1.5	1.5	1.5
> 6	> 4	7.5	6.0	5.5	4.5

Table 9.29 Adjustment (f_{LW}) for Lane Width

Lane Width (ft)	Reduction in FFS, f_{LW} (mi/h)
12	0.0
11	1.9
10	6.6

Table 9.30   Adjustment (f_{LC}) for Lateral Clearance

Four-Lane Highways		Six-Lane Highways	
Total Lateral Clearance (ft)	Reduction in FFS (mi/h)	Total Lateral Clearance (ft)	Reduction in FFS (mi/h)
12	0.0	12	0.0
10	0.4	10	0.4
8	0.9	8	0.9
6	1.3	6	1.3
4	1.8	4	1.7
2	3.6	2	2.8
0	5.4	0	3.9

Table 9.31   Adjustment for Median Type

Median Type	Reduction in FFS (mi/h)
Undivided highways	1.6
Divided highways (including TWLTLs)	0.0

Table 9.32   Access-Point Density Adjustment

Access Points/Mile	Reduction in FFS (mi/h)
0	0.0
10	2.5
20	5.0
30	7.5
≥ 40	10.0

Table 9.33 Level-of-Service Criteria for Basic Freeway Segments

Criteria	LOS				
	A	*B*	*C*	*D*	*E*
FFS = 75 mi/h					
Maximum density (pc/mi/ln)	11	18	26	35	45
Maximum speed (mi/h)	75.0	74.8	70.6	62.2	53.3
Maximum v/c	0.34	0.56	0.76	0.90	1.00
Maximum service flow rate (pc/h/ln)	820	1350	1830	2170	2400
FFS = 70 mi/h					
Maximum density (pc/mi/ln)	11	18	26	35	45
Maximum speed (mi/h)	70.0	70.0	68.2	61.5	53.3
Maximum v/c	0.32	0.53	0.74	0.90	1.00
Maximum service flow rate (pc/h/ln)	770	1260	1770	2150	2400
FFS = 65 mi/h					
Maximum density (pc/mi/ln)	11	18	26	35	45
Maximum speed (mi/h)	65.0	65.0	64.6	59.7	52.2
Maximum v/c	0.30	0.50	0.71	0.89	1.00
Maximum service flow rate (pc/h/ln)	710	1170	1680	2090	2350
FFS = 60 mi/h					
Maximum density (pc/mi/ln)	11	18	26	35	45
Maximum speed (mi/h)	60.0	60.0	60.0	57.6	51.1
Maximum v/c	0.29	0.47	0.68	0.88	1.00
Maximum service flow rate (pc/h/ln)	660	1080	1560	2020	2300
FFS = 55 mi/h					
Maximum density (pc/mi/ln)	11	18	26	35	45
Maximum speed (mi/h)	55.0	55.0	55.0	54.7	50.0
Maximum v/c	0.27	0.44	0.64	0.85	1.00
Maximum service flow rate (pc/h/ln)	600	990	1430	1910	2250

Table 9.34 Adjustments for Lane Width

Lane Width (ft)	Reduction in Free-Flow Speed, f_{LW} (mi/h)
12	0.0
11	1.9
10	6.6

Table 9.35 Adjustments (f_{LC}) for Right-Shoulder Lateral Clearance

| Right-Shoulder Lateral Clearance (ft) | Reduction in Free-Flow Speed, f_{LC} (mi/h) | | | |
| | Lanes in One Direction | | | |
	2	3	4	≥ 5
≥ 6	0.0	0.0	0.0	0.0
5	0.6	0.4	0.2	0.1
4	1.2	0.8	0.4	0.2
3	1.8	1.2	0.6	0.3
2	2.4	1.6	0.8	0.4
1	3.0	2.0	1.0	0.5
0	3.6	2.4	1.2	0.6

Table 9.36 Adjustments (f_N) for Number of Lanes

Number of Lanes (One Direction)	Reduction in Free-Flow Speed, f_N (mi/h)
≥ 5	0.0
4	1.5
3	3.0
2	4.5

Table 9.37 Adjustments (f_{ID}) for Interchange Density

Interchanges per Mile	Reduction in Free-Flow Speed f_{ID} (mi/h)
≥ 0.50	0.0
0.75	1.3
1.00	2.5
1.25	3.7
1.50	5.0
1.75	6.3
2.00	7.5

CHAPTER 10

Capacity and Level of Service at Signalized Intersections

The level of service at any intersection on a highway has a significant effect on the overall operating performance of that highway. Thus, improvement of the level of service at each intersection usually results in an improvement of the overall operating performance of the highway. An analysis procedure that provides for the determination of capacity or level of service at intersections is therefore an important tool for designers, operation personnel, and policy makers. Factors that affect the level of service at intersections include the flow and distribution of traffic, the geometric characteristics, and the signalization system.

A major difference between consideration of level of service on highway segments and level of service at intersections is that only through-flows are used in computing the levels of service at highway segments (see Chapter 9), whereas turning flows are significant when computing the levels of service at signalized intersections. The signalization system, which includes the allocation of time among the conflicting movements of traffic and pedestrians at the intersection, is also an important factor. For example, the distribution of green times among these conflicting flows significantly affects both the capacity and operation of the intersection. Other factors such as lane widths, traffic composition, grade, and speed also affect the level of service at intersections in a similar manner as for highway segments.

This chapter presents procedures for determining the capacity and level of service at signalized intersections, based on results of some recent research projects and field observations as presented in the *Highway Capacity Manual*.

DEFINITIONS OF SOME COMMON TERMS

Most of the terms commonly used in capacity and level of service analyses of signalized intersections were defined in Chapter 8. However, some additional terms need to be defined and others need to be redefined to understand their use in this chapter.

- **Permitted turning movements** are those made within gaps of an opposing traffic stream or through a conflicting pedestrian flow. For example, when a right turn is made while pedestrians are crossing a conflicting crosswalk, the right turn is a permitted turning movement. Similarly, when a left turn is made between two consecutive vehicles of the opposing traffic stream, the left turn is a permitted turn. The suitability of permitted turns at a given intersection depends on the geometric characteristics of the intersection, the turning volume, and the opposing volume.
- **Protected turns** are those turns protected from any conflicts with vehicles in an opposing stream or pedestrians on a conflicting crosswalk. A permitted turn takes more time than a similar protected turn and will use more of the available green time.
- **Change and clearance interval** is the sum of the "yellow" and "all-red" intervals (given in seconds) that are provided between phases to allow vehicular and pedestrian traffic to clear the intersection before conflicting movements are released.
- **Geometric conditions** is a term used to describe the roadway characteristics of the approach. They include the number and width of lanes, grades, and the allocation of the lanes for different uses, including the designation of a parking lane.
- **Signalization conditions** is a term used to describe the details of the signal operation. They include the type of signal control, the phasing sequence, the timing, and an evaluation of signal progression on each approach.
- **Flow ratio** (v/s) is the ratio of the actual flow rate or projected demand v on an approach or lane group to the saturation flow rate s.
- **Lane group** consists of one or more lanes that have a common stop line, carry a set of traffic streams, and whose capacity is shared by all vehicles in the group.

CAPACITY AT SIGNALIZED INTERSECTIONS

The capacity at a signalized intersection is given for each lane group and is defined as the maximum rate of flow for the subject lane group that can go through the intersection under prevailing traffic, roadway, and signalized conditions. Capacity is given in vehicles per hour (veh/h), but it is based on the flow during a peak 15-min period. The capacity of the intersection as a whole is not considered; rather, emphasis is placed on providing suitable facilities for the major movements of the intersections. Capacity is therefore meaningfully applied only to major movements or approaches of the intersection. Note also that in comparison with other locations such as freeway segments, the capacity of an intersection approach is not as strongly correlated with the level of service. It is therefore necessary that both the level of service and the capacity be analyzed separately when signalized intersections are being evaluated.

The concept of a saturation flow or saturation flow rate s is used to determine the capacity of a lane group. The saturation flow rate is the maximum flow rate on the approach or lane group that can go through the intersection under prevailing traffic and roadway conditions when 100 percent effective green time is available. The saturation flow rate is given in units of veh/h of effective green time.

The capacity of an approach or lane group is given as

$$c_i = s_i(g_i/C)$$

(10.1)

where

c_i = capacity of lane group i (veh/h)

s_i = saturation flow rate for lane group or approach i

(g_i/C) = green ratio for lane group or approach i

g_i = effective green for lane group i or approach i

C = cycle length

The ratio of flow to capacity (v/c) is usually referred to as the *degree of saturation* and can be expressed as

$$(v/c)_i = X_i = \frac{v_i}{s_i(g_i/C)} \tag{10.2}$$

where

X_i = (v/c) ratio for lane group or approach i

v_i = actual flow rate or projected demand for lane group or approach i (veh/h)

s_i = saturation flow for lane group or approach i (veh/h/g)

g_i = effective green time for lane group i or approach i (sec)

It can be seen that when the flow rate equals capacity, X_i equals 1.00; when flow rate equals zero, X_i equals zero.

When the overall intersection is to be evaluated with respect to its geometry and the total cycle time, the concept of critical volume-to-capacity ratio (X_c) is used. The critical (v/c) ratio is usually obtained for the overall intersection but considers only the critical lane groups or approaches, which are those lane groups or approaches that have the maximum flow ratio (v/s) for each signal phase. For example, in a two-phase signalized intersection, if the north approach has a higher (v/s) ratio than the south approach, more time will be required for vehicles on the north approach to go through the intersection during the north-south green phase, and the phase length will be based on the green time requirements for the north approach. The north approach will therefore be the critical approach for the north-south phase. As shown in Chapter 8, the critical v/c ratio for the whole intersection is given as

$$X_c = \sum_i (v/s)_{ci} \frac{C}{C - L} \tag{10.3}$$

where

X_c = critical v/c ratio for the intersection

$\sum_i(v/s)_{ci}$ = summation of the ratios of actual flows to saturation flow for all critical lanes, groups, or approaches

C = cycle length (sec)

L = total lost time per cycle computed as the sum of the lost time, (t_ℓ), for each critical signal phase, $L = \sum_i t_\ell$

Equation 10.3 can be used to estimate the signal timing for the intersection if this is unknown and a critical (v/c) ratio is specified for the intersection. Alternatively, this equation can be used to obtain a broader indicator of the overall sufficiency of the intersection by substituting the maximum permitted cycle length for the jurisdiction and determining the resultant critical (v/c) ratio for the intersection. When the critical (v/c) ratio is less than 1.00, the cycle length provided is adequate for all critical movements to go through the intersection if the green time is proportionately distributed to the different phases. That is, for the assumed phase sequence, all movements in the intersection will be provided with adequate green times if the total green time is proportionately divided among all phases. If the total green time is not properly allocated to the different phases, it is possible to have a critical (v/c) ratio of less than 1.00, but with one or more individual oversaturated movements within a cycle.

LEVEL OF SERVICE AT SIGNALIZED INTERSECTIONS

The procedures can be used for either a detailed or operational evaluation of a given intersection or a general planning estimate of the overall performance of an existing or planned signalized intersection. At the design level of analysis, more input data are required for a direct estimate of the level of service to be made. It is also possible at this level of analysis to determine the effect of changing signal timing.

The procedures presented here for the operational evaluation are those given in the 2000 edition of the *Highway Capacity Manual.* These procedures deal with the computation of the level of service at the intersection approaches and the level of service at the intersection as a whole. See the "Additional Readings" section at the end of this chapter for other methods of determining capacity and level of service at signalized intersections.

Control delay is used to define the level of service at signalized intersections, since delay not only indicates the amount of lost travel time and fuel consumption, it is also a measure of the frustration and discomfort of motorists. Control or signal delay, which is that portion of total delay that is attributed to the control facility, is computed to define the level of service at the signalized intersection. This includes the delay due to the initial deceleration, queue move-up time, stopped time, and final acceleration. Delay, however, depends on the red time, which in turn depends on the length of the cycle. Reasonable levels of service can therefore be obtained for short cycle lengths, even though the (v/c) ratio is as high as 0.9. To the extent that signal coordination reduces delay, different levels of service may also be obtained for the same (v/c) ratio when the effect of signal coordination changes.

OPERATION ANALYSIS

The procedure at the operation level of analysis can be used to determine the capacity or level of service at the approaches of an existing signalized intersection or the overall level of service at an existing intersection. The procedure can also be used in the detailed design of a given intersection. In using the procedure to analyze an existing signal, operational data such as phasing sequence, signal timing, and geometric details (lane widths, number of lanes) are known. The procedure is used to determine the level of service at which the intersection is performing in terms of control or signal delay. In using the procedure for detailed design, the operational data usually are not known and therefore have to be computed or assumed. The delay and level of service are then determined.

The LOS criteria are given in terms of the average control delay per vehicle during an analysis period of 15 min. Six levels of service are prescribed. The criteria for each are described below and are shown in Table 10.1.*

- **Level of service A** describes that level of operation at which the average delay per vehicle is 10.0 sec or less. At level of service A, vehicles arrive mainly during the green phases, resulting in only a few vehicles stopping at the intersection. Short cycle lengths may help in obtaining low delays.
- **Level of service B** describes that level of operation at which delay per vehicle is greater than 10 sec but not greater than 20 sec. At level of service B, the number of vehicles stopped at the intersection is greater than that for level of service A, but progression is still good and the cycle length may also be short.
- **Level of service C** describes that level of operation at which delay per vehicle is greater than 20 sec and up to 35 sec. At level of service C, many vehicles go through the intersection without stopping, but a significant number of vehicles are stopped. In addition, some vehicles at an approach will not clear the intersection during the first cycle (cycle failure). The higher delay may be due to the significant number of vehicles arriving during the red phase (fair progression) and/or relatively long cycle lengths.
- **Level of service D** describes that level of operation at which the delay per vehicle is greater than 35 sec but not greater than 55 sec. At level of service D, more vehicles are stopped at the intersection, resulting in the longer delay. The number of individual cycles failing is now noticeable. The longer delay at this level of service is due to a combination of two or more of several factors that include long cycle lengths, high (v/c) ratios, and unfavorable progression.
- **Level of service E** describes that level of operation at which the delay per vehicle is greater than 55 sec but not greater than 80 sec. At level of service E, individual cycles frequently fail. This long delay, which is usually taken as the limit of acceptable delay by many agencies, generally indicates high (v/c) ratios, long cycle lengths, and poor progression.
- **Level of service F** describes that level of operation at which the delay per vehicle is greater than 80 sec. This long delay is usually unacceptable to most motorists. At level of service F, *oversaturation* usually occurs—that is, arrival flow rates are greater than the capacity at the intersection. Long delay can also occur as a result of poor progression and long cycle lengths. Note that this level of service can occur when approaches have high (v/c) ratios, which are less than 1.00, but also have many individual cycles failing.

It should be emphasized once more that, in contrast to other locations, the level of service at a signalized intersection does not have a simple one-to-one relationship with capacity. For example, at freeway segments, the (v/c) ratio is 1.00 at the upper limits of level of service E. At the signalized intersection, however, it is possible for the delay to be unacceptable at level of service F although the (v/c) ratio is less than 1.00 and even as low as

---

*For convenience in looking up values, all tables referenced in this chapter are gathered in an appendix at the end of the chapter.

0.75. When long delays occur at such (v/c) ratios, it may be due to a combination of two or more of the following conditions:

- Long cycle lengths
- Green time is not properly distributed, resulting in a longer red time for one or more lane groups—that is, there is one or more disadvantaged lane groups
- A poor signal progression, which results in a large percentage of the vehicles on the approach arriving during the red phase

It is also possible to have short delays at an approach when the (v/c) ratio equals 1.00—that is, saturated approach—which can occur if the following conditions exist:

- Short cycle length
- Favorable signal progression, resulting in a high percentage of vehicles arriving during the green phase

Clearly, level of service F does not necessarily indicate that the intersection, approach, or lane group is oversaturated, nor can it be automatically assumed that the demand flow is below capacity for a level of service range of A to E. It is therefore imperative that both the capacity and level of service analyses be carried out when a signalized intersection is to be evaluated fully.

Methodology of Operation Analysis Procedure

The tasks involved in an operational analysis are presented in the flow chart shown in Figure 10.1. The tasks have been divided into five modules: (1) input parameters; (2) lane grouping and demand flow rate; (3) saturation flow rate; (4) capacity analysis v/c; and (5) level-of-service module. Each of these modules will be discussed in turn, including a detailed description of each task involved.

Input Parameters

This module involves the collection and presentation of the data that will be required for the analysis. The tasks involved are

- Identifying and recording the geometric characteristics
- Identifying and specifying the traffic conditions
- Specifying the signalized conditions

Table 10.2 gives the input data required for each analysis lane group.

Specifying Geometric Characteristics.   The physical configuration of the intersection is obtained in terms of the number of lanes, lane width, grade, movement by lane, parking locations, lengths of storage bays, and so forth and is recorded on the appropriate form, shown in Figure 10.2. In cases where the physical configuration of the intersection is unknown, the planning level of analysis (see page 461) may be used to determine a suitable configuration, or state and local policies and/or guidelines can be used. If no guidelines are available, guidelines given in Chapter 8 may be used.

Specifying Traffic Conditions.   This phase involves the recording of bicycle, pedestrian, and vehicular hourly volumes on the appropriate cell of the form shown on Figure 10.2.

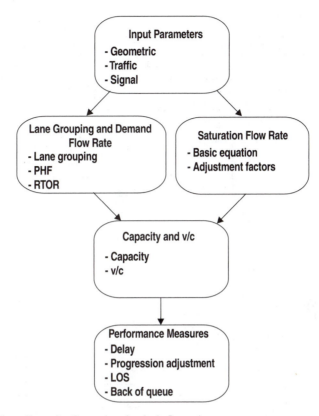

Figure 10.1   Flow Chart for Operation Analysis Procedure

SOURCE: *Highway Capacity Manual,* Special Report 209, Fourth Edition, Transportation Research Board, National Research Council, Washington, D.C., 2000. Used with permission.

Pedestrian and bicycle volumes are recorded such that those that conflict with a given stream of right-turning vehicles are in the same direction as the conflicting right-turning vehicles. For example, pedestrians on the north crosswalk will conflict with the westbound (WB) right-turning vehicles and should be recorded in the WB row of the form. Similarly, pedestrians on the east crosswalk will conflict with the northbound (NB) right-turning vehicles and should be recorded in the NB row of the form. The traffic volumes are the flow rates (equivalent hourly volumes) for the analysis period, which is usually taken as 15 min ($T = 0.25$). This flow rate may also be computed from the hourly volumes and the peak-hour factors. Control delay is significantly influenced by the length of the analysis period where v/c is greater than 0.9. Therefore, when v/c is greater than 0.9 and the 15-min flow rate is relatively constant for periods longer than 15 min, the analysis period (T) in hours should be the length of time the flow remains relatively constant. In cases of oversaturation ($v/c > 1$) in which the flow rate remains relatively constant, the analysis period should be extended to cover the period of oversaturation. However, when the resulting analysis period is longer than 15 min, and different flow rates are observed during

INPUT WORKSHEET

General Information

Analyst _____
Agency or Company _____
Date Performed _____
Analysis Time Period _____

Site Information

Intersection _____
Area Type ☐ CBD ☐ Other
Jurisdiction _____
Analysis Year _____

Intersection Gemoetry

grade=_____

Street

grade=_____

Show North Arrow

grade=_____

Street

grade=_____

☐▷	= Pedestrian Button
⊢——⊣	= Lane Width
↑	= Through
↱	= Right
↰	= Left
↑↱	= Through + Right
↰↑	= Left + Through
↰↱	= Left + Right
↰↑↱	= Left + Through + Right

Volume and Timing Input

	EB			WB			NB			SB		
	LT	TH	RT[1]	LT	TH	RT[1]	LT	TH	RT[1]	LT	TH	RT[1]
Volume, V (veh/h)												
% heavy vehicles, % HV												
Peak-hour factor, PHF												
Pretimed (P) or actuated (A)												
Start-up lost time, l_1 (s)												
Extension of effective green time, e (s)												
Arrival type, AT												
Approach pedestrian volume,[2] v_{ped} (p/h)												
Approach bicycle volume,[2] v_{bic} (bicycles/h)												
Parking (Y or N)												
Parking maneuvers, N_m (maneuvers/h)												
Bus stopping, N_B (buses/h)												
Min. timing for pedestrians,[3] G_p (s)												

Signal Phasing Plan

D I A G R A M	Ø1	Ø2	Ø3	Ø4	Ø5	Ø6	Ø7	Ø8
Timing	G = Y =	G = Y =	G = Y =	G = Y =	G = Y =	G = Y =	G = Y =	G = Y =

↗ Protected turns   ⁻ ⁼ ⁻ ↗ Permitted turns Pedestrian   Cycle length, C = _____ s

Notes

1. RT volumes, as shown, exclude RTOR.
2. Approach pedestrian and bicycle volumes are those that conflict with right turns from the subject approach.
3. Refer to Equation 16-2.

Figure 10.2 Input Worksheet for Operation Level of Analysis

SOURCE: *Highway Capacity Manual,* Special Report 209, Fourth Edition, Transportation Research Board, National Research Council, Washington, D.C., 2000. Used with permission.

sub-periods of equal length within the longer analysis period, a special multiple-period analysis should be conducted. Details of traffic volume should include the percentage of heavy vehicles (%HV) in each movement, where heavy vehicles are defined as all vehicles having more than four tires on the pavement. In recording the number of buses, only buses that stop to pick up or discharge passengers on either side of the intersection are included. Buses that go through the intersection without stopping to pick up or discharge passengers are considered heavy vehicles.

The level of coordination between the lights at the intersection being studied and those at adjacent intersections is a critical characteristic, and it is determined in terms of the type of vehicle arrival at the intersection. Six arrival types (AT) have been identified.

- **Arrival Type 1,** which represents the worst condition of arrival, is a dense platoon containing over 80 percent of the lane group volume arriving at the beginning of the red phase.
- **Arrival Type 2,** which, while better than Type 1, is still considered unfavorable, is either a dense platoon arriving in the middle of the red phase or a dispersed platoon containing 40 to 80 percent of the lane group volume arriving throughout the red phase.
- **Arrival Type 3,** which usually occurs at isolated and noninterconnected inter-sections, is characterized by highly dispersed platoons and entails the random arrival of vehicles in which the main platoon contains less than 40 percent of the lane group volume. Arrivals at coordinated intersections with minimum benefits of progression may also be described by this arrival type.
- **Arrival Type 4,** which is usually considered a favorable platoon condition, is either a moderately dense platoon arriving in the middle of a green phase or a dispersed platoon containing 40 to 80 percent of the lane group volume arriving throughout the green phase.
- **Arrival Type 5,** which represents the best condition of arrival that usually occurs, is a dense platoon containing over 80 percent of the lane group volume arriving at the start of the green phase.
- **Arrival Type 6,** which represents exceptional progression quality, is a very dense platoon progressing through several closely spaced intersections with very low traffic from the side streets.

It is necessary to determine, as accurately as possible, the type of arrival for the inter-section being considered, since both the estimate of delay and the determination of the level of service will be significantly affected by the arrival type used in the analysis. Field observation is the best way to determine the arrival type, although time-space diagrams for the street being considered could be used for an approximate estimation. In using field observations, the percentage of vehicles arriving during the green phase is determined and the arrival type is then obtained for the platoon ratio for the approach. The HCM gives the platoon ratio as

$$R_p = P(C/g) \qquad (10.4)$$

where

   R_p = platoon ratio
   P = proportion of all vehicles in the movement arriving during the green phase
   C = cycle length (sec)
   g = effective green time for the movement (sec)

The arrival type is obtained from Table 10.3, which gives a range of platoon ratios and progression quality for each arrival type.

The number of parking maneuvers at the approach is another factor that influences capacity and level of service of the approach. A parking maneuver is when a vehicle enters or leaves a parking space. The number of parking maneuvers adjacent to an analysis lane group is given as the number of parking maneuvers per hour (N_m) that occur within 250 ft upstream of the intersection.

Specifying Signalized Conditions.   Details of the signal system should be specified, including a phase diagram and the green, yellow, and cycle lengths. The phasing scheme at an intersection determines which traffic stream or streams are given the right of way at the intersection, and therefore has a significant effect on the level of service. A poorly designed phasing scheme may result in unnecessary delay. Different phasing plans for pretimed and actuated signals are illustrated in Chapter 8.

Lane Grouping and Demand Flow Rate

Three main tasks are involved in this module: identifying the different lane groups, adjusting hourly volumes to peak 15-min flow rates using the PHF, and adjusting for a right turn on red (RTOR). Figure 10.3 shows a worksheet that can be used for this module.

Identifying the Different Lane Groups.   The lane groups at each approach must be identified, as the HCM methodology considers each approach at the intersection and individual lane groups on each approach. The guidelines given in Chapter 8 to establish lane groups are also used here. However, when a lane group includes a lane that is shared by left turning and straight through vehicles (shared lane), it is necessary to determine whether the shared lane is operating as a de facto left turn lane. The shared lane is considered a de facto left lane, if the computed proportion of left turns in the shared lane is 1.0 (i.e., 100 percent).

Adjustment of Hourly Volumes.   Earlier, we saw that the analysis for level of service requires that flow rates be based on the peak 15-min flow rate. It is therefore necessary to convert hourly volumes to 15-min flow rates by dividing the hourly volumes by the peak hour factor (PHF). Note that although not all movements of an approach may peak at the same time, dividing all hourly volumes by a single PHF assumes that the peaking occurs for all movements at the same time, which is a conservative assumption.

Adjustment for Right-Turn-on-Red (RTOR).   The right-turn volume in a lane group may be reduced by the volume of vehicles turning right during the red phase. This reduction is done in terms of hourly volume and prior to the conversion to flow rates. It is also recommended to consider field data on the number of vehicles turning right during the red phase when existing intersections are being considered. When a future intersection is being considered, it is suggested that the total right-turn volumes be used in the analysis,

VOLUME ADJUSTMENT AND SATURATION FLOW RATE WORKSHEET

General Information

Project Description_____

Volume Adjustment

	EB			WB			NB			SB		
	LT	TH	RT	LT	TH	RT	LT	TH	RT	LT	TH	RT
Volume, V (veh/h)												
Peak-hour factor, PHF												
Adjusted flow rate, v_p = V/PHF (veh/h)												
Lane group												
Adjusted flow rate in lane group, v (veh/h)												
Proportion[1] of LT or RT (P_{LT} or P_{RT})		-			-			-			-	

Saturation Flow Rate

Base saturation flow, s_o (pc/h/ln)												
Number of lanes, N												
Lane width adjustment factor, f_w												
Heavy-vehicle adjustment factor, f_{HV}												
Grade adjustment factor, f_g												
Parking adjustment factor, f_p												
Bus blockage adjustment factor, f_{bb}												
Area type adjustment factor, f_a												
Lane utilization adjustment factor, f_{LU}												
Left-turn adjustment factor, f_{LT}												
Right-turn adjustment factor, f_{RT}												
Left-turn ped/bike adjustment factor, f_{Lpb}												
Right-turn ped/bike adjustment factor, f_{Rpb}												
Adjusted saturation flow, s (veh/h) $s = s_o$ N f_w f_{HV} f_g f_p f_{bb} f_a f_{LU} f_{LT} f_{RT} f_{Lpb} f_{Rpb}												

Notes

1. P_{LT} = 1.000 for exclusive left-turn lanes, and P_{RT} = 1.000 for exclusive right-turn lanes. Otherwise, they are equal to the proportions of turning volumes in the lane group.

Figure 10.3 Volume Adjustment and Saturation Flow Rate Worksheet

SOURCE: *Highway Capacity Manual,* Special Report 209, Fourth Edition, Transportation Research Board, National Research Council, Washington, D.C., 2000. Used with permission.

since it is very difficult to estimate the number of right-turning vehicles that move on the red phase. An exception to this is when an exclusive left-turn phase for the cross street "shadows" the right-turn-lane movement. For example, an eastbound exclusive left-turn phase will "shadow" the southbound right-turning vehicles. In such a case, the shadowing left-turn volume per lane may be used as the volume of right-turning vehicles that move on the red phase and can be deducted from the right-turn volumes. Also, right turns that are free-flowing and not controlled by the signal are not included in the analysis.

Saturation Flow Rate

This module provides for the computation of a saturation flow rate for each lane group. The saturation flow rate is defined as the flow rate in veh/h that the lane group can carry if it has the green indication continuously, that is, if $g/C = 1$.

Base Equation for Saturation Flow Rate.   The saturation flow rate (s) depends on an ideal saturation flow (s_o), which is usually taken as 1900 passenger cars per hour of green time per lane. This ideal saturation flow is then adjusted for the prevailing conditions to obtain the saturation flow for the lane group being considered. The adjustment is made by introducing factors that correct for the number of lanes, lane width, the percent of heavy vehicles in the traffic, approach grade, parking activity, local buses stopping within the intersection, area type, lane utilization factor, and right and left turns. The HCM gives the saturation flow as

$$s = (s_o)(N)(f_w)(f_{HV})(f_g)(f_p)(f_a)(f_{bb})(f_{Lu})(f_{RT})(f_{LT})(f_{Lpb})(f_{Rpb}) \qquad (10.5)$$

where

s = saturation flow rate for the subject lane group, expressed as a total for all lanes in lane group under prevailing conditions (veh/h/g)

s_o = ideal saturation flow rate per lane, usually taken as 1900 (veh/h/ln)

N = number of lanes in lane group

f_w = adjustment factor for lane width

f_{HV} = adjustment factor for heavy vehicles in the traffic stream

f_g = adjustment factor for approach grade

f_p = adjustment factor for the existence of parking lane adjacent to the lane group and the parking activity on that lane

f_a = adjustment factor for area type (for CBD, 0.90; for all other areas, 1.00)

f_{bb} = adjustment factor for the blocking effect of local buses stopping within the intersection area

f_{Lu} = adjustment factor for lane utilization

f_{RT} = adjustment factor for right turns in the lane group

f_{LT} = adjustment factor for left turns in the lane group

f_{Lpb} = pedestrian adjustment factor for left turn movements

f_{Rpb} = pedestrian adjustment factor for right turn movements

Adjustment Factors.   Although the necessity for using some of these adjustment factors was presented in Chapter 8, the basis for using each of them is given again here to facilitate comprehension of the material. The equations which are used to determine the factors are given in Table 10.4.

- **Lane Width Adjustment Factor, f_w.** This factor depends on the average width of the lanes in a lane group. It is used to account for both the reduction in saturation flow rates when lane widths are less than 12 ft and the increase in saturation flow rates when lane widths are greater than 12 ft. When lane widths are 16 ft or greater, such lanes may be divided into two narrow lanes of 8 ft each. Lane width factors should not be computed for lanes less than 8 ft wide. See Table 10.4 for the equation used to compute these factors.

- **Heavy Vehicle Adjustment Factor, f_{HV}.** The heavy vehicle adjustment factor is related to the percentage of heavy vehicles in the lane group. This factor corrects for the additional delay and reduction in saturation flow due to the presence of heavy vehicles in the traffic stream. The additional delay and reduction in saturation flow are due mainly to the difference between the operational capabilities of the heavy vehicles and passenger cars and the additional space taken up by heavy vehicles. In this procedure, heavy vehicles are defined as any vehicle that has more than four tires touching the pavement. A passenger-car equivalent (E_t) of two is used for each heavy vehicle. This factor is computed by using the appropriate equation given in Table 10.4.

- **Grade Adjustment Factor, f_g.** This factor is related to the slope of the approach being considered. It is used to correct for the effect of slopes on the speed of vehicles, including both passenger cars and heavy vehicles, since passenger cars are also affected by grade. This effect is different for up-slope and down-slope conditions; therefore, the direction of the slope should be taken into consideration. This factor is computed by using the appropriate equation given in Table 10.4.

- **Parking Adjustment Factor, f_p.** On-street parking within 250 ft upstream of the stop line of an intersection causes friction between parking and nonparking vehicles, which results in a reduction of the maximum flow rate that the adjacent lane group can handle. This effect is corrected for by using a parking adjustment factor on the base saturation flow. This factor depends on the number of lanes in the lane group and the number of parking maneuvers per hour. The equation given in Table 10.4 for the parking adjustment factor indicates that the higher the number of lanes in a given lane group, the less effect parking has on the saturation flow; the higher the number of parking maneuvers, the greater the effect. In determining these factors it is assumed that each parking maneuver (either in or out) blocks traffic on the adjacent lane group for an average duration of 18 sec. It should be noted that when the number of parking maneuvers per hour is greater than 180 (equivalent to more than 54 min), a practical limit of 180 should be used. This adjustment factor should be applied only to the lane group immediately adjacent to the parking lane. When parking occurs on both sides of a single lane group, the sum of the number of parking maneuvers on both sides should be used.

- **Area Type Adjustment Factor, f_a.** The general types of activities in the area at which the intersection is located have a significant effect on speed and therefore on saturation volume at an approach. For example, because of the complexity of intersections located in areas with typical central business district characteristics, such as narrow sidewalks, frequent parking maneuvers, vehicle blockades, narrow streets, and high pedestrian activities, these intersections operate less efficiently than intersections at other areas. This is corrected for by using the area type adjustment factor f_a, which is 0.90 for a central business district (CBD) and 1.0 for all other areas. It should be noted, however, that 0.90 is not automatically used for all areas designated as CBDs, nor should it be limited only to CBDs. It should be used for locations that exhibit the characteristics referred to earlier, that result in significant impact on the intersection capacity.

- **Bus Blockage Adjustment Factor, f_{bb}.** When buses have to stop on a travel lane to discharge or pick up passengers, some of the vehicles immediately behind the bus will also have to stop. This results in a decrease in the maximum volume that can be handled by that lane. This effect is corrected for by using the bus blockage adjustment factor, which is related to the number of buses in an hour that stop on the travel lane, within 250 ft upstream or downstream from the stop line, to pick up or discharge passengers, as well as the number of lanes in the lane group. This factor is also computed using the appropriate equation given in Table 10.4.

- **Lane Utilization Adjustment Factor, f_{Lu}.** The lane utilization factor is used to adjust the ideal saturation flow rate to account for the unequal utilization of the lanes in a lane group. This factor is also computed using the appropriate equation given in Table 10.4. It is given as

$$ f_{Lu} = \frac{v_g}{v_{gl}N} \tag{10.6} $$

where

v_g = unadjusted demand flow rate for lane group veh/h
v_{gl} = unadjusted demand flow rate on the single lane in the lane group with the highest volume
N = number of lanes in the lane group

It is recommended that actual field data be used for computing f_{Lui}. Values shown in Table 10.5 can, however, be used as default values when field information is not available.

- **Right-Turn Adjustment Factor, f_{RT}.** This factor accounts for the effect of geometry as other factors are used to account for pedestrians and bicycles using the conflicting crosswalk. It depends on the lane from which the right turn is made, (i.e., exclusive or shared lane) and the proportion of right-turning vehicles on the shared lane. This factor is also computed using the appropriate equation given in Table 10.4.

- **Left-Turn Adjustment Factor, f_{LT}.** This adjustment factor is used to account for the fact that left-turn movements take more time than through movements. The values of this factor also depend on the type of phasing (protected, permitted, or protected-plus-permitted), the type of lane used for left turns (exclusive or shared lane), the proportion of left-turn vehicles using a shared lane, and the opposing flow rate when there is a permitted left-turn phase. The left turns can be made under any one of the following conditions:

- **Case 1:** Exclusive lanes with protected phasing
- **Case 2:** Exclusive lanes with permitted phasing
- **Case 3:** Exclusive lanes with protected-plus-permitted phasing
- **Case 4:** Shared lane with protected phasing

- **Case 5:** Shared lane with permitted phasing
- **Case 6:** Shared lane with protected-plus-permitted phasing
- **Case 7:** Single-lane approaches with permitted left turns

Cases 1 through 6 are for multilane approaches. Case 7 is for single-lane approaches in which either the subject approach and/or the opposing approach consists of a single lane. The methodology for computing the left-turn factors for the multilane approaches is first presented. These computations take into account:

- The portion of the effective green time in seconds during which left turns cannot be made because they are blocked by the clearance of an opposing saturated queue of vehicles, or g_q
- The portion of the effective green time in seconds that expires before a left-turning vehicle arrives, or g_f
- The portion of the effective green time in seconds during which left turns filter through the opposing unsaturated flow (after the opposing queue clears), or g_u

The appropriate left-turn adjustment factor is determined through the following computations for the different cases below:

Case 1—Exclusive Left-Turn Lane with Protected Phasing. As shown in Table 10.4, a left-turn factor of $f_{LT1} = 0.95$ is used.

Case 2A—Exclusive Left-Turn Lane with Permitted Phasing (Multilane permitted left turns opposed by a multilane approach). The left-turn factor f_{LT2A} is computed from the expression

$$f_{LT2A} = \left(\frac{g_u}{g}\right)\left[\frac{1}{1+P_L(E_{L1}-1)}\right] \quad (f_{min} \le f_{LTA} \le 1.00), \quad (10.7)$$

$$f_{min} = 2(1+P_L)/g$$

$$P_L = \left[1+\frac{(N-1)g}{g_u/(E_{L1}+4.24)}\right] \quad (10.8)$$

$$g_u = g-g_q \quad g_q \ge 0 \text{ or}$$

$$g_u = g \quad g_q < 0$$

$$g_q = \frac{v_{olc}qr_o}{0.5-[v_{olc}(1-qr_o)/g_o]}-t_l, v_{olc}(1-qr_o)/g_o \le 0.49 \quad (10.9)$$

where
$$qr_o = \text{opposing queue ratio}$$
$$= \max\left[1-R_{po}\left(\frac{g_o}{C}\right),0\right]$$
$$R_{po} = \text{opposing platoon ratio (see Table 10.3)}$$
$$v_{olc} = \text{adjusted opposing flow per lane, per cycle}$$
$$= \frac{v_oC}{3600N_of_{LU_o}}, \text{ veh/C /ln}$$

f_{Lu_o} = opposing lane utilization factor (determined from Eq. 10.6 or Table 10.5)

g_u = portion of the effective green time during which left turns filter through the opposing saturated flow (sec)

g_q = portion of the effective green time during which left turns cannot be made because they are blocked by the clearance of an opposing saturated queue of vehicles

C = cycle length (sec)

g = effective permitted green time for left-turn lane group (sec)

g_o = opposing effective green time (sec)

N = number of lanes in exclusive left-turn group

N_o = number of lanes in opposing approach

t_l = lost time for left-turn lane group

v_o = adjusted flow rate for opposing approach, (veh/h)

E_{L1} = through car equivalent for permitted left turns (see Table 10.6)

Case 2B—Exclusive Left-Turn Lane with Permitted Phasing (Multilane permitted left turns opposed by a single lane approach). The left-turn factor f_{LTB} is computed from the equation

$$f_{LT2B} = \left(\frac{g_u}{g}\right)\left[\frac{1}{1+(E_{L1}-1)}\right] + \left[\frac{g_{diff}\big/g}{1+(E_{L2}-1)}\right], \text{ where } (f_{min} \le f_{LTB} \le 1) \quad (10.10)$$

$$f_{min} = \frac{4}{g} \quad (10.11)$$

$$E_{L2} = \max\left[\left(1-P_{THo}^n\right)/P_{LTo}, 1.0\right]$$

$$g_{diff} = \max\left[g_q, 0\right] \quad \text{(when opposing left turn volume is 0, } g_{diff} \text{ is zero)}$$

$$n = \max\left[(g_q/2), 0\right]$$

$$P_{THo} = 1 - P_{Lto}$$

$$g_u = g - g_q \quad \text{if } g_q \ge 0 \text{ or}$$

$$g_u = g \quad \text{if } g_q < 0$$

$$v_{olc} = v_o C / 3600 f_{Luo} \quad \text{(veh/hr)} \quad (10.12)$$

$$g_q = 4.943 v_{olc}^{0.762} q_{ro}^{1.061} - t_L \quad (10.13)$$

$$q_{ro} = \max\left[1 - R_{po}(g_o/C), 0\right] \quad (10.14)$$

g = effective permitted green time for left-turn lane group (sec)

P_{LTo} = proportion of opposing left-turn volume in opposing flow

g_u = proportion of the effective green time during which left turns filter through the opposing saturated flow (sec)

g_q = proportion of the effective green time during which left turns cannot be made because they are blocked by the clearance of an opposing saturated queue of vehicles

q_{ro} = opposing queue ratio

g_o = opposing effective green time (sec)

v_o = adjusted flow rate for opposing flow (veh/h)

t_L = lost time for left turn lane group

E_{L1} = through car equivalent for permitted left turns (see Table 10.6)

C = cycle length (sec)

R_{po} = opposing platoon ratio (determined from Eq. 10.6)

Since the proportion of left turns on an exclusive left turn lane is 1, then

$$f_{LT2}(\min) = \frac{2(1+1)}{g} = \frac{4}{g}$$

where

g = effective green time for the lane group (sec)

$f_{LT2}(\min)$ = practical minimum value for left-turn adjustment factor for exclusive lanes–permitted left turns and assuming an approximate average headway of 2 sec per vehicle in an exclusive lane

Case 3—Exclusive Lane with Protected-Plus-Permitted Phasing.
In determining the left-turn factor for this case, the protected portion of the phase is separated from the permitted portion, and each portion is analyzed separately. That is, the protected portion of the phase is considered a protected phase with a separate lane group, and the permitted portion is considered a permitted phase with its own separate lane group. The left-turn factor for the protected portion is then obtained as 0.95, and the left-turn factor for the permitted phase is computed from the appropriate equation. However, care should be taken to compute the appropriate values of G, g, g_f, and g_q, as discussed later. Doing so may result in different saturation flow rates for the two portions. A method for estimating delay in such cases is defined later.

Case 4—Shared Lane with Protected Phasing.
In this case, the left-turn factor f_{LT4} is either obtained directly from Table 10.13 or computed from the expression

$$f_{LT4} = \frac{1.0}{1.0 + 0.5P_{LT}} \tag{10.15}$$

where P_{LT} is the proportion of left turns in the lane group.

Case 5A—Shared Lane with Permitted Phasing (Permitted left turns opposed by multilane approach).
In this case, the resultant effect on the entire lane group should be considered. The left-turn factor f_{LT5A} is computed from the expression

$$f_{LT5A} = \frac{f_{m5A} + 0.91(N-1)}{N} \tag{10.16}$$

where f_{m5} is the left-turn adjustment factor for the lane from which permitted left turns are made. This value is computed from the expression

$$f_{m5} = \frac{g_{f5A}}{g} + \frac{g_u}{g}\left[\frac{1}{1 + P_L(E_{L1} - 1)}\right] f_{\min} \le fm5A \le 1.00 \qquad (10.17)$$

where

g = effective permitted green time for the left-turn lane group (sec)

g_{f5A} = portion of the effective green time that expires before a left-turning vehicle arrives (sec)

g_u = portion of the effective green time during which left-turning vehicles filter through the opposing flow (sec)

$\quad = g - g_q$ when $g_q \ge g_{f5}$

$\quad = g - g_{f5}$ when $g_q < g_{f5}$

$\quad\quad$ where g_q is obtained from Eq. 10.9

E_{L1} = through-car equivalent for each turning vehicle, as obtained from Table 10.6

P_L = proportion of left turns from shared lane(s)

The value of g_{f5A} is calculated from

$$g_{f5A} = G\exp(-0.882\, LTC^{0.717}) - t_L \qquad (10.18)$$

and the proportion of left turns from shared lanes, P_L, is calculated from

$$P_L = P_{LT}\left[1 + \frac{(N-1)g}{g_{f5A} + \dfrac{g_u}{E_{L1}} + 4.24}\right] \qquad (10.19)$$

where

G = actual green time (sec)

LTC = left turns per cycle = $v_{LT}C/3600 \qquad (10.20)$

C = cycle length (sec)

t_L = lost time per phase (sec)

P_{LT} = proportion of left turns in the lane group

N = number of lanes in the lane group

E_{L1} = through-car equivalent for each turning vehicle, as obtained from Table 10.6

Also, note that in order to account for sneakers, the practical minimum value of f_{m5A} is estimated as $2(1 + P_L)/g$.

Case 5B—Shared Lane with Permitted Phasing (Permitted left turns opposed by a single-lane approach).
In this case the left-turn factor is computed from the expression

$$f_{LT5B} = \left[\frac{g_f}{g}\right] + \left[\frac{g_u/g}{1+P_{LT}(E_{L1}-1)}\right] + \left[\frac{g_{diff}/g}{1+P_{LT}(E_{L2}-1)}\right] \tag{10.21}$$

$$(f_{min} \le f_{LT5B} \le 1)$$

$$f_{min} = 2(1+P_{LT})/g \tag{10.22}$$

$$g_{diff} = \max[(g_q - g_f),0] \text{ (when left turn volume is 0, } g_{diff} \text{ is 0)} \tag{10.23}$$

$$E_{L2} = \max[(1-P_{THo}^n)/P_{LTo, 1.0}] \tag{10.24}$$

$$P_{THo} = 1 - P_{LTo}$$

$$g_u = g - g_q \quad \text{if } g_q \ge 0 \text{ or}$$

$$g_u = g - g_f \quad \text{if } g_q < 0$$

$$g_q = 4.943 v_{olc}^{0.762} q_{ro}^{1.061} - t_L \quad (g_q \le g) \tag{10.25}$$

$$v_{olc} = v_o C / 3600 \quad (\text{veh/h}) \tag{10.26}$$

$$q_{ro} = \max[1 - R_{po}(g_o/C),0] \tag{10.27}$$

$$g_f = G\left[e^{0.860(LTC^{0.629})}\right] - t_L \quad (g_f \le g) \tag{10.28}$$

$$LTC = v_{LT}C / 3600 \tag{10.29}$$

where

g = effective permitted green time for left-turn lane group (sec)
G = total actual green for left-turn lane group (sec)
P_{LTo} = proportion of opposing left-turn volume in opposing flow
g_u = proportion of the effective green time during which left turns filter through the opposing saturated flow (sec)
g_q = proportion of the effective green time during which left turns cannot be made because they are blocked by the clearance of an opposing saturated queue of vehicles
q_{ro} = opposing queue ratio
g_o = opposing effective green time (sec)
v_o = adjusted flow rate for opposing flow (veh/h)
v_{LT} = adjusted left-turn flow rate
P_{LT} = proportion of left turn volume in left-turn lane group
t_L = lost time for left-turn lane group
E_{L1} = through car equivalent for permitted left turns (see Table 10.6)
C = cycle length (sec)
R_{po} = opposing platoon ratio (determined from Eq. 10.6)

Case 6—Shared Lane with Protected-Plus-Permitted Phasing. In determining the left-turn factor for this case, the protected portion of the phase is separated from the permitted portion and each portion is analyzed separately. The protected portion is considered as a protected phase and Eq. 10.15 is used to determine the left-turn factor for this portion.

The left-turn factor for the permitted portion is determined by using the procedure for case 5A or 5B, depending on whether the left turns are opposed by a multilane approach or a single-lane approach. Which means that for an opposing multilane approach Eq. 10.16 is used and for an opposing single lane approach Eq. 10.21 is used. However, the appropriate values for G, g, g_f and g_q must be computed as discussed later.

Case 7—Single-Lane Approaches with Permitted Left Turns.   Three different conditions exist under this case: case 7A, a single-lane approach opposed by a single-lane approach; case 7B, a single-lane approach opposed by a multilane approach; and case 7C, a multilane approach opposed by a single-lane approach.

In case 7A (single-lane approach opposed by a single-lane approach), the left-turn adjustment f_{LT7A} factor is computed from

$$f_{LT7A} = \frac{g_{f7A}}{g} + \frac{g_{diff}}{g}\left[\frac{1}{1 + P_{LT}(E_{L2} - 1)}\right] + \frac{g_{u7A}}{g}\left[\frac{1}{1 + P_{LT}(E_{L1} - 1)}\right] \quad (10.30)$$

where

g_{diff} = $\max[(g_q - g_{f7A}), 0]$ (when no opposing left turns are present g_{diff} is zero)

g_{f7A} = $G \exp(-0.860 LTC^{0.629}) - t_L$ (sec) $0 \le g_{f7A} \le g$ (10.31)

g_{q7A} = $4.943 v_{olc}^{0.762} qr_o^{1.061} - t_L$ (sec) $0 \le g_{q7A} \le g$)   (10.32)

g_{u7A} = $g - g_{q7A}$ when $g_{q7A} \ge g_{f7A}$

g_{u7A} = $g - g_{f7A}$ when $g_{q7A} < g_{f7A}$

g = effective green time (sec)

G = actual green time for the permitted phase (sec)

P_{LT} = proportion of left turns in the lane group

LTC = left turns per cycle

     = $v_{LT}C/3600$

v_{LT} = adjusted left-turn flow rate (veh/h)

C = cycle length (sec)

t_L = lost time for subject left-turn lane group (sec)

v_{olc} = adjusted opposing flow rate per lane per cycle (veh/l/c)

     = $v_oC/(3600 \, f_{Luo})$, veh/l/c

v_o = adjusted opposing flow rate (veh/h)

qr_o = opposing queue ratio, that is, the proportion of opposing flow rate originating in opposing queues

     = $1 - R_{po}(g_o/C)$

R_{po} = platoon ratio for the opposing flow, obtained from Table 10.3, based on the arrival type of the opposing flow

g_o = effective green time for the opposing flow (sec)

E_{L2} = $(1 - P_{THO}^n)/P_{LTO}$; $E_{L2} \ge 1.0$

P_{LTO} = proportion of left turns in opposing single-lane approach

P_{THO} = proportion of through and right-turning vehicles in opposing single-lane approach computed as $(1 - P_{LTO})$

n = maximum number of opposing vehicles that can arrive during $(g_{q7A} - g_{f7A})$, computed as $(g_{q7A} - g_{f7A})/2$ with $n \geq 0$

E_{L1} = through-car equivalent for each left-turning vehicle, obtained from Table 10.14

For case 7B (single-lane approach opposed by a multilane approach), gaps are not opened in the opposing flow by opposing left-turning vehicles blocking opposing through movements. The single-lane model therefore does not apply and the multilane models in Eqs. 10.16 and 10.17 apply but $f_{LT} = f_M$; however, the single-lane model for g_f is used. That is,

$$f_{LT7B} = \frac{g_{f7B}}{g} + \frac{g_{u7B}}{g} \left[\frac{1}{1 + P_L \, (E_{L1} - 1)} \right] \tag{10.33}$$

g = effective green time for the lane group (sec)

g_{f7B} = portion of the effective green time that expires before a left-turning vehicle arrives (sec)

= G exp $(-0.860 LTC^{0.629}) - t_L$

g_{u7B} = portion of the effective green time during which left-turning vehicles filter through the opposing flow (sec)

= $g - g_{q7B}$ when $g_q \geq g_{f7B}$

= $g - g_{f7B}$ when $g_q < g_{f7B}$

where g_q is obtained from Eq. 10.13, and P_L is obtained from Eq. 10.19.

Note that the practical minimum value of f_{LT7B} may be estimated as $2(1 + P_L)/g$.

For case 7C (multilane approach opposed by a single-lane approach), the single-lane model given in Eq. 10.30 applies with two revisions. First, g_{f7A} is replaced by g_{f7C}, where

$$g_{f7C} = G \exp \left(-0.882 LTC^{0.717}\right) - t_L \tag{10.34}$$

Second, P_{LT} should be replaced by an estimated P_L that accounts for the effect of left turns on the other lanes of the lane group from which left turns are not made. These substitutions give the left turn factor (f_{LT7C}) as

$$f_{LT7C} = \left[f_{m7C} + 0.91(N - 1)/N \right] \tag{10.35}$$

$$f_{m7C} = \frac{g_{f7C}}{g} + \frac{g_{diff}}{g} \left[\frac{1}{1 + P_L \, (E_{L2} - 1)} \right] + \frac{g_u}{g} \left[\frac{1}{1 + P_L \, (E_{L1} - 1)} \right] \tag{10.36}$$

$$P_L = P_{LT} \left[1 + \frac{(N - 1)g}{g_{f7C} + \dfrac{g_u}{E_{L1}} + 4.24} \right] \tag{10.37}$$

$$E_{L2} = \frac{\left(1 - P_{THO}^{n}\right)}{P_{LTO}}$$ (10.38)

where

$$g_{q7C} = 4.943 v_{olC}^{0.762} qr_{o}^{1.061} - t_{L}, \quad 0.0 \leq g_{q} \leq g$$ (10.39)

$$g_{diff} = max\left[(g_{q7C} - g_{f7C}), \, 0\right]$$

$$g_{u} = g - g_{f7C} \text{ when } g_{q7C} \geq g_{f7C}$$ (10.40)

$$= g - g_{q7C} \text{ when } g_{q7C} < g_{f7C}$$

g = effective green time (sec)

$$g_{f7C} = G \exp(-0.882 LTC^{0.717}) - t_{L}$$

G = actual green time for the permitted phase (sec)

LTC = left turns per cycle

$\quad = v_{LT} C / 3600$

v_{LT} = adjusted left-turn flow rate (veh/h)

C = cycle length (sec)

t_{L} = lost time per phase (sec)

v_{olc} = adjusted opposing flow rate per lane per cycle (veh/l/c)

$\quad = v_{o} C / (3600 N_{o}) f_{Lo}$

qr_{o} = opposing queue ratio, that is, the proportion of opposing flow rate originating in opposing queues

$\quad = 1 - R_{po}(g_{o}/C)$

R_{po} = platoon ratio for the opposing flow, obtained from Table 10.3, based on the opposing flow

g_{o} = effective green time for the opposing flow (sec)

P_{LT} = proportion of left turns in the lane group

N = number of lanes in the lane group

f_{s} = left-turn saturation factor ($f_{s} \geq 0$)

$\quad = (875 - 0.625 v_{o}) / 1000$

v_{o} = adjusted opposing flow rate (veh/h)

P_{LTO} = proportion of left turns in opposing single-lane approach

P_{THO} = proportion of through and right-turning vehicles in opposing single-lane approach

$\quad = (1 - P_{LTO})$

n = maximum number of opposing vehicles that can arrive during $(g_{q7C} - g_{f7C})$ computed as $(g_{q7C} - g_{f7C})/2, \, n \geq 0$

E_{L1} = through-car equivalent for each left-turning vehicle, obtained from Table 10.6

Note, however, that when the subject approach is a dual left-turn lane, $f_{LT7C} = f_{m7C}$, and Eq. 10.35 does not apply.

The worksheets shown in Figures 10.4 and 10.5 may be used to compute the left-turn factors for multilane and single-lane opposing approaches, respectively.

To compute the appropriate values for G, g, g_{f}, g_{q}, and g_{u} for protected-plus-permitted-left-turn phases, one can use the models presented in the previous section when the left

SUPPLEMENTAL WORKSHEET FOR PERMITTED LEFT TURNS OPPOSED BY MULTILANE APPROACH

General Information

Project Description _____

Input

	EB	WB	NB	SB
Cycle length, C (s)				
Total actual green time for LT lane group,[1] G (s)				
Effective permitted green time for LT lane group,[1] g (s)				
Opposing effective green time, g_o (s)				
Number of lanes in LT lane group,[2] N				
Number of lanes in opposing approach, N_o				
Adjusted LT flow rate, v_{LT} (veh/h)				
Proportion of LT volume in LT lane group,[3] P_{LT}				
Adjusted flow rate for opposing approach, v_o (veh/h)				
Lost time for LT lane group, l_L				

Computation

	EB	WB	NB	SB
LT volume per cycle, $LTC = v_{LT}C/3600$				
Opposing lane utilization factor, f_{LUo} (refer to Volume Adjustment and Saturation Flow Rate Worksheet)				
Opposing flow per lane, per cycle $v_{olc} = \dfrac{v_o C}{3600 N_o f_{LUo}}$ (veh/C/ln)				
$g_f = G[e^{-0.882(LTC^{0.717})}] - l_L,\ g_f \le g$ (except for exclusive left-turn lanes)[1, 4]				
Opposing platoon ratio, R_{po} (refer to Exhibit 16-11)				
Opposing queue ratio, $qr_o = \max[1 - R_{po}(g_o/C), 0]$				
$g_q = \dfrac{v_{olc}qr_o}{0.5 - [v_{olc}(1 - qr_o)/g_o]} - l_L,\ v_{olc}(1 - qr_o)/g_o \le 0.49$ (note case-specific parameters)[1]				
$g_u = g - g_q$ if $g_q \ge g_f$, or $g_u = g - g_f$ if $g_q < g_f$				
E_{L1} (refer to Exhibit C16-3)				
$P_L = P_{LT}\left[1 + \dfrac{(N-1)g}{(g_f + g_u/E_{L1} + 4.24)}\right]$ (except with multilane subject approach)[5]				
$f_{min} = 2(1 + P_L)/g$				
$f_m = [g_f/g] + [g_u/g]\left[\dfrac{1}{1 + P_L(E_{L1} - 1)}\right],\ (f_{min} \le f_m \le 1.00)$				
$f_{LT} = [f_m + 0.91(N - 1)]/N$ (except for permitted left turns)[6]				

Notes

1. Refer to Figure 10.6 for case-specific parameters and adjustment factors.
2. For exclusive left-turn lanes, N is equal to the number of exclusive left-turn lanes. For shared left-turn lanes, N is equal to the sum of the shared left-turn, through, and shared right-turn (if one exists) lanes in that approach.
3. For exclusive left-turn lanes, $P_{LT} = 1$.
4. For exclusive left-turn lanes, $g_f = 0$, and skip the next step. Lost time, l_L, may not be applicable for protected–permitted case.
5. For a multilane subject approach, if $P_L \ge 1$ for a left-turn shared lane, then assume it to be a de facto exclusive left-turn lane and redo the calculation.
6. For permitted left turns with multiple exclusive left-turn lanes $f_{LT} = f_m$.

Figure 10.4 Supplemental Worksheet for Permitted Left Turns Where Approach Is Opposed by Multilane Approach

SOURCE: *Highway Capacity Manual,* Special Report 209, Fourth Edition, Transportation Research Board, National Research Council, Washington, D.C., 2000. Used with permission.

SUPPLEMENTAL WORKSHEET FOR PERMITTED LEFT TURNS OPPOSED BY SINGLE-LANE APPROACH

General Information

Project Description _____

Input

	EB	WB	NB	SB
Cycle length, C (s)				
Total actual green time for LT lane group,[1] G (s)				
Effective permitted green time for LT lane group,[1] g (s)				
Opposing effective green time, g_o (s)				
Number of lanes in LT lane group,[2] N				
Adjusted LT flow rate, v_{LT} (veh/h)				
Proportion of LT volume in LT lane group, P_{LT}				
Proportion of LT volume in opposing flow, P_{LTo}				
Adjusted flow rate for opposing approach, v_o (veh/h)				
Lost time for LT lane group, t_L				

Computation

LT volume per cycle, $LTC = v_{LT}C/3600$				
Opposing flow per lane, per cycle, $v_{olc} = v_oC/3600$ (veh/C/ln)				
Opposing platoon ratio, R_{po} (refer to Exhibit 16-11)				
$g_f = G[e^{-0.860(LTC^{0.629})}] - t_L \quad g_f \le g$ (except exclusive left-turn lanes)[3]				
Opposing queue ratio, $qr_o = \max[1 - R_{po}(g_o/C), 0]$				
$g_q = 4.943 v_{olc}^{0.762} qr_o^{1.061} - t_L \quad g_q \le g$				
$g_u = g - g_q$ if $g_q \ge g_f$, or				
$g_u = g - g_f$ if $g_q < g_f$				
$n = \max[(g_q - g_f)/2, 0]$				
$P_{THo} = 1 - P_{LTo}$				
E_{L1} (refer to Exhibit C16-3)				
$E_{L2} = \max[(1 - P_{THo}^n)/P_{LTo}, 1.0]$				
$f_{min} = 2(1 + P_{LT})/g$				
$g_{diff} = \max[g_q - g_f, 0]$ (except when left-turn volume is 0)[4]				
$f_{LT} = f_m = [g_f/g] + \left[\dfrac{g_u/g}{1 + P_{LT}(E_{L1} - 1)}\right] + \left[\dfrac{g_{diff}/g}{1 + P_{LT}(E_{L2} - 1)}\right]$ $(f_{min} \le f_m \le 1.00)$				

Notes

1. Refer to Figure 10.6 for case-specific parameters and adjustment factors.
2. For exclusive left-turn lanes, N is equal to the number of exclusive left-turn lanes. For shared left-turn lanes, N is equal to the sum of the shared left-turn, through, and shared right-turn (if one exists) lanes in that approach.
3. For exclusive left-turn lanes, $g_f = 0$, and skip the next step. Lost time, t_L, may not be applicable for protected-permitted case.
4. If the opposing left-turn volume is 0, then $g_{diff} = 0$.

Figure 10.5 Supplemental Worksheet for Permitted Left Turns Where Approach Is Opposed by Single-Lane Approach

SOURCE: *Highway Capacity Manual,* Special Report 209, Fourth Edition, Transportation Research Board, National Research Council, Washington, D.C., 2000. Used with permission.

turn can move only in a permitted phase. When the left turns can be made during protected-plus-permitted phases, the protected portion of the phase is separated from the permitted portion, and each portion is treated separately. Two left-turn factors are then determined: one for the protected phase and another for the permitted phase. The left-turn factor for the protected phase is determined as discussed earlier, and that for the permitted phase is obtained from the appropriate model, but with modified values of $G, g, g_f,$ and g_q (which are denoted as $G^\star, g^\star, g_f^\star, g_q^\star$). Figure 10.6 shows the equation for obtaining

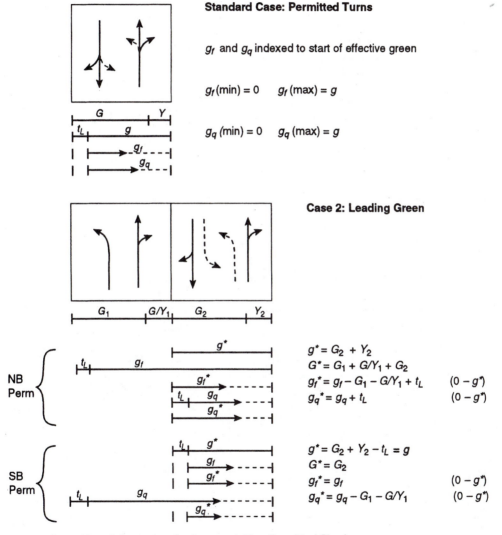

Figure 10.6 Green Time Adjustments for Protected-Plus-Permitted Phasing

SOURCE: *Highway Capacity Manual,* Special Report 209, Fourth Edition, Transportation Research Board, National Research Council, Washington, D.C., 2000. Used with permission.

Continued

Case 3: Lagging Green

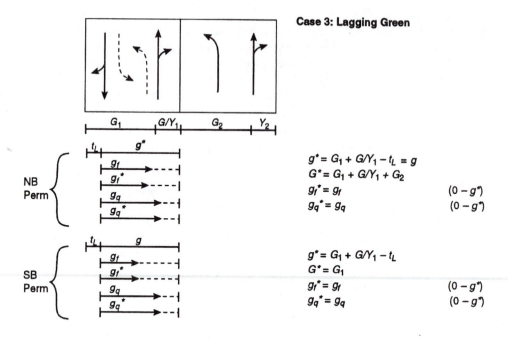

$g^* = G_1 + G/Y_1 - t_L = g$
$G^* = G_1 + G/Y_1 + G_2$
$g_f^* = g_f$ $(0 - g^*)$
$g_q^* = g_q$ $(0 - g^*)$

$g^* = G_1 + G/Y_1 - t_L$
$G^* = G_1$
$g_f^* = g_f$ $(0 - g^*)$
$g_q^* = g_q$ $(0 - g^*)$

Case 4: Leading and Lagging Green

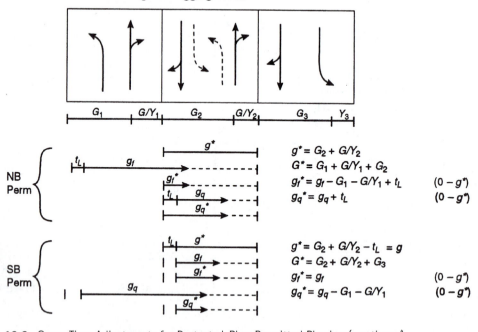

$g^* = G_2 + G/Y_2$
$G^* = G_1 + G/Y_1 + G_2$
$g_f^* = g_f - G_1 - G/Y_1 + t_L$ $(0 - g^*)$
$g_q^* = g_q + t_L$ $(0 - g^*)$

$g^* = G_2 + G/Y_2 - t_L = g$
$G^* = G_2 + G/Y_2 + G_3$
$g_f^* = g_f$ $(0 - g^*)$
$g_q^* = g_q - G_1 - G/Y_1$ $(0 - g^*)$

Figure 10.6 Green Time Adjustments for Protected-Plus-Permitted Phasing (*continued*)

SOURCE: *Highway Capacity Manual,* Special Report 209, Fourth Edition, Transportation Research Board, National Research Council, Washington, D.C., 2000. Used with permission.

Case 5: LT Phase with Leading Green

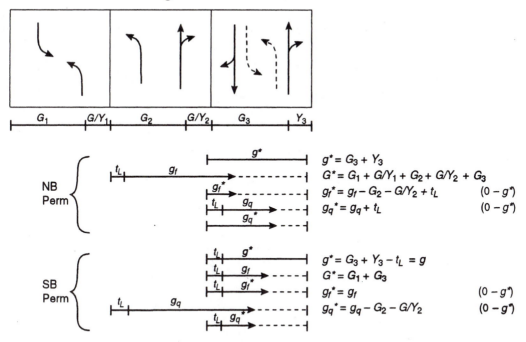

Figure 10.6 Green Time Adjustments for Protected-Plus-Permitted Phasing (*continued*)

SOURCE: *Highway Capacity Manual,* Special Report 209, Fourth Edition, Transportation Research Board, National Research Council, Washington, D.C., 2000. Used with permission.

$G^\star, g^\star, g_f^\star$, and $g_q^\star$ for the different cases. For example, Case 2 shows the case with an exclusive left-turn lane and a leading green G, which is followed by a period G/Y_1 during which the left-turn movement is given the yellow indication and the through movement is still given the green ball indication. This is then followed by a period G_2, during which both the NB and SB traffic streams have first the green ball indication and then a full yellow indication, Y_2. This results in an effective green time for the NB permitted phase $g^\star$ of $G_2 + Y_2$, and for the SB direction, $g^\star$ of $G_2 + Y_2 - t_L$. The reason for this is that there is no lost time for the NB traffic during the permitted-left-turn phase since the movement was initiated during the protected portion of the phase, and the lost time is assessed there. This results in different effective green times for NB and SB traffic streams. Similarly, if the NB left turns are made from a shared lane, g_f would be computed from the total green time of $G_1 + G/Y_1 + G_2$, which includes the leading phase green time. However, in computing the appropriate g_f, only the portion of g_f that applies to the permitted phase should be used. This results in $g_f^\star$ being $g_f - G_1 - G/Y_1 + t_L$. Also, in computing the appropriate $g_q^\star$ for the NB traffic stream, it is noted that this should be the portion of the NB permitted green phase ($g^\star$) that is blocked by the clearance of the opposing queue. However, the permitted NB phase does not account for the lost time, and $g_q^\star$ is obtained as

$g_q + t_L$. Similar considerations are used to obtain the modified equations for $g^\star$, $G^\star$, $g_f^\star$, and $g_q^\star$ for the different cases shown.

Pedestrian and Bicycle Adjustment Factors. These factors are included in the saturation flow equation to account for the reduction in the saturation flow rate resulting from the conflicts between automobiles and pedestrians, and bicycles. The specific zones within the intersection where these conflicts occur are shown in Figure 10.7. The parameters required for the computation for these factors as presented by the HCM are shown in Table 10.4. The flow chart shown in Figure 10.8 illustrates the procedure. The procedure can be divided into the following four main tasks:

- Determine average pedestrian occupancy, OCC_{pedg}
- Determine relevant conflict zone occupancy, OCC_r
- Determine permitted phase pedestrian-bicycle adjustment factors for turning movements A_{pBT}

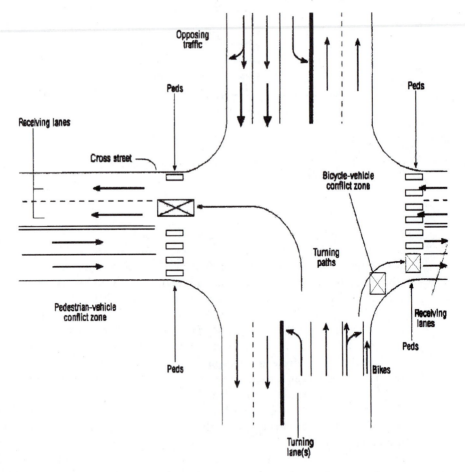

Figure 10.7   Conflict Zone Locations

SOURCE: *Highway Capacity Manual,* Special Report 209, Fourth Edition, Transportation Research Board, National Research Council, Washington, D.C., 2000. Used with permission.

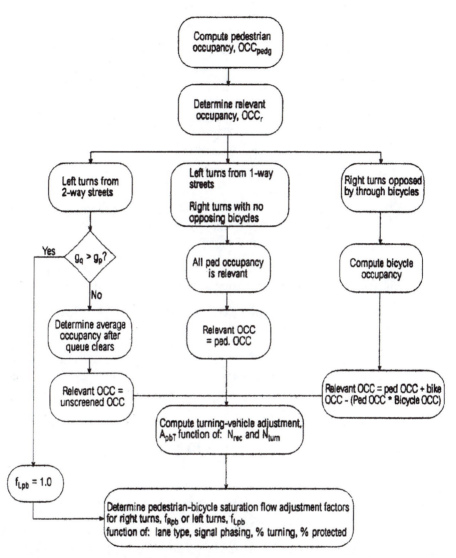

Figure 10.8   Outline of Computational Procedure for f_{Rpb} and f_{Lpb}

SOURCE: *Highway Capacity Manual,* Special Report 209, Fourth Edition, Transportation Research Board, National Research Council, Washington, D.C., 2000. Used with permission.

- Determine saturation flow adjustment factors for turning movements (f_{Lpb} for left-turn movements and f_{Rpb} for right-turn movements)

Determine Average Pedestrian Occupancy (OCC_{pedg}).   In this task, the pedestrian flow rate is first computed from the pedestrian volume using Eq. 10.41, and the average pedestrian occupancy is then computed from the pedestrian flow rate using Eq. 10.42 or 10.43.

$$v_{pedg} = v_{ped}\left(\frac{C}{g_p}\right), \left(v_{pedg} \leq 5000\right) \qquad (10.41)$$

$$OCC_{pedg} = v_{pedg} / 2000, \left(v_{pedg} \leq 1000, \text{ and } OCC_{pedg} \leq 0.5\right) \qquad (10.42)$$

$$OCC_{pedg} = 0.4 + v_{pedg} / 10,000, \left(1000 < v_{pedg} \leq 5000, \text{ and } 0.5 < OCC_{pedg} \leq 0.9\right) \quad (10.43)$$

where

v_{pedg} = pedestrian flow rate
v_{ped} = pedestrian volume
g_p = pedestrian green walk + flashing don't walk time (sec) (if pedestrian signal timing is unknown, g_p may be assumed to be equal to g (sec))
C = cycle length (sec)

Note that in using Eqs. 10.42 and 10.43, it is not necessary to compute v_{pedg} from Eq. 10.41 if pedestrian flow rate is collected directly from the field.

Determine Relevant Conflict Zone Occupancy (OCC$_r$).   Two conditions influence the computation of this factor. These are (1) right-turning bicycles and automobiles weave to the right before reaching the stop line, and (2) left-turn movements are made from a one-way street. When the first condition exists, the bicycle/automobile interaction within the intersection is eliminated; the bicycle volume should therefore be ignored and only the impact of pedestrians should be considered. For both of these conditions, the HCM gives the relevant conflict zone occupancy as

$$OCC_r = OCC_{pedg} \qquad (10.44)$$

where

OCC_{pedg} = average pedestrian occupancy (obtained from Eq. 10.42 or 10.43)

When bicycle interaction is also expected within the intersection, the bicycle flow rate (v_{bicg}) is first computed from the bicycle volume using Eq. 10.45, and the bicycle conflict zone occupancy (OCC_{bicg}) is then determined from the bicycle flow rate using Eq. 10.46. The relevant conflict zone occupancy (OCC_T) is then computed from the pedestrian occupancy (OCC_{pedg}) and the bicycle occupancy (OCC_{bicg}) using Eq. 10.47.

$$v_{bicg} = v_{bic}\left(C/g\right) \qquad \left(v_{bicg} \leq 1900\right) \qquad (10.45)$$

$$OCC_{bicg} = 0.02 + \left(v_{bicg}/2700\right) \qquad \left(v_{bicg} < 1900 \text{ and } OCC_{bicg} \leq 0.72\right) \qquad (10.46)$$

$$OCC_r = OCC_{pedg} + OCC_{bicg} - \left(OCC_{pedg}\right)\left(OCC_{bicg}\right) \qquad (10.47)$$

where

v_{bicg} = bicycle flow rate (bicycles per hr)
v_{bic} = bicycle volume

As in the previous case, when bicycle flow rates are collected directly in the field these values should be used and Eq. 10.45 is not used.

When left turns are made from an approach on a two-way street, a comparison of the opposing queue clearance (g_q) and the pedestrian green time (g_p) is first made to determine whether g_q is less or greater than g_p. If $g_q \geq g_p$, then the pedestrian green time is used entirely by the opposing queue and the pedestrian adjustment factor for left-turn movements (f_{Lpb}) is 1.0. However, if $g_q < g_p$, then the pedestrian occupancy after the opposing queue clears (OCC_{pedu}) is determined from the average pedestrian occupancy (OCC_{pedg}) using Eq. 10.48. The relevant conflict zone occupancy (OCC_r) is then determined from OCC_{pedu} using Eq. 10.49. Equation 10.49 is based on the fact that left-turning vehicles can go through the intersection only after the opposing queue has cleared, and that accepted gaps in the opposing flow v_o must be available for the left-turning vehicles.

$$OCC_{pedu} = OCC_{pedg}\left[1 - 0.5(g_q / g_p)\right] \qquad (10.48)$$

$$OCC_r = OCC_{pedu}\left[e^{-(5/3600)v_o}\right] \qquad (10.49)$$

where

OCC_{pedu} = pedestrian occupancy after the opposing queue clears
OCC_{pedg} = average pedestrian occupancy
g_q = opposing queue clearing time (sec)
q_p = pedestrian green walk + flashing DON'T WALK (sec) (if the pedestrian signal timing is unknown, g_p may be assumed to be equal to the effective green (g))

Determine Permitted Phase Pedestrian-Bicycle Adjustment Factors for Turning Movement (A_{pbT}): Two conditions are considered in the determination of the A_{pbT}. These are (1) the number of turning lanes (N_{turn}) is the same as the number of the cross-street receiving lanes (N_{rec}) and (2) the number of the turning lanes is less than the number of the cross-street receiving lanes.

When N_{turn} is equal to N_{rec}, the proportion of the time the conflict zone is occupied is the adjustment factor, as it is unlikely that the turning vehicles will be able to move around the pedestrians or bicycles. The permitted phase pedestrian-bicycle adjustment factor (A_{pbT}) is therefore obtained from Eq. 10.50.

$$A_{pbT} = 1 - OCC_r \qquad (N_{turn} = N_{rec}) \qquad (10.50)$$

When N_{turn} is less than the N_{rec}, the impact of pedestrian and bicycles on the saturation flow is reduced, as it is more likely that the turning vehicles will be able to move around the pedestrians and the bicycles. The A_{pbT} in this case is obtained from Eq. 10.51.

$$A_{pbT} = 1 - 0.6(OCC_r) \qquad (N_{trun} < N_{rec}) \qquad (10.51)$$

where
N_{turn} = the number of turning lanes
N_{rec} = the number of receiving lanes

It is recommended that actual field observation be carried out to determine the number of turning lanes (N_{turn}) and the number of receiving lanes (N_{rec}). The reason for this is that at some intersections, left-turns are illegally made deliberately from an outer lane or the receiving lane is blocked by vehicles that are double parked, making it difficult for the turning vehicles to make a proper turn. Simply reviewing the intersection plans and noting the striping cannot identify these conditions.

Determine Saturation Flow Adjustment Factors for Turning Movements (f_{Lpb} for left turn movements, and f_{Rpb} for right-turn movements).   These factors depend on the A_{pbT} and the proportion of the turning flow that uses the protected phase. The pedestrian–bicycle adjustment factor for left-turns (f_{Lpb}) is obtained from Eq. 10.52, and that for right turns (f_{Rpb}) is obtained from Eq. 10.53.

$$f_{Lpb} = 1.0 - P_{LT}(1 - A_{pbT})(1 - P_{LTA}) \qquad (10.52)$$

where
P_{LT} = proportion of left turn volumes (used only for left turns made from a single lane approach or for shared lanes)
A_{pbT} = permitted phase pedestrian-bicycle adjustment factor for turning movements (obtained from Eq. 10.50 or 10.51)
P_{LTA} = the proportion of left turns using protected phase (used only for protected/permissive phases)

$$f_{Rpb} = 1.0 - P_{RT}(1 - A_{pbT})(1 - P_{RTA}) \qquad (10.53)$$

where
P_{RT} = proportion of right-turn volume (used only for right turns made from single-lane approach or for shared turning lanes)
A_{pbT} = permitted phase pedestrian-bicycle adjustment factor for turning movements (obtained from Eq. 10.50 or 10.51)
P_{RTA} = the proportion of right turns using protected phase (used only for protected/permissive phases)

Figure 10.9 shows a supplemental worksheet that can be used to carry out these procedures.

Field Determination of Saturation Flow.   An alternative to the use of adjustment factors is to determine directly the saturation flow in the field. It was shown in Chapter 8 that the saturation flow rate is the maximum discharge flow rate during the green time. This flow rate is usually achieved 10 to 14 seconds after the start of the green phase, which is usually the time the fourth, fifth or sixth passenger car crosses the stop line. Therefore,

SUPPLEMENTAL WORKSHEET FOR PEDESTRIAN-BICYCLE EFFECTS ON PERMITTED LEFT TURNS AND RIGHT TURNS

General Information

Project Description _____

Permitted Left Turns

	EB	WB	NB	SB
	_↗	_↙	↖	↓
Effective pedestrian green time,[1,2] g_p (s)				
Conflicting pedestrian volume,[1] v_{ped} (p/h)				
$v_{pedg} = v_{ped} (C/g_p)$				
$OCC_{pedg} = v_{pedg}/2000$ if ($v_{pedg} \leq 1000$) or $OCC_{pedg} = 0.4 + v_{pedg}/10,000$ if ($1000 < v_{pedg} \leq 5000$)				
Opposing queue clearing green,[3,4] g_q (s)				
Effective pedestrian green consumed by opposing vehicle queue, g_q/g_p; if $g_q \geq g_p$ then $f_{Lpb} = 1.0$				
$OCC_{pedu} = OCC_{pedg} [1 - 0.5(g_q/g_p)]$				
Opposing flow rate,[3] v_o (veh/h)				
$OCC_r = OCC_{pedu} [e^{-(5/3600)v_o}]$				
Number of cross-street receiving lanes,[1] N_{rec}				
Number of turning lanes,[1] N_{turn}				
$A_{pbT} = 1 - OCC_r$ if $N_{rec} = N_{turn}$				
$A_{pbT} = 1 - 0.6(OCC_r)$ if $N_{rec} > N_{turn}$				
Proportion of left turns,[5] P_{LT}				
Proportion of left turns using protected phase,[6] P_{LTA}				
$f_{Lpb} = 1.0 - P_{LT}(1 - A_{pbT})(1 - P_{LTA})$				

Permitted Right Turns

	EB	WB	NB	SB
	_↘	↙_	↗	↙
Effective pedestrian green time,[1,2] g_p (s)				
Conflicting pedestrian volume,[1] v_{ped} (p/h)				
Conflicting bicycle volume,[1,7] v_{bic} (bicycles/h)				
$v_{pedg} = v_{ped}(C/g_p)$				
$OCC_{pedg} = v_{pedg}/2000$ if ($v_{pedg} \leq 1000$), or $OCC_{pedg} = 0.4 + v_{pedg}/10,000$ if ($1000 < v_{pedg} \leq 5000$)				
Effective green,[1] g (s)				
$v_{bicg} = v_{bic}(C/g)$				
$OCC_{bicg} = 0.02 + v_{bicg}/2700$				
$OCC_r = OCC_{pedg} + OCC_{bicg} - (OCC_{pedg})(OCC_{bicg})$				
Number of cross-street receiving lanes,[1] N_{rec}				
Number of turning lanes,[1] N_{turn}				
$A_{pbT} = 1 - OCC_r$ if $N_{rec} = N_{turn}$				
$A_{pbT} = 1 - 0.6(OCC_r)$ if $N_{rec} > N_{turn}$				
Proportion of right turns,[5] P_{RT}				
Proportion of right turns using protected phase,[8] P_{RTA}				
$f_{Rpb} = 1.0 - P_{RT}(1 - A_{pbT})(1 - P_{RTA})$				

Notes

1. Refer to Input Worksheet.
2. If intersection signal timing is given, use Walk + flashing Don't Walk (use G + Y if no pedestrian signals). If signal timing must be estimated, use (Green Time – Lost Time per Phase) from Quick Estimation Control Delay and LOS Worksheet.
3. Refer to supplemental worksheets for left turns.
4. If unopposed left turn, then $g_q = 0$, $v_o = 0$, and $OCC_r = OCC_{pedu} = OCC_{pedg}$.
5. Refer to Volume Adjustment and Saturation Flow Rate Worksheet.
6. Ideally determined from field data; alternatively, assume it equal to $(1 - \text{permitted phase } f_{LT})/0.95$.
7. If $v_{bic} = 0$ then $v_{bicg} = 0$, $OCC_{bicg} = 0$, and $OCC_r = OCC_{pedg}$.
8. P_{RTA} is the proportion of protected green over the total green, $g_{prot}/(g_{prot} + g_{perm})$. If only permitted right-turn phase exists, then $P_{RTA} = 0$.

Figure 10.9 Supplemental Worksheet for Pedestrian-Bicycle Effects

saturation flow rates are computed starting with the headway after the fourth vehicle in the queue.

Two people are needed to carry out the procedure, with one being the timer equipped with a stopwatch, and the other the recorder equipped with a push-button event recorder or a notebook computer with appropriate software. The form shown in Figure 10.10 is used to record the data. It is suggested that the general information section of the form shown in Figure 10.10 be completed and other details such as area type, width, and grade of the lane being evaluated be measured and recorded. An observation point is selected at the intersection such that a clear vision of the traffic signals and the stop line is maintained. A reference point is selected to indicate when a vehicle has entered the intersection. This reference point is usually the stop line such that all vehicles that cross the stop line are considered as having entered the intersection. The following steps are then carried out for each cycle and for each lane.

Step 1.   The timer starts the stopwatch at the beginning of the green phase and notifies the recorder.

Step 2.   The recorder immediately notes the last vehicle in the stopped queue and describes it to the timer and also notes which vehicles are heavy vehicles and which vehicles turn left or right.

Step 3.   The timer then counts aloud each vehicle in the queue as its rear axle crosses the reference point (that is, "one," "two," "three," and so on). Note that right- or left-turning vehicles that are yielding to either pedestrians or opposing vehicles are not counted until they have gone through the opposing traffic.

Step 4.   The timer calls out the times that the fourth, tenth, and last vehicles in the queue cross the stop line, and these are noted by the recorder.

Step 5.   In cases where queued vehicles are still entering the intersection at the end of the green phase, the number of the last vehicle at the end of the green phase is identified by the timer and told to the recorder so that number can be recorded.

Step 6.   The width of the lane and the slope of the approach are then measured and recorded together with any unusual occurrences that might have affected the saturation flow.

Step 7.   Since the flow just after the start of the green phase is less than saturation flow, the time considered for calculating the saturation flow is that between the time the rear axle of the fourth car crosses the reference point (t_4) and the time the rear axle of the last vehicle queued at the beginning of the green crosses the same reference point (t_n). The saturation flow is then determined from Eq. 10.54.

$$\text{saturation flow} = \frac{3600}{(t_4 - t_n)/(n - 4)} \tag{10.54}$$

where n is the number of the last vehicle surveyed. The data recorded on heavy vehicles, turning vehicles, and approach geometrics can be used in the future if adjustment factors are to be applied.

FIELD SATURATION FLOW RATE STUDY WORKSHEET

General Information		Site Information		
Analyst _____		Intersection _____		
Agency or Company _____		Area Type	☐ CBD	☐ Other
Date Performed _____		Jurisdiction _____		
Analysis Time Period _____		Analysis Year _____		

Lane Movement Input

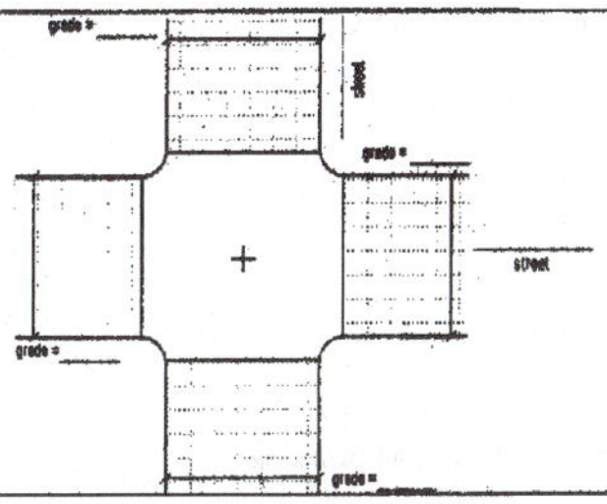

Movements Allowed
☐ Through
☐ Right turn
☐ Left turn

Identify all lane movements and the lane studied

Input Field Measurement

Veh. in queue	Cycle 1			Cycle 2			Cycle 3			Cycle 4			Cycle 5			Cycle 6		
	Time	HV	T	Time	HV	T	Time	HV	T	Time	HV	T	Time	HV	T	Time	HV	T
1																		
2																		
3																		
4																		
5																		
6																		
7																		
8																		
9																		
10																		
11																		
12																		
13																		
14																		
15																		
16																		
17																		
18																		
19																		
20																		
End of saturation																		
End of green																		
No. veh. > 20																		
No. veh. on yellow																		

Gloassary and Notes

HV = Heavy vehicles (vehicles with more than 4 tires on pavement)
T   = Turning vehicles (L = Left, R = Right)
Pedestrians and buses that block vehicles should be noted with the time that they block traffic, for example,
P12 = Pedestrians blocked traffic for 12 s
B15 = Bus blocked traffic for 15 s

Figure 10.10 Field Sheet for Direct Observation of Prevailing Saturation Flow Ratio

SOURCE: *Highway Capacity Manual,* Special Report 209, Fourth Edition, Transportation Research Board, National Research Council, Washington, D.C., 2000. Used with permission.

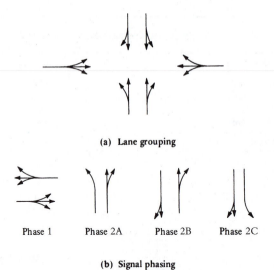

(a) Lane grouping

Phase 1   Phase 2A   Phase 2B   Phase 2C

(b) Signal phasing

Figure 10.11 Illustrative Example for Determining Critical Lane Group

Capacity and v/c Analysis Module

In this module, results of the computations carried out in the previous modules are used to determine the important capacity variables, which include

- Flow ratios for the different lane groups
- Capacities for the different lane groups
- (v/c) ratios for the different lane groups
- The critical (v/c) ratio for the overall intersection

The adjusted demand volume obtained for each lane group in the volume adjustment module is divided by the saturation flow for the appropriate lane group determined in the saturation flow module to obtain the flow ratio (v_i/s_i) for that lane group. The capacity of each lane group is then determined by using Eq. 10.1.

$$c_i = s_i(g_i/C)$$

Using Eq. 10.2, the volume-to-capacity (v/c) ratio is then computed for each lane group.

$$X_i = (v_i/c_i)$$

Similarly, using Eq. 10.3, the critical volume-to-capacity ratio X_c is then computed for the intersection.

$$X_c = \sum_i (v/s)_{ci} \frac{C}{C - L}$$

The identification of the critical lane group for each green phase is necessary before the critical volume-to-capacity ratio (X_c) can be determined for the intersection. This identification is relatively simple when there are no overlapping phases, because the lane group with the maximum flow ratio (v_i/s_i) during each green phase is the critical lane group for that phase. When the phases overlap, however, identification of the critical lane group is not as simple. The basic principle used in this case is that the critical (v/c) ratio for the intersection is based on the combinations of lane groups that will use up the largest amount of the capacity available. This is demonstrated by Figure 10.11, which shows a phasing system that provides for exclusive left-turn lanes in the south and north approaches and overlapping phases.

There are only two lane groups during Phase 1—that is, EBLT/TH/RT and WBLT/TH/RT. The critical lane group is simply selected as the lane group with the greater (v_i/s_i) ratio. The three other phases, however, include overlapping movements, and the critical lane group is not as straightforward to identify. It can be seen that the NBTH/RT lane group moves during Phases 2A and 2B and therefore overlaps with the SBTH/RT lane group, which moves during Phases 2B and 2C, while the NBLT lane group moves only during Phase 2A and the SBLT lane group moves only during Phase 2C. The NBTH/RT lane group can therefore be critical for the sum of Phases 2A and 2B, whereas the SBLT lane group can be critical for Phase 2C.

In determining the critical lane group, any one phase or portion of a phase can have only one critical lane group. If a critical lane group has been determined for the sum of phases i and j, no other lane group can be critical for either phase i or j or for any combination of phases that includes phase i or j. Note also that in determining the signal timing for any intersection—that is, for design of the intersection—the critical lane group is used.

Two possibilities therefore exist for the (v/s) ratios for the overlapping Phases 2A, 2B, and 2C:

1. NBTH/RT + SBLT
2. SBTH/RT + NBLT

The critical lane flow ratio for the intersection is therefore the maximum flow ratio of the following:

EBLT/TH/RT + NBTH/RT + SBLT
EBLT/TH/RT + SBTH/RT + NBLT
WBLT/TH/RT + NBTH/RT + SBLT
WBLT/TH/RT + SBTH/RT + NBLT

It is also necessary to determine the total lost time L before the critical (v/c) ratio can be computed. The general rule is that it is assumed that a lost time of t_L occurs when a movement is initiated. Therefore, the total lost time L for each possible critical movement identified above is $3t_L$, where t_L is the lost time per phase. Thus, in general, $L = nt_L$, where n is the number of movements in the critical path through the signal cycle. Note that the lost time for each phase (t_L) is the sum of the start-up lost times (L_i) and the yellow-plus and red interval (y) minus the extension time (e).

The computation of the critical volume-to-capacity ratio (X_c) completes the definition of the capacity characteristics of the intersection. As stated earlier, these characteristics must be evaluated separately and in conjunction with the delay and levels of service that are determined in the next module. Following are some key points that should be kept in mind when the capacity characteristics are being evaluated:

1. When the critical (v/c) ratio is greater than 1.00, the geometric and signal characteristics are inadequate for the critical demand flows at the intersection. The operating characteristics at the intersection may be improved by increasing the cycle length, changing the cycle phase, and/or changing the roadway geometrics.
2. When there is a large variation in the (v/c) ratio for the different critical lane groups but the critical (v/c) ratio is acceptable, the green time is not proportionately distributed, and reallocation of the green time should be considered.
3. A protected left-turn phase should be considered when the use of permitted left turns results in drastic reductions of the saturation flow rate for the appropriate lane group.
4. When the critical v/c ratio approaches 1.0, it is quite likely that the existing signal and geometric characteristics will not be adequate for an increased demand flow rate. Consideration should therefore be given to changing either the signal timing and/or the roadway geometrics.
5. If the (v/c) ratios are above acceptable limits and protected turning phases have been included for the turning movements with high flows, then changes in roadway geometrics will be required to reduce the (v/c) ratios.

The computation required for this module may be carried out in the format shown in Figure 10.12. Note that in row 2 of Figure 10.12, "Phase Type," when left turns are made from exclusive lanes during a protected permissive phase, the protected phase should be represented by a separate column. The protected phase is considered to be the primary phase and is designated as "P," the permitted phase is considered as the secondary phase and is designated "S," and the column containing the total flows is designated as "T." However, certain quantities (such as lane group capacity) should be computed as the sum of the primary and secondary phase flows. It should also be noted that both lane groups with shared left-turn lanes and lane groups with only protected or permitted phasing have only a primary phase.

Performance Measures Module

The results obtained from the volume adjustment, saturation flow rate, and capacity analysis modules are now used in this module to determine the average control time delay per vehicle in each lane group and hence the level of service for each approach and the intersection as a whole. The computation first involves the determination of the uniform, incremental, and residual delays.

Uniform Delay. The uniform delay is that delay that will occur in a lane group if vehicles arrive with a uniform distribution and if saturation does not occur during any cycle. It is based on the first term of the Webster delay model discussed in Chapter 8. Uniform delay is determined as

$$d_{1i} = 0.50C \frac{(1 - g_i/C)^2}{1 - (g_i/C)[\min(X_i, 1.0)]} \tag{10.55}$$

CAPACITY AND LOS WORKSHEET

General Information

Project Description _____

Capacity Analysis

Phase number													
Phase type													
Lane group													
Adjusted flow rate, v (veh/h)													
Saturation flow rate, s (veh/h)													
Lost time, t_L (s), $t_L = l_1 + Y - e$													
Effective green time, g (s), $g = G + Y - t_L$													
Green ratio, g/C													
Lane group capacity,[1] $c = s(g/C)$, (veh/h)													
v/c ratio, X													
Flow ratio, v/s													
Critical lane group/phase (√)													
Sum of flow ratios for critical lane groups, Y_c $Y_c = \sum$ (critical lane groups, v/s)													
Total lost time per cycle, L (s)													
Critical flow rate to capacity ratio, X_c $X_c = (Y_c)(C)/(C - L)$													

Lane Group Capacity, Control Delay, and LOS Determination

	EB			WB			NB			SB		
Lane group												
Adjusted flow rate,[2] v (veh/h)												
Lane group capacity,[2] c (veh/h)												
v/c ratio,[2] X = v/c												
Total green ratio,[2] g/C												
Uniform delay, $d_1 = \frac{0.50\, C\,[1 - (g/C)]^2}{1 - [\min(1, X)g/C]}$ (s/veh)												
Incremental delay calibration,[3] k												
Incremental delay,[4] d_2 $d_2 = 900T[(X - 1) + \sqrt{(X - 1)^2 + \frac{8kIX}{cT}}\,]$(s/veh)												
Initial queue delay, d_3 (s/veh)												
Uniform delay, d_1 (s/veh)												
Progression adjustment factor, PF												
Delay, $d = d_1(PF) + d_2 + d_3$ (s/veh)												
LOS by lane group												
Delay by approach, $d_A = \frac{\sum(d)(v)}{\sum v}$ (s/veh)												
LOS by approach												
Approach flow rate, v_A (veh/h)												
Intersection delay, $d_I = \frac{\sum(d_A)(v_A)}{\sum v_A}$ (s/veh)												

Notes

1. For permitted left turns, the minimum capacity is $(1 + P_L)(3600/C)$.
2. Primary and secondary phase parameters are summed to obtain lane group parameters.
3. For pretimed or nonactuated signals, k = 0.5. Otherwise, refer to Table 10.1.
4. T = analysis duration (h); typically T = 0.25, which is for the analysis duration of 15 min.
   I = upstream filtering metering adjustment factor; I = 1 for isolated intersections.

Figure 10.12 Capacity Analysis Worksheet

SOURCE: *Highway Capacity Manual,* Special Report 209, Fourth Edition, Transportation Research Board, National Research Council, Washington, D.C., 2000. Used with permission.

where

d_{1i} = uniform delay (sec/vehicle) for lane group i
C = cycle length (sec)
g_i = effective green time for lane group i (sec)
X_i = (v/c) ratio for lane group i

Note that a special procedure is used to determine uniform delay for protected-plus-permitted left-turn operation. This procedure is described later in this chapter.

Incremental Delay.   The incremental delay takes into consideration that the arrivals are not uniform but random and that some cycles will overflow (random delay) as well as delay caused by sustained periods of oversaturation. It is given as

$$d_{2i} = 900T\left[(X_i - 1) + \sqrt{(X_i - 1)^2 + \frac{8k_i I_i X_i}{c_i T}}\,\right] \qquad (10.56)$$

where

d_{2i} = incremental delay (sec/vehicle) for lane group i
c_i = capacity of lane group i (veh/h)
T = duration of analysis period (hr)
k_i = incremental delay factor that is dependent on controller settings (see Table 10.7)
I_i = upstream filtering metering adjustment factor accounts for the effect of filtered arrivals from upstream signals (for isolated intersections, $I = 1$; for nonisolated intersections see Table 10.8)
X_i = v/c ratio for lane group i

Residual Demand Delay.   This delay occurs as a result of an initial unmet demand Q_b vehicles at the start of the analysis period T. That is, a residual event of length Q_b exists at the start of the analysis period. In computing this residual demand, one of the following five cases will apply:

- **Case 1:** $Q_b = 0$, analysis period is unsaturated
- **Case 2:** $Q_b = 0$, analysis period is saturated
- **Case 3:** $Q_b > 0$ and Q_b can be fully served during analysis period T, i.e., unmet demand Q_b and total demand in period T (qT) should be less than capacity cT, i.e., $Q_b + qT < cT$
- **Case 4:** $Q_b > 0$, but Q_b is decreasing, i.e., demand in time T, (qT) is less than the capacity cT
- **Case 5:** $Q_b > 0$, and demand in time T exceeds capacity cT

Residual demand is obtained from Eq. 10.57 as

$$d_{3i} = \frac{1800Q_{bi}(1 + u_i)t_i}{c_i T} \qquad (10.57)$$

where
Q_{bi} = initial unmet demand at the start of period T_i vehicles for lane group i
c_i = adjusted lane group capacity veh/h
T = duration of analysis period (h)
t_i = duration of unmet demand in T for lane group i (h)
u_i = delay parameter for lane group i

and

$$t_i = 0 \ if \ Q_b = 0, \ else \ t_i = \min\left[T_i, \frac{Q_{bi}}{c_i(1 - \min(1 - X_i))} \right] \qquad (10.58)$$

$$u_i = 0 \ if \ t < T, \ else \ u_i = 1 - \frac{c_i T}{Q_{bi}[1 - \min(1, X_i)]} \qquad (10.59)$$

where
X_i is the degree of saturation (v/c) for lane group i.

Total Control Delay.   The total control delay for lane group i is given as:

$$d_i = d_{1i} PF + d_{2i} + d_{3i} \qquad (10.60)$$

where
d_i = the average control delay per vehicle for a given lane group
PF = uniform delay adjustment factor for quality of progression (see Table 10.9)
d_{1i} = uniform control delay component assuming uniform arrival
d_{2i} = incremental delay component for lane group i, no residual demand at the start of the analysis period T
d_{3i} = residual demand delay for lane group i

It should be noted that the delay adjustment factor (PF) accounts for the effect of quality of signal progression at the intersection. The adjustment factor for controller type (k) accounts for the ability of actuated controllers to adjust timing from cycle to cycle. The adjustment factor for quality of progression (PF) accounts for the positive effect that good signal progression has on the flow of traffic through the intersection and depends on the arrival type. It has a value of one for isolated intersections (arrival type 3). The six different arrival types were defined earlier under "Specifying Traffic Conditions." Table 10.9 gives values for PF.

When Q_b is greater than zero, which is for Cases 3, 4 and 5, it is necessary to evaluate the uniform control delay for two periods; (i) the period when oversaturation queue exists, i.e., $X \geq 1$ and (ii) the period of undersaturation when $X < 1$. A value of $X = 1$ is used to determine the portion of the uniform control delay during the oversaturated period (t) using Eq. 10.55, and the actual value of X is used to find the portion of the uniform control delay during the undersaturated period ($T - t$). The value of d_1 is then obtained as the sum of the weighted values of the delay for each period as shown in Eq. 10.61.

$$d_1 = \frac{d_s t}{T} + \frac{d_u (PF)(t - T)}{T} \tag{10.61}$$

where

d_s = saturated delay (d_1 evaluated for $X = 1$) (hr)
d_u = undersaturated delay (d_1 evaluated for actual X value) (hr)
T = analysis period (hr)

It is obvious from Eq. 10.61 that when the over-saturated period is as long as the analysis period, (Cases 4 and 5), i.e., $T = t$, the d_u term drops off and the uniform delay is obtained directly from Eq. 10.55 using $X = 1$. However, when left turns are made from exclusive left-turn lanes with a protected-permissive phase, a special procedure, described in the following paragraphs, is used to estimate d_s and d_u.

Special Procedure for Uniform Delay with Protected-Plus-Permitted Operations.   The uniform delay given by Eq. 10.55 is based on the queue storage as a function of time. When there is only a single green phase per cycle for a given lane group, the variation of the queue storage with time can be represented as a triangle. When left turns are allowed to proceed on both protected and permitted phases, the variation of queue storage is no longer a simple triangle but rather a more complex polygon. In order to determine the area of this complex polygon, representing the uniform delay, it is necessary to determine the proper values of the arrival and discharge rates during the different intervals. If the protected phase is considered the "primary" phase and the permitted phase is considered the secondary phase, then the following quantities must be known to compute the uniform delay.

- The arrival rate q_a (vehicle/sec), assumed to be uniform throughout the cycle.
- The saturation flow rate S_p (vehicle/sec) for the primary phase.
- The saturation flow rate S_s (vehicle/sec) for the unsaturated portion of the secondary phase. (The unsaturated portion begins when the queue of opposing vehicles has not been served.)
- The effective green time g (sec) for the primary phase in which the left-turn traffic has the green arrow.
- The green time g_q (sec) during which the permitted left turns cannot be made because they are blocked by the clearance of an opposing queue. (This interval begins at the start of the permitted green and ends when the queue of the opposing through vehicles is completely discharged.)
- The green time g_u (sec) during which the permitted left turns can filter through gaps in the opposing flow. This green period starts at the end of g_q.
- The red time r during which the signal is effectively red for the left turns.

The equations for determining these queue lengths depend on whether the phasing system is protected and permitted (leading) or permitted and protected (lagging). Shown below are the three conditions under the leading phase system and two under the lagging phase system.

- Protected and Permitted (Leading Left Turns)
  Condition 1. No queue remains at the end of a protected or permitted phase. Note that this does not occur for uniform delays if $v/c < 1$.

Condition 2. A queue remains at the end of the protected phase but not at the end of the permitted phase.

Condition 3. No queue remains at the end of the protected phase, but there is a queue at the end of the permitted phase.

- Permitted and Protected (Lagging Left Turns)

Condition 4. No queue remains at the end of the permitted phase. This means that there will also be no queue at the end of the protected phase, since all left-turning vehicles will be cleared during the protected phase.

Condition 5. A queue remains at the end of the permitted phase. Note also that if v/c < 1, this queue will be cleared during the protected phase.

These different queue lengths and the uniform delay for any one of the five conditions are determined from the equations given in Figure 10.13.

Approach Delay.   Having determined the average stopped delay for each lane group, we can now determine the average stopped delay for any approach as the weighted average of the stopped delays of all lane groups on that approach. The approach delay is given as

$$d_A = \frac{\sum\limits_{i=1}^{n_A} (d_{ia} v_i)}{\sum\limits_{i=1}^{n_A} v_i} \tag{10.62}$$

where

d_A = delay for approach A (sec/vehicle)

d_{ia} = adjusted delay for lane group i on approach A (sec/vehicle)

v_i = adjusted flow rate for lane group i (veh/h)

n_A = number of lane groups on approach A

The level of service of approach A can then be determined from Table 10.1.

Intersection Delay.   The average intersection stopped delay is found in a manner similar to the approach delay. In this case, the weighted average of the delays at all approaches is the average stopped delay at the intersection. The average intersection delay is therefore given as

$$d_I = \sum_{A=1}^{A_n} \frac{d_A v_A}{\sum\limits_{A} v_A} \tag{10.63}$$

where

d_I = average stopped delay for the intersection (sec/vehicle)

d_A = adjusted delay for approach A (sec/vehicle)

v_A = adjusted flow rate for approach A (veh/h)

A_n = number of approaches at the intersection

SUPPLEMENTAL UNIFORM DELAY WORKSHEET FOR LEFT TURNS FROM EXCLUSIVE LANES WITH PROTECTED AND PERMITTED PHASES

General Information

Project Description _____

v/c Ratio Computation

	EB	WB	NB	SB
Cycle length, C (s)				
Protected phase eff. green interval, g (s)				
Opposing queue effective green interval, g_q (s)				
Unopposed green interval, g_u (s)				
Red time, r (s) $r = C - g - g_q - g_u$				
Arrival rate, q_a (veh/s) $q_a = \dfrac{v}{3600 \cdot \max[X,\, 1.0]}$				
Protected phase departure rate, s_p (veh/s) $s_p = \dfrac{s}{3600}$				
Permitted phase departure rate, s_s (veh/s) $s_s = \dfrac{s(g_q + g_u)}{(g_u \cdot 3600)}$				
If leading left (protected + permitted) v/c ratio, $X_{perm} = \dfrac{q_a(g_q + g_u)}{s_s\, g_u}$				
If lagging left (permitted + protected) v/c ratio, $X_{perm} = \dfrac{q_a(r + g_q + g_u)}{s_s\, g_u}$				
If leading left (protected + permitted) v/c ratio, $X_{prot} = \dfrac{q_a(r + g)}{s_p\, g}$				
If lagging left (permitted + protected) v/c ratio, X_{prot} is N/A				

Uniform Queue Size and Delay Computations

	EB	WB	NB	SB
Queue at beginning of green arrow, Q_a				
Queue at beginning of unsaturated green, Q_u				
Residual queue, Q_r				
Uniform delay, d_1				

Uniform Queue Size and Delay Equations

	Case	Q_a	Q_u	Q_r	d_1
If $X_{perm} \le 1.0$ & $X_{prot} \le 1.0$	1	$q_a r$	$q_a g_q$	0	$[0.50/(q_a C)][r Q_a + Q_a^2/(s_p - q_a) + g_q Q_u + Q_u^2/(s_s - q_a)]$
If $X_{perm} \le 1.0$ & $X_{prot} > 1.0$	2	$q_a r$	$Q_r + q_a g_q$	$Q_a - g(s_p - q_a)$	$[0.50/(q_a C)][r Q_a + g(Q_a + Q_r) + g_q(Q_r + Q_u) + Q_u^2/(s_s - q_a)]$
If $X_{perm} > 1.0$ & $X_{prot} \le 1.0$	3	$Q_r + q_a r$	$q_a g_q$	$Q_u - g_u(s_s - q_a)$	$[0.50/(q_a C)][g_q Q_u + g_u(Q_u + Q_r) + r(Q_r + Q_a) + Q_a^2/(s_p - q_a)]$
If $X_{perm} \le 1.0$ (lagging lefts)	4	0	$q_a(r + g_q)$	0	$[0.50/(q_a C)][(r + g_q)Q_u + Q_u^2/(s_s - q_a)]$
If $X_{perm} > 1.0$ (lagging lefts)	5	$Q_u - g_u(s_s - q_a)$	$q_a(r + g_q)$	0	$[0.50/(q_a C)][(r + g_q)Q_u + g_u(Q_u + Q_a) + Q_a^2/(s_p - q_a)]$

Figure 10.13 Supplemental Uniform Delay Worksheet for Left Turns from Exclusive Lanes with Primary and Secondary Phases

SOURCE: *Highway Capacity Manual,* Special Report 209, Fourth Edition, Transportation Research Board, National Research Council, Washington, D.C., 2000. Used with permission.

The level of service for the intersection is then determined from Table 10.1, using the average control delay for the intersection. The computation required to determine the different levels of service may be organized in the format shown in Figure 10.12.

It should be emphasized again that short or acceptable delays do not automatically indicate adequate capacity. Both the capacity and the delay should be considered in the evaluation of any intersection. Where long and unacceptable delays are determined, it is necessary to find the cause of the delay. For example, if (v/c) ratios are low and delay is long, the most probable cause for the long delay is that the cycle length is too long and/or the progression (arrival type) is unfavorable. Delay can therefore be reduced by improving the arrival type, by coordinating the intersection signal with the signals at adjacent intersections, and/or by reducing the cycle length at the intersection. When delay is long but arrival types are favorable, the most probable cause is that the intersection geometrics are inadequate and/or the signal timing is improperly designed.

Example 10.1 Computing Level of Service at a Signalized Intersection Using the Operation Level of Analysis

Figure 10.14 shows peak hour volumes, pedestrian volumes, geometric layout, traffic mix, and signal timings at an isolated pretimed signalized intersection. If the intersection does not exhibit central business district characteristics, and studies have shown that the PHF at the intersection is 0.85, what is the expected level of service if curb parking is allowed on each approach? Use the operation evaluation level of analysis. There is no all-red phase and the amber time is 3 sec.

Solution: To demonstrate the application of the operational level of analysis procedure, the problem will be solved by systematically going through each of the modules discussed.

The tasks in the input module are to identify the geometric, traffic, and signalization conditions.

- **Geometric Conditions.** Since this is an existing intersection, the actual geometrics are shown in Figure 10.14. Parking is permitted on three approaches, and the number of parking maneuvers is recorded for each approach as shown.
- **Traffic Conditions.** Figure 10.14 shows volumes at the intersection and other traffic conditions, including the PHF, %HV, the number of buses stopped per hour, and the number of conflicting pedestrians per hour.
- **Signalization Condition.** There is no pedestrian push button since the intersection is pretimed. It is also necessary to estimate the minimum time required for pedestrians to safely cross each street. Using Eq. 8.12, the minimum green times required for pedestrian crossings are

$$G_p = 3.2 + \frac{L}{S_p} + (0.27 N_{ped}) \qquad \text{for } W_e \leq 10 \text{ ft}$$

INPUT WORKSHEET

General Information

Analyst _____

Agency or Company _____

Date Performed _____

Analysis Time Period _____

Site Information

Intersection _____

Area Type   ☐ CBD   ☒ Other

Jurisdiction _____

Analysis Year _____

Intersection Geometry

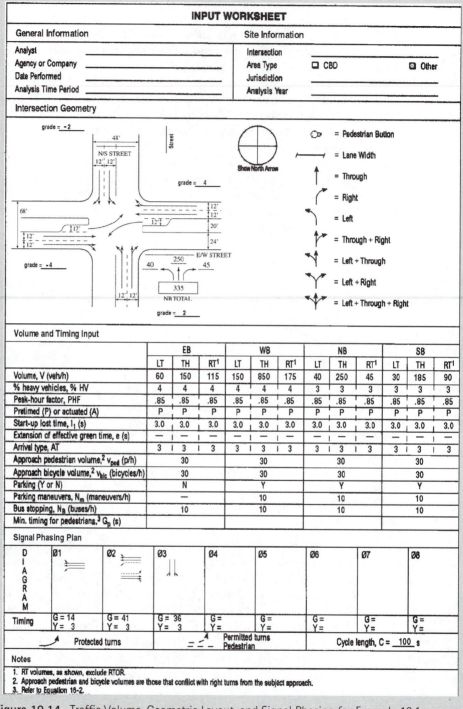

Volume and Timing Input

	EB			WB			NB			SB		
	LT	TH	RT[1]	LT	TH	RT[1]	LT	TH	RT[1]	LT	TH	RT[1]
Volume, V (veh/h)	60	150	115	150	850	175	40	250	45	30	185	90
% heavy vehicles, % HV	4	4	4	4	4	4	3	3	3	3	3	3
Peak-hour factor, PHF	.85	.85	.85	.85	.85	.85	.85	.85	.85	.85	.85	.85
Pretimed (P) or actuated (A)	P	P	P	P	P	P	P	P	P	P	P	P
Start-up lost time, l_1 (s)	3.0	3.0	3.0	3.0	3.0	3.0	3.0	3.0	3.0	3.0	3.0	3.0
Extension of effective green time, e (s)	—	—	—	—	—	—	—	—	—	—	—	—
Arrival type, AT	3	3	3	3	3	3	3	3	3	3	3	3
Approach pedestrian volume,[2] v_{ped} (p/h)	30			30			30			30		
Approach bicycle volume,[2] v_{bic} (bicycles/h)	30			30			30			30		
Parking (Y or N)	N			Y			Y			Y		
Parking maneuvers, N_m (maneuvers/h)	—			10			10			10		
Bus stopping, N_B (buses/h)	10			10			10			10		
Min. timing for pedestrians,[3] G_p (s)												

Signal Phasing Plan

DIAGRAM	Ø1	Ø2	Ø3	Ø4	Ø5	Ø6	Ø7	Ø8
Timing	G = 14, Y = 3	G = 41, Y = 3	G = 36, Y = 3	G = , Y =	G = , Y =	G = , Y =	G = , Y =	G = , Y =

Protected turns ↗

Permitted turns ↗ / Pedestrian

Cycle length, C = __100__ s

Notes

1. RT volumes, as shown, exclude RTOR.
2. Approach pedestrian and bicycle volumes are those that conflict with right turns from the subject approach.
3. Refer to Equation 16-2.

Figure 10.14 Traffic Volume, Geometric Layout, and Signal Phasing for Example 10.1

SOURCE: *Highway Capacity Manual,* Special Report 209, Fourth Edition, Transportation Research Board, National Research Council, Washington, D.C., 2000. Used with permission.

where

G_p = minimum green time
L = crosswalk length
S_p = average speed of pedestrian (ft/s)
W_e = effective crosswalk width
3.2 = pedestrian start up time
N_{ped} = number of pedestrians crossing during an interval

Number of pedestrians crossing/hr = 30
Number of intervals/hr = 3600/100 = 36
Average number of pedestrians crossing/interval = 30/36 = 0.83

For E-W direction

$$G_p = 3.2 + 48/4 + 0.27 \times 0.83 = 15.42 \text{ sec}$$

For the N-S direction

$$G_p = 3.2 + 68/4 + 0.27 \times 0.83 = 20.42 \text{ sec}$$

These results allow pedestrians to cross the whole width of road in one attempt. There is also no coordination since the lights are pretimed. Arrivals will therefore be assumed to be random, and arrival type 3 is recorded in Figure 10.14.

The existing signal phases are shown in Figure 10.14, under Phasing, and the phase lengths are recorded as shown in the diagram.

The computations required for the volume adjustment module are carried out using the volume adjustment worksheet shown in Figure 10.15.

- *Flow Rate.* The volumes are adjusted to the peak 15-min flow using the PHF, and are recorded in row 6 of Figure 10.15. For example,

$$\text{EBLT flow rate} = \frac{60}{0.85} = 71 \text{ veh/h}$$

- *Lane Groups.* The lane groups selected are shown in Figure 10.15. The EB and WB left lanes are exclusive left-turn lanes and are therefore considered to be a single-lane group, and the through and right lanes also are made into a single group. Since the lanes on the north and south approaches are all served by the same green phase and there are no exclusive left- or right-turn lanes, all lanes in the north approach are made into a single-lane group, and all lanes in the south approach are made into a single-lane group. WB left turns are protected plus permitted, whereas EB left turns are permitted.
- *Adjust Volume.* The total adjusted flow rate in each lane group is determined and recorded in row 7 of Figure 10.15. The proportions of left turns and right turns are also computed for each lane group. For example, the proportion of EB through and right-turn flow rate that turns right is 135/1017 = 0.13.

VOLUME ADJUSTMENT AND SATURATION FLOW WORKSHEET

Volume Adjustment

	Eastbound			Westbound			Northbound			Southbound		
	L	T	R	L	T	R	L	T	R	L	T	R
Volume, V	60	750	115	150	850	175	40	250	45	30	185	90
PHF	0.85	0.85	0.85	0.85	0.85	0.85	0.85	0.85	0.85	0.85	0.85	0.85
Adj flow	71	882	135	176	1000	206	47	294	53	35	218	106
No. Lanes	1	2	0	1	2	0	0	2	0	0	2	0
Lane group	L	TR		L	TR			LTR			LTR	
Adj flow	71	1017		176	1206			394			359	
Prop LTs	1.000			1.000				0.119			0.097	
Prop RTs		0.133			0.171			0.135			0.295	

Saturation Flow Rate (see Table 10.4 to determine the adjustment factors)

	Eastbound			Westbound			Northbound			Southbound		
LG	L	TR		L	TR			LTR			LTR	
So	1900	1900		1900	1900			1900			1900	
Lanes	1	2	0	1	2	0	0	2	0	0	2	0
fW	1.000	1.000		1.000	1.000			1.000			1.000	
fHV	0.962	0.962		0.962	0.962			0.971			0.971	
fG	1.020	1.020		0.980	0.980			0.990			1.010	
fP	1.000	1.000		1.000	0.925			0.925			0.925	
fBB	1.000	0.980		1.000	0.980			0.980			0.980	
fA	1.00	1.00		1.00	1.00			1.00			1.00	
fLU	1.00	0.95		1.00	0.95			0.95			0.95	
fRT		0.980			0.974			0.980			0.956	
fLT	0.217	1.000		0.950	1.000			0.872			0.889	
Sec.				0.122								
fLpb	0.996	1.000		0.999	1.000			0.998			0.999	
fRpb		0.993			0.993			0.993			0.984	
S	403	3379		1700	2985			2663			2679	

Figure 10.15 Volume Adjustment Worksheet for Example 10.1

The computations carried out in the saturation flow rate module are also shown in Figure 10.15. The ideal saturation flow is taken as 1900 passenger cars per hour of green per lane and recorded as shown. The adjustment factors are then computed and recorded as shown.

- **_Lane Width Adjustment Factor, f_w._** The lane width adjustment factors are computed from the equation given in Table 10.4, i.e., $f_w = 1 + \dfrac{(w - 12)}{30}$ $w \geq 8.0$.

  For 12-ft lanes: $f_w = 1.00$

- **Heavy Vehicle Adjustment Factor, f_{HV}.** These factors are based on the percentage of heavy vehicles in the traffic stream on each approach and are computed from the equation given in Table 10.4, which is

$$f_{HV} = \frac{100}{100 + \%HV(E_T - 1)}$$

where
    $\%HV$ = percent of heavy vehicle for lane group volume
    E_T = 2 pc/HV

For example, the f_{HV} for the EB left turn lane group is

$$f_{HV} = \left(\frac{100}{100 + 4(v - 1)} \right)$$
$$= 0.962$$

- **Grade Adjustment Factor, f_g.** These factors are also computed from the appropriate equation given in Table 10.4, which is

$$f_g = 1 - \left(\frac{\%G}{200} \right) \qquad -6 \leq \%G \leq +10$$

$$\text{negative is downhill}$$

where
    $\%G$ = percent grade on a lane group approach

For example f_g for the EB approach

$$f_g = 1 - \left(\frac{-4}{200} \right)$$
$$= 1.02$$

the grade adjustment factors for the other approaches are obtained as

    WB approach + 4 percent: $f_g = 0.98$
    NB approach + 2 percent: $f_g = 0.99$
    SB approach − 2 percent: $f_g = 1.01$

- **Parking Adjustment Factor, f_p.** The parking adjustment factors are also computed from the appropriate equation in Table 10.4. In this case, there are 10 parking maneuvers per hour at each approach. Each of the lane groups affected by the parking maneuvers has two lanes. The parking adjustment factor for each lane group affected by parking is therefore 0.925 as shown below. Note that only the right-lane group on each approach is affected by parking.

The parking adjustment factors are obtained from the equation given in Table 10.4, which is

$$f_p = \left(\frac{N - 0.1 - \left(\frac{18 N_m}{3600} \right)}{N} \right)$$

where

N = number of lanes in the lane group $0 \leq N_m \leq 180$
N_m = number of parking maneuvers/h $f_p \geq 0.050$
                                    $f_p = 1.00$ for no parking

For example, for the NB left, through, and right turn lane group

$$f_g = \left(\frac{2 - 0.1 - \left(\frac{18 \times 10}{3600} \right)}{2} \right)$$

$$= 0.925$$

- **Bus Blockage Adjustment Factor, f_{bb}.** These factors depend on the number of bus blockages per hour and the number of lanes in the lane group affected. The factors are computed from the appropriate equation in Table 10.4 and recorded in Figure 10.15. Since there are 10 buses stopping at each approach per hour and since each of the lane groups affected consists of 2 lanes, $f_{bb} = 0.98$ as shown below. Note that the left-turn lane groups are not affected by stopping buses.

$$f_{bb} = \left[\frac{N - \left(\frac{14.4 N_B}{3600} \right)}{N} \right] \qquad 0 \leq N_B \leq 250, \, f_{bb} \geq 0.050$$

where

N = number of lanes in lane group
N_B = number of buses stopping

$$f_{bb} = \left[\frac{2 - \frac{14.4 \times 10}{3600}}{2} \right]$$

$$= 0.98$$

- *Area Type Adjustment Factor, f_a.* The intersection is not located within a business district; therefore, the area type adjustment factor is 1.00 for all approaches.
- *Right-Turn Lane Adjustment Factor, f_{RT}.* Adjustment is required only for lane groups with right-turning vehicles. All other lane groups will have an f_{RT} of 1.0. The factors are computed from the appropriate equation in Table 10.4.

    For example, the EBTH/RT lane group is considered a shared lane. The factor is computed from the expression

$$f_{RT} = 1.0 - (0.135)P_{RT}, \qquad (f_{RT} \geq 0.050)$$

where

P_{RT} = proportion of right turns in the group

$$f_{RT} = 1.0 - (0.150)0.133 = 0.980$$

- *Left-Turn Lane Adjustment Factor, f_{LT}.* The left-turn adjustment factor for the protected phase of the WB approach is obtained directly from Table 10.4 as 0.95, since the left turns on this approach are made from exclusive left-turn lanes with protected phasing. However, it is necessary to use the special procedure both for the other approaches, since the left turns at all of these are made during a permitted phase, and also for the permitted portion on the WB approach. The computation required for each of these is shown in Figures 10.16 and 10.17.

    For example, in computing the left-turn adjustment factor for the left turns during the permitted phase of the WB approach (left-turn lanes from exclusive lanes with permitted phasing), Eq. 10.7 is used

$$f_{LT2A} = \left(\frac{g_u}{g} \right) \left[\frac{1}{1 + P_L(E_{L1} - 1)} \right]$$

where

$g_u = g - g_q$

$$g_q = \frac{v_{olc}\, qr_o}{[0.5 - v_{olc}(1 - qr_o)/g_o]} - t_L$$

$v_{olc} = v_o C/[3600\,(N_o)(f_{Luo})]$
f_{LW} = lane utilization factor = 0.95 (see Table 10.5)
v_{olc} = $(1017 \times 100)/(3600 \times 2)0.95 = 14.87$ (veh/l/c)
v_o = 1062 (veh/h) (note that opposing left turns are on an exclusive lane and therefore are not included)
N_o = 2 (note that opposing left-turn lane is exclusive and therefore not included)
$qr_o = 1 - R_{po}(g_o/C)$
R_{po} = platoon ratio for the opposing flow

SUPPLEMENTAL WORKSHEET FOR PERMITTED LEFT TURNS
*** For Use Where the Subject Approach is Opposed by a Multilane Approach ***
SUPPLEMENTAL PERMITTED LT WORKSHEET
for exclusive lefts

Input		EB	WB	NB	SB
Cycle length, C	100.0 sec				
Total actual green time for LT lane group, G (s)		41.0	58.0		
Effective permitted green time for LT lane group, g(s)		41.0	44.0		
Opposing effective green time, go (s)		58.0	41.0		
Number of lanes in LT lane group, N		1	1		
Number of lanes in opposing approach, No		2	2		
Adjusted LT flow rate, VLT (veh/h)		71	176		
Proportion of LT in LT lane group, PLT		1.000	1.000		
Proportion of LT in opposing flow, PLTo		0.00	0.00		
Adjusted opposing flow rate, Vo (veh/h)		1206	1017		
Lost time for LT lane group, tL		3.00	3.00		
Computation					
LT volume per cycle, LTC = VLTC/3600		1.97	4.89		
Opposing lane util. factor, fLUo		0.95	0.95	0.95	0.95
Opposing flow, Volc = VoC/[3600(no)fLUo] (veh/ln/cyc)		17.63	14.87		
gf = [exp(- a * (LTC ** b)))] -tl, gf<=g		0.0	0.0		
Opposing platoon ratio, Rpo (see Table 10.3)		1.00	1.00		
Opposing Queue Ratio, qro=Max[1-Rpo(go/C), 0]		0.42	0.59		
gq, (See Eq. 10.9)		2.88	24.97		
gu=g-gq if gq<=gf, or = g-gf if gq<gf		38.12	19.03		
n=Max(gq-gf)/2,0)		1.44	12.49		
PTHo=1-PLTo		1.00	1.00		
PL*=PLT[1+(N-1)g/(gf+gu/EL1+4.24)]		1.00	1.00		
EL1 (See Table 10.6)		4.29	3.55		
EL2=Max ((1-Ptho**n)/(Plto, 1.0)					
fmin=2 (1+PL)/g or fmin=2(1+Pl)/g		0.10	0.09		
gdiff=max (gq-gf, 0)		0.00	0.00		
fm=[gf/g]+[gu/g][1+PL(EL1-1)], (fmin=fmin;max=1.00)		0.22	0.12		
flt=fm=[gf/g]+[gu/g]/[1+PL(EL1-1)]+[gdiff/f]/ [1+PL(EL2-1)], (fmin<=fm<=1.00) or flt={fm+0.91(N-1)]/N**					
Left-turn adjustment, fLT		0.217	0.122		

*   For special case of single-lane approach opposed by multilane approach, see text.
** If Pl>=1 for shared left-turn lanes with N>1, then assume de-facto left-turn lane and redo calculations.
*** For permitted left-turns with multiple exclusive left-turn lanes, flt=fm. For special case of multilane approach opposed by single-lane approach or when gf>gq, see text.

Figure 10.16 Supplemental Worksheet for Computation of Left-Turn Adjustment Factors for Permitted Left Turns of Example 10.1

SUPPLEMENTAL WORKSHEET FOR PERMITTED LEFT TURNS
*** For Use Where the Subject Approach is Opposed by a Multilane Approach ***

	EB	WB	NB	SB
Cycle length, C 100.0 sec				
Total actual green time for LT lane group, G (s)			36.0	36.0
Effective permitted green time for LT lane group, g(s)			36.0	36.0
Opposing effective green time, go (s)			36.0	36.0
Number of lanes in LT lane group, N			2	2
Number of lanes in opposing approach, No			2	2
Adjusted LT flow rate, VLT (veh/h)			47	35
Proportion of LT in LT lane group, PLT			0.119	0.097
Proportion of LT in opposing flow, PLTo			0.10	0.12
Adjusted opposing flow rate, Vo (veh/h)			359	394
Lost time for LT lane group, tL			3.00	3.00
Computation				
LT volume per cycle, LTC = VLTC/3600			1.31	0.97
Opposing lane util. factor, fLUo	0.95	0.95	0.95	0.95
Opposing flow, Volc = VoC/[3600(no)fLUo] (veh/ln/cyc)			5.25	5.76
gf = [exp(- a * (LTC ** b))] -tl, gf<=g			9.4	12.2
Opposing platoon ratio, Rpo (see Table 10.3)			1.00	1.00
Opposing Queue Ratio, qro=Max[1-Rpo(go/C), 0]			0.64	0.64
gq, (See Eq. 10.9)			4.51	5.33
gu=g-gq if gq<=gf, or = g-gf if gq<gf			26.62	23.83
n=Max(gq-gf)/2,0)			0.00	0.00
PTHo=1-PLTo			0.90	0.88
PL*=PLT[1+(N-1)g/(gf+gu/EL1+4.24)]			0.28	0.22
EL1 (See Table 10.6)			2.04	2.11
EL2=Max ((1-Ptho**n)/(Plto, 1.0)			1.00	1.00
fmin=2 (1+PL)/g or fmin=2(1+Pl)/g			0.07	0.07
gdiff=max (gq-gf, 0)			0.00	0.00
fm=[gf/g]+[gu/g][1+PL(EL1-1)], (fmin=fmin;max=1.00)			0.83	0.87
flt=fm=[gf/g]+[gu/g]/[1+PL(EL1-1)]+[gdiff/f]/ [1+PL(EL2-1)], (fmin<=fm<=1.00) or flt={fm+0.91(N-1)]/N**				
Left-turn adjustment, fLT	0.217	0.122	0.872	0.889

  *   For special case of single-lane approach opposed by multilane approach, see text.
  ** If Pl>=1 for shared left-turn lanes with N>1, then assume de-facto left-turn lane and
       redo calculations.
 *** For permitted left-turns with multiple exclusive left-turn lanes, flt=fm. For special
       case of multilane approach opposed by single-lane approach or when gf>gq, see text.

Figure 10.17 Supplemental Worksheet for Computation of Left-Turn Adjustment Factors for Permitted Left Turns of Example 10.1

SOURCE: *Highway Capacity Manual,* Special Report 209, Fourth Edition, Transportation Research Board, National Research Council, Washington, D.C., 2000. Used with permission.

= 1.00 (for arrival type 3; see Table 10.3)

g_o = effective green time for opposing flow

= $G_o + Y_o - t_L$ = 41 sec

For the permissive phase, the lost time is zero because the lost time has occurred at the beginning of the movement, i.e., at the start of the protected phase.

$$qr_o = 1 - R_{po}(g_o/C)$$

$$= 1 - 1(41/100)$$

$$= 0.59$$

$$g_q = \frac{14.87 \times 0.59}{0.5 - 14.87(1 - 0.59)/41} - 0$$

$$= 24.97 \text{ (sec)}$$

$$g_u = g - g_q$$

$$= 44 - 24.97$$

$$= 19.03 \text{ (sec)}$$

$$f_{LT} = \frac{g_u}{g}\left[\frac{1}{1 + P_L(E_{L1} - 1)}\right]$$

where

$$P_L = \left[1 + \frac{(N-1)g}{\left(g_u / E_{L1}\right) + 4.24}\right]$$

and

N = number of lanes in exclusive left turn lane group

E_{L1} = through car equivalent for permitted left turns

In this case since $N = 1$, $P_L = 1$

To determine E_{L1}, we should first determine v_{oe} (see Table 10.6)

where

$$v_{oe} = v_o/f_{Luo} = 1017/.95 = 1071$$

Using Table 10.6 and extrapolating, we obtain

$$E_{L1} = 3.55$$

Substituting for g_w, g, P_L, and E_{L1} in the equation for f_{LT}, we obtain

$$f_{LT} = \left(\frac{19.03}{44}\right)\left[\frac{1}{1+1(3.55-1)}\right]$$
$$= 0.128$$

Also note that for the NB and SB left turns, Eq. 10.15 is used because these are shared lanes with permitted phasing.

- **Pedestrian and Bicycle Adjustment Factors.**   These factors account for the reduction in the saturation flow rate as a result of the conflicts between the automobiles and pedestrians and bicycles. These factors are based on the pedestrian and bicycle flow rates. The procedure is illustrated by considering the EB approach. Let us first determine the pedestrian-bicycle left-turn adjustment factor (f_{Lpb}). Note that in this case since the opposing queue clearing green time, which has been calculated earlier as 2.88 sec, is less than the effective pedestrian green time, which is 41 sec, then $f_{Lpb} \neq 1$. We determine f_{Lpb} as follows:

  Determine pedestrian flow rate using Eq. 10.41

$$v_{pedg} = v_{ped}\left(C\Big/g_p\right), \left(v_{pedg} \leq 5000\right)$$

where

v_{pedg} = pedestrian flow rate
v_{ped} = pedestrian volume
g_p = pedestrian green walk + flashing don't walk time (sec) (if pedestrian signal timing is unknown g_p may be assumed to be equal to g (sec))
C = cycle length (sec)
v_{pedg} = 30(100/41) = 73

Determine average pedestrian occupancy from Eq. 10.42

$$OCC_{pedg} = v_{ped}/2000, \left(v_{pedg} \leq 1000, and\ OCC_{pedg} \leq 0.5\right)$$
$$OCC_{pedg} = 73/2000, \left(v_{pedg} \leq 1000, and\ OCC_{pedg} \leq 0.5\right)$$
$$= 0.037$$

Since the right-turning vehicles do not weave to the right before the stop line, there is also bicycle/automobile interaction.

Determine bicycle flow rate (f_{bicg}) from Eq. 10.45

$$v_{bicg} = v_{bic}(C/g) \qquad (v_{bicg} \leq 1900)$$
$$v_{bicg} = 30(100/41) \qquad (v_{bicg} \leq 1900)$$
$$= 73$$

Determine the pedestrian occupancy after the opposing queue (OCC_{pedu} for the left turns) from Eq. 10.48

$$OCC_{pedu} = OCC_{pedu}[1 - 0.5(g_q/g_p)]$$
$$= 0.037(1 - 0.5(2.88/41))$$
$$= 0.35$$

Determine the relevant conflict zone occupancy (OCC_r) using Eq. 10.49

$$OCC_r = OCC_{pedu}\left[e^{-(5/3600)v_o}\right]$$
$$= 0.35\left[e^{-(5/3600)1206}\right]$$
$$= 0.007$$

Determine the permitted phase adjustment factor using Eq. 10.51. Note that in this case $N_{turn} < N_{rec}$. If $N_{turn} = N_{rec}$, then Eq. 10.50 should be used

$$A_{pbt} = 1 - 0.6(OCC_r)$$
$$= 1 - 0.6(0.007)$$
$$= 0.996$$

Determine the pedestrian/bicycle saturation flow adjustment factor for left turns from Eq. 10.52

$$f_{Lpb} = 1.0 - P_{LT}(1 - A_{pbT})(1 - P_{LTA})$$

where
A_{pbT} = permitted phase pedestrian-bicycle adjustment factor for turning movements (obtained from Eq. 10.50 or 10.51)
P_{LTA} = the proportion of left turns using protected phase (used only for protected/permissive phases)
f_{Lpb} = $1.0 - 1(1 - 0.996)(1 - 0)$

Note that in this case, $P_{LT} = 1$ as the EB left turn lane is an exclusive lane and $P_{LTA} = 0$ as the EB left turn traffic moves only during a permitted phase.

$$f_{Lpb} = 0.996$$

Let us now determine the pedestrian-bicycle adjustment factor for right-turning movements.

Determine the bicycle conflict zone occupancy (OCC_{bicg}) from Eq. 10.46,

$$OCC_{bicg} = 0.02 + (v_{bicg}/2700) \qquad (v_{bicg} < 1900 \text{ and } OCC_{bicg} \leq 0.72)$$
$$= 0.02 + (73/2700)$$
$$= 0.047$$

Determine the relevant conflict zone for pedestrians and bicycles (OCC_r) using Eq. 10.47

$$OCC_r = OCC_{pedg} + OCC_{bicg} - (OCC_{pedg})(OCC_{bicg})$$
$$= 0.037 + 0.047 - (0.037)(0.047)$$
$$= 0.082$$

Determine the permitted phase pedestrian-bicycle adjustment factor A_{pbt}

$$A_{pbt} = 1 - 0.6(OCC_r)$$
$$= 1 - 0.6(0.082)$$
$$= 0.951$$

Determine the pedestrian-bicycle saturation flow adjustment factor for right turns using Eq. 10.53.

$$f_{Rpb} = 1.0 - P_{RT}(1 - A_{pbT})(1 - P_{RTA})$$
$$= 1.0 - 1(1 - 0.951)(1 - 0)$$
$$= 0.993$$

The computation for the other movements follows the same procedure and is shown in Figure 10.18.

The adjusted saturated flow is then determined using Eq. 10.5 and the factors listed in Figure 10.15. For example, the adjusted saturation flow rate for the EB left turn group is

$$(1900)(1)(0.962)(1.020)(1.00)(1.00)(1.00)(1.00)(1.00)(0.217)(0.996) = 403 \text{ veh/h/g}$$

SUPPLEMENTAL PEDESTRIAN-BICYCLE EFFECTS WORKSHEET

Permitted Left Turns	EB	WB	NB	SB
Effective pedestrian green time, gp (s)	41.0	44.0	36.0	36.0
Conflicting pedestrian volume, Vped (p/h)	30	30	30	30
Pedestrian flow rate, Vpedg (p/h)	73	68	83	83
OCCpedg	0.037	0.034	0.042	0.042
Opposing queue clearing green, gq (s)	2.88	24.97	4.51	5.33
Eff. ped. green consumed by opp. veh. queue, gq/gp	0.070	0.568	0.125	0.148
OCCpedu	0.035	0.024	0.039	0.038
Opposing flow rate, Vo (veh/h)	1206	1017	359	394
OCCr	0.007	0.006	0.024	0.022
Number of cross-street receiving lanes, Nrec	2	2	2	2
Number of turning lanes, Nturn	1	1	1	1
ApbT	0.996	0.996	0.986	0.987
Proportion of left turns, PLT	1.000	1.000	0.119	0.097
Proportion of left turns using protected phase, PLTA	0.000	0.800	0.000	0.000
Left-turn adjustment, fLpb	0.996	0.999	0.998	0.999
Permitted Right Turns				
Effective pedestrian green time, gp (s)	41.0	58.0	36.0	36.0
Conflicting pedestrian volume, Vped (p/h)	30	30	30	30
Conflicting bicycle volume, Vbic (bicycles/h)	30	30	30	30
Vpedg	73	51	83	83
OCCpedg	0.037	0.026	0.042	0.042
Effective green, g (s)	41.0	58.0	36.0	36.0
Vbicg	73	52	83	83
OCCbicg	0.047	0.039	0.051	0.051
OCCr	0.082	0.064	0.090	0.090
Number of cross-street receiving lanes, Nrec	2	2	2	2
Number of turning lanes, Nturn	1	1	1	1
ApbT	0.951	0.962	0.946	0.946
Proportion right-turns, PRT	0.133	0.171	0.135	0.295
Proportion right-turns using protected phase, PRTA	0.000	0.000	0.000	0.000
Right turn adjustment, fRpb	0.993	0.993	0.993	0.984

Figure 10.18 Supplemental Pedestrian-Bicycle Effects Worksheet for Example 10.1

- *Capacity and v/c module.* In the capacity analysis module, we compute the capacity and the (v/c) ratio for each lane group and the aggregate results. The computations required are carried out in Figure 10.19. The adjusted flow rates and adjusted saturation flow rates are extracted from Figure 10.15 and recorded in Figure 10.19, together with the appropriate lane group as shown. The flow ratios are then computed and recorded in column five of the figure. Since this is an analysis

CAPACITY ANALYSIS WORKSHEET
Capacity Analysis and Lane Group Capacity

Appr/Mvmt	Lane Group	Adj Flow Rate (v)	Adj Sat Flow Rate (s)	Flow Ratio (v/s)	Green Ratio (g/C)	Lane Group Capacity (c)	Lane Group v/c Ratio
Eastbound							
Prot							
Perm							
Left	L	71	403	0.18	0.41	165	0.43
Prot							
Perm							
Thru	TR	1017	3379	0.30	0.41	1385	0.73
Right							
Westbound							
Prot		176	1700	0.10	0.140	238	0.74
Perm			225			99	0.00
Left	L	176			0.58	337	0.52
Prot							
Perm							
Thru	TR	1206	2985	0.40	0.58	1731	0.70
Right							
Northbound							
Prot							
Perm							
Left							
Prot							
Perm							
Thru	LTR	394	2663	0.15	0.36	959	0.41
Right							
Southbound							
Prot							
Perm							
Left							
Prot							
Perm							
Thru	LTR	359	2679	0.13	0.36	964	0.37
Right							

Sum of flow ratios for critical lane groups, Y_c = Sum (v/s) = 0.55
Total lost time per cycle, L = 9.00 sec
Critical flow rate to capacity ratio, $X_c = (Y_c)(C)/(C-L) = 0.61$

Figure 10.19 Worksheet for the Capacity Analysis Module for Example 10.1

SOURCE: *Highway Capacity Manual,* Special Report 209, Fourth Edition, Transportation Research Board, National Research Council, Washington, D.C., 2000. Used with permission.

problem, the signal times are known and the g/C ratios are determined, assuming a lost time of 3 sec per phase. The amber time is given as 3 sec per phase.

$$NB: \quad (g/C) = \frac{36-3+3}{100} = 0.36$$

$$EB: \quad (g/C) = \frac{41-3+3}{100} = 0.41$$

Figure 10.20 is used to carry out the required computations for the level of service module. The uniform delay d_1 and the incremental delay d_2 are calculated for each phase using the appropriate equations. Note that this is an isolated intersection, therefore the value of I is 1. Also note that since there is no residual demand, d_3 is zero.

$$d_1 = 0.50C \frac{(1-g/C)^2}{1-(g/C)[\min (X, 1.0)]}$$

and

$$d_2 = 900T\left[(X-1)+\sqrt{(X-1)^2 + \frac{8kIx}{cT}}\right]$$

Since the intersection is isolated, type 3 arrival is assumed, which gives a PF of 1.0 from Table 10.9. The total delay for each lane group is obtained by first applying the delay adjustment factor for control type PF from Table 10.9 to the d_1, and using the appropriate value of k in d_2 from Table 10.7. In this case, PF is 1, and k is 0.5. Then the level of service is obtained from Table 10.1. The average delay for each approach is found by determining the weighted average of the delays of the lane groups in that approach. Similarly, the average delay for the intersection is found by determining the weighted delay for all approaches. For example,

$$\text{EB approach delay} = \frac{(71)(29.1)+(1017(28.4))}{71+1017} = 28.4 \text{ sec}$$

$$\text{intersection delay} = \frac{28.4(71+1017)+17.5(176+1206)+25.3(394)+(24.8(359))}{71+1017+176+1206+394+359}$$

$$= 23.0 \text{ sec}$$

A summary of the intersection performance is given in Figure 10.21.

The computations required to solve Example 10.1 were carried out manually to facilitate a detailed description of each step of the procedure. Note, however, that computer programs are available that can be used to carry out these computations. For example, the HCS (developed by McTrans Center, at the University of Florida) is a microcomputer program that can be used to determine the level of service of a signalized intersection at both the "planning" and the "operation and design" levels of analysis. This software package can be used in solving this example or any similar problem.

LOS MODULE WORKSHEET

Control Delay and LOS Determination

Appr/ Lane Grp	Ratios v/c	Ratios g/C	Unf Del dl	Prog Adj Fact	Lane Grp Cap	Incremental Factor k	Del d2	Res Del d3	Lane Group Delay	Lane Group LOS	Approach Delay	Approach LOS
Eastbound												
L	0.43	0.41	21.1	1.000	165	0.50	8.0	0.0	29.1	C		
TR	0.73	0.41	24.9	1.000	1385	0.50	3.5	0.0	28.4	C	28.4	C
Westbound												
L	0.52	0.58	14.6	1.000	337	0.50	5.7	0.0	20.2	C		
TR	0.70	0.58	14.8	1.000	1731	0.50	2.3	0.0	17.1	B	17.5	B
Northbound												
LTR	0.41	0.36	24.0	1.000	959	0.50	1.3	0.0	25.3	C	25.3	C
Southbound												
LTR	0.37	0.36	23.7	1.000	964	0.50	1.1	0.0	24.8	C	24.8	C

Intersection delay = 23.0 (sec/veh)

Intersection LOS = C

A summary of the Intersection Performance is given in Figure 10.21.

Figure 10.20 Worksheet for the Level-of-Service Module for Example 10.1

SOURCE: *Highway Capacity Manual,* Special Report 209, Fourth Edition, Transportation Research Board, National Research Council, Washington, D.C., 2000. Used with permission.

Intersection Performance Summary

Appr/ Lane Grp	Lane Group Capacity	Adj Sat Flow Rate (s)	Ratios v/c	Ratios g/C	Lane Group Delay	Lane Group LOS	Approach Delay	Approach LOS
Eastbound								
L	165	403	0.43	0.41	29.1	C		
TR	1385	3379	0.73	0.41	28.4	C	28.4	C
Westbound								
L	337	1700	0.52	0.58	20.2	C		
TR	1731	2985	0.70	0.58	17.1	B	17.5	B
Northbound								
LTR	959	2663	0.41	0.36	25.3	C	25.3	C
Southbound								
LTR	964	2679	0.37	0.36	24.8	C	24.8	C

Intersection delay = 23.0 (sec/veh)

Intersection LOS = C

Figure 10.21 Intersection Performance Summary

PLANNING ANALYSIS

The planning level of analysis presented here can be used to determine the required geometries of an intersection for a given demand flow or to estimate its operational status during the planning stage. The methodology uses the sum of the critical lane volumes (X_{cm}) to determine whether the intersection will operate at "under capacity," "near capacity," "at capacity," or "over capacity" as shown in Table 10.10. The main data requirements are the traffic volumes and the lane configuration of each approach at the intersection. Figure 10.22 shows the planning input worksheet, on which the input data are entered. Figure 10.23 is the lane volume worksheet, used to determine the critical lane volume at each approach. It is necessary that a separate worksheet be completed for each approach. However, if it is necessary to obtain an estimate of the level of service of the intersection based on the stopped delay, this can be done by establishing a signal timing plan for the intersection and incorporating it in additional steps of the methodology. This is done by using the signal operations worksheet shown in Figure 10.24. The procedure for this planning analysis consists of the following steps:

Step 1.   After completing the input worksheet, the lane volumes for each movement are determined using Figure 10.23. When necessary, the formula required to compute the value for each item is given. In using these formulas it should be noted that the numbers in square brackets represent the values for the item numbers shown. For example, the value for item 5 (LT lane volume), given as [1]/([3]×[4]), is obtained by dividing the value entered in item 1 by the product of the values entered in items 3 and 4. Also, in this analysis the following should be noted:
- Enter only exclusive lanes for turning movements
- Include shared lanes with through lanes
- Right-turn volumes on shared lanes are ultimately added to the through volumes
- Left-turn volumes on shared lanes should be converted to equivalent through vehicles using factors given in Table 10.11, and the proportion of the shared lane occupied by this volume should be deducted from through lane capacity
- Shared lanes that carry left turns with no opposing flow should be treated as shared right-turn lanes
- It is essential that the actual left-turn treatment for the intersection be used in the analysis

Step 2.   Determine whether the left-turn movement at the subject approach is protected or permitted.

Step 3.   Determine the sum of the critical volumes.

Step 4.   Using the signal operation worksheet shown in Figure 10.24, select an appropriate phase plan that will both provide the desired degree of left-turn protection and accommodate the left-turn volume balance.

Step 5.   Determine the probable capacity from Table 10.10.

If it is desired to estimate the level of service based on the stopped delay, the following additional steps should be carried out.

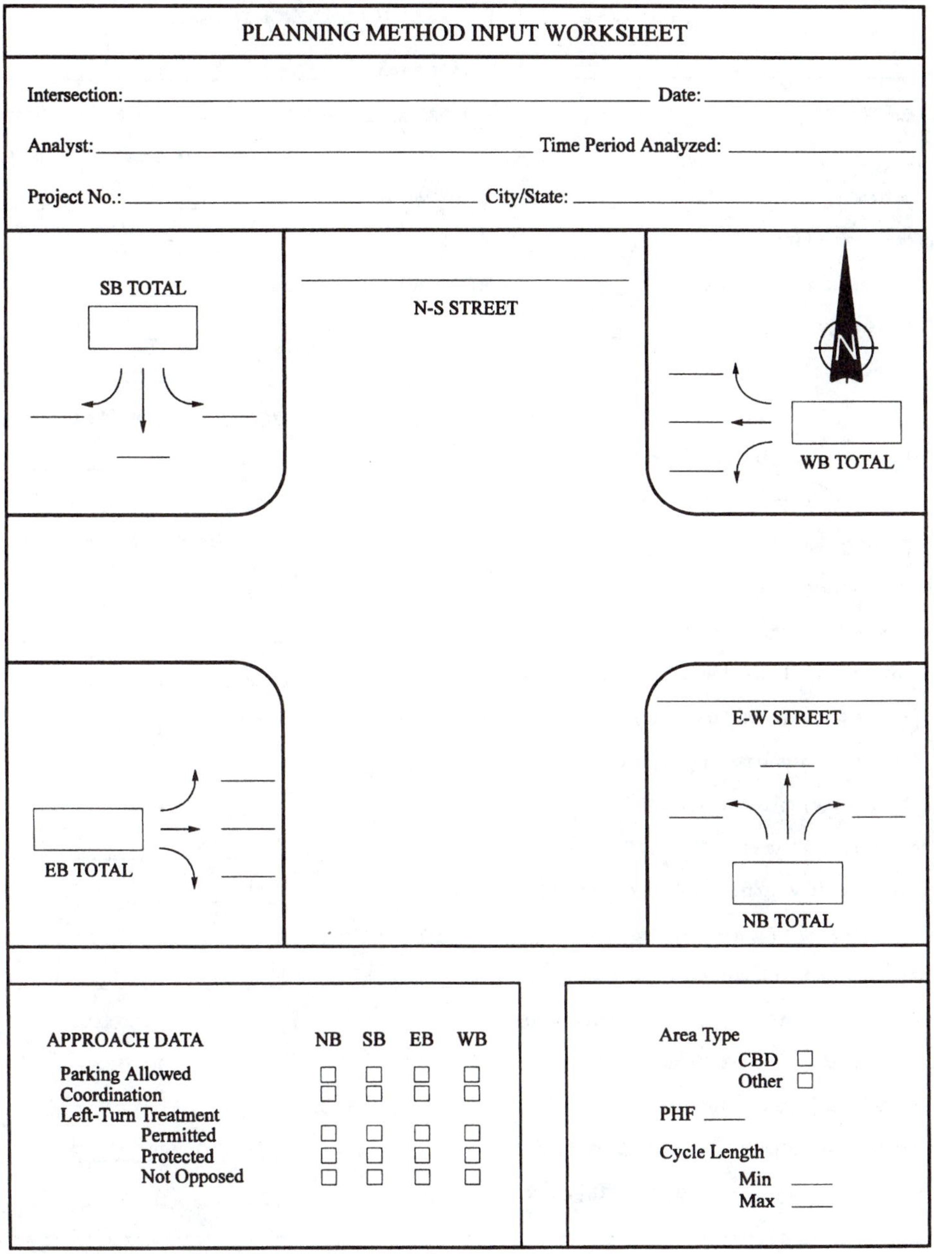

PLANNING METHOD INPUT WORKSHEET

Intersection:_____ Date:_____

Analyst:_____ Time Period Analyzed:_____

Project No.:_____ City/State:_____

SB TOTAL

N-S STREET

N

WB TOTAL

E-W STREET

EB TOTAL

NB TOTAL

APPROACH DATA	NB	SB	EB	WB
Parking Allowed	☐	☐	☐	☐
Coordination	☐	☐	☐	☐
Left-Turn Treatment				
Permitted	☐	☐	☐	☐
Protected	☐	☐	☐	☐
Not Opposed	☐	☐	☐	☐

Area Type
CBD ☐
Other ☐

PHF ____

Cycle Length
Min ____
Max ____

Figure 10.22 Planning Method Input Worksheet

SOURCE: *Highway Capacity Manual,* Special Report 209, Fourth Edition, Transportation Research Board, National Research Council, Washington, D.C., 2000. Used with permission.

PLANNING METHOD LANE VOLUME WORKSHEET

Location: _____   Direction _____

Left Turn Movement	**Right Turn Movement**	**Exclusive**	**Shared**
		RT Lane	**RT Lane**

1. LT volume _____   6. RT volume _____   _____

2. Opposing mainline volume _____   7. RT Lanes _____   1

3. No of exclusive LT lanes _____   8. RT adjustment factor _____   _____

4. LT adjustment factor _____   RT lane vol: [9] _____   [10] _____

   (See instructions)

            Cross product: [2] * [1] _____ --->   **Permitted**   **Protected**   **Not Opposed**

5. LT lane volume: [1] / ([3] * [4])   0   _____   _____

Through Movement

11. Through volume   _____   _____   _____

12. Parking adjustment factor   _____   _____   _____

13. No. of through lanes including shared lanes   _____   _____   _____

----------- Exclusive LT lane computations -------------------

14. Total approach volume: ([10] + [11]) /[12]   _____   _____   _____

16. Left turn equivalence: (Table 10.6)   _____   XXXXXXXXX   XXXXXXXXX

18. Through lane volume: [14] / [13]   _____   _____   _____

19. Critical lane volume: (See instructions)   _____   _____   _____

----------- Shared LT lane computations ----------------------

14. Total approach volume: (See instructions)   _____   _____   _____

15. Proportion of left turns in the lane group   _____   XXXXXXXXX   XXXXXXXXX

16. Left turn equivalence: (Table 10.6)   _____   XXXXXXXXX   XXXXXXXXX

17. Left turn adjustment factor: _____   _____   1.0

18. Through lane volume: [14] / ([13] * [17])   _____   _____   _____

19. Critical lane volume: Max([5][9],[18])   _____   _____   _____

Left Turn Check (if [16] > 8)

20. Permitted left turn sneaker capacity: 7200 / C_{max} _____   XXXXXXXXX   XXXXXXXXX

Figure 10.23 Planning Method Lane Volume Worksheet

SOURCE: *Highway Capacity Manual,* Special Report 209, Fourth Edition, Transportation Research Board, National Research Council, Washington, D.C., 2000. Used with permission.

PLANNING METHOD SIGNAL OPERATIONS WORKSHEET

Phase Plan Selection from Lane Volume Worksheets		EASTBOUND	WESTBOUND	NORTHBOUND	SOUTHBOUND

Critical Through-RT lane volume: [19]   _____   _____   _____   _____

LT lane volume: [5]   _____   _____   _____   _____

Left turn protection: (Perm, Prot, N/O)   _____   _____   _____   _____

Dominant left turn: (Indicate by '*')   _____   _____   _____   _____

Selection Criteria based on the specified left turn treatment:	Plan 1:	Perm	Perm	Perm	Perm
		Perm	N/O	Perm	N/O
		N/O	Perm	N/O	Perm
	Plan 2a:	Perm	Prot	Perm	Prot
	Plan 2b:	Prot	Perm	Prot	Perm
* Indicates the dominant left turn	Plan 3a:	*Prot	Prot	*Prot	Prot
for each opposing pair	Plan 3b:	Prot	*Prot	Prot	*Prot
	Plan 4:	N/O	N/O	N/O	N/O

Phase plan selected (1 to 4)   _____   _____

Min. cycle [C_{min}] _____   Max cycle [C_{max}] _____   [PHF] (From Input Worksheet) _____

Phasing Plan From Table 10.19   ------- EAST-WEST -------   ------- NORTH-SOUTH -----

	Note	Value	Phase 1	Phase 2	Phase 3	Phase 1	Phase 2	Phase 3
Movement codes			____	____	____	____	____	____
Critical Phase Volume [CV]			____	____	____	____	____	____
Critical Sum [CS]	1	____						
Lost time/phase [PL]			____	____	____	____	____	____
Lost time/cycle [TL]	2	____						
CBD adjustment [CBD]	3	____						
Critical v/c ratio [X_{cm}]	4	____						
Intersection status	5	____						

Optional Timing Plan Computation

	Note	Value						
Reference Sum [RS]	6	____						
Cycle length [CYC]	7	____						
Green time	8		____	____	____	____	____	____

Notes
1. Critical sum = Sum of critical phase volumes [CV's] for all phases.
2. Lost time/cycle = Sum of all lost times/phase, [PL's].
3. CBD adjustment = .9 within CBD, 1.0 elsewhere.
4. Critical v/c ratio = CS /((1-[TL]/C_{max}) * 1900 * [CBD] * [PHF]).
5. Status: (See instructions).
6. Reference Sum = 1710 * [PHF] * [CBD].
7. Cycle length = [TL] / (1-(Min([CS],[RS]) / [RS])), Subject to [C_{min}] and [C_{max}].
8. Green time = ([CYC]-[TL]) * ([CV]/[CS]) + [PL].

Figure 10.24 Planning Method Signal Operations Worksheet

SOURCE: *Highway Capacity Manual,* Special Report 209, Fourth Edition, Transportation Research Board, National Research Council, Washington, D.C., 2000. Used with permission.

Step 6. Assuming a 90 percent degree of saturation, determine the cycle length adequate for the observed volumes.

Step 7. Proportionately assign green times to the conflicting base, using the phasing plan selected in step 5 and the cycle length determined in step 6.

Step 8. Use the operational level of analysis to estimate the level of service at this intersection.

In this text, only steps 1 through 5 will be presented, because in practice, planning analyses usually do not include the estimation of the level of service based on stop delay.

Steps 1 through 3 are executed for each approach by using the lane volume worksheet (Figure 10.23), and steps 4 and 5 are executed by using the signal operations worksheet (Figure 10.24). Although most of the items on the lane volume worksheet are self-explanatory, it is necessary to explain some of the items used in steps 1, 2, and 3.

In determining the left-turn volume to be recorded for item 1, it should be noted that when left turns are made from exclusive left-turn lanes during a protected-plus-permitted phase, the left-turn volume is reduced by two vehicles per phase to account for sneakers. The number of cycles per hour will be required for this, but unfortunately in most cases the cycle length will not be known. It is suggested that the maximum cycle length that was entered in Figure 10.22 be used. However, it should be noted that in order to avoid unreasonably short green time for the protected left-turn phase, this adjustment should not result in a left-turn volume less than four vehicles per cycle.

The opposing mainline volume in item 2 and the number of exclusive left turn lanes in item 3 take the same definitions and have the same guidelines for their determination as those given in the procedure for operational analysis.

The left-turn adjustment factor in item 4 should be determined only for the protected left turns from exclusive turn lanes or for left turns without any opposing flow. However, this factor should be corrected for dual lanes using the lane utilization factor of 0.97 from Table 10.5, thus obtaining a reduced left-turn adjustment factor of 0.92. In cases where the nonexistence of an opposing flow is due to a T intersection or a one-way street, then the interference of pedestrians should be considered and a factor of 0.85 used for a single lane. This factor is adjusted also when dual lanes exist, using a lane utilization factor of 0.885 to obtain a factor of 0.75. The product of [1] and [2] shown below item 4 is used for comparison purposes to determine if a protected phase should be assumed. This is discussed further under item 20.

In item 5, the left-turn volume per lane is determined. Note that if there are no exclusive left-turn lanes, then the value for item 5 will be the actual left-turn volume entered in item 1. Also, note that if the left turns are made during a permitted phase, then the value for item 5 is automatically zero.

The right-turn volume to be recorded in item 6 includes right turns on shared through and right-turn lanes or on exclusive right-turn lanes, minus the volume of right turns made during the red phase, as discussed earlier for operational analysis.

Item 7 is self-explanatory. The value for the right-turn adjustment factor to be recorded for item 8 is 0.85 for a single or shared lane. For dual right-turn lanes, this factor is adjusted by the lane utilization factor of 0.885 (from Table 10.5) to obtain a right-turn factor of 0.75.

In item 9, the right-lane volume is obtained as [6]/([7]*[8]), except that if there is no exclusive right-turn lane, a value of 1 is used for [7] and the resultant computation is recorded for item 10.

Item 11 is self-explanatory. The parking adjustment factor to be recorded for item 12 is obtained from Table 10.4, assuming 20 parking maneuvers per hour if parking is expected. If parking will not be allowed, the parking adjustment factor is 1; if parking will be permitted, the factors are 0.8, 0.9, and 0.933 for one, two, and three lanes, respectively. Item 13 is also self-explanatory.

Items 14 through 19 should be considered separately for exclusive and shared left-turn lanes. For exclusive left-turn lanes, the expression given in item 14 for the total approach volume takes into consideration that the through volume should be adjusted for parking and that only the total of the shared-lane right-turn volumes and the through volumes is included. Note that the left-turn volumes are not included since they are on an exclusive left-turn lane and therefore not part of the through-volume lane group. Item 15 is not applicable for exclusive left-turn lanes. For item 16, although the through-car equivalent for permitted left turns is not used for the case of an exclusive left-turn phase, it should nevertheless be determined from Table 10.6 because it will be used later in item 20. Item 17 is not applicable to exclusive left-turn lanes. The computation for item 18 is self- explanatory; this value will be used to determine the critical lane volumes. The critical lane volume for item 19 is usually that obtained in item 18, except when there is no opposing flow to the left-turn volume or when the right turns are made from an exclusive right-turn lane, and when any one of these movements is higher than the through movement. In situations where both conditions exist, the critical volume is max([5],[9],[18]).

For shared left-turn lanes, in computing the value to be recorded for item 14, the same expression for item 14 under the exclusive left-turn lane is used, that is, ([10] + [11])/[12]; however, when the left turn is not opposed, [11] should be replaced by ([11] + [5]). This latter situation means that the total through volume is taken as the sum of the through volume (which is recorded in item 11) and the unopposed left-turn volumes. Item 15 is self-explanatory. The value for item 16 is the through-car equivalent for permitted left turns, obtained from Table 10.6. This factor is used to compute the shared-lane left-turn adjustment factor, f_{LT}, using the appropriate formula in Table 10.4. However, when the left turns are protected, this item is not used.

In computing the value for item 17, the shared left-turn adjustment factor is computed for the appropriate evaluation in Table 10.4. This is a reduction factor that will be used in item 18 to take into consideration the effect of the left-turning vehicles waiting for a gap in the opposing traffic to make the turn. Note that this factor is 1 for lanes that are not opposed, since in that case the left-turns can always turn on arrival at the intersection. Item 18 is self-explanatory and gives the equivalent through volume per lane. The critical lane volume in item 19 is the maximum of either the value for item 18 or the right-turn volume from an exclusive right-turn lane, as computed in item 9.

Item 20 is used to check for both the exclusive left-turn and shared left-turn cases whether a protected left-turn phase is required if one or more of the left turns has been designated as permitted, i.e., when protected phase is provided. It is recommended that the protected left-turn phases be used when either of the following exists:

- The product ([2] × [1]) exceeds an adopted threshold.
- The left-turn equivalent recorded for item 16 is greater than 3.5, and the left-turn volume is greater than two vehicles per cycle (i.e., [1] > $7200/C_{max}$). This is because under such conditions, it is very likely that the subject left turn will not have adequate capacity without a protected phase.

In step 4, the phase plan is selected from the six alternatives given on the signal operations worksheet (Figure 10.24). Note that the phase plan deals with only one street at a time, so the phasing plan for the whole intersection consists of two phase plans. A brief description of each phase plan is necessary. In plan 1 there is no provision made for any protected left turns, which means all through right and left turns on both approaches of the street under consideration are made during the same phase. This requires that left-turning vehicles yield to the opposing through movements. In plans 2a and 2b, provision is made for a protected left-turn phase for only one approach on the street being considered. The only difference between plans 2a and 2b is in terms of which approach of the street being considered has the protected left turns. The phase plans will therefore consist of two phases. The first phase allows movement of the protected left turns and the through movement on the same approach. In the second phase, the two through movements proceed. In plans 3a and 3b, both sets of opposing left turns on a given street are protected. This plan consists of three phases. During the first phase, the two opposing left turns proceed, followed by the second phase in which the higher left-turn movement will continue with the through movement on the same approach. In the third phase, the two through movements will proceed. Again, the only difference between plans 3a and 3b is in terms of which approach has the dominant left-turn volume. Phase plan 4 is usually referred to as the "split phase" operation. It consists of two phases, with the through and left-turn movements from one of the two approaches moving during each phase. The table given in Figure 10.24 shows the criteria for selecting the appropriate phase plan.

The critical phase volume (CV) and the lost time for each phase are determined in step 5 and entered in Figure 10.24. Table 10.12 gives a summary of the appropriate choice for the CV and lost time for each of the six phase plans. The critical volume-to-capacity ratio X_{cm} is then determined by the ratio of the critical sum CS (the sum of critical phase volumes) to the sum of the critical lane volumes, CL, that can be accommodated at the maximum cycle length. The sum of the critical lane volumes is computed using the formula

$$CL = \left(1 - \frac{T_L}{C_{max}}\right)(1900)(CBD)(PHF) \tag{10.64}$$

where
$\quad\quad T_L$ = total lost time (sec)
$\quad\quad C_{max}$ = maximum cycle length (sec)
$\quad\quad CBD$ = CBD correction factor (0.9 for CBD, 1.0 for non-CBD)

The ratio X_{cm} is then used to obtain the probable capacity from Table 10.10.

Example 10.2   Computing Capacity at a Signalized Intersection Using the Planning
Level of Analysis

Figure 10.25 shows the design volumes and the layout of a new signalized intersection being planned for projected traffic growth in 10 years at the intersection of
10th and Main Streets in a CBD area. Using the planning level of analysis and assuming a PHF of 0.85, determine whether the capacity of the proposed design will be
adequate.

Solution: The computations required are in the lane volume and signal operations worksheets shown in Tables 10.13 and 10.14. The procedure will be illustrated
by going through that for the eastbound approach. The left-turn volume for item 1
is 200 veh/h, as obtained from Figure 10.25. The opposing flow is the sum of the
westbound through and right-turn volumes, that is, $(190 + 735) = 925$. Note that
in this case, there is no exclusive right-turn lane; therefore, the total right-turn volume is considered as part of the opposing flow. The number of exclusive left-turn lanes
is one (item 3). Since the left turns are made from exclusive left-turn lanes, the
left-turn adjustment factor is 0.95 (item 4). The value of item 4 is then adjusted by the
left- turn adjustment factor to obtain $200/0.95 = 211$ (item 5). Similar procedures are
carried out for right-turn movements in items 6 through 10 and for the through
movements in items 11 through 18. Note that the right-turn factor obtained is 0.85 because the right turns are made from a shared lane. Also note that since no parking is
allowed, the parking adjustment factor is 1 and the total through movement is the
sum of the right-turn adjustment volume on the shared right and through lane and
the through volume, i.e., $(212 + 475) = 687$. This is inserted for item 14. The volume per lane is therefore $687/2 = 344$, since this total volume is served by two lanes.
The critical lane volume for the through and right lanes on this approach is therefore 344. This procedure is repeated for each of the other approaches, as shown in
Table 10.13.

The most appropriate plan for the east-west road is plan 3a, since the left turns on
both approaches on this road are made from exclusive left-turn lanes and should therefore have protected phases, with the eastbound left-turn movement being the higher
left-turn volume. Similarly, phase plan 3b is selected for the north-south street, since the
southbound left-turn volume is the dominant volume.

The critical phase volume, CV, for each phase is shown on the signal operations
worksheet (Table 10.14), from which the critical sum is obtained as $105 + 106 + 480 +
116 + 68 + 488 = 1363$. The total lost time is 12 sec (assuming 3 sec per phase; see Table
10.12). The critical lane volume that can be accommodated at the maximum cycle length
is computed from Eq. 10.64:

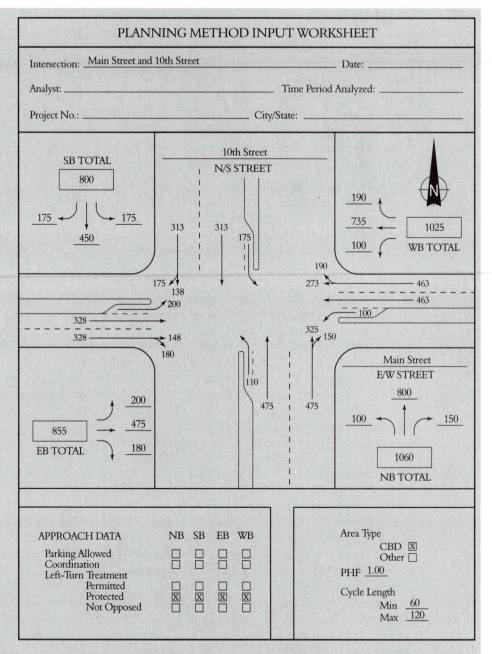

Figure 10.25 Data and Solution for Example 10.2

SOURCE: *Highway Capacity Manual,* Special Report 209, Fourth Edition, Transportation Research Board, National Research Council, Washington, D.C., 2000. Used with permission.

$$CL = \left(1 - \frac{T_L}{C_{\max}}\right)(1900)(CBD)(PHF)$$

$$= \left(1 - \frac{12}{120}\right)(1900)(0.9)(0.85)$$

$$= 1462 \text{ veh/h}$$

$$X_{cm} = \frac{CV}{CL}$$

$$= \frac{1363}{1462}$$

$$= 0.93$$

Using Table 10.10 and X_{cm} = 0.93, the intersection will operate at near capacity.

Example 10.3   Determining a Suitable Layout for a Signalized Intersection Using the Planning Level of Analysis

Figure 10.26 shows the estimated maximum flows at a signalized intersection planned for a CBD in an urban area. Using these data, we have to determine a suitable layout for the intersection that will provide a condition of under capacity. Due to buildings adjacent to Riverbend Drive, only two lanes can be provided on each approach of this road. Assume a PHF of 1.00.

Solution: In this example, it is necessary to determine whether the intersection geometry will be adequate for the volumes shown in Figure 10.26, such that the intersection will operate at "under capacity." If this is not achieved, then the intersection geometry should be modified until the desired operational level is obtained. Tables 10.15 and 10.16 show the solution for the trial layout of the intersection, which shows that the intersection will operate "at capacity." The operational level is therefore not achieved, and the layout of the intersection is modified by including exclusive left-turn lanes at the east and west approaches for Main Street, as shown in Figure 10.27. The analysis is repeated (Tables 10.17 and 10.18), and the results now indicate that the modified layout of the intersection will provide for an operational level of "under capacity." This problem illustrates how the planning level of analysis can be used to obtain the general intersection layout for a required level of operation.

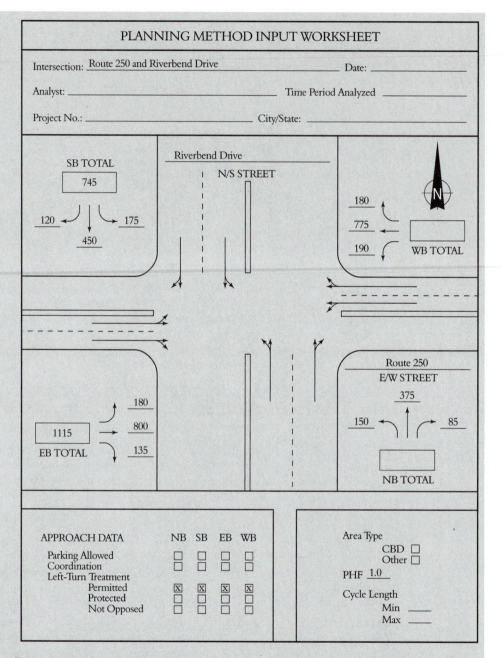

Figure 10.26 Data for Example 10.3

SOURCE: *Highway Capacity Manual,* Special Report 209, Fourth Edition, Transportation Research Board, National Research Council, Washington, D.C., 2000. Used with permission.

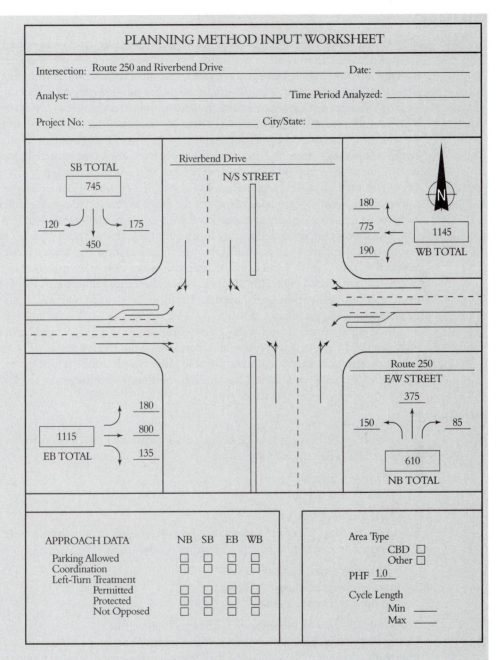

Figure 10.27   Revised Layout for Example 10.3

SOURCE: *Highway Capacity Manual,* Special Report 209, Fourth Edition, Transportation Research Board, National Research Council, Washington, D.C., 2000. Used with permission.

SUMMARY

The complexity of a signalized intersection requires that several factors be considered when its quality of operation is being evaluated. In particular, the geometric characteristics, prevailing traffic conditions, and signal characteristics must be used in determining the level of service at the intersection. This requirement makes the determination and improvement of level of service at a signalized intersection much more complex than at other locations of a highway. For example, in Chapter 9, the primary characteristics used in determining capacity and level of service at segments of highways are those of geometry of the highway and traffic composition. Since the geometry is usually fixed, the capacity of a highway segment can be improved by improving the highway geometry if consideration is given to some variations over time in traffic composition. However, this cannot be done easily at signalized intersections because of the added factor of the green time allocation to the different traffic streams, which has a significant impact on the operation of the intersection.

The procedures presented in this chapter for the determination of delay and level of service at a signalized intersection take into consideration the most recent results of research in this area. In particular, the procedures not only take into consideration such factors as traffic mix, lane width, and grades but also factors inherent in the signal system itself, such as whether the signal is pretimed, semiactuated, or fully actuated and the impact of pedestrians and bicycles.

In using the procedures for design, remember that the cycle length has a significant effect on the delay at an intersection and therefore on the level of service. Thus, it is useful first to determine a suitable cycle and suitable phase lengths, using one of the methods discussed in Chapter 8, and then to use this information to determine the level of service.

PROBLEMS

10-1 A two-phase signal system is to be designed for an isolated intersection with a peak hour factor of 0.95. The critical lane volumes are:

Phase A = 550 /h
Phase B = 600 veh/h

The total lost time L is 14 sec.
Using the *Highway Capacity Manual* procedure, determine an appropriate cycle length for the intersection that will satisfy the following conditions:

minimum cycle length = 45 sec
maximum cycle length = 120 sec
maximum critical v/c ratio (X_c) = 0.85

The saturation flow per lane for each phase is 1900 veh/h.

10-2 Figure 10.28 shows peak hour volumes and other traffic and geometric characteristics of an intersection located in the CBD of a city. Determine the overall level of service at the intersection. Use operation analysis.

10-3 Figure 10.29 shows an isolated intersection at Third Street and Ellis Avenue outside the CBD of a city. Third Street is one way in the NB direction, with parking allowed on either side. Ellis Avenue is a major arterial with a separate left-turn lane on the EB approach. Traffic volume counts at the intersection have shown that the growth rate over the past three years is 4 percent per annum, and it is predicted that this rate will continue for the

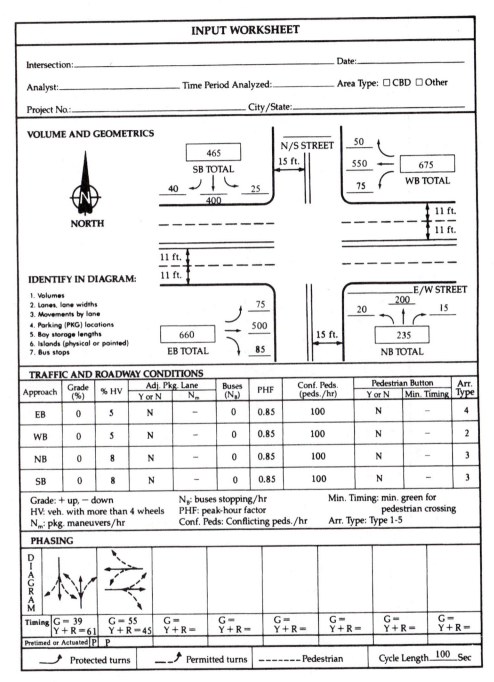

Figure 10.28 Characteristics of the Intersection in Problem 10-2

SOURCE: *Highway Capacity Manual,* Special Report 209, Fourth Edition, Transportation Research Board, National Research Council, Washington, D.C., 2000. Used with permission.

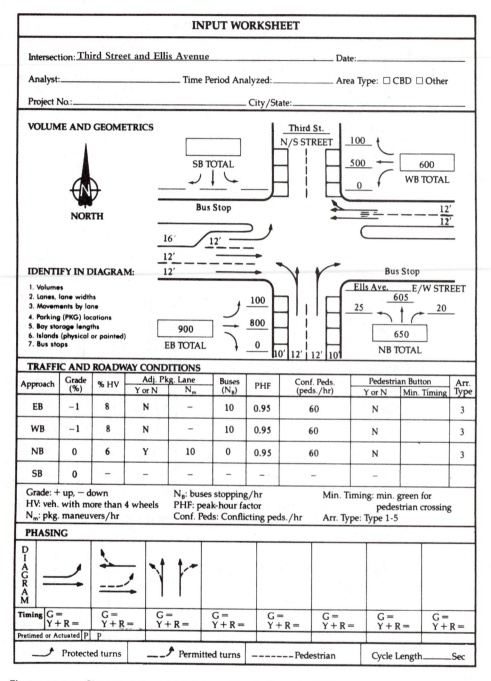

INPUT WORKSHEET

Intersection: Third Street and Ellis Avenue _____ Date:_____

Analyst:_____ Time Period Analyzed:_____ Area Type: ☐ CBD ☐ Other

Project No.:_____ City/State:_____

VOLUME AND GEOMETRICS

NORTH

Third St.
N/S STREET

SB TOTAL

100
500 ←
0

600
WB TOTAL

Bus Stop

12'
12'

16'   12'
12'
12'

IDENTIFY IN DIAGRAM:

1. Volumes
2. Lanes, lane widths
3. Movements by lane
4. Parking (PKG) locations
5. Bay storage lengths
6. Islands (physical or painted)
7. Bus stops

100
800
0

900
EB TOTAL

Bus Stop

Ells Ave.   E/W STREET
605
25   20

650
NB TOTAL

10' 12' 12' 10'

TRAFFIC AND ROADWAY CONDITIONS

Approach	Grade (%)	% HV	Adj. Pkg. Lane		Buses (N_B)	PHF	Conf. Peds. (peds./hr)	Pedestrian Button		Arr. Type
			Y or N	N_m				Y or N	Min. Timing	
EB	−1	8	N	−	10	0.95	60	N		3
WB	−1	8	N	−	10	0.95	60	N		3
NB	0	6	Y	10	0	0.95	60	N		3
SB	0	−	−	−	−	−	−	−		

Grade: + up, − down
HV: veh. with more than 4 wheels
N_m: pkg. maneuvers/hr

N_B: buses stopping/hr
PHF: peak-hour factor
Conf. Peds: Conflicting peds./hr

Min. Timing: min. green for pedestrian crossing
Arr. Type: Type 1-5

PHASING

D I A G R A M

Timing | G = Y + R = | G = Y + R = | G = Y + R = | G = Y + R = | G = Y + R = | G = Y + R = | G = Y + R = | G = Y + R =

Pretimed or Actuated | P | P

Protected turns   Permitted turns   ------- Pedestrian   Cycle Length_____ Sec

Figure 10.29 Characteristics of the Intersection in Problem 10-3

SOURCE: *Highway Capacity Manual,* Special Report 209, Fourth Edition, Transportation Research Board, National Research Council, Washington, D.C., 2000. Used with permission.

INPUT WORKSHEET

Intersection: **3rd Street and K Street** _____ Date: _____

Analyst: _____ Time Period Analyzed: _____ Area Type: ☐ CBD ☐ Other

Project No.: _____ City/State: _____

VOLUME AND GEOMETRICS

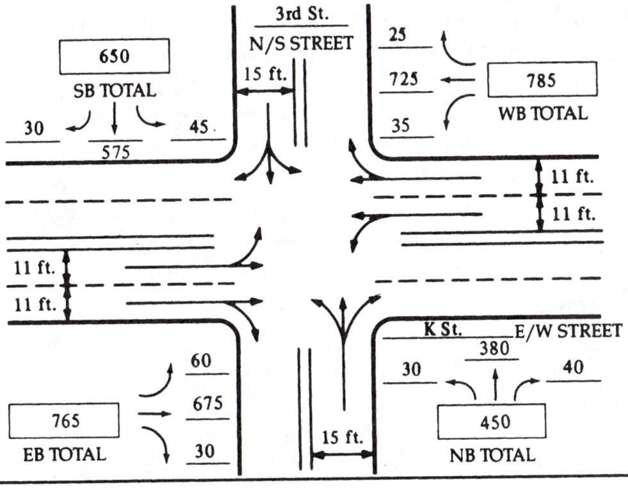

NORTH

3rd St.
N/S STREET
15 ft.

650
SB TOTAL

30   ↙ ↓ ↘   45
575

25
725
35
785
WB TOTAL

11 ft.
11 ft.

11 ft.
11 ft.

IDENTIFY IN DIAGRAM:

1. Volumes
2. Lanes, lane widths
3. Movements by lane
4. Parking (PKG) locations
5. Bay storage lengths
6. Islands (physical or painted)
7. Bus stops

60
675
765
EB TOTAL
30

K St.   E/W STREET
380
30   ↑   40
450
NB TOTAL
15 ft.

TRAFFIC AND ROADWAY CONDITIONS

Approach	Grade (%)	% HV	Adj. Pkg. Lane Y or N	Adj. Pkg. Lane N_m	Buses (N_B)	PHF	Conf. Peds. (peds./hr)	Pedestrian Button Y or N	Pedestrian Button Min. Timing	Arr. Type
EB	−1	4	N	−	0	0.95	60	N		3
WB	−1	4	N	−	0	0.95	60	N		3
NB	0	6	N	−	0	0.95	60	N		3
SB	0	6	N	−	0	0.95	60	N		3

Grade: + up, − down N_B: buses stopping/hr   Min. Timing: min. green for
HV: veh. with more than 4 wheels   PHF: peak-hour factor   pedestrian crossing
N_m: pkg. maneuvers/hr   Conf. Peds: Conflicting peds./hr   Arr. Type: Type 1-5

PHASING

DIAGRAM

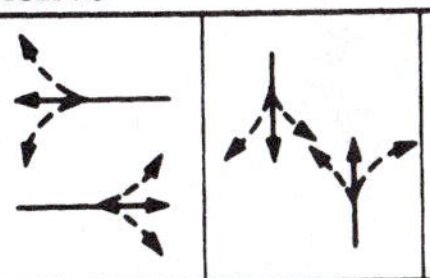

Timing	G = 30 Y + R = 40	G = 34 Y + R = 36	G = Y + R =	G = Y + R =	G = Y + R =	G = Y + R =	G = Y + R =	G = Y + R =
Pretimed or Actuated	P	P						

⤴ Protected turns   ⤴ Permitted turns   ------- Pedestrian   Cycle Length_____ Sec

Figure 10.30 Characteristics of the Intersection in Problem 10-4

SOURCE: *Highway Capacity Manual,* Special Report 209, Fourth Edition, Transportation Research Board, National Research Council, Washington, D.C., 2000. Used with permission.

next five years. The existing peak hour volumes and traffic characteristics are shown in Figure 10.29. Determine a suitable timing of the existing three-phase signal that will be suitable for the traffic volumes in five years. Assume that, except for the traffic volumes, characteristics will remain the same. Also determine the LOS at which the intersection will operate. A critical (v/c) ratio of 0.85 or lower is required at the intersection.

10-4   Figure 10.30 on page 477 shows projected peak hour volumes in five years and other traffic and geometric characteristics for the intersection of 3rd and K Streets in the CBD of an urban area. Determine whether the existing signal timing will be suitable for the projected demand if the level of service at each approach must be D or better. If the existing system will not satisfy this requirement, make suitable changes to the phasing and/or signal timing that will achieve the level of service requirement of D. Determine the critical volume and the intersection overall level of service, using your phasing and/or signal timing. A critical (v/c) ratio of 0.85 or lower is required at the intersection.

10-5   Repeat Problem 10-4, assuming that the intersection is located outside the CBD. There is adequate right of way available for significant geometric improvement, and the level of service on each approach should be C or better.

10-6   Figure 10.31 shows the estimated maximum flows at an isolated signalized intersection planned for an urban area outside the CBD. Using the planning level of analysis, determine a suitable intersection layout and phasing sequence that will provide a condition of under capacity.

10-7   It is estimated that traffic will grow at a rate of 4 percent per annum at the intersection described in Problem 10-6. Using the planning level of analysis, determine under what conditions the intersection will be operating after 10 years of operation if your recommended geometric layout is adopted.

10-8   Using your solution for Problem 10-6 and the operation analysis procedure, determine the overall level of service at the intersection if it is located within the CBD. All approaches have 0 percent grades and 3 percent heavy vehicles in the traffic stream. Parking is not allowed, and there is no bus stop on any approach. The PHF at each approach is 0.90. There are 60 conflicting pedestrians at each approach, and there are no pedestrian buttons. A critical (v/c) ratio of 0.85 or lower is desired. Use a four-phase signal (one phase for each approach).

10-9   Figure 10.31 shows existing peak hour volumes at an isolated four-leg intersection, with each approach consisting of two 10-ft-wide lanes. A signal system is to be installed at the

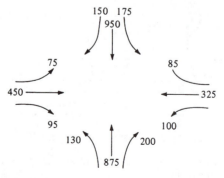

Figure 10.31   Isolated Four-Legged Intersection for Problems 10-6 and 10-9

intersection in three years. Traffic growth is projected at 3 percent per annum. Using the planning level of analysis, determine whether it will be necessary to improve the intersection geometry before the installation of the signals.

10-10 Using the peak hour volumes given for Problem 10-9, determine a suitable intersection layout if the growth rate is 4 percent per annum, the signal system is to be installed in five years, and a condition of under capacity is required.

REFERENCES

Highway Capacity Manual, Special Report 209, 4th ed., Transportation Research Board, National Research Council, Washington, D.C., 2000.

ADDITIONAL READINGS

Akcelik, R. *Traffic Signals: Capacity and Timing Analysis,* Research Report Number 123, Australian Road Research Board, Nunawading, Victoria, Australia, 1982.

Akcelik, R. (ed.), *Signalized Intersection Capacity and Timing Guide,* Signalized Intersection Capacity Workshop Papers—Internal Report AIR 1094-1, Australian Road Research Board, Nunawading, Victoria, Australia, 1979.

Interim Materials on Highway Capacity, Transportation Research Circular 212, National Research Council, Transportation Research Board, Washington, D.C., January 1980.

JHK & Associates and the Traffic Institute, Northwestern University, *Development of an Improved Highway Capacity Manual,* NCHRP 3-28(2), National Research Council, Transportation Research Board, Washington, D.C., 1979.

JHK & Associates, in cooperation with the Traffic Institute, Northwestern University, *NCHRP Signalized Intersections Capacity Method,* National Research Council, Transportation Research Board, Washington, D.C., May 1982.

McInerney, H.B., and S.G. Peterson, "Intersection Capacity Measurement Through Critical Movement Summations: A Planning Tool," *Traffic Engineering* 41 (June 1971).

Miller, A.J., "Australian Road Capacity Guide," *Australian Road Research Bulletin* 3 (1981).

Reilly, W.R., C.C. Gardner, and J.H. Kell, *A Technique for Measurement of Delay at Intersections,* Federal Highway Administration Report No. RD-76-135/137, U.S. Department of Transportation, Federal Highway Administration, Washington, D.C., November 1976.

Shenk, E., W.R. McShane, J.M. Ulerio, and R.P. Roess, *Highway Capacity Software User's Manual,* U.S. Department of Transportation, Federal Highway Administration, Washington, D.C., 1986.

CHAPTER 10 APPENDIX: TABLES

All tables in this appendix except Table 10.2 and Tables 10.13 through 10.18 are reprinted from *Highway Capacity Manual,* Special Report 209, Third Edition, copyright by the Transportation Research Board, National Research Council, Washington, D.C., 2000. Used with permission.

Table 10.2 is reprinted from C. J. Messer, "Guidelines for Signalized Left-Turn Treatments, *Implementation Package FHWA-IP-814,* U.S. Department of Transportation, Federal Highway Administration, Washington, D.C., 1981.

Table 10.1   Level-of-Service Criteria for Signalized Intersections

Level of Service	Control Delay Per Vehicle (sec)
A	≤ 10.0
B	> 10.0 and ≤ 20.0
C	> 20.0 and ≤ 35.0
D	> 35.0 and ≤ 55.0
E	> 55.0 and ≤ 80.0
F	> 80.0

Table 10.2   Input Data Needs for Each Analysis Lane Group

Type of Condition	Parameter
Geometric conditions	Area Type Number of lanes, N Average lane width, (ft) Grade, G (%) Existence of exclusive LT or RT lanes Length of storage bay, LT or RT lane, L_s (ft) Parking
Traffic conditions	Demand volume by movement, V (veh/h) Base saturation flow rate, s_o (pc/h/ln) Peak-hour factor, PHF Percent heavy vehicles, HV (%) Approach pedestrian flow rate, v_{ped} (p/h) Local buses stopping at intersection, N_B (buses/h) Parking activity, N_m (maneuvers/h) Arrival type, AT Proportion of vehicles arriving on green, P Approach speed, S_A (mi/h)
Signalization conditions	Cycle length, C (s) Green time, G (s) Yellow-plus-all-red-change-and-clearance interval (intergreen), Y (s) Actuated or pretimed operation Pedestrian push-button Minimum pedestrian green, G_p (s) Phase plan Analysis period, T (h)

Table 10.3 Relationship Between Arrival Type and Platoon Ratio (R_p)

Arrival Type	Range of Platoon Ratio (R_p)	Default Value (R_p)	Progression Quality
1	≤ 0.50	0.333	Very poor
2	> 0.50 and ≤ 0.85	0.667	Unfavorable
3	> 0.85 and ≤ 1.15	1.000	Random arrivals
4	> 1.15 and ≤ 1.50	1.333	Favorable
5	> 1.50 and ≤ 2.00	1.667	Highly favorable
6	> 2.00	2.000	Exceptional

Table 10.4 Adjustment Factors for Saturation Flow Rates[a]

Factor	Formula	Definition of Variables	Notes
Lane width	$f_w = 1 + \dfrac{(W-12)}{30}$	W = lane width (ft)	$W \geq 8.0$ If $W > 16$, a two-lane analysis may be considered
Heavy vehicles	$f_{HV} = \dfrac{100}{100 + \%HV(E_T - 1)}$	$\% HV$ = percent heavy vehicles for lane group volume	E_T = 2.0 pc/HV
Grade	$f_g = 1 - \dfrac{\%G}{200}$	$\% G$ = percent grade on a lane group approach	$-6 \leq \% G \leq +10$ Negative is downhill
Parking	$f_p = \dfrac{N - 0.1 - \dfrac{18N_m}{3600}}{N}$	N = number of lanes in lane group N_m = number of parking maneuvers/h	$0 \leq N_m \leq 180$ $f_p = \geq 0.050$ $f_p = 1.000$ for no parking
Bus blockage	$f_{bb} = \dfrac{N - \dfrac{14.4N_B}{3600}}{N}$	N = number of lanes in lane group N_B = number of buses stopping/h	$0 \leq N_B \leq 250$ $f_{bb} = \geq 0.050$
Type of area	$f_a = 0.900$ in CBD $f_a = 1.000$ in all other areas		
Lane utilization	$f_{LU} = v_g/(v_{g1}N)$	v_g = unadjusted demand flow rate for the lane group, veh/h v_{g1} = unadjusted demand flow rate on the single lane in the lane group with the highest volume N = number of lanes in the lane group	
Left turns	Protected phasing: Exclusive lane: $F_{LT} = 0.95$ Shared lane: $f_{LT} = \dfrac{1}{1.0 + 0.05P_{LT}}$	P_{LT} = proportion of LTs in lane group	See pages 415 through 427 for non-protected phasing alternatives
Right turns	Exclusive lane: $f_{RT} = 0.85$ Shared lane: $f_{RT} = 1.0 - (0.15)P_{RT}$ Single lane $f_{RT} = 1.0 - (0.135)P_{RT}$	P_{RT} = proportion of RTs in lane group	$f_{RT} = \geq 0.050$
Pedestrian-bicycle blockage	LT adjustment: $f_{Lpb} = 1.0 - P_{LT}(1 - A_{pbT})$ $(1 - P_{LTA})$ RT adjustment: $f_{Rpb} = 1.0 - P_{RT}(1 - A_{pbT})$ $(1 - P_{RTA})$	P_{LT} = proportion of LTs in lane group A_{pbT} = permitted phase adjustment P_{LTA} = proportion of LT protected green over total LT green P_{RT} = proportion of RTs in lane group P_{RTA} = proportion of RT protected green over total RT green	See pages 428 to 432 for step-by-step procedure

[a] The table contains formulas for all adjustment factors. However, for situations in which permitted phasing is involved, either by itself or in combination with protected phasing, separate tables are provided, as indicated in this exhibit.

Table 10.5 Default Lane Utilization Factors

Lane Group Movements	No. of Lanes in Lane Group	Percent of Traffic in Most Heavily Traveled Lane	Lane Utilization Factor (f_{Lu})
Through or shared	1	100.0	1.000
	2	52.5	0.952
	3[a]	36.7	0.908
Exclusive left turn	1	100.0	1.000
	2[a]	51.5	0.971
Exclusive right turn	1	100.0	1.000
	2[a]	56.5	0.885

[a]If lane group has more lanes than number shown in this table, it is recommended that surveys be made or the largest f_{Lu}-factor shown for that type of lane group be used.

Table 10.6 Through-Car Equivalents, E_{L1}, for Permitted Left Turns

Type of Left-Turn Lane	Effective Opposing Flow, $v_{oe} = v_o/f_{LU_o}$						
	1	200	400	600	800	1000	1200[a]
Shared	1.4	1.7	2.1	2.5	3.1	3.7	4.5
Exclusive	1.3	1.6	1.9	2.3	2.8	3.3	4.0

Notes:

[a]Use formula for effective opposing flow more than 1200; v_{oe} must be > 0.

$E_{L1} = s_{HT}/s_{LT} - 1$ (shared)

$E_{L1} = s_{HT}/s_{LT}$ (exclusive)

$$s_{LT} = \frac{v_{oe}e^{\left(\frac{-v_{oe}t_c}{3600}\right)}}{1 - e^{\left(\frac{-v_{oe}t_f}{3600}\right)}}$$

where

E_{L1} = through-car equivalent for permitted left turns

s_{HT} = saturation flow of through traffic (veh/h/ln) - 1900 veh/h/ln

s_{LT} = filter saturation flow of permitted left turns (veh/h/ln)

t_c = critical gap = 4.5 s

t_f = follow-up headway = 4.5 s (shared), 2.5 s (exclusive)

Table 10.7 Recommended k Values for Lane Groups Under Actuated and Pretimed Control

Unit Extension (UE) (sec)	Degree of Saturation (X)					
	≤ 0.50	0.60	0.70	0.80	0.90	≥ 1.0
≤ 2.0	0.04	0.13	0.22	0.32	0.41	0.50
2.5	0.08	0.16	0.25	0.33	0.42	0.50
3.0	0.11	0.19	0.27	0.34	0.42	0.50
3.5	0.13	0.20	0.28	0.35	0.43	0.50
4.0	0.15	0.22	0.29	0.36	0.43	0.50
4.5	0.19	0.25	0.31	0.38	0.44	0.50
5.0^1	0.23	0.28	0.34	0.39	0.45	0.50
Pretimed or nonactuated movement	0.50	0.50	0.50	0.50	0.50	0.50

Note: For a given UE and its k_{min} value at $X = 0.5$: $k = (1 - 2k_{min})(X - 0.5) + k_{min}$, $k \geq k_{min}$, $k \leq 0.5$.
[1]For UE > 5.0, extrapolate to find k, keeping $k \leq 0.5$.

Table 10.8 Recommended I-Values for Lane Groups with Upstream Signals

	Degree of Saturation at Upstream Intersection, X_u						
	0.40	0.50	0.60	0.70	0.80	0.90	≥ 1.0
I	0.922	0.858	0.769	0.650	0.500	0.314	0.090

Note: $I = 1.0 - 0.91 X_u^{2.68}$ and $X_u \leq 1.0$.

Table 10.9 Progression Adjustment Factor (PF)

Progression Adjustment Factor (PF) $PF = (1 - P)f_p/(1 - g/C)$ (see Note)

Green Ratio (g/C)	Arrival Type (AT)					
	AT-1	AT-2	AT-3	AT-4	AT-5	AT-6
0.20	1.167	1.007	1.000	1.000	0.833	0.750
0.30	1.286	1.063	1.000	0.986	0.714	0.571
0.40	1.445	1.136	1.000	0.895	0.555	0.333
0.50	1.667	1.240	1.000	0.767	0.333	0.000
0.60	2.001	1.395	1.000	0.576	0.000	0.000
0.70	2.556	1.653	1.000	0.256	0.000	0.000
Default, f_p	1.00	0.93	1.00	1.15	1.00	1.00
Default, R_p	0.333	0.667	1.000	1.333	1.667	2.000

Note: 1. Tabulation is based on default values of f_p and R_p.
2. $P = R_p \, g/C$ (may not exceed 1.0).
3. PF may not exceed 1.0 for AT-3 through AT-6.
4. For arrival type see Table 10.3.

Table 10.10 Intersection Status Criteria for Signalized Intersection Planning Analysis

Critical v/c Ratio (X_{cm})	Relationship to Probable Capacity
$X_{cm} \leq 0.85$	Under capacity
$0.85 < X_{cm} \leq 0.95$	Near capacity
$0.95 < X_{cm} \leq 1.00$	At capacity
$X_{cm} > 1.00$	Over capacity

Table 10.11 Shared-Lane Left-Turn Adjustment Computations for Planning-Level Analysis

Permitted Left Turn

Lane groups with two or more lanes:
$$[17] = \{[13] - 1 + e^{(-[13] \cdot [1] \cdot [16]/600)}\}/[13]$$
Subject to a minimum value that applies at very low left-turning volumes when some cycles will have no left-turn arrivals:
$$[17] = \{[13] - 1 + e^{(-[1] \cdot C_{max}/3600)}\}/[13]$$
Lane groups with only one lane for all movements:
$$[17] = e^{-(0.02 \cdot ([16] + 10 \cdot [15]) \cdot [1] \cdot C_{max}/3600)}$$

Protected-Plus-Permitted Left Turn
(One Direction Only)

If $[2] < 1220$
$$[17] = 1/\{1 + [(235 + 0.435 \cdot [2]) \cdot [15]]/(1400 - [2])\}$$
If $[2] \geq 1220$
$$[17] = 1/(1 + 4.525 \cdot [15])$$

Table 10.12 Phase Plan Summary for Planning Analysis

Phase Plan	Phase No.	Lost Time	East-West		North-South	
			Movement Code	Critical Sum	Movement Code	Critical Sum
1	1	3	EWT	Max(ET, EL, WT, WL)	NST	Max(NT, NL, ST, SL)
2a	1	3	WTL	WL	STL	SL
	2	3	EWT	Max(WT-WL, ET)	NST	Max(ST-SL, NT)
2b	1	3	ETL	EL	NTL	NL
	2	3	EWT	Max(ET-EL, WT)	NST	Max(NT-NL, ST)
3a	1	3	EWL	WL	NSL	SL
	2	0	ETL	EL-WL	NTL	NL-SL
	3	3	EWT	Max(WT, ET-(EL-WL))	NST	Max(ST, NT-(NL-SL))
3b	1	3	EWL	EL	NSL	NL
	2	0	WTL	WL-EL	STL	SL-NL
	3	3	EWT	Max(ET, WT-(WL-EL))	NST	Max(NT, ST-(SL-NL))
4	1	3	ETL	Max(ET, EL)	NTL	Max(NT, NL)
	2	3	WTL	Max(WT, WL)	STL	Max(ST, SL)

Note: EWT = eastbound and westbound through; ETL = eastbound through and left; WTL = westbound through and left; NST = northbound and southbound through; STL = southbound through and left; NTL = northbound through and left; ET = eastbound through; EL = eastbound left; WT = westbound through; WL = westbound left; NT = northbound through; NL = northbound left; ST = southbound through; SL = southbound left.

Table 10.13 Lane Volume Computations for Example 10.2

Type of Movement	East-bound	West-bound	North-bound	South-bound
Left Turn Movement				
1. LT volume	200	100	110	175
2. Opposing mainline volume	925	655	625	950
3. Number of exclusive LT lanes	1	1	1	1
	Prot	Prot	Prot	Prot
4. LT adjustment factor	.95	.95	.95	.95
5. LT lane volume	211	105	116	184
Right Turn Movement				
Right Lane Configuration (E=Excl, S=Shrd)	S	S	S	S
6. RT volume	180	190	150	175
7. Exclusive lanes	N/A	N/A	N/A	N/A
8. RT adjustment factor	.85	.85	.85	.85
9. Exclusive RT lane volume	0	0	0	0
10. Shared lane volume	212	224	176	206
Through Movement				
11. Through volume	475	735	800	450
12. Parking adjustment factor	1	1	1	1
13. No. of through lanes including shared	2	2	2	2
14. Total approach volume	687	959	976	656
15. Proportion of left turns in lane group	0	0	0	0
16. Left turn equivalence	N/A	N/A	N/A	N/A
17. LT adj. factor	N/A	N/A	N/A	N/A
18. Through lane volume	344	480	488	328
19. Critical lane volume	344	480	488	328
Left Turn Check (if [16] > 8)				
20. Permitted left turn sneaker capacity: $7200/C_{max}$	N/A	N/A	N/A	N/A

Table 10.14 Signal Operations Worksheet for Example 10.2

Phase Plan Selection from Lane Volume Worksheet		*East-bound*	*West-bound*	*North-bound*	*South-bound*
Critical through-RT vol: [19]		344	480	488	328
LT lane volume: [5]		211	105	116	184
Left-turn protection: (P/U/N)		P	P	P	P
Dominant left turn: (Indicate by '★')		★			★
Selection criteria based on the	Plan 1:	U	U	U	U
specified left-turn protection	Plan 2a:	U	P	U	P
	Plan 2b:	P	U	P	U
	Plan 3a:	★P	P	★P	P
	Plan 3b:	P	★P	P	★P
	Plan 4:	N	N	N	N
			3a		3b

Min. cycle (C_{min}): 70
Max. cycle (C_{max}): 120

Timing Plan	*Value*	*East-West*			*North-South*		
		Ph 1	*Ph 2*	*Ph 3*	*Ph 1*	*Ph 2*	*Ph 3*
Movement codes		EWL	ETL	EWT	NSL	STL	NST
Critical phase volume [CV]		105	106	480	116	68	488
Critical sum [CS]	1363						
CBD adjustment [CBD]	.9						
Reference sum [RS]	1462						
Lost time/phase [P_L]		3	0	3	3	0	3
Lost time/cycle [T_L]	12						
Cycle length [CYC]	120						
Green time		11.3	8.4	41	12.2	5.4	41.7
Critical *v/c* ratio [X_{cm}]	0.93						
Status	Near capacity						

★Indicates the dominant left turn for each opposing pair.

Table 10.15 Lane Volume Computations for First Run of Example 10.3

Type of Movement	East-bound	West-bound	North-bound	South-bound
Left Turn Movement				
1. LT volume	180	190	150	175
2. Opposing mainline volume	1145	1115	745	610
3. Number of exclusive LT lanes	0	0	0	0
	206100	211850	111750	106750
	S	S	S	S
	Perm	Perm	Perm	Perm
4. LT adjustment factor	N/A	N/A	N/A	N/A
5. LT lane volume	N/A	N/A	N/A	N/A
Right Turn Movement				
Right Lane Configuration (E=Excl, S=Shrd)	S	S	S	S
6. RT volume	135	180	85	120
7. Exclusive lanes	N/A	N/A	N/A	N/A
8. RT adjustment factor	.85	.85	.85	.85
9. Exclusive RT lane volume	0	0	0	0
10. Shared lane volume	159	212	100	141
Through Movement				
11. Through volume	800	775	375	450
12. Parking adjustment factor	1	1	1	1
13. No. of through lanes including shared	2	2	2	2
14. Total approach volume	959	987	475	591
15. Proportion of left turns in lane group	N/A	N/A	N/A	N/A
16. Left turn equivalence	16	16	5.34	3.72
17. LT adj. factor	.5	.5	.53	.56
18. Through lane volume	957	985	444	530
19. Critical lane volume	957	985	444	530
Left Turn Check (if [16] > 8)				
20. Permitted left turn sneaker capacity: 7200/Cmax	60	60	N/A	N/A

Table 10.16 Signal Operations Worksheet for First Run of Example 10.3

Phase Plan Selection from Lane Volume Worksheet		East-bound	West-bound	North-bound	South-bound
Critical through-RT vol: [19]		957	985	444	530
LT lane volume: [5]		N/A	N/A	N/A	N/A
Left turn protection: (P/U/N)		U	U	U	U
Dominant left turn: (indicate by '★')			★		★
Selection criteria based on the specified	Plan 1:	U	U	U	U
left-turn protection	Plan 2a:	U	P	U	P
	Plan 2b:	P	U	P	U
	Plan 3a:	★P	P	★P	P
	Plan 3b:	P	★P	P	★P
	Plan 4:	N	N	N	N
			1		1

Min. cycle (C_{min}): 60
Max. cycle (C_{max}): 120

Timing Plan	Value	East-West			North-South		
		Ph 1	Ph 2	Ph 3	Ph 1	Ph 2	Ph 3
Movement codes		EWG			NSG		
Critical phase volume [CV]		985	0	0	530	0	0
Critical sum [CS]	1515						
CBD adjustment [CBD]	.9						
Reference sum [RS]	1462						
Lost time/phase [P_L]		3	0	0	3	0	0
Lost time/cycle [T_L]	6						
Cycle length [CYC]	120						
Green time		77.1	0	0	42.9	0	0
Critical v/c ratio [Xcm]	0.98						
Status	At capacity						

*Indicates the dominant left turn for each opposing pair.

Table 10.17 Lane Volume Computation for Revised Layout for Example 10.3

	East-bound	West-bound	North-bound	South-bound
Left Turn Movement				
1. LT volume	180	190	150	175
2. Opposing mainline volume	955	935	745	610
3. Number of exclusive LT lanes	1	1	0	0
	171900	177650	111750	106750
	E	E	S	S
	Perm	Perm	Perm	Perm
4. LT adjustment factor	1.0	1.0	N/A	N/A
5. LT lane volume	N/A	N/A	N/A	N/A
Right Turn Movement				
Right Lane Configuration (E=Excl, S=Shrd)	S	S	S	S
6. RT volume	135	180	85	120
7. Exclusive lanes	N/A	N/A	N/A	N/A
8. RT adjustment factor	.85	.85	.85	.85
9. Exclusive RT lane volume	0	0	0	0
10. Shared lane volume	159	212	100	141
Through Movement				
11. Through volume	800	775	375	450
12. Parking adjustment factor	1	1	1	1
13. No. of through lanes including shared	2	2	2	2
14. Total approach volume	959	987	475	591
15. Proportion of left turns in lane group	0	0	N/A	N/A
16. Left turn equivalence	5.73	5.52	5.34	3.72
17. LT adj. factor	N/A	N/A	.53	.56
18. Through lane volume	480	494	444	530
19. Critical lane volume	480	494	444	530
Left Turn Check (if [16] > 8)				
20. Permitted left turn sneaker capacity: 7200/Cmax	N/A	N/A	N/A	N/A

Table 10.18 Signal Operations Worksheet for Revised Layout for Example 10.3

		East-bound	West-bound	North-bound	South-bound
Phase Plan Selection from Lane Volume Worksheet					
Critical through-RT vol: [19]		480	494	444	530
LT lane volume: [5]		N/A	N/A	N/A	N/A
Left-turn protection: (P/U/N)		U	U	U	U
Dominant left turn: (indicate by '★')					★
Selection criteria based on the specified	Plan 1:	U	U	U	U
left turn protection	Plan 2a:	U	P	U	P
	Plan 2b:	P	U	P	U
	Plan 3a:	★P	P	★P	P
	Plan 3b:	P	★P	P	★P
	Plan 4:	N	N	N	N
		1		1	

Min. cycle (C_{min}): 60
Max. cycle (C_{max}): 120

Timing Plan	Value	East-West			North-South		
		Ph 1	*Ph 2*	*Ph 3*	*Ph 1*	*Ph 2*	*Ph 3*
Movement codes		EWG			NSG		
Critical phase volume [CV]		494	0	0	530	0	0
Critical sum [CS]	1024						
CBD adjustment [CBD]	1						
Reference sum [RS]	1624						
Lost time/phase [PL]		3	0	0	3	0	0
Lost time/cycle [TL]	6						
Cycle length [CYC]	60						
Green time		29.1	0	0	30.9	0	0
Critical *v/c* ratio [Xcm]	0.60						
Status	Under capacity						

★Indicates the dominant left turn for each opposing pair.

P A R T   3

Transportation Planning

The process of transportation planning involves the elements of situation and problem definition, search for solutions and performance analysis, as well as evaluation and choice of project. The process is useful for describing the effects of a proposed transportation alternative and for explaining the benefits to the traveler of a new transportation system and its impacts on the community. The highway and traffic engineer is responsible for developing forecasts of travel demand, conducting evaluations based on economic and noneconomic factors, and identifying alternatives for short-, medium-, and long-range purposes.

CHAPTER 11

The Transportation Planning Process

This chapter explains how decisions to build transportation facilities are made and highlights the major elements of the process. Transportation planning has become institutionalized; federal guidelines, regulations, and requirements for local planning are often driving forces behind existing planning methods.

The formation of the nation's transportation system has been evolutionary, not the result of a grand plan. The system now in place is the product of many individual decisions to build or improve its various parts, such as bridges, highways, tunnels, harbors, railway stations, and airport runways. Most of these transportation facilities were selected for construction or improvement because those involved concluded that the project would result in overall improvement.

Among the factors believed to justify a transportation project are improvements in traffic flow and safety, savings in energy consumption and travel time, economic growth, and increased accessibility. Some transportation projects, however, may be selected for other reasons—for example, to stimulate employment in a particular region, to compete with other cities or states for prestige, to attract industry, to respond to pressures from a political constituency, or to gain personal benefit from a particular route location or construction project. In some instances, transportation projects are *not* selected for construction because of opposition from those who would be adversely affected. For example, a new highway may require the taking of residential property, or the construction of an airport may introduce undesirable noise due to low-flying planes. Whatever the reason for selecting or rejecting a transportation project, a specific process led to the conclusion to build or not to build.

The process for planning transportation systems is a rational one that intends to furnish unbiased information about the effects that the proposed transportation project will have on the community and on its expected users. For example, if noise or air pollution is a concern, the process will examine and estimate how much additional noise or air pollution will occur if the transportation facility is built. Usually cost is a major factor, and so the process will include estimates of the construction, maintenance, and operating costs.

The process must be flexible enough to be applicable to any transportation project or system, because the kinds of problems that transportation engineers work on will vary over time. Transportation has undergone considerable change in emphasis over a 200-year period; such modes as canals, railroads, highways, air, and public transit have each been dominant at one time or another. Thus, the activities of transportation engineers have varied considerably during this period, depending on society's needs and concerns. Examples of societal concerns include energy conservation, traffic congestion, environmental impacts, safety, security, efficiency, productivity, and community preservation.

The transportation planning process is not intended to furnish a decision or to give a single result that must be followed, although it can do so in relatively simple situations. Rather, the process is intended to give the appropriate information to those who will be responsible for deciding whether the transportation project should go forward.

BASIC ELEMENTS OF TRANSPORTATION PLANNING

The transportation planning process comprises seven basic elements, which are interrelated and not necessarily carried out sequentially. The information acquired in one phase of the process may be helpful in some earlier or later phase, so there is a continuity of effort that finally results in a decision. The elements in the process are:

- Situation definition
- Problem definition
- Search for solutions
- Analysis of performance
- Evaluation of alternatives
- Choice of project
- Specification and construction

These elements are described and illustrated in Figure 11.1, using a scenario involving the feasibility of constructing a new bridge.

Situation Definition

The first step in the planning process is *situation definition*, which involves all of the activities required to understand the situation that gave rise to the perceived need for a transportation improvement. In this phase, the basic factors that created the present situation are described, and the scope of the system to be studied is delineated. The present system is analyzed and its characteristics are described. Information about the surrounding area, its people, and their travel habits may be obtained. Previous reports and studies that may be relevant to the present situation are reviewed and summarized. Both the scope of the study and the domain of the system to be investigated are delineated.

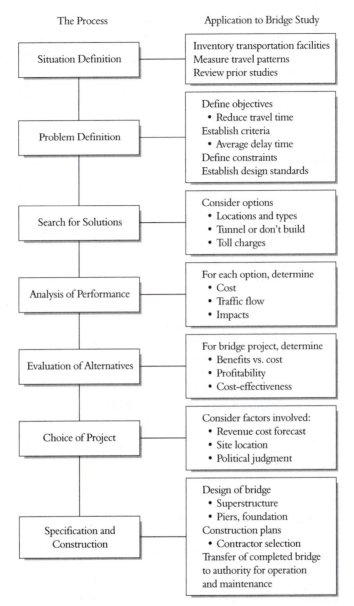

Figure 11.1 Basic Elements in the Transportation Planning Process Applied to Consider the Feasibility of a New Bridge

In the example, in which a new bridge is being considered, situation definition involves developing a description of the present highway and transportation services in the region; measuring present travel patterns and highway traffic volumes; reviewing prior studies, geological maps, and soil conditions; and delineating the scope of the study and the area affected. A public hearing might also be held to obtain citizen input. The situation will then be described in an environmental impact statement (EIS), a report that documents all alternatives and their impacts.

Problem Definition

The purpose of this step is to describe the problem in terms of the objectives to be accomplished by the project and to translate those objectives into criteria that can be quantified. Objectives are statements of purpose, such as to reduce traffic congestion; to improve safety; and to maximize net highway-user benefits. Criteria are the measures of effectiveness that can be used to measure how effective a proposed transportation project will be in meeting the stated objectives. For example, the objective "to reduce traffic congestion" might use "travel time" as the measure of effectiveness. The characteristics of an acceptable system should be identified, and specific limitations and requirements should be noted. Also, any pertinent standards and restrictions that the proposed transportation project must conform to should be understood.

An objective for the bridge project might be to reduce travel congestion on other roads or to reduce travel time between certain areas. The criterion used to measure how well these objectives are achieved is average delay or average travel time. Constraints placed on the project might be physical limitations, such as the presence of other structures, topography, or historic buildings. Design standards for bridge width, clearances, loadings, and capacity should also be noted.

Search for Solutions

In this phase of the planning process, consideration is given to a variety of ideas, designs, locations, and system configurations that might provide solutions to the problem. This is the *brainstorming* stage, in which many options may be proposed for later testing and evaluation. The alternatives can be suggested by any group or organization. In fact, the planning study may have been originated to determine the feasibility of a particular project or idea. The transportation engineer has a variety of options available in any particular situation, and any or all may be considered in this idea-generating phase. Among the options that might be used are different types of transportation technology or vehicles, various system or network arrangements, and different methods of operation. This phase also includes preliminary feasibility studies, which might narrow the range of choices to those that appear most promising. Some data gathering, field testing, and cost estimating may be necessary at this stage to determine the practicality and financial feasibility of the alternatives being proposed.

In the case of the bridge project, a variety of options may be considered, including different locations and bridge types. The study should also include the option of not building the bridge and might also consider what other alternatives are available, such as a tunnel. Operating policies should be considered, including various toll charges and methods of collection.

Analysis of Performance

The purpose of *performance analysis* is to estimate how each of the proposed alternatives would perform under present and future conditions. The criteria identified in the previous steps are calculated for each transportation option. Included in this step is a determination of the investment cost of building the transportation project, as well as annual costs for maintenance and operation. This element also involves the use of mathematical models for estimating travel demand. The number of persons or vehicles that will use the system is determined, and these results, expressed in vehicles or persons/hour, serve as the basis for project design. Other information about the use of the system, such as trip length, travel by time of day, and vehicle occupancy, are also determined and used in calculating user benefits for various criteria or measures of effectiveness. Environmental effects of the transportation project, such as noise and air pollution levels and acres of land required, are estimated. These nonuser impacts are calculated in situations where the transportation project could have significant impacts on the community or as required by law.

This task is sometimes referred to as the *transportation planning process*, but it is really the systems analysis process that integrates system supply on a network with travel demand forecasts to show equilibrium travel flows. The forecasting-model system and related network simulation are discussed in Chapter 12.

To analyze the performance of the new bridge project, first prepare preliminary cost estimates for each location being considered. Then compute estimates of the traffic that would use the bridge, given various toll levels and bridge widths. The average trip length and average travel time for bridge users would be determined and compared with existing or no-build conditions. Other impacts, such as land required, visual effects, noise levels, and air or water quality changes, would also be computed.

Evaluation of Alternatives

The purpose of the evaluation phase is to determine how well each alternative will achieve the objectives of the project as defined by the criteria. The performance data produced in the analysis phase are used to compute the benefits and costs that will result if the project is selected. In cases where the results cannot be reduced to a single monetary value, a weighted ranking for each alternative might be produced and compared with other proposed projects. For those effects that can be described in monetary terms, the benefit-cost ratio (described in Chapter 13) for each project is calculated to show the extent to which the project would be a sound investment. Other economic tests might also be applied, including the net present worth of benefits and costs. In complex situations where there are many criteria, expressed in both monetary and nonmonetary terms, the results might simply be shown in a cost-effectiveness matrix (for example, cost versus number of homes displaced) that allows the person or group making the decision to understand how each alternative performs for each of the criteria and at what cost. The results can be plotted to provide a visual comparison of each alternative and its performance.

In the evaluation of the bridge project, first determine the benefits and costs and compute the benefit-cost ratio. If the result is positive, the evaluation of alternative sites requires additional comparison of factors, both for engineering and economic feasibility and for environmental impact. A cost-effectiveness matrix that compares the cost of each alternative with its effectiveness in achieving certain goals will further assist in the evaluation.

Choice of Project

The final project selection is made after considering all the factors involved. In a simple situation, for example, where the project has been authorized and is in the design phase, a single criterion (such as cost) might be used and the chosen project would be the one with the lowest cost. With a more complex project, however, many factors have to be considered, and selection is based on how the results are perceived by those involved in decision making. If the project involves the community, it may be necessary to hold additional public hearings. A bond issue or referendum may be required. Perhaps none of the alternatives will meet the criteria or standards, and additional investigations will be necessary. The transportation engineer who furnishes a recommendation may have developed a strong opinion as to which alternative to select. If the engineer is not careful, such bias could result in the early elimination of promising alternatives or the presentation to decision makers of inferior projects. If the engineer is acting professionally and ethically, he or she will perform the task such that the information necessary to make an informed choice is available and that every feasible alternative is considered.

In deciding whether or not to build the proposed bridge, decision makers would look carefully at the revenue-cost forecasts and would probably select the project that appeared to be most financially sound. The site location would be selected based on a careful study of the factors involved. The information gathered in the earlier phases would be used, together with engineering judgment and political considerations, to arrive at a final project selection.

Specification and Construction

Once the transportation project has been selected, a detailed design phase is begun, in which each of the components of the facility is specified. For a transportation facility, this involves its physical location, geometric dimensions, and structural configuration. Design plans are produced that can be used by contractors to estimate the cost of building the project. When a construction firm is selected, these plans will be the basis on which the project will be built.

For the bridge project, once a decision to proceed has been made, a design is produced that includes the type of superstructure, piers and foundations, roadway widths and approach treatment, as well as appurtenances such as tollbooths, traffic signals, and lighting. These plans are made available to contractors, who prepare bids for the construction of the bridge. If a bid does not exceed the amount of funds available and the contractor is acceptable to the client, the project proceeds to the construction phase. Upon completion, the new bridge is turned over to the local transportation authority for operation and maintenance.

Example 11.1   Planning the Relocation of a Rural Road

To illustrate the transportation planning process a situation that involves a rural road relocation project is described. Each of the activities that are part of the project is discussed in terms of the seven-step planning process described previously. This project includes both a traffic analysis and an environmental assessment, and is typical of those conducted by transportation consultants or metropolitan transportation organizations. (This example is based on a study completed by the engineering firm, Edwards and Kelsey.)

Step 1.   **Situation definition.** The project is a proposed relocation or reconstruction of 3.3 mi of US 1A located in the coastal town of Harrington, Maine. The town center, a focal point of the project, is located near the intersection of highways US 1 and US 1A, on the banks of the Harrington River, an estuary of the Gulf of Maine. (See Figure 11.2.) The town of Harrington has 553 residents, of whom 420 live within the study area and 350 live in the town center. The population has been declining in recent years; many young people have left because of the lack of employment opportunities. Most of the town's industry consists of agriculture or fishing, so a realignment of the road that damages the environment would also affect the town's livelihood. There are 10 business establishments within the study area; 20 percent of the town's retail sales are tourism-related. The average daily traffic is 2620 vehicles per day, of which 69 percent represent through traffic and 31 percent represent local traffic.

Step 2.   **Problem definition.** The Maine Department of Transportation wishes to improve US 1A, primarily to reduce the high accident rate on this road in the vicinity of the town center. The problem is caused by a narrow bridge that carries the traffic on US 1A into the town center, the poor horizontal and vertical alignment of the road within the town center, and a dangerous intersection where US 1A and US 1 meet. The accident rate on US 1A in the vicinity of the town center is four times the statewide

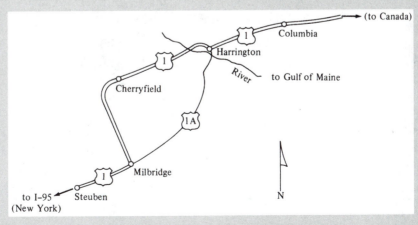

Figure 11.2 Location Map for Highways U.S. 1 and U.S. 1A

rate. A secondary purpose of the proposed relocation is to improve the level of service for through traffic by increasing the average speed on the relocated highway.

The measures of effectiveness for the project will be the accident rate, travel time, and construction cost. Other aspects that will be considered are the effects that each alternative would have on a number of businesses and residences that would be displaced, the changes in noise levels and air quality, and the changes in natural ecology. The criteria that will be used to measure these effects will be the number of businesses and homes displaced, noise levels and air quality, and the acreage of salt marsh and trees affected.

Step 3. **Search for solutions.** The Department of Transportation has identified four alternative routes, as illustrated in Figure 11.3, in addition to the present route—alternative 0—referred to as the null or "do-nothing" alternative. All routes begin at the same location—3 mi southwest of the center of Harrington—and end at a common point northeast of the town center. The alternatives are as follows:

- *Alternative 1:* This road bypasses the town to the south on a new location across the Harrington River. The road would have two lanes, each 12-ft wide, and 8-ft shoulders. A new bridge would be constructed about 1/2 mi downstream from the old bridge.
- *Alternative 2:* This alternative would use the existing US 1A into town, but with improvements in the horizontal and vertical alignment throughout its length and the construction of a new bridge. The geometric specifications would be the same as for alternative 1.

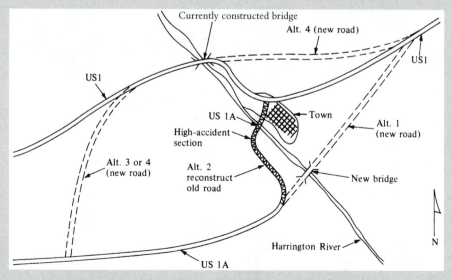

Figure 11.3 Alternative Routes for Highway Relocation

- *Alternative 3:*   This new road would merge with US 1 west of Harrington, and then continue through town. It would use the Route 1 bridge, which was recently constructed. Geometric specifications are the same as those for the other alternatives.
- *Alternative 4:*   This road would merge with US 1 and use the Route 1 bridge, as in alternative 3. However, it would bypass the town center on a new alignment.

Step 4.   **Analysis of performance.** The measures of effectiveness are calculated for each of the alternatives. The results of these calculations are shown in Table 11.1 for alternatives 1 through 4 and for the null alternative. The relative ranking of each alternative is presented in Table 11.2. For example, the average speed on the existing road is 25 mi/h, whereas for alternatives 1 and 4, the speed is 55 mi/h, and for alternatives 2 and 3, the speed is 30 mi/h. Similarly, the accident rate, which is now four times the statewide average, would be reduced to 0.6 for alternative 4, and 1.2 for alternative 1. The project cost ranges from $1.18 million for alternative 3 to $1.58 million for alternative 2. Other items that are calculated include the number of residences displaced, the volume of traffic within the town both now and in the future, air quality, noise, lost taxes, and acreage of trees removed.

Step 5.   **Evaluation of alternatives.** Each of the alternatives is compared with the others to assess the improvements that would occur based on a given

Table 11.1   Measures of Effectiveness for Rural Road Alternatives

	Alternatives				
Criteria	*0*	*1*	*2*	*3*	*4*
Speed (mi/h)	25	55	30	30	55
Distance (mi)	3.7	3.2	3.8	3.8	3.7
Travel time (min)	8.9	3.5	7.6	7.6	4.0
Accident factor[a]	4	1.2	3.5	2.5	0.6
Construction cost ($ million)	0	1.50	1.58	1.18	1.54
Residences displaced	0	0	7	3	0
City traffic					
Present	2620	1400	2620	2520	1250
Future (20 years)	4350	2325	4350	4180	2075
Air quality ($\mu g/m^3$ CO)	825	306	825	536	386
Noise (dBA)	73	70	73	73	70
Tax loss	None	Slight	High	Moderate	Slight
Trees removed (acres)	None	Slight	Slight	25	28
Runoff	None	Some	Some	Much	Much

[a] Relative to statewide average for this type of facility.

Table 11.2   Ranking of Alternatives

Criterion/Alternative	Alternatives				
	0	1	2	3	4
Travel time	4	1	3	3	2
Accident factor[a]	5	2	4	3	1
Cost ($ millions)	1	3	5	2	4
Residences displaced	1	1	3	2	1
Air quality	4	1	4	3	2
Noise	2	1	2	2	1
Tax loss	1	2	4	3	2
Trees removed (acres)	1	2	2	3	4
Increased runoff	1	2	2	3	3

Note: 1 = highest; 5 = lowest.
[a] Relative to statewide average for this type of facility.

criterion. In this example, we consider the following measures of effectiveness and their relationship to project cost:

- *Travel time:*   Every alternative improves the travel time. The best is alternative 1, followed by alternative 4. Alternatives 2 and 3 are equal, but neither reduces travel time significantly (Figure 11.4).
- *Accident factor:*   The best accident record will occur with alternative 4, followed by alternatives 1, 3, and 2 (Figure 11.5).
- *Cost:*   The least costly alternative is simply to do nothing, but the dramatic potential improvements in travel time and safety would indicate that the proposed project should probably be undertaken. Alternative 3 is lowest in cost at $1.18 million. Alternative 2 is highest in cost, would not be as safe as alternative 3, and would produce the same travel time. Thus, alternative 2 would be eliminated. Alternative 1 would cost $0.32 million more than alternative 3, but would reduce the accident factor by 1.3 and travel time by 4.1 min. Alternative 4 would cost $0.04 million more than alternative 1 and would increase travel time, but would decrease the accident factor. These cost-effectiveness values are shown in Figures 11.4 and 11.5. They indicate that alternatives 1 and 4 are both more attractive than alternatives 2 and 3 because the former would produce significant improvements in travel time and accidents.

The Department of Transportation could also make a cost-benefit analysis that compares the increased cost for each alternative with the economic benefits, including travel-time savings, accident reduction, and a reduction in maintenance. Procedures for economic evaluation are described in Chapter 13.

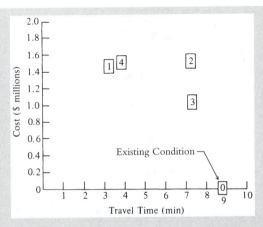

Figure 11.4 Travel Time Between West Harrington and U.S. 1 Versus Cost

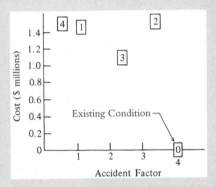

Figure 11.5 Accident Factor (relative to statewide average) Versus Cost

- *Displaced residences:* Three residences would be displaced if alternative 3 were selected; seven residences would be displaced if alternative 2 were selected. No residences would have to be removed if alternative 1 or 4 were selected.
- *Air quality in the town:* Alternative 1 would produce the highest air quality, followed by alternatives 4, 3, and 2. The air quality improvement would result from removing a significant amount of the slow-moving through traffic from the center of the city to a high-speed road where most of the pollution would be dispersed.
- *Noise levels:* Noise levels are lower for alternatives 1 and 4.
- *Tax losses:* Tax losses would be slight for alternatives 1 and 4, moderate for alternative 3, and high for alternative 2.
- *Trees removed:* Alternatives 3 and 4 would eliminate 25 and 28 acres of trees, respectively. Alternative 1 would result in slight losses; alternative 2, no loss.

- *Runoff:* There would be no runoff for alternative 0, some for alternatives 1 and 2, and a considerable amount for alternatives 3 and 4.

Step 6:   **Choice of project.** From strictly a cost point of view, the Department of Transportation would select alternative 3 since it results in travel time and safety improvements at the lowest cost. However, if additional funds are available, then alterative 1 or 4 would be considered. Since alternative 1 is lower in cost than alternative 4 and is equal or better than alternative 3 for each criterion related to community impacts, this alternative would be the one most likely to be selected. In the selection process, each alternative would be reviewed. Also, comments would be received from citizens and elected officials to assist in the design process so that environmental and community effects would be minimized.

Step 7:   **Specifications and construction.** The choice has been made, and alternative 1, a bypass south of Harrington, has been ranked of sufficiently high priority so that it will be constructed. This alternative involves building both a new bridge across the Harrington River and a new road connecting US 1A with US 1. The designs for the bridge and road will be prepared. Detailed estimates of the cost to construct will be made, and the project will be announced for bid. The construction company that produces the lowest bid and can meet other qualifications will be awarded the contract, and the road will be built. Upon completion, the road will be turned over to the Department of Transportation, who will be responsible for its maintenance and operation. Follow-up studies will be conducted to determine how successful the road was in meeting its objectives; where necessary, modifications will be made to improve its performance.

INSTITUTIONALIZATION OF TRANSPORTATION PLANNING

This transportation planning process is based on the systems approach to problem solving and is quite general in its structure. The process can be applied to many cases for transportation decision making, such as intercity high-speed rail feasibility studies, airport location, port and harbor development, and urban transportation systems. The most common application is in urban areas, where it has been mandated by law since 1962, when the Federal Aid Highway Act required that all transportation projects in urbanized areas with populations of 50,000 or more be based on a transportation planning process that was continuous, comprehensive, and coordinated with other modes.

Because the urban transportation planning process provides an institutionalized and formalized planning structure, it is important to identify the environment in which the transportation planner works. The forecasting modeling process that has evolved is presented in Chapter 12 to provide an illustration of the methodology. The planning process used at other problem levels is a variation of this basic approach.

Transportation Planning Organization

In carrying out the urban transportation planning process, several committees are created, that represent various community interests and viewpoints. These committees are the policy committee, the technical committee, and the citizens' advisory committee.

Policy Committee

The policy committee is composed of elected or appointed officials, such as the mayor and director of public works, who represent the governing bodies or agencies that will be affected by the results. This committee makes the basic policy decisions and acts as a board of directors for the study. They will decide on management aspects of the study as well as key issues of financial and political nature.

Technical Committee

The technical committee is composed of the engineering and planning staffs who are responsible for carrying out the work and evaluating the technical aspects of the project. This group will make the necessary evaluations and cost comparisons for each project alternative, and will supervise the technical details of the entire process. Typically, the technical committee will include highway, transit, and traffic engineers, as well as other specialists in land-use planning, economics, and computers.

Citizens' Advisory Committee

The citizens' advisory committee is composed of a cross section of the community and may include representatives from labor, business, and the League of Women Voters, as well as interested citizens and members of other community interest groups. This committee's function is to express community goals and objectives, to suggest alternatives, and to react to proposed alternatives. Through this committee structure, an open dialogue is produced between the policy makers, the technical staff, and the community. It is expected that when a selection is made and recommendations are produced by the study, a consensus of all interested parties has been reached. Although this is not always possible, the role of a citizens' advisory committee should be to increase communications and, it is hoped, result in plans that reflect community interests.

URBAN TRANSPORTATION PLANNING

Urban transportation planning involves the evaluation and selection of highway or transit facilities to serve present and future land uses. For example, the construction of a new shopping center, airport, or convention center will require additional transportation services. Also, new residential development, office space, and industrial parks will generate additional traffic, requiring the creation or expansion of roads and transit services.

The process must also consider other proposed developments and improvements that will occur within the planning period. The urban transportation planning process has been enhanced through the efforts of the Federal Highway Administration and the Federal

Transit Administration of the U.S. Department of Transportation by the preparation of manuals and computer programs that assist in organizing data and forecasting travel flows.

Urban transportation planning is concerned with two separate time horizons. The first is a short-term emphasis intended to select projects that can be implemented within a one- to three-year period. These projects are designed to provide better management of existing facilities by making them as efficient as possible. The second time horizon deals with the long-range transportation needs of an area and identifies the projects to be constructed over a 20-yr period.

Short-term projects involve programs such as traffic signal timing to improve flow, car and van pooling to reduce congestion, park-and-ride fringe parking lots to increase transit ridership, and transit improvements.

Long-term projects involve programs such as adding new highway elements, additional bus lines or freeway lanes, rapid transit systems and extensions, or access roads to airports or shopping malls.

The urban transportation planning process can be carried out in terms of the procedures outlined previously and is usually described as follows. Figure 11.6 illustrates the comprehensive urban area transportation planning process.

Inventory of Existing Travel and Facilities

This is the data-gathering activity in which urban travel characteristics are described for each defined geographic unit or traffic zone within the study area. Inventories and surveys are made to determine traffic volumes, land uses, origins and destinations of travelers, population, employment, and economic activity. Inventories are made of existing transportation facilities, both highway and transit. Capacity, speed, travel time, and traffic volume are determined. The information gathered is summarized by geographic areas called *traffic zones* and for the existing highway and transit system.

The size of the zone will depend on the nature of the transportation study, and it is important that the number of zones be adequate for the type of problem being investigated. Often census tracts or census enumeration districts are used as traffic zones because population data are easily available by this geographic designation.

Establishment of Goals and Objectives

The urban transportation study is carried out to develop a program of highway and transit projects that should be completed in the future. Thus, a statement of goals, objectives, and standards is prepared that identifies deficiencies in the existing system, desired improvements, and what is to be achieved by the transportation improvements.

For example, if a transit authority is considering the possibility of extending an existing rail line into a newly developed area of the city, its objectives for the new service might be to maximize its revenue from operations, maximize ridership, promote development, and attract the largest number of auto users so as to relieve traffic congestion.

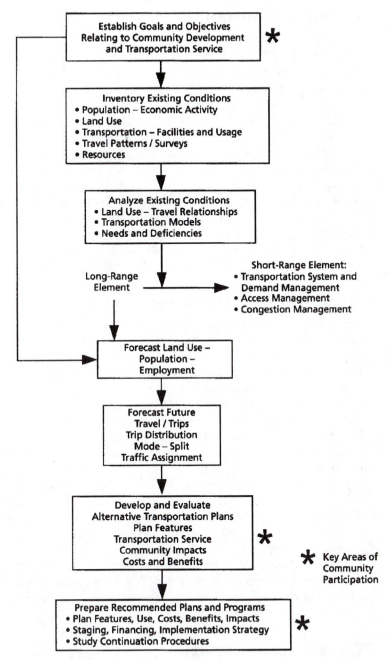

Figure 11.6 Comprehensive Urban Area Transportation Planning Process

SOURCE: Redrawn from *Transportation Planning Handbook,* Institute of Transportation Engineers, 2nd Edition, Prentice-Hall, 1999.

Generation of Alternatives

In this phase of the urban transportation planning process, the alternatives to be analyzed will be identified. It may also be necessary to analyze the travel effects of different land-use plans and to consider various lifestyle scenarios. The transportation options available to the urban transportation planner include various technologies, network configurations, vehicles, operating policies, and organizational arrangements.

In the case of a transit line extension, the technologies could be rail rapid transit or bus. The network configuration could be defined by a single line, two branches, or a geometric configuration such as a radial or grid pattern. The guideway, which represents a homogeneous section of the transportation system, could be varied in length, speed, waiting time, capacity, and direction. The intersections, which represent the end points of the guideway, could be a transit station or the line terminus. The vehicles could be singly driven buses or multicar trains. The operating policy could involve 10-min headways during peak hours and 30-min headways during off-peak hours, or other combinations. The organizational arrangements could be private or public. These and other alternatives would be considered in this phase of the planning process.

Estimation of Project Cost and Travel Demand

This activity in the urban transportation planning process involves two separate tasks. The first is to determine the project cost and the second is to estimate the amount of traffic expected in the future. The estimation of facility cost is relatively straightforward, whereas the estimation of future traffic flows is a complex undertaking requiring the use of mathematical models and computers.

Future travel is determined by forecasting future land use in terms of the economic activity and population that the land use in each traffic zone will produce. With the land-use forecasts established in terms of number of jobs, residents, auto ownership, income, and so forth, the traffic that this land use will produce can be determined. This is carried out in a four-step process that includes the determination of the number of trips generated, the origin and destination of trips, the mode of transportation used by each trip (for example, auto, bus, rail), and the route taken by each trip. The urban traffic forecasting process thus involves four distinct activities: *trip generation, trip distribution, modal split,* and *network assignment,* as illustrated in Figure 11.7. This forecasting process is described in Chapter 12.

When the travel forecasting process is completed, the highway and transit volumes on each link of the system will be estimated. The actual amount of traffic is not known until it occurs. These results can be compared with the present capacity of the system to determine the operating level of service.

Evaluation of Alternatives

This phase of the process is similar in concept to what was described earlier but can be complex in practice because of the conflicting objectives and diverse groups that will be affected by an urban transportation project.

Among the groups that could be affected are the traveling public (user), the highway or transit agencies (operator), and the nontraveling public (community). Each of these

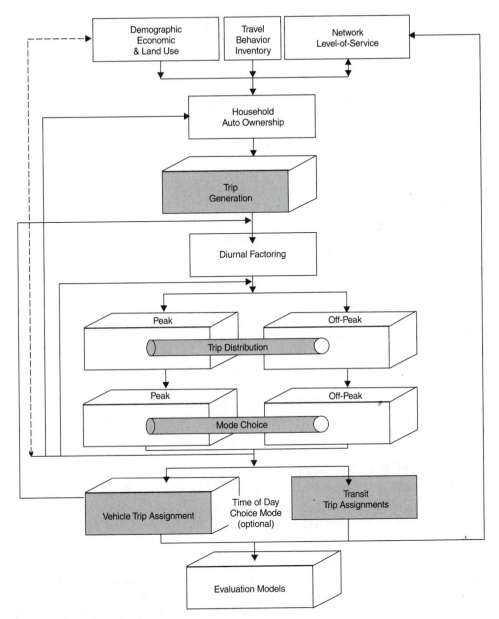

Figure 11.7 Travel Demand Model Flowchart

SOURCE: *Transportation Planning Handbook,* Institute of Transportation Engineers, 2nd Edition, Prentice-Hall, 1999.
ITE SOURCE: Parsons, Brinkerhoff, Quade and Douglas.

groups will have different objectives and viewpoints concerning how well the system performs. The traveling public wants to improve speed, safety, and comfort; the transportation agency wishes to minimize cost; and the community wants to preserve its lifestyle and minimize any adverse impacts.

The purpose of the evaluation process is to identify feasible alternatives in terms of cost and traffic capacity, to estimate the effects of each alternative in terms of the objectives expressed, and to assist in identifying those alternatives that will serve the traveling public and be acceptable to the community.

Choice of Project

Selection of a project will be based on a process that will ultimately involve elected officials and the public. Quite often funds to build an urban transportation project, such as a subway system, may involve a public referendum. In other cases, a vote by a state legislature may be required before funds are committed. A multiyear program will then be produced that outlines the projects to be carried out over the next 20 years. With approval in hand, the project can proceed to the specification and construction phase.

FORECASTING TRAVEL

To accomplish the objectives and tasks of the urban transportation planning process, a technical effort referred to as the *urban transportation forecasting process* is carried out to analyze the performance of various alternatives. There are four basic elements and related tasks in the process. These are (1) data collection (or inventories), (2) analysis of existing conditions and calibration of forecasting techniques, (3) forecast of future travel demand, and (4) analysis of the results. These elements and related tasks are described in the following sections.

Defining the Study Area

Prior to collecting and summarizing the data, it is usually necessary to delineate the study area boundaries and to further subdivide the area into traffic zones for data tabulation and analysis. The selection of these zones is based on the following criteria:

1. Socioeconomic characteristics should be homogeneous.
2. Intrazonal trips should be minimized.
3. Physical, political, and historical boundaries should be utilized where possible.
4. Zones should not be created within other zones.
5. The zone system should generate and attract approximately equal trips, households, population, or area.
6. Zones should use census tract boundaries where possible.

An illustration of analysis zones for a transportation study is shown in Figure 11.8.

Data Collection

The data collection phase provides information about the city and its people that will serve as the basis for developing travel demand estimates. The data include information about

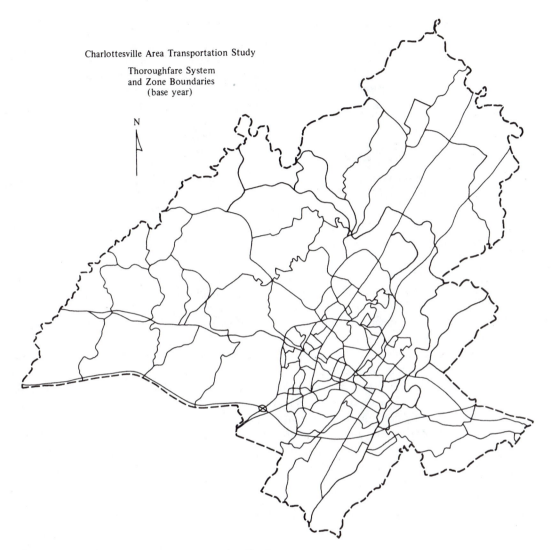

Charlottesville Area Transportation Study

Thoroughfare System
and Zone Boundaries
(base year)

N

Figure 11.8 Analysis Zones for Transportation Study

SOURCE: Modified from the Virginia Department of Transportation.

economic activity (employment, sales volume, income, etc.), land use (type, intensity), travel characteristics (trip and traveler profile), and transportation facilities (capacity, travel speed, etc.). This phase may involve surveys and can be based on previously collected data.

Population and Economic Data

Once a zone system for the study area is established, population and socioeconomic forecasts prepared at a regional or statewide level are used. These are allocated to the study area, and then the totals are distributed to each zone. This process can be accomplished by using either a ratio technique or small-area land-use allocation models.

The population and economic data usually will be furnished by the agencies responsible for planning and economic development, whereas providing travel and transportation data is the responsibility of the traffic engineer. For this reason, we will focus our attention on the data required to describe travel characteristics and the transportation system.

Transportation Inventories

Transportation system inventories involve a description of the existing transportation services; the available facilities and their condition; location of routes and schedules; maintenance and operating costs; system capacity and existing traffic; volumes, speed, and delay; and property and equipment. The types of data collected about the current system will depend on the specifics of the problem.

For a highway planning study, the system would be classified functionally into categories that reflect their principal use. These are the major arterial system, minor arterials, collector roads, and local service (see Chapter 16). Physical features of the road system would include number of lanes, pavement and approach width, traffic signals, and traffic control devices. Street and highway capacity would be determined, including capacity of intersections. Traffic volume data would be determined for intersections and highway links. Travel times along the arterial highway system would also be determined.

A computerized network of the existing street and highway system is produced. The network consists of a series of links, nodes, and centroids. A *link* is a portion of the highway system that can be described by its capacity, lane width, and speed. A *node* is the end point of a link and represents an intersection or location where a link changes direction, capacity, width, or speed. A *centroid* is the location within a zone where trips are considered to begin and end. Coding of the network requires information from the highway inventory in terms of link speeds, length, and capacities. The network is then coded to locate zone centroids, nodes, and the street system. A portion of a link-node map is illustrated in Figure 11.9.

For a transit planning study, the inventory includes present routes and schedules, including headways, transfer points, location of bus stops, terminals, and parking facilities. Information about the bus fleet, such as its number, size, and age, would be identified. Maintenance facilities and maintenance schedules would be determined, as would the organization and financial condition of the transit companies furnishing service in the area. Other data would include revenue and operating expenses.

The transportation facility inventories provide the basis for establishing the networks that will be studied to determine present and future traffic flows. Data needs can include the following items:

- Public streets and highways
    —Rights of way
    —Roadway and shoulder widths
    —Locations of curbed sections
    —Locations of structures such as bridges, overpasses, underpasses, and major culverts
    —Overhead structure clearances

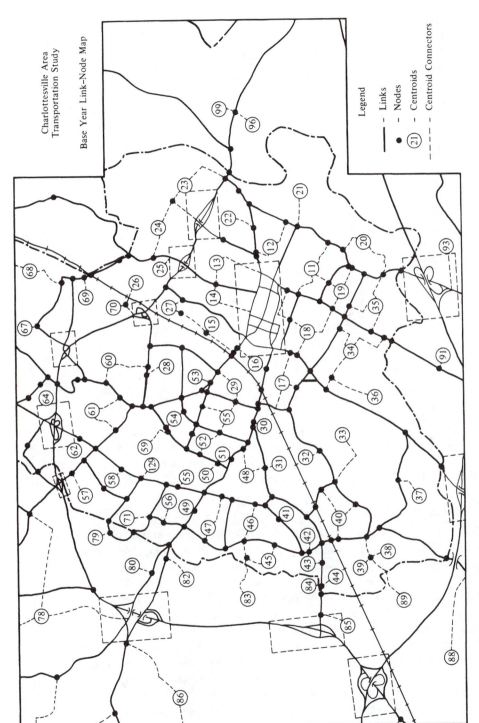

Figure 11.9 Link-Node Map for Highway System

SOURCE: Virginia Department of Transportation.

—Railroad crossings
—Location of critical curves or grades
—Identification of routes by governmental unit having maintenance jurisdiction
—Functional classification
—Street lighting
• Land use and zoning controls
• Traffic generators
—Schools
—Parks
—Stadiums
—Shopping centers
—Office complexes
• Laws, ordinances, and regulations
• Traffic control devices
—Traffic signs
—Signals
—Pavement markings
• Transit system
—Routes by street
—Locations and lengths of stops and bus layover spaces
—Location of off-street terminals
—Change of mode facilities
• Parking facilities
• Traffic volumes
• Travel times
• Intersection and roadway capacities

In many instances, the data will already have been collected and is available in the files of city, county, or state offices. In other instances, some data may be more essential than others. A careful evaluation of the data needs should be undertaken prior to the study.

Travel Surveys

Travel surveys are obtained to establish a complete understanding of the travel patterns within the study area. For single projects, such as a highway project, it may be sufficient to use traffic counts on existing roads or, for transit, counts of passengers riding the present system. However, to understand why people travel and where they wish to go, origin-destination (O-D) survey data can be useful. The O-D survey asks questions about each trip that is made on a specific day, such as where the trip begins and ends, the purpose of the trip, the time of day, and the vehicle involved (auto or transit), and about the person making the trip: age, sex, income, vehicle owner, and so on. The O-D survey may be completed as a home interview, or people may be asked questions while riding the bus or when stopped at a roadside interview station. Sometimes the information is requested by telephone or by return postcard. O-D surveys are rarely made in communities where these data have been collected previously. Due to the high cost of O-D surveys, prior data are updated and U.S. Census travel-to-work data are used. Figure 11.10 illustrates a home interview origin-destination survey form.

Travel Day ________ and Date

Sample Number

NORTH CENTRAL TEXAS
COUNCIL OF GOVERNMENTS
HOME INTERVIEW SURVEY

Section I: Household Data

A. Sample Address ________ (House Number, Street Name, Apt. No.) ________ (City/Town) ________ (County) ________ (Zip Code)

B. Structure Type ________

C. Number of People Living at this Address ________

D. Number of People Age 5 and Over Living at this Address ________

E. Number of Out-of-Area Visitors Staying at this Address ________

F. Number of Passenger Cars, Vans, and Pickups Available for Use ________

G. Household Income: (Do Not Ask Until Interview Is Complete) ________

Section II: Data on Persons Age 5 and Over

A	B	C	D	E	F	G	H	I	J	K
Person Number	✓ If Interviewed	Relation To Head	Age	Sex	Licensed to Drive?	Occupation	Industry	Worked on Travel Day?	Made Trips While at Work?	Made Other Trips on Travel Day?
01		Head 1		1 M 2 F	1 YES 2 NO			1 YES 2 NO 3 Worked at Home	1 YES 2 NO	1 YES 2 NO
02				1 M 2 F	1 YES 2 NO			1 YES 2 NO 3 Worked at Home	1 YES 2 NO	1 YES 2 NO
03				1 M 2 F	1 YES 2 NO			1 YES 2 NO 3 Worked at Home	1 YES 2 NO	1 YES 2 NO
04				1 M 2 F	1 YES 2 NO			1 YES 2 NO 3 Worked at Home	1 YES 2 NO	1 YES 2 NO
05				1 M 2 F	1 YES 2 NO			1 YES 2 NO 3 Worked at Home	1 YES 2 NO	1 YES 2 NO
06				1 M 2 F	1 YES 2 NO			1 YES 2 NO 3 Worked at Home	1 YES 2 NO	1 YES 2 NO
07				1 M 2 F	1 YES 2 NO			1 YES 2 NO 3 Worked at Home	1 YES 2 NO	1 YES 2 NO
08				1 M 2 F	1 YES 2 NO			1 YES 2 NO 3 Worked at Home	1 YES 2 NO	1 YES 2 NO
09				1 M 2 F	1 YES 2 NO			1 YES 2 NO 3 Worked at Home	1 YES 2 NO	1 YES 2 NO
10				1 M 2 F	1 YES 2 NO			1 YES 2 NO 3 Worked at Home	1 YES 2 NO	1 YES 2 NO

Age Codes

1	5 - 10	6	36 - 45
2	11 - 15	7	46 - 55
3	16 - 20	8	56 - 65
4	21 - 25	9	65 - OVER
5	26 - 35	0	UNKNOWN

Relation Codes

1	HEAD	6	GRANDCHILD
2	SPOUSE	7	OTHER RELATIVE
3	SON	8	UNRELATED
4	DAUGHTER	9	OUT-OF-AREA VISITORS
5	GRANDPARENT	0	UNKNOWN

Section III: Trip Summary

A. Total Vehicular Trips Reported ________

B. Persons Age 5 and Over Making Trips ________

C. Persons Age 5 and Over Not Making Trips ________

D. Complete or Incomplete Interview Code ________

Section IV: Administrative

A. Household Telephone Number ________

B. Interviewer ________

C. Telephone Contacts (If Any):

Date	Time	Purpose/Outcome

D. Personal Contacts In Household:

Date	Time	Talked To/Comments

E. Completed Interview Submitted:

Date: ________ By: ________

I Certify That All Information On This Form Is Correct And True.

________ Signature of Interviewer

F. If Interview Submitted Incomplete

Interviewer's Reason: ________

________ Date ________ Initials

Supervisor's Comments ________

________ Date ________ Initials

G. First Edit: Fail Pass

________ Date ________ Initials

H. Final Edit: Fail Pass

________ Date ________ Initials

I. Coding Complete

________ Date ________ Initials

Figure 11.10 Travel Behavior Inventory: Home Interview Survey

SOURCE: North Central Texas Council of Governments.

O-D data are compared with other sources to assure the accuracy and consistency of the results. Among the tests used are crosschecks that compare the number of dwelling units or the trips per dwelling unit observed in the survey with published data. Screenline checks (see Chapter 4) can be made to compare the number of reported trips that cross a defined boundary, such as a bridge or two parts of a city, with the number actually observed. For example, the number of cars observed crossing one or more bridges might be compared with the number estimated from the surveys. It is also possible to assign trips to the existing network to compare how well the data replicate actual travel. If the screenline crossings are significantly different from those produced by the data, it is possible to make adjustments in the O-D results so that conformance with the actual conditions is assured.

Following the O-D checking procedure, a set of *trip tables* is prepared that shows the number of trips between each zone in the study area. These tables can be subdivided by trip purpose, truck trips, and taxi trips. Tables are also prepared that list the socioeconomic characteristics for each zone and the travel time between zones. Examples of trip tables are shown in Chapter 12.

Calibration

Calibration is concerned with establishing mathematical relationships that can be used to estimate future travel demand. Usually, analysis of the data will reveal the effect on travel demand of factors such as land use, socioeconomic characteristics, or transportation system factors.

Example 11.2   Estimating Trips per Day Using Multiple Regression

A multiple regression analysis shows the following relationship for the number of trips per household:

$$T = 0.82 + 1.3P + 2.1A$$

where

T = number of trips per household per day
P = number of persons per household
A = number of autos per household

If a particular zone contains 250 households with an average of 4 persons and 2 autos for each household, determine the average number of trips per day in that zone.

Solution:

   Step 1.   Calculate the number of trips per household.

$$T = 0.82 + 1.3P + 2.1A$$
$$= 0.82 + (1.3 \times 4) + (2.1 \times 2)$$
$$= 10.22 \text{ trips/household/day}$$

> **Step 2.**   Determine the number of trips in the entire zone.
>
> $$N = T \times (\text{number of households})$$
> $$= 10.22 \times 250$$
> $$= 2555 \text{ trips/day}$$

Other mathematical formulas establish the relationships for trip length, percentage of trips by auto or transit, or the particular travel route selected.

Travel forecasts are made by applying the relationships developed in the calibration process. These formulas rely upon estimates of future land use, socioeconomic characteristics, and transportation conditions.

Forecasting can be summarized in a simplified way by indicating the task that each step in the process is intended to perform. These tasks are as follows (and are described more fully in Chapter 12).

1. **Population and economic analysis** determines the magnitude and extent of activity in the urban area.
2. **Land use analysis** determines where the activities will be located.
3. **Trip generation** determines how many trips each activity will produce or attract.
4. **Trip distribution** determines the origin or destination of trips that are generated at a given activity.
5. **Modal split** determines which mode of transportation will be used to make the trip.
6. **Traffic assignment** determines which route on the transportation network will be used when making the trip.

Computers are used extensively in the urban transportation planning process. Over the years, a package of programs has been developed by the Federal Highway Administration (FHWA) and the Federal Transit Administration (FTA). This package is called the Urban Transportation Planning System (UTPS).

A microcomputer version of the UTPS strategy was developed by the FHWA. It is referred to as the Quick Response System (QRS) because it is not data intensive and uses a set of general models that are transferable. The techniques have been computerized, and various versions of the original UTPS program have been developed for microcomputers under such acronyms as MINUTP, TRANPLAN, and MICROTRIPS. The Center for Microcomputers in Transportation maintains a complete file of transportation planning software. Among these are QRSII for Windows and TRIPS, which are comprehensive transportation planning software packages.

Environmental Impact Statements

Federal and/or state regulations may require that the environmental impacts of proposed projects be assessed. These impacts may include effects on air quality, noise levels, water quality, wetlands, and the preservation of historic sites of interest. The analytical process through which these effects are identified can take one of three forms, depending on the scope of the proposed project: a full *environmental impact statement* (EIS), a simpler *environmental assessment* (EA), or a cursory checklist of requirements known as a *categorical exclusion*

(CE). Regardless of the level of detail, the purpose of this environmental review process is "to assure that all potential effects (positive as well as adverse) are addressed in a complete manner so that decision-makers can understand the consequences" of the proposed project. Once an agency, such as a state department of transportation, completes a draft EIS, the public is given an opportunity to comment, and a project cannot proceed until the revised, final EIS has been accepted by the approporiate federal and state regulatory agencies.

Elements of an EIS.   Although the entire environmental review process is beyond the scope of this text, examination of some of the common elements of an EIS illustrate its role in transportation planning.

The project's *purpose and need* articulate why the project is being undertaken: is it to improve safety, to increase capacity in response to expected future traffic growth, or is it a deficient link in a region's comprehensive transportation network? The purpose and need section should include relevant AADT projections, crash rates, and a description of existing geometric conditions.

The *alternatives* to the proposed project, such as the do-nothing case, should be described, as well as any criteria that have been used to eliminate alternatives from further consideration. For example, if a second bridge crossing over a body of water is being considered, then alternatives could be to change the location, to widen an existing bridge, to improve ferry service, or to do nothing. Criteria that prevent further consideration, such as the presence of an endangered species at what would have been another potential location or the permanent loss of several acres of wetlands, are given in this section.

The *environmental effects* of the proposed project, such as water quality (during construction and once construction is complete), soil, wetlands, and impacts on plant and animal life, especially endangered species should be analyzed. Note that environmental effects also include the impact on communities such as air quality, land use, cultural resources, and noise.

Example of Analytic Procedures Used in an EIS.   The level of detail in a full EIS can be staggering, given the amount of analysis required to answer some seemingly simple questions: What is the noise level? How much will automobile emissions increase? A number of tools are provided by regulatory agencies that can answer some of these questions. For example, the Environmental Protection Agency (EPA) has developed an emissions-based model (known as MOBILE) that may be used to evaluate emissions impacts of alternatives. The results of this model, compared with observed carbon monoxide concentrations, can be used to determine the relative impact of different project alternatives on the level of carbon monoxide.

One approach for quantifying noise impacts is to use the Federal Highway Administration's Transportation Noise Model (TNM), a computer program that forecasts noise levels as a function of traffic volumes and other factors. The method by which noise impacts are assessed can vary by regulatory agency: for example, the FHWA will permit one to use the L_{10} descriptor, which is "the percentile noise level that is exceeded for ten percent of the time." A more common noise descriptor is L_{eq}, which is the average noise intensity over time. A variant of this descriptor may be used by the U.S. Department of Housing and Urban Development (HUD), where the L_{eq} for each hour is determined but then a 10 dB "penalty" is added to the values from 10 p.m. to 7 a.m. Since noise is proportional to traffic speed, the impact of this last type of descriptor is to favor projects that would not necessarily result in high speeds in close proximity to populated areas during the evening hours. Thus it would be reasonable for an EIS to use current and forecasted volume counts (automobiles, medium trucks, and heavy trucks), speeds, and directional split in order to compare noise effects for the current conditions, future conditions with the proposed project, and future conditions assuming any other alternatives, which at a minimum should include the no-build option.

SUMMARY

Transportation projects are selected for construction based on a variety of factors and considerations. The transportation planning process is useful when it can assist decision makers and others in the community to select a course of action for improving transportation services.

The seven-step planning process is a useful guide for organizing the work necessary to develop a plan. The seven steps are (1) situation definition, (2) problem definition, (3) search for solutions, (4) analysis of performance, (5) evaluation of alternatives, (6) choice of project, and (7) specification and construction. Although the process does not produce a single answer, it assists the transportation planner or engineer in carrying out a logical procedure that will result in a solution to the problem. The process also is valuable as a means of describing the effects of each course of action and for explaining to those involved how the new transportation system will benefit the traveler and what its impacts will be on the community.

The urban transportation planning process is an institutionalized version of the basic seven-step approach to planning for any new transportation project. The elements of the urban transportation planning process are (1) inventory of existing travel and facilities, (2) establishment of goals and objectives, (3) generation of alternatives, (4) estimation of project costs and travel demand, (5) evaluation of alternatives, and (6) choice of project. An understanding of the elements of urban transportation planning is essential to place in perspective the analytical processes for estimating travel demand. It is the forecasting model system to which the next chapter is devoted.

PROBLEMS

11-1 Explain why the transportation planning process is not intended to furnish a decision or give a single result.

11-2 Describe the steps that an engineer should follow if he or she were asked to determine the need for a grade-separated railroad grade crossing that would replace an at-grade crossing of a two-lane highway with a rail line.

11-3 Describe the basic steps in the transportation planning process.

11-4 Select a current transportation problem in your community or state with which you are familiar or interested. Briefly describe the situation and the problem involved. Indicate the options available and the major impacts of each option on the community.

11-5 You have been asked to evaluate a proposal that tolls be increased on roads and bridges within your state. Describe the general planning and analysis process that you would use in carrying out this task.

11-6   You have been asked to prepare a study that will result in a decision regarding improved transportation between an airport and the city it serves. The city has a population of 500,000, but the airport also serves a region of approximately 3 million people. It is anticipated that the region will grow to approximately double its size in 15 years and that air travel will increase by 150 percent. Briefly outline the elements necessary to undertake the study, and show these in a flow diagram.

11-7   What caused transportation planning to become institutionalized in urban areas, and what does the process need to be based on?

11-8   Urban transportation planning is concerned with two separate time horizons. Briefly describe each and provide examples of the types of projects that can be categorized in each horizon.

11-9   What is the purpose of performing inventories and surveys for each defined geographic unit or traffic zone within a study area?

11-10   What are the basic four elements that make up the urban transportation forecasting process?

11-11   In the data collection phase of the urban transportation forecasting process, what type of information should the data reveal for a traffic analysis zone?

11-12   List four ways of obtaining origin-destination information. Which method would produce the most accurate results?

11-13   Define the following terms: (a) link, (b) node, (c) centroid, and (d) network.

11-14   Draw a link-node map of the streets and highways within the boundaries of your immediate neighborhood or campus. For each link, show travel times and distances (to nearest 0.1 mile).

REFERENCES

Center for Microcomputers in Transportation, University of Florida, Transportation Research Center, 2000.

Cohn, L. F., Harris, R. A., and Lederer, P. R., "Environmental and Energy Considerations" Chapter 8.

Computer Programs for Urban Transportation, U.S. Department of Transportation, Federal Highway Administration, Washington, D.C., April 1977.

Forecasting Inputs to Transportation Planning, NCHRP Report 266, Transportation Research Board, National Research Council, Washington, D.C., 1983.

Levinson, H. S., and Jurasin, R. R., "Transportation Planning Studies," Chapter 5, *Transportation Planning Handbook,* 2nd ed., Institute of Transportation Engineers, 1999.

Institute of Transportation Engineers. *Transportation Planning Handbook,* 2nd ed., Washington, D.C., 1999.

Manheim, M. L., *Fundamentals of Transportation Systems Analysis,* MIT Press, Cambridge, Mass., 1979.

Microcomputers in Transportation, Software Source Book, U.S. Department of Transportation, Washington, D.C., February 1986.

U.S. Department of Transportation Fifth Coast Guard District. *Parallel Crossing of Chesapeake Bay Draft Environmental Impact Statement/4(f) Statement,* Portsmouth, Va., 1993.

ADDITIONAL READINGS

Blanchard, B. S., and Fabrycky, W. J., *Systems Engineering and Analysis,* Prentice-Hall, Englewood Cliffs, N.J., 1981.

Information Requirements for Transportation Economic Analysis, Conference Proceedings 21, Transportation Research Board, National Research Council, Washington, D.C., 2000.

Meyer, Michael D., and E. J. Miller, *Urban Transportation Planning: A Decision-Oriented Approach,* McGraw-Hill, New York, 2000.

Toward a Sustainable Future, Special Report 251, Transportation Research Board, National Research Council, Washington, D.C., 1997.

CHAPTER 12

Forecasting Travel Demand

Travel demand is expressed as the number of persons or vehicles per unit time that can be expected to travel on a given segment of a transportation system under a set of given land-use, socioeconomic, and environmental conditions. Forecasts of travel demand are used to establish the vehicular volume on future or modified transportation system alternatives. The methods for forecasting travel demand can range from a simple extrapolation of observed trends to a sophisticated computerized process involving extensive data gathering and mathematical modeling. The travel demand forecasting process is as much an art as it is a science because judgments are required concerning the various parameters—that is, population, car ownership, and so forth—that provide the basis for a travel forecast. The methods used in forecasting demand will depend on the availability of data as described in Chapter 11 and on specific constraints on the project, such as availability of funds and project schedules. Sources and techniques for transportation studies are given in the *Transportation Planning Handbook.* Survey techniques and in-depth descriptions of available data can be found in the "Additional Readings" section at the end of this chapter.

DEMAND FORECASTING APPROACHES

There are two basic demand forecasting situations in transportation planning. The first involves travel demand studies for urban areas, and the second deals with intercity travel demand. Urban travel demand forecasts, when first developed in the 1950s and 1960s, required that extensive databases be prepared using home interview and/or road-side interview surveys. The information gathered provided useful insight concerning the characteristics of the trip maker, such as age, sex, income, auto ownership, and so forth, the land use at each end of the trip, and the mode of travel. Travel data could then be aggregated by zone and/or be used at a more disaggregated level—that is, household or individual—to formulate relationships between variables and to calibrate models.

In the intercity case, data generally are aggregated to a greater extent than for urban travel forecasting, such as city population, average city income, and travel time or travel cost between city pairs. The U.S. statistical data base, particularly for intercity travel, has been weak since the late 1970s. A positive shift in the availability of travel data has come about as a result of the creation of the Bureau of Transportation Statistics, of the U.S. DOT following the ISTEA legislation. This chapter describes the urban travel forecasting process. The underlying concepts may also be applied to intercity travel demand.

The databases that were established in urban transportation studies during the 1955–1970 period were used for the calibration and testing of models for trip generation, distribution, modal choice, and traffic assignment. These data collection and calibration efforts involved a significant investment of money and personnel resources, and consequent studies were based on updating the existing database and using models that had been developed previously.

Factors Influencing Travel Demand

Three factors affect the demand for urban travel: the location and intensity of land use; the socioeconomic characteristics of people living in the area; and the extent, cost, and quality of available transportation services. These factors are incorporated in most travel forecasting procedures.

Land-use characteristics are a primary determinant of travel demand. The amount of traffic generated by a parcel of land depends on how the land is used. For example, shopping centers, residential complexes, and office buildings produce different traffic generation patterns.

Socioeconomic characteristics of the people living within the city also influence the demand for transportation. Lifestyles and values affect how people decide to use their resources for transportation. For example, a residential area consisting of high-income, white-collar workers will generate more trips by automobile per person than a residential area populated primarily by retirees.

The availability of transportation facilities and services, referred to as the *supply,* also affects the demand for travel. Travelers are sensitive to the level of service provided by alternative transportation modes. When deciding whether to travel at all or which mode to use, they consider attributes such as travel time, cost, convenience, comfort, and safety.

Sequential Steps for Travel Forecasting

Prior to the technical task of travel forecasting, the study area must be delineated into a set of traffic zones that form the basis for analysis of travel movements within, into, and out of the urban area. (See Figure 11.8 and the discussion in Chapter 11 for further details.) The set of zones can be aggregated into larger units, called *districts,* for certain analytical techniques or analyses that work at such levels. Land-use estimates are also developed.

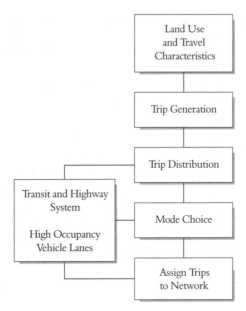

Figure 12.1 Travel Forecasting Process

Travel forecasting is solely within the domain of the transportation planner and is an integral part of site development and traffic engineering studies as well as areawide transportation planning. Techniques that represent the state-of-the-practice of each task are described in this chapter to introduce the topic and to illustrate how demand forecast can be determined. Variations of each forecasting technique can be readily found in the literature.

The approach most commonly used to forecast travel demand is the "four-step process" of trip generation, trip distribution, modal choice, and traffic assignment. Simultaneous model structures have also been used in practice, particularly to forecast intercity travel. Figure 12.1 illustrates the sequential travel forecasting process.

TRIP GENERATION

Trip generation is the process of determining the number of trips that will begin or end in each traffic zone within a study area. Since the trips are determined without regard to destination, they are referred to as *trip ends.* Each trip has two ends, and these are described in terms of trip purpose, or whether the trips are either produced by a traffic zone or attracted to a traffic zone. For example, a home-to-work trip would be considered to have a trip end produced in the home zone and attracted to the work zone. Trip generation analysis has two functions: (1) to develop a relationship between trip end production or attraction and land use, and (2) to use the relationship developed to estimate the number of trips generated at some future date under a new set of land-use conditions. To illustrate the process we will consider two methods: cross classification and rates based on activity units. Another commonly used method is regression analysis, which has been applied to estimate both productions and attractions. This method is used infrequently because it relies on zonal aggregated data. Trip generation methods that use a disaggregated analysis, based on individual sample units such as persons, households, income, and vehicle units, are preferred. The application of multiple regression analysis was illustrated in Example 11.2 (Chapter 11).

Cross Classification

Cross classification is a technique developed by the Federal Highway Administration (FHWA) to determine the number of trips that begin or end at the home. Home-based trip generation is a very useful number because it can represent a significant proportion of all trips. The first step is to develop a relationship between socioeconomic measures and trip production. The two variables most commonly used are average income and auto ownership. Other variables that could be considered are household size and stage in the household life cycle. The relationships are developed based on income data and results of O-D surveys. Figure 12.2 illustrates the variation in average income within a zone.

Example 12.1   Developing Trip Generation Curves from Household Data

A travel survey produced the data shown in Table 12.1. Twenty households were interviewed. The table shows the number of trips produced per day for each of the households (numbered 1–20), as well as the corresponding annual household income and the number of automobiles owned. Based on the data provided, develop a set of curves showing the number of trips per household versus income and auto ownership.

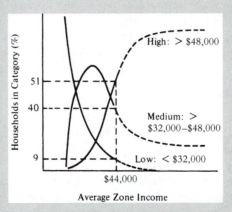

Figure 12.2 Average Zonal Income Versus Households in Income Category

SOURCE: Modified from *Computer Programs for Urban Transportation Planning*, U.S. Department of Transportation, Washington, D.C., April 1977.

Table 12.1 Survey Data Showing Trips per Household, Income, and Auto Ownership

Household Number	Trips Produced per Household	Household Income ($1000s)	Autos per Household
1	2	16	0
2	4	24	0
3	10	68	2
4	5	44	0
5	5	18	1
6	15	68	3
7	7	38	1
8	4	36	0
9	6	28	1
10	13	76	3
11	8	72	1
12	6	32	1
13	9	28	2
14	11	44	2
15	10	44	2
16	11	52	2
17	12	60	2
18	8	44	1
19	8	52	1
20	6	28	1

Solution:

Step 1. From the information in Table 12.1, produce a matrix that shows the number and percentage of households as a function of auto ownership and income grouping (see Table 12.2). The numerical values in each cell represent the number of households observed in each combination of income–auto ownership category. The value in parentheses is the percentage observed at each income level. In actual practice, the sample size would be at least 25 data points per cell to insure statistical accuracy. The data shown in Table 12.2 are used to develop relationships between the percent of households in each auto ownership category by household income, as illustrated in Figure 12.3.

Step 2. A second table produced from the data in Table 12.1 shows the average number of trips per household versus income and cars owned. The results, shown in Table 12.3, are illustrated in Figure 12.4, which depicts the relationship between trips per household per day by income and auto ownership. The table indicates that for a given income, trip generation increases with the number of cars owned. Similarly, for a given car ownership, trip generation increases with the rise in income.

Table 12.2   Number and Percent of Household in Each Income Category Versus Car Ownership

Income ($1000s)	Autos Owned			Total
	0	1	2+	
24	2(67)	1(33)	0(0)	3(100)
24–36	1(25)	3(50)	1(25)	5(100)
36–48	1(20)	2(40)	2(40)	5(100)
48–60		1(33)	2(67)	3(100)
>60		1(25)	3(75)	4(100)
Total	4	8	8	20

Note: Values in parentheses are percent of automobiles owned at each income range.

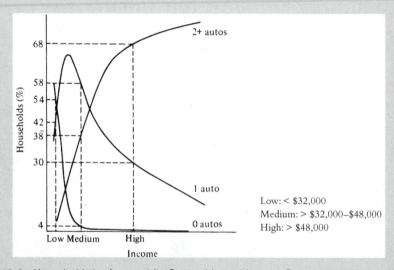

Figure 12.3   Households by Automobile Ownership and Income Category

SOURCE: Modified from *Computer Programs for Urban Transportation Planning,* U.S. Department of Transportation, Washington, D.C., April 1977.

Step 3.   As a further refinement, additional O-D data (not shown in Table 12.1) can be used to determine the percentage of trips by each trip purpose for each income category. These results are shown in Figure 12.5, wherein three trip purposes are used: home-based work (HBW), home-based other (HBO), and non-home based (NHB). The terminology refers to the origination of a trip as either at the home or not at the home.

Table 12.3 Average Trips per Household Versus Income and Car Ownership

Income ($1000s)	Autos Owned		
	0	*1*	*2+*
≤24	3	5	—
24–36	4	6	9
36–48	5	7.5	10.5
48–60	—	8.5	11.5
>60	—	8.5	12.7

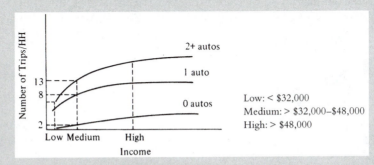

Figure 12.4 Trips per Household per Day by Auto Ownership and Income Category

SOURCE: Modified from *Computer Programs for Urban Transportation Planning,* U.S. Department of Transportation, Washington, D.C., April 1977.

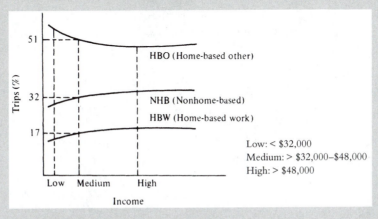

Figure 12.5 Trips by Purpose and Income Category

SOURCE: Modified from *Computer Programs for Urban Transportation Planning,* U.S. Department of Transportation, Washington, D.C., April 1977.

The trip generation model that has been developed based on survey data can now be used to estimate the number of home- and nonhome-based trips for each trip purpose.

Example 12.2   Computing Trips Generated in a Suburban Zone

Consider a zone that is located in a suburban area of a city. The population and income date for the zone are as follows:

Number of dwelling units: 60
Average income per dwelling unit: $44,000

Determine the number of trips per day generated in this zone for each trip purpose, assuming that the characteristics depicted in Figures 12.2–12.5 apply in this situation. The problem is solved in four basic steps.

Solution:

Step 1. Determine the percentage of households in each economic category. These results can be obtained by analysis of census data for the area. A typical plot of average zonal income versus income distribution is shown in Figure 12.2. For an average zonal income of $44,000, the following distribution is observed:

Income ($)	Households (%)
Low (under 32,000)	9
Medium (32,000–48,000)	40
High (over 48,000)	51

Step 2. Determine the distribution of auto ownership per household for each income category. A typical curve showing percent of households, at each income level, that own zero, one, or two+ autos is shown in Figure 12.3, and the results are listed in Table 12.4.

Table 12.4 shows that 58 percent of medium-income families own one auto per household. Also, from the previous step we know that a zone, with an average income of $44,000, contains 40 percent of households in the medium-income category. Thus, we can calculate that of the 60 households in that zone, there will be $60 \times 0.40 \times 0.58 = 14$ medium-income households that own one auto.

Step 3. Determine the number of trips per household per day for each income–auto ownership category. A typical curve showing the relationship between trips per household, household income, and auto ownership is shown in Figure 12.4. The results are listed in Table 12.5. The table shows that a medium-income household owning one auto will generate eight trips per day.

Table 12.4 Percentage of Households in Each Income Category Versus Auto Ownership

	Cars/Household		
Income	0	1	2+
Low	54	42	4
Medium	4	58	38
High	2	30	68

Table 12.5 Number of Trips per Household per Day

	Autos/Household		
Income	0	1	2+
Low	1	6	7
Medium	2	8	13
High	3	11	15

Step 4: Calculate the total number of trips per day generated in the zone. This is done by computing the number of households in each income–auto ownership category, multiplying this result by the number of trips per household, as determined in step 3, and summing the result. Thus,

$$P_{gh} = HH \times I_g \times A_{gh} \times (P_H)_{gh} \tag{12.1}$$

$$P_T = \sum_{g}^{3} \sum_{h}^{3} P_{gh} \tag{12.2}$$

where

HH = number of households in the zone

I_g = percentage of households in zone with income level g (low, medium, or high)

A_{gh} = percentage of households in income level g with h autos per household (h = 0, 1, or 2+)

P_{gh} = number of trips per day generated in the zone by householders with income level g and auto ownership h

$(P_H)_{gh}$ = number of trips per day produced in a household at income level g and auto ownership h

P_T = total number of trips generated in the zone

The calculations are shown in Table 12.6. For a zone with 60 households and an average income of \$44,000, the number of trips generated is 666 auto trips per day.

Table 12.6 Number of Trips per Day Generated by Sixty Households

	Income, Auto Ownership	Total Trips by Income Group
$60 \times 0.09 \times 0.54 \times 1 = 3$ trips	L, 0	
$60 \times 0.09 \times 0.42 \times 6 = 14$ trips	L, 1	
$60 \times 0.09 \times 0.04 \times 7 = 2$ trips	L, 2+	19
$60 \times 0.40 \times 0.04 \times 2 = 2$ trips	M, 0	
$60 \times 0.40 \times 0.58 \times 8 = 111$ trips	M, 1	
$60 \times 0.40 \times 0.38 \times 13 = 119$ trips	M, 2+	232
$60 \times 0.51 \times 0.02 \times 3 = 2$ trips	H, 0	
$60 \times 0.51 \times 0.30 \times 11 = 101$ trips	H, 1	
$60 \times 0.51 \times 0.68 \times 15 = 312$ trips	H, 2+	415
Total = 666 trips		666

Step 5. Determine the percentage of trips by trip purpose. As a final step, we can calculate the number of trips that are HBW, HBO, and NHB. If these percentages are 17, 51, and 32, respectively (see Figure 12.5), for the medium-income category, then the number of trips from the zone for the three trip purposes are $232 \times 0.17 = 40$ HBW, $232 \times 0.51 = 118$ HBO, and $232 \times 0.32 = 74$ NHB. (Similar calculations would be made for other income groups.) The final result, which is left for the reader to verify, is obtained by using the following percentages: low income, 15, 55, and 30, and high income, 18, 48, and 34. These yield 118 HBW, 327 HBO, and 221 NHB trips.

We have illustrated how trip generation values are determined and used to calculate the number of trips in each zone. Values for each income or auto ownership category can be developed using survey data or published statistics compiled for other cities. Figure 12.6 shows the trip generation rate per vehicle for single-family detached housing. The average rate is 6.27, and the range of rates is 2.69–9.38. This table is one of over one thousand, furnished by the Institute of Transportation Engineers, for 10 different land uses, including port and terminal, industrial, agricultural, residential, lodging, institutional, medical, office, retail, and services. An extensive amount of useful trip generation data is also available in *Quick Response Urban Travel Estimation Techniques and Transferrable Parameters*. The ITE *Transportation Planning Handbook* (1999) also provides trip generation rates.

Rates Based on Activity Units

The preceding section illustrated how trip generation is determined for residential zones where the basic unit is the household. Trips generated at the household end are referred to

Single-Family Detached Housing (210)

Average Vehicle Trip Ends vs: Vehicles
On a: Weekday
Number of Studies: 133
Average Number of Vehicles: 274
Directional Distribution: 50% entering, 50% exiting

Trip Generation per Vehicle

Average Rate	Range of Rates	Standard Deviation
6.27	2.69–9.38	2.84

Data Plot and Equation

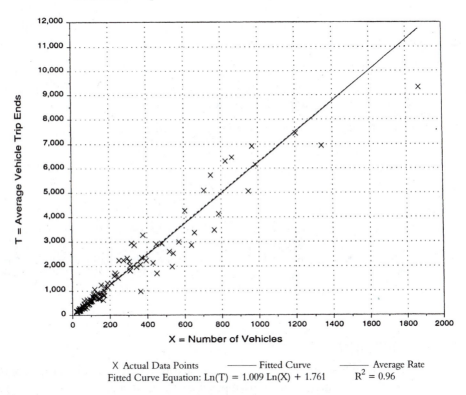

X Actual Data Points ———— Fitted Curve ———— Average Rate
Fitted Curve Equation: $\text{Ln}(T) = 1.009\,\text{Ln}(X) + 1.761$ $R^2 = 0.96$

Figure 12.6 Trip Generation Characteristics

SOURCE: *Trip Generation,* 5th ed., Institute of Transportation Engineers, Washington, D.C., 1991.

as *productions,* and they are *attracted* to zones for purposes such as work, shopping, visiting friends, and medical trips. Thus, an activity unit can be described by measures such as square feet of floor space or number of employees. Trip generation rates for attraction zones can be determined from survey data or are tabulated in some of the sources listed in the "References" and "Additional Readings" sections at the end of this chapter. Trip attraction rates are illustrated in Table 12.7.

Table 12.7 Trip Generation Rates by Trip Purpose and Employee Category

	Attractions per Household	*Attractions per Nonretail Employee*	*Attractions per Downtown Retail Employee*	*Attractions per Other Retail Employee*
HBW	—	1.7	1.7	1.7
HBO	1.0	2.0	5.0	10.0
NHB	1.0	1.0	3.0	5.0

Example 12.3 Computing Trips Generated in an Activity Zone

A commercial center in the downtown contains several retail establishments and light industries. Employed at the center are 220 retail and 650 nonretail workers. Determine the number of trips per day attracted to this zone.

Solution: Use the trip generation rates listed in Table 12.7:

HBW: $(220 \times 1.7) + (650 \times 1.7) = 1479$
HBO: $(220 \times 5.0) + (650 \times 2.0) = 2400$
NHB: $(220 \times 3.0) + (650 \times 1.0) = 1310$
Total = 5189 trips/day

Note that three trip purposes are given in Table 12.7: home-based work (HBW), home-based other (HBO), and nonhome based (NHB). For example, for HBO trips, there are 5.0 attractions per downtown retail employee (in trips per day), and 2.0 attractions per nonretail employee.

TRIP DISTRIBUTION

Trip distribution is a process by which the trips generated in one zone are allocated to other zones in the study area. For example, if the trip generation analysis results in an estimate of 200 HBW trips in zone 10, then the trip distribution analysis would determine how many of these trips would be made between zone 10 and all the other zones.

Several basic methods are used for trip distribution. Among these are the gravity model, growth factor models, and intervening opportunities. The gravity model is preferred because it uses the attributes of the transportation system and land-use characteristics and has been calibrated extensively for many urban areas. The gravity model has achieved virtually universal use because of its simplicity, its accuracy, and its support from the U.S. Department of Transportation. Growth factor models, which were used more widely in the 1950s and 1960s, require that the origin-destination matrix be known for the base (or current) year, as well as an estimate of the number of future trip ends in each zone. The intervening opportunities model and other models are available but not widely used in practice.

Gravity Model

The most widely used and documented trip distribution model is the *gravity model,* which states that the number of trips between two zones is directly proportional to the number of trip attractions generated by the zone of destination and inversely proportional to a function of time of travel between the two zones.

Mathematically, the gravity model is expressed as

$$T_{ij} = P_i \left[\frac{A_j F_{ij} K_{ij}}{\sum_j A_j F_{ij} K_{ij}} \right] \tag{12.3}$$

where

T_{ij} = number of trips that are produced in zone i and attracted to zone j
P_i = total number of trips produced in zone i
A_j = number of trips attracted to zone j
F_{ij} = a value which is an inverse function of travel time
K_{ij} = socioeconomic adjustment factor for interchange ij

The values of P_i and A_j have been determined in the trip generation process. The sum of P_i's for all zones must equal the sum of A_j's for all zones. K_{ij} values are used when the estimated trip interchange must be adjusted to ensure that it agrees with the observed trip interchange.

The values for F_{ij} are determined by a calibrating process in which trip generation values as measured in the O-D survey are distributed using the gravity model. After each distribution process is completed, the percentage of trips in each trip length category produced by the gravity model is compared with the percentage of trips recorded in the O-D survey. If the percentages do not agree, then the F_{ij} factors that were used in the distribution process are adjusted and another gravity model trip distribution is performed. The calibration process is continued until the trip length percentages are in agreement.

Figure 12.7 illustrates F values for calibrations of a gravity model. (Normally this curve is a semilog plot.) F values can also be determined using travel time values and an inverse relationship between F and t. For example, the relationship for F might be in the form t^{-1}, t^{-2}, e^{-t}, and so forth, since F values decrease as travel times increase.

The socioeconomic factor K_{ij} is used to make adjustments of trip distribution values between zones where differences between estimated and actual values are significant. The K value is referred to as the "socioeconomic factor" since it accounts for variables other than travel time. The values for K are determined in the calibration process, but it is used judiciously when a zone is considered to possess unique characteristics.

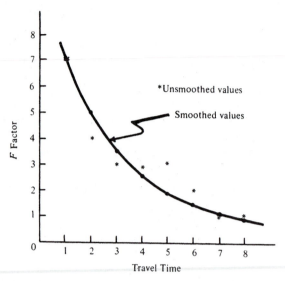

Figure 12.7 Calibration of *F* Factors

SOURCE: *Computer Programs for Urban Transportation Planning*, U.S. Department of Transportation, Washington, D.C., April 1977.

Example 12.4   Use of Calibrated F Factors and Iteration

To illustrate the application of the gravity model, consider a study area consisting of three zones. The data have been determined as follows. The number of productions and attractions have been computed for each zone by methods described in the section on trip generation, and the average travel times between each zone have been determined. Both are shown in Tables 12.8 and 12.9. Assume K_{ij} is the same unit value for all zones. Finally, the F values have been calibrated as previously described and are shown in Table 12.10 for each travel time increment. Note that the intrazonal travel time for zone 1 is larger than those of most of the other interzone times because of the geographical characteristics of the zone and the lack of access within the area. This zone could represent conditions in a congested downtown area.

Solution:

- Determine the number of trips between each zone using the gravity model formula and the given data. (Note: F_{ij} is obtained by using the travel times in Table 12.9 and selecting the correct F value from Table 12.10. For example, travel time is 2 min between zones 1 and 2. The corresponding F value is 52.)

Table 12.8 Trip Productions and Attractions for a Three-Zone Study Area

Zone	1	2	3	Total
Trip Productions	140	330	280	750
Trip Attractions	300	270	180	750

Table 12.9 Travel Time Between Zones (min)

Zone	1	2	3
1	5	2	3
2	2	6	6
3	3	6	5

Use equation 12.3.

$$T_{ij} = P_i \left(\frac{A_j F_{ij} K_{ij}}{\sum_{j=1}^{n} A_j F_{ij} K_{ij}} \right) \qquad K_{ij} = 1 \text{ for all zones}$$

$$T_{1\text{-}1} = 140 \times \frac{300 \times 39}{(300 \times 39) + (270 \times 52) + (180 \times 50)} = 47$$

$$T_{1\text{-}2} = 140 \times \frac{270 \times 52}{(300 \times 39) + (270 \times 52) + (180 \times 50)} = 57$$

$$T_{1\text{-}3} = 140 \times \frac{180 \times 50}{(300 \times 39) + (270 \times 52) + (180 \times 50)} = 36$$

$$P_1 = 140$$

- Make similar calculations for zones 2 and 3.

$$T_{2\text{-}1} = 188; \quad T_{2\text{-}2} = 85; \quad T_{2\text{-}3} = 57; \quad P_2 = 330$$

$$T_{3\text{-}1} = 144; \quad T_{3\text{-}2} = 68; \quad T_{3\text{-}3} = 68; \quad P_3 = 280$$

- Summarize the results as shown in Table 12.11. Note that the sum of the productions in each zone is equal to the number of productions given in the problem statement. However, the number of attractions estimated in the trip distribution

Table 12.10 Travel Time Versus Friction Factor

Time (min)	F
1	82
2	52
3	50
4	41
5	39
6	26
7	20
8	13

Note: *F* values were obtained from the calibration process.

Table 12.11 Zone-to-Zone Trips: First Iteration

Zone	1	2	3	P
1	47	57	36	140
2	188	85	57	330
3	144	68	68	280
Computed *A*	379	210	161	750
Given *A*	300	270	180	750

phase differs from the number of attractions given. For zone 1, the correct number is 300, whereas the computed value is 379. Values for zone 2 are 270 versus 210, and for zone 3 they are 180 versus 161.

- Calculate the adjusted attraction factors according to the formula

$$A_{jk} = \frac{A_j}{C_{j(k-1)}} A_{j(k-1)} \tag{12.4}$$

where
A_{jk} = adjusted attraction factor for attraction zone (column) j, iteration k.
$A_{jk} = A_j$ when $k = 1$
C_{jk} = actual attraction (column) total for zone j, iteration k
A_j = desired attraction total for attraction zone (column) j
j = attraction zone number, $j = 1, 2, \ldots, n$
n = number of zones
k = iteration number, $k = 1, 2, \ldots, m$
m = number of iterations

- To produce a mathematically correct result, repeat the trip distribution computations using modified attraction values, so that the numbers attracted will be increased or reduced as required. For zone 1, for example, the estimated attractions were too great. Therefore, the new attraction factors are adjusted downward by multiplying the original attraction value by the ratio of the original to estimated attraction values.

$$\text{zone 1:} \quad A_{12} = 300 \times \frac{300}{379} = 237$$

$$\text{zone 2:} \quad A_{22} = 270 \times \frac{270}{210} = 347$$

$$\text{zone 3:} \quad A_{32} = 180 \times \frac{180}{161} = 201$$

- Apply the gravity model formula (Eq. 12.3) in each iteration to calculate zonal trip interchanges using the adjusted attraction factors obtained from the preceding iteration. In practice, the gravity model formula thus becomes

$$\left(T_{ijk} = \frac{P_i A_{jk} F_{ij} K_{ij}}{\sum\limits_{j=1}^{n} A_{jk} F_{ij} K_{ij}} \right)_p$$

where T_{ijk} is the trip interchange between i and j for iteration k, and $A_{jk} = A_j$ when $k = 1$. Subscript j goes through one complete cycle every time k changes, and i goes through one complete cycle every time j changes. The above formula is enclosed in parentheses and subscripted to indicate that the complete process is performed for each trip purpose.
- Perform a second iteration using the adjusted attraction values.

$$T_{1-1} = 140 \times \frac{237 \times 39}{(237 \times 39) + (347 \times 52) + (201 \times 50)} = 34$$

$$T_{1-2} = 140 \times \frac{347 \times 52}{(237 \times 39) + (347 \times 52) + (201 \times 50)} = 68$$

$$T_{1-3} = 140 \times \frac{201 \times 50}{(237 \times 39) + (347 \times 52) + (201 \times 50)} = 37$$

$$P_1 = 140$$

- Make similar calculations for zones 2 and 3.

$$T_{2-1} = 153; \quad T_{2-2} = 112; \quad T_{2-3} = 65; \quad P_2 = 330$$

$$T_{3-1} = 116; \quad T_{3-2} = 88; \quad T_{3-3} = 76; \quad P_3 = 280$$

- List the results as shown in Table 12.12. Note that, in each case, the sum of the attractions is now much closer to the given value. The process will be continued until there is a reasonable agreement (within 5 percent) between the A that is estimated using the gravity model and the values that are furnished in the trip generation phase.

Table 12.12 Zone to Zone Trips: Second Iteration

Zone	1	2	3	P
1	34	68	38	140
2	153	112	65	330
3	116	88	76	280
Computed A	303	268	179	750
Given A	300	270	180	750

Growth Factor Models

Trip distribution can also be computed when the only data available are the origins and destinations between each zone for the current or base year and the trip generation values for each zone for the future year. This method was widely used when O-D data were available but the gravity model and calibrations for F factors had not yet become operational. Growth factor models are used primarily to distribute trips between zones in the study area and zones in cities external to the study area. Since they rely upon an existing O-D matrix, they cannot be used to forecast traffic between zones where no traffic currently exists. Further, the only measure of travel friction is the amount of current travel. Thus, the growth factor method cannot reflect changes in travel time between zones, as does the gravity model.

The most popular growth factor model is the *Fratar method*, which is a mathematical formula that proportions future trip generation estimates to each zone as a function of the product of the current trips between the two zones T_{ij} and the growth factor of the attracting zone G_j. Thus,

$$T_{ij} = (t_i G_i) \frac{t_{ij} G_j}{\sum_x t_{ix} G_x} \qquad (12.5)$$

where

T_{ij} = number of trips estimated from zone i to zone j
t_i = present trip generation in zone i
G_x = growth factor of zone x
$T_i = t_i G_i$ = future trip generation in zone i
t_{ix} = number of trips between zone i and other zones x
t_{ij} = present trips between zone i and zone j
G_j = growth factor of zone j

To illustrate the application of the growth factor method, consider the following example.

Example 12.5   Forecasting Trips Using the Fratar Model

A study area consists of four zones (A, B, C, and D). An O-D survey indicates that the number of trips between each zone is as shown in Table 12.13. Planning estimates for the area indicate that in five years the number of trips in each zone will increase by the growth factor shown in Table 12.14, and that trip generation will be increased to the amounts shown in the last column of the table. Determine the number of trips between each zone for future conditions.

Solution: Using the Fratar formula (Eq. 12.5), we can calculate the number of trips between zones A and B, A and C, A and D, and so forth. Note that we obtain two values for each zone pair (that is, T_{AB} and T_{BA}). These values are averaged, yielding a value $\overline{T}_{AB} = (T_{AB} + T_{BA})/2$.

The calculations are as follows:

$$T_{ij} = (t_i G_i) \frac{t_{ij} G_i}{\sum\limits_{x} t_{ix} G_x}$$

$$T_{AB} = 600 \times 1.2 \frac{400 \times 1.1}{(400 \times 1.1) + (100 \times 1.4) + (100 \times 1.3)} = 446$$

$$T_{BA} = 700 \times 1.1 \frac{400 \times 1.2}{(400 \times 1.2) + (300 \times 1.4)} = 411$$

$$\overline{T}_{AB} = \frac{T_{AB} + T_{BA}}{2} = \frac{446 + 411}{2} = 428$$

Similar calculations yield

$$\overline{T}_{AC} = 141; \quad \overline{T}_{AD} = 124; \quad \overline{T}_{BC} = 372; \quad \overline{T}_{CD} = 430$$

Table 12.13   Present Trip Generation and Growth Factors

Zone	Present Trip Generation (trips/day)	Growth Factor	Trip Generation in Five Years
A	600	1.2	720
B	700	1.1	770
C	700	1.4	980
D	400	1.3	520

Table 12.14 Present Trips Between Zones

Zone	A	B	C	D
A	—	400	100	100
B	400	—	300	—
C	100	300	—	300
D	100	—	300	—
Total	600	700	700	400

Table 12.15 First Estimate of Trips Between Zones

Zone	A	B	C	D	Estimated Total Trip Generation	Actual Trip Generation
A	—	428	141	124	693	720
B	428	—	372	—	800	770
C	141	372	—	430	943	980
D	124	—	430	—	554	520
Totals	693	800	943	554		

Table 12.16 Growth Factors for Second Iteration

Zone	Estimated Trip Generation	Actual Trip Generation	Growth Factor
A	693	720	1.04
B	800	770	0.96
C	943	980	1.04
D	554	520	0.94

The results of the preceding calculations have produced the first estimate (or iteration) of future trip distribution and are shown in Table 12.15. Note, however, that the totals for each zone do not equal the values of future trip generation as stated earlier. For example, the trip generation in zone A is estimated as 693 trips, whereas the correct value is 720 trips. Similarly, the estimate for zone B is 800 trips, whereas the desired value is 770 trips.

We now proceed with a second iteration in which the input data are the numbers of trips between zones as previously calculated. Also, new growth factors are computed as the ratio of the trip generation expected to occur in five years and the trip generation estimated in the preceding calculation. The values are given in Table 12.16.

The calculations for the second iteration are left to the reader and can be repeated as many times as needed until the estimate and actual trip generation values are close in agreement.

MODE CHOICE

Mode choice is that aspect of the demand analysis process that determines the number (or percentage) of trips between zones that are made by automobile and by transit. The selection of one mode or another is a complex process that depends on factors such as the traveler's income, the availability of transit service or auto ownership, and the relative advantages of each mode in terms of travel time, cost, comfort, convenience, and safety. Mode choice models attempt to replicate the relevant characteristics of the traveler, the transportation system, and the trip itself, such that a realistic estimate of the number of trips by each mode for each zonal pair is obtained. A discussion of the many mode choice models is beyond the scope of this chapter and the interested student should refer to other primary sources cited.

Types of Mode Choice Models

Since public transportation is an important factor, primarily in larger cities, mode choice calculations may involve only distinguishing trip interchanges as either auto or transit. Depending on the level of detail required, three types of transit estimating procedures are used: (1) direct generation of transit trips, (2) use of trip end models, and (3) trip interchange modal split models.

For *direct generation* methods, transit trips are generated directly, by estimating either total person trips or auto driver trips. Figure 12.8 is a graph that illustrates the relationship between transit trips per day per 1000 population and persons per acre versus auto ownership. As density of population increases, it can be expected that transit riding will also increase for a given level of auto ownership.

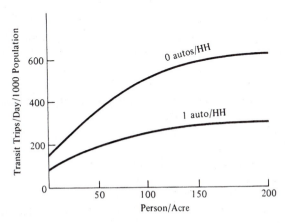

Figure 12.8 Number of Transit Trips by Population Density and Automobile Ownership per Household

Example 12.6   Estimating Mode Choice by Direct Trip Generation

Determine the number of transit trips per day in a zone which has 5000 people living on 50 acres. The auto ownership is 40 percent of zero autos per household and 60 percent of one auto per household.

Solution:

Calculate the number of persons per acre: $5000/50 = 100$. Then determine the number of transit trips per day per 1000 persons (from Figure 12.8) to calculate the total of all transit trips per day for the zone.

zero autos/HH: 510 trips/day/1000 population
one auto/HH: 250 trips/day/1000 population

$$\text{Total Transit Trips: } (0.40)(510)(5) + (0.60)(250)(5) =$$

$$1020 + 750 = 1770 \text{ transit trips per day}$$

Note that this method assumes that the attributes of the system are not important. Factors such as travel time, cost, and convenience are not considered. These so-called predistribution models apply when transit service is poor and riders are "captive," or when transit service is excellent and "choice" clearly favors transit. When highway and transit modes "compete" for auto riders, then system factors are considered.

Trip end models determine the percentage of total person or auto trips that will use transit. The estimates are made prior to the trip distribution phase and are based on land-use or socioeconomic characteristics of the zone. They do not incorporate the quality of service. The procedure follows:

1. Generate total person trip productions and attractions by trip purpose.
2. Compute the urban travel factor.
3. Determine the percentage of these trips by transit using a mode choice curve (see Figure 12.9).
4. Apply auto occupancy factors.
5. Distribute transit and auto trips separately.

The mode choice model shown in Figure 12.9 is based on two factors: households per auto and persons per square mile. The product of these variables is called the urban travel factor (UTF). Percentage of travel by transit will increase in an S curve fashion as the UTF increases.

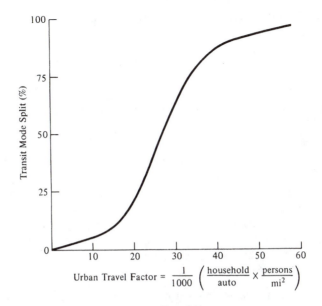

Figure 12.9 Transit Mode Split Versus Urban Travel Factor

Example 12.7   *Estimating Trip Productions by Transit*

The total number of trip productions in a zone is 10,000 trips/day. The number of house-holds per auto is 1.80, and there are 15,000 persons/square mile living there. Determine the percent of residents who can be expected to use transit.

Solution: Compute the urban travel factor:

$$UTF = \frac{1}{1000}\left(\frac{HH}{auto}\right)\left(\frac{persons}{sq\ mi}\right)$$

$$= \frac{1}{1000} \times 1.80 \times 15{,}000 = 27.0$$

Enter Figure 12.9. Transit mode split = 45 percent.

Trip interchange models incorporate system level of service variables such as relative travel time, relative travel cost, economic status of the trip maker, and relative travel service. The following describes one such model, the QRS method. This model illustrates the basic elements that are considered in estimating mode choice.

The QRS method is based on the following relationship:

$$MS_a = \frac{I_{ijt}^{-b}}{I_{ijt}^{-b} + I_{ija}^{-b}} \times 100 \ \ or \ \ \frac{I_{ija}^{b}}{I_{ijt}^{b} + I_{ija}^{b}} \times 100 \tag{12.6}$$

$$MS_t = (1 - MS_a) \times 100 \tag{12.7}$$

where

MS_t = proportion of trips between zone i and j using transit
MS_a = proportion of trips between zone i and j using auto
I_{ijm} = a value referred to as the *impedance* of travel of mode m, between i and j, which is a measure of the total cost of the trip. [Impedance = (in-vehicle time min) + (2.5 × excess time min) + (3 × trip cost, \$/income earned/min).]
b = an exponent, which depends on trip purpose
m = t for transit mode; a for auto mode

Note that in-vehicle time is time spent traveling in the vehicle, and excess time is time spent traveling but not in the vehicle (waiting, walking, and so forth).

The impedance value is determined for each zone pair and represents a measure of the expenditure required to make the trip by either auto or transit. The data required for estimating mode choice include (1) distance between zones by auto and transit, (2) transit fare, (3) out-of-pocket auto cost, (4) parking cost, (5) highway and transit speed, (6) exponent values, b, (7) median income, and (8) excess time, which includes the time required to walk to a transit vehicle and time waiting or transferring. Assume that the time worked per year is 120,000 min.

Example 12.8   Computing Mode Choice Using the QRS Model

To illustrate the application of the QRS method, assume that the data shown in Table 12.17 have been developed for travel between a suburban zone S and a downtown zone D. Determine the percent of work trips by auto and transit. An exponent value of 2.0 is used for work travel. Median income is \$12,000 per year.

Table 12.17   Travel Data Between Two Zones, S and D

	Auto	Transit
Distance	10 mi	8 mi
Cost per mile	\$0.15	\$0.10
Excess time	5 min	8 min
Parking cost	\$1.50 (or 0.75/trip)	—
Speed	30 mi/h	20 mi/h

Solution: Use Eq. 12.6.

$$MS_a = \frac{I_{ija}^b}{I_{ijt}^b + I_{ija}^b}$$

$$I_{SDa} = \left(\frac{10}{30} \times 60\right) + (2.5 \times 5) + \left\{ \frac{3 \times [(1.50/2) + 0.15 \times 10]}{12,000/120,000} \right\} = 20 + 12.5 + 67.5$$

$$= 100.0 \text{ equivalent min}$$

$$I_{SDt} = \left(\frac{8}{20} \times 60\right) + (2.5 \times 8) + \left[\frac{3 \times (8 \times 0.10)}{12,000/120,000}\right] = 24 + 20 + 24$$

$$= 68 \text{ equivalent min}$$

$$MS_a = \frac{(68)^2}{(68)^2 + (100)^2} \times 100 = 31.6 \text{ percent}$$

$$MS_t = (1 - 0.316) \times 100 = 68.4 \text{ percent}$$

Thus the mode choice of travel by transit between zones S and D is 68.4 percent, and by highway the value is 31.6 percent. These percentages are applied to the estimated trip distribution values to determine the number of trips by each mode. If, for example, the number of work trips between zones S and D were computed to be 500, then the number by auto would be $500 \times 0.316 = 158$, and by transit the number of trips would be $500 \times 0.684 = 342$.

Logit Models

An alternative approach used in transportation demand analysis is to consider the relative utility of each mode as a summation of each modal attribute. Then the choice of a mode is expressed as a probability distribution. For example, assume that the utility of each mode is

$$U_x = \sum_{i=1}^{n} a_i X_i \tag{12.8}$$

where
    U_x = utility of mode x
    n = number of attributes
    X_i = attribute value (time, cost, and so forth)
    a_i = coefficient value for attribute i (negative, since the values are disutilities)

Then if 2 modes, auto (A) and transit (T), are being considered, the probability of selecting the auto mode A can be written as

$$P(A) = \frac{e^{U_A}}{e^{U_A} + e^{U_T}} \tag{12.9}$$

This form is called the *logit model* and provides a convenient way to compute mode choice. Choice models are utilized within the urban transportation planning process but they are also used in transit marketing studies and in directly estimating travel demand. An illustration of a logit curve is shown in Figure 12.10.

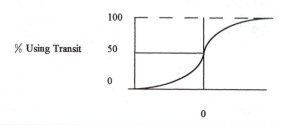

Figure 12.10 Modal Choice for Transit Versus Automobile

Example 12.9 Use of Logit Model to Compute Mode Choice

The utility functions for auto and transit are as follows:

$$\text{Auto: } U_A = -0.46 - 0.35T_1 - 0.08\,T_2 - 0.005C$$

$$\text{Transit: } U_T = -0.07 - 0.05T_1 - 0.15T_2 - 0.005C$$

where
T_1 = total travel time (minutes)
T_2 = waiting time (minutes)
C = cost (cents)

The travel characteristics between two zones are as follows:

	Auto	Transit
T_1	20	30
T_2	8	6
C	320	100

Solution: Use the logit model to determine the percent of travel in the zone by auto and transit.

$$U_x = \sum_{i=1}^{n} a_i x_i$$

$$U_A = -0.46 - (0.35 \times 20) - (0.08 \times 8) - (0.005 \times 320) = -9.70$$

$$U_B = -0.07 - (0.35 \times 30) - (0.08 \times 6) - (0.005 \times 100) = -11.55$$

Using Equation 12.9 yields

$$P_A = \frac{e^{U_A}}{e^{U_A} + e^{U_T}} = \frac{e^{-9.70}}{e^{-9.7} + e^{-11.55}} = 0.86$$

$$P_T = \frac{e^{U_T}}{e^{U_A} + e^{U_T}} = \frac{e^{-11.55}}{e^{-9.7} + e^{-11.55}} = 0.14$$

TRAFFIC ASSIGNMENT

The final step in the transportation forecasting process is to determine the actual street and highway routes that will be used and the number of automobiles and buses that can be expected on each highway segment. The procedure used to determine the expected traffic volumes is known as *traffic assignment*. Up to this point, we know the number of trips by transit and auto that will travel between zones. We now assign these trips to a logical highway route and sum up the results for each highway segment. Our result is what we have been seeking: a forecast of the average daily or peak hour traffic volumes that will occur on the urban transportation system that serves the study area.

To carry out a trip assignment, the following data are required: First, we need to know how many trips will be made from one zone to another (this information was determined in the trip distribution phase). Second, we need to know the available highway or transit routes between zones and how long it will take to travel on each route. Third, we need a decision rule (or algorithm) that states the criteria by which motorists or transit users will select a route.

Basic Approaches

Three basic approaches can be used for traffic assignment purposes: (1) diversion curves, (2) minimum time path (all-or-nothing) assignment, and (3) minimum time path with capacity restraint. The *diversion curve* method is similar in approach to a mode choice curve. The traffic between two routes is determined as a function of relative travel time or cost. Figure 12.11 illustrates a diversion curve based on travel time ratio.

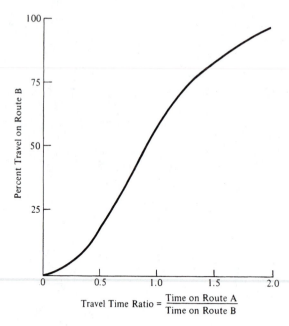

Figure 12.11 Travel Time Ratio Versus Percentage of Travel on Route B

The *minimum time path* method assigns all trips to those links that comprise the shortest time path between the two zones. *Capacity restraint* is a refinement of the minimum path method in that, after all traffic has been assigned to a link, the travel times on each link are adjusted, based on the capacity of the link and the number of trips on each link. The capacity restraint method requires repeated assignments and adjustments of travel time until a balance is achieved.

Minimum Path Algorithm

The traffic assignment process is illustrated using the minimum path algorithm. This method is selected because it is commonly used, generally produces accurate results, and adequately demonstrates the basic principles involved.

The minimum path assignment is based on the theory that a motorist or transit user will select the quickest route between any O-D pair. In other words, the traveler will always select the route that represents minimum travel time. Thus, to determine which route that will be, it is necessary to find the shortest route from the zone of origin to all other destination zones. The results can be depicted as a tree, referred to as a *skim tree*. All trips from that zone are assigned to links on the skim tree. Each zone is represented by a node in the network, which represents the entire area being examined. To determine the minimum path, a procedure is used that finds the shortest path without having to test all possible combinations.

The algorithm that will be used in the example below is to connect all nodes from the home (originating) node and keep all paths as contenders until one path to the same node is a faster route than others, at which juncture those links on the slower path are eliminated.

The general mathematical algorithm that describes the process is to select paths that minimize the expression

$$\sum_{\text{all } ij} V_{ij} T_{ij} \qquad (12.10)$$

where

V_{ij} = volume on link i, j
T_{ij} = travel time on link i, j
i, j = adjacent nodes

Example 12.10 Finding Minimum Paths in a Network

To illustrate the process of path building, consider the following 16-node network with travel times on each link shown for each node (zone) pair:

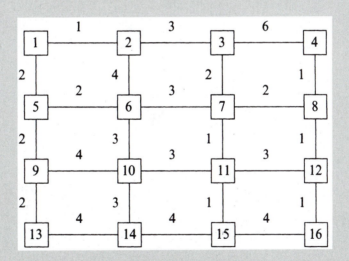

The link and node network is representative of the road and street system. We wish to determine the shortest travel path from node 1 (home node) to all other zones.

Solution: To determine minimum time paths from node 1 to all other nodes, proceed as follows:

Step 1.   Determine the time to nodes connected to node 1. Time to node 2 is 1 min. Time to node 5 is 2 min. Times are noted near nodes in diagram.

Step 2. From the node closest to the home node (node 2 is the closest to home node 1), make connections to nearest nodes. These are nodes 3 and 6. Write the cumulative travel times at each node.

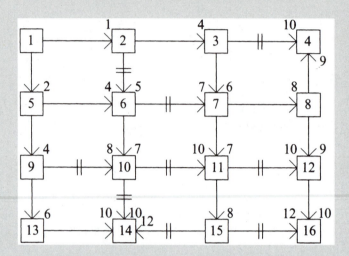

Step 3. From the node that is now closest to the home node (node 5), make connections to the nearest nodes (node 6 and 9). Write the cumulative travel times at each node.

Step 4. Time to node 6 via node 5 is shorter than that via zone 2. Therefore, link 2–6 is deleted.

Step 5. Three nodes are equally close to the home node (nodes 3, 6, and 9). Select the lowest-numbered node (node 3) and add corresponding links to nodes 4 and 7.

Step 6. Of the three equally close nodes, node 6 is the next closest to the home node. Connect to zone 7 and 10. Eliminate link 6–7.

Step 7. Building proceeds from node 9 to nodes 10 and 13. Eliminate link 9–10.

Step 8. Build from node 7.

Step 9. Build from node 13.

Step 10. Build from node 10, and eliminate link 10–11.

Step 11. Build from node 11.

Step 12. Build from node 8, and eliminate link 11–12.

Step 13. Build from node 15, and eliminate link 14–15.

Step 14. Build from node 12, and eliminate link 15–16.

To find the minimum path from any node to node 1, follow the path backwards. Thus, for example, the links on the minimum path from zone 1 to zone 11 are 7–11, 3–7, 2–3, and 1–2. This process is then repeated for the other 15 zones to produce the skim trees for each of the zones in the study area. Figure 12.12 illustrates the skim tree produced for zone 1.

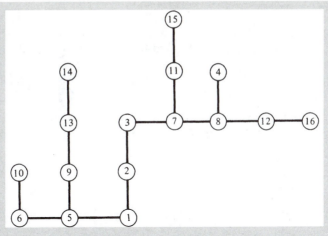

Figure 12.12 Minimum Path Tree for Zone 1

Example 12.11   Network Loading Using Minimum Path Method

The links that are on the minimum path for each of the nodes connecting node 1 are shown in Table 12.18. Also shown are the number of auto trips between zone 1 and all other zones. From these results, the number of trips on each link is determined. To illustrate, link 1–2 is used by trips from node 1 to nodes 2, 3, 4, 7, 8, 11, 12, 15, and 16. Thus, the trips between these node pairs are assigned to link 1–2 as illustrated in Table 12.18. The volumes are 50, 75, 80, 60, 30, 80, 25, 20, and 85 for a total of 505 trips on link 1–2 from node 1.

Table 12.18 Links on Minimum Path for Trips from Node 1

From	To	Trips	Links on the Minimum Path
1	2	50	1–2
	3	75	1–2, 2–3
	4	80	1–2, 2–3, 3–7, 7–8, 4–8
	5	100	1–5
	6	125	1–5, 5–6
	7	60	1–2, 2–3, 3–7
	8	30	1–2, 2–3, 3–7, 7–8
	9	90	1–5, 5–9
	10	40	1–5, 5–6, 6–10
	11	80	1–2, 2–3, 3–7, 7–11
	12	25	1–2, 2–3, 3–7, 7–8, 8–12
	13	70	1–5, 5–9, 9–13
	14	60	1–5, 5–9, 9–13, 13–14
	15	20	1–2, 2–3, 3–7, 7–11, 11–15

Solution: We can now calculate the number of trips that should be assigned to each link of those that have been generated in node 1 and distributed to nodes 2 through 16 (Table 12.19). A similar process of network loading would be completed for all other zone pairs. Calculations for traffic assignment, as well as for other stops in the forecasting model system, can be performed using the microcomputer program TRIPS.

Table 12.19 Assignment of Trips from Node 1 to Links on Highway Network

Link	Trips on Link	
1–2	50, 75, 80, 60, 30, 80, 25, 20, 85 =	505
2–3	75, 80, 60, 30, 80, 25, 20, 85 =	455
1–5	100, 125, 90, 40, 70, 60 =	485
5–6	125, 40 =	165
7–8	80, 30, 25, 85 =	220
4–8	80 =	80
5–9	90, 70, 60 =	220
6–10	40 =	40
7–11	80, 20 =	100
8–12	25, 85 =	110
9–13	70, 60 =	130
11–15	20 =	20
12–16	85 =	85
13–14	60 =	60

Capacity Restraint.   A modification of the process just described is known as *capacity restraint.* The number of trips assigned to each link is compared with the capacity of the link to determine the extent to which link travel times have been reduced. Using relationships between volume and travel time (or speed) similar to those derived in Chapter 6, it is possible to recalculate the new link travel time. A reassignment is then made based on these new values. The iteration process continues until an equilibrium balance is achieved.

The speed-volume relationship most commonly used in computer programs, which was developed by the U.S. Department of Transportation, is depicted by Figure 12.13 and expressed in the following formula:

$$t = t_o \left[1 + 0.15 \left(\frac{V}{C} \right)^4 \right] \tag{12.11}$$

where
t = travel time on the link
t_o = free-flow travel time
V = volume on the link
C = capacity of the link

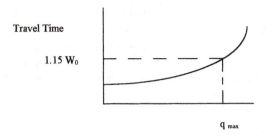

Figure 12.13 Travel Time Versus Vehicle Volume

Example 12.12 Computing Capacity Restrained Travel Times

In Example 12.11, the volume on link 1–5 was 485 and the travel time was 2 minutes. If the capacity of the link is 500, determine the link travel time that should be used for the next traffic assignment iteration.

Solution:

$$t_1 = t_o \left[1 + 0.15 \left(\frac{V}{C} \right)^4 \right]$$

$$t_{1-5} = 2 \left[1 + 0.15 \left(\frac{485}{500} \right)^4 \right]$$

$$= 2.27 \ \text{min}$$

The process of calculating the travel demand for an urban transportation system is now completed. The results of this work will be used to determine where improvements will be needed in the system, to make economic evaluations of project priority, and to assist in the geometric and pavement design phases. In actual practice, the calculations are carried out by computer, because the process becomes computationally more intensive as the number of zones increases.

OTHER METHODS FOR FORECASTING DEMAND

This chapter has described how travel demand is forecasted by using the four-step procedure of trip generation, distribution, mode choice, and traffic assignment. There are many variations within each of these steps, and the interested reader should refer to the references cited for additional details. Furthermore, there are other methods that can be used to forecast demand. Some of these are listed below.

Trend Analysis

This approach to demand estimation is based simply on the extrapolation of past trends. For example, to forecast the amount of traffic on a rural road, traffic count data from previous years are plotted versus time. Then, to compute the volume of traffic at some future date, the *trend line* is extrapolated forward, or else the average growth rate is used. Often a mathematical expression is developed using statistical techniques, and quite often a semilog relationship is used. Although simple in application, trend line analysis has the disadvantage that future demand estimates are based on extrapolations of the past, and thus no allowance is made for changes that may be time-dependent.

Demand Elasticity

Travel demand can also be determined if the relationship between demand and a key service variable (such as travel cost) is known. If V is the volume (demand) at a given service level X, then the elasticity of demand, $E(V)$, is the percent change in volume divided by the percent change in service level, or

$$E(V) = \frac{\% \, \Delta \, \text{in} \, V}{\% \, \Delta \, \text{in} \, X}$$

$$E(V) = \frac{\Delta V/V}{\Delta X/X} = \frac{X}{V} \frac{\Delta V}{\Delta X}$$

Example 12.13   Forecasting Transit Ridership Reduction Due to Increase in Fares

To illustrate the use of demand elasticity, a rule of thumb in the transit industry states that for each 1 percent increase in fares, there will be one-third of 1 percent reduction in ridership. If current ridership is 2000/day at a fare of 30¢, what will the ridership be if the fare is increased to 40¢?

Solution: In this case $E(V) = 1/3$, and

$$\frac{1}{3} = \frac{X}{V} \frac{\Delta V}{\Delta X} = \frac{30}{2000} \times \frac{\Delta V}{10} \quad or \quad \frac{30 \Delta V}{(2000)(10)}$$

$$\Delta V = 222$$

This means that new ridership is $2000 - 222 = 1778$ passengers/day.

SUMMARY

The process of forecasting travel demand is necessary to determine the number of persons or vehicles that will use a new transportation system or component. The methods used to forecast demand include extrapolation of past trends, elasticity of demand, and relating travel demand to socioeconomic variables.

Urban travel demand forecasting is a complex process because demand for urban travel is influenced by the location and intensity of land use, the socioeconomic characteristics of the population, and the extent, cost, and quality of transportation services.

Forecasting urban travel demand involves a series of tasks. These include population and economic analysis, land-use forecasts, trip generation, trip distribution, mode choice, and traffic assignment. The development of computer programs to calculate the elements within each task has greatly simplified implementation of the demand forecasting process.

Travel demand forecasts are also required for completing an economic evaluation of various system alternatives. This topic is described in the next chapter.

PROBLEMS

12-1   Identify and briefly describe the two basic demand forecasting situations in transportation planning.

12-2   What are the three factors that affect the demand for urban travel?

12-3   Define the following terms: (a) home-based work (HBW) trips, (b) home-based other (HBO) trips, (c) nonhome-based (NHB) trips, (d) production, (e) attractions, (f) origin, and (g) destination.

12-4   The following cross-classification data have been developed for Jeffersonville Transportation Study Area.

($000) Income	HH (%) High	HH (%) Med	HH (%) Low	Autos/HH (%) 0	Autos/HH (%) 1	Autos/HH (%) 2	Autos/HH (%) 3	Trip Rate/Auto 0	Trip Rate/Auto 1	Trip Rate/Auto 2	Trip Rate/Auto 3+	Trips (%) HBW	Trips (%) HBO	Trips (%) NHB
10	0	30	70	48	48	4	0	2.0	6.0	11.5	17.0	38	34	28
20	0	50	50	4	72	24	0	2.5	7.5	12.5	17.5	38	34	28
30	10	70	20	2	53	40	5	4.0	9.0	14.0	19.0	35	34	31
40	20	75	5	1	32	52	15	5.5	10.5	15.5	20.5	27	35	38
50	50	50	0	0	19	56	25	7.5	12.0	17.0	22.0	20	37	43
60	70	30	0	0	10	60	30	8.0	13.0	18.0	23.0	16	40	44

Develop the family of cross-classification curves, and determine the number of trips produced (by purpose) for a traffic zone containing 500 houses with an average household income of $35,000. (Use high = 55,000; medium = 25,000; low = 15,000.)

12-5   A person travels to work in the morning and returns home in the evening. How many productions and attractions are generated in the work and residence zones?

12-6   The Federal Highway Administration's method for estimating trip productions and attractions is based on the use of cross classification. Describe and illustrate by sketches the procedures required for (a) trip production and (b) trip attraction.

12-7   The following socioeconomic data have been collected for the Jeffersonville Transportation Study (JTS).

Population = 72,173
Area = 70 square miles
Registered vehicles = 26,685

Single-family housing units = 15,675
Apartment units = 7567
Retail employment = 5502
Nonretail employment = 27,324
Student attendance = 28,551 by zone of attendance
Average household income = $17,500
Total traffic zones = 129

The results of the cross-classification analysis are as follows:

total trips produced for study area = 282,150 trips/day

Home-to-work	13 percent	(36,680)
Home-to-nonwork	62 percent	(174,933)
Nonhome trips	25 percent	(70,537)

The attraction rates for the study area have been developed using the following assumptions:

100 percent of home-to-work trips go to employment locations.
Home-to-nonwork trips are divided into the following types:

Visit friends	10 percent
Shopping	60 percent
School	10 percent
Nonretail employment	20 percent

Nonhome trips are divided into the following types:

Other employment area (nonretail)	60 percent
Shopping	40 percent

Determine the number of home-to-work, home-to-nonwork, and nonhome-based trips attracted to a zone with the following characteristics:

Population = 1920
Dwelling units = 800
Retail employment = 50
Nonretail employment = 820
School attendance = 0

12-8   A small town has been divided into three traffic zones. An origin-destination survey was conducted earlier this year and yielded the number of trips between each zone, as shown in the table below. Travel times between zones were also determined. Provide a trip distribution calculation using the gravity model for two iterations. Assume $K_{ij} = 1$.

The following table shows the number of productions and attractions in each zone:

Zone	1	2	3	Total
Productions	250	450	300	1000
Attractions	395	180	425	1000

The survey's results for the zones' travel time in minutes was as follows:

Zone	1	2	3
1	6	4	2
2	2	8	3
3	1	3	5

The following table shows travel time versus friction factor:

Time (minutes)	Friction Factor
1	82
2	52
3	50
4	41
5	39
6	26
7	20
8	13

12-9 The Jeffersonville Transportation Study Area has been divided into four large districts (traffic zones). The following data have been collected for those districts. Provide a trip distribution calculation using the gravity model for two iterations. Assume $K_{ij} = 1$.

District	Production	Attractions	Travel Times (min)			
			1	2	3	4
1	3400	2800	4	11	15	10
2	6150	6500	11	6	6	9
3	3900	2550	15	6	6	11
4	2800	4400	10	9	11	4

Travel Time	F_{ij}
1	2.0
4	1.6
6	1.0
9	0.9
10	0.86
11	0.82
12	0.80
15	0.68
20	0.49

12-10 The following table shows the productions and attractions used in the first iteration of a trip distribution procedure and the productions and attractions that resulted. Determine the number of productions and attractions that should be used for each zone in the second iteration.

	1	2	3	4
P	100	200	400	600
A	300	100	200	700
P^1	100	200	400	600
A^1	250	150	300	600

12-11 The Jeffersonville Transportation Study area has been divided into four large districts (traffic zones). The following data have been compiled:

District	Productions	Attractions	Travel Time (min) 1	2	3	4
1	1000	1000	5	8	12	15
2	2000	700	8	5	10	8
3	3000	6000	12	10	5	7
4	2200	500	15	8	7	5

Travel Time	F_{ij}
1	2.00
5	1.30
6	1.10
7	1.00
8	0.95
10	0.85
12	0.80
15	0.65

After the first iteration the trip table was

District	1	2	3	4	P_s
1	183	94	677	46	1000
2	256	244	1372	128	2000
3	250	186	2404	160	3000
4	180	183	1657	180	2200
As	869	707	6110	514	8200

Complete the second iteration.

12-12   For the travel pattern in Figure 12.14, develop a Fratar distribution for two iterations.

12-13   What data are required in order to use (a) the gravity model and (b) the Fratar model?

12-14   Determine the minimum path for nodes 1, 3, and 9 in Figure 12.15. Sketch the final trees.

12-15   Assign the vehicle trips shown in the O-D trip table to the network, shown in Figure 12.16 using the all-or-nothing assignment technique. Make a list of the links in the network and indicate the volume assigned to each. Calculate the total vehicle minutes of travel. Show the minimum path and assign traffic for each of the five nodes.

	Trips Between Zones				
From/To	1	2	3	4	5
1	0	100	100	200	150
2	400	0	200	100	500
3	200	100	0	100	150
4	250	150	300	0	400
5	200	100	50	350	0

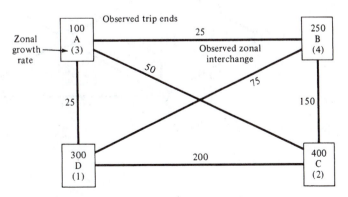

Figure 12.14 Travel Pattern

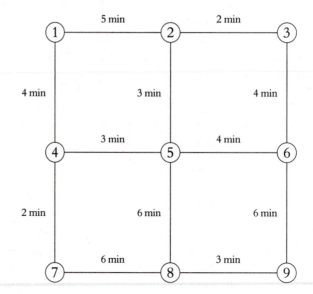

Figure 12.15 Link Node Network

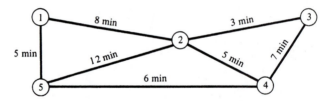

Figure 12.16 Highway Network

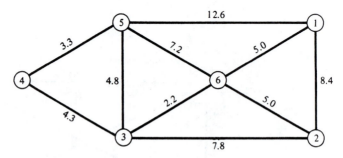

Figure 12.17 Link Travel Times

12-16   Figure 12.17 represents travel times on the links connecting six zonal centroids. Determine the minimum path from each zone to each other zone. Use the all-or-nothing trip assignment method to determine the total trips for each link after all of the trips from the following two-way trip table have been loaded onto the network.

Zone	\multicolumn{6}{c}{Trips Between Zones}
	1	2	3	4	5	6
1	0	1000	1100	400	1000	1300
2	—	0	1050	700	1100	1200
3	—	—	0	1200	1150	1600
4	—	—	—	0	800	400
5	—	—	—	—	0	700
6	—	—	—	—	—	0

REFERENCES

Application of Disaggregate Travel Demand Models, NCHRP Report 253, Transportation Research Board, National Research Council, Washington, D.C., 1982.

Application of New Travel Demand Forecasting Techniques to Transportation Planning: A Study of Individual Choice Modes, U.S. Department of Transportation, Federal Highway Administration, Washington, D.C., 1977.

Caldwell III, L. C., and Demetsky, M. J., *An Evaluation of the Transferability of Cross Classification Trip Generation Models,* VHTRC 78-R39, Virginia Highway and Transportation Research Council, Charlottesville, Va., 1978.

Center for Microcomputers in Transportation, University of Florida, Transportation Research Center, 1997.

Chang, C. L. and Meyers, D. T., "Transportation Models," Chapter 6, *Transportation Planning Handbook,* 2nd ed., Institute of Transportation Engineers, 1999.

Computer Programs for Urban Transportation Planning, U.S. Department of Transportation, Federal Highway Administration, Washington, D.C., 1977.

Cosmis Corporation, *Traffic Assignment,* U.S. Department of Transportation, Federal Highway Administration, Washington, D.C., 1974.

Edwards, Jr., J. D., editor, *Transportation Planning Handbook,* 2nd ed., Institute of Transportation Engineers, Prentice Hall, Englewood Cliffs, N.J., 1999.

Guidelines for Trip Generation Analysis, U.S. Department of Transportation, Federal Highway Administration, Washington, D.C., April 1973.

Kanafani, A., *Transportation Demand Analysis,* McGraw-Hill, New York, 1983.

Manheim, M. L., *Fundamentals of Transport Systems Analysis,* MIT Press, Cambridge, Mass., 1979.

Meyer, M. D., and Miller, E. J., *Urban Transportation Planning,* 2nd Edition, McGraw-Hill, New York, 2000.

Mode Choice Forecasting Methodology, Development and Calibration of the Washington Mode Choice Models, Technical Report 8, R.H. Pratt Associates, Inc., Washington, D.C., June 1973, Appendix II.

New Approaches to Understanding Travel Behavior, NCHRP Report 250, Transportation Research Board, National Research Council, Washington, D.C., 1982.

Pisarski, A. E., "Intercity Passenger Travel," *Transportation Planning Handbook,* 2nd ed., Chapter 11, Institute of Transportation Engineers, Prentice Hall, Englewood Cliffs, N.J., 1999.

Quick Response Urban Travel Estimation Techniques and Transferable Parameters, Users Guide, NCHRP Report No. 187, National Research Council, Transportation Research Board, Washington, D.C., 1978.

Shunk, G. A., "Urban Transportation Systems," Chapter 4, *Transportation Planning Handbook,* Institute of Transportation Engineers, 1992.

Stopher, P. R., and Meyberg, A. H., *Urban Transportation Modeling and Planning,* Lexington Books, D.C. Heath, Lexington, Mass., 1975.

Traffic Assignment and Distribution for Small Urban Areas, U.S. Department of Commerce, Bureau of Public Roads, Washington, D.C., September 1965.

UTPS Network Development Manual, U.S. Department of Transportation, Washington, D.C., 1974.

ADDITIONAL READINGS

An Introduction to Urban Development Models and Guidelines for Their Use in Urban Transportation Planning, U.S. Department of Transportation, Federal Highway Administration, Washington, D.C., 1975.

Bass, P., and Dresser, G. B., *Traffic Forecasting Requirements by Project Type,* Report No. FHWA/TX-95/1235-8, Texas Transportation Institute in cooperation with the Texas Department of Transportation and the Federal Highway Administration, U.S. Department of Transportation, August 1994.

Bell, M. G. H., and Y. Iida, *Transportation Network Analysis,* John Wiley & Sons, New York, 1997.

Black, J., *Urban Transport Planning,* Johns Hopkins University Press, Baltimore, Md., 1981.

Calibrating and Testing a Gravity Model for Any Size Urban Area, U.S. Department of Transportation, Federal Highway Administration, Washington, D.C., 1975.

Cohen, H. S., "Review of Empirical Studies of Induced Traffic," Appendix B of *Expanding Metropolitan Highways: Implications for Air Quality and Energy Use,* Special Report 245, Committee for Study of Impacts of Highway Capacity Improvements on Air Quality and Energy Use, Transportation Research Board, National Research Council, National Academy Press, Washington, D.C., 1995.

Consequences of Small Sample O-D Data Collection in the Transportation Planning Process, U.S. Department of Transportation, Federal Highway Administration, Washington, D.C., 1976.

Forecasting Impacts to Transportation Planning, NCHRP Report No. 266, Transportation Research Board, National Research Council, Washington, D.C., 1983.

Guidelines for Designing Travel Surveys for Statewide Transportation Planning, U.S. Department of Transportation, Federal Highway Administration, Washington, D.C., 1976.

Harvey, G. W., "Transportation Pricing and Travel Behavior," in *Curbing Gridlock: Peak-Period Fees to Reduce Congestion,* Special Report 242, Vol. 2, Committee for Study on Urban Transportation Congestion Pricing, Transportation Research Board, National Research Council, National Academy Press, Washington, D.C., 1994.

Keller, C. R., and Mehra, J., *Site Impact Traffic Evaluation Handbook,* U.S. Department of Transportation, Federal Highway Administration, Washington, D.C., January 1985.

Oppenheim, N., *Urban Travel Demand Modeling,* John Wiley & Sons, New York, 1995.

Parking Generation, 2nd ed., Institute of Transportation Engineers, Washington, D.C., 1987.

Report of the Conference on Economic and Demographic Methods for Projections of Population, American Statistical Association, Washington, D.C., 1977.

Trip Generation, 6th ed., Institute of Transportation Engineers, Washington, D.C., 1997.

CHAPTER 13

Evaluating Transportation Alternatives

In the previous chapter we described methods and techniques for establishing the demand for transportation services under a given set of conditions. The results of this process furnish the necessary input data to prepare an evaluation of the relative worth of alternative projects. In this chapter we describe various ways in which transportation project evaluations are carried out.

BASIC ISSUES IN EVALUATION

The basic concept of an evaluation is simple and straightforward, but the actual process itself can be complex and involved. A transportation project is usually proposed because of a perceived problem or need. For example, a project to improve safety at a railroad grade crossing may be based on citizen complaints about accidents or time delays at the crossing site. In most instances, there are many ways to solve the problem, and each solution or alternative will result in a unique outcome in terms of project cost and results. In the railroad grade crossing example, one solution would be to install gates and flashing lights; another solution would be to construct a grade-separated overpass. These two solutions are quite different in terms of their costs and effectiveness. The first solution will be less costly than the second, but it will also be less effective in reducing accidents and delays.

A transportation improvement can be viewed as a mechanism for producing a result desired by society at a price. The question is, will the benefits of the project be worth the cost? In some instances, the results may be confined to the users of the system (as in the case of the grade crossing), whereas in other instances, those affected may include persons in the community who do not use the system.

Prior to beginning an analysis to evaluate a transportation alternative, the engineer or planner should consider a number of basic questions and issues. These will assist in determining the proper approach to be taken, what data are needed, and what analytical techniques should be used. These issues are discussed in the following paragraphs.

Objectives of Evaluation

What information is needed for project selection? The objective of an evaluation is to furnish the appropriate information about the outcome of each alternative so that a selection can be made. The evaluation process should be viewed as an activity in which information relevant to the selection is available to the person or group who will make a decision. An essential input in the process is to know what information will be important in making a project selection. In some instances, a single criterion may be paramount (such as cost); in other cases, there may be many objectives to be achieved. The decision maker may wish to have the relative outcome of each alternative expressed as a single number, whereas at other times, it may be more helpful to see the results individually for each criteria and each alternative.

There are many methods and approaches for preparing a transportation project evaluation, and each one can be useful when correctly applied. In this chapter we first describe the considerations in selecting an evaluation method and discuss issues that are raised in the evaluation process. We then describe two classes of evaluation methods that are based on a single measure of effectiveness: the first reduces all outcomes to a monetary value, and the second reduces all outcomes to a numerical relative value. Finally, we discuss evaluation as a fact-finding process in which all outcomes are reported separately in a matrix format so that the decision maker has complete information about the project outcome. We also discuss how this information can be used in public forums for citizen input and how the decision process can be extended to include public participation.

Evaluations can also be made after a project is completed to determine if the outcomes for the project are as had been anticipated. Post facto evaluation can be very helpful in formulating information useful for evaluating similar projects elsewhere or in making modifications in original designs. In the final section of this chapter, we discuss the issues in post facto evaluations and illustrate the results of evaluations for completed projects.

Identifying Project Stakeholders

Who will use the information, and what are their viewpoints? A transportation project can affect a variety of groups in different ways. In some instances, only one or a few groups are involved; in other cases, many factions have an interest. Examples of groups that could be affected by a transportation project include the system users, transportation management, labor, citizens in the community, business, and local, state, and national governments. Each of these groups will be concerned with something different, and the viewpoint that each represents will influence the evaluation process itself.

For smaller, self-contained projects, those groups with something to gain or lose by the project—the *stakeholders*—will usually be limited to the system users and transportation management. For larger, regional-scale transportation projects, the number and variety of stakeholders will increase because the project will affect many groups in addition to the users and management. For example, a major project could increase business in the downtown area, or expanded construction activity could trigger an economic boom in the area. The project might also require the taking of land, or it could create other environmental effects. Thus, if the viewpoint is that of an individual traveler or business, the analysis can be made on narrow economic grounds. If the viewpoint is the community at large, then the analysis must consider a wider spectrum of concerns.

If the viewpoint is that of a local community, then the transfer of funds by grants from the state or federal government would not be considered to be a cost, whereas increases in land values within the area would be considered a benefit. However, if the viewpoint were expanded to a regional or state level, these grants and land value increases would be viewed as costs to the region or as transfers of benefits from one area to another. Thus, a clear definition of whose viewpoint is being considered in the evaluation is necessary if proper consideration is to be given to how these groups are either positively or negatively affected by each proposed alternative.

Selecting and Measuring Evaluation Criteria

What are the relevant criteria, and how should these be measured? A transportation project is intended to accomplish one or more goals and objectives, which are made operational as criteria. The numerical or relative results for each criteria are called *measures of effectiveness*. For example, in a railroad grade crossing problem, if the goal is to reduce accidents, the criteria can be measured as the number of accidents expected to occur for each of the alternatives considered. If another goal is to reduce waiting time, the criteria could be the number of minutes per vehicle consumed at the grade crossing. Non-quantifiable criteria can also be used and expressed in a relative scale, such as high, medium, and low.

Criteria selection is a basic element of the evaluation process because the measure used becomes the basis on which each project is compared. Thus, it is important that the criteria be related as closely as possible to the stated objective. To use a nontransportation example for illustration, if the objective of a course is to learn traffic and highway engineering, then a relevant criterion to measure results is exam grades, whereas a less relevant criterion is the number of class lectures attended. Both are measures of class performance, but the first is more relevant in measuring how well one achieved the stated objective.

Criteria not only must be relevant to the problem but also must have other attributes as well. They should be easy to measure and sensitive to changes made in each alternative. Also, it is advisable to limit the number of criteria to those that will be most helpful in reaching a decision in order to keep the analysis manageable for both the engineer who is doing the work and the person(s) who will act on the result. Too much information can be confusing and counterproductive and, rather than being helpful, could create uncertainty and could encourage a decision on political or other nonquantitative basis. Some examples of criteria used in transportation evaluation are listed in Table 13.1.

Measures of Effectiveness

How are measures of effectiveness used in the evaluation process itself? One approach is to convert each measure of effectiveness to a common unit, and then, for each alternative,

Table 13.1   Criteria for Evaluating Transportation Alternatives

- Capital Costs
  —Construction
  —Right of way
  —Vehicles
- Maintenance Costs
- Facility Operating Costs
- Travel Time Cost
  —Total hours and cost of system travel
  —Average door-to-door speed
  —Distribution of door-to-door speeds
- Vehicle Operating Costs
- Accident Costs

compute the summation for all measures. A common unit is money, and it may be possible to make a transformation of the relevant criteria to equivalent dollars and then compare each alternative from an economic point of view. For example, if the cost of an accident is known and the value of travel time can be determined, then for the railroad grade crossing problem, it would be possible to compute a single number that would represent the total cost involved for each alternative, since construction, maintenance, and operating costs are already known in dollar terms, and the accident and time costs can be computed using conversion rates.

A second approach is to convert each measure of effectiveness to a numerical score. For example, if a project alternative does well in one criterion, it is given a high score; if it does poorly in another criterion, it is given a low score. A single number can be calculated that represents the weighted average score of all the measures of effectiveness that were considered. This approach is similar to calculating grades in a course. The instructor establishes both a set of criteria to measure a student's performance (for example, homework, midterms, finals, class attendance, and a term paper) and weights for each criterion. The overall measure of the student's performance is the weighted sum of the outcome for each measure of effectiveness. Measures of effectiveness should be independent of each other if a summation procedure (such as adding grades) is to be used in the evaluation. If the criteria are correlated, then adding up the weighted scores will bias the outcome. (This would suggest, for example, that homework grades should not be included in a student's final grade since they may correlate with midterm results.)

A third way is to identify the measures of effectiveness for each alternative in a matrix form, with no attempt made to combine them. This approach furnishes the maximum amount of information without prejudging either how the measures of effectiveness should be combined or their relative importance.

Evaluation Procedures and Decision Making

How well will the evaluation process assist in making a decision? The decision maker typically needs to know what the costs of the project will be; in many instances, this alone will determine the outcome. Another question may be, do the benefits justify the expenditure of funds for transportation, or would the money be better spent elsewhere? The decision maker will also want to know if the proposed project is likely to produce the stated results—that is, how confident can we be of the predicted outcomes?

It may be necessary to carry out a sensitivity analysis that shows a range of values rather than a single number. Also, evaluations of similar projects elsewhere may provide clues to the probable success of the proposed venture. The decision maker also may wish to know if all the alternatives have been considered and how they compare with the one being recommended. Are there other ways to accomplish the objective, such as using management and traffic control strategies, that would eliminate the need for a costly construction project? It may be that providing separate bus and carpool lanes results in significant increases in the passenger-carrying capacity of a freeway, thus eliminating the need to build additional highway lanes.

The decision maker may want to know the cost to highway users as the result of travel delays during construction. Also of interest may be the length of time necessary to finish the project, since public officials are often interested in seeing work completed during their administration. The source of funds for the project and other matters dealing with its implementation will also be of concern. Thus, in addition to the fairly straightforward problem of evaluation based on a selected set of measurable criteria, the transportation engineer must be prepared to answer any and all questions about the project and its implications.

The evaluation process requires that the engineer have all appropriate facts about a proposed project and be able to convey these in a clear and logical manner to facilitate decision making. In addition to the formal numerical summaries of each project, the engineer also must be prepared to answer other questions about the project that relate to its political and financial feasibility. In the final analysis, the selection itself will be based on a variety of factors and considerations that reflect all of the inputs that a decision maker receives from the appropriate source.

EVALUATION BASED ON ECONOMIC CRITERIA

To begin the discussion of economic evaluation, it is helpful to consider the relationship between the supply and demand for transportation services. Consider a particular transportation project, such as a section of roadway or a bridge. Further, assume that we can calculate the cost involved for a motorist to travel on the facility. (These costs would include fuel, tolls, travel time, maintenance, and other actual or perceived out-of-pocket expenses.) Using methods described in Chapter 12 we can calculate the traffic volumes (or demand) for various values of user cost. As explained in Chapter 2, as the cost of using the facility decreases, the number of vehicles per day will increase. This relationship is shown schematically in Figure 13.1, and represents the demand curve for the facility for a particular group of motorists. A demand curve could shift upward or downward and have a different slope for users with different incomes or for various trip purposes. If the curve moved upward, it would indicate a greater willingness to pay, reflecting perhaps a group with a higher income. If the slope approached horizontal, it would indicate that demand is elastic, that is, that a small change in price would result in a large change in volume. If the slope approached vertical, it would indicate that the demand is inelastic, that is, that a large change in price has little effect on demand. As an example, the price of gasoline is said to be

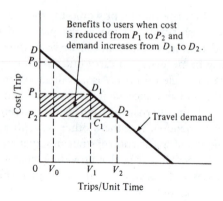

Figure 13.1   Demand Curve for Travel on a Given Facility

inelastic because people seem to drive equally as often after gas prices increase as before the increase.

Consider that the cost to travel on this facility is P_1. Then the number of trips per unit time will be V_1, and the total cost for all users over a given period per hour or day will be $(P_1) \times (V_1)$. This amount can be shown graphically as the area $0P_1D_1V_1$. As can be seen from the demand curve, all but the last user would have been willing to pay more than the actual price. For example, V_0 users would have been willing to pay P_0 to use the facility, whereas they paid the lesser amount P_1. The area under the demand curve $0DD_1V_1$ is the amount that the V_1 users would be willing to pay. If we subtract the area $0P_1D_1V_1$, which is the amount the users actually paid, we are left with the area P_1DD_1. This triangular area is referred to as a *consumer surplus* and represents the value of the economic benefit for the current users of the facility.

Suppose the price for using the facility is reduced to P_2 because of various improvements that have been made to the facility. The total user cost is now equal to $(P_2) \times (V_2)$, and the consumer surplus is the triangular area between the demand curve and P_2D_2 or P_2DD_2. The net benefit of the project is the net increase in consumer surplus, or area P_2DD_2 minus area P_1DD_1, and is represented by the shaded trapezoidal area $P_2P_1D_1D_2$. That area is made up of two parts. The first is the reduction in total cost paid by the original travelers, V_1, and is represented by the rectangular area $P_2P_1D_1C_1$. The second is the consumer surplus earned by the new users $V_2 - V_1$ and is represented by the triangular area $C_1D_1D_2$.

We now have a theoretical basis to calculate the net benefits to users of an improved transportation facility, which can then be compared with the improvement cost. For a linear demand curve, the formula for user benefits is

$$B_{2,1} = \frac{1}{2}(P_1 - P_2)(V_1 + V_2) \qquad (13.1)$$

where

$B_{2,1}$ = net benefits to transport users
P_1 = user cost of unimproved facility

P_2 = user cost of improved facility
V_1 = volume of travel on unimproved facility
V_2 = volume of travel on improved facility

As we have noted earlier, it is not practical to develop demand curves, but rather the four-step process described in Chapter 12 is used. In these instances, the value for the volume that is used in economic calculations is taken to be the number of trips that will occur on the improved facility. Equation 13.2, which replaces the term $1/2(V_1 + V_2)$ of Eq. 13.1 with V_2, has been commonly used in highway engineering studies:

$$B_{2,1} = (P_1 - P_2)(V_2) \tag{13.2}$$

This formula will overstate benefits unless demand is inelastic, that is, if the demand curve is vertical (i.e., $V_1 = V_2$).

To consider the economic worth of improving this transportation facility, we calculate the cost of the improvement and compare it with the cost of maintaining the facility in its present condition (the do-nothing alternative). One approach is to consider the difference in costs, to compare this with the difference in benefits, and then to select the project if the net increase in benefits exceeds the net increase in costs. Another approach is to consider the total costs of each alternative, including user and facility costs, and then to select the project that has the lowest total cost. Thus, to carry out an economic evaluation, it is necessary to develop the elements of cost for both the facility and the users. These include facility costs for construction, maintenance, and operation and user costs for travel time, accidents, and vehicle operations. The elements of cost are discussed in the following section.

Elements of Cost

The cost of a transportation facility improvement includes two components: *first cost* and *continuing costs.* Since we are concerned with cost differences, those costs that are common to both projects can be excluded. The first cost for a highway or transit project may include engineering design, right of way, and construction. Each transportation project is unique, and the specifics of the design will dictate what items will be required and at what cost. Continuing costs include maintenance, operation, and administration. These are recurring costs that will be incurred over the life of the facility and are usually based on historical data for similar projects. For example, if one alternative involves the purchase of buses, then the first (or capital) cost is the price of the bus, and the operating and maintenance cost will be known from manufacturer data or experience.

Expenses for administration or other overhead charges are usually excluded in an economic evaluation because they will be incurred regardless of whether or not the project is selected. Other excluded costs are those that have already been incurred. These are known as *sunk costs* and as such are not relevant to the decision of what to do in the future since these expenditures have already been made. For most capital projects, a service life must be determined and a salvage value estimated. *Salvage value* is the worth of an asset at the end of its service life. For example, a transit bus costing $150,000 may be considered to have a service life of 12 years and a salvage value of $20,000, and a concrete pavement may have a service life of 15 years and no salvage value. Suggested service lives for various facilities can be obtained from various transportation organizations, such as the American Association of State Highway and Transportation Officials and the American Public Transit Association.

Three commonly used measures of user costs are included in a transportation project evaluation: costs for vehicle operation, travel time costs, and costs of accidents. These costs are sometimes referred to as *benefits,* the implication being that the improvements to a transportation facility will reduce the cost for the users—that is, lower the perceived price, as shown on the demand curve—and result in a user benefit. It is simpler to consider these items in terms of relative cost because the data are needed in this format for purposes of an economic evaluation. The interactions between road user costs and highway geometric and operational factors are illustrated in Figure 13.2.

Vehicle Operating Costs

User costs for motor vehicle operation are significant items in a highway project evaluation. For example, a road improvement that eliminates grades, curves, and traffic signals as well as shortening the route can result in major cost reductions to the motorist. Agencies, such as the U.S. Department of Transportation and various vehicle manufacturers, furnish data about vehicle costs on highways or at intersections.

Travel Time Costs

One of the most important reasons for making transportation improvements is to increase speed or to reduce travel delay. In the world of trade and commerce, time is equivalent to money. For example, business ventures that furnish overnight delivery of small packages

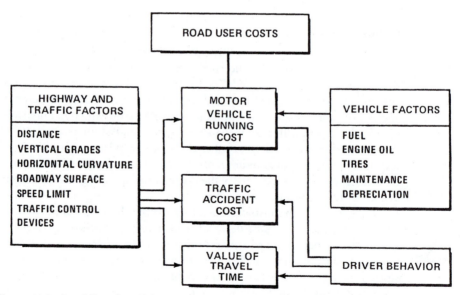

Figure 13.2 Road User Cost Factors

SOURCE: *Highway Engineering Economy,* U.S. Department of Transportation, Federal Highway Administration, April 1983, p. 28.

have grown and flourished. Transoceanic airline service replaced steamships because airplanes reduced the time to cross the ocean from 6–9 days to 6–9 hours. The tunnel between Great Britain and France (called the Chunnel), which opened in 1994, is intended to shorten trip time by replacing ferries with rail.

The method of handling travel time savings in an economic analysis has stirred considerable debate. No one disagrees that time savings have an economic value, but the question is, how should these be converted to dollar amounts (if at all)? One problem is that a typical stream of traffic contains both private and commercial vehicles, each of which values time savings quite differently. Time savings for a trucking firm can be translated directly into savings in labor cost by using an hourly rate for labor and equipment. Personal travel, on the other hand, is made for a variety of reasons; some are work related but many are not (shopping, school, social, recreation). Time saved in traveling to and from work can be related to wages earned, but time saved in other pursuits has little if any economic basis for conversion.

The value of time saved also depends on the length of trip and family income. If time savings are small—less than 5 minutes—they will not be perceived as significant and therefore have little value. If time savings are above a threshold level where they will make a noticeable difference in total travel time—over 15 minutes—then they could have significant economic value. The American Association of State Highway and Transportation Officials (AASHTO) recommends that for an average income of $20,000, time savings for trips less than 5 minutes are worth only $0.27 per hour, whereas the value increases by a factor of 18, to $2.32/hour, for trips greater than 15 minutes. The problem in using this type of time value data is that it is often difficult and costly to develop trip length and income distributions at the level of detail required.

The apparent monetary savings from even small travel time reductions can be quite large. For example, if a highway project that will carry average daily traffic (ADT) of 50,000 autos saves only 2 minutes per traveler, and the value of time for the average motorist is estimated conservatively at $5.00/hour, the total minimum annual savings is $50,000 \times (2/60) \times 365 \times 5 = \$3,041,667$. At 10 percent interest, these savings could justify spending a total of almost $26 million for a 20-year project life. Clearly, this result is an exaggeration of the actual benefits received. Although travel time does represent an economic benefit, the conversion to a dollar value is always open to question. AASHTO has used the average value approach. Others would argue that time savings should be credited only for commercial uses or stated simply in terms of actual value of number of hours saved.

The Texas Transportation Institute (TTI) has developed a computerized tool for analyzing benefits and costs of highway improvements. Values are furnished for travel time at $9.75/person/hour for auto and $22.53 for semitrailers. The TTI program, Micro Bencost, is a menu-driven program that can perform an economic analysis for projects such as capacity increases, bypasses, intersection/interchange development, pavement rehabilitation, bridges, additional safety, and highway-railroad grade crossings. The user's manual furnishes default values for user and facility cost parameters.

Accident Costs

Loss of life, injury, and property damage incurred in a transportation accident is a continuing national concern. Following every major air tragedy is an extensive investigation, and following the investigation, expenditures of funds are often authorized to improve the nation's air navigation system. Similarly, the 55 mi/h highway speed limit imposed by Congress following the oil crisis of 1973–1974 was retained long after the crisis was ended because it was credited with saving lives on the nation's highways. (In 1995, Congress removed all federal restrictions on speed limits.) It has been well established that the accident rate/million vehicle-miles is substantially lower on limited-access highways than on four-lane undivided roads. Reflecting the economic costs of accidents requires both an estimate of the number and type of accidents that are likely to occur over the life of the facility and an estimate of the value of each occurrence. Property damage and injury-related accidents can be valued using insurance data. The value of a human life is "priceless," but in economic terms, measures such as future earnings have been used. There is no simple numerical answer to the question, "What is the value of a human life or the cost of an accident?" although everyone would agree that economic value does exist. Published data vary widely, and the most prudent course, if an economic value is desired, is to select a value that appears most appropriate for the given situation.

Economic Evaluation Methods

An economic evaluation of a transportation project is completed using one of the following methods: present worth (PW), equivalent uniform annual cost (EUAC), benefit-cost ratio (BCR), or internal rate of return (ROR). Each method, when correctly used as shown in Example 13.1, will produce the same results. The reason for selecting one over the other is preference for how the results will be presented. Since transportation projects are usually built to serve traffic over a long period of time, it is necessary to consider the time-dependent value of money over the life of a project.

Present worth (PW) is the most straightforward of the methods, since it represents the current value of all the costs that will be incurred over the lifetime of the project. The general expression for present worth of a project is

$$PW = \sum_{n=0}^{N} \frac{C_n}{(1+i)^n} \tag{13.3}$$

where
C_n = facility and user costs incurred in year n
N = service life of the facility (in years)
i = rate of interest

Net present worth (NPW) is the present worth of a given cash flow that has both receipts and disbursements. The use of an interest rate in an economic evaluation is common practice because it represents the cost of capital. Money spent on a transportation project is no longer available for other investments. Therefore, a minimal value of interest rate is the rate that would have been earned if the money were invested elsewhere. For example, if $1000 were deposited in a bank at 8 percent interest, its value in 5 years would be $1000(1 + 0.08) = \$1469.33$. Thus the PW of having $1469.33 in 5 years at 8 percent interest is equal to $1000, and the opportunity cost is 8 percent. Discount rates can be higher or lower, depending on risk of investment and economic conditions.

It is helpful to use a cash flow diagram to depict the costs and revenues that will occur over the lifetime of a project. Time is plotted as the horizontal axis and money as the vertical axis, as illustrated in Figure 13.3. Using Eq. 13.4, we can calculate the NPW of the project, which is

$$\text{NPW} = \sum_{n=0}^{N} \frac{R_n}{(1+i)^n} + \frac{S}{(1+i)^n} - \sum_{n=0}^{N} \frac{M_n + O_n + U_n}{(1+i)^n} - C_o \qquad (13.4)$$

where

C_o = initial construction cost
n = a specific year
M_n = maintenance cost in year n
O_n = operating cost in year n
U_n = user cost in year n
S = salvage value
R_n = revenues in year n
N = service life, years

In this manner we have converted a time stream of costs and revenues into a single number: its NPW. The term $1/(1+i)^n$ is known as the present worth factor of a single payment and is written as $P/F - i - N$, where P is the present value given the future amount F, and N is the years of service life.

Equivalent uniform annual worth (EUAW) is a conversion of a given cash flow to a series of equal annual amounts. If the amounts are considered to occur at the end of the interest period, then the formula is

$$\text{EUAW} = \text{NPW}\left[\frac{i(1+i)^N}{(1+i)^N - 1} \right] = \text{NPW}(A/P - i - N) \qquad (13.5)$$

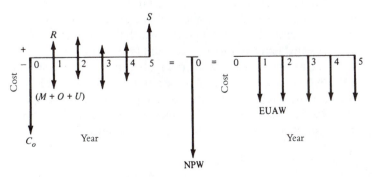

Figure 13.3 Typical Cash Flow Diagram for a Transportation Alternative and Equivalence as Net Present Worth of Annual Cost

Similarly,

$$\text{NPW} = \text{EUAW}\left[\frac{(1 + i)^N - 1}{i(1 + i)^N}\right] = \text{EUAW}(P/A - 1 - N) \qquad (13.6)$$

where
  EUAW = equivalent uniform annual worth
  NPW = net present worth
    i = interest rate, expressed as a decimal
   N = number of years

The term in the brackets in Eq. 13.5 is referred to as the capital recovery factor and represents the amount necessary to repay $1 if N equal payments are made at interest rate i. For example, if a loan is made for $5000 to be repaid in equal monthly payments over a 5-year period at 1 percent/month, then the amount is

$$\text{EUAW} = 5000\left[\frac{0.01(1 + 0.01)^{60}}{(1 + 0.01)^{60} - 1}\right] = 5000(0.02225) = 111.25$$

Thus, 60 payments of $111.25 would repay a $5000 debt, including both principal and interest. The NPW of a cash flow is converted to an EUAW by multiplying the NPW by the capital recovery factor.

The inverse of the capital recovery factor is the present worth factor for a uniform series, as stated in Eq. 13.6. Thus, the present value of 60 payments of $111.25, at 1 percent per month, is

$$\text{NPW} = 111.25\left[\frac{(1 + 0.01)^{60} - 1}{0.01(1 + 0.01)^{60}}\right] = 111.25(44.96) = 5000$$

Formula solutions for values of i and N that convert a monetary value from a future to a present time period $(P/F - i - N)$ and from a present time period to equal end-of-period payments $(A/P - i - N)$ are tabulated in textbooks on engineering economics. Table 13.2 lists values of single-payment present worth factors (P/F) and capital recovery factors (A/P) for a selected range of interest rates and time periods.

The *benefit-cost ratio* (BCR) is a ratio of the present worth of net project benefits and net project costs. This method is used in situations where it is desired to show the extent to which an investment in a transportation project will result in a benefit to the investor. To do this, it is necessary to make project comparisons to determine how the added investment compares with the added benefits. The formula for BCR is

$$\text{BCR}_{2/1} = \frac{B_{2/1}}{C_{2/1}}$$

Table 13.2 Present Worth and Capital Recovery Factors

N	$i = 3$		$i = 5$		$i = 10$		$i = 15$	
	(P/F)	(A/P)	(P/F)	(A/P)	(P/F)	(A/P)	(P/F)	(A/P)
1	0.9709	1.0300	0.9524	1.0500	0.9091	1.1000	0.8696	1.1500
2	0.9426	0.5226	0.9070	0.5378	0.8264	0.5762	0.7561	0.6151
3	0.9151	0.3535	0.8638	0.3672	0.7513	0.4021	0.6575	0.4380
4	0.8885	0.2690	0.8227	0.2820	0.6830	0.3155	0.5718	0.3503
5	0.8626	0.2184	0.7835	0.2310	0.6209	0.2638	0.4972	0.2983
10	0.7414	0.1172	0.6139	0.1295	0.3855	0.1627	0.2472	0.1993
15	0.6419	0.0838	0.4810	0.0963	0.2394	0.1315	0.1229	0.1710
20	0.5537	0.0672	0.3769	0.0802	0.1486	0.1175	0.0611	0.1598
25	0.4776	0.0574	0.2953	0.0710	0.0923	0.1102	0.0304	0.1547
30	0.4120	0.0510	0.2314	0.0651	0.0573	0.1061	0.0151	0.1523
35	0.3554	0.0465	0.1813	0.0611	0.0356	0.1037	0.0075	0.1511
40	0.3066	0.0433	0.1420	0.0583	0.0221	0.1023	0.0037	0.1506
45	0.2644	0.0408	0.1113	0.0563	0.0137	0.1014	0.0019	0.1503
50	0.2281	0.0389	0.0872	0.0548	0.0085	0.1009	0.0009	0.1501

where

$B_{2/1}$ = reduction in user and operation costs between higher cost alternative 2
      and lower cost alternative 1, expressed as PW or EUAW

$C_{2/1}$ = increase in facility costs, expressed as PW or EUAW

If the BCR is 1 or greater, then the higher cost alternative is economically attractive. If the BCR is less than 1, this alternative is discarded.

Correct application of the BCR method requires that costs for each alternative be converted to PW or EUAW values. The proposals must be ranked in ascending order of capital cost, including the do-nothing alternative, which usually has little if any initial cost. Incremental BCRs are calculated for pairs of projects, beginning with the lowest cost alternative. If the higher cost alternative yields a BCR less than 1, it is eliminated and the next higher cost alternative is compared with the lower cost alternative. If the higher cost alternative yields a BCR equal to or greater than 1, it is retained and the lower cost alternative is eliminated. This process continues until every alternative has been compared. The alternative selected is the one with the highest initial cost and a BCR of 1 or more with respect to lower cost alternatives and a BCR less than 1 when compared with all higher cost projects.

The *internal rate-of-return* (ROR) method determines the interest rate at which the PW of reductions in user and operation costs $B_{2/1}$ equals the PW of increases in facility costs $C_{2/1}$. If the ROR exceeds the interest rate (referred to as minimum attractive rate of return), the higher cost project is retained. If the ROR is less than the interest rate, the higher priced project is eliminated. The procedure for comparison is similar to that used in the BCR method.

Example 13.1   Illustration of Economic Analysis Methods

The Department of Traffic is considering three improvement plans for a heavily traveled intersection within the city. The intersection improvement is expected to achieve three goals: improve travel speeds, increase safety, and reduce operating expenses for motorists. The annual dollar value of savings compared with existing conditions for each criterion as well as additional construction and maintenance costs is shown in Table 13.3. If the economic life of the road is considered to be 50 years and the discount rate is 3 percent, which alternative should be selected? Solve the problem using the four methods for economic analysis.

Solution:

- Compute the NPW of each project.

$$(P/A - 3 - 50) = \frac{(1+i)^N - 1}{i(1+i)^N} = \frac{(1+0.03)^{50} - 1}{0.03(1+0.03)^{50}} = 25.729$$

$$\text{NPW}_\text{I} = -185{,}000 + (-1500 + 5000 + 3000 + 500)(P/A - 3 - 50)$$

$$= -185{,}000 + (7000)(25.729) = -185{,}000 + 180{,}103$$

$$= -4897$$

$$\text{NPW}_\text{II} = -220{,}000 + (-2500 + 5000 + 6500 + 500)(P/A - 3 - 50)$$

$$= -220{,}000 + (9500)(25.729) = -220{,}000 + 244{,}425$$

$$= +24{,}465$$

$$\text{NPW}_\text{III} = -310{,}000 + (-3000 + 7000 + 6000 + 2800)(P/A - 3 - 50)$$

$$= -310{,}000 + (12{,}800)(25.729) = -310{,}000 + 329{,}331$$

$$= +19{,}331$$

The project with the highest NPW is alternative II.

Table 13.3   Cost and Benefits for Improvement Plans with Respect to Existing Conditions

Alternative	Construction Cost	Annual Savings in Accidents	Annual Travel Time Benefits	Annual Operating Savings	Annual Additional Maintenance Cost
I	$185,000	$5000	$3000	$ 500	$1500
II	220,000	5000	6500	500	2500
III	310,000	7000	6000	2800	3000

- Solve by the EUAW method. Note $(A/P - 3 - 50) = 1/25.729 = 0.03887$.

$$\text{EUAW}_I = -185,000(A/P - 3 - 50) - 1500 + 5000 + 3000 + 500$$

$$= -185,000(0.03887) + 7000 = -7190 + 7000$$

$$= -190$$

$$\text{EUAW}_{II} = -220,000(A/P - 3 - 50) - 2500 + 5000 + 6500 + 500$$

$$= -220,000(0.03887) + 9500 = -8551 + 9500$$

$$= +949$$

$$\text{EUAW}_{III} = -310,000(0.03887) - 3000 + 7000 + 6000 + 2800$$

$$= -12,050 + 12,800$$

$$= +750$$

The project with the highest EUAW is alternative II, which is as expected since EUAW = NPW(0.03887).

- Solve by the BCR method.

1. Compare the BCR of alternative I with respect to do-nothing (DN).

$$\text{BCR}_{I/DN} = \frac{180,103}{185,000} = 0.97$$

Since $\text{BCR}_{I/DN}$ is less than 1, we would not build alternative I.

2. Compare BCR of alternative II with respect to DN.

$$\text{BCR}_{II/DN} = \frac{244,425}{220,000} = 1.11$$

Since BCR > 1, we would select alternative II over DN.

3. Compare BCR of alternative III with respect to alternative II.

$$\text{BCR} = \frac{(329,331) - (244,425)}{(310,000) - (220,000)} = \frac{84,906}{90,000} = 0.94$$

Since BCR is less than 1, we would not select alternative III, and we reach the same conclusion as previously, which is to select alternative II.

- Solve by the ROR method. In this situation, we solve for the value of interest rate for which NPW = 0.

1. Compute ROR for alternative I versus DN. (Recall that all values are with respect to existing conditions.)

$$NPW = 0 = -185,000 + (-1500 + 5000 + 3000 + 500) \times (P/A - i - 50)$$

$$(P/A - i - 50) = 185,000/7000$$

$$(P/A - i - 50) = 26.428$$

$$i = 2.6 \text{ percent}$$

Since the ROR is lower than 3 percent we discard alternative I.

2. Compute ROR for alternative II versus DN.

$$NPW = 0 = -220,000 + (-2500 + 5000 + 6500 + 500) \times (P/A - i - 50)$$

$$(P/A - i - 50) = 220,000/9500$$

$$(P/A - i - 50) = 23.16$$

$$i = 3.6 \text{ percent}$$

Since ROR is greater than 3 percent, select alternative II over DN.

3. Compute ROR for alternative III versus alternative II.

$$NPW = 0 = -(310,000 - 220,000) + (12,800 - 9500) \times (P/A - i - 50)$$

$$(P/A - i - 50) = 90,000/3300$$

$$(P/A - i - 50) = 27.27$$

$$i = 2.7 \text{ percent}$$

Since the increased investment in alternative III yields an ROR less than 3 percent, we do not select it but again pick alternative II.

The preceding example illustrates the basic procedures used in an economic evaluation. Four separate methods were used, each producing the same result. The PW or EUAC method is simplest to understand and apply and is recommended for most purposes when the economic lives of each alternative are equal. The BCR gives less information to the decision maker and must be carefully applied if it is to produce the correct answer. (For example, the alternative with the highest BCR with respect to the do-nothing case is not necessarily the best.) The ROR method requires more calculations but does provide additional information. For example, the highest ROR in the preceding problem was 3.6 percent. This says that if the minimum attractive ROR were greater than 3.6 percent (say 5 percent), none of the projects would be economically attractive.

EVALUATION BASED ON MULTIPLE CRITERIA

Many problems associated with economic methods limit their usefulness. Among these are

- Converting criteria values directly into dollar amounts.
- Choosing the appropriate value of interest rate and service life.
- Distinguishing between the user groups that benefit from a project and those that pay.
- Failing to distinguish between groups that benefit and those that lose.
- Considering all costs, including external costs.

For these reasons, economic evaluation methods should be used either in narrowly focused projects or as one of many inputs in larger projects. The next section discusses evaluation methods that seek to include measurable criteria that are not translated only into monetary terms.

Rating and Ranking

Numerical scores are helpful in comparing the relative worth of alternatives in cases where criteria values cannot be transformed into monetary amounts. The basic equation is as follows:

$$S_i = \sum_{j=1}^{N} K_j V_{ij} \tag{13.7}$$

where

S_i = total value of score of alternative i
K_j = weight placed on criteria j
V_{ij} = relative value achieved by criteria j for alternative i

The application of this method is illustrated by the following example.

Example 13.2   Evaluating Light-Rail Transit Alternatives Using the Rating and Ranking Method

A transportation agency is considering the construction of a light-rail transit line from the center of town to a growing suburban region. The transit agency wishes to examine five alternative alignments, each of which has advantages and disadvantages in terms of cost, ridership, and service provided. The alternatives differ in length of the line, location, types of vehicles used, seating arrangements, operating speeds, and numbers of stops. Estimated values achieved by each criterion for each of the five alternatives are shown in Table 13.4. The agency wants to evaluate each alternative using a ranking process. Determine which project should be selected.

Table 13.4 Estimated Values for Measures of Effectiveness

		Alternatives				
Number	Measure of Effectiveness	I	II	III	IV	V
1	Annual return on investment (%)	13.0	14.0	11.0	13.5	15.0
2	Daily ridership (1000s)	25	23	20	18	17
3	Passengers seated in peak hour (%)	25	35	40	50	50
4	Length of line (mi)	8	7	6	5	5
5	Auto drivers diverted (1000s)	3.5	3.0	2.0	1.5	1.5

Solution:

Step 1. Identify the goals and objectives of the project. The transit agency has determined that five major objectives should be achieved by the new transit line.

1. Net revenue generated by fares should be as large as possible with respect to the capital investment.
2. Ridership on the transit line should be maximized.
3. Service on the system should be comfortable and convenient.
4. The transit line should extend as far as possible to promote development and accessibility.
5. The transit line should divert as many auto users as possible during the peak hour in order to reduce highway congestion.

Step 2. Develop the alternatives that will be tested. In this case five alternatives have been identified as feasible candidates. These vary in length from 5 to 8 miles. The alignment, the amount of the system below, at, and above grade, vehicle size, headways, number of trains, and other physical and operational features of the line are determined in this step.

Step 3. Define an appropriate measure of effectiveness for each objective. For the objectives listed in step 1, the following measures of effectiveness are selected:

Objective	Measure of Effectiveness
1	Net annual revenue divided by annual capital cost
2	Total daily ridership
3	Percent of riders seated during the peak hour
4	Miles of extension into the corridor
5	Number of auto drivers diverted to transit

Step 4.   Determine the relative weight for each objective. This step requires a subjective judgment on the part of the group making the evaluation and will vary among individuals and vested interests. One approach is to allocate the weights on a 100-point scale (just as would be done in developing final grade averages for a course). Another approach is to rank each objective in order of importance and then use a formula of proportionality to obtain relative weights. In this example, the objectives are ranked as shown in Table 13.5. The weighting factor is determined by assigning the value n to the highest ranked alternative, $n - 1$ to the next highest, and so forth and computing a relative weight as

$$K_j = \frac{W_j}{\sum\limits_{j=1}^{n} W_j} \tag{13.8}$$

where

K_j = weighting factor of objective j

W_j = relative weight for objective j

The resulting values for each objective are shown in Table 13.5. Objective 1, which is to generate revenue, will be worth 30 points, whereas objective 5, which is to divert auto drivers, is weighted 12 points. Other weighting methods, such as by ballot or group consensus, could be used. It is not necessary to use weights that total 100 as any range of values can be selected, and the final results normalized to 100 at the end of the process.

Step 5.   Determine the value of each measure of effectiveness. In this step the measures of effectiveness are calculated for each alternative. Techniques for demand estimation, as described in Chapter 12, are used to obtain daily and hourly ridership on the line. Cost estimates are developed based on the length of line, number of vehicles and stations, right of way

Table 13.5   Ranking and Weights for Each Objective

Objective	Ranking	Relative Weight (W_j)	Weighting Factor* $(\times 100)$
1	1	5	30
2	2	4	24
3	3	3	17
4	3	3	17
5	4	2	12
Total		17	100

*Rounded to whole numbers to equal 100.

costs, electrification, and so forth. Revenues are computed, and ridership volumes during the peak hour are estimated. In some instances, forecasts are difficult to make, so a best or most likely estimate is produced. Since it is the comparative performance of each alternative that is of interest, relative values of effectiveness measures can be used.

Step 6. Compute a score and ranking for each alternative. The score for each alternative is computed by considering each measure of effectiveness and awarding the maximum score to the alternative with the highest value and a proportionate amount to the other alternatives. Consider the first criterion, return on investment. Table 13.4 shows that alternative V achieves the highest value and is awarded 30 points. The value for alternative I is calculated as (13/15)(30) = 26. The results are shown in Table 13.6. (An alternative approach is to award the maximum points to the highest valued alternative and zero points to the lowest.)

Table 13.6 Point Score for Candidate Transit Lines

Measure of Effectiveness	Alternatives				
	I	*II*	*III*	*IV*	*V*
1	26.0	28.0	22.0	27.0	30.0
2	24.0	22.1	19.2	17.3	16.3
3	8.5	11.9	13.6	17.0	17.0
4	17.0	14.9	12.8	10.6	10.6
5	12.0	10.3	6.9	5.1	5.1
Total	87.5	87.2	74.5	77.0	79.0

The total point score indicates that the ranking of the alternatives in order of preference is I, II, V, IV, and III. Alternatives I and II are clearly superior to the others and are very similar in ranking. These two will bear further investigation prior to making a decision.

Ranking and rating evaluation is an attractive approach because it can accommodate a wide variety of criteria and can incorporate various viewpoints. Reducing all inputs to a single number is a convenient way to rate the alternatives. The principal disadvantage is that the dependence on a numerical outcome masks the major issues underlying the selection and the tradeoffs involved.

Another problem with ranking methods is that the mathematical form for the rating value (Eq. 13.7) is a summation of the products of the criteria weight and the relative value. For this mathematical operation to be correct, the scale of measurement must be a constant interval (for example, temperature). If the ranking values are ordinal (such as the numbering of a sports team), the ranking formula cannot be used.

Results also could be changed by revising the ranking of the objectives and their relative weights.

Finally, there is the problem of communicating the results to decision makers, since the interpretation is often difficult to visualize. People think in concrete terms and are able to judge alternatives only when they are presented realistically, rather than as numerical values.

The next section describes a more general and comprehensive approach to evaluation that furnishes information for decision making but stops short of computing numerical values for each alternative.

Cost Effectiveness

Cost effectiveness attempts to be comprehensive in its approach while using the best attributes of economic evaluation. In this method, the criteria that reflect the goals of the project are listed separately from project costs. Thus, the project criteria are considered to be measures of its effectiveness, and the costs are considered as the investment required if that effectiveness value is to be achieved. This approach uses data from economic analysis but permits other intangible effects, such as environmental consequences, which are measured as well. The following example illustrates the use of the cost effectiveness method.

Example 13.3   Evaluating Metropolitan Transportation Plans Using Cost Effectiveness

Five alternative system plans are being considered for a major metropolitan area. They are intended to provide added capacity, improved levels of service, and reductions in travel time during peak hours. Plan A retains the status quo with no major improvements, plan B is an all-rail system, plan C is all highways, plan D is a mix of rail transit and highways, and plan E is a mix of express buses and highways. An economic evaluation has been completed for the project, with the results shown in Table 13.7.

Plan B, the all-rail system, and plan D, the combination rail and highway system, have incremental BCRs of less than 1, whereas plan C, all highways, and plan E, highways and express buses, have incremental BCRs greater than 1. These results would suggest that the highway–bus alternative (plan E) is preferable to the highway–rail transit

Table 13.7   Benefit-Cost Comparisons for Highway and Transit Alternatives

Plan Comparisons	Annual Cost Difference ($ million)	Annual Savings ($ million)	BCR
A versus B	28.58	21.26	0.74
A versus C	104.14	116.15	1.12
C versus D	22.66	17.16	0.76
C versus E	16.73	19.75	1.18

SOURCE: Adapted from *Alternative Multimodal Passenger Transportation Systems,* NCHRP Report 146, Transportation Research Board, National Research Council, Washington, D.C., 1973.

alternatives (plans B and D). To examine these options more fully, noneconomic impacts have been determined for each and are displayed as an evaluation matrix in Table 13.8. Among the measures of interest are numbers of persons and businesses displaced, number of fatal and personal injury accidents, emissions of carbon monoxide and hydrocarbons, and average travel speeds by highway and transit.

An examination of Table 13.8 yields several observations. In terms of number of transit passengers carried, plan E ranks highest, followed by plan B. The relationship between annual cost and transit passengers carried is illustrated in Figure 13.4. This cost effectiveness analysis indicates that plan B produces a significant increase in transit passengers over plan A. Although plans C, D, and E are much more costly, they do not produce many more transit riders for the added investment.

Community impacts are reflected in the number of homes and businesses displaced and the extent of environmental pollution. Figure 13.5 illustrates the results for number of businesses displaced, and Figure 13.6 depicts the results for emissions of hydrocarbons.

In terms of businesses displaced versus transit passengers carried, Plans C and D require considerable disruption with very little increase in transit patronage over plan B. Plan B is clearly preferred if the impact on the community is to be minimized. On the

Table 13.8 Measure of Effectiveness Data for Alternative Highway–Transit Plans

| Measure of Effectiveness | Plan A | Plan B | Plan C | Plan D | Plan E |
	Null	All Rail	All Highway	Rail and Highway	Bus and Highway
Persons displaced	0	660	8000	8000	8000
Businesses displaced	0	15	183	183	183
Annual total fatal accidents	159	158	137	136	134
Annual total personal injuries	6767	6714	5596	5544	5517
Daily emissions of carbon monoxide (tons)	2396	2383	2233	2222	2215
Daily emissions of hydrocarbons (tons)	204	203	190	189	188
Average door-to-door auto trip speed (mi/h)	15.9	16.2	21.0	21.2	21.5
Average door-to-door transit trip speed (mi/h)	6.8	7.6	6.8	7.6	7.8
Annual transit passengers (millions)	154.2	161.7	154.2	161.7	165.2
Total annual cost ($ millions)	2.58	31.16	106.72	129.38	123.44
Interest rate, percent	8.0	8.0	8.0	8.0	8.0

SOURCE: Adapted from *Alternative Multimodal Passenger Transportation Systems,* NCHRP Report 146, Transportation Research Board, National Research Council, Washington, D.C., 1973.

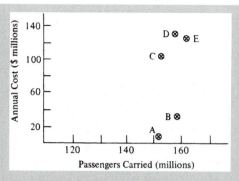

Figure 13.4 Relationship Between Annual Cost and Passengers Carried

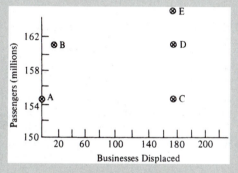

Figure 13.5 Relationship Between Passengers Carried and Businesses Displaced

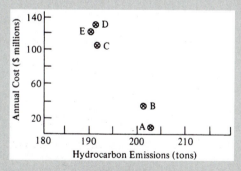

Figure 13.6 Annual Cost Versus Hydrocarbon Emissions

other hand, plan C, which is considerably more costly than plan B, results in a significant reduction in pollution levels, whereas the other two plans, D and E, although more expensive than C, have little further impact on pollution levels.

The items described are but a few of the many relationships that could be examined. They do, however, illustrate the cost-effectiveness procedure and the various conflicting tradeoffs that can result. One conclusion that seems evident is that, although the BCRs for plans B and D are less than 1, these plans bear further investigation since they produce several environmentally and socially beneficial effects and attract more transit ridership. A sensitivity analysis of the benefit-cost study would show that if the interest rate were reduced to 4 percent or the value of travel time were increased by $0.30 per hour, the rail transit plan, plan B, would have a BCR greater than 1.

The cost effectiveness approach does not yield a recommended result, as do economic methods or ranking schemes. However, it is a valuable tool because it defines more fully the impacts of each course of action and helps to clarify the issues. With more complete information, one should come to a better decision. Rather than closing out the analysis, the approach opens it up and permits a wide variety of factors to be considered.

Evaluation as a Fact-Finding Process

The preceding discussion of economic and rating methods for evaluation has illustrated the technique and application of these approaches. These so-called rational methods are inadequate when the transportation alternatives create a large number of impacts on a wide variety of individuals and groups. Under these conditions, the evaluation process is primarily one of fact finding to provide the essential information from which a decision can be made. The evaluation procedure for complex projects is illustrated in Figure 13.7 and should include four activities that follow the development and organization of basic data

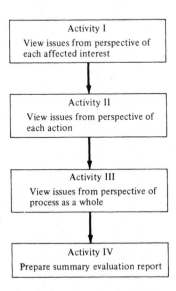

Figure 13.7 Evaluation Procedure for Complex Projects

SOURCE: Redrawn from *Transportation Decision Making*, NCHRP Report 156, Transportation Research Board, National Research Council, Washington, D.C., 1975.

and the identification of the major problems or issues that must be addressed in the evaluation process. The activities are as follows:

Activity 1. View the issues from the perspective of each affected interest group. The thrust of this activity is to view the consequences of proposed alternatives as they affect particular groups and individuals and as those groups perceive them. For each interest group, the information should be examined and a statement prepared that indicates how that group will react to each alternative. This step can be considered as a means of understanding where each of the stakeholders are coming from, what their biases, likes, and dislikes are, what problems they represent, and so forth.

Activity 2. View the issues from the perspective of each alternative. The purpose of this activity is to describe each alternative in terms of its advantages and disadvantages. Each proposed project is discussed from the point of view of community concern, feasibility, equity, and potential acceptability.

Activity 3. View the issues from the perspective of the process as a whole. In this step, all of the alternatives and each of the issues (criteria) are examined together to see if patterns develop from which general statements can be made. For example, there may be one or more alternatives that prove to have so many disadvantages that they can be eliminated. There may be several groups who share the same viewpoint or who are in fierce opposition. Compromise solutions may emerge that will reconcile conflicts, or there may be popular alternatives that generate little controversy.

Activity 4. Summarize the results of the previous activities. In this activity, the result of the evaluation is documented for use by decision makers and other interested individuals. The report should include a description of the alternatives, the advantages and disadvantages of each alternative, areas of conflict and of agreement, and identification of the alternatives that have the greatest potential for success in accomplishing major objectives and achieving public acceptance.

The above process is comprehensive and goes beyond a simple listing of criteria and alternatives. One essential feature of the approach is that it requires the analyst to furnish values of each measure of effectiveness for each alternative, without attempting to reduce the results to a single numerical value. The information is then used to make judgments about the relative merits of each alternative.

Trade-Off and Balance-Sheet Approaches

According to Lockwood and Wagner, in any evaluation process there are several important conditions that must be met if the difference among alternative projects is to be adequately considered:

1. All alternatives should be evaluated in a framework of common objectives. Measures of effectiveness should be derived from the objectives covering all impact areas.
2. The incidence and timing of impacts on groups and areas should be identified for all impact categories.
3. Standards or accepted impact significance thresholds for measures of effectiveness should be indicated where accepted or required by law.
4. All measures of effectiveness should be treated at an equal level of detail and appropriate scale.

5. Uncertainties or probabilities or both should be expressed for each impact category.
6. A sensitivity analysis should be conducted to describe variations in results for alternatives when values of key parameters are changed.

Balance-sheet or trade-off approaches satisfy these criteria in contrast to economic or weighting schemes, which provide little in the way of comparative information. These approaches display the impacts of plan alternatives to various groups. The method is based on the viewpoint that individuals or groups that review the data will first introduce their own sets of values and weights, and then reach a judgment based on the merits of each alternative using all the data in a disaggregated fashion. Each impact category or goal may have more than one measure of effectiveness. To determine the cost effectiveness of each impact category in a balance-sheet framework, it is useful to compare proposals with the do-nothing alternative. In this way, the positive and negative impacts of not constructing a new project are fully understood.

Evaluation of Completed Projects

The material on evaluation discussed thus far has dealt with the evaluation of plans, and the focus has been to answer the "what if" questions in sufficient detail so that a good decision will be made. Another form of evaluation is to examine the results of a project after it has been implemented to determine (1) how effective it has been in accomplishing its objectives, (2) what can be learned that is useful for other project decisions, (3) what changes should be made to improve the current situation, or (4) if the project should be continued or abandoned.

The subject of postevaluation of transportation projects is closely related to the more general topic of experimental design. If, for example, the effects of a particular medical treatment are to be determined, two population groups with similar characteristics are selected, one serving as the control group and the other as the experimental group. Then the treatment is applied to the experimental group but not to the control group. Differences are measured before and after treatment for both groups, and the net effect is considered to be the result of the treatment itself.

Example 13.4   Evaluating the Effect of Bus Shelters on Transit Ridership

A transit authority wishes to evaluate the effectiveness of new bus shelters on transit ridership as well as acceptance by the community. A series of new shelters were built along one bus route but not on the other lines.

Solution: Bus ridership has been measured before and after the shelters had been installed on the test line and on a control line where nothing new had been added. Both lines serve similar neighborhoods. The ridership results are shown in Table 13.9. The line with new shelters increased ridership by 13.3 percent, whereas the line without shelters increased by only 2.5 percent. It should be stressed that only in the absence of any other factors can we conclude that the effect of the new shelters was to increase ridership by $(13.3 - 2.5) = 10.8$ percent.

Table 13.9 Transit Ridership

	Before	After	Change (%)
Line A: new shelters	1500	1700	13.3
Line B: no shelters	1950	2000	2.5

Another tool for evaluation of completed projects is to conduct a survey of users of the new facility. The questionnaire can probe in greater depth why riders use the facility and to what extent the project improvement influenced their choice. In the bus shelter example, a survey of bus riders would inquire if passengers were longtime bus riders or are new riders. If new riders, the survey would ask the reasons for riding to find out how many of the new riders considered the new shelters a factor. (Other reasons for riding could be that the rider is new in the neighborhood, his or her car is being repaired, gasoline prices have just gone up, and so forth.) The survey would also ask the old riders to comment on the new shelters. Thus, the survey would corroborate the before-and-after ridership data as well as furnish additional information about the riders themselves and how they reacted to the new project. This information would be useful in deciding whether or not to implement a bus shelter program for the entire city.

In the transportation field, it is difficult to achieve an experimental design with a well-defined control and experimental group because (1) transportation projects influence a wide range of outcomes, (2) implementation times are very long and therefore funds are not available to gather before data, (3) a control group is difficult to identify, and (4) changes that occur over a long time period are difficult to connect with a single event, such as a new transportation system. An example of a postevaluation for a major transportation project is the Bay Area Rapid Transit impact study. The approach to postevaluation in this instance was to predict the effect on the region of measures such as air pollution, noise, travel time, and so forth without the rail transit system and then measure the actual amounts with the rapid transit system in place. In this approach, a control group was impossible to obtain but in its place a forecast was made of conditions in the region if a rapid transit system had not been constructed. The obvious difficulty with this method is that it depends on the accuracy of the forecast and must take into account all the changes that have occurred in the region that might have an effect on the impact measures of interest.

Another type of transportation project postevaluation is to make comparisons between different systems or technologies that serve similar travel markets. These comparisons can be helpful to decision makers in other localities because they furnish useful information about what happened in an actual situation. A postevaluation study can be useful because it can consider the actual results for many variables, whereas a preproject evaluation of a mode or technology tends to focus primarily on cost factors or is based on hypothetical situations that require many questionable assumptions.

Example 13.5   Comparing the Effectiveness of Bus and Rail Transit

Compare the effectiveness of rail and bus based on the experience with a rail transit line serving downtown Philadelphia and a suburb of New Jersey with an express bus line connecting downtown Washington, D.C., with the Virginia suburbs. The rail line, known as the Lindenwold Line, serves 12 stations with 24-hour service per day, whereas the busway, known as the Shirley Highway, extends for 11 miles, with no stations along the way and with bus service provided on exclusive lanes only during the peak hour. Both systems serve relatively low-density, auto-oriented residential areas with heavy travel during the peak hours.

Solution: To determine the relative effectiveness, a comparative analysis of each project was made after they had been in operation for several years. Measures of effectiveness were considered from the viewpoint of the passenger, the operator, and the community. Data were collected for each system and for each measure of effectiveness. A detailed evaluation for each parameter was prepared that both described how each system performed and discussed its advantages and disadvantages. To illustrate, consider the evaluation of one service parameter—*reliability*—expressed as schedule adherence. The variance from scheduled travel times may result from traffic delays, vehicle breakdowns, or adverse weather conditions. It depends mostly on the control that the operator has over the entire system. By far the most significant factor for reliability is availability of exclusive rights of way.

- *Lindenwold:* That year 99.15 percent of all trains ran less than 5 min late, and the following year the figure was 97 percent.
- *Shirley:* Surveys conducted over a 4-day period indicated that 22 percent arrived before schedule time, 32 percent were more than 6 minutes late, and only 46 percent arrived at the scheduled time within a 5-minute period.
- *Comparison:* The Lindenwold Line (rail) is superior to the Shirley Highway (bus) with respect to reliability.

A summary of the comparative evaluations of the two systems is shown in Table 13.10.

Table 13.10   Comparative Evaluation of Completed Rail and Bus Transit

Measure of Effectiveness	Lindenwold (Rail)	Shirley (Bus)	Higher Rated System
Availability	Good	Poor	Rail
Absolute travel time	Very good	Good	Rail
Reliability	Very good	Poor	Rail
Comfort	Good	Poor	Rail
Convenience	Good	Fair	Rail
Safety and Security	Very good	Good	Rail
Area coverage	Good	Very good	Bus
Frequency	Very good	Very poor	Rail

continued

Table 13.10   Comparative Evaluation of Completed Rail and Bus Transit (*Continued*)

Measure of Effectiveness	Lindenwold (Rail)	Shirley (Bus)	Higher Rated System
Investment cost	Very poor	Fair	Bus
Operating cost	Good	Fair	Rail
Capacity	Good	Poor	Rail
Passenger attraction	Very good	Good	Rail
System impact	Very good	Good	Rail

SOURCE: Adapted from V. R. Vuchic and R. M. Stanger, "Lindenwold Rail Line and Shirley Busway: A Comparison," *Highway Research Record* 459, Transportation Research Board, National Research Council, Washington, D.C., 1973.

A detailed analysis of the results would indicate that each system has advantages and disadvantages. The principal reasons why the rail system appears more attractive than the bus is because it provides all-day service, is simpler to understand and use, and produces a higher quality of service.

SUMMARY

The evaluation process for selecting a transportation project has been described. Various methods have been presented that, when used in the proper context, can assist a decision maker in making a selection. The most important attribute of an evaluation method is its ability to correctly describe the outcomes of a given alternative. The evaluation process begins with a statement of the goals and objectives of the proposed project, and these are converted into measures of effectiveness. Evaluation methods differ by the way in which measures of effectiveness are considered.

Economic evaluation methods require that each measure of effectiveness be converted into dollar units. Numerical ranking methods require that each measure of effectiveness be translated to an equivalent score. Both methods produce a single number to indicate the total worth of the project. Cost effectiveness methods require only that each measure of effectiveness be displayed in matrix form, and it is the task of the analysts to develop relationships between various impacts and the costs involved. For projects with many impacts that will influence a wide variety of individuals and groups, the evaluation process is essentially one of fact finding, and the projects must be considered from the viewpoint of the stakeholders and community. The reasons for selecting a project will include many factors in addition to simply how the project performs. A decision maker must consider issues such as implementation, schedules, financing, and legal and political matters.

When a project has been completed and has been in operation for some time, a postevaluation can be a useful means to examine the effectiveness of the results. To conduct a postevaluation, it is necessary to separate the effect of the project on each measure of effectiveness from other influencing variables. A standard procedure is the use of a control group for comparative purposes, but this is usually not possible for most transportation projects. A typical procedure is to compare the results with a forecast of the region without the project in place. Postevaluations can also be used to compare alternative modes and technologies, using a wide range of measures of effectiveness.

The usefulness of an evaluation procedure is its effectiveness in assisting decision makers to arrive at a solution that will best accomplish the intended goals.

PROBLEMS

13-1 What is the main objective of conducting a transportation project evaluation?

13-2 Describe four basic issues that should be considered prior to selection of an evaluation procedure.

13-3 List the basic criteria used for evaluating transportation alternatives. What units are used for measurement?

13-4 Average demand on a rural roadway ranges from zero to 500 veh/day when the cost per trip goes from $1.50 to zero.

(a) Calculate the net user benefits per year if the cost decreases from $1.00 to $0.75/trip (assume a linear demand function).

(b) Compare the value calculated in (a) with the benefits as calculated in typical highway studies.

13-5 A ferry is currently transporting 250 veh/day at a cost of $1.25/vehicle. The ferry can attract 500 more veh/day when the cost/veh is $0.75. Calculate the net user benefits/year if the cost /veh decreases from $1.10 to $0.95.

13-6 What are the two components of the cost of a transportation facility improvement? Describe each.

13-7 Estimate the average unit costs for (a) operating a standard vehicle on a level roadway, (b) travel time for a truck company, (c) single-vehicle property damage, (d) personal injury, and (e) fatality.

13-8 Derive the equation to compute the equivalent annual cost given the capital cost of a highway, such that $A = (A/P) \times P$, where A/P is the capital recovery factor. Compute the equivalent annual cost if the capital cost of a transportation project is $100,000, annual interest $= 10$ percent, and $n = 15$ years.

13-9 A highway project is expected to cost $1,500,000 initially. The annual operating and maintenance cost after the first year is $2000 and will increase by $250 each year for the next 10 years. At the end of the fifth year, the project must be resurfaced at a cost of $300,000.

(a) Calculate the present worth of costs for this project if the annual interest rate is 8 percent.

(b) Convert the value obtained in (a) to equivalent uniform annual costs.

13-10 Three transportation projects have been proposed to increase the safety in and around a residential neighborhood. Each project consists of upgrading existing street signing to highly retroreflective sheeting to increase visibility. The table below shows the initial construction costs, annual operating costs, the useful life of the sheeting, and the salvage values for each alternative. Assume that the discount rate is 10 percent. Calculate the present worth for each alternative and determine the preferred project based on the economic criteria.

Alternative	Initial Construction Costs ($)	Annual Operating Costs ($)	Useful Life (Years)	Salvage Value ($)
I	10,000	2000	10	2500
II	12,000	1600	10	3000
III	5,000	2500	5	500

13-11 Two designs have been proposed for a short span bridge in a rural area, as shown in the table below. The first proposal is to construct the bridge in two phases (Phase I now and Phase II in 25 years). The second alternative is to construct it in one phase. Assuming that the annual interest rate is 4 percent, determine which alternative is preferred using present worth analysis.

Alternative	Construction Costs (Dollars)	Annual Maintenance Costs (Dollars)	Service Period (Years)
I (Phase I)	14,200,000	75,000	1–50
I (Phase II)	12,600,000	25,000	25–50
II	22,400,000	100,000	1–50

13-12 Three designs have been proposed to improve traffic flow at a major intersection in a heavily traveled suburban area. The first alternative involves improved traffic signaling. The second alternative includes traffic signal improvements and intersection widening for exclusive left turns. The third alternative includes extensive reconstruction, including a grade separation structure. The construction costs, as well as annual maintenance and user costs, are listed in the following table for each alternative. Determine which alternative is preferred based on economic criteria if the analysis period is 20 years and the annual interest rate is 15 percent. Show that the result is the same using the present worth, equivalent annual cost, benefit-cost ratio, and rate of return methods.

Alternative	Capital Cost ($)	Annual Maintenance ($)	Annual User Cost ($)	Salvage Value ($)
Present condition	—	15,000	500,000	—
Traffic signals	440,000	10,000	401,000	15,000
Intersection widening	790,000	9,000	350,000	11,000
Grade separation	1,250,000	8,000	301,000	—

13-13 A road is being proposed to facilitate a housing development on a scenic lake. Two alternatives have been suggested. One of the roadway alignments is to go around the lake and slightly impact a wetland. The second alternative will also go around the lake and will significantly impact two wetlands. The following table shows the anticipated costs for each alternative. Assuming that the annual interest rate is 7 percent, determine which alternative is preferred using equivalent annual cost analysis.

Alternative	First Cost ($)	Annual Maintenance ($)	Service Life (Years)	Salvage Value ($)	Annual Wetland Rehab. Costs ($)	Annual Roadway Lighting Costs ($)
I	75,000	3000	15	45,000	7500	1500
II	125,000	2000	15	25,000	2500	2500

13-14   The light-rail transit line described in this chapter is being evaluated by another group of stakeholders. In examining the objectives, they place the following rankings on each as follows:

Objective	Ranking
1	5
2	3
3	1
4	2
5	4

Using the revised information, determine the weighted score for each alternative and comment on your result.

13-15   You have been hired as a consultant to a medium-size city to develop and implement a procedure for evaluating whether or not to build a highway bypass around the CBD. Write a short report describing your proposal and recommendation as to how the city should proceed with this process.

13-16   The following data have been developed for four alternative transportation plans for a high-speed transit line that will connect a major airport with the downtown area of a large city. Prepare an evaluation report for these proposals by considering the cost effectiveness of each attribute. Show your results in graphical form and comment on each proposal.

Rail Alternatives

Measure of Effectiveness	Existing Service	Plan A	Plan B	Plan C	Plan D
Persons displaced	0	264	3200	3200	3200
Businesses displaced	0	23	275	275	275
Average door-to-door trip speed (mi/h)	10.2	38	45	46	48
Annual passengers (millions)	118.6	124.4	118.6	124.4	127.0
Annual cost (millions)	—	16.4	20.2	23.8	22.7

13-17   A new carpool lane has replaced one lane of an existing six-lane highway. During peak hours the lane is restricted to cars carrying three or more passengers. After five months of operation, the carpool lane handles 800 autos/hour, whereas the existing lanes are operating at capacity levels of 1500 veh/h/lane at an occupancy rate of 1.2. How would you determine if the new carpool lane is successful or if the lane should be open to all traffic?

REFERENCES

Frye, F. R., *Alternative Multimodal Passenger Transportation Systems,* NCHRP Report 146, Transportation Research Board, National Research Council, Washington, D.C., 1973.

Lockwood, S. C., and Wagner, F. A., *Methodological Framework for the TSM Planning Process,* TRB Special Report 172, Transportation Research Board, National Research Council, Washington, D.C., 1973.

Manheim, M .L., et al., *Transportation Decision Making,* NCHRP Report 156, Transportation Research Board, National Research Council, Washington, D.C., 1975.

Texas Transportation Institute, *Micro Bencost Users Manual,* National Cooperative Highway Research Program, Transportation Research Board, National Research Council, Washington, D.C., 1993.

Vuchic, V. R., and Stanger, R. M., "Lindenwold Rail Line and Shirley Busway: A Comparison," *Highway Research Record* 459, Transportation Research Board, National Research Council, 1973.

ADDITIONAL READINGS

A Manual on User Benefit Analysis of Highway and Bus Transit Improvements, American Association of State Highway and Transportation Officials, Washington, D.C., 1978.

Highway Engineering Economy, Federal Highway Administration, Washington, D.C., 1983.

Manheim, M. L., *Fundamentals of Transportation Systems Analysis,* MIT Press, Cambridge, Mass., 1979.

Microcomputer Evaluation of Highway User Benefits, Final report for NCHRP 7-12, Texas Transportation Institute, Texas A & M University System, College Station, Tex., October 31, 1993.

Riggs, J. L., *Engineering Economics,* McGraw Hill, New York, 1996.

Rutherford, G. Scott, *Multimodal Evaluation in Passenger Transportation,* NCHRP Synthesis 201, Transportation Research Board, National Research Council, National Academy Press, Washington, D. C., 1994.

Thomas, E. N., and Schofer, J. L., *Strategies for the Evaluation of Alternative Transportation Plans,* NCHRP Report 96, Transportation Research Board, National Research Council, Washington, D.C., 1970.

Weisbrod, G., and Weisbrod, B., *Assessing the Economic Impact of Transportation Projects: How to Choose the Appropriate Technique for Your Project,* Transportation Research Board, National Research Council, National Academy Press, Washington, D.C., October 1997.

CHAPTER 14

Transportation Systems Management

In the preceding chapters we described the process by which transportation plans are developed and evaluated. The context for that planning process is the need to establish long-range plans for the construction of transportation facilities, such as highways and mass transit systems, as well as to evaluate short-range improvements. In many situations, the major transportation facilities are already built, and the problem faced by transportation engineers is to operate these systems at their most productive and efficient levels. The term used to describe this operational planning process is *transportation systems management* (TSM). This chapter first describes various TSM strategies that can be used to improve transportation performance and then illustrates how these alternatives are evaluated.

The basic objective of TSM is to create more efficient use of existing facilities through improved management and operation of vehicles and the roadway. For example, a highway that carries 1500 autos per day with only one person per vehicle could increase its carrying capacity to 6000 persons per day with four persons per car. Thus, ridesharing could represent an effective TSM action. Other TSM actions include park-and-ride, separate lanes for high occupancy vehicles, bicycle and pedestrian facilities, traffic signal improvements to increase capacity and reduce delay at intersections, and motorist information.

The term *transportation demand management* (TDM) is also used to describe actions intended to reduce the volume of highway travel. In this chapter TSM refers to all strategies that contribute to the efficient use of existing highway facilities, both on the "supply" side (i.e., changing the facility itself) the "demand" side (changing the way people travel) and the use of information technology to improve safety and efficiency.

The TSM planning process entails defining the problem, establishing system objectives and criteria, deriving measures of effectiveness, identifying TSM actions, estimating costs and impacts, selecting TSM projects, and post facto monitoring of project results. The process is similar to that used for long-range planning but is less data intensive and often relies on field tests or before-and-after studies for evaluation. Since most TSM projects are relatively less costly than major construction and are easily implemented, it is often possible to try out an idea as a demonstration project and, if not successful, discontinue the service or method when the trial period is over. If the idea proves successful, modifications might improve it further, and it may also be implemented in other areas of the city or state.

Often, TSM projects are initiated to accomplish a nontransportation objective. For example, the Environmental Protection Agency has been mandated to ensure that air quality standards are maintained. The U.S. Department of Transportation is concerned with highway safety, energy conservation, cost effectiveness, reduction in peak hour travel, and minimizing transit operating expenses. These and other objectives can be addressed through TSM by examining many diverse ways to move people.

TSM STRATEGIES

According to the Institute of Transportation Engineers, TSM strategies can be classified into three basic categories: creating efficient use of road space (managing transportation supply), reducing vehicle use in congested areas (managing transportation demand and land use), and providing transit service. Within each category are a variety of strategies and tactics that can be used. These are described in the following sections.

TSM Actions to Create Efficient Use of Road Space

The actions in this group are intended to improve the flow of traffic, usually without altering the total number of vehicles that use the roadway during an average day. These TSM project improvements are intended to reduce travel time and delay for motorists, pedestrians, and transit users. Techniques used include traffic operations and traffic signalization, improvements for vehicle travel, bicycle and pedestrian projects, priorities in roadway assignments for high-occupancy vehicles, traffic restrictions in residential and congested areas, parking management, altering work schedules, and improving the coordination between modes.

Traffic Operations Improvements

TSM strategies in this category are similar to those a traffic engineer carries out in normal operations. These include (1) widening intersections, (2) creating one-way streets, (3) installing separate lanes for right and left turns, and (4) restricting turning movements (especially left turns).

Traffic Signalization Improvements

Many tools and techniques can be used to improve traffic signal performance. Usually, improvements are made that will benefit all traffic flow, but signals can also be used to favor high occupancy vehicles, buses, and carpools, or to create a safe environment in residential areas. Traffic signals can be improved by the following actions:

- Physically improving local intersections
- Coordinating signal timing for arterial roadways

- Computerizing areawide signal coordination in downtown grid networks
- Controlling traffic for freeways, including changeable message signs
- Metering ramps
- Television monitoring of traffic

Improvements for Pedestrians and Bicycles

These TSM strategies encourage nonmotorized travel by making walking and bicycling safer and more pleasant. Among the most effective means are widening sidewalks; providing lighting, benches, and pedestrian malls; building grade separations, such as underpasses or overpasses, to avoid conflicts at intersections; building bikeways; and installing pedestrian controls at intersections.

Special Roadway Designations

This TSM strategy is based on the idea that if one or more lanes of traffic are reserved for high-occupancy vehicles, then the person-carrying capacity/lane will be increased, use of public transportation will be stimulated, energy will be conserved, and service reliability will be improved. Roadways can be specially designated in a variety of ways, including the following:

- Reserving the curb lane of an arterial for buses only
- Reserving an entire street, usually in the CBD, for the exclusive use of buses
- Reserving a lane in the opposite direction of traffic (contraflow lane) when traffic is heavy in one direction and light in the other direction
- Setting up reversible freeway lanes for heavy inbound morning traffic and outbound evening traffic
- Providing bypass lanes on freeway ramps for use by buses or high-occupancy vehicles
- Providing bus or carpool lanes on freeways (referred to as diamond lanes)
- Permitting use of highway shoulders during peak hours
- Using the center lane for left turns only, usually for all vehicles in off-peak hours and for buses only in peak hours

Vehicle Restrictions in Pedestrian Areas

In instances when traffic volume is excessive or when pedestrian-vehicular conflicts exist, it may be necessary to restrict the area for use by autos and truck traffic. This strategy can also improve air quality and safety. As with any TSM strategy that restricts or limits use, its success will often depend on whether or not the restrictions cause severe hardship on those no longer permitted to use the facility. For this reason, traffic restriction strategies must also be accompanied by other improvements such as transit, parking, traffic operations, and marketing. Types of vehicle restriction include the following:

- Peak hour pricing in which tolls or charges are placed on vehicles that use the area during congested hours
- Auto-restricted areas that limit the use of autos within the CBD. (One technique is to divide the CBD into zones and restrict travel across zone boundaries. Transit is permitted into the CBD using street rights of way that define the zones.)

- Pedestrian malls that restrict streets to pedestrian traffic only or sometimes pedestrian and transit traffic
- Vehicle restrictions in residential areas, including stop signs, street closings, speed humps, diagonal diverters at intersections, and truck restrictions

Parking Management

Parking availability is a means of regulating the flow of traffic. When parking becomes more scarce and expensive, the number of vehicles entering the area will be reduced. Some people will shift to transit or carpools; others will park at the periphery of the city, where space is available at more reasonable rates. The TSM strategies for parking include the following:

- Curb parking restrictions to reduce the amount of on-street parking
- Off-street parking restrictions, such as pricing differentials to discourage all-day parking, the elimination of free parking, and parking subsidies
- Preferential parking for cars and vanpools to serve as an incentive for ridesharing
- Parking rate charges designed to encourage ridesharing or to limit vehicular traffic
- Parking information for motorists regarding space availability

Work-Schedule Management

Traffic congestion can be reduced if peak period flows can be spread out over a longer period of time. If journey to work traffic is normally concentrated between the hours of 7:30 a.m. and 8:30 a.m., then a reduction in traffic density would be possible if instead it were spread from 7:00 a.m. to 9:00 a.m. This TSM strategy is controlled by employers and will depend on work schedules within the organizations. The following approaches are used:

- Staggered work hours that require an adjustment in the beginning and ending work times
- Flextime that permits an employee to begin and end the work day within a flexible time range
- A four-day work week, based on 10 hours/day
- Substitution of communications for transportation (which has become more prevalent as computers are used at home)

Intermodal Coordination

Most TSM strategies require that the traveler change modes at some point. For this reason, coordination between modes must be considered, including improvements between transit, carpools, autos, walking, and bicycles. Many possibilities exist for intermodal coordination, such as park-and-ride facilities to assist the transfer between autos and rail or bus transit, and transit interface improvements that simplify the transfer between modes by eliminating fare collection or locating bus stops to minimize walking.

Relationship Between TSM Actions

This section has listed a variety of actions intended to improve the efficiency of road space. For example, better traffic signal timing should result in fewer delays and higher average speeds. These actions may have secondary effects by encouraging more or less vehicular travel as motorists adjust to the changed level of service. Thus, if a road has been improved by traffic signalization, more people may decide to use it; or if a freeway lane is dedicated to exclusive use by buses, then some people may shift from auto to transit. The TSM actions described in the following section are intended primarily to reduce the volume of vehicles and in this way save energy and reduce congestion. These actions will also result in more efficient use of road space. Thus, all TSM actions are interrelated and should result in both efficient use of road space and less congestion.

TSM Actions to Reduce Vehicle Volume

Techniques available to reduce vehicle volume on streets and highways include increasing vehicle occupancy through ridesharing, imposing economic disincentives on auto users, encouraging the use of travel by other modes, and reduction of truck traffic. These techniques are also referred to as transportation demand management (TDM).

Increasing Vehicle Occupancy Through Ridesharing

Every day, throughout the nation, millions of people drive to work alone. Average car occupancy in most cities is less than 1.5 persons/vehicle. If many of these motorists would rideshare, fewer autos would be driven, resulting in less traffic congestion. Methods of ridesharing include the following:

- Carpools
- Vanpools
- Subscription bus service
- Shared-ride taxi service

Carpools involve 2 to 6 persons who share driving to work, usually on a rotating basis. Vanpools are prearranged, usually by the employer, and involve 8 to 15 persons who each pay a prorated share of vehicle and operating costs. Subscription bus services are usually provided by a transit company to prearranged groups of 30 to 40 persons who commute to work from the same general area. The success of ridesharing depends on how well the proposed strategy compares with the auto in terms of travel time, cost, and convenience. Other TSM strategies, such as reserved parking for ridesharing and separate lanes for high occupancy vehicles, provide additional incentives. Carpooling is the strategy that has the greatest potential because it is easily implemented, requiring very little in the way of organization or new vehicles. Many transportation agencies provide carpool matching and information services that assist people in communities to find rides.

Discouraging Auto Use by Economic Means

The use of tolls and other user charges was mentioned earlier as a means of restricting autos in congested areas. This strategy, known as *congestion* or *value pricing,* can also be used to reduce or eliminate traffic on congested streets or highways. The approaches that could be used include the following:

- Increased parking fees in downtown areas
- Special licenses required for parking in restricted areas
- Charges for each trip based on vehicle metering of travel
- Tolls on highways and bridges

The two most common methods are parking charges and tolls. (Tolls are used not only to limit demand but also to raise revenue.) Other economic disincentives, such as congestion toll pricing, in which higher tolls are charged during peak periods, or drastic increases in parking charges, have not been widely accepted in the United States because they are politically unpopular and are usually opposed by the motoring public as well as by the business community.

Encouraging Travel by Means Other Than Auto

If motorists would leave the car at home and travel to work by bus, train, bicycle, or on foot, then traffic congestion on streets and highways would be reduced. For this to occur, the service provided by the other modes must be perceived as comparable or better than driving alone. The following methods are intended to make these modes look more attractive:

- Restrictions in auto travel such that transit becomes more competitive, for example, tolls, auto-restricted zones, freeway ramp metering, and parking elimination
- Provision of fringe parking lots in outlying areas connected with high-speed rapid transit or express bus service
- Provision of special facilities such as downtown distributors or bikeways in areas where autos are prohibited

Park-and-ride facilities have proven to be an effective means of encouraging transit use. The provision of parking lots at outlying locations is relatively easy to implement, and they can be built at low cost. For a park-and-ride facility to be successful, a high level of transit service must be provided, including preferential treatment on some highways. The total cost to the motorist should be lower than the cost by auto. Parking lots should be constructed so that they are compatible with the neighborhood and located in an area where travel demand and traffic congestion are sufficient to warrant service.

Reduction of Truck Traffic in Congested Areas

The conflict between auto traffic and truck traffic on city streets can create major traffic congestion and delay problems. Curbside loading and unloading, double parking, and increased truck sizes can result in significant interference to the smooth flow of traffic. Solutions proposed include the following:

- Arrange deliveries during off-peak hours
- Limit the size of trucks permitted for downtown delivery
- Create separate truck-only streets
- Streamline truck deliveries and scheduling operations
- Limit truck traffic to designated routes

Attempts to restrict truck operations either in time or space are usually difficult to achieve because of the large number of trucking firms involved, difficulties in arranging deliveries and pickups during nonworking hours, and resistance by local businesses and unions. To a great extent, trucks already schedule themselves outside of peak hours so that they can avoid slow speeds during hours of congestion. The most effective strategy to accommodate trucking needs and reduce congestion is by building off-street loading areas and special truck lanes.

Improving Transit Service

The principal goal of TSM is to reduce traffic congestion. Thus, if economical, reliable, and fast transit service is available, it is likely that some auto users will switch to transit service. Any of the TSM actions that improve the flow of traffic should also improve transit service. However, if service is to be truly competitive with the automobile, special treatment is required. If transit is to be effective, it should travel on its own right of way, have flexible routing patterns, and be "hassle free" and easy to use. Techniques used to improve transit include express bus services, shuttle services from fringe parking areas to downtown, internal circulation in low-density areas, improved flexibility in route scheduling and dispatching, simplified fare collection procedures, park-and-ride facilities, shelters, bus stop signs, bus fleet modernization, and improved passenger information services. Carrying out a transit improvement program requires an integrated approach that incorporates each of these actions where appropriate so that the net effect is a system that operates as effectively and efficiently as possible.

TSM ACTION PROFILES

A convenient way to examine various TSM strategies is to describe the problem that each action addresses, the conditions necessary for application of the action, potential problems of implementation, and potential evaluation factors. Table 14.1 lists a variety of TSM actions that could be considered for use in solving a particular transportation problem.

To illustrate how each action can be described, the attributes of three TSM strategies are profiled in this section. These are alternative work hours, travel on freeway shoulders during peak periods, and two-way left-turn lanes. Similar profiles of each TSM strategy listed in Table 14.1 are described in the reports *Simplified Procedures for Evaluating Low-Cost TSM Projects* and *A Toolbox for Alleviating Traffic Congestion*.

Staggered Work Hours

Problems Addressed
- Traffic congestion develops consistently during peak commuting periods in or near an employment center.
- Congestion is expected to increase if business expansion or plans are approved and implemented.
- Transit vehicles serving an employment center are consistently overcrowded during parts of or all of the peak commuting periods.

Conditions for Application
- *Slack capacity:* There should be periods of low traffic volume within the hours that work trips will be staggered. If traffic congestion is uniform, this strategy will not be effective.

Table 14.1   TSM Actions Available for Solving Transportation Problems

Staggered work hours
Flexible work hours
Increased peak period tolls
Toll discounts for carpools during peak hours
Residential parking permits
Neighborhood traffic barriers
Park-and-ride lots along transit routes
On-street parking bans during peak hours
Parking reserved for short-term use only
Increased parking rates
Parking rate fines and time-limit adjustments
Expanded off-street parking
Freeway ramp control
Travel on freeway shoulders during peak periods
One-way streets
Reversible lanes
Two-way left-turn lanes
New street segments
Signal phases for left turns
Employer-based carpool matching programs
Employer vanpool programs
Freeway lanes reserved for buses and carpools
Priority freeway access/egress for buses and carpools
Arterial street lanes reserved for express buses or carpools
Shuttle buses or vans
Bus transfer stations
Expanded regular-route bus service
Pedestrian-only streets
Community transit services

- *Employment concentration:* The industrial center should contain about 40 percent of employment in a 1–2 mi radius. Application in large firms with 1000 or more employees are more effective than many small firms.
- *Transit availability:* Transit routes should serve the center with service convenient to changes in work schedules.

Implementation Problems

- Some businesses and labor organizations may not cooperate.
- A reduction may occur in vanpools or carpools when work times are shifted.

Evaluation Factors

- Changes in vehicle volumes entering and leaving the employment center
- Changes in peak operation speeds or levels of service

- Changes in average loads on transit routes
- Changes in transit ridership and revenue

Travel on Freeway Shoulders

Problems Addressed

- Traffic congestion develops consistently on a freeway or in a freeway corridor during peak commuting periods.
- Localized traffic congestion develops on a freeway as a result of a lane drop.

Conditions for Application

- *Geometric conditions:* There should be wide shoulders (minimum 10 ft) and horizontal and vertical alignment for safe speed on shoulder lanes.
- *Location:* This strategy should be implemented on freeways where most drivers are commuters and thus familiar with use of shoulders for moving traffic during peak periods.
- *Traffic composition:* If more than 3 percent of traffic stream is composed of trucks, this strategy may cause traffic flow disruptions and accidents due to lane-change movements.

Implementation Problems

- The shoulder is not available for breakdowns and traffic enforcement.
- Use of the shoulder will attract travelers who would normally use other routes, thus limiting the effectiveness of this strategy.
- This action could simply shift the problem to another location, such as the ramps or points where the shoulder terminates.

Evaluation Factors

- Increases in traffic volumes
- Changes in peak period speeds or levels of service
- Changes in traffic volumes on parallel facilities
- Increased cost of freeway operations for ramp modifications, maintenance, and enforcement
- Changes in crash rate

Two-Way Left-Turn Lanes

Problems Addressed

- Left turns on arterial streets at midblock locations interfere with through traffic and cause delays.
- Left turns are also a major contributor to crashes, especially rear-end collisions.

Conditions for Application

- Since a separate lane is designated solely for left turns, sufficient roadway width is required. Streets with 4 lanes, each 12 ft wide, can be restriped with lanes 9½ ft wide, thus creating an additional turning lane 10 ft wide. It should be understood, however, that narrowing lanes could result in some compromise in safety, as 12-ft lanes are preferred. Lane narrowing should be permitted only when sufficient right of way is unavailable. Speed limit reductions should be posted along the stretch where lanes have been reduced in width.

- Left-turn movements should be distributed along the arterials to avoid long queues. This strategy is most appropriate where strip development exists. At major shopping centers, separate left-turn signals are recommended.
- Distances between traffic signals should be sufficient to allow left-turn storage lanes at intersections. If signals are placed closer than 1000 ft apart, less than half of the lane may be available for two-way operation.

Evaluation Factors
- Change in peak operating speed along arterial street
- Change in number of accidents involving left turns
- Cost to modify intersections, maintain pavement markings, and install traffic signs

Each of the TSM strategies described involve both advantages and disadvantages. Although a strategy may solve one problem, it may create others. Also, the project may be accepted by some groups but opposed by others. It is helpful to understand the likely effects as well as the limitations of each TSM strategy before the project is selected. Thus, profiles of each TSM action are helpful in deciding if a given action should be used in a given situation.

Example 14.1   TSM Improvements for an Urban Arterial Corridor

The following example illustrates how the planning process is used to select TSM improvements. The situation involves an urban arterial corridor in a large metropolitan area.

- *Existing Conditions.* The arterial corridor studied (Lee Street) is 2 mi in length and connects two freeways, as illustrated in Figure 14.1. Traffic volumes increase toward the CBD, going north. Railroad tracks, which parallel the road on the east side for its entire length, limit traffic access from that direction. The width of the arterial is 60 ft, and the arterial is striped for four lanes throughout its length. Bus service consists of five routes that provide headways between 6 and 30 min. No special or preferential bus facilities exist in the corridor, and all routes terminate in the downtown area. At the peak load point along the arterial, there are 25 buses per hour.
- *Objectives.* Improvements to the corridor are intended to accomplish the following objectives: (1) to improve travel time for transit users, (2) to improve comfort and convenience, (3) to increase capacity, and (4) to minimize capital and operating costs.

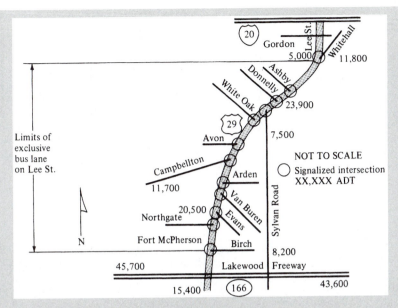

Figure 14.1 Urban Arterial Corridor

SOURCE: Redrawn from Charles M. Abrams, et al., *Measures of Effectiveness for TSM Strategies,* U.S. Department of Transportation, Washington, D.C., 1979.

Solution:

- **Proposed TSM Actions.** The TSM program being evaluated consists of the following elements:

    1. *Exclusive Bus Lanes in Peak Hours.* The arterial will be striped as five lanes. During the morning peak, one northbound lane will be reserved for buses only; during the evening peak, one southbound lane will be reserved for buses only. During off-peak hours, the center lane will be used for two-way left turns. See Figure 14.2.
    2. *Express Bus Service.* Two of the bus routes will operate express bus service.
    3. *Park-and-Ride Facilities.* Park-and-ride lots will be built at two locations, one at Campbellton Road and the other near the Lakewood Freeway (Figure 14.1).
    4. *Reduced Headways on Existing Bus Routes.* The headways on two bus routes will be reduced from 20 to 12 min. Three new bus routes will be added and will stop at park-and-ride lots every 8 min.

- **Selection of Measures of Effectiveness.** For each objective, a measure of effectiveness is used that best represents the improvement. These are (1) point-to-point travel time, (2) frequency of bus service, (3) volume-to-capacity ratio, and (4) equivalent annual cost of each action.
- **Estimating Travel Demand.** Since this is a short-term planning study, the data and modeling techniques available from the local planning agency will be used. The Urban Transportation Planning System (UTPS) is selected because of its ability

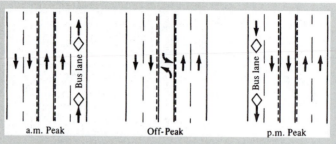

Figure 14.2 Bus Headway and Lane Designations

SOURCE: Redrawn from Charles M. Abrams, et al., *Measures of Effectiveness for TSM Strategies,* U.S. Department of Transportation, Washington, D.C., 1979.

to perform mode choice as well as transit and highway assignment. The key procedure used to evaluate shifts in transit use is the logit form of the mode choice model, which is expressed as

$$P_t(x) = \frac{U_t(x_i b)^B}{U_a(x_i b)^B + U_t(x_i b)^B}$$

(14.1)

$$P_a(x) = 1 - P_t(x)$$

(14.2)

where
$P_t(x)$ = probability of an individual using transit for a trip
$P_a(x)$ = probability of an individual using auto for a trip
$U_t(x_i b)$ = utility function incorporating appropriate costs x_i of using transit and the coefficients b_i of the various costs
$U_a(x_i b)$ = utility function, similar to $U_t(x_i b)$, for auto
B = constant that varies by trip purpose

The mode choice model is calibrated for three trip purposes: home-based work, home-based other, and nonhome based. The variables used are transit in-vehicle time, 2.5 × (walk + wait time), driving time to transit stop, out-of-pocket cost, transit fare, and parking cost.

For the proposed TSM actions, the highway network was modified to simulate planned conditions, such that where lanes were added, link capacities and speeds were increased. Trip table data were available, but it was necessary to update these to present conditions. A Fratar model was used for this purpose because growth in the region was generally stable over the period and zonal population and employment data were available to develop growth factors.

- **Impact Evaluation.** The measures of effectiveness were calculated by determining the differences between the existing conditions and those with the TSM actions in place. The results of the impact evaluation are as follows:

1. *Travel Time.* Overall, the proposed project is estimated to reduce peak period person hours of travel by about 2 percent in the corridor. The reduction is due mainly to the increase in express bus service and improved highway travel. Door-to-door transit travel times are substantially reduced from an average of 48.3 min to 40.8 min, whereas auto travel is reduced only from 29.2 min to 28.9 min.

2. *Comfort and Convenience.* Frequency of bus service was selected as the measure of effectiveness, and this would increase substantially if the proposed actions were placed in effect.

3. *Capacity of the Transportation System.* The volume-to-capacity ratios at selected locations decreased slightly with the proposed TSM actions in place. This is due mainly to a modest (2 percent) reduction in highway travel. There was a significant demand for park-and-ride facilities, which is estimated to serve 1000 vehicles at maximum accumulation.

4. *Equivalent Annual Cost.* The capital cost of the proposed action is $3.9 million, which includes 33 new buses, 1000 parking spaces, and 2.5 mi of bus lane. At an interest rate of 8 percent and a 10-yr life, the equivalent annual cost is $0.58 million. Operating and maintenance costs are estimated to increase by $1.1 million. In addition, the operating deficits for the transit authority will increase by $0.95 million.

- *Project Analysis and Selection.* The proposed TSM actions in this corridor contribute to the achievement of the objectives; however, the effectiveness of the strategy is quite modest. Improvements in auto traffic were about 2 percent, whereas the increases in transit patronage and levels of service were more substantial. On the other hand, the annual cost of the project is estimated to be approximately $2.63 million. Whether the project will be selected will depend on the values for each measure of effectiveness developed in the analysis, as well as the relative importance of each objective, availability of funds, and other projects that have been proposed for the region.

Example 14.2   Using Simplified Procedures for Low-Cost TSM Projects

The preceding example illustrated how the transportation planning process, which typically is used for long-range studies, can be applied to the evaluation of TSM actions. The procedure is still somewhat cumbersome, however, since it depends on an extensive database, the use of calibrated models for trip distribution and mode choice, and a link-node network for traffic assignment. Many TSM projects are low-cost improvements and to evaluate them by using such complex procedures is both impractical and costly. The following examples illustrates an evaluation procedure intended for low-cost projects. Additional details of this approach are described in *Simplified Procedures for Evaluating Low-Cost TSM Projects*. The procedure follows four basic steps: (1) analyze problems and their setting, (2) identify and screen candidate solutions, (3) design, analyze, and evaluate solutions, and (4) recommend an action plan. The following problem involves a travel corridor in a small city and for this reason transit does not play a major role.

Solution:

Step 1. Analyze Problems and Setting

A 2-mi corridor connects the CBD of a small town with a residential district and a freeway interchange. The artery, Lisbon Street, is a two-lane road that carries 5000–8000 veh/d. (See Figure 14.3.) Strip development borders the road and also serves as frontage for residential development.

The residents and merchants in the community have complained about delays and unsafe driving conditions. In response, the county engineering office conducted field observations of traffic movements at intersections, travel time runs, and speed and delay studies, and collected crash data. In addition, the physical condition of the road was noted, including surface, curbs, signals, markings, and signs.

The results of this data-gathering phase were then compared with a set of preestablished deficiency criteria, shown in Table 14.2. The performance measures noted in the field were then plotted, showing the observed value at each intersection or along the road itself. These results are shown in Figure 14.4 for three performance measures: peak hour volume-to-capacity ratio, average peak travel speed, and crash rate factor.

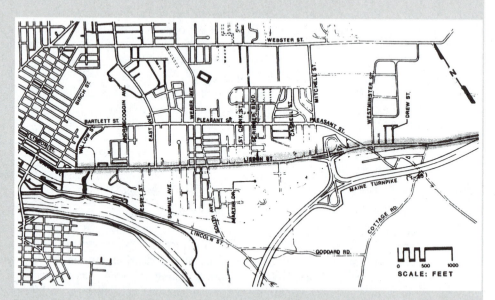

Figure 14.3 Lisbon Street Corridor

SOURCE: J. H. Batchelder, et al., *Simplified Procedures for Evaluating Low-Cost TSM Projects: User's Manual,* NCHRP Report 263, Transportation Research Board, National Research Council, Washington, D.C., 1983.

Table 14.2 Conditions That Establish Deficiency of Roadway Segments

- Supply Capacity

  Lane width under 11 ft

  Signs and pavement marking not in conformance with *Manual of Uniform Traffic Control Devices* (MUTCD)

  Traffic control not in conformance with ITE/AASHTO standards

  Curb radii less than 20 ft

  Sight distance not in conformance with MUTCD

- Service Quality

  Intersection level of service lower than C, or critical lane volumes greater than 1200 cars/hr

  Road segment level of service lower than C, or average lane volume greater than 1200 cars/hr

  Peak average travel speed less than 20 mi/h

- Crashes

  Three-year average crash rate exceeds station average with 95 percent confidence

  Average rates

     Signalized intersections: 1.22/million entering vehicles

     Nonsignalized intersections: 0.39/million entering vehicles

     Road segments: 3.74/million vehicles

SOURCE: J. H. Batchelder et al., *Simplified Procedures for Evaluating Low-Cost TSM Projects: User's Manual,* NCHRP Report 263, Transportation Research Board, National Research Council, Washington, D.C., 1983.

The profiles both show locations along the corridor where problems exist and suggest possible relationships among the problems. Notes on the margin of the existing condition profile indicate the probable causes of each deficiency. For example, low travel speeds between Summit Street and South Street are due to faulty traffic signal operation and curb parking near intersections.

The problems are summarized on a site map, Figure 14.5, by road segment. In this case, there are four problem areas along the corridor. Other constraints are noted; for example, no other feasible parallel routes exist, so improvements must be limited to the narrow corridor along Lisbon Street.

Step 2.   Identify and Screen Candidate Solutions

As is typically the case in many communities, funds to correct street and highway problems are limited. In this instance, solutions must be restricted to those not requiring the taking of right of way or the purchase of new traffic control equipment. Also, any solution that is likely to create opposition and delay in implementation would be eliminated. Thus, solutions are limited to traffic and parking controls. Ideas such as ridesharing, transit, or physical improvements are discarded as being impractical, ineffective, or too costly.

Possible solutions are identified with the aid of a set of TSM Action Identification Tables and TSM Action Profiles. The TSM Action

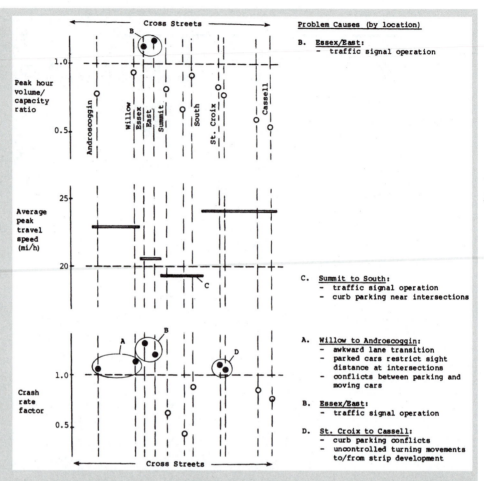

Figure 14.4 Observed Conditions Along Lisbon Street Corridor

SOURCE: J. H. Batchelder, et al., *Simplified Procedures for Evaluating Low-Cost TSM Projects: User's Manual,* NCHRP Report 263, Transportation Research Board, National Research Council, Washington, D.C., 1983.

Identification Tables are intended to assist the engineer in identifying general approaches and specific types of TSM actions that may be applicable in solving short-term transportation problems. Tables have been prepared in the *Simplified Procedures Manual* for six problem types: (1) isolated intersections or street segments, (2) corridors, (3) residential communities, (4) employment centers, (5) commercial centers, and (6) regional, state, and national problems. For each problem type, a subset of problems is identified. The tables are organized around five factors: (1) location and scale of the problem, (2) nature of the problem, (3) underlying deficiency

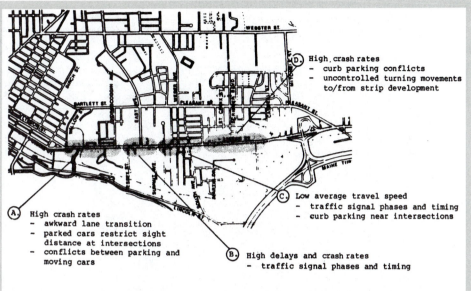

Figure 14.5 Problem Summary for Lisbon Street Corridor

SOURCE: J. H. Batchelder, et al., *Simplified Procedures for Evaluating Low-Cost TSM Projects: User's Manual*, NCHRP Report 263, Transportation Research Board, National Research Council, Washington, D.C., 1983.

in transportation system, (4) strategy or approach to the solution, and (5) location or scale of action.

In the Lisbon Street example, the problem falls under the category of "Isolated Intersections or Street Segments," and the subset of this problem includes: (1) vehicle flow conflicts and crashes and (2) traffic congestion. Table 14.3 shows the Action Identification Table for these two problems. Other problem solution ideas would be obtained from the Action Identification Table for corridors.

The Action Identification Tables and Action Profiles (described earlier) assist the engineer in identifying candidate solutions. A list of possible actions is prepared, together with notes or comments pertaining to their relevance to the problem being studied. For the Lisbon Street example, a list of 20 possible action improvements was developed. Comments

Table 14.3   Action Identification: Isolated Problems at Intersections or on Street Segments

Underlying Transportations Deficiencies	Corrective Strategies	Actions Applied at Intersections	Actions Applied on Street Segments
Problem 1: Vehicle Flow Conflicts and Accidents			
Turning and parking movements inhibit traffic flow →1A.	*Reduce delays and conflicts* by separating flows.	—New signals —Signal phasing and timing changes —Turn lanes —Striping —Traffic police —Islands	—Two-way left-turn lanes —Expanded off-street parking or loading areas —On-street parking restrictions or removal
1B.	*Reduce delays and conflicts* by diverting movements.	—Left-turn prohibitions —Jug-handles —On-street parking restrictions near intersections	—Medians —Side street and curb cut closures —On-street parking restrictions or removal —Expanded off-street parking or loading areas
Inadequate sight or stopping distances → 1C.	*Increase time available for driver reaction.*	—New signals or stop signs —Signal phasing and timing changes —Warning devices	—Speed restriction —On-street parking restrictions or removal —Warning devices
Problem 2: Traffic Congestion			
Turning and parking movements inhibit traffic flow →2A&B.	Same as strategies 1A and 1B above		
Inadequate capacity to handle peak traffic volumes →2C.	*Reduce delays* by adding capacity.	—New lanes —Signal timing	

SOURCE: Adapted from J.H. Batchelder et al., *Simplified Procedures for Evaluating Low-Cost TSM Projects: User's Manual,* NCHRP Report 263, Transportation Research Board, National Research Council, Washington, D.C., 1983.

concerning their appropriateness were prepared and a recommendation made as to whether or not the action should be pursued (based on a judgment assessment of feasibility and effectiveness). The action screening worksheet is shown in Figure 14.6. The process resulted in a recommendation to pursue 12 of the 20 candidate actions. These were combined into two initial action packages. The first set of actions included minimum improvements required to solve the problem at the lowest cost. The second set included widening the roadway along one segment of the corridor.

CANDIDATE STRATEGIES/ACTIONS	NOTES/COMMENTS	PURSUE?
• Separate traffic flows at intersections to reduce crashes and delays		
- new traffic signals	- major intersections already signalized	No
- separate turn phases	- significant turning movements at 5 intersections	Yes
- left-turn lanes	- ditto; adequate right-of-way, but need to remove parking	Yes
- striping/islands	- probably useful at some locations	Yes
• Separate traffic flows on street segments to reduce delays		
- two-way left turn lanes	- inadequate right-of-way	No
• Divert turning movements to reduce crashes and delays		
- left-turn prohibitions	- added traffic on side streets and driver confusion rule this out	No
- street and curb-cut closures	- no alternative entrances to many abutting properties; entrances to some side streets could be closed but this would hinder access to off-street parking and residences; must be done selectively	Maybe
- medians	- added travel distances and increased turns at intersections (including illegal U-turns) would disrupt traffic more than existing midblock turns	No
• Move parking from street to reduce crashes and delays		
- remove parking near intersections	- adequate vacancy rate in nearby spaces along most of corridor, so little opposition expected from merchants	Yes
- remove parking in midblock	- in most locations, adequate off-street or side street parking exists near stores; may need to retain parking in some blocks of Lisbon St.	Yes
- add off-street parking	- not needed, vacant spaces can absorb cars currently parking on Lisbon St.	No
• Increase time for driver reaction to avoid crashes		
- new signals, signs or warning devices	- most intersections meet signal and signing standards, but some signing could be improved	Yes
- parking removal near intersections	- should be effective in improving sight distance at problem	Yes
• Add capacity at intersections to reduce delays		
- new lanes	- can use width of parking lanes to add turning or through lanes if curb radii are increased	Yes
- signal timing	- retiming signals will add effective capacity to at least two intersections	Yes
• Add capacity on street segments to reduce delays		
- new lane	- existing parking lanes can be safely used as travel lanes if curb radii are increased	Yes
• Add street capacity to corridor to reduce delays		
- new segments or lanes on parallel streets	- not feasibile in residential neighborhoods	No
• Improve use of existing corridor capacity to reduce delays		
- signal coordination	- physically interconnecting Essex/East St. signals should be considered; manual coordination probably adequate elsewhere	Yes
- reversible lanes	- inadequate right-of-way; balanced traffic	No
- one-way pairs	- too much local traffic; no good route for other half of pair	No

Figure 14.6 Candidate TSM Actions for Lisbon Street Corridor

SOURCE: J. H. Batchelder, et al., *Simplified Procedures for Evaluating Low-Cost TSM Projects: User's Manual,* NCHRP Report 263, Transportation Research Board, National Research Council, Washington, D.C., 1983.

The initial actions are shown in Figure 14.7. Essentially, the improvement program involves removal of parking to open up lanes for travel or turning movements and to modify signal operations.

Step 3. Design, Analyze, and Evaluate Solutions

An analysis plan is developed that will permit an evaluation of each criterion considered to be important for the project. The criteria selected are as follows:

- Travel lanes at least 11 ft wide
- Signs, signals, and markings conform to MUTCD
- Turning radii at least 20 ft
- Adequate separation between turning and through traffic
- Adequate buffer between curb travel lanes and sidewalk flows
- Safe crossing time during each signal cycle
- Level of service C or higher
- Peak travel times 20 mi/h or higher
- Parking provided within 300 ft of stores
- Budget limited to $200,000

The analysis plan for the project is shown in Figure 14.8. It addresses each of the criteria posed. Since this is a short-term action, factors such as traffic growth and changing travel patterns need not be considered. The techniques used to carry out the analysis plan are listed in Figure 14.9. Many of the traffic engineering analysis methods required to solve this problem

<u>Package #1</u>: improvements requiring only a small amount of new traffic control equipment and no new right-of-way

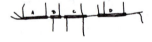

<u>Package #2</u>: improvements requiring only a small amount of new traffic control equipment and/or right-of-way

- All sections:
  - increase curb radii
  - remove parking near major intersections

- Section A (Willow to Androscoggin):
  - add turn and travel lanes through restriping

- Section B (Essex to East):
  - connect traffic signals
  - add turn lanes and phases

- Section C (Summit to South):
  - restripe for 2 approach lanes at intersections

- Section D (St. Croix to Cassell):
  - remove parking
  - restripe for 2 lanes in each direction

- All actions included in Package #1

- Section A:
  - widen to add travel lanes

Figure 14.7 Action Packages for Further Evaluation

SOURCE: J. H. Batchelder, et al., *Simplified Procedures for Evaluating Low-Cost TSM Projects: User's Manual*, NCHRP Report 263, Transportation Research Board, National Research Council, Washington, D.C., 1983.

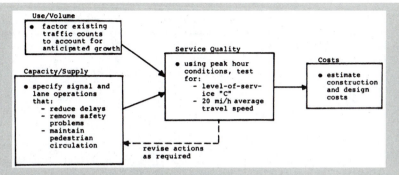

Figure 14.8 Analysis Plan for Corridor Evaluation

SOURCE: J. H. Batchelder, et al., *Simplified Procedures for Evaluating Low-Cost TSM Projects: User's Manual,* NCHRP
Report 263, Transportation Research Board, National Research Council, Washington, D.C., 1983.

- **Service Quality**

  - Traffic delays at intersections - Use
    critical lane method to test specific
    lane and signal options for achievement
    of level-of-service C during peak hour.

  - Average travel speed - Intersection
    improvements and parking restrictions
    should be sufficient to meet 20 mi/h
    standard. Proposed phasing sequences
    should be checked for interference.

  - Parking availability - Field survey showed
    vacancy rates along side streets and in
    adjacent lots are adequate to accomodate
    any reductions in on-street spaces.

  - Pedestrian access and circulation -
    Existing sidewalks and crosswalks are
    sufficient to accommodate increased flows.
    Anticipated pedestrian and turning traffic
    volumes do not warrant phase separations.
    Signal timing must continue to allow
    adequate crossing time for pedestrians.

- **Safety**

  Examine crash records to
  identify correctible safety problems.

- **Supply/Capacity**

  Specify different signal phasings and lane
  configurations and operations for analysis.

- **Financial**

  Prepare cost estimates for engineering and
  construction/implementation of recommended
  actions. Changes in on-going street
  maintenance costs are expected to be
  minimal.

- **Public Costs**

  - Crashes - Intersection safety improve-
    ments identified in safety analysis
    should sufficiently reduce crash rates.

  - Air quality - No problems exist or are
    anticipated in the corridor.

Figure 14.9 Techniques Required to Complete Analysis

SOURCE: J. H. Batchelder, et al., *Simplified Procedures for Evaluating Low-Cost TSM Projects: User's Manual,* NCHRP
Report 263, Transportation Research Board, National Research Council, Washington, D.C., 1983.

have been described in earlier chapters of this textbook. Table 14.4 lists
the methods suggested for carrying out travel time estimation procedures.
Method selection aids for computing other variables can be found in a
guide from the Transportation Research Board.

The principal elements of this problem required the analysis and
design of seven intersections and three short segments of roadway. The
engineering analyses required to solve the problem were examination of

Table 14.4 Methods Selection Table Travel Time Estimation

Spot or Segment Analysis	*Appropriate Techniques*
Vehicular delays at intersections	
Check for acceptable level of service	Use of capacity adequacy techniques cited in the Methods Selection Table for supply/capacity estimation
Estimate average travel time, delay, or queue duration	Apply analytical techniques and worksheets
	Apply simulation models of signalized intersection operations (mainly applicable to complex, high-volume intersections)
Set signal timing to minimize delays	Apply Webster's equation or similar procedure
Vehicular travel time along roadway segment	
Check for acceptable level of service	Use capacity adequacy techniques cited in the Methods Selection Table for supply/capacity estimation
Estimate average travel time	Use curves relating speed to volume-to-capacity ratio for different types of roadways
Travel time along transit route segment	
Estimate route running times	Apply an analytical procedure to calculate route segment running times
Estimate operating speeds	Use average operating speeds observed under similar conditions

SOURCE: Adapted from J.H. Batchelder et al., *Simplified Procedures for Evaluating Low-Cost TSM Projects: User's Manual,* NCHRP Report 263, Transportation Research Board, National Research Council, Washington, D.C., 1983.

crash records and safety conditions and the use of critical lane analysis and signal timing equations to test for operation at level of service C or better. Similar techniques were applied to the remainder of the corridor and final checks were made of signal phasing and timing to assure that no conflicts existed. Finally, the results were summarized and used in estimating costs.

Step 4. Recommend an Action Plan

The major constraints on the project are the requirement that costs not exceed $200,000 and that the project be implemented over a 3 to 5 yr period. Accordingly, the recommended improvements were grouped in order of priority such that projects selected first addressed the most significant problems and second, could be completed within a 2 to 3 yr period. Other projects were identified that should be implemented but are deferred subject to availability of funds. The recommended action plan, shown in Figure 14.10, consists of the following projects. Detailed diagrams of each improvement are also prepared as illustrated in Figure 14.11.

Priority 1 ($55,000)

- Section A
  —Restripe for one eastbound and two westbound lanes
  —Paint transition zone to start of curb parking
  —Increase curb radii
- Section B
  —Change lane-use designations (left-turn lanes)
  —Change traffic signal phases and timing
  —Increase curb radii

Priority 2 ($60,000)

- Section C
  —Relocate traffic signal heads
  —Prohibit parking near intersections
  —Restripe for two approach lanes
  —Increase curb radii

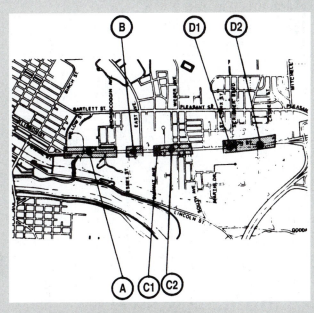

Figure 14.10 Recommended Action Plan

SOURCE: J. H. Batchelder, et al., *Simplified Procedures for Evaluating Low-Cost TSM Projects: User's Manual,* NCHRP Report 263, Transportation Research Board, National Research Council, Washington, D.C., 1983.

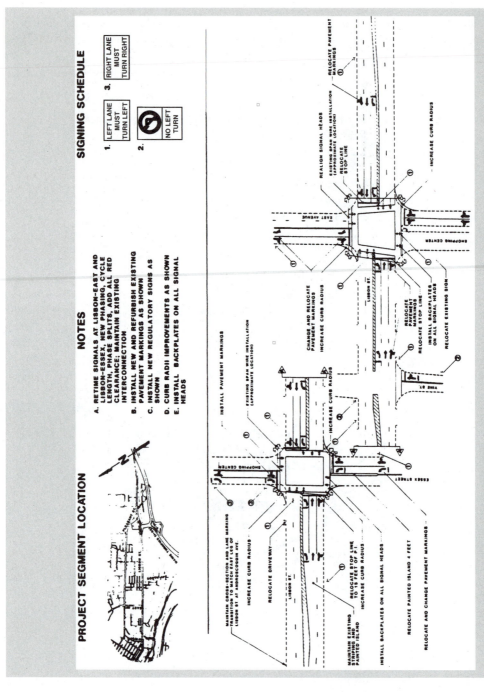

Figure 14.11   Example of a Project Improvement Diagram for Lisbon Street Corridor

SOURCE: J. H. Batchelder, et al., *Simplified Procedures for Evaluating Low-Cost TSM Projects: User's Manual*, NCHRP Report 263, Transportation Research Board, National Research Council, Washington, D.C., 1983.

- Section D
  —Remove parking and restripe for two lanes in each direction
  —Upgrade traffic signal hardware
  —Increase curb radii

Priority 3 ($75,000)

- Section C2
  —Prohibit parking near intersections
  —Restripe for two approach lanes
  —Increase curb radii
- Section D2
  —Remove parking and restripe for two lanes in each direction
  —Increase curb radii

TSM AND INTELLIGENT TRANSPORTATION SYSTEMS (ITS)

The term intelligent transportation systems (ITS) refers to the use of information technology including computers, electronics, and communications to improve traffic operations. ITS represents a twenty-first century tool than can be used to implement many of the TSM strategies described in this chapter. ITS projects for TSM are operational improvements to existing transportation facilities. The approaches used in the highway field are focused on three broad categories of ITS strategies. They are: (1) advanced traveler information systems (ATIS), (2) advanced traffic management systems (ATMS), and (3) advanced vehicle control systems (AVCS). Each of these ITS techniques addresses a distinct element of the highway system and is intended to achieve reductions in congestion and pollution and to increase safety. Transit and commercial vehicle applications of ITS, referred to as commercial vehicle operations (CVO) and advanced public transportation systems (APTS) are also available and may be used in some TSM projects (such as ridesharing and vehicle scheduling). The three major applications of ITS in highway and traffic engineering are highlighted and described in the following sections. Additional information is available from the references cited at the end of the chapter.

Advanced Traveler Information Systems (ATIS)

The purpose of ATIS is to furnish the motorist with up-to-the-minute information about highway conditions in the region where a trip is being planned or executed. Information is provided in real time about congestion, bottlenecks caused by accidents or construction, and weather to assist the motorist in making informed decisions regarding the best departure time and route. Information may also include alternate routes to avoid heavy congestion. Information is provided for pre-trip planning and while the motorist is en route. Pre-trip information can include parking availability at destination sites, toll facilities and charges, and route alternates. Motorists are most interested in information regarding unexpected incidents, weather and roadway conditions, travel times and delays, and construction or maintenance activity. For the visitor to the area, additional information can include cultural and recreational sites and route guidance.

Advanced Traffic Management Systems (ATMS)

The purpose of ATMS is to improve the efficiency and effectiveness of the existing highway system while minimizing environmental effects. Among the applications of ATMS are urban traffic control, incident detection, and freeway or corridor management. Urban traffic control utilizes ITS technologies by using sensors in the roadway that furnish data to a central control facility where traffic is monitored. Variable message signs are used to provide motorists with information about road conditions ahead and alternative routes if needed. Ramp metering controls traffic flow that enters a freeway to assure smooth and continuous flow. ITS, to be effective, requires constant surveillance of existing traffic, as well as flow parameters, including volume, speed occupancy, and travel times. Surveillance technologies include loop detectors, video camera, image processing, acoustics, radar, and microwave technologies. Surveillance also includes road and weather information including fog, rain, snow, and pavement condition. Incident detection is an important element of ATMS, as the ability to detect an incident and to quickly remove vehicles from the roadway can have a major impact on congestion and delay. ATMS is also used to collect tolls electronically, thus eliminating long queues at manually operated booths.

Advanced Vehicle Control Systems (AVCS)

The purpose of AVCS is to enhance the control of the vehicle through electronics and computer technology, thus increasing safety by minimizing the potential for crashes and the damage they cause. Among the components of AVCS are rear-end collision avoidance systems, intelligent cruise control, and protection from unsafe maneuvers such as lane changing. The ultimate goal of AVCS is to develop a fully automated highway system in which the driver is on "automatic pilot" while monitoring the situation. A "hands-off, feet-off" technology is feasible only in limited corridors and under carefully controlled conditions, and its widespread use is still many years away. Many of the AVCS technologies are the product of innovations in automobile technology. A gradual evolution in the way people drive will occur as new technological options become less costly and are included in the "standard" packages of vehicle options. Over time, both the roadway and the automobile will become "smarter."

Applications of ITS Technologies

There are four major areas where ITS technologies are enhancing travel demand and transportation systems management. These are (1) providing real-time traveler information, (2) parking management, (3) dynamic pricing of highway capacity, and (4) telecommuting.

Real-time traveler information, as previously described, provides the motorist with up-to-the-minute information about travel conditions and choices. The availability of this information can influence a traveler to change routes or mode, from highway to transit, thus reducing congestion. As the saying goes, "knowledge is power," and the benefits to the motorist in receiving accurate and current information are improvements in travel time and safety.

Parking management is intended to provide the motorist with information about the availability of parking spaces within large structures or vast surface lots. Everyone has had the frustrating experience of circling round and round in a vain attempt to locate a vacant parking space. Sometimes the only way to find one is to follow someone who is approaching a parked car. Electronic parking information and guidance systems can serve as an effective means to eliminate the uncertainty in finding a space. These systems, which are more common in Western Europe than in the United States, serve as an effective TSM tool by dispersing parking demand, guiding tourists in unfamiliar areas, and effectively utilizing existing parking facilities.

Dynamic pricing of highway capacity, known as congestion or value pricing, has the potential to reduce peak traffic loads on highways by charging higher prices during congested periods. Peak-hour pricing is common practice in the telephone and utility industries. However, prior to the availability of ITS technologies, no practical means existed for charging for highway transportation on the basis of miles driven and time of day. With electronic toll collection, the possibility now exists to price highway service based on peak and off-peak travel demand.

The potential for telecommuting is being realized as employees are able to work at home or other remote sites, thus reducing travel demand at congested commercial locations. Technologies such as video-conferencing, high-speed voice and data links, and portable computer equipment have minimized the need for many workers to be at a centralized employment location. As the nation moves toward a service and information–based economy, remoteness from the main office is no longer a barrier to effective communication and worker productivity. The extent to which telecommuting will play a significant role in traffic reduction will depend to a great extent on the willingness of industry's leadership to adapt their practices to a new style, one in which the worker is no longer under direct supervision or management.

SUMMARY

Transportation systems management (TSM) is that aspect of transportation engineering that seeks to improve the existing street and highway system by making changes in its operations. A wide variety of changes can be made both to the system itself (supply) and how the system is used (demand). The term transportation demand management (TDM) is often used to reflect the latter set of actions.

This chapter has described the many TSM techniques available for consideration in solving a given transportation problem. Actions can be selected to improve the use of road space, reduce vehicle use, and improve transit service. Each action has its advantages and disadvantages, and these should be fully understood before it is recommended.

A TSM planning process should be followed when evaluating any action. The process includes identification of the problem, establishment of a set of objectives and criteria, identification of candidate TSM actions, design and evaluation of solutions, preparation of a recommended action plan, and, after the project is completed, an evaluation of its actual performance.

Carrying out a TSM analysis and design requires techniques and skills of traffic engineering and highway planning. Among these are signal timing, capacity analysis, speed and accident studies, pavement striping and signing, and parking regulations. The application of these topics has been described in previous chapters, and the techniques can be applied in the design and testing of TSM alternatives.

PROBLEMS

14-1   For the following situations, indicate if a TSM project is an appropriate solution. For each situation where TSM would be selected, describe two ways in which the problem could be solved.
(a)   Traffic is congested on downtown city streets.
(b)   A suburban employment center lacks necessary parking facilities.
(c)   A freeway during peak morning and evening periods has a level of service D.
(d)   Traffic crashes increase along an arterial street.

14-2   Describe two situations with which you are familiar in which TSM actions have been undertaken to accomplish the following objectives:
(a)   Create efficient use of road space
(b)   Reduce vehicle volume
(c)   Improve transit service

14-3   Prepare a TSM action profile for the following actions: (a) increased peak period tolls, (b) one-way streets, and (c) employer vanpool programs. In your profile, consider the problem addressed, implementation problems, and evaluation factors.

14-4   Select a transportation corridor in your community for which traffic congestion is a problem. Prepare a report discussing how you would proceed to develop an improvement program. In your report include the following items.
(a)   Description of existing conditions
(b)   Objectives to be accomplished
(c)   Proposed TSM actions to be investigated
(d)   Measures of effectiveness that best represent the improvement
(e)   Travel demand data required
(f)   Impact analysis and project selection

14-5   What TSM strategies would improve the parking situation at your college or university or at a neighborhood parking facility? What is the likelihood that these actions would be implemented by the appropriate administration?

14-6   A proposal has been made to convert one lane (in each direction) of a six-lane freeway for use by carpools only. The lane would extend for 10 mi. Currently, the lanes operate at capacity in morning and evening peak hours, with an average of 1.2 persons/car. It is expected that about 20 percent of drivers will carpool with an average of 3 persons/auto. If the jam density is 124 veh/m and the mean free speed is 58 mi/h, what will be the travel time savings under these conditions? (Assume linear relationships between speed and density.) Peak hours are 7–9 a.m. and 4–6 p.m.

14-7   A separate bus lane is proposed that will replace a single lane of a freeway with a capacity of 2000 veh/h. If auto occupancy is 1.5 persons/vehicle, what are the fewest bus trips/hour required if each bus were to carry 50 passengers? What would be the average headway for each bus? Comment on the advantages and disadvantages of this proposal.

14-8   A large corporation located in a suburban area wishes to limit the need for parking by encouraging its employees to vanpool to work. The vans will be driven by an employee who rides free, and each van will accommodate 10 riders who live an average of 10 mi from the work site. A van costs $18,000, and maintenance and operating costs are $0.50/ mile.
(a)   Determine the monthly charge for each passenger, assuming the life of a van is 8 years and interest rates are 10 percent.

(b)   If 50 vans are in use, how many fewer parking spaces are needed?

(c)   If a parking space costs $4000 to construct, how much could the company spend for the vanpool program and still break even?

14-9   A corridor in an urban area is being investigated to determine the effectiveness of several proposed TSM actions to improve travel conditions. The proposed actions are park-and-ride lots, improved signal timings and phasings, one-way streets, and alternative work schedules. What measures of effectiveness could be used to evaluate the proposed improvements?

14-10   Discuss the differences between TSM studies and long-range planning studies.

14-11   Discuss the four phases for developing a TSM plan and the effort required in each step.

14-12   Identify the three broad categories of ITS intended to improve operations on existing highways. Briefly describe strategies available in each category.

REFERENCES

Abrams, C. M., and Renzo, J. D., *Measures of Effectiveness for TSM Strategies,* U.S. Department of Transportation, Federal Highway Administration, Washington, D.C., 1979.

American Association of State Highway Officials, *Guide for Development of Bicycle Facilities,* Washington, D.C., 1999.

Batchelder, J. H., et al., *Simplified Procedures for Evaluating Low-Cost TSM Projects: User's Manual,* NCHRP Report 263, Transportation Research Board, National Research Council, Washington, D.C., 1983.

Mason, Jr., J. M., et al., *Training Aid for Applying NCHRP Report 263,* NCHRP Report 283, Transportation Research Board, National Research Council, Washington, D.C., 1986.

Meyer, M. D., editor, *A Toolbox for Alleviating Traffic Congestion,* 2nd ed., Institute of Transportation Engineers, Washington, D.C., 1997.

Institute of Transportation Engineers, *Intelligent Transportation System Primer,* Washington, D.C., 2000.

ADDITIONAL READINGS

Edwards, Jr., J. D., editor, Transportation Planning Handbook, 2nd ed., Institute of Transportation Engineers, Washington, D.C., 1999.

Federal Highway Administration, *Transportation Management for Corridors and Activity Centers: Opportunities and Experiences, Final Report,* Technology Sharing Program, Washington, D.C., 1986.

New York State Department of Transportation, *Energy Impacts of Transportation Systems Management Actions,* Urban Mass Transportation Administration, Washington, D.C., 1981.

Sussman, Joseph, *Introduction to Transportation Systems,* Artech House, Boston-London, 2000.

Transportation Research Board, *Transportation Demand Management and Ride Sharing,* TRR 1564, National Research Council, Washington, D.C., 1996.

Urbitran Associates, *Transportation System Management: Implementation and Impacts,* Urban Mass Transportation Administration, Washington, D.C., 1982.

Voorhees, Alan M., Inc., *TSM: An Assessment of Impacts,* Urban Mass Transportation Administration, Washington, D.C., 1978.

P A R T 4

Location, Geometrics, and Drainage

Highway location involves the acquisition of data concerning the terrain upon which the road will traverse and the economical siting of an alignment. To be considered are factors of earthwork, geologic conditions, and land use. Geometric design principles are used to establish the horizontal and vertical alignment, including consideration of the driver, the vehicle, and roadway characteristics. Design of parking and terminal facilities must be considered as they form an integral part of the total system. Since the new highway will alter existing patterns of surface and subsurface flow—and be influenced by it—careful attention to the design of drainage facilities is required.

CHAPTER 15

Highway Surveys and Location

Selecting the location of a proposed highway is an important initial step in its design. The decision to select a particular location is usually based on topography, soil characteristics, environmental factors such as noise and air pollution, and economic factors. The data required for the decision process are usually obtained from different types of surveys, depending on the factors being considered. Most engineering consultants and state agencies presently involved in highway locations use computerized techniques to process the vast amounts of data that are generally handled in the decision process. These techniques include remote sensing, which uses aerial photographs for the preparation of maps, and computer graphics, which is a combination of the analysis of computer-generated data with a display on a computer monitor. In this chapter we present a brief description of the current techniques used in highway surveys to collect and analyze the required data, and the steps involved in the procedure for locating highways. We also cover earthwork computations and mass diagrams, since an estimate of the amount of earthwork associated with any given location is required for an economic evaluation of the highway at that location. The result of the economic evaluation aids in the decision to accept or reject that location.

TECHNIQUES FOR HIGHWAY SURVEYS

Highway surveys usually involve measuring and computing horizontal and vertical angles, vertical heights (elevations), and horizontal distances. The surveys are then used to prepare base maps with contour lines (that is, lines on a map connecting points that have the same elevation) and longitudinal cross sections. Highway surveying techniques have been revolutionized due to the rapid development of electronic equipment and computers. Surveying techniques can be grouped into three general categories:

- Ground surveys
- Remote sensing
- Computer graphics

Ground Surveys

Ground surveys are the basic location technique for highways. The *transit* is used for measuring angles in both vertical and horizontal planes, the *level* for measuring changes in elevation, and the *tape* for measuring horizontal distances. *Electronic distance measuring* (EDM) *devices* and computers are used in contemporary surveying. A summary of survey equipment follows. For greater detail, refer to the surveying texts cited at the end of this chapter.

The Transit

The transit consists of a telescope with vertical and horizontal cross hairs, a graduated arc or vernier for reading vertical angles, and a graduated circular plate for reading horizontal angles. The telescope is mounted so that it can rotate vertically about a horizontal axis, and the amount of rotation can be directly obtained from the graduated arc. Two vertical arms support the telescope on its horizontal axis, with the graduated arc attached to one of the arms. The arms are attached to a circular plate, which can rotate horizontally with reference to the graduated circular plate, thereby providing a means for measuring horizontal angles. An example of the engineer's transit is shown in Figure 15.1a.

Measurement of Horizontal Angles. A horizontal angle, such as angle *XOY* in Figure 15.1b, can be measured by setting the engineer's transit directly over the point *O*. The telescope is then turned toward point *X* and viewed first over its top and then by sighting through the telescope to obtain a sightline of *OX*, using the horizontal tangent screws provided. The graduated horizontal scale is then set to zero and locked into position, using the clamp screws provided. The telescope is rotated in the horizontal plane until the line of sight is *OY*. The horizontal angle (α) is read from the graduated scale.

Measurement of Vertical Angles. The transit measures vertical angles by using the fixed vertical graduated arc or vernier and a spirit level attached to the telescope. When a transit is used to determine the vertical angle to a point, the angle obtained is the angle of elevation or the angle of depression from the horizontal. The transit is set up in a way similar to that for measuring horizontal angles, ensuring that the longitudinal axis is in the horizontal position. The telescope is approximately sighted at the point. The tangent screw of the telescope is then used to set the horizontal cross hair at the point. The vertical angle is obtained from the graduated arc or the vertical vernier.

The Level

The essential parts of a level are the telescope, with vertical and horizontal cross hairs, a level bar, a spindle, and a leveling head. The level bar on which the telescope is mounted is rigidly fixed to the spindle. The level tube is attached to the telescope or the level bar so that it is parallel to the telescope. The spindle is fitted into the leveling head in such a way that allows the level to rotate about the spindle as an axis, with the leveling head attached to a tripod. The level also carries a bubble that indicates whether the level is properly centered. The centering of the bubble is done by using the leveling screws provided.

(a) Typical engineer's transit

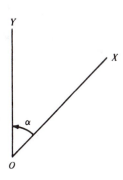

(b) Measurement of a horizontal angle using a transit

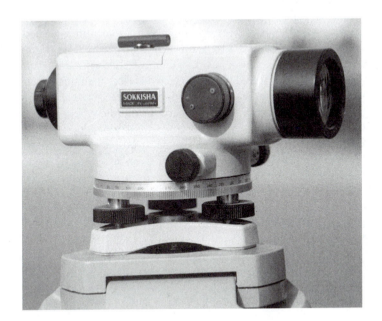

(c) Automatic level

(d) Wooden leveling rod

Figure 15.1 Typical Surveying Equipment

SOURCE: Photographs by Ed Deasy, Virginia Transportation Research Council, Charlottesville, Va. Used with permission.

The most common types of levels are the Wye level and the self-leveling or automatic levels, shown in Figure 15.1c. The specific features of these levels are that the telescope of the Wye level is permanently fixed to the level bar and that the self-leveling level automatically levels itself.

A leveling rod is required to measure changes in elevation. These are rods of rectangular cross sections, graduated in feet or meters so that differences in heights can be measured. Figure 15.1d shows a typical leveling rod, graduated in feet.

The difference in elevation between two points, A and B, is obtained by setting the level at a third point. The rod is then placed at the first point, say, point A, and the level is sighted at the rod. The reading indicated by the cross hair is recorded. The level rod is transferred to point B, the level is again sighted at the rod, and again the reading indicated by the cross hair is recorded. The difference between these readings gives the difference in elevation between points A and B. If the elevation of one of the points is known, the elevation of the other point can then be determined.

Measuring Tapes

Tapes are used for direct measurement of horizontal distances. They are available in several materials, but the types used for engineering work usually are made of steel or a woven nonmetallic or metallic material. They are available in both U.S. and metric units.

Electronic Distance Measuring Devices

Several EDM devices are available, which may have slightly different advantages, but work on the same general principle. An EDM device consists mainly of a transmitter located at one end of the distance to be measured and a reflector at the other end. The transmitter sends either a light beam, a low-power laser beam, or a high-frequency radio beam in the form of very short waves, which are picked up and reflected back to the transmitter. The difference in phase between the transmitted and reflected waves is measured electronically and used to determine the distance between the transmitter and the reflector. Figure 15.2a shows the TOPCON DM-A2 electronic distance measuring device. This equipment can measure distances up to about 3300 ft in average atmospheric conditions. When used in conjunction with the slope reduction calculator, it can also measure slope and height differences. Special features permit the operator to automatically change the display from slope to horizontal distance. Units can also be changed from meters to feet. More recent EDM devices can be mounted on the framework of a transit, which permits the determination of vertical and horizontal angles. The result is that distances and directions can be determined from a single instrument setup. Figure 15.2b shows a set up of TOPCON DM-A2 mounted on a transit for angular and linear measurement. Figure 15.2c shows a geodetic total station, consisting of an EDM and a theodolite in a single instrument.

Remote Sensing

Remote sensing is the measurement of distances and elevations by using devices located above the earth, such as airplanes or orbiting satellites using Global Positioning Systems. For a discussion of GPS and Land/Geographic Information Systems see *The Surveying Handbook*, R. C. Brinker (ed), 1995. The most commonly used remote sensing method is

(a) TOPCON DM-A2 electronic distance measuring device

Figure 15.2 Typical Electronic Surveying Instruments

aerial photography, usually referred to as *photogrammetry*. Photogrammetry is the process by which distances between objects and other information are obtained indirectly, by taking measurements from aerial photographs of the objects. This process is fast and cheap for large projects but can be very expensive for small projects. The break-even size for which photogrammetry can be used varies between 30 and 100 acres, depending on the circumstances of the specific project. The successful use of the method depends on the type of terrain. Difficulties will arise when it is used for terrain with the following characteristics:

- Areas of thick forest, such as tropical rain forests, that completely cover the ground surface
- Areas that contain deep canyons or tall buildings, which may conceal the ground surface on the photographs
- Areas that photograph as uniform shades, such as plains and some deserts

The most common uses of photogrammetry in highway engineering are the identification of suitable locations for highways and the preparation of base maps with contours of 2- or 5-ft intervals. In both of these uses, the first task is to obtain the aerial photographs of the area if none is available. A brief description of the methodology used to obtain and interpret the photographs is presented below. Readers interested in a more detailed treatment of the subject should refer to sources listed in the "Additional Readings" section at the end of this chapter.

Obtaining and Interpreting Aerial Photographs

The photographs are taken from airplanes, with the axis of the camera at a near vertical position. The axis should be exactly vertical, but this position is usually difficult to

(b) TOPCON DM-A2 mounted on a theodolite for both angular and linear distance measurements

Figure 15.2 Typical Electronic Surveying Instruments (*continued*)

obtain because the motion of the aircraft may cause some tilting of the camera up to a maximum of about 5°, although on average this value is about 1°. Photographs taken this way are defined as vertical aerial photographs and are used for highway mapping. In some cases, however, the axis of the camera may be intentionally tilted so that a greater area will be covered by a single photograph. Photographs of this type are known as oblique photographs.

Vertical aerial photographs are taken in a square format, usually with dimensions 9 in. × 9 in., so that the area covered by each photograph is also a square. The airplane flies over the area to be photographed in parallel runs such that any two adjacent photographs overlap both in the direction of flight and in the direction perpendicular to flight. The overlap in the direction of flight is the forward or end overlap and provides an overlap of about 60 percent for any two consecutive photographs. The overlap in the direction

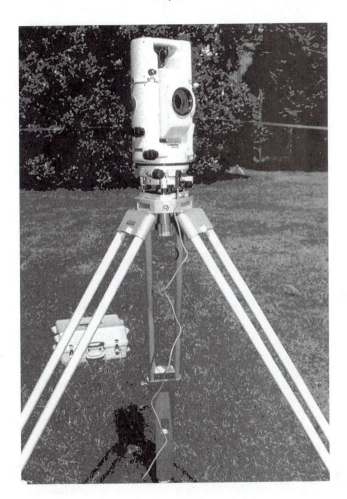

(c) TOPCON GTS-2B geodetic total station

Figure 15.2 Typical Electronic Surveying Instruments (*continued*)

SOURCE: Photographs by Ed Deasy, Virginia Transportation Research Council, Charlottesville, Va., with permission of TOPCON, Paramus, N.J.

perpendicular to flight is the side overlap and provides an overlap of about 25 percent between consecutive flight lines, as shown in Figure 15.3. This ensures that each point on the ground is photographed at least twice, which is necessary for obtaining a three-dimensional view of the area. Figure 15.4 shows a set of consecutive vertical aerial photographs, usually called stereopairs. Certain features of the photographs are worth pointing out. The four marks A, B, C, and D that appear in the middle of each side of the photographs are known as *fiducial marks*. The intersection of the lines joining opposite fiducial marks gives the geometric center of the photograph, which is also called the *principal point*. It is also necessary to select a set of points on the ground that can be easily

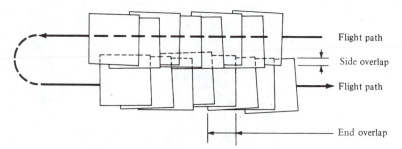

Figure 15.3 End and Side Overlaps in Aerial Photography

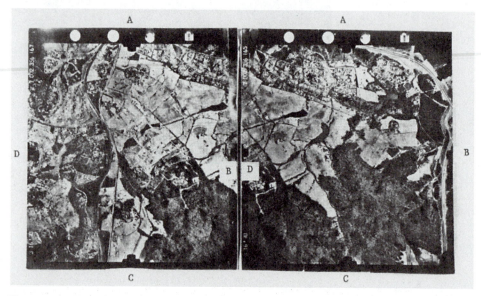

Figure 15.4 Set of Stereopairs

SOURCE: Photographs by Ed Deasy, Virginia Transportation Research Council, Charlottesville, Va. Used with permission.

identified on the photographs as *control points.* These control points are used to set elevations on the maps developed from the aerial photographs.

The information on the aerial photographs is then used to convert these photographs into maps. The instruments used for this process are known as *stereoscopes* or *stereo-plotters,* and they vary from a simple mirror stereoscope, shown in Figure 15.5, to more complex types such as the stereo-plotter shown in Figure 15.6. All of these stereoscopes use the principle of *stereoscopy,* which is the ability to see objects in three dimensions when these objects are viewed by both eyes. When a set of stereopairs is properly placed under a stereoscope, so that an object on the left photograph is viewed by the left eye and the same object on the right photograph is viewed by the right eye, the observer perceives the object in

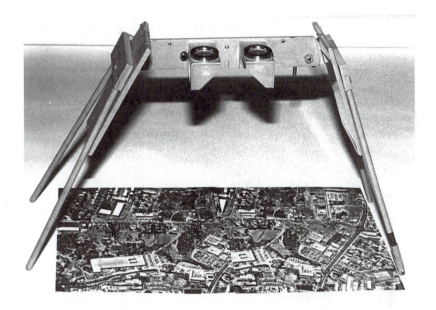

Figure 15.5 Mirror Stereoscope

SOURCE: Photograph by Nicholas J. Garber.

Figure 15.6 Planicomp CIVO-Analytical Stereo-Plotter

SOURCE: Photograph by Sarath C. Joshua, with permission of Carl Zeiss, Inc., Thornwood, N.Y.

three dimensions and therefore sees the area in the photograph as if it were in three dimensions. This technique can be used to produce maps that show the features on the ground and the topography of the area.

Scale at a Point on a Vertical Photograph

Figure 15.7 shows a schematic of a single photograph taken of points M and N. The camera is located at O, and it is assumed that the axis is vertical. H is the flying height above the datum XX, and h_m and h_n are the elevations of M and N. If the images are recorded on plane FF, the scale at any point such as M'' on the photograph can be found. For example, the scale at M'' is given as OM''/OM.

Since triangles OMK and $OM''N''$ are similar,

$$\frac{OM''}{OM} = \frac{ON''}{OK}$$

but $ON'' = f$ the focal length of camera, and $OK = H - h_m$. The scale at M'' is $f/(H - h_m)$. Similarly, the scale at P'' is $f/(H - h_p)$, and the scale at N'' is $f/(H - h_n)$. In general, the scale at any point is given as

$$S = \frac{f}{H - h} \tag{15.1}$$

where
  S = scale at a given point on the photograph
  f = focal length of camera lens
  H = flying height (ft)
  h = elevation of point at which the scale is being determined (ft)

Thus, even when the axis of the camera is perfectly vertical, the scale on the photograph varies from point to point and depends on the elevation of the point at which the scale is to be determined. One cannot therefore refer to the scale of an aerial photograph but rather can refer to the scale at a point on an aerial photograph. However, an approxi-

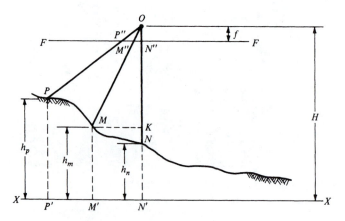

Figure 15.7 Schematic of a Single Aerial Photograph

mate overall scale may be obtained by determining the average height (h_{av}) of different points of the photograph and by substituting h_{av} for h in Eq. 15.1.

Example 15.1   Computing Elevation of a Point from an Aerial Photograph

The elevations of two points A and B on an aerial photograph are 500 ft and 565 ft, respectively. The scales at these points on the photograph are 1:10,000 and 1:9870, respectively. Determine the elevation at point C, if the scale at C is 1:8000.

Solution:

Use Eq. 15.1 to solve for focal length, f,

$$S = \frac{f}{H - h} \tag{15.1}$$

$$\frac{1}{10,000} = \frac{f}{H - 500} \tag{Point A}$$

$$\frac{1}{9870} = \frac{f}{H - 565} \tag{Point B}$$

$$H - 500 = 10,000\,f$$
$$-H + 565 = -9870\,f$$
$$65 = 130\,f$$
$$f = 0.5 \text{ ft}$$

Since $f = 0.5$ ft and $H - 500 = 10,000 \times 0.5$, $H = 5500$ ft.
Elevation at Point C

$$\frac{1}{8000} = \frac{0.5}{5500 - h_C}$$
$$h_C = 5500 - 4000 = 1500 \text{ ft}$$

Example 15.2   Computing Flying Height of an Airplane Used in Aerial Photography

Vertical photographs are to be taken of an area whose mean ground elevation is 2000 ft. If the scale of the photographs is approximately 1:15,000, determine the flying height. Focal length of the camera lens is 7.5 in.

Solution:

$$f = 7.5 \text{ in} = 0.625 \text{ ft}$$
$$S = \frac{1}{15,000} = \frac{0.625}{H - 2000} \tag{15.1}$$
$$H = 11,375 \text{ ft}$$

Distances and Elevations from Aerial Photographs

Figure 15.8 is a schematic of a stereopair of vertical photographs taken on a flight. D is the orthogonal projection to the datum plane of a point A located on the ground. The height AD is therefore the elevation of point A. O and O' are the camera positions for the two photographs, with OO' being the air base (the distance between the centers of two consecutive aerial photographs). Images of A on the photographs are a' and a'', whereas d' and d'' are the images of D. The lines $a'd'$ and $a''d''$ represent the radial shift, which is caused by the elevation of point A above point D on the datum plane. This radial shift is known as the parallax, and it is an important parameter when elevations and distances are to be obtained from aerial photographs. The parallax at any point of an aerial photograph can be determined by the use of an instrument called the parallax bar, shown in Figure 15.9. Two identical dots on the bar known as half-marks appear as a single floating dot above the three-dimensional view (terrain model) when stereoscopic viewing is attained.

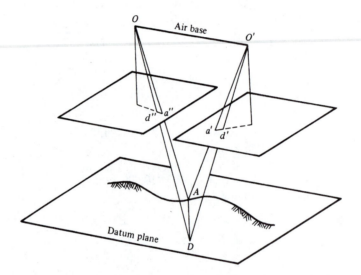

Figure 15.8 Schematic of a Stereopair of Vertical Photographs

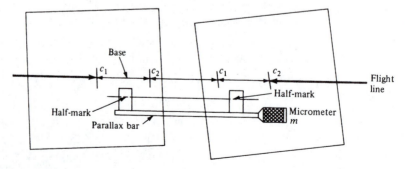

Figure 15.9 Parallax Bar

The distance between the half-marks can be changed by turning the micrometer, which results in the single dot moving up and down over the terrain model. The micrometer is moved until the dot appears to be just touching the point on the model at which the parallax is required. The distance between the half-marks as read by the micrometer is the parallax at that point. It can be shown that

$$\frac{H - h}{f} = \frac{B}{p} \tag{15.2}$$

where

H = flying height (ft)
h = elevation of a given point (ft)
p = parallax at the given point (ft)
B = air base (ft)
f = focal length of the camera (ft)

The X and Y coordinates of any point can also be found from the parallax by using the expression for scale given in Eq. 15.1.

$$S = \frac{f}{H - h}$$

From Eq. 15.2,

$$H = \frac{fB}{p} + h$$

Substituting for H in Eq. 15.1 gives

$$S = \frac{p}{B} \tag{15.3}$$

By establishing a coordinate system in the terrain with the x axis being parallel to the photographic axis of flight, the coordinates of any point are given as

$$X = \frac{x}{S} = \frac{xB}{p} \tag{15.4}$$

and

$$Y = \frac{y}{S} = \frac{yB}{p} \tag{15.5}$$

where X and Y are ground coordinates of the point, and x and y are coordinates of the image of the point on the left photograph.

The horizontal distance between any two points A and B on the ground then can be obtained from Eq. 15.6.

$$D = \sqrt{(X_A - X_B)^2 + (Y_A - Y_B)^2}$$ (15.6)

Example 15.3   Computing the Elevation and Distance Between Two Points Using an Aerial Photograph

The scale at the image of a control point on an aerial photograph is 1:15,000, and the elevation of the point is 1300 ft. The focal length of the camera lens is 6 in. Determine the elevations of points B and C and the horizontal ground distance between them for

	B	C
x coordinate on axis of left photograph	6 in.	7 in.
y coordinate on axis of left photograph	3 in.	4 in.
Scale on photograph	1:17,500	1:17,400

the following data. Also determine the parallax at B if the air base is 200 ft.

Solution:

For the control point,

$$\frac{1}{15,000} = \frac{0.5}{H - 1300} \quad \text{or} \quad H = 8800 \text{ ft}$$

Thus, the elevations at B and C are found:

$$\frac{1}{17,500} = \frac{0.5}{8800 - h_B} \quad \text{or} \quad h_B = 50 \text{ ft}$$

$$\frac{1}{17,400} = \frac{0.5}{8800 - h_C} \quad \text{or} \quad h_C = 100 \text{ ft}$$

$X_B = 0.5/(1/17,500) = 8750$ ft (from Eq. 15.4)
$Y_B = 0.25/(1/17,500) = 4375$ ft (from Eq. 15.5)
$X_C = (7.0/12.0)/(1/17,400) = 10,150$ ft
$Y_C = (4.0/12)/(1/17,400) = 5800$ ft

The distance between B and C, using Eq. 15.6, is

$$D = \sqrt{[(8750 - 10,150)^2 + (4375 - 5800)^2]}$$
$$= \sqrt{(1,960,000 + 2,030,625)}$$
$$= 1997.65 \text{ ft}$$

The parallax at B is

$$p_B = \frac{0.5 \times 200}{8750} = 0.01143 \text{ ft (from Eq. 15.4)},$$

$$p_B = \frac{0.25 \times 200}{4375} = 0.01143 \text{ ft (from Eq. 15.5)}$$

Computer Graphics

Computer graphics, when used for highway location, is usually the combination of photogrammetry and computer techniques. The information obtained from photogrammetry is stored in a computer, which is linked with a stereo-plotter and computer monitor. With input of the appropriate command, the terrain model or the horizontal and vertical alignments are obtained and displayed on the monitor. It is therefore easy to change some control points and obtain a new alignment of the highway on the screen, permitting the designer to immediately see the effects of any changes made.

Figure 15.10 shows the setup of a Galileo stereo digitizer and the Intergraph/Intermap workstation, which has good computer graphics capabilities. A typical workstation is controlled by system software that covers four main areas of design work:

- Preparatory work
- Restitution and plotting

Figure 15.10 Galileo Stereo Digitizer Intergraph/Intermap Work Station

SOURCE: Photograph by Sarath C. Joshua, with permission of Galileo Corporation of America, East Chester, N.Y.

- Data transfer
- Systems calibration and diagnosis

The software for preparatory work is used for the input of control data, the determination of the calibration of camera characteristics, and the preparation of manuscripts. The restitution and plotting software programs are used for measuring data from photographs, automatically locating fiducial marks, digital plotting, and on-line computing of horizontal and vertical angles, vertical and slope distances, and radii and centers of circles. The data transfer programs store and check all data and either print them out or put them in a form suitable for use by other programs. The systems calibration and diagnosis programs are used for calibration of the measuring system and interactive checking of all the important functions of the optomechanical units, the operating elements, control electronics, and the digital plotting table.

PRINCIPLES OF HIGHWAY LOCATION

The basic principle for locating highways is that roadway elements such as curvature and grade must blend with each other to produce a system that provides for the easy flow of traffic at the design capacity, while meeting design criteria and safety standards. The highway should also cause a minimal disruption to historic and archeological sites and to other land-use activities. Environmental impact studies are therefore required in most cases before a highway location is finally agreed upon.

The highway location process involves four phases:

- Office study of existing information
- Reconnaissance survey
- Preliminary location survey
- Final location survey

Office Study of Existing Information

The first phase in any highway location study is the examination of all available data of the area in which the road is to be constructed. This phase is usually carried out in the office prior to any field or photogrammetric investigation. All the available data are collected and examined. These data can be obtained from existing engineering reports, maps, aerial photographs, and charts, which are usually available at one or more of the state's departments of transportation, agriculture, geology, hydrology, and mining. The type and amount of data collected and examined depend on the type of highway being considered, but in general, data should be obtained on the following characteristics of the area:

- Engineering, including topography, geology, climate, and traffic volumes
- Social and demographic, including land use and zoning patterns
- Environmental, including types of wildlife; location of recreational, historic, and archeological sites; and the possible effects of air, noise, and water pollution
- Economic, including unit costs for construction and the trend of agricultural, commercial, and industrial activities

Preliminary analysis of the data obtained will indicate whether any of the specific sites should be excluded from further consideration because of one or more of the above characteristics. For example, if it is found that a site of historic and archeological importance is located within an area being considered for possible route location, it may be immediately decided that any route that traverses that site should be excluded from further consideration. At the completion of this phase of the study, the engineer will be able to select general areas through which the highway can traverse.

Reconnaissance Survey

The object of this phase of the study is to identify several feasible routes, each within a band of a limited width of a few hundred feet. When rural roads are being considered, there is often little information available on maps or photographs, and therefore aerial photography is widely used to obtain the required information. Feasible routes are identified by a stereoscopic examination of the aerial photographs, taking into consideration factors such as

- Terrain and soil conditions
- Serviceability of route to industrial and population areas
- Crossing of other transportation facilities, such as rivers, railroads, and other highways
- Directness of route

Control points between the two endpoints are determined for each feasible route. For example, a unique bridge site with no alternative may be taken as a primary control point. The feasible routes identified are then plotted on photographic base maps.

Preliminary Location Survey

During this phase of the study, the positions of the feasible routes are set as closely as possible by establishing all the control points and determining preliminary vertical and horizontal alignments for each. Preliminary alignments are used to evaluate the economic and environmental feasibility of the alternative routes.

Economic Evaluation

Economic evaluation of each alternative route is carried out to determine the future effect of investing the resources necessary to construct the highway.

The benefit-cost ratio method described in Chapter 13 is used for this evaluation. Factors usually taken into consideration include road user costs, construction costs, maintenance costs, road user benefits, and any disbenefits, which may include adverse impacts due to dislocation of families, businesses, and so forth. The results obtained from the economic evaluation of the feasible routes provide valuable information to the decision maker. For example, these results will provide information on the economic resources that will be gained or lost if a particular location is selected. This information is also used to aid the policy maker in determining whether the highway should be built, and if so, what type of highway it should be.

Environmental Evaluation

Construction of a highway at any location will have a significant impact on its surroundings. A highway is therefore an integral part of the local environment and must be considered as such. This environment includes plant, animal, and human communities and encompasses social, physical, natural, and man-made variables. These variables are interrelated in a manner that maintains equilibrium and sustains the lifestyle of the different communities. The construction of a highway at a given location may result in significant changes in one or more variables, which in turn may offset the equilibrium and result in significant adverse effects on the environment. This may lead to a reduction of the quality of life of the animals and/or human communities. It is therefore essential that the environmental impact of any alignment selected be fully evaluated.

Federal legislation has been enacted that sets forth the requirements of the environmental evaluation required for different types of projects. In general, the requirements call for the submission of environmental impact statements for many projects. These statements as described in Chapter 11 should include

- A detailed description of alternatives
- The probable environmental impact, including the assessment of positive and negative effects
- An analysis of short-term impact as differentiated from long-term impact
- Any secondary effects, which may be in the form of changes in the patterns of social and economic activities
- Probable adverse environmental effects that cannot be avoided if the project is constructed
- Any irreversible and irretrievable resources that have been committed

In cases where an environmental impact study is required, it is conducted at this stage to determine the environmental impact of each alternative route. Such a study will determine the negative and/or positive effects the highway facility will have on the environment. For example, the construction of a freeway at grade through an urban area may result in an unacceptable noise level for the residents of the area (negative impact), or the highway facility may be located so that it provides better access to jobs and recreation centers (positive impact). Public hearings are also held at this stage to provide an opportunity for constituents to give their views on the positive and negative impacts of the proposed alternatives.

The best alternative, based on all the factors considered, is then selected as the preliminary alignment of the highway.

Final Location Survey

The final location survey is a detailed layout of the selected route. The final horizontal and vertical alignments are determined, and the final positions of structures and drainage channels are located. The method used is to set out the points of intersections (PI) of the straight portions of the highway and fit a suitable horizontal curve between these. This is usually a trial-and-error process until, in the designer's opinion, the best alignment is obtained, taking both engineering and aesthetic factors into consideration. Splines and curve templates are available that can be used in this process. The *spline* is a flexible plastic guide that can be bent into different positions and is used to lay out different curvilinear alignments, from which the most suitable is selected. *Curve templates* are transparencies giving circular curves, three-center compound curves, and spiral curves of different radii and different standard scales. Figure 15.11 shows circular curve templates, and Figure 15.12 shows three-centered curve templates. The spline is used first to obtain a hand-fitted smooth curve that fits in with the requirements of grade, cross sections, curvature, and drainage. The hand-fitted curve is then changed to a more defined curve by using the standard templates.

The availability of computer-based techniques has significantly enhanced this process since a proposed highway can be displayed on a monitor, enabling the designer to have a driver's eye view of both the horizontal and vertical alignments of the road. The designer can therefore change either or both alignments until the best alignment is achieved.

Detailed design of the vertical and horizontal alignments is then carried out to obtain both the deflection angles for horizontal curves and the cuts or fills for vertical curves and straight sections of the highway. The design of horizontal and vertical curves is presented in Chapter 16.

Location of Recreational and Scenic Routes

The location process of recreational and scenic routes follows the same steps as discussed earlier, but the designer of these types of roads must be aware of their primary purpose. For example, although it is essential for freeways and arterial routes to be as direct as possible, a circuitous alignment may be desirable for recreational and scenic routes to provide access to recreational sites such as lakes or campsites or to provide special scenic views. The designer must realize, however, the importance of adopting adequate design standards, as

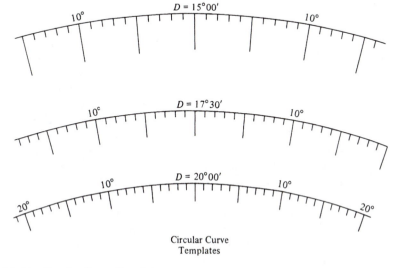

Figure 15.11 Circular Curve Template

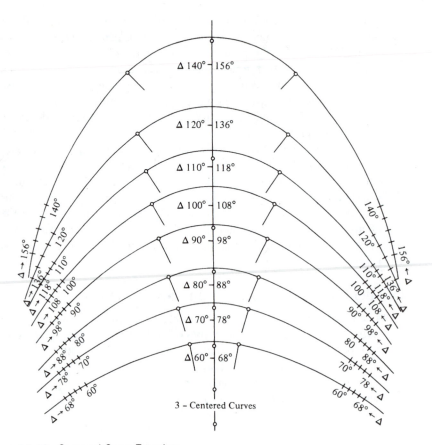

Figure 15.12 Centered Curve Template

given in Chapter 16. Three additional factors should be considered in the location of recreational and scenic routes:

1. Design speeds are usually low, and therefore special provisions should be made to discourage fast driving, for example, by providing a narrower lane width.
2. Location should be such that the conflict between the driver's attention on the road and the need to enjoy the scenic view is minimized. This can be achieved by providing turn-outs with wide shoulders and adequate turning space at regular intervals, or by providing only straight alignments when the view is spectacular.
3. Location should be such that minimum disruption is caused to the area.

Special guides are available for the location and design of recreational and scenic routes. The reader is referred to *A Policy on Geometric Design of Highways and Streets*, published by the American Association of State Highway and Transportation Officials, for more information on this topic.

Location of Highways in Urban Areas

Urban areas usually present complex conditions that must be considered in the highway location process. In addition to factors discussed under office study and reconnaissance survey, other factors that significantly influence the location of highways in urban areas include

- Connection to local streets
- Right-of-way acquisition
- Coordination of the highway system with other transportation systems
- Adequate provisions for pedestrians

Connection to Local Streets

When the location of an expressway or urban freeway is being planned, it is important that adequate thought be given to which local streets should connect with on- and off-ramps to the expressway or freeway. The main factor to consider is the existing travel pattern in the area. The location should enhance the flow of traffic on the local streets. The techniques of traffic assignment, as discussed in Chapter 12, can be used to determine the effect of the proposed highway on the volume and traffic flow on the existing streets. The location should provide for adequate sight distances at all ramps. Ramps should not be placed at intervals that will cause confusion or increase the crash potential on the freeway or expressway.

Right-of-Way Acquisition

One factor that significantly affects the location of highways in urban areas is the cost of acquiring right of way. This cost is largely dependent on the predominant land use in the right of way of the proposed highway. Costs tend to be much higher in commercial areas, and landowners in these areas are often unwilling to give up their land for highway construction. Thus, freeways and expressways in urban areas have been placed on continuous elevated structures in order to avoid the acquisition of rights of way and the disruption of commercial and residential activities. This method of design has the advantage of minimal interference with existing land-use activities, but it is usually objected to by occupiers of adjacent land because of noise or for aesthetic reasons. The elevated structures are also very expensive to construct and therefore do not completely eliminate the problem of high costs.

Coordination of the Highway System
with Other Transportation Systems

Many urban planners now realize the importance of a balanced transportation system and strive toward a fully integrated system of highways and public transportation. This integration should therefore be taken into consideration during the location process of an urban highway. Several approaches have been considered, but the main objective is to provide new facilities that will increase the overall level of service of the transportation system in the urban area. In Washington, D.C., for example, park-and-ride facilities have been provided at transportation terminals to facilitate the use of the subway system, and exclusive bus lanes have been used to reduce the travel time of express buses during the peak hour.

Another form of transportation system integration is the multiple use of rights of way by both highways and transit agencies. In this case the right of way is shared between them, and bus or rail facilities are constructed either in the median or alongside the freeway. Examples include the WMATA rail system in the median of Interstate 66 in Northern Virginia (see Figure 15.13), Congress Street and Dan Ryan Expressway in Chicago, and some sections of Bay Area Rapid Transit in San Francisco.

Adequate Provisions for Pedestrians

Providing adequate facilities for pedestrians should be an important factor in deciding the location of highways, particularly for highways in urban areas. Pedestrians are an integral part of any highway system but are more numerous in urban areas than in rural areas. Therefore, special attention must be given to the provision of adequate pedestrian facilities in planning and designing urban highways. Facilities that should be provided include sidewalks, crosswalks, traffic control features, curb cuts, and ramps for the handicapped. In heavily congested urban areas, the need for grade-separated facilities, such as overhead bridges and/or subways, may have a significant effect on the final location of the highway. Although vehicular traffic demands in urban areas are of primary concern in deciding the location of highways in these areas, the provision of adequate pedestrian facilities must also be of concern because pedestrians are an indispensable vital component of the urban area.

Figure 15.13 Washington Metropolitan Area Transit Authority Rail System in the Median of Route 66, Northern Virginia

SOURCE: Photograph by Ed Deasy, Virginia Transportation Research Council, Charlottesville, Va. Used with permission.

Principles of Bridge Location

The basic principle for locating highway bridges is that the highway location should determine the bridge location, not the reverse. When the bridge is located first, in most cases the resulting highway alignment is not the best. The general procedure for most highways, therefore, is to first determine the best highway location and thus determine the bridge site. In some cases, this will result in skewed bridges, which are more expensive to construct, or in locations where foundation problems exist. When serious problems of this nature occur, all factors such as highway alignments, construction costs of the bridge deck and its foundation, and construction cost of bridge approaches should be considered in order to determine a compromise route alignment that will give a suitable bridge site. This will include the economic analysis of the benefits and costs, as discussed in Chapters 11 and 13.

A detailed report should be obtained for the bridge site selected to determine whether there are any factors that make the site unacceptable. This report should include accurate data on soil stratification, the engineering properties of each soil stratum at the location, the crushing strength of bedrock, and water levels in the channel or waterway.

When the waterway to be crossed requires a major bridge structure, however, it is necessary to first identify a narrow section of the waterway with suitable foundation conditions for the location of the bridge and then determine acceptable highway alignments that cross the waterway at that section. This will significantly reduce the cost of bridge construction in many situations.

Effect of Terrain on Route Location

One factor that significantly influences the selection of a highway location is the terrain of the land, which in turn affects the laying of the grade line. The primary factor that the designer considers on laying the grade line is the amount of earthwork that will be necessary for the selected grade line. One method to reduce the amount of earthwork is to set the grade line as closely as possible to the natural ground level. This is not always possible, especially in undulating or hilly terrain. The least overall cost may also be obtained if the grade line is set such that there is a balance between the excavated volume and the volume of embankment. Another factor that should be considered in laying the grade line is the existence of fixed points, such as railway crossings, intersections with other highways, and in some cases existing bridges, which require that the grade be set to meet them. When the route traverses flat or swampy areas, the grade line must be set high enough above the water level to facilitate proper drainage and to provide adequate cover to the natural soil. The height of the grade line is usually dictated by the expected floodwater level. Grade lines should also be set such that the minimum sight distance requirements as discussed in Chapter 3 are obtained. The criteria for selecting maximum and minimum grade lines are presented in Chapter 16. In addition to these guidelines, the amount of earthwork associated with any grade line influences the decision on whether the grade line should be accepted or rejected.

The following sections describe how a highway grade is established that minimizes earth moving and maximizes the use of native soil. Figure 15.16 on page 668 is an example of a highway plan and the proposed vertical alignment. The solid line is the vertical projection of the centerline of the road profile. The dotted lines represent points along the terrain a distance of 55 feet from the centerline. The circles and triangles are points along the terrain that are 85 feet from the centerline of the road. This information can be used to plot cross-sections that depict the shape of the roadway when completed. The final grade line is adjusted until the amount of excess cut or fill has been minimized. If there is an excess of cut material then it must be removed and stored at another location. If there is an excess of fill then material must be purchased and delivered to the site. Thus an ideal situation occurs when there is a balance between the amount of cut and fill.

Computing Earthwork Volumes

One of the major objectives in selecting a particular location for a highway is to minimize the amount of earthwork required for the project. Therefore, the estimation of the amount of earthwork involved for each alternative location is required at both the preliminary and final stages.

To determine the amount of earthwork involved for a given grade line, cross sections are taken at regular intervals along the grade line. The cross sections are usually spaced 50 ft apart, although this distance is sometimes increased for preliminary engineering. These cross sections are obtained by plotting the natural ground levels and proposed grade profile of the highway along a line perpendicular to the grade line to indicate areas of excavation and areas of fill. Figure 15.14 shows three types of cross sections. When the computation is done manually, the cross sections are plotted on standard cross-section paper, usually to a scale of 1 in. to 10 ft for both the horizontal and vertical directions. The areas of cuts and

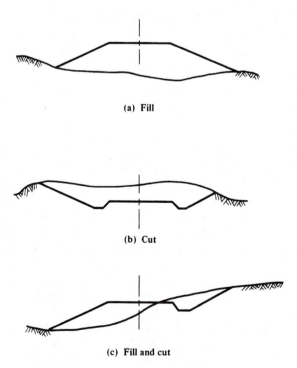

(a) Fill

(b) Cut

(c) Fill and cut

Figure 15.14 Types of Cross Sections

fills at each cross section are then determined by the use of a planimeter or by any other suitable method. Surveying books document the different methods for area computation. The volume of earthwork is then computed from the cross-sectional areas and the distances between the cross sections.

A common method of determining the volume is that of average end areas. This procedure is based on the assumption that the volume between two consecutive cross sections is the average of their areas multiplied by the distance between them, as given in Eq. 15.7,

$$V = \frac{L}{54}(A_1 + A_2)$$ (15.7)

where
$\quad V$ = volume (yd^3)
$\quad A_1$ and A_2 = end areas (ft^2)
$\quad L$ = distance between cross sections (ft)

The average end area method has been found to be sufficiently accurate for most earthwork computations since cross-sections are taken 50–100 feet apart, and minor irregularities tend to cancel each other. When greater accuracy is required, such as in situations where the grade line moves from a cut to a fill section, the volume may be considered as a pyramid or other geometric shapes.

It is common practice in earthwork construction to move suitable materials from cut sections to fill sections to reduce to a minimum the amount of material borrowed from borrow pits. When the materials excavated from cut sections are compacted at the fill sections, they fill less volume than was originally occupied. This phenomenon is referred to as *shrinkage* and should be accounted for when excavated material is to be reused as fill material. The amount of shrinkage depends on the type of material. Shrinkages of up to 50 percent have been observed for some soils. However, shrinkage factors used are generally between 1.10 and 1.25 for high fills and between 1.20 and 1.25 for low fills. These factors are applied to the fill volume in order to determine the required quantity of fill material.

Example 15.4   Computing Fill and Cut Volumes using the Average End Area Method

A roadway section is 2000 ft long (20 stations). The cut and fill volumes are to be computed between each station. Table 15.1 lists the station numbers (Column 1) and lists the end area values (ft^2) between each station that are in Cut (Column 2) and that are in Fill (Column 3). Material in a fill section will consolidate (known as shrinkage), and for this road section, is 10 percent. (For example, if 100 yd^3 of net fill is required, the total amount of fill material that is supplied by a cut section is 100 + (0.10 × 100) = 100 + 10 = 110 ft^3.)

Determine the net volume of cut and fill that is required between station 0 and station 1.

Table 15.1 Computation of Fill and Cut Volumes and Mass Haul Diagram Ordinate

	End Area (ft^2)		Volume (yd^3)				Net Volume (4–7)		
1	2	3	4	5	6	7	8	9	10
					Shrinkage 10 percent	Total Fill (5 + 6)			Mass Diagram Ordinate
Station	Cut	Fill	Total Cut	Fill			Fill (−)	Cut (+)	
0	3	18							0
			9	126	13	139	130		
1	2	50							−130
			7	272	27	299	292		
2	2	97							−422
			11	420	42	462	451		
3	4	130							−873
			22	335	34	369	347		
4	8	51							−1220
			89	178	18	196	107		
5	40	45							−1327
			157	120	12	132		25	
6	45	20							−1302
			231	46	5	51		180	
7	80	5							−1122
			374	13	1	14		360	
8	122	2							−762
			467	4	0	4		463	
9	130	0							−299
			500	0	0	0		500	
10	140	0							201
			444	6	1	7		437	
11	100	3							638
			333	61	6	67		266	
12	80	30							904
			287	93	9	102		185	
13	75	20							1089
			231	130	13	143		88	
14	50	50							1177
			130	241	24	265	135		
15	20	80							1042
			56	333	33	366	310		
16	10	100							732
			19	407	41	448	429		
17	0	120							303
			6	444	44	488	482		
18	3	120							−179
			80	315	31	346	266		
19	40	50							−445
			130	148	15	163	33		
20	30	30							−478

Solution:

$$V_{Cut} = \frac{100(A_{0C} + A_{1C})}{54} = \frac{100(3+2)}{54} = 9.25 \text{ yd}^3$$

$$V_{Fill} = \frac{100(A_{0F} + A_{1F})}{54} = \frac{100(18+50)}{54} = 125.9 \text{ yd}^3$$

Shrinkage = 125.9 (0.10) = 13 yd^3
Total fill volume = 126 + 13 = 139 yd^3
The cut and fill volume between station 0 + 00 and 1 + 00 is shown in Column 4 & 7

Cut: 9 yd^3 (Column 4)

Fill: 126 yd^3 (Column 5) +

Shrinkage: 13 yd^3 (Column 6)

Total Fill required: 139 yd^3 (Column 7)

Net Volume between stations 0-1 = Total Cut – Total Fill = 9 – 139 = –130 yd^3 (Column 8)

Note: Net fill volumes are negative (–) (Column 8) and net cut volumes are positive (+) (Column 9).

Similar calculations are performed between all other stations, from station 1 + 00 to 20 + 00, to obtain the remaining Cut or Fill values shown in Columns 2-9.

Computing Ordinates of the Mass Haul Diagram.   The mass haul diagram is a series of connected lines that depicts the *net* accumulation of cut or fill between any two stations. The ordinate of the mass diagram is the net accumulation in yd^3 from an arbitrary starting point. Thus, the difference in ordinates between any two stations represents the net accumulation of cut or fill between these stations. If the first station of the roadway is considered to be the starting point, then the net accumulation at this station is zero.

Example 15.5   Computing Mass Haul Diagram Ordinates

Use the data obtained in Example 15.4 to determine the net accumulation of cut or fill beginning with station 0 + 00. Plot the results.

Solution: Columns 8 and 9 show the net cut and fill between each station. To compute the mass haul diagram ordinate between station X and $X + 1$, add the *net* accumulation from Station X (the first station) to the net cut or fill volume (Column 8 or 9) between stations X and $X + 1$. Enter this value in Column 10.

Station 0 + 00 Mass Diagram Ordinate = 0

Station 1 + 00 Mass Diagram Ordinate = 0 – 130 = –130 yd^3

Station 2 + 00 Mass Diagram Ordinate = –130 – 292 = –422 yd^3

Station 3 + 00 Mass Diagram Ordinate = –422 – 451 = –873 yd^3

Station 4 + 00 Mass Diagram Ordinate = –873 – 347 = –1220 yd^3

Station 5 + 00 Mass Diagram Ordinate = –1220 – 107 = –1327 yd^3

Station 6 + 00 Mass Diagram Ordinate = –1327 + 25 = –1302 yd^3

Station 7 + 00 Mass Diagram Ordinate = –1302 + 180 = –1122 yd^3

Continue the calculation process for the remaining 13 stations to obtain the values shown in Column 10 of Table 15.1. A plot of the results is shown in Figure 15.15.

Interpretation of the Mass Haul Diagram.   Inspection of Figure 15.15 and Table 15.1 reveals the following characteristics:

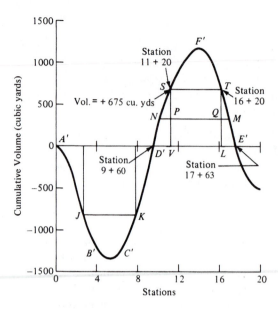

Figure 15.15 Mass Haul Diagram for Computation Shown in Table 15.1

1. When the mass haul diagram slopes downward (negative), the preceding section is in fill and when the slope is upward (positive) the preceding section is in cut.
2. The difference in mass haul diagram ordinates between any two stations represents the *net accumulation* between the two stations (cut or fill). For example the net accumulation between station 6 + 00 and 12 + 00 is (−1302) + (904) = −398 yd³ (fill).
3. A horizontal line on the mass haul diagram defines the locations where the net accumulation between these two points is zero. These are referred to as "balance points," because there is a balance in cut and fill volumes between these points. In Figure 15.15 the "*x*" axis represents a balance between points A' and D' and a balance between points D' and E'. Beyond point E' the mass haul diagram indicates a fill condition for which there is no compensating cut. The maximum value is the ordinate at station 20 + 00 of −478 yd³. For this section imported material (called borrow) will have to be purchased and transported from an off-site location.
4. Other horizontal lines can be drawn connecting portions of the mass haul diagram. For example lines J-K and S-T, which are each five stations long, depict a balance of cut and fill between stations at points J and K and S and T.

Example 15.6 Computing Balance Point Stations

Compute the value of balance point stations for the mass haul diagram in Figure 15.15 for the following situations:

    (a) The x axis
    (b) The horizontal distance S-T, which measures 500 ft

Solution: Balance points are computed by interpolation using the even stations where the ordinates change from cut to fill (or vice versa)

Balance point D' occurs between Station 9 + 00 and 10 + 00 (since ordinate values are −299 and +201)

Assuming that the mass haul diagram ordinate changes linearly between stations, by similar triangles we can write

$$\text{Station of the Balance Point } D' = (9 + 00) + [299/(299 + 201)](100) = 9 + 60$$

Similarly,

$$\text{Station of the Balance Point } E' = (17 + 00) + [303/(303 + 179)](100) = 17 + 63$$

(b) To determine the balance point stations for line ST it is necessary to draw the mass haul diagram to a larger scale than depicted in the textbook, and to read the station for one of the points directly from the diagram. Using this technique, station 11 + 20 was measured for points and from this value the station for point T is computed as

$$(11 + 20) + (5 + 00) = \text{Station } 16 + 20$$

Computing Overhaul Payments.   Contractors are compensated for the cost of earthmoving in the following manner. Typically, the contract price will include a stipulated maximum distance that earth will be moved without the client incurring additional charges. If this distance is exceeded, then the contract stipulates a unit price add-on quoted in additional station-yd^3 of material moved. The maximum distance for which there is no charge is called free haul. The extra distance is called overhaul.

Example 15.7   Computing Overhaul Payment

The free-haul distance in a highway construction contract is 500 ft and the overhaul price is $11/yd^3 station. For the mass haul diagram shown in Figure 15.15, determine the extra compensation that must be paid to a contractor to balance the cut and fill between station 9 + 60 (D) and station 17 + 63 (E).

Step 1.   Determine the number of cubic yards of overhaul.
The overhaul volume will occur between stations 9 + 60 and 11 + 20, and between stations 16 + 20 and 17 + 63. The overhaul value is obtained by interpolation between stations 11 + 00 and 12 = 00 or by reading the value from the mass haul diagram.
By interpolation the value is:

$$\text{Overhaul} = \text{Ordinate at station } 11 + (\text{difference in ordinates at 12 and 11}) (20/100)$$
$$= 638 + (904 - 638)(0.2) = 638 + 53 = 691 \text{ yd}^3$$

This overhaul value should equal the value at station 16 + 20
By interpolation the value is

$$732 - (732 - 320)(0.2) = 649 \text{ yd}^3$$

Since the values are not equal, use the average (670 yd^3)
Or measure the overhaul from a larger scale diagram to obtain a value of 675 yd^3. This value is selected for the calculation of contractor compensation.

Step 2. Determine the overhaul distance.
The method of moments is used is to compute the weighted average of the overhaul distances from the balance line to the station where free haul begins.
Beginning with stations 9 + 60 to 10 + 00, the volume moved is 201 yd^3, and the average distance to the free haul station (11 + 20) is (10 + 00 − 9 + 60)/2 + 100 + 20 = 140 ft.
From stations 10 + 00 to 11 + 00, the volume moved is (638 − 201) = 437 yd^3, and the distance moved to the free haul line is (11 + 00 − 10 + 00)/2 + 20 = 70 ft.
From station 11 + 00 to station 11 + 20, the volume moved is 675 − 638 = 37 yd^3 and the average distance is 10 ft.
Overhaul distance moved between station 9 + 60 and 11 + 20:

$$\{(201)(140) + (437)(70) + (37)(10)\} \div 675 = 59,100 \div 675 = 87.6 \text{ ft}$$

Similarly, we compute the overhaul distance between the balance point at station 17 + 63 and the beginning of free haul at station 16 + 20.
Beginning with stations 17 + 63 to 17 + 00, the volume moved is 303 yd^3, and the average distance to the free haul station (16 + 20) is (17 + 63 − 17 + 00) ÷ 2 + (17 + 00 − 16 + 20) = 111.5 ft.
From stations 17 + 00 to 16 + 20, the volume moved is (675 − 303) = 372 yd^3, and the distance moved to the free haul line is (17 + 00 − 16 + 20) ÷ 2 = 40 ft.
Overhaul average distance moved between station 16 + 20 and 17 + 63:

$$\{(303)(111.5) + (372)(40)\} \div 675 = 48,664.5 \div 675 = 72.1 \text{ ft}$$

Total overhaul distance = 87.6 + 72.1 = 159.7 ft

Step 3. Compute overhaul cost due to the contractor.

Overhaul cost = contract price ($/yd^3 station) × overhaul (yd^3) × stations
= 11 × 675 × (0.876 + 0.721) = $11,858

Computer programs are now available that can be used to compute cross-sectional areas and volumes directly from the elevations given at the cross sections. Some programs will also compute the ordinate values for a mass diagram and determine the overhaul, if necessary.

PREPARATION OF HIGHWAY PLANS

Once the final location of the highway system is determined, it is then necessary to provide the plans and specifications for the facility. The plans and specifications of a highway are the instructions under which the highway is constructed. They are also used for the preparation of engineer's estimates and contractor's bids. When a contract is let out for the construction of a highway, the plans and specifications are part of the contract documents and are therefore considered legal documents. The plans are drawings that contain all details necessary for proper construction, whereas the specifications give written instructions on quality and type of materials. Figure 15.16 (top) shows an example of a highway plan. Figure 15.16 (bottom) shows the vertical alignment, sometimes referred to as the *profile,* indicating the natural ground surface and the center line of the road, with details of vertical curves. The horizontal alignment is usually drawn to a scale of 1 in. to 100 ft, although in some cases, the scale of 1 in. to 50 ft is used to provide greater detail. In drawing the vertical alignment the horizontal scale used is the same as that of the horizontal alignment, but the vertical scale is exaggerated five to ten times. The vertical alignment may also give estimated earthwork quantities at regular intervals, usually at 100 ft stations.

Most state agencies require consultants to prepare final design drawings on standard sheets 36 in. × 22 in. These drawings are then usually reduced to facilitate easy handling in the field during construction. Other drawings showing typical cross sections and specific features such as pipe culverts and concrete box culverts are also provided. Standard drawings of some of these features that occur frequently in highway construction have been provided by some states and can be obtained directly from the highway agencies. Consultants may not have to produce them as part of their scope of work.

SUMMARY

The selection of a suitable location for a new highway requires information obtained from highway surveys. These surveys can be carried out by either conventional ground methods or use of electronic equipment and computers. A brief description of some of the more commonly used methods of surveys has been presented to introduce the reader to these techniques.

A detailed discussion of the four phases of the highway location process has been presented to provide the reader with the information required and the tasks involved in selecting the location of a highway. The computation of earthwork volumes is also presented since the amount of earthwork required for any particular location may significantly influence the decision to either reject or select that location. Note, however, that the final selection of a highway location, particularly in an urban area, is not now purely in the hands of the engineers. The reason is that citizen groups, with interest in the environment and historical preservation, can be extremely vocal in opposing highway locations that, in their opinion, conflict with their objectives. Thus, in selecting a highway location, the engineer must take into consideration the environmental impact of the road on its

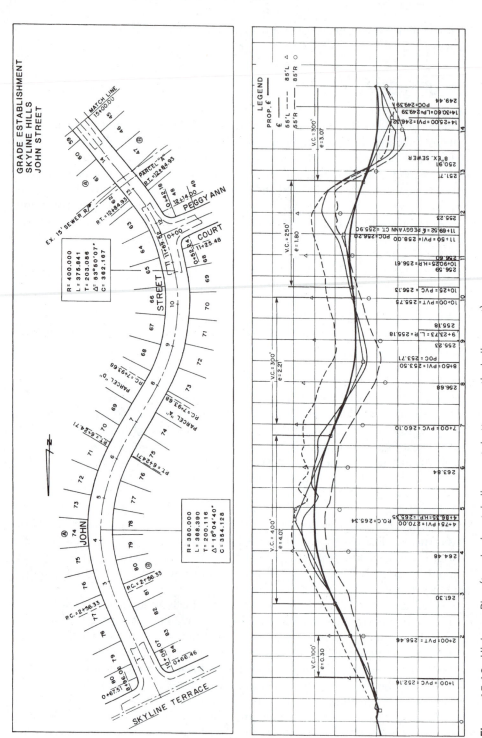

Figure 15.16 Highway Plan (top: horizontal alignment; bottom: vertical alignment)

SOURCE: Courtesy of Byrd, Tallamy, McDonald and Lewis, Fairfax, Va.

surroundings. For many types of projects, federal regulations also require environmental impact statements, which require detailed analyses.

PROBLEMS

15-1   What are the three elements that highway surveys usually involve?

15-2   Briefly describe the use of each of the following instruments in conventional ground surveys: (a) the transit, (b) the level, (c) measuring tapes, and (d) electronic distance measuring devices.

15-3   Briefly compare the factors that should be considered in locating an urban freeway with those for a rural freeway.

15-4   The slope distance between two points A and B on a highway is 150 feet. If the difference in elevation between these is 15 feet, what is the horizontal distance between these points?

15-5   Describe how each of the following could be used in highway survey location: (a) aerial photogrammetry, (b) computer graphics, and (c) conventional survey techniques.

15-6   A photograph is to be obtained at a scale of 1:10,000 by aerial photogrammetry. If the focal length of the camera to be used is 6.5 inches, determine the height at which the aircraft should be flown if the average elevation of the terrain is 950 feet.

15-7   The distance in the x direction between two control points on a vertical aerial photograph is 4.5 inches. If the distance between these same two points is 3.6 inches on another photograph having a scale of 1:24,000, determine the scale of the first vertical aerial photograph. If the focal length of the camera is 6.0 inches and the average elevation at these points is 100 feet, determine the flying height from which each photograph was taken.

15-8   The scale at the image of a well-defined object on an aerial photograph is 1 : 24,000, and the elevation of the object is 1500 feet. The focal length of the camera lens is 6.5 inches. If the air base (B) is 250 feet, determine the elevation of the two points A and C and the distance between them if the coordinates of A and C are as given below.

	A	C
x coordinate	5.5 inch	6.5 inch
y coordinate	3.5 inch	5.0 inch
Scale of photograph	1:13,000	1:17,400

15-9   A vertical photo has an air base of 200 ft. Stereoscopic measurements of parallex at a point representing the top of a 200 ft tower is 0.278 in. The camera focal length is 6.5 in. Photos were taken at an elevation of 7500 ft. Determine the elevation of the *base* of the tower.

15-10   The length of a runway at a national airport is 7500 ft long and at elevation 1,500 ft above sea level. The airport was recently expanded to include another runway used primarily for corporate aircraft. It is desired to determine the length of this runway whose elevation is 1800 ft. An aerial photograph was taken of the airport. Measurements on the photograph for the national airport runway are 4.80 in and for the corporate runway, 3.4 in. The camera focal length is 6 in. Determine the length of the corporate runway.

15-11   Using an appropriate diagram, discuss the importance of the side and forward overlaps in aerial photography.

15-12	Briefly discuss the factors that are of specific importance in the location of scenic routes.
15-13	Under what conditions would you prefer borrowing new material from a borrow pit for a highway embankment over using material excavated from an adjacent section of the road?
15-14	Using the data given in Table 15.1, determine the total overhaul cost if the free haul is 700 feet and the overhaul cost is $7.50 per cubic yard station. Stations of the free haul lines are 1 + 80 and 8 + 80 and 10 + 20 and 17 + 20.
15-15	The following table shows the stations and ordinates for a mass diagram. The free-haul distance is 600 ft. Overhaul cost is $15 per station yard.

Station	Ordinate (yd^3)
0 + 00	0
1 + 00	45
2 + 00	60
2 + 20	90
4 + 00	120
6 + 00	140
7 + 00	110
8 + 20	90
9 + 00	82
10 + 00	60
10 + 30	0

(a) Use the method of movements to compute the additional cost that must be paid to the contractor.

(b) Sketch the ground profile if the finished grade of this roadway section is level (0 percent).

ADDITIONAL READINGS

A Policy on Geometric Design of Highways and Streets, American Association of State Highway and Transportation Officials, Washington, D.C., 2001.

Brinker, R.C. (editor), *The Surveying Handbook,* Chapman and Hall, December, 1995.

Davis, R.E., F.S. Foote, J.M. Anderson, and E.M. Makhail, *Surveying Theory and Practice,* McGraw-Hill, New York, 1981.

Mence, C.F., and D.W. Gibson, *Route Surveying,* 5th ed., Harper & Row, New York, 1980.

CHAPTER 16

Geometric Design of Highway Facilities

The geometric design of highway facilities deals with the proportioning of the physical elements of highways, such as vertical and horizontal curves, lane widths, cross sections, and parking bays. The characteristics of driver, pedestrian, vehicle, and road, as discussed in Chapter 3, serve as the basis for determining the physical dimensions of these elements. For example, lengths of vertical curves or radii of circular curves are determined such that the minimum stopping sight distance is provided on the curve for the design speed of the highway. The basic objective in geometric design of highways, however, is to produce a smooth-flowing, crash-free facility. This can be achieved by having a consistent design standard along the highway that satisfies the characteristics of the drivers and vehicles.

The American Association of State Highway and Transportation Officials (AASHTO) plays a very important role in the development of guidelines and standards used in highway geometric design. The membership of this association consists of representatives from all state highway and transportation departments and the Federal Highway Administration (FHWA). The association has several technical committees that consider suggested standards from individual states. When a standard is approved by the required majority, it is declared as adopted and is accepted by all members of the association. *A Policy on Geometric Design of Highways and Streets,* published by AASHTO, gives all of these standards for geometric design of highways. In this chapter, the current standards used for design, as recommended by AASHTO, are presented, and their relationships to the characteristics of the driver, pedestrian, vehicle, and road are noted. The principles and theories used in the design of both the horizontal and vertical alignments are also presented.

HIGHWAY FUNCTIONAL CLASSIFICATION

Highways are classified according to their respective functions in terms of the character of the service they are providing. This classification system facilitates the systematic development of highways and the logical assignment of highway responsibilities among different jurisdictions. Highways and streets are primarily described as rural or urban roads, depending on the area in which they are located. This primary classification is essential since urban and rural areas have fundamentally different characteristics, particularly those related to type of land use and population density, which significantly influence travel patterns. Following the primary classification, highways are then classified separately for urban and rural areas under the following categories:

- Principal arterials
- Minor arterials
- Major collectors
- Minor collectors
- Local roads and streets

Freeways are not listed as a separate functional class since they are generally classified as part of the principal arterial system. Note, however, that freeways have unique geometric criteria that require special consideration during design.

Functional System of Urban Roads

Urban roads are all highway facilities within urban areas. Urban areas are usually designated by state and local officials and usually have populations of 5000 or more, although some states use other values. For example, the Virginia Department of Transportation uses a population of 3500 to define an urban area. Urban areas are further subdivided into urbanized areas with populations of 50,000 or more and small urban areas with populations between 5000 and 50,000. Urban roads are functionally classified into principal arterials, minor arterials, collectors, and locals. Figure 16.1 is a schematic of the different classes.

Urban Principal Arterial System

This system of highways serves the major activity centers of the urban area and consists mainly of the highest-traffic-volume corridors. It carries a high proportion of the total vehicle-miles of travel within the urban area and carries most trips with origin or destination within the urban area. The system also serves trips that bypass the central business districts (CBDs) of urbanized areas. All controlled-access facilities are within this system, although controlled access is not necessarily a condition for a highway to be classified as an urban principal arterial. Highways within this system are further divided into three subclasses based mainly on the type of access to the facility: (1) interstate, with fully controlled access and grade-separated interchanges; (2) expressways, which have controlled access but may also include at-grade intersections; and (3) other principal arterials (with partial or no controlled access).

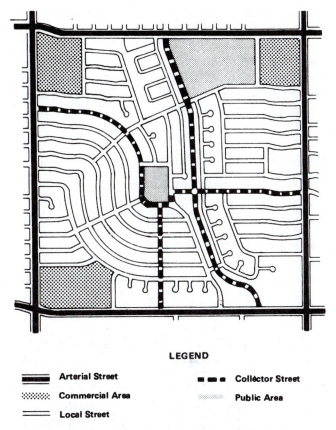

LEGEND

▬▬▬ **Arterial Street**   ■ ■ ● **Collector Street**

⋰⋰⋰ **Commercial Area**   ░░░ **Public Area**

▭▭▭ **Local Street**

Figure 16.1 Schematic of the Functional Classes of Suburban Roads

SOURCE: *A Policy on Geometric Design of Highways and Streets,* American Association of State Highway and Transportation Officials, Washington, D.C., 2001. Used with permission.

Urban Minor Arterial System

Streets and highways that interconnect with and augment the urban primary arterials are classified as urban minor arterials. This system serves trips of moderate length and places more emphasis on land access than does the primary arterial system. All arterials not classified as primary are included in this class. Although highways within this system may serve as local bus routes and may connect communities within the urban areas, they do not normally go through identifiable neighborhoods. The spacing of minor arterial streets in fully developed areas is usually not less than 1 mi, but the spacing can be 2 to 3 mi in suburban fringes.

Urban Collector Street System

The main purpose of streets within this system is to collect traffic from local streets in residential areas or in CBDs and convey it to the arterial system. Thus, collector streets usually go through residential areas and facilitate traffic circulation within residential, commercial, and industrial areas.

Urban Local Street System

This system consists of all other streets within the urban area that are not included in the three systems described earlier. The primary purposes of these streets are to provide access to abutting land and to the collector streets. Through traffic is deliberately discouraged on these streets.

Functional System of Rural Roads

Highway facilities outside urban areas form the rural road system. These highways are categorized as principal arterials, minor arterials, major collectors, minor collectors, and locals. Figure 16.2 is a schematic of the system.

Rural Principal Arterial System

This system consists of a network of highways that serves most of the interstate trips and a substantial amount of intrastate trips. Virtually all highway trips between urbanized areas

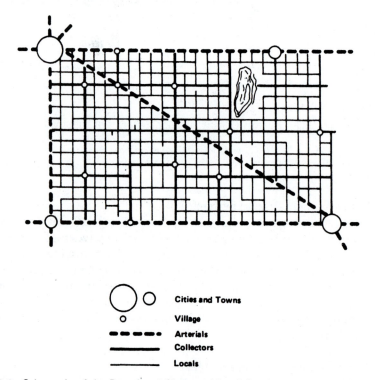

Figure 16.2 Schematic of the Functional Classes of Rural Roads

SOURCE: *A Policy on Geometric Design of Highways and Streets,* American Association of State Highway and Transportation Officials, Washington, D.C., 2001. Used with permission.

and a high percentage of the trips between small urban areas with populations of 25,000 or more are made on this system. The system is further divided into freeways (which are divided highways with fully controlled access and no at-grade intersections) and other principal arterials consisting of all principal arterials not classified as freeways.

Rural Minor Arterial System

This system of roads augments the principal arterial system in the formation of a network of roads that connects cities, large towns, and other traffic generators, such as large resorts. Travel speeds on these roads are usually as high as those on the principal arterial system.

Rural Collector System

Highways within this system carry traffic primarily within individual counties, and trip distances are usually shorter than those on the arterial roads. This system of roads is subdivided into major collector roads and minor collector roads.

Rural Major Collector System

Routes under this system carry traffic primarily to and from county seats and large cities that are not directly served by the arterial system. The system also carries the main intracounty traffic.

Rural Minor Collector System

This system consists of routes that collect traffic from local roads and convey it to other facilities. One important function of minor collector roads is that they provide linkage between rural hinterland and locally important traffic generators such as small communities.

Rural Local Road System

This system consists of all roads within the rural area not classified within the other systems. These roads serve trips of relatively short distances and connect adjacent lands with the collector roads.

FACTORS INFLUENCING HIGHWAY DESIGN

Highway design is based on several design standards and controls, which in turn depend on

- Functional classification of the highway being designed
- Expected traffic volume and vehicle mix
- Design speed
- Topography of the area in which the highway will be located
- Level of service to be provided
- Available funds
- Safety
- Social and environmental factors

These factors are often interrelated. For example, design speed depends on functional classification, and functional classification, to a certain extent, depends on expected traffic volume. The design speed may also depend on the topography, particularly in cases where limited funds are available. In general, however, the principal factors used to determine the standards to which a particular highway will be designed are the level of service to be provided, the expected traffic volume, the design speed, and the design vehicle. These factors coupled with the basic characteristics of drivers, vehicles, and road, as discussed in Chapter 3, are used to determine standards for the geometric characteristics of the highway, such as cross sections and horizontal and vertical alignments. For example, appropriate geometric standards should be selected to maintain a desired level of service for a known proportional distribution of different types of vehicles.

HIGHWAY DESIGN STANDARDS

Selection of the appropriate set of geometric design standards is the first step in the design of any highway. This is essential because no single set of geometric standards can be used for all highways. For example, geometric standards that may be suitable for a scenic mountain road with low average daily traffic (ADT) are inadequate for a freeway carrying heavy traffic. The characteristics of the highway should therefore be considered in selecting the geometric design standards.

Design Hourly Volume

The design hourly volume (DHV) is the projected hourly volume that is used for design. This volume is usually taken as a percentage of the expected ADT on the highway. Figure 16.3 shows the relationship between traffic hourly volumes as a percentage of ADT and the number of hours in one year with higher volumes. This relationship was computed from the analysis of traffic count data over a wide range of volumes and geographic conditions. For example, Figure 16.3 shows that an hourly volume equal to 12 percent of the ADT is exceeded at 85 percent of locations during 20 hours in the entire year. A close examination of this curve also shows that between 0 and about 25 highest hours, a small increase in the number of hours results in a significant reduction in the percentage of ADT, whereas a relatively large increase in number of hours at the right of the 30th-highest hour results in only a slight decrease in the percentage of ADT. This characteristic of the curve has led to the conclusion that it is uneconomical to select a DHV greater than that which will be exceeded during only 29 hours in a year. The 30th-highest hourly volume is therefore usually selected as the DHV. Experience has also shown that the 30th-highest hourly volume as a percentage of ADT varies only slightly from year to year, even when significant changes of ADT occur. It has also been shown that, excluding rural highways with unusually high or low fluctuation in traffic volume, the 30th-highest hourly volume for rural highways is usually between 12 percent and 18 percent of the ADT, with the average being 15 percent.

Note, however, that the 30th-highest hourly volume should not be indiscriminately used as the DHV, particularly on highways with unusual or high seasonal fluctuation in the traffic flow. Although the percentage of annual average daily traffic (AADT) represented by the 30th-highest hourly volume on such highways may not be significantly

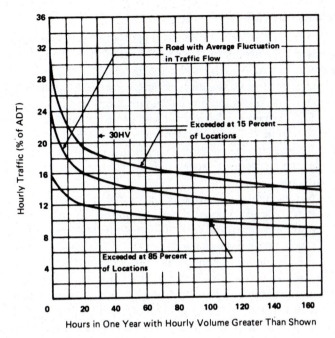

Figure 16.3 Relationship Between Hourly Volume (two way) and Annual Average Daily Traffic on Rural Roads

SOURCE: *A Policy on Geometric Design of Highways and Streets,* American Association of State Highway and Transportation Officials, Washington, D.C., 2001. Used with permission.

different from those on most rural roads, this criterion may not be applicable, since the seasonal fluctuation results in a high percentage of high-volume hours and a low percentage of low-volume hours. For example, economic consideration may not permit the design to be carried out for the 30th-highest hourly volume, but at the same time, the design should not be such that severe congestion occurs during peak hours. A compromise is to select a DHV that will result in traffic operating at a somewhat slightly lower level of service than that which normally exists on rural roads with normal fluctuations. For this type of rural road, therefore, it is desirable to select 50 percent of the volume that occurs for only a few peak hours during the design year as the DHV, even though this may not be equal to the 30th-highest hourly volume. This may result in some congestion during the peak hour, but the capacity of the highway normally will not be exceeded.

The 30th-highest hourly volume may also be used as the DHV for urban highways. It is usually determined by applying between 8 percent and 12 percent to the ADT. However, other relations may be used for highways with seasonal fluctuation in traffic flow much different from that on rural roads. One alternative is to use the average of the highest afternoon peak hour volume for each week in the year as the DHV.

Design Speed

Design speed is defined by AASHTO as "a selected speed to determine the various geometric features of the roadway." Design speed depends on the functional class of the highway, the topography of the area in which the highway is located, and the land use of the adjacent area. For highway design, topography is generally classified into three groups: level terrain, rolling terrain, and mountainous terrain.

Level terrain is relatively flat, and horizontal and vertical sight distances are generally long or can be achieved without much construction difficulty or major expense.

Rolling terrain has natural slopes that often rise above and fall below the highway grade, with occasional steep slopes that restrict the normal vertical and horizontal alignments.

Mountainous terrain has sudden changes in ground elevation in both the longitudinal and transverse directions, thereby requiring frequent hillside excavations to achieve acceptable horizontal and vertical alignments. The criteria for describing different topographies were presented in Chapter 9.

It is important that the design speed selected not be significantly different from the speed at which motorists will expect to drive. For example, a low design speed should not automatically be selected for a rural collector road, because when such a road is located in an area of flat topography, motorists will tend to drive at high speeds. The average trip length on the highway is another factor that should be considered in selecting the design speed. In general, highways with longer average trips should be designed for higher speeds.

Design speeds range from 20 mi/h to 70 mi/h, with intermediate values at 10 mi/h increments. Design elements show no significant difference when increments are less than 10 mi/h but do show very large differences with increments of 15 mi/h or higher. In general, however, freeways are designed for 60–70 mi/h, whereas design speeds for other arterial roads range from 30 mi/h to 60 mi/h. Tables 16.1 and 16.2 give recommended values for minimum design speeds for different classes of highway.

Note that a design speed is selected to achieve a desired level of operation and safety on the highway. It is one of the first parameters selected in the design process because several other design variables are determined from it.

Table 16.1   Minimum Design Speeds for Rural Collector Roads

	Design Speed (mi/h) for Specified Design Volume (veh/day)		
Type of Terrain	*0–400*	*400 to 2000*	*Over 2000*
Level	40	50	60
Rolling	30	40	50
Mountainous	20	30	40

SOURCE: Adapted from *A Policy on Geometric Design of Highways and Streets,* American Association of State Highway and Transportation Officials, Washington, D.C., 2001. Used by permission.

Table 16.2 Minimum Design Speeds for Various Functional Classifications

Class		Speed (mi/h)					
		20	30	40	50	60	70
Rural Principal Arterial	Min. 50 mi/h for freeways			x	x	x	x
Rural Minor Arterial				x	x	x	x
Rural Collector Road	DHV over 400			x	x	x	
	DHV 20–400			x	x	x	
	DHV 100–200		x	x	x		
	Current ADT over 400		x	x	x		
	Current ADT under 400	x	x	x			
Rural Local Road	DHV over 400			x	x	x	
	DHV 200 to 400			x	x	x	
	DHV 100 to 200			x	x	x	
	Current ADT over 400			x	x	x	
	Current ADT 250 to 400	x	x	x			
	Current ADT 50 to 250	x	x				
	Current ADT under 50	x	x				
Urban Principal Arterial	Minimum 50 mi/h for freeways			x	x	x	x
Urban Minor Arterial				x	x	x	x
Urban Collector Street				x	x	x	
Urban Local Street		x	x				

SOURCE: *Road Design Manual,* Virginia Department of Transportation, Richmond, Va., 2000.

Design Vehicle

The design vehicle is that vehicle selected to represent all vehicles on the highway. Its weight, dimensions, and operating characteristics will be used to establish the geometric standards of the highway, such as radii at intersections and radii of turning roadways. The different classes of vehicles and their dimensions were discussed in Chapter 3. The vehicle type selected as the design vehicle is the largest that is likely to use the highway with considerable frequency. The selected design vehicle is then used to determine critical design features such as radii at intersections and radii of turning roadways.

The following guidelines are given by AASHTO for the selection of a design vehicle:

- When a parking lot or a series of parking lots are the main traffic generators, the passenger car may be used.
- For the design of intersections at local streets and park roads, a single-unit truck may be used.
- At intersections of state highways and city streets that serve buses with relatively few large trucks, a city transit bus may be used.

- At intersections of highways and low-volume county highways or township/local roads with less than 400 ADT, either an 84-passenger large school bus 40 ft long or a 65-passenger conventional bus 36 ft long may be used. The selection of either of these will depend on the expected usage of the facility.
- At intersections of freeway ramp terminals and arterial crossroads, and at intersections of state highways and industrialized streets that carry high volumes of traffic, the minimum size of the design vehicle should be WB-20.

Cross-Section Elements

The principal elements of a highway cross section consist of the travel lanes, shoulders, and medians (for some multilane highways). Marginal elements include median and roadside barriers, curbs, gutters, guard rails, sidewalks, and side slopes. Figure 16.4 shows a typical cross section for a two-lane highway, and Figure 16.5 shows that for a multilane highway.

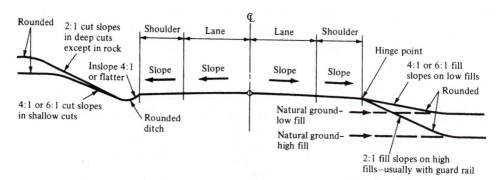

Figure 16.4 Typical Cross Section for Two-Lane Highways

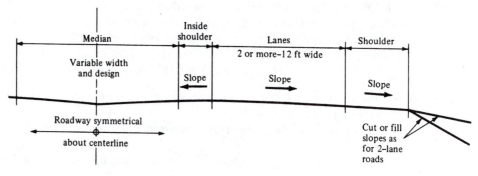

Figure 16.5 Typical Cross Section for Multilane Highways (half section)

Width of Travel Lanes

Travel lane widths usually vary from 9 ft to 12 ft. Most arterials have 12-ft travel lanes since the extra cost for constructing 12-ft lanes over 10-ft lanes is usually offset by the lower maintenance cost for shoulders and pavement surface, resulting in the reduction of wheel concentrations at the pavement edges. On two-lane, two-way rural roads, lane widths of 10 ft or 11 ft may be used, but two factors must be considered when selecting a lane width less than 12 ft wide. The first is that it has been shown by Garber and others that when pavement surfaces are less than 22 ft, the crash rates for large trucks tend to increase. Second, as discussed in Chapters 9 and 10, the capacity of a highway significantly decreases as the lane width is reduced from 12 ft. Lane widths of 10 ft are therefore used only on low-speed facilities. Lanes that are 9 ft wide are used occasionally in urban areas if traffic volume is low and there are extreme right-of-way constraints.

Shoulders

Figures 16.4 and 16.5 show the sections of the highway cross section designated as shoulders. The shoulder is contiguous with the traveled lane and provides an area along the highway for vehicles to stop, particularly during an emergency. In some cases, bicycles can be accommodated by the shoulder. Shoulders are also used to laterally support the pavement structure. The shoulder width is known as either *graded* or *usable,* depending on the section of the shoulder being considered. The graded shoulder width is the whole width of the shoulder measured from the edge of the traveled pavement to the intersection of the shoulder slope and the plane of the side slope. The usable shoulder width is that part of the graded shoulder that can be used to accommodate parked vehicles. The usable width is the same as the graded width when the side slope is equal to or flatter than 4 : 1 (horizontal : vertical), as the shoulder break is usually rounded to a width between 4 ft and 6 ft, thereby increasing the usable width.

When a vehicle stops on the shoulder, it is desirable for it to be at least 1 ft and preferably 2 ft from the edge of the pavement. Based on this, AASHTO recommends that usable shoulder widths of at least 10 ft and preferably 12 ft be used on highways having a large number of trucks and on highways with high traffic volumes and high speeds. However, it may not always be feasible to provide this minimum width, particularly when the terrain is difficult or when traffic volume is low. A minimum shoulder width of 2 ft may therefore be used on the lowest type of highways, but 6- to 8-ft widths should preferably be used. However, when pedestrians and bicyclists are permitted to use the shoulder, the minimum width should be 4 ft. The width for usable shoulders within the median for divided arterials having two lanes in each direction, however, may be reduced to 3 ft, since drivers rarely use the median shoulder for stopping on these roads. The usable median shoulder width for divided arterials with three or more lanes in each direction should be at least 8 ft, since drivers in difficulty on the lane next to the median often find it difficult to maneuver to the outside shoulder.

It is essential that all shoulders be flush with the edge of the traveled lane and be sloped to facilitate the drainage of surface water on the traveled lanes. Recommended slopes are 2 percent to 6 percent for bituminous and concrete-surfaced shoulders, and 4 percent to 6 percent for gravel or crushed-rock shoulders.

Medians

A median is the section of a divided highway that separates the lanes in opposing directions. The width of a median is the distance between the edges of the inside lanes, including the median shoulders. The functions of a median include

- Providing a recovery area for out-of-control vehicles
- Separating opposing traffic
- Providing stopping areas during emergencies
- Providing storage areas for left-turning and U-turning vehicles
- Providing refuge for pedestrians
- Reducing the effect of headlight glare
- Providing temporary lanes and cross-overs during maintenance operations

Medians can either be raised, flush, or depressed. Raised medians are frequently used in urban arterial streets because they facilitate the control of left-turn traffic at intersections by using part of the median width for left-turn-only lanes. Some disadvantages associated with raised medians include possible loss of control of the vehicle by the driver if the median is accidentally struck, and the casting of a shadow from oncoming headlights, which results in drivers having difficulty seeing the curb.

Flush medians are commonly used on urban arterials. They can also be used on freeways, but with a median barrier. To facilitate drainage of surface water, the flush median should be crowned. The practice in urban areas of converting flush medians into two-way left-turn lanes is rather popular, since this helps to increase the capacity of the urban highway while providing some of the features of a median.

Depressed medians are generally used on freeways and are more effective in draining surface water. A side slope of 6 : 1 is suggested for depressed medians, although a slope of 4 : 1 may be adequate.

Median widths vary from a minimum of 4 ft to 80 ft or more. Median widths should be as wide as possible but should be balanced with the other elements of the cross section and the cost involved. In general, the wider the median, the more effective it is in providing safe operating conditions. A minimum width of 10 ft is recommended for four-lane urban freeways, which is adequate for two 4-ft shoulders, and a 2-ft median barrier. A minimum of 22 ft, preferably 26 ft, is recommended for six or more lanes of freeway.

Median widths for urban collector streets, however, vary from 2 ft to 40 ft, depending on the median treatment. For example, when the median is a paint-striped separation, 2- to 4-ft medians are required. For narrow raised or curbed areas, 2- to 6-ft medians are required, and for curbed sections, 16- to 40-ft medians are required. The larger width is necessary for curbed sections because it provides space for protecting vehicles crossing an intersection. It can also be used for landscape treatment.

Median and Roadside Barriers

AASHTO defines a median barrier as a longitudinal system used to prevent an errant vehicle from crossing the portion of a divided highway separating the traveled ways for traffic in opposite directions. Roadside barriers, on the other hand, protect vehicles from obstacles or slopes on the roadside. They may also be used to shield pedestrians and property from the traffic stream. The provision of median barriers must be considered when traffic volumes are high and when access to multilane highways and other highways is only partially controlled. However, when the median of a divided highway has physical characteristics that may create unsafe conditions, such as a sudden lateral drop-off or obstacles, the provision of a median barrier should be considered regardless of the traffic volume or the median width. Roadside barriers should be provided whenever conditions exist on the side of the road that warrant protection of vehicles. For example, when the slope of an embankment is high or when there is a roadside object such as a view of an overhead bridge, the provision of a roadside barrier is warranted. Figures 16.6 and 16.7 show typical roadside and median barriers currently in use. For additional information on selecting, locating, and designing median and roadside barriers, see the AASHTO *Roadside Design Guide.*

Curbs and Gutters

Curbs are raised structures made of either Portland cement concrete or bituminous concrete (rolled asphalt curbs) that are used mainly on urban highways to delineate pavement edges and pedestrian walkways. Curbs are also used to control drainage, improve aesthetics, and reduce right of way. Curbs can be generally classified as either vertical or sloping. Vertical curbs, which may be vertical or nearly vertical, range from 6 to 8 in high,

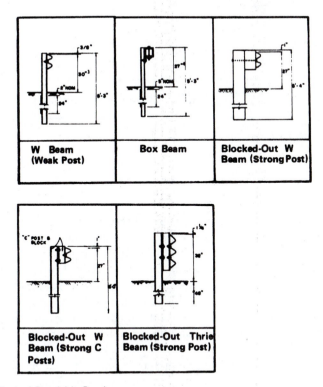

Figure 16.6 Typical Roadside Barriers

SOURCE: *Roadside Design Guide,* American Association of State Highway and Transportation Officials, Washington, D.C., 2000. Used with permission.

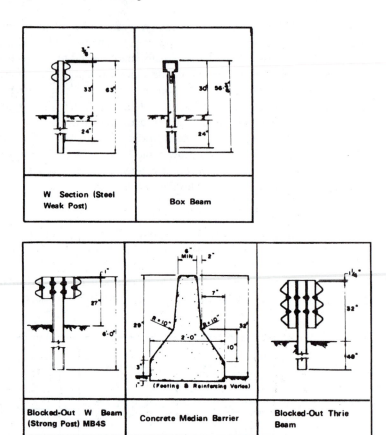

Figure 16.7 Typical Median Barriers

SOURCE: *Roadside Design Guide,* American Association of State Highway and Transportation Officials, Washington, D.C., 1996. Used with permission.

with steep sides, and are designed to prevent vehicles from leaving the highway. Sloping curbs are designed so that vehicles can cross them if necessary. Figure 16.8 shows some typical highway curbs. Both vertical and sloping curbs may be designed separately or as integral parts of the pavement. In general, vertical curbs should not be used in conjunction with traffic barriers, such as bridge railings or median and roadside barriers, because they could contribute to vehicles rolling over the traffic barriers. Vertical curbs should also be avoided on highways with design speeds greater than 40 mi/h, because at such speeds it is usually difficult for drivers to retain control of the vehicle after an impact with the curb.

Gutters or drainage ditches are usually located on the pavement side of a curb to provide the principal drainage facility for the highway. They are sloped to prevent any hazard to traffic, and they usually have cross slopes of 5 percent to 8 percent and are 1 to 6 ft wide. Gutters can be designed as V-type sections or as broad, flat, rounded sections.

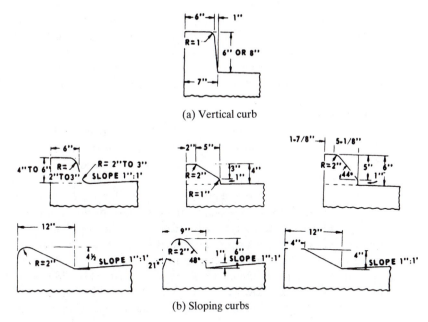

(a) Vertical curb

(b) Sloping curbs

Figure 16.8 Typical Highway Curbs

SOURCE: *A Policy on Geometric Design of Highways and Streets,* American Association of State Highway and Transportation Officials, Washington, D.C., 2001. Used with permission.

Guard Rails

Guard rails are longitudinal barriers placed on the outside of sharp curves and at sections with high fills. Their main function is to prevent vehicles from leaving the roadbed. They are installed at embankments higher than 8 ft and when shoulder slopes are greater than 4 : 1. Research has been done on the design of guard rails, particularly in the treatment of the end sections. The main objective of these research activities has been to select the best materials and shapes for guard rails. Shapes commonly used include the W beam and the box beam. Research has also led to the development of the weak post system, which provides for the post to collapse on impact, with the rail deflecting and absorbing the energy due to impact.

Sidewalks

Sidewalks are usually provided on roads in urban areas, but very seldom are they provided in rural areas. However, the provision of sidewalks in rural areas should be evaluated during the planning process to determine if any sections of the road require them. For example, rural high-speed highways may require sidewalks at areas with high pedestrian concentrations, such as areas adjacent to schools, industrial plants, and local businesses. Generally, sidewalks should be provided when pedestrian traffic is high along main or high-speed roads in either rural or urban areas. When no shoulders are provided on arterials, sidewalks are necessary even when pedestrian traffic is low. In urban areas, sidewalks should also be provided along both sides of collector streets that serve as pedestrian access to schools, parks, shopping centers, and transit stops, and along collector streets in commercial areas. Sidewalks should have a minimum clear width of 4 ft in residential areas and a range of 4 to 8 ft in commercial areas.

To encourage pedestrian use of sidewalks during all weather conditions, they should have all-weather surfaces; otherwise, pedestrians will tend to use the traffic lanes.

Cross Slopes

Pavements on straight sections of two-lane and multilane highways without medians are sloped from the middle downward to both sides of the highway. This provides a cross slope, whose cross section can be either curved or plane or a combination of the two. The parabola is generally used for the curved cross section. In this case, the highest point of the pavement (the crown) is slightly rounded, with the cross slope increasing toward the pavement edge. Plane cross slopes consist of uniform slopes at both sides of the crown. The curved cross section has one advantage in that the slope increases outward to the pavement edge, thereby enhancing the flow of surface water away from the pavement. One major disadvantage is that they are difficult to construct.

The cross slopes on divided highways are provided by either crowning the pavement in each direction, as shown in Figure 16.9(a), or by sloping the entire pavement in one direction, as shown in Figure 16.9(b). The advantage of draining the pavement in each direction separately is that surface water is quickly drained away from the traveled roadway during heavy rain storms, whereas the disadvantage is that more drainage facilities, such as inlets and underground drains, are required. This method is therefore mainly used at areas with heavy rain and snow.

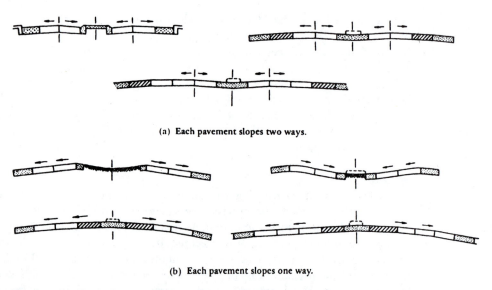

(a) **Each pavement slopes two ways.**

(b) **Each pavement slopes one way.**

Figure 16.9 Basic Cross Slope Arrangements for Divided Highways

SOURCE: *A Policy on Geometric Design of Highways and Streets,* American Association of State Highway and Transportation Officials, Washington, D.C., 2001. Used with permission.

In determining the rate of cross slope for design, two conflicting factors should be considered. Although a steep cross slope is required for drainage purposes, it may be undesirable in that vehicles will tend to drift to the edge of the pavement, particularly under icy conditions. Recommended rates of cross slopes are 1.5 percent to 2 percent for high-type pavements and 2 percent to 6 percent for low-type pavements. High-type pavements have wearing surfaces that can adequately support the expected traffic load without visible distress due to fatigue and are not susceptible to weather conditions. Low-type pavements are used mainly for low-cost roads and have wearing surfaces ranging from untreated loose material to surface-treated earth.

Side Slopes

Side slopes are provided on embankments and fills to provide stability for earthworks. They also serve as a safety feature by providing a recovery area for out-of-control vehicles. When being considered as a safety feature, the important sections of the cross slope are the hinge point, the foreslope, and the toe of the slope, as shown in Figure 16.10. The hinge point is potentially hazardous because it may cause vehicles to jump into the air while crossing it, resulting in loss of control by the driver. Rounding the hinge point enhances the control of the vehicle by the driver. The foreslope is the area that serves principally as a recovery area, where vehicle speeds can be reduced and other recovery maneuvers taken to regain control of the vehicle. The gradient of the foreslope should therefore not be high. Slopes of 3 : 1 (horizontal : vertical) or flatter are generally used for high embankments. This can be increased only when conditions at the site dictate it. The recommended values for foreslopes and backslopes are given in Table 16.3. To facilitate the safe movement of vehicles from the foreslope to the backslope, the toe of slope is rounded up as shown in Figure 16.10.

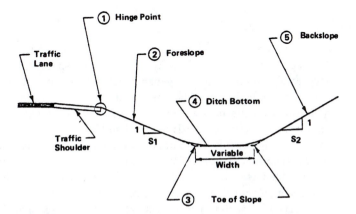

Figure 16.10 Designation of Roadside Regions

SOURCE: *A Policy on Geometric Design of Highways and Streets,* American Association of State Highway and Transportation Officials, Washington, D.C., 2001. Used with permission.

Table 16.3 Guide for Earth Slope Design

	Earth Slope, for Type of Terrain		
Height of Cut or Fill (ft)	Flat or Rolling	Moderately Steep	Steep
0–4	6 : 1	6 : 1	4 : 1
4–10	4 : 1	4 : 1	2 : 1★
10–15	4 : 1	2.50 : 1	1.75 : 1★
15–20	2 : 1★	2 : 1★	1.75 : 1★
Over 20	2 : 1★	2 : 1★	1.75 : 1★

★Slopes 2 : 1 or steeper should be subject to a soil stability analysis and should be reviewed for safety.
SOURCE: *Roadside Design Guide,* American Association of State Highway and Transportation Officials, Washington, D.C., 1996. Used by permission.

Right of Way

The right of way is the total land area acquired for the construction of the highway. Its width should be enough to accommodate all the elements of the highway cross section, any planned widening of the highway, and any public utility facilities that will be installed along the highway. In some cases, however, the side slopes may be located outside the right of way on easement areas. The right of way for two-lane urban collector streets should be between 40 and 60 ft, whereas the desirable minimum for two-lane arterials is 84 ft. Right-of-way widths for undivided four-lane arterials vary from 64 ft to 108 ft, whereas for divided arterials, they range from about 120 ft to 300 ft, depending on the numbers of lanes and whether frontage roads are included. The minimum right-of-way widths for freeways depend on the number of lanes and the existence of a frontage road.

DESIGN OF THE ALIGNMENT

The vertical and horizontal layouts of the highway make up the alignment. The design of the alignment depends primarily on the design speed selected for the highway. The least costly alignment is one that generally takes the form of the natural topography. Often this is not possible, however, because the designer has to adhere to certain standards that may not exist on the natural topography. It is important that the alignment of a given section has consistent standards to avoid sudden changes in the vertical and horizontal layout of the highway. It is also important that both horizontal and vertical alignments be designed to complement each other, since this will result in a safer and more attractive highway. One factor that should be considered to achieve this is the proper balancing of the grades of tangents with curvatures of horizontal curves and the location of horizontal and vertical curves with respect to each other. For example, a design that achieves horizontal curves with large radii at the expense of steep or long grades is a poor design. Similarly, if sharp horizontal curves are placed at or near the top of pronounced crest vertical curves or at or near the bottom of a pronounced sag vertical curve, this will create hazardous sections of the highway. It is important that this coordination of the vertical and horizontal alignments be considered at the early stages of preliminary design.

Vertical Alignment

The vertical alignment of a highway consists of straight sections of the highway known as grades, or tangents, connected by vertical curves. The design of the vertical alignment therefore involves the selection of suitable grades for the tangent sections and the design of the vertical curves. The topography of the area through which the road traverses has a significant impact on the design of the vertical alignment.

Grades

The effect of grade on the performance of heavy vehicles was discussed in Chapter 3, where it was shown that the speed of a heavy vehicle can be significantly reduced if the grade is steep and/or long. In Chapter 9, we noted that steep grades affect not only the performance of heavy vehicles but also the performance of passenger cars. In order to limit the effect of grades on vehicular operation, the maximum grade on any highway should be selected judiciously.

The selection of maximum grades for a highway depends on the design speed and the design vehicle. It is generally accepted that grades of 4 to 5 percent have little or no effect on passenger cars, except for those with high weight/horsepower ratios, such as those found in compact and subcompact cars. As the grade increases above 5 percent, however, speeds of passenger cars decrease on upgrades and increase on downgrades.

Grade has a greater impact on trucks than on passenger cars. Extensive studies have been conducted, and results (some of which were presented in Chapter 9) have shown that truck speed may increase up to 5 percent on downgrades and decrease by 7 percent on upgrades, depending on the percent and length of the grade.

The impact of grades on recreational vehicles is more significant than that for passenger cars, but it is not as critical as that for trucks. However, it is very difficult to establish maximum grades for recreational routes, and it may be necessary to provide climbing lanes on steep grades when the percentage of recreational vehicles is high.

Maximum grades have been established, based on the operating characteristics of the design vehicle on the highway. These vary from 5 percent for a design speed of 70 mi/h to between 7 and 12 percent for a design speed of 30 mi/h, depending on the type of highway. Table 16.4 gives recommended values of maximum grades. Note that these recommended maximum grades should not be used frequently, particularly when grades are long and the traffic includes a high percentage of trucks. On the other hand, when grade lengths are less than 500 ft and roads are one-way in the downgrade direction, maximum grades may be increased by up to 2 percent, particularly on low-volume rural highways.

Minimum grades depend on the drainage conditions of the highway. Zero-percent grades may be used on uncurbed pavements with adequate cross slopes to laterally drain the surface water. When pavements are curbed, however, a longitudinal grade should be provided to facilitate the longitudinal flow of the surface water. It is customary to use a minimum of 0.5 percent in such cases, although this may be reduced to 0.3 percent on high-type pavement constructed on suitably crowned, firm ground.

Table 16.4 Recommended Maximum Grades

| | Rural Collectors[a] Design Speed (mi/h) | | | | | | | | |
Type of Terrain	20	25	30	35	40	45	50	55	60
	Grades (%)								
Level	7	7	7	7	7	7	6	6	5
Rolling	10	10	9	9	8	8	7	7	6
Mountainous	12	11	10	10	10	10	9	9	8

| | Urban Collectors[a] Design Speed (mi/h) | | | | | | | | |
Type of Terrain	20	25	30	35	40	45	50	55	60
	Grades (%)								
Level	9	9	9	9	9	8	7	7	6
Rolling	12	12	11	10	10	9	8	8	7
Mountainous	14	13	12	12	12	11	10	10	9

| | Rural Arterials Design Speed (mi/h) | | | | | | | | |
Type of Terrain	40	45	50	55	60	65	70	75	80
	Grades (%)								
Level	5	5	4	4	3	3	3	3	3
Rolling	6	6	5	5	4	4	4	4	4
Mountainous	8	7	7	6	6	5	5	5	5

| | Rural and Urban Freeways[b] Design Speed (mi/h) | | | | | | |
Type of Terrain	50	55	60	65	70	75	80
	Grades (%)						
Level	4	4	3	3	3	3	3
Rolling	5	5	4	4	4	4	4
Mountainous	6	6	6	5	5	—	–

| | Urban Arterials Design Speed (mi/h) | | | | | | |
	30	35	40	45	50	55	60
	Grades (%)						
Level	8	7	7	6	6	5	5
Rolling	9	8	8	7	7	6	6
Mountainous	11	10	10	9	9	8	8

[a]Maximum grades shown for rural and urban conditions of short lengths (less than 500 ft) and on one-way downgrades may be up to 2 percent steeper.

[b]Grades that are 1 percent steeper than the value shown may be used for extreme cases in urban areas where development precludes the use of flatter grades and for one-way downgrades, except in mountainous terrain.

SOURCE: Adapted from *A Policy on Geometric Design of Highways and Streets,* American Association of State Highway and Transportation Officials, Washington, D.C., 2001. Used by permission.

Vertical Curves

Vertical curves are used to provide a gradual change from one tangent grade to another so that vehicles may run smoothly as they traverse the highway. These curves are usually parabolic in shape. The expressions developed for minimum lengths of vertical curves are therefore based on the properties of a parabola. They are classified as crest vertical curves or sag vertical curves. The different types of vertical curves are shown in Figure 16.11.

The main criteria used for designing vertical curves are

- Provision of minimum stopping sight distance
- Adequate drainage
- Comfortable in operation
- Pleasant appearance

The first criterion is the only criterion associated with crest vertical curves, whereas all four criteria are associated with sag vertical curves.

Crest Vertical Curves.   Two conditions exist for the minimum length of crest vertical curves. These are (1) when the sight distance is greater than the length of the vertical curve, and (2) when the sight distance is less than the length of the vertical curve. Let us first consider the case of the sight distance being greater than the length of the vertical curve. Figure 16.12 shows this condition. This figure schematically presents a vehicle on the grade at C with the driver's eye at height H_1 and an object of height H_2 located at D. If

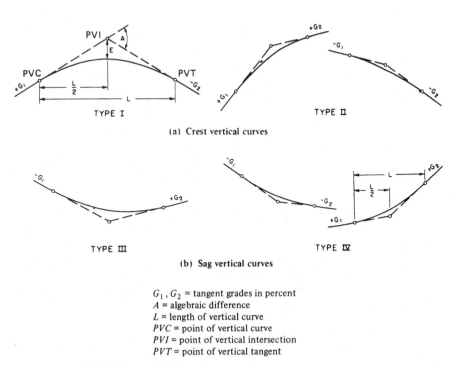

(a) Crest vertical curves

(b) Sag vertical curves

G_1, G_2 = tangent grades in percent
A = algebraic difference
L = length of vertical curve
PVC = point of vertical curve
PVI = point of vertical intersection
PVT = point of vertical tangent

Figure 16.11 Types of Vertical Curves

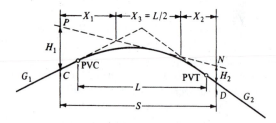

L = length of vertical curve (ft)
S = sight distance (ft)
H_1 = height of eye above roadway surface (ft)
H_2 = height of object above roadway surface (ft)
G_1 = slope of first tangent
G_2 = slope of second tangent
PVC = point of vertical curve
PVT = point of vertical tangent

Figure 16.12 Sight Distance on Crest Vertical Curve ($S > L$)

this object is seen by the driver, the line of sight is PN and the sight distance is S. Note that the line of sight is not necessarily horizontal, but in calculating the sight distance, the horizontal projection is considered.

From the properties of the parabola,

$$X_3 = \frac{L}{2}$$

The sight distance S is then given as

$$S = X_1 + \frac{L}{2} + X_2 \tag{16.1}$$

X_1 and X_2 can be found in terms of the grades G_1 and G_2 and their algebraic difference A. The minimum length of the vertical curve for the required sight distance is obtained as

$$L_{min} = 2S - \frac{200\left(\sqrt{H_1} + \sqrt{H_2}\right)^2}{A} \quad (\text{for } S > L) \tag{16.2}$$

It has been the practice to assume that the height of the driver, H_1, is 3.75 ft, and that the height of the object is 0.5 ft. Because of the increasing number of compact automobiles on the nation's highways, the height of the driver's eye is now taken as 3.5 ft. Also, AASHTO now recommends an object height of 2.0 ft, which is equivalent to the taillight height of a passenger car. With this assumption, Eq. 16.2 becomes

$$L_{min} = 2S - \frac{2158}{A} \quad (\text{for } S > L) \tag{16.3}$$

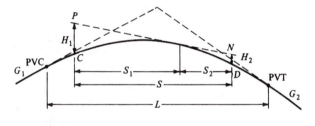

L = length of vertical curve (ft)
S = sight distance (ft)
H_1 = height of eye above roadway surface (ft)
H_2 = height of object above roadway surface (ft)
G_1 = slope of first tangent
G_2 = slope of second tangent
PVC = point of vertical curve
PVT = point of vertical tangent

Figure 16.13 Sight Distance on Crest Vertical Curve ($S < L$)

When the sight distance is less than the length of the crest vertical curve, the configuration shown in Figure 16.13 applies. Also, the properties of a parabola can be used to show that the minimum length of the vertical curve is given as

$$L_{min} = \frac{AS^2}{200\left(\sqrt{H_1} + \sqrt{H_2}\right)^2} \quad \text{(for } S < L) \tag{16.4}$$

Substituting 3.5 ft for H_1 and 2.0 ft for H_2 gives

$$L_{min} = \frac{AS^2}{2158} \text{ (for } S < L) \tag{16.5}$$

Example 16.1   Minimum Length of a Crest Vertical Curve

A crest vertical curve is to be designed to join a +3 percent grade with a −3 percent grade at a section of a two-lane highway. Determine the minimum length of the curve if the design speed of the highway is 60 mi/h and $S < L$. Assume that the perception-reaction time is 2.5 sec., and the deceleration rate for braking (a) is 11.2 ft/sec².

Solution:

- Use Eq. 3.32 from Chapter 3 to determine the SSD required for the design conditions. (Since the grade changes constantly on a vertical curve, the worst-case value for G is used to determine the braking distance.)

$$SSD = 1.47ut + \cfrac{u^2}{30\left\{\left(\cfrac{a}{32.2}\right)-G\right\}}$$

$$= 1.47 \times 60 \times 2.5 + \cfrac{60^2}{30\left\{\cfrac{11.2}{32.2}-0.03\right\}}$$

$$= 220.50 + 377.56$$

$$= 598.1 \text{ ft}$$

- Use Eq. 16.5 to obtain the minimum length of vertical curve:

$$L_{\min} = \frac{AS^2}{2158}$$

$$= \frac{6 \times 598.1^2}{2158}$$

$$= 994.3 \text{ ft}$$

Example 16.2   Maximum Safe Speed on a Crest Vertical Curve

An existing vertical curve on a highway joins a $+4.4$ percent grade with a 4.4 percent grade. If the length of the curve is 275 ft, what is the maximum safe speed on this curve? Take $a = 11.2$ ft/sec^2 and perception-reaction time $= 2.5$ sec. Also assume $S < L$.

Solution:

- Determine the SSD using the length of the curve and Eq. 16.5.

$$L_{\min} = \frac{AS^2}{2158}$$

$$275 = \frac{8.8 \times S^2}{2158}$$

$$S = 259.69 \text{ ft}$$

- Now determine the maximum safe speed for this sight distance from Eq. 3.32 from Chapter 3.

$$259.69 = 1.47 \times 2.5u + \cfrac{u^2}{30\left\{\cfrac{11.2}{32.2}-0.04\right\}}$$

from which we obtain the quadratic equation

$$u^2 - 33.50u - 2367.02 = 0$$

- Solving the quadratic equation to find the maximum safe speed,

$$u = 34.70 \text{ mi/h}$$

The maximum safe speed for the SSD available is therefore 34.7 mi/h. However, if a speed limit is to be posted to satisfy this condition, 30 mi/h will be used since speed limits are usually set at 5 mi/h increments.

Sag Vertical Curves.   The selection of the minimum length of a sag vertical curve is usually controlled by the following four different criteria: (1) sight distance provided by the headlight, (2) rider comfort, (3) control of drainage, and (4) general appearance.

The headlight sight distance requirement is based on the fact that as a vehicle is driven on a sag vertical curve at night, the position of the headlight and the direction of the headlight beam dictate the stretch of highway ahead that is lighted—and therefore the distance that can be seen by the driver. Figure 16.14 is a schematic of the situation when $S > L$. The headlight is located at a height H above the ground, and the headlight beam is inclined upward at an angle β to the horizontal. The headlight beam intersects the road at D, thereby restricting the available sight distance to S. The values used for H and β are usually 2 ft and 1°, respectively. Using the properties of the parabola, it can be shown that

$$L = 2S - \frac{200(H + S \tan \beta)}{A} \quad (\text{for } S > L) \qquad (16.6)$$

Substituting 2 ft for H and 1° for β makes Eq. 16.6 become

$$L = 2S - \frac{(400 + 3.5S)}{A} \quad (\text{for } S > L) \qquad (16.7)$$

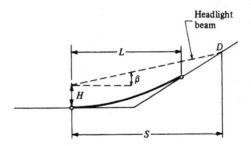

Figure 16.14 Headlight Sight Distance on Sag Vertical Curves ($S>L$)

Similarly, for the condition when $S < L$, it can be shown that

$$L = \frac{AS^2}{200(H + S \tan \beta)} \quad (\text{for } S < L) \tag{16.8}$$

and substituting 2 ft for H and 1° for β gives

$$L = \frac{AS^2}{400 + 3.5S} \quad (\text{for } S < L) \tag{16.9}$$

To provide a safe condition on a sag vertical curve, the curve must be of such a length that it will make the light beam sight distance S be at least equal to the SSD. The SSDs for the appropriate design speeds are therefore used for S when Eqs. 16.8 and 16.9 are used to compute minimum lengths of sag vertical curves.

The comfort criterion for the design of sag vertical curves takes into consideration that when a vehicle traverses a sag vertical curve, both the gravitational and centrifugal forces act in combination, resulting in a greater effect than on a crest vertical curve where these forces act in opposition to each other. Several factors such as weight carried, body suspension of the vehicle, and tire flexibility affect comfort due to change in vertical directions. This makes it difficult for comfort to be measured directly. However, it is generally accepted that a comfortable ride will be provided if the radial acceleration is not greater than 1 ft/sec². An expression that has been used for the comfort criterion is

$$L = \frac{Au^2}{46.5} \tag{16.10}$$

where u is the design speed in mi/h and L and A are the same as previously used. This length is usually about 75 percent of that obtained from the headlight sight distance requirement.

The drainage criterion for sag vertical curves is essential when the road is curbed. This criterion is different from the other criteria in that a maximum length requirement is given, rather than a minimum length. The requirement usually specified to satisfy this criterion is that a minimum grade of 0.35 percent be provided within 50 ft of the level point of the curve. It has been observed that the maximum length for this criterion is usually greater than the minimum lengths for the other criteria for speeds up to 60 mi/h and about the same for a speed of 70 mi/h.

The criterion of general appearance is usually satisfied by the use of a rule of thumb expressed as

$$L = 100A \tag{16.11}$$

where L is the minimum length of the sag vertical curve. Experience has shown, however, that longer curves are frequently necessary for high-type highways if the general appearance of these highways is to be improved.

Example 16.3 Minimum Length of a Sag Vertical Curve

A sag vertical curve is to be designed to join a −3 percent grade to a +3 percent grade. If the design speed is 40 mi/h, determine the minimum length of the curve that will satisfy all criteria. Take $a = 11.2$ ft/sec^2 and perception-reaction time = 2.5 sec.

Solution:

- Find the stopping sight distance.

$$SSD = 1.47ut + \frac{u^2}{30\left(\dfrac{11.2}{32.2} - G\right)}$$

$$= 1.47 \times 40 \times 2.5 + \frac{40^2}{30(0.35 - .03)} = 147.0 + 166.67$$

$$= 313.67 \text{ ft}$$

- Determine whether $S < L$ or $S > L$ for the headlight sight distance criterion. For $S > L$,

$$L = 2S - \frac{(400 + 3.5S)}{A}$$

$$= 2 \times 313.67 - \frac{400 + 3.5 \times 313.67}{6}$$

$$= 377.70 \text{ ft}$$

(This condition is not appropriate since $313.67 < 377.70$ and therefore $S \not> L$.) For $S < L$, then,

$$L = \frac{AS^2}{400 + 3.5S}$$

$$= \frac{6 \times 313.67^2}{400 + 3.5 \times 313.67}$$

$$= 394.12 \text{ ft}$$

(This condition applies.)
- Determine minimum length for the comfort criterion.

$$L = \frac{Au^2}{46.5}$$

$$= \frac{6 \times 40^2}{46.5} = 206.5 \text{ ft}$$

- Determine minimum length for the general appearance criterion.

$$L = 100A$$
$$= 100 \times 6 = 600 \text{ ft}$$

The minimum length to satisfy all criteria is 600 ft.

The solution of Example 16.3 demonstrates a theoretical analysis to determine the minimum length of a sag vertical curve that satisfies the criteria discussed. AASHTO, however, suggests that the headlight criterion seems to be the most logical for general use, and therefore it uses this criterion to establish minimum lengths of sag vertical curves.

Elevations on Vertical Curves.   The properties of a parabola are again used to determine the elevation on the curve at regular intervals after determining the length of the curve.

Crest Vertical Curves.   The expressions for the minimum lengths of the crest vertical curves are given in Eq. 16.3 and Eq. 16.5, which are repeated here:

$$L = \frac{AS^2}{2158} \quad \text{(for } S < L) \tag{16.12}$$

and

$$L = 2S - \frac{2158}{A} \quad \text{(for } S > L) \tag{16.13}$$

Equation 16.12 can be written as

$$L = KA \tag{16.14}$$

where K is the length of the vertical curve per percent change in A. K can therefore be used as a simple and convenient way to establish design control for crest vertical curves.

Table 16.5 gives values for K based on stopping sight distance requirements. The use of K as a design control is convenient, since the value for any design speed will represent all combinations of A and L for that speed. Similarly, K values can be computed for the case where the sight distance is less than the vertical curve.

In using Eq. 16.13, it has been found that the minimum lengths obtained for the case of S greater than L are not practical design values and are generally not used. The common practice for this condition is for individual states to set minimum limits, which range from 100 ft to 325 ft. Alternatively, minimum lengths can be set at three times the design speed, which gives values that are directly dependent on the design speed.

Having determined the minimum length of the crest vertical curve, the elevations of the curve at regular intervals can then be determined. This is done by considering the

Table 16.5 Design Controls for Crest Vertical Curves Based on Stopping Sight Distance

		Rate of Vertical Curvature, K^a	
Design Speed (mi/h)	Stopping Sight Distance (ft)	Calculated	Design
15	80	3.0	3
20	115	6.1	7
25	155	11.1	12
30	200	18.5	19
35	250	29.0	29
40	305	43.1	44
45	360	60.1	61
50	425	83.7	84
55	495	113.5	114
60	570	150.6	151
65	645	192.8	193
70	730	246.9	247
75	820	311.6	312
80	910	383.7	384

[a]Rate of vertical curvature, K, is the length of curve per percent algebraic difference in intersecting grades (A). $K = L/A$.
SOURCE: Adapted from *A Policy on Geometric Design of Highways and Streets,* American Association of State Highway and Transportation Officials, Washington, D.C., 2001. Used by permission.

properties of the parabola. Consider a crest vertical curve in the form of a parabola shown in Figure 16.15.

From the properties of a parabola, $Y = ax^2$, where a is a constant. The rate of change of slope is

$$\frac{d^2Y}{dx^2} = 2a$$

but

$$T_1 = T_2 = T$$
$$L = 2T$$

where L is the length of the curve in feet. (Note that the length of the vertical curve is the horizontal projection of the curve and not the length along the curve.)

If the total change in slope is A, then

$$2a = \frac{A}{100L}$$

and

$$a = \frac{A}{200L}$$

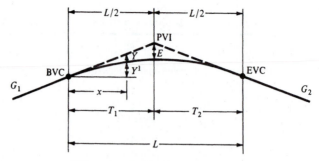

PVI = point of vertical intersection.
BVC = beginning of vertical curve (same point as PVC)
EVC = end of vertical curve (same point as PVT)
E = **external distance**
G_1, G_2 = grades of tangents (%)
L = length of curve
A = algebraic difference of grades, $G_1 - G_2$

Figure 16.15 Layout of a Crest Vertical Curve for Design

The equation of the curve can therefore be written as

$$Y = \frac{A}{200L} x^2 \tag{16.15}$$

When $x = L/2$, the external distance E from the point of vertical intersection (PVI) to the curve is determined by substituting $L/2$ for x in Eq. 16.15:

$$E = \frac{A}{200L} \left(\frac{L}{2} \right)^2 = \frac{AL}{800} \tag{16.16}$$

Since stations are given in 100-ft intervals, E can be given as

$$E = \frac{AN}{8} \tag{16.17}$$

where N is the length of the curve in stations. The vertical offset Y at any point on the curve can also be given in terms of E. Substituting $800E/L$ for A in Eq. 16.15 will give

$$Y = \left(\frac{x}{L/2} \right)^2 E \tag{16.18}$$

The location of the high point of a crest vertical curve is frequently of interest to the designer because of drainage requirements. The distance between the beginning of the vertical curve (BVC) and the high point can be determined by considering the expression for Y^1 in ft (see Figure 16.15):

$$
\begin{aligned}
Y^1 &= \frac{G_1 x}{100} - Y \\
&= \frac{G_1 x}{100} - \frac{A}{200L} x^2 \\
&= \frac{G_1 x}{100} - \left(\frac{G_1 - G_2}{200L}\right) x^2
\end{aligned} \tag{16.19}
$$

Differentiating Eq. 16.19 and equating it to zero will give the value of x for the high point on the curve:

$$
\frac{dY^1}{dx} = \frac{G_1}{100} - \left(\frac{G_1 - G_2}{100L}\right) x = 0 \tag{16.20}
$$

Therefore,

$$
x_{high} = \frac{100L}{(G_1 - G_2)} \frac{G_1}{100} = \frac{LG_1}{(G_1 - G_2)}
$$

where x_{high} = distance in ft from BVC to the turning point—that is, the point with the highest elevation on the curve.

Similarly, it can be shown that the difference in elevation between the BVC and the turning point (Y^1_{high}) can be obtained by substituting the expression x_{high} for x in Eq. 16.19 to obtain

$$
Y^1_{high} = \frac{LG_1^2}{200(G_1 - G_2)} \tag{16.21}
$$

The design of a crest vertical curve will generally proceed in the following manner:

Step 1. Determine the minimum length of curve to satisfy sight distance requirements.

Step 2. Determine from the layout plans the station and elevation of the PVI—that is, the point where the grades intersect.

Step 3. Compute the elevations of the BVC and end of vertical curve (EVC).

Step 4. Compute offsets Y from the tangent to the curve at equal distances, usually 100 ft apart, beginning with the first whole station after the BVC.

Step 5. Compute elevations on curve.

Example 16.4   Design of Crest Vertical Curve

A crest vertical curve joining a +3 percent and a −4 percent grade is to be designed for 75 mi/h. If the tangents intersect at station (345 + 60.00) at an elevation of 250 ft, determine the stations and elevations of the BVC and EVC. Also, calculate the elevations of intermediate points on the curve at the whole stations.

A sketch of the curve is shown in Figure 16.16.

Solution: For a design speed of 75 mi/h, $K = 312$, from Table 16.5.

$$\text{minimum length} = 312 \times [3 - (-4)] = 2184 \text{ ft}$$

$$\text{station of BVC} = (345 + 60) - \left(\frac{21 + 84}{2}\right) = 334 + 68$$

$$\text{station of EVC} = (334 + 68) + (21 + 84) = 356 + 52$$

$$\text{elevation of BVC} = 250 - \left(0.03 \times \frac{2184}{2}\right) = 217.24 \text{ ft}$$

The remainder of the computation is efficiently done using the format shown in Table 16.6.

Sag Vertical Curves.   The design of sag vertical curves takes the same format as for crest vertical curves. The headlight sight distance requirement is used for design purposes since stopping sight distances based on this requirement are generally within the limits of practical accepted values. The expressions developed earlier for minimum length of sag vertical curves based on this requirement are

$$L = 2S - \frac{400 + 3.5S}{A} \quad (\text{for } S > L) \tag{16.22}$$

and

$$L = \frac{AS^2}{400 + 3.5S} \quad (\text{for } S < L) \tag{16.23}$$

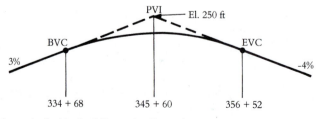

Figure 16.16   Layout of a Vertical Curve for Example 16.4

Table 16.6   Elevation Computations for Example 16.4

Station	Distance from BVC (x) (ft)	Tangent Elevation (ft)	Offset $\left(y = \dfrac{Ax^2}{200L} \right)$ ft	Curve Elevation (Tangent Elevation − Offset) ft
BVC 334 + 68	0	217.24	0	217.24
335 + 00	32	$217.24 + \dfrac{32}{100} \times 3 = 218.20$	0.02	218.18
336 + 00	132	221.20	0.28	220.92
337 + 00	232	224.20	0.86	223.34
338 + 00	332	227.20	1.77	225.43
339 + 00	432	230.20	2.99	227.21
340 + 00	532	233.20	4.54	228.66
341 + 00	632	236.20	6.40	229.80
342 + 00	732	239.20	8.59	230.61
343 + 00	832	242.20	11.09	231.11
344 + 00	932	245.20	13.92	231.28
345 + 00	1032	248.20	17.07	231.13
346 + 00	1132	251.20	20.54	230.66
347 + 00	1232	254.20	24.32	229.88
348 + 00	1332	257.20	28.43	228.77
349 + 00	1432	260.20	32.86	227.34
350 + 00	1532	263.20	37.61	225.59
351 + 00	1632	266.20	42.68	223.52
352 + 00	1732	269.20	48.07	221.13
353 + 00	1832	272.20	53.79	218.41
354 + 00	1932	275.20	59.82	215.38
355 + 00	2032	278.20	66.17	212.03
356 + 00	2132	281.20	72.84	208.36
EVC 356 + 52.00	2184	282.76	76.44	206.32

The design control for sag vertical curves is also conveniently expressed in terms of K rate for all values of A. The computed values of K are shown in Table 16.7.

Computation of the elevations at different points on the curve takes the same form as that for the crest vertical curve. In this case, however, the offset Y is added to the appropriate tangent elevation to obtain the curve elevation since the formation elevation of the curve is higher than the tangent.

Example 16.5   Design of Sag Vertical Curve

A sag vertical curve joins a −3 percent grade and a +3 percent grade. If the PVI of the grades is at station (435 + 50) and has an elevation of 235 ft, determine the station and elevation of the BVC and EVC for a design speed of 70 mi/h. Also compute the elevation on the curve at 100-ft intervals. Figure 16.17 shows a layout of the curve.

Table 16.7   Design Controls for Sag Vertical Curves Based on Stopping Sight Distance

Design Speed (mi/h)	Stopping Sight Distance (ft)	Rate of Vertical Curvature, K^a	
		Calculated	Design
15	80	9.4	10
20	115	16.5	17
25	155	25.5	26
30	200	36.4	37
35	250	49.0	49
40	305	63.4	64
45	360	78.1	79
50	425	95.7	96
55	495	114.9	115
60	570	135.7	136
65	645	156.5	157
70	730	180.3	181
75	820	205.6	206
80	910	231.0	231

[a]Rate for vertical curvature, K, is the length of curve (ft) per percent algebraic difference intersecting grades (A). $K = L/A$

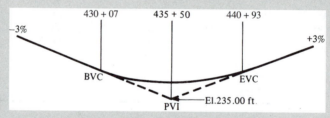

Figure 16.17   Layout for a Sag Vertical Curve for Example 16.5

Solution: For a design speed of 70 mi/h, $K = 181$, using the higher rounded value in Table 16.7.

$$\text{length of curve} = 181 \times 6 = 1086 \text{ ft}$$
$$\text{station of BVC} = (435 + 50) - (5 + 43) = 430 + 07$$
$$\text{station of EVC} = (435 + 50) + (5 + 43) = 440 + 93$$
$$\text{elevation of BVC} = 235 + 0.03 \times 543 = 251.29$$
$$\text{elevation of EVC} = 235 + 0.03 \times 543 = 251.29$$

The computation of the elevations is shown in Table 16.8.

Table 16.8 Elevation Computations for Example 16.5

Station	Distance from BVC (x) (ft)	Tangent Elevation (ft)	Offset $\left(y = \dfrac{Ax^2}{200L} \right)$ (ft)	Curve Elevation (Tangent Elevation + Offset) (ft)
BVC 430 + 07	0	251.29	0	251.29
431 + 00.00	93	248.50	0.24	248.74
432 + 00.00	193	245.50	1.03	246.53
433 + 00.00	293	242.50	2.37	244.87
434 + 00.00	393	239.50	4.27	243.77
435 + 00.00	493	236.50	6.71	243.21
436 + 00.00	593	233.50	9.71	243.21
437 + 00.00	693	230.50	13.27	243.77
438 + 00.00	793	227.50	17.37	244.87
439 + 00.00	893	224.50	22.03	246.53
440 + 00.00	993	221.50	27.24	248.74
EVC 440 + 93.00	1086	218.71	32.58	251.29

Horizontal Alignment

The horizontal alignment consists of straight sections of the road, known as tangents, connected by horizontal curves. The curves are usually segments of circles, which have radii that will provide for a smooth flow of traffic along the curve. It was shown in Chapter 3 that the minimum radius of a horizontal curve depends on the design speed u of the highway, the superelevation e, and the coefficient of side friction f_s. This relationship was shown to be

$$R = \frac{u^2}{15(e + f_s)} \tag{16.24}$$

Also, when a vehicle is being driven around a horizontal curve, an object located near the inside edge of the road may interfere with the view of the driver, which will result in a reduction of the driver's sight distance ahead of her. When such a situation exists, it is necessary to design the horizontal curve such that the available sight distance is at least equal to the safe stopping distance. Figure 16.18 is a schematic of this situation.

Consider a vehicle at point A and an object at point T. The line of sight that will just permit the driver to see the object is the chord AT. However, the horizontal distance traveled by the vehicle from point A to the object is arc AT. If 2θ (in degrees) is the angle subtended at the center of the circle by arc AT, then

$$\frac{2R\theta\pi}{180} = S$$

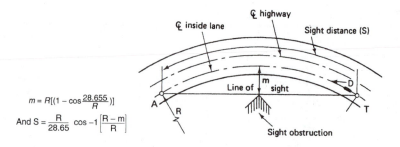

$$m = R[(1 - \cos\frac{28.655}{R})]$$

$$\text{And } S = \frac{R}{28.65} \cos{-1}\left[\frac{R-m}{R}\right]$$

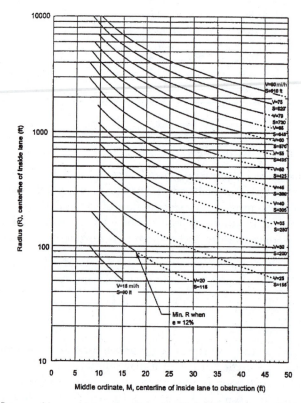

Figure 16.18 Range of Lower Values for Stopping Sight Distances on Horizontal Curves

SOURCE: Adapted from *A Policy on Geometric Design of Highways and Streets,* American Association of State Highways and Transportation Officials, Washington, D.C., 2001. Used by permission.

where

R = radius of horizontal curve

S = sight distance = length of arc AT

$$\theta = \frac{28.65}{R} S$$

However,

$$\frac{R - m}{R} = \cos \theta$$

$$\frac{R - m}{R} = \cos \left(\frac{28.65}{R} S \right)$$

$$m = R \left(1 - \cos \frac{28.65}{R} S \right) \tag{16.25}$$

Equation 16.25 can be used to determine m, R, or S, depending on the information known.

Example 16.6   Location of Object Near a Horizontal Curve

A horizontal curve having a radius of 800 ft forms part of a two-lane highway that has a posted speed limit of 35 mi/h. If the highway is flat at this section, determine the minimum distance a large billboard can be placed from the center line of the inside lane of the curve, without reducing the required SSD. Assume perception-reaction time of 2.5 sec, and $a = 11.2$ ft/sec^2.

Solution:

- Determine the required SSD.
  Use Equation 3.24

$$(1.47 \times 35 \times 2.5) + \frac{(35)^2}{30(0.35)} = 245.29 \text{ ft}$$

- Determine m
  Use Eq. 16.25.

$$m = 800 \left[1 - \cos \left(\frac{28.65}{800} (245.29) \right) \right] = 800(1 - 0.988) \text{ ft}$$

$$= 9.4 \text{ ft}$$

The design of the horizontal alignment therefore entails the determination of the minimum radius, the determination of the length of the curve, and the computation of the horizontal offsets from the tangents to the curve to facilitate the setting out of the curve. In some cases, to avoid a sudden change from a tangent with infinite radius to a curve of finite radius, a curve with radii varying from infinite value to the radius of the circular curve is placed between the circular curve and the tangent. Such a curve is known as a *spiral* or *transition curve.*

Horizontal Curves

There are four types of horizontal curves: simple, compound, reversed, and spiral. Computations required for each of them are presented in turn.

Simple Curves. Figure 16.19 shows a layout sketch of a simple horizontal curve. As stated earlier, the curve is a segment of a circle with radius R. The point at which the curve begins is known as the *point of curve* (PC), and the point at which it ends is known as the *point of tangent* (PT). The point at which the two tangents intersect is known as the *point of intersection* (PI) or vertex (V). The simple circular curve is described either by its radius—for example, 200-ft-radius curve—or by the degree of the curve. Two definitions exist for the degree of the curve: the arc and the chord.

The arc defines the curve in terms of the angle subtended at the center by a circular arc 100 ft in length [Figure 16.20(a)]. This means that for a 2° curve, for example, an arc of 100 ft will be subtended by an angle of 2° at the center. If θ is the angle in radians subtended at the center by an arc of a circle, the length of that arc is given by $R\theta = L$. If D_a° is the angle in degrees subtended at the center by an arc of length L, then

$$\theta = \frac{\pi D_a^\circ}{180} \text{ (rad)}$$

$$\frac{R\pi D^\circ}{180} = 100$$

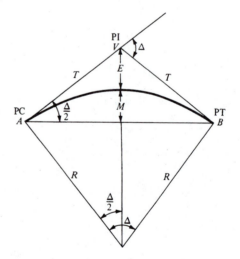

R = radius of circular curve	PC = point of curve
T = tangent length	PT = point of tangent
Δ = deflection angle	PI = point of intersection
M = middle ordinate	E = external distance

Figure 16.19   Layout of a Simple Horizontal Curve

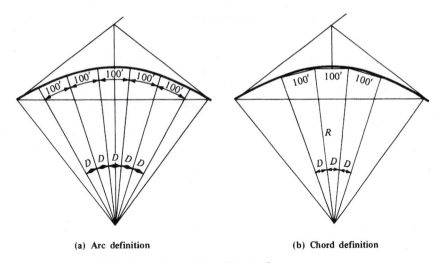

(a) Arc definition   **(b) Chord definition**

Figure 16.20 Arc and Chord Definitions for a Circular Curve

Therefore,

$$R = \frac{180 \times 100}{\pi D_a^\circ}$$

giving

$$R = \frac{5729.6}{D_a^\circ} \tag{16.26}$$

The radius of the curve can therefore be determined if the degree of curve is known.

The chord defines the curve in terms of the angle subtended at the center by a chord of 100 ft [see Figure 16.20(b)], and the radius is given as

$$R = \frac{50}{\sin D_c^\circ/2} \tag{16.27}$$

where D_c is the angle in degrees subtended at the center by a 100-ft chord. The arc definition is commonly used for highway work, and the chord definition is commonly used for railway work.

Formulas for Simple Circular Curves. Referring to Figure 16.19 and using the properties of a circle, the two tangent lengths AV and BV are equal and designated as T. The angle Δ formed by the two tangents is known as the deflection angle. We can therefore obtain the tangent length as

$$T = R \tan \frac{\Delta}{2} \tag{16.28}$$

The length C of the chord AB, which is known as the *long chord,* is given as

$$C = 2R \, \sin \, \frac{\Delta}{2} \tag{16.29}$$

The external distance E is the distance from the point of intersection to the curve on a radial line and is given as

$$E = R \, \sec \, \frac{\Delta}{2} - R$$

$$= R \left(\frac{1}{\cos \dfrac{\Delta}{2}} - 1 \right) \tag{16.30}$$

The middle ordinate M is the distance between the midpoint of the long chord and the midpoint of the curve and is given as

$$M = R - R \, \cos \frac{\Delta}{2}$$

$$= R \left(1 - \cos \frac{\Delta}{2} \right) \tag{16.31}$$

The length of the curve L is given as

$$L = \frac{R\Delta\pi}{180} \tag{16.32}$$

Setting Out of a Simple Horizontal Curve.   Simple horizontal curves are usually set out in the field by staking out points on the curve using deflection angles measured from the tangent at the point of curve (PC) and the lengths of the chords joining consecutive whole stations. Figure 16.21 is a schematic of the procedure involved. The first deflection angle VAp determined to the first whole station on the curve, which is usually less than a station away from the PC, is equal to $\delta_1/2$ (properties of a circle).

The next deflection angle VAq is

$$\frac{\delta_1}{2} + \frac{D}{2}$$

and the next angle VAv is

$$\frac{\delta_1}{2} + \frac{D}{2} + \frac{D}{2} = \frac{\delta_1}{2} + D$$

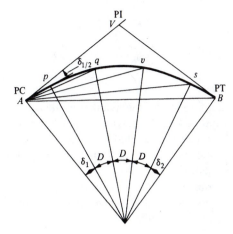

Figure 16.21　Deflection Angles on a Simple Circular Curve

and the next angle VAs is

$$\frac{\delta_1}{2} + \frac{D}{2} + \frac{D}{2} + \frac{D}{2} = \frac{\delta}{2} + \frac{3D}{2}$$

and the last angle VAB is

$$\frac{\delta_1}{2} + \frac{D}{2} + \frac{D}{2} + \frac{D}{2} + \frac{\delta_2}{2} = \frac{\delta_1}{2} + \frac{\delta_2}{2} + \frac{3D}{2} = \frac{\Delta}{2}$$

To set out the horizontal curve, it is necessary to determine δ_1 and δ_2. The length of the first arc, l_1, is related to δ_1 as

$$l_1 = \frac{R\pi}{180}\,\delta_1 \tag{16.33}$$

giving

$$R = \frac{l_1 \times 180}{\delta_1 \pi}$$

From Eq. 16.32,

$$R = \frac{180L}{\Delta\pi}$$

Therefore,

$$\frac{l_1}{\delta_1} = \frac{L}{\Delta} = \frac{l_2}{\delta_2}$$

In setting out a simple horizontal curve in the field, the PC and PT (point of tangent) are first located and staked. Deflection angles from the PC to each whole station are then computed. A transit is mounted over the PC and each whole station is located on the ground using the appropriate deflection angle, and the chord distance is measured from the preceding station.

Note that the lengths l_1 and l_2 are measured along the curve, and that the corresponding chord lengths should be calculated, particularly when circular curves are sharp. These chord lengths can be calculated from

$$C_1 = 2R \, \sin \frac{\delta_1}{2} \qquad (16.34)$$

$$C_D = 2R \, \sin \frac{D}{2}$$

$$C_2 = 2R \, \sin \frac{\delta_2}{2}$$

where C_1, C_D, and C_2 are the first, intermediate, and last chords, respectively.

Example 16.7   Design of a Horizontal Curve

The deflection angle of a 4° curve is 55°25′. If the PC is located at station (238 + 44.75), determine the length of the curve and the station of PT. Also determine the deflection angles and chords for setting out the curve at whole stations from the PC. Figure 16.22 shows a layout of the curve.

Solution:

$$\text{radius of curve} = \frac{5729.6}{D} = \frac{5729.6}{4}$$
$$\approx 1432.4 \text{ ft}$$

$$\text{length of curve} = \frac{R\Delta\pi}{180} = \frac{1432.4 \times 55.4167\pi}{180}$$
$$= 1385.42 \text{ ft}$$

Therefore, station at PT is equal to station (238 + 44.75) + (13 + 85.42) stations = 252 + 30.17. The distance between PC and the first station is 239 − (238 + 44.75) stations = 55.25 ft.

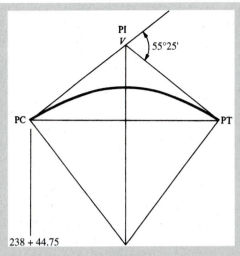

Figure 16.22 Layout of Curve for Example 16.7

$$\frac{\delta_1}{\Delta} = \frac{l_1}{L}$$

$$\frac{\delta_1}{55.4167} = \frac{55.25}{1385.42}$$

Therefore,

$$\delta_1 = 2.210°$$

$$C_1 = 2 \times 1432.4 \, \sin\left(\frac{2.210}{2}\right) = 55.25 \text{ ft}$$

The first deflection angle to station 239 is $\delta_1/2 = 1.105° = 1°6'18''$. Similarly,

$$l_2 = (252 + 30.17) - (252 + 00.00) = 30.17 \text{ ft}$$

$$\frac{\delta_2}{2} = \frac{30.17}{1385.42} \times \frac{55.4167}{2} = 0.6034°$$

$$= 36°12''$$

$$C_2 = 2 \times 1432.4 \, \sin(0.6034°)$$

$$= 30.17 \text{ ft}$$

The other deflection angles are computed in Table 16.9.

Table 16.9   Computations of Deflection Angles and Chord Lengths for Example 16.7

Station	Deflection Angle	Chord Length (ft)
PC 238 + 44.75	0	0
239	1°6′18″	55.25
240	3°6′18″	99.98
241	5°6′18″	99.98
242	7°6′18″	99.98
243	9°6′18″	99.98
244	11°6′18″	99.98
245	13°6′18″	99.98
246	15°6′18″	99.98
247	17°6′18″	99.98
248	19°6′18″	99.98
249	21°6′18″	99.98
250	23°6′18″	99.98
251	25°6′18″	99.98
252	27°6′18″	99.98
PT 252 + 30.17	27°42′30″	30.17

$$D = 4°$$

$$C_D = 2 \times 1432.42 \ \sin\left(\frac{4}{2}\right)$$

$$= 99.98 \ \text{ft}$$

Note that the deflection angle to PT is half the deflection angle of the tangents. This serves as a check of the computation. Also, because the curve is flat, the chord lengths are approximately equal to the arc lengths.

Compound Curves.   Compound curves consist of two or more curves in succession, turning in the same direction, with any two successive curves having a common tangent point. Figure 16.23 shows a typical layout of a compound curve, consisting of two simple curves. These curves are used mainly in obtaining desirable shapes of the horizontal alignment, particularly at at-grade intersections, ramps of interchanges, and highway sections in difficult topographic conditions. To avoid abrupt changes in the alignment, the radii of any two consecutive simple curves that form a compound curve should not be widely different. AASHTO recommends that the ratio of the flatter radius to the sharper radius at intersections should not be greater than 2:1 as drivers can adjust to sudden changes in curvature and speed. The maximum desirable ratio recommended by AASHTO for interchanges is 1.75:1, although 2:1 may be used.

To provide smooth transition from a flat curve to a sharp curve, and to facilitate a reasonable deceleration rate on a series of curves of decreasing radii, the length of each

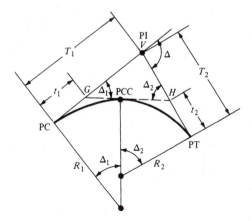

R_1, R_2 = radii of simple curves forming compound curve
Δ_1, Δ_2 = deflection angles of simple curves
Δ = deflection angle of compound curve
t_1, t_2 = tangent lengths of simple curves
T_1, T_2 = tangent lengths of compound curve
PCC = point of compound curve
 PI = point of intersection
PC = point of curve
PT = point of tangent

Figure 16.23 Layout of a Compound Curve

curve should not be too short. Minimum lengths based on the radius of each curve as recommended by AASHTO are given in Table 16.10.

Figure 16.23 shows seven variables—R_1, R_2, Δ_1, Δ_2, Δ, T_1, and T_2—six of which are independent, since $\Delta = \Delta_1 + \Delta_2$. Several solutions can be developed for the compound curve, but only the vertex triangle method is presented here, since this is the method frequently used in highway design.

In Figure 16.23, R_1 and R_2 are usually known.

$$\Delta = \Delta_1 + \Delta_2 \tag{16.35}$$

$$t_1 = R_1 \ \tan \ \frac{\Delta_1}{2} \tag{16.36}$$

$$t_2 = R_2 \ \tan \ \frac{\Delta_2}{2} \tag{16.37}$$

$$\frac{\overline{VG}}{\sin \Delta_2} = \frac{\overline{VH}}{\sin \Delta_1} = \frac{t_1 + t_2}{\sin(180 - \Delta)} = \frac{t_1 + t_2}{\sin \Delta} \tag{16.38}$$

$$T_1 = VG + t_1 \tag{16.39}$$

$$T_2 = VH + t_2 \tag{16.40}$$

Table 16.10 Lengths of Circular Arc for a Compound Intersection Curve When Followed by a Curve of One-Half Radius or Preceded by a Curve of Double Radius

Length of Circular Arc (ft)	Radius (ft)						
	100	150	200	250	300	400	500 or more
Minimum	40	50	60	80	100	120	140
Desirable	60	70	90	120	140	180	200

SOURCE: Adapted from *A Policy on Geometric Design of Highways and Streets*, American Association of State Highway and Transportation Officials, Washington, D.C., 2001. Used by permission.

where

R_1 and R_2 = radii of simple curves forming the compound curve
Δ_1 and Δ_2 = deflection angles of simple curves
t_1 and t_2 = tangent lengths of simple curves
T_1 and T_2 = tangent lengths of compound curves
Δ = deflection angle of compound curve

To be able to lay out the curve, it is necessary to know the value of at least one other variable. Usually, Δ_1 or Δ_2 can be obtained from the layout plans. Equations 16.35 through 16.40 can then be used to solve for Δ_1 or Δ_2, VG, VH, t_1, t_2, T_1, and T_2. Deflection angles can then be determined for each simple curve in turn.

Example 16.8   Design of a Compound Curve

Figure 16.24 shows a compound curve that is to be set out at an intersection. If the point of compound curve (PCC) is located at station (565 + 35), determine the deflection angles for setting out the curve.

Solution:

$$t_1 = 500 \ \tan \ \frac{34}{2} = 152.87 \text{ ft}$$

$$t_2 = 350 \ \tan \ \frac{26}{2} = 80.80 \text{ ft}$$

For length of horizontal curve of 500 ft radius,

$$L = R\Delta_1 \ \frac{\pi}{180} = 500 \times \frac{34\pi}{180}$$
$$= 296.71 \text{ ft}$$

For length of horizontal curve of 350 ft radius,

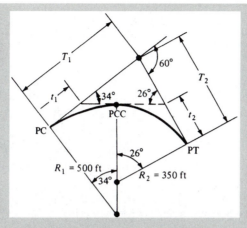

Figure 16.24   Compound Curve for Example 16.8

$$L = 350 \, \frac{26\pi}{180}$$
$$= 158.82 \, (\text{ft})$$

Therefore, station of PC is equal to $(565 + 35.00) - (2 + 96.71) = 562 + 38.29$, and station of PT is equal to $(565 + 35.00) + (1 + 58.82) = 566 + 93.82$.

For curve of 500 ft radius,

$$\frac{D}{2} = \frac{5729.6}{2 \times 500} = 5°43'47'' \qquad (\text{from Eq. 16.26})$$

$$l_1 = (563 + 00) - (562 + 38.29) = 61.71 \, (\text{ft})$$

$$\frac{\delta_1}{l_1} = \frac{\Delta}{L}$$

$$\frac{\delta_1}{2} = \frac{61.71 \times 34}{2 \times 296.71} = 3°32'8''$$

$$l_2 = (565 + 35.00) - (565 + 00) = 35 \, (\text{ft})$$

$$\frac{\delta_2}{2} = \frac{35 \times 34}{2 \times 296.71} = 2°0'19''$$

For curve of 350 ft radius,

$$\frac{D}{2} = \frac{5729.6}{2 \times 350} = 8°11'7''$$

$$l_1 = (566 + 00) - (565 + 35.00) = 65 \, \text{ft}$$

$$\frac{\delta_1}{2} = \frac{65 \times 26}{2 \times 158.82} = 5°19'16''$$

$$l_2 = (566 + 93.82) - (566 + 00) = 93.82 \text{ ft}$$

$$\frac{\delta_2}{2} = \frac{93.82 \times 26}{2 \times 158.8} = 7°40'44''$$

For computation of the deflection angles, see Table 16.11.

Note that deflection angles for the 350 ft radius curve are turned from the common tangent with the transit located at PCC. Also note that with the relatively flat simple curves, the calculated lengths of the chords are almost equal to the corresponding arc lengths.

Table 16.11   Computations for Example 16.8

| | *500 ft Radius Curve* | |
Station	Deflection Angle	Chord Length (ft)
PC 562 + 38.29	0	0
563	3°32'8''	61.66'
564	9°15'55''	99.84'
565	14°59'42''	99.84'
PCC 565 + 35.00	17°00'00''	35.00'

| | *350 ft Radius Curve* | |
Station	Deflection Angle	Chord Length (ft)
PCC 565 + 35.00	0	0
566	5°19'16''	64.9'
PT 566 + 93.82	13°00'00''	93.5'

Reverse Curves.   Reverse curves usually consist of two simple curves with equal radii turning in opposite directions with a common tangent. They are generally used to change the alignment of a highway. Figure 16.25 shows a reverse curve with parallel tangents. If d and D are known, it is necessary to determine Δ_1, Δ_2, and R to set out the curve:

$$\Delta = \Delta_1 = \Delta_2$$

$$\text{angle OWX} = \frac{\Delta_1}{2} = \frac{\Delta_2}{2}$$

$$\text{angle OYZ} = \frac{\Delta_1}{2} = \frac{\Delta_2}{2}$$

Therefore, line *WOY* is a straight line, and hence

$$\tan \frac{\Delta}{2} = \frac{d}{D}$$

$$d = R - R \cos \Delta_1 + R - R \cos \Delta_2$$

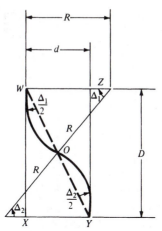

R = radius of simple curves
Δ_1, Δ_2 = deflection angles of simple cu
d = distance between parallel tangents
D = distance between tangent points

Figure 16.25 Geometry of a Reverse Curve with Parallel Tangents

$$= 2R(1 - \cos \Delta)$$

$$R = \frac{d}{2(1 - \cos \Delta)}$$

If d and R are known,

$$\cos \Delta = 1 - \frac{d}{2R}$$

$$D = d \cot \frac{\Delta}{2}$$

Reverse curves are seldom recommended because sudden changes to the alignment may result in drivers finding it difficult to keep in their lanes. When it is necessary to reverse the alignment, a preferable design consists of two simple horizontal curves, separated by a sufficient length of tangent between them, to achieve full superelevation. Alternatively, the simple curves may be separated by an equivalent length of spiral, which is described in the next section.

Transition (Spiral) Curves.   Transition curves are placed between tangents and circular curves or between two adjacent circular curves having substantially different radii. The use of transition curves provides a vehicle path that gradually increases or decreases the radial force as the vehicle enters or leaves a circular curve. The degree of a transition curve placed between a tangent and a circular curve varies from zero at the tangent end to the degree of the circular curve at the curve end. When placed between two circular curves, the degree varies from that of the first circular curve to that of the second circular curve.

The expression given in Eq. 16.41 is used by some highway agencies to compute the minimum length of a transition curve.

$$L = \frac{3.15u^3}{RC}$$

(16.41)

where

L = minimum length of curve (ft)

u = speed (mi/h)

R = radius of curve (ft)

C = rate of increase of radial acceleration (ft/sec^2/sec)

C is an empirical factor that indicates the level of comfort and safety involved. Values used for C in highway engineering vary from 1 to 3.

The results of a recent study, however, indicate that under operational conditions, the most desirable length of a spiral curve is approximately the length of the natural spiral path used by drivers as they traverse the curve. Table 16.12 gives AASHTO recommended lengths of spiral curves based on this consideration.

The computation for setting out the transition curve is beyond the scope of this book. In fact, many highway agencies do not use transition curves, since drivers will usually guide their vehicles into circular curves gradually. A practical alternative for determining the minimum length of a spiral is to use the length required for superelevation runoff (see following section).

Superelevation Runoff

In Chapter 3 we showed that highway pavements on circular curves should be superelevated to counteract the effect of centrifugal force. The length of highway section required to achieve a full superelevated section from a section with adverse crown removed, or vice versa, is known as the *superelevation runoff*. This length depends on the

Table 16.12 Desirable Lengths of Spiral Curves

Design Speed (mi/h)	Spiral Length (ft)
15	44
20	59
25	74
30	88
35	103
40	117
45	132
50	147
55	161
60	176
65	191
70	205
75	220
80	235

SOURCE: Adapted from *A Policy on Geometric Design of Highways and Streets,* American Association of State Highway and Transportation Officials, Washington, D.C., 2001. Used by permission.

design speed, the rate of superelevation, and the pavement width. AASHTO recommends that when spiral curves are used in transition design, the *superelevation runoff* should be achieved over the length of the spiral curve. Based on this, it is recommended that the length of the spiral curve should be the length of the *superelevation runoff*. Table 16.13 shows recommended lengths for the *superelevation runoff*.

Attainment of Superelevation

It is essential that the change from a crowned cross section to a superelevated one be achieved without causing any discomfort to motorists or creating unsafe conditions. One of four methods can be used to achieve this change on undivided highways:

1. A crowned pavement is rotated about the profile of the center line.
2. A crowned pavement is rotated about the profile of the inside edge.
3. A crowned pavement is rotated about the profile of the outside edge.
4. A straight cross-slope pavement is rotated about the profile of the outside edge.

   Figure 16.26(a) is a schematic of method 1. This is the most commonly used method since the distortion obtained is less than those obtained with the other methods. The procedure used is first to raise the outside edge of the pavement relative to the centerline, until the outer half of the cross section is horizontal (point B). The outer edge is then raised by an additional amount to obtain a straight cross section. Note that the inside edge is still at its original elevation, as indicated at point C. The whole cross section is then rotated as a unit about the centerline profile until the full superelevation is achieved at point E. Figure 16.26(b) illustrates method 2, where the centerline profile is raised with

Table 16.13   Tangent Runout Length for Spiral Curve Transition Design

Design Speed (mi/h)	Tangent Runout Length (ft)				
	Superelevation Rate (%)				
	2	4	6	8	10
15	44	—	—	—	—
20	59	30	—	—	—
25	74	37	25	—	—
30	88	44	29	—	—
35	103	52	34	26	—
40	117	59	39	29	—
45	132	66	44	33	—
50	147	74	49	37	—
55	161	81	54	40	—
60	176	88	59	44	—
65	191	96	64	48	38
70	205	103	68	51	41
75	220	110	73	55	44
80	235	118	78	59	47

SOURCE: Adapted from *A Policy on Geometric Design of Highways and Streets,* American Association of State Highway and Transportation Officials, Washington, D.C., 2001. Used by permission.

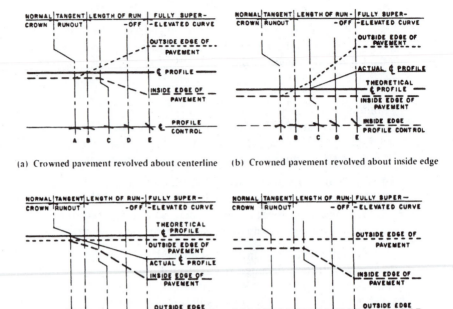

(a) Crowned pavement revolved about centerline

(b) Crowned pavement revolved about inside edge

(c) Crowned pavement revolved about outside edge

(d) Straight cross slope pavement revolved about outside edge

(e) Angular breaks appropriately rounded (dotted lines)

Figure 16.26 Diagrammatic Profiles Showing Methods of Attaining Superelevation for a Curve to the Right

SOURCE: *A Policy on Geometric Design of Highways and Streets,* American Association of State Highway and Transportation Officials, Washington, D.C., 2001. Used with permission.

respect to the inside pavement edge to obtain half the required change, while the remaining half is achieved by raising the outside pavement edge with respect to the profile of the centerline. Method 3, demonstrated by Figure 16.26(c), is similar to method 2. The only difference is that the change is effected below the outside edge profile. Figure 16.26(d) illustrates method 4, which is used for sections of straight cross slopes.

Note that straight lines have been conveniently used to illustrate the different methods, but in practice, the angular breaks are appropriately rounded by using short vertical curves, as shown in Figure 16.26(e).

Superelevation is achieved on divided highways by using one of three methods. Method 1 involves superelevating the whole cross section, including the median, as a plane section. The rotation in most cases is done about the centerline of the median. This method is used only for highways with narrow medians and moderate super-elevation rates, since large differences in elevation can occur between the extreme pavement edges if the median is wide. Method 2 involves rotating each pavement separately around the median edges, while keeping the median in a horizontal plane. This method is used mainly for pavements with median widths of 30 ft or less, although it can be used for any median, because by keeping the median in the horizontal plane, the difference in elevation between the extreme pavement edges does not exceed the pavement superelevation. In method 3, the two pavements are treated separately, resulting in variable elevation differences between the median edges. This method generally is used on pavements with median widths of 40 ft or greater. The large difference in elevation between the extreme pavement edges is avoided by providing a compensatory slope across the median.

SPECIAL FACILITIES FOR HEAVY VEHICLES ON STEEP GRADES

Recent statistics indicate a continual increase in the annual number of vehicle-miles of large trucks on the nation's highways over the past few years. We noted in both Chapter 3 and Chapter 9 that large trucks have different operating characteristics than those of passenger cars, and that this difference increases as the grade of a section of highway increases. These factors make it necessary to consider the provision of special facilities on sections of highways with steep grades carrying high volumes of heavy vehicles. The two facilities normally considered are climbing lanes and emergency escape ramps.

Climbing Lanes

A climbing lane is an extra lane in the upgrade direction for use by heavy vehicles whose speeds are significantly reduced by the grade. A climbing lane eliminates the need for drivers of other vehicles using the normal upgrade lane to reduce their speed, which would be necessary if they encounter a slow-moving heavy vehicle just ahead on the same lane. In the past, the provision of climbing lanes was not common because of the added cost of construction. The increasing rate of crashes directly associated with the reduction in speed of heavy vehicles on steep sections of two-lane highways and the significant reduction of the capacity of these sections when heavy vehicles are present have made it necessary to seriously consider the provision of climbing lanes on these highways. The basic condition that suggests the use of a climbing lane is when a grade is longer than its critical length. The critical length is that length that will cause a speed reduction of the heavy vehicle by 10 mi/h or more. This is important as the amount by which a truck's speed reduces on a grade depends on the length of the grade. For example, Figure 9.6 shows that the speed of a truck entering a grade of 5 percent at 55 mi/h will be reduced to about 43 mi/h for a grade length of 1000 ft and to about 27 mi/h for a grade length of 6000 ft. The length of the climbing lane will depend on the physical characteristics of the grade, but a general guideline is that the climbing lane should be long enough to facilitate the heavy vehicle's rejoining the main traffic stream without causing a hazardous condition. The climbing lane is provided only if, in addition to the critical length requirement, the upgrade traffic flow rate is greater than 200 veh/h and the upgrade truck flow is higher than 20 veh/h.

Climbing lanes are frequently not used on multilane highways, since relatively faster moving vehicles can pass the slower moving vehicles by using one of the other lanes. In addition, it is usually difficult to justify the provision of climbing lanes based on capacity since the multilane highways usually have adequate capacity to carry the traffic demand, including the normal percentage of slow-moving vehicles, without becoming too congested.

Emergency Escape Ramps

An emergency escape ramp is one provided on the downgrade of a highway for the use of a driver who has lost control of the vehicle because of brake failure. The objective is to provide a lane that diverges away from the main traffic stream while the uncontrolled vehicle's speed is gradually reduced and the vehicle is eventually brought to rest. The four basic types of design commonly used are shown in Figure 16.27. The sandpile shown in

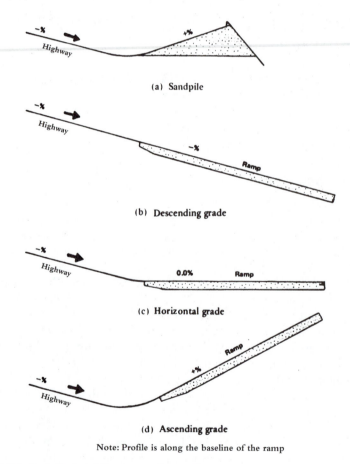

(a) Sandpile

(b) Descending grade

(c) Horizontal grade

(d) Ascending grade

Note: Profile is along the baseline of the ramp

Figure 16.27   Basic Types of Emergency Escape Ramps

SOURCE: *A Policy on Geometric Design of Highways and Streets,* American Association of State Highway and Transportation Officials, Washington, D.C., 2001. Used with permission.

Figure 16.27(a) has a pile of loose dry sand at the end of the escape ramp on which the vehicle is brought to rest. The sandpile provides an increased rolling resistance, and it is placed with an upgrade to use the influence of gravity in bringing the vehicle to a stop. Sandpiles are usually not longer than 400 ft. The descending and horizontal grades shown respectively in Figures 16.27(b) and (c) do not use this influence of gravity in stopping the vehicle but instead rely mainly on the increased rolling resistance provided by the arresting bed, which is made up of loose aggregates. These escape ramps can therefore be very long. The ascending-grade ramp shown in Figure 16.27(d) combines the effect of gravity and the increased rolling resistance by providing both the upgrade and the arresting bed. Thus, these ramps are relatively shorter than the descending or horizontal grade ramps.

BICYCLE FACILITIES

Use of the bicycle as a viable mode of transportation is now recognized, particularly in urban areas. The bicycle has therefore become an important factor that is considered in the design of urban highways, with the result that a wide variety of bicycle-related projects have been implemented.

Although most of the existing highway and street systems can be used to carry bicycles, it is often necessary to carry out improvements to these facilities to guarantee the safety of bicycle riders. Guides for the development of new bicycle facilities have been published by AASHTO.

Following is a brief description of the design for bicycle lanes and bicycle paths.

Bicycle Lanes

A bicycle (bike) lane is that part of the street or highway specifically reserved for the exclusive or preferential use of bicycle riders. Bicycle lanes can be delineated by striping, signing, or pavement markings. These lanes should always be one-way with traffic. Figure 16.28 shows typical bicycle lane cross sections. The minimum width under ideal conditions is 4 ft but should be increased when conditions demand it. For example, when bike lanes are located on a curbed urban street with parking allowed at both sides, as shown in Figure 16.28(a), a minimum width of 5 ft is required. Note that when parking is provided, the bicycle lane should always be located between the parking lane and the traveled lanes for automobiles, as shown in Figure 16.28(a). In cases where parking is not allowed, as in Figure 16.28(b), a minimum bike lane of 5 ft should be provided, because cyclists tend to ride away from the curb to avoid hitting it. In cases where the highway has no curb or gutter, the bicycle lane should be located between the shoulders and the travel lanes of the motor vehicles. A minimum bicycle lane of 4 ft could be provided if additional maneuvering width is provided by the shoulders.

Special consideration should be given to pavement markings at intersections, since through movement of bicycles will tend to conflict with right-turning movement of motor vehicles, and left-turning movement of bicycles will tend to conflict with through movement of motor vehicles. Pavements should therefore be striped to encourage both cyclists and motorists to commence their turning maneuvers in a merging configuration some distance away from the intersection. Figure 16.29 shows alternative ways for achieving this, as suggested by AASHTO.

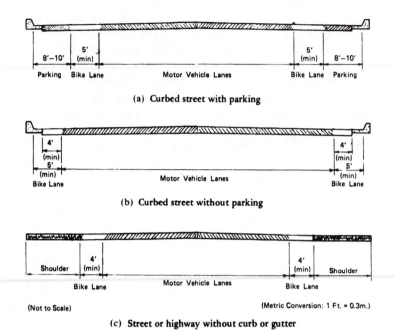

(a) Curbed street with parking

(b) Curbed street without parking

(Not to Scale)   (Metric Conversion: 1 Ft. = 0.3m.)

(c) Street or highway without curb or gutter

Figure 16.28 Typical Bicycle Lane Cross Sections

SOURCE: *Guide for Development of Bicycle Facilities,* American Association of State Highway and Transportation Officials, Washington, D.C., 1999. Used with permission.

Bicycle Paths

Bicycle (bike) paths are physically separated from automobile traffic and are used exclusively by cyclists. The design criteria for bicycle paths are somewhat similar to those for highways, but some of these criteria are governed by bicycle operating characteristics, which are significantly different from those of automobiles. Designers should be aware of those bicycle operating characteristics that are similar to and different from automobile characteristics. Important design considerations for a safe bicycle path include the path width, the design speed, the horizontal alignment, and the vertical alignment.

Width of Bicycle Paths

A typical layout of a bicycle path on a separate right of way is shown in Figure 16.30. The minimum width recommended by AASHTO is 10 ft for two-way paths, and 5 ft for one-way paths. Under certain conditions, the width for the two-way path may be reduced to 8 ft. These conditions include

- Low bicycle volumes at all times
- Vertical and horizontal alignments properly designed to provide frequent safe passing conditions
- Only occasional use of path by pedestrians
- No loaded maintenance vehicles on path

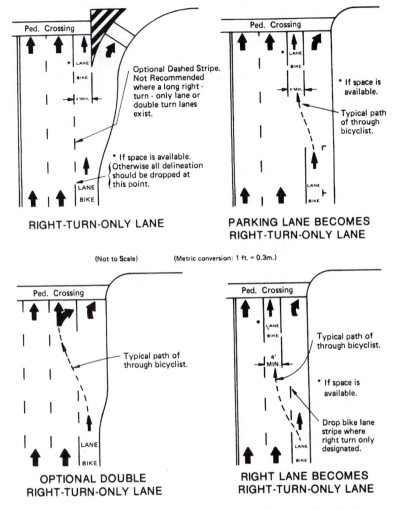

Figure 16.29 Channelization of Bicycle Lanes Approaching Motor Vehicle Right-Turn-Only Lanes

SOURCE: *Guide for Development of Bicycle Facilities,* American Association of State Highway and Transportation Officials, Washington, D.C., 1999. Used with permission.

Just as shoulders are provided on highways, it is necessary to provide a uniform graded area at least 2 ft wide at both sides of the path.

When bicycle paths are located adjacent to a highway, a wide area must be provided between the highway and the bicycle path to facilitate the independent use of the bicycle path. In cases where this area is less than 5 ft wide, physical barriers, such as dense shrubs or a fence, should be used to separate the two facilities.

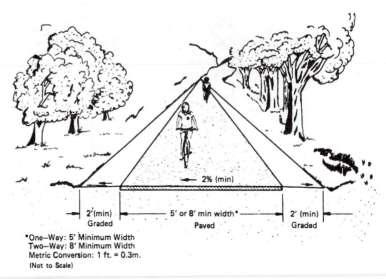

Figure 16.30 Bicycle Path on Separated Right of Way

SOURCE: *Guide for Development of Bicycle Facilities,* American Association of State Highway and Transportation Officials, Washington, D.C., 1999. Used with permission.

Design Speed

AASHTO recommends that a minimum design speed of 20 mi/h be used for paved paths, but this should be increased to 30 mi/h for grades greater than 4 percent or where strong prevailing tail winds exist. The minimum design speed can also be reduced to 15 mi/h on unpaved bicycle paths, since riding speeds tend to be lower on these paths.

Horizontal Alignment

Similar criteria exist for bicycle paths as for highways, in that the minimum radius of horizontal curves depends on the velocity and the superelevation and is given as

$$R = \frac{u^2}{15(e + f_s)}$$

where

    R = minimum radius of the curve (ft)
    u = design speed (mi/h)
    e = rate of superelevation
    f_s = coefficient of side friction

The factors to be considered here are the superelevation rate and the coefficient of side friction. Superelevation rates for bicycle paths vary from a minimum of 2 percent, which will facilitate surface drainage, to a maximum of 5 percent. Higher superelevation rates may result in some cyclists finding it difficult to maneuver around the curve. Coefficients of side friction used for design vary from 0.3 to 0.22. It is suggested that for unpaved bicycle paths, values varying from 0.15 to 0.11 should be used.

Vertical Alignment

The design of the vertical alignment for a bicycle path is similar to that for a highway, in that the selection of suitable grades and the design of appropriate vertical curves are required.

Grades should not exceed 5 percent and should be lower when possible, particularly on long inclines. This is necessary to enable cyclists to easily ascend the grade and to prevent excessive speed while descending the grade.

The vertical curves are designed to provide the minimum length as dictated by the stopping sight distance requirement. Figure 16.31 shows minimum stopping sight distances for different design speeds, and Figure 16.32 gives minimum lengths for vertical curves. The minimum stopping sight distances given in Figure 16.31 are based on a reaction time of 2.5 sec, which is the value normally used for bicyclists.

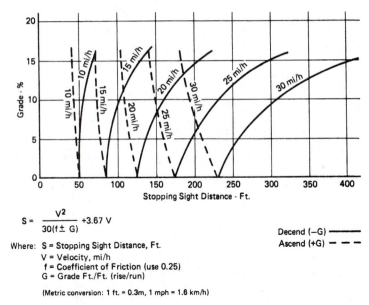

$$S = \frac{V^2}{30(f \pm G)} + 3.67V$$

Where: S = Stopping Sight Distance, Ft.
V = Velocity, mi/h
f = Coefficient of Friction (use 0.25)
G = Grade Ft./Ft. (rise/run)

Decend (−G) ——————
Ascend (+G) — — —

(Metric conversion: 1 ft. = 0.3m, 1 mph = 1.6 km/h)

Figure 16.31 Stopping Sight Distance for Bicycles

SOURCE: *Guide for Development of Bicycle Facilities,* American Association of State Highway and Transportation Officials, Washington, D.C., 1999. Used with permission.

PARKING FACILITIES

The geometric design of parking facilities mainly involves the dimensioning and arranging of parking bays to provide safe and easy access without seriously restricting the flow of traffic on the adjacent traveling lanes. Guidelines for the design of on-street and off-street parking facilities are presented.

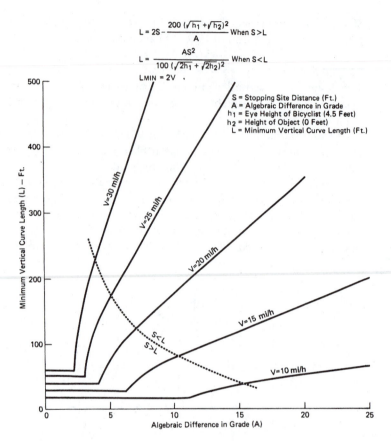

Figure 16.32 Minimum Lengths of Vertical Curves for Bicycles

SOURCE: *Guide for Development of Bicycle Facilities,* American Association of State Highway and Transportation Officials, Washington, D.C., 1999. Used with permission.

Design of On-Street Parking Facilities

On-street parking facilities may be designed with the parking bays parallel to the curb or inclined to the curb. Figure 16.33 shows the angles of inclination commonly used for curb parking and the associated dimensions for automobile parking. It can be seen that the number of parking bays that can be fitted along a given length of curb increases as the angle of inclination increases up to 90°. The higher the inclination angle, however, the greater the encroachment of the parking bays on the traveling pavement of the highway. Parking bays inclined at angles to the curb severely interfere with the movement of traffic on the highway, with the result that crash rates tend to be higher on sections of roads with angle parking than on those with parallel parking. It should be noted that the dimensions given in Figure 16.33 are for automobiles. When parking bays are to be provided for trucks and other types of vehicles, the dimensions should be determined on the basis of the dimensions of the specific vehicle being considered.

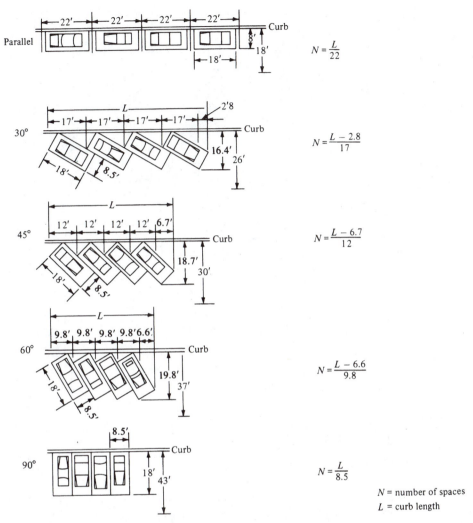

Figure 16.33   Street Space Used for Various Parking Configurations

SOURCE: Redrawn from R. H. Weant, H. S. Levinson, G. Mogren, *Parking*, Eno Foundation for Highway Traffic Control, Inc., Saugatuck, Conn., 1990.

Design of Off-Street Parking Facilities—Surface Car Parks

The primary aim in designing off-street parking facilities is to obtain the maximum possible number of spaces. Figures 16.34 and 16.35 show different types of layouts that can be used in a surface lot. The most important consideration is that the layout should be such that parking a vehicle involves only one distinct maneuver, without the necessity to reverse. The different layouts also indicate that the parking space is most efficiently used when the parking bays are inclined at 90° to the direction of traffic flow. The use of

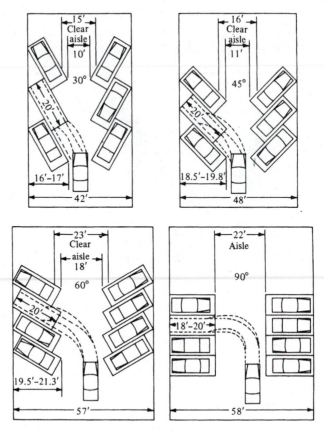

Figure 16.34 Parking Stall Layout

the herringbone layout (Figure 16.35) facilities traffic circulation since it provides for one-way flow of traffic on each aisle.

Design of Off-Street Parking Facilities—Garages

Parking garages basically consist of several platforms, supported by columns, which are placed in such a way to facilitate an efficient arrangement of parking bays and aisles. Access ramps connect each level with the one above. The gradient of these ramps is usually not greater than 1 in 10 on straight ramps and 1 in 12 on the centerline of curved ramps. It is recommended that the radius of curved ramps measured to the end of the outer curve should not be less than 70 ft, and that the maximum superelevation should be 0.15 ft/ft. It is also recommended that the width should not be less than 16 ft for curved ramps and 9 ft for straight ramps. These ramps can be one-way or two-way, although one-way ramps are preferable. When two-way ramps are used, the lanes must be clearly marked and perhaps physically divided, particularly at curves and turning points. This helps to avoid head-on collisions that may occur as drivers cut corners or swing wide at bends.

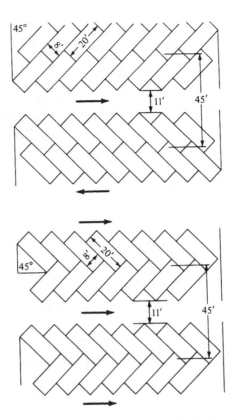

Figure 16.35 Herringbone Layout of Parking Stalls in an On-Surface Lot

In some cases, the platforms are connected by elevators into which cars are driven or mechanically placed. These elevators then lift the car to the appropriate level for parking. The vehicle is then either removed by an attendant who eventually parks it or it is mechanically removed and then parked by that attendant.

The size of the receiving area is an important factor in garage design and depends on whether the cars are owner-parked (self-parking) or attendant-parked. When cars are self-parked, very little or no reservoir space is required, since drivers need pause only for a short time to pick up a ticket from a machine or to be given one by an attendant. When cars are parked by attendants, the driver must stop and leave the vehicle. The attendant then enters and drives the vehicle to the parking bay. Thus, a reservoir space must be provided that will accommodate temporary storage for entering vehicles. The size of this reservoir space depends on the ratio of the rate of storage of the vehicles to the rate of arrival of the vehicles. The rate of storage must take into consideration the time required to transfer the vehicle from its driver to the attendant. The number of temporary storage bays can be determined by applying the queuing theory presented in Chapter 5. Figure 16.36 shows the size of reservoir space required for different ratios of storage rate to arrival rate, with overloading occurring only 1 percent of the time.

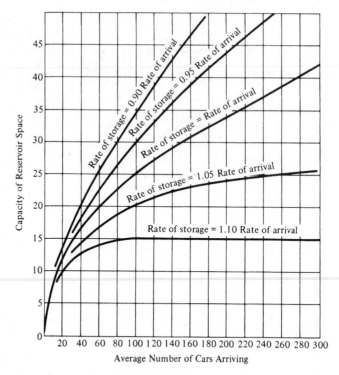

Figure 16.36   Reservoir Space Required If Facility Is Overloaded Less Than One Percent of Time

SOURCE: Redrawn from E. R. Ricker, *Traffic Design of Parking Garages,* Eno Foundation for Highway Traffic Control, Inc., Saugatuck, Conn., 1957.

COMPUTER USE IN GEOMETRIC DESIGN

Computer programs are now available to carry out all of the designs discussed in this chapter. Most highway agencies have developed programs suitable for their individual hardware systems, which may not easily be used by another agency with a noncompatible system. AASHTO regularly publishes an index of available software packages. The province of Manitoba, Canada, for example, has developed a software package known as the Roadway Design System. This package will undertake the complete design process from preliminary highway location to the detailed design of vertical and geometric alignments. A comprehensive system used in all phases of highway design is the COGO System, which is used for roadway surveying and geometric design. Several programs have also been developed for use with microcomputers, offering even small consulting firms the opportunity to use computer-aided methods in their design work. An example is Survey 4, developed by the Simplicity Company.

SUMMARY

This chapter presents the fundamental principles of highway geometric design, together with the necessary formulas required for the design of the vertical and horizontal alignments of the highway. The basic characteristics of the driver, pedestrian, vehicle, and road are used to determine the geometric characteristics of any highway or bikeway. The fundamental characteristic on which several design standards are based is the design speed, which depends on the type of facility being considered.

The material presented in this chapter provides the mathematical tools required by an engineer engaged in the geometric design of highways. Since any motor vehicle on a highway has to be parked at one time or another, the design of parking bays and garages is therefore included.

A brief introduction to available computer programs for geometric design of highways is also presented. The intention is to let the reader know of the opportunities available in computer-aided design in highway engineering, rather than the presentations of different computer programs, since many of those currently in use may soon become outdated as new programs become available.

PROBLEMS

16-1   A rural collector highway located in a mountainous terrain is to designed to carry a design volume of 800 veh/day, 98 percent of which are passenger cars. Determine the following: (a) a suitable design speed, (b) lane and usable shoulder widths, (c) maximum desirable grade, (d) minimum radius of horizontal curvature, (e) minimum length of crest vertical curves, and (f) minimum length of sag vertical curves, for the maximum grade.

16-2   Repeat Problem 16-1 for an urban freeway in rolling terrain.

16-3   With the aid of an appropriate diagram, describe a suitable method you will use to achieve superelevation on a six-lane rural freeway (three lanes in each direction) with a 50-ft-wide median.

16-4   A rural collector is to be constructed in an area of rolling terrain where the ADT is 650 veh/day. For this situation determine the following design values.

  (a)   Minimum design speed
  (b)   Recommended lane width
  (c)   Preferable shoulder width
  (d)   Maximum grade
  (e)   Minimum length of curve based on design speed
        Crest
        Sag
  (f)   Maximum superelevation
  (g)   Maximum degree of curve ($f_s = 0.15$)
        (Degrees-minutes-seconds)

16-5   A +2 percent grade on an arterial highway intersects with a −1 percent grade at station (535 + 24.25) at an elevation of 300 ft. If the design speed of the highway is 65 mi/h, determine the stations and elevations of the BVC and EVC, the elevation of each 100-ft station, and the high point. Use L=KA.

16-6   Determine the minimum length of a crest vertical curve if the grades are +4% and −2%. Design speed is 70 mi/h. State assumptions used in solving this problem.

16-7   Determine the minimum length of a sag vertical curve if the grades are −4% and +2%. Design speed is 70 mi/h. State assumptions used. Consider the following criteria:
(a)   Stopping sight distance
(b)   Comfort
(c)   General appearance

16-8   A +g_1 grade is connected by a vertical curve to a −g_2 grade ($g_2 < g_1$). If the rate of change of gradient of the curve is constant, show that the height y of any point of the curve above the BVC is given as

$$100y = g_1x - \frac{(g_1 - g_2)x^2}{2L}$$

16-9   A sag vertical curve connects a −1.5 percent grade with a +2.5 percent grade on a rural arterial highway. If the criterion selected for design is the minimum stopping sight distance, and the design speed of the highway is 70 mi/h, compute the elevation of the curve at 100-ft stations if the grades intersect at station (475 + 00) at an elevation of 300 ft. Also determine the elevation and station of the low point.

16-10   A horizontal curve is designed for a two-lane road in mountainous terrain. The following data are known:

Deflection angle: 40 degrees
Tangent length: 436.76 feet
Station of PI: 2700 + 10.65
$f_s = 0.12$
$e = 0.08$

Determine the following:
(a)   Design speed
(b)   Station of the PC
(c)   Station of the PT
(d)   Deflection angle and chord length to the first even 100 ft. station.

16-11   A building is located 19 feet from the centerline of the inside lane of a curved section of highway whose radius is 400 feet. The road is level and $e = 0.10$. Determine the posted speed limit (to the nearest 5 mi/h) considering the following conditions: (1) stopping sight distance and (2) curve radii.

16-12   A circular curve connects two tangents that deflect at an angle of 48°. The point of intersection of the tangents is located at station (948 + 67.32). If the design speed of the highway is 60 mi/h, determine both the point of the tangent and the deflection angles to whole stations for laying out the curve. (Select appropriate values for e and f.)

16-13   To determine the radius of an existing curve, two chords xy and yz of 135 ft each were marked out on the curve. The perpendicular distance between y and the chord xz was found to be 15 ft. Determine the radius of the curve and the angle you will set out from xz to get a line to the center of the curve.

16-14   A compound circular curve having radii of 600 ft and 450 ft is to be designed to connect two tangents deflecting by 75°. If the deflection angle of the first curve is 45° and the PCC is located at station (675 + 35.25), determine the deflection angles and the corresponding chord lengths for setting out the curve.

16-15   An arterial road is to be connected to a frontage road by a reverse curve with parallel tangents. The distance between the centerline of the two roads is 60 ft. If the PC of the curve is located at station (38 + 25.31) and the deflection angle is 25°, determine the station of PT.

16-16   A parking lot is to be designed for a shopping center. If the area available is 400 ft × 500 ft, design a suitable layout for achieving each of the following objectives—that is, one layout for each objective:

(a)   Provide the maximum number of spaces.

(b)   Provide the maximum number of spaces while facilitating traffic circulation by providing one-way flow on each aisle.

16-17   Describe the factors that must be taken into account in the design of bicycle paths.

16-18   Determine the minimum radius of horizontal curves on a bicycle path if the average speed of bicycles is 20 mi/h and the minimum superelevation is 2% to facilitate drainage. If the total change in grade is 10%, determine the minimum length of the vertical curve.

16-19   Describe the four methods by which superelevation can be attained on a curved section of roadway.

REFERENCES

A Policy on Geometric Design of Highways and Streets, American Association of State Highway and Transportation Officials, Washington, D.C., 2001.

Computer System Index, American Association of State Highway and Transportation Officials, Washington, D.C.

Garber, N. J., Patel, S., and Kalaputapu, R., *The Effect of Trailer Width and Length on Large-Truck Accidents,* Virginia Transportation Research Council Report No. VTRC 93-R11, 1993.

Guide for Development of Bicycle Facilities, American Association of State Highway and Transportation Officials, Washington, D.C., 1999.

Roadside Design Guide, American Association of State Highway and Transportation Officials, Washington, D.C., 1996.

ADDITIONAL READINGS

GPS for GIS Data Collection, U.S. Army Topographic Engineering Center, Ft. Belvoir, Va., 1997.

Kavanagh, Barry F., *Surveying with Construction Applications,* 3rd ed., Prentice-Hall, Englewood Cliffs, N.J., 1997.

Van Sickle, Jan, *GPS for Land Surveyors,* Ann Arbor Press, Chelsea, Mich., 1996.

CHAPTER 17

Highway Drainage

Provision of adequate drainage is an important factor in the location and geometric design of highways. Drainage facilities on any highway or street should adequately provide for the flow of water away from the surface of the pavement to properly designed channels. Inadequate drainage will eventually result in serious damage to the highway structure. In addition, traffic may be slowed by accumulated water on the pavement, and accidents may occur as a result of hydroplaning and loss of visibility from splash and spray. The importance of adequate drainage is recognized in the amount of highway construction dollars allocated to drainage facilities. About 25 percent of highway construction dollars are spent for erosion control and drainage structures, such as culverts, bridges, channels, and ditches.

The highway engineer is concerned primarily with two sources of water. The first source, surface water, is that which occurs as rain or snow. Some of this is absorbed into the soil, and the remainder remains on the surface of the ground and should be removed from the highway pavement. Drainage for this source of water is referred to as *surface drainage*. The second source, ground water, is that which flows in underground streams. This may become important in highway cuts or at locations where a high water table exists near the pavement structure. Drainage for this source of water is referred to as *subsurface drainage*.

In this chapter, we present the fundamental design principles for surface and subsurface drainage facilities. The principles of hydrology necessary for understanding the design concepts are also included, together with a brief discussion on erosion prevention.

SURFACE DRAINAGE

Surface drainage encompasses all means by which surface water is removed from the pavement and right of way of the highway or street. A properly designed highway surface drainage system should effectively intercept all surface and watershed runoff and direct this water into adequately designed channels and gutters for eventual discharge into the natural waterways. Water seeping through cracks in the highway riding surface and shoulder areas into underlying layers of the pavement may result in serious damage to the highway pavement. The major source of water for this type of intrusion is surface runoff. An adequately designed surface drainage system will therefore minimize this type of damage. The surface drainage system for rural highways should include adequate transverse and longitudinal slopes on both the pavement and shoulder to ensure positive runoff, and longitudinal channels (ditches), culverts, and bridges to provide for the discharge of the surface water to the natural waterways. Storm drains and inlets are also provided on the median of divided highways in rural areas. In urban areas, the surface drainage system also includes adequate longitudinal and transverse slopes, but the longitudinal drains are usually underground pipe drains, designed to carry both surface runoff and ground water. Curbs and gutters may also be used in urban and rural areas to control street runoff, although they are more frequently used in urban areas.

Transverse Slopes

The main objective for providing slopes in the transverse direction is to facilitate the removal of surface water from the pavement surface in the shortest possible time. This is achieved by crowning the surface at the center of the pavement, thereby providing cross slopes on either side of the centerline or providing a slope in one direction across the pavement width. The different ways in which this is achieved were discussed in Chapter 16. Shoulders, however, are usually sloped to drain away from the pavement, except on highways with raised narrow medians. The need for high cross slopes to facilitate drainage is somewhat in conflict with the need for relatively flat cross slopes for driver comfort. Selection of a suitable cross slope is therefore usually a compromise between the two requirements. It has been determined that cross slopes of 2 percent or less do not significantly affect driver comfort, particularly with respect to the driver's effort in steering. Recommended values of cross slopes for different shoulders and types of highway pavement were presented in Chapter 16.

Longitudinal Slopes

A minimum gradient in the longitudinal direction of the highway is required to obtain adequate slope in the longitudinal channels, particularly at cut sections. Slopes in longitudinal channels should generally not be less than 0.2 percent for highways in very flat terrain. Although zero percent grades may be used on uncurbed pavements with adequate cross slopes, a minimum of 0.5 percent is recommended for curbed pavements. This may be reduced to 0.3 percent on suitably crowned high-type pavements constructed on firm ground.

Longitudinal Channels

Longitudinal channels (ditches) are constructed along the sides of the highway to collect the surface water that runs off from the pavement surface, subsurface drains, and other areas of the highway right of way. When the highway pavement is located at a lower level than the adjacent ground, such as in cuts, water is prevented from flowing onto the pavement by constructing a longitudinal drain (intercepting drain) at the top of the cut to intercept the water. The water collected by the longitudinal ditches is then transported to a drainage channel and next to a natural waterway or retention pond.

Curbs and Gutters

As pointed out in Chapter 16, curbs and gutters can be used to control drainage in addition to other functions, which include preventing the encroachment of vehicles on adjacent areas and delineating pavement edges. Some typical shapes for curbs and gutters were presented in Chapter 16. Curbs and gutters are used more frequently in urban areas, particularly in residential areas, where they are used in conjunction with storm sewer systems to control street runoff. When it is necessary to provide relatively long continuous sections of curbs in urban areas, the inlets to the storm sewers must be adequately designed for both size and spacing so that the impounding of large amounts of water on the pavement surface is prevented.

HIGHWAY DRAINAGE STRUCTURES

Drainage structures are constructed to carry traffic over natural waterways that flow below the right of way of the highway. These structures also provide for the flow of water below the highway, along the natural channel, without significant alteration or disturbance to its normal course. One of the main concerns of the highway engineer is to provide an adequate size structure, such that the waterway opening is sufficiently large to discharge the expected flow of water. Inadequately sized structures can result in water impounding, which may lead to failure of the adjacent sections of the highway due to embankments being submerged in water for long periods.

The two general categories of drainage structures are major and minor. Major structures are those with clear spans greater than 20 feet, whereas minor structures are those with clear spans of 20 feet or less. Major structures are usually large bridges, although multiple-span culverts may also be included in this class. Minor structures include small bridges and culverts.

Major Structures

It is beyond the scope of this book to discuss the different types of bridges. Emphasis is therefore placed on selecting the span and vertical clearance requirements for such structures. The bridge deck should be located above the high water mark. The clearance above the high water mark depends on whether the waterway is navigable. If the waterway is navigable, the clearance above the high water mark should allow the largest ship using the channel to pass underneath the bridge without colliding with the bridge deck. The clearance height, type, and spacing of piers also depend on the probability of ice jams and the extent to which floating logs and debris appear on the waterway during high water.

An examination of the banks on either side of the waterway will indicate the location of the high water mark, since this is usually associated with signs of erosion and debris deposits. Local residents, who have lived near and observed the waterway during flood stages over a number of years, can also give reliable information on the location of the high water mark. Stream gauges that have been installed in the waterway for many years can also provide data that can be used to locate the high water mark.

Minor Structures

Minor structures, consisting of short-span bridges and culverts, are the predominant type of drainage structures on highways. Although openings for these structures are not designed to be adequate for the worst flood conditions, they should be large enough to accommodate the flow conditions that might occur during the normal life expectancy of the structure. Provision should also be made for preventing clogging of the structure due to floating debris and large boulders rolling from the banks of steep channels.

Culverts are made of different materials and in different shapes. Materials used to construct culverts include concrete (reinforced and unreinforced), corrugated steel, and corrugated aluminum. Other materials may also be used to line the interior of the culvert to prevent corrosion and abrasion or to reduce hydraulic resistance. For example, asphaltic concrete may be used to line corrugated metal culverts. The different shapes normally used in culvert construction include circular, rectangular (box), elliptical, pipe arch, metal box, and arch. Figure 17.1(a) shows a corrugated metal arch culvert and Figure 17.1(b) shows a rectangular (box) culvert.

(a) Corrugated metal arch culvert (ARMCO)

Figure 17.1 Different Types of Culverts

SOURCE: J. M. Normann, R. J. Houghtalen, W. J. Johnston, *Hydraulic Design of Highway Culverts,* Report No. FHWA-IP-85-15, U.S. Department of Transportation, Office of Implementation, McLean, Va., September 1985.

(b) Rectangular (box) culvert

Figure 17.1 Different Types of Culverts (*continued*)

SOURCE: Photograph by Lewis Woodson, Virginia Transportation Research Council, Charlottesville, Va. Used with permission.

SEDIMENT AND EROSION CONTROL

Continuous flow of surface water over shoulders, side slopes, and unlined channels often results in soil eroding from the adjacent areas of the pavement. Erosion can lead to conditions that are detrimental to the pavement structure and other adjacent facilities. For example, soil erosion of shoulders and side slopes can result in failure of embankment and cut sections, and soil erosion of highway channels often results in the pollution of nearby lakes and streams. Prevention of erosion is an important factor when highway drainage is being considered, both during construction and when the highway is completed. The methods used to prevent erosion and control sediment are now briefly discussed.

Intercepting Drains

Provision of an intercepting drain at the top of a cut helps to prevent erosion of the side slopes of cut sections, since the water is intercepted and prevented from flowing freely down the side slopes. The water is collected and transported in the intercepting drain to paved spillways that are placed at strategic locations on the side of the cut. The water is then transported through these protected spillways to the longitudinal ditches alongside the highway.

Curbs and Gutters

Curbs and gutters can be used to protect unsurfaced shoulders on rural highways from eroding. They are placed along the edge of the pavement such that surface water is prevented from flowing over and eroding the unpaved shoulders. Curbs and gutters can also be used to protect embankment slopes from erosion when paved shoulders are used. In this case, the curbs and gutters are placed on the outside edge of the paved shoulders, and the surface water is then directed to paved spillways located at strategic positions and transported to the longitudinal drain at the bottom of the embankment.

Turf Cover

Using a firm turf cover on unpaved shoulders, ditches, embankments, and cut slopes is an efficient and economic method of preventing erosion when slopes are flatter than 3 to 1. The turf cover is commonly developed by sowing suitable grasses immediately after grading. The two main disadvantages of using turf cover on unpaved shoulders are that turf cover cannot resist continued traffic and it loses its firmness under conditions of heavy rains.

Slope and Channel Linings

When the highway is subjected to extensive erosion, more effective preventive action than any of those already described is necessary. For example, when cut and embankment side slopes are steep and are located in mountainous areas subjected to heavy rain or snow, additional measures should be taken to prevent erosion. A commonly used method is to line the slope surface with rip-rap or hand-placed rock.

Channel linings are also used to protect longitudinal channels from eroding. The lining is placed along the sides and in the bottom of the drain. Protective linings can be categorized into two general groups: flexible and rigid. Flexible linings include dense-graded bituminous mixtures and rock rip-rap, whereas rigid linings include Portland cement concrete and soil cement. Rigid linings are much more effective in preventing erosion under severe conditions, but they are more expensive and, because of their smoothness, tend to create high unacceptable velocities at the end of the linings. When the use of rigid lining results in high velocities, a suitable energy dissipator must be placed at the lower end of the channel to prevent excessive erosion. The energy dissipator is not required if the water discharges into a rocky stream or into a deep pool. Detailed design procedures for linings are presented later in the chapter under "Channel Design."

Erosion Control During Construction

During highway construction, special precautions are required to control erosion and sediment accumulation. Among the techniques used are: sediment basins, check dams, silt fence/filter barriers, brush barriers, diversion dikes, slope drains, and dewatering basins.

Sediment basins are required when runoff from a drainage area that is greater than 3 acres flows across a disturbed area. The basin allows sediment-laden runoff to pond where the sediment settles in the bottom of the basin. *Check dams* are used to slow the velocity of a concentrated flow of water and are made of local materials such as rock, logs, or straw bales. A *silt fence* is a fabric, often reinforced with wire mesh. *Brush barriers* are made of construction spoil material from the construction site, often combined with filter fabric. A *diversion dike* is an earthen berm that diverts water to a sediment basin. *Slope drains* are used to convey water down a slope, avoiding erosion before a permanent drainageway is constructed. *Dewatering basins* are detention areas to which sediment-laden water is pumped.

HYDROLOGIC CONSIDERATIONS

Hydrology is the science that deals with the characteristics and distribution of water in the atmosphere, on the earth's surface, and in the ground. The basic phenomenon in hydrology is the cycle that consists of precipitation occurring onto the ground in the form of water, snow, hail, and so forth and returning to the atmosphere in the form of vapor. It is customary in hydrology to refer to all forms of precipitation as rainfall, with precipitation usually measured in terms of the equivalent depth of water that is accumulated on the ground surface.

Highway engineers are primarily concerned with three properties of rainfall: the rate of fall, known as *intensity;* the length of time for a given intensity, known as *duration;* and the probable number of years that will elapse before a given combination of intensity and duration will be repeated, known as *frequency.* The U.S. Weather Bureau has a network of automatic rainfall instruments that collect data on intensity and duration over the entire country. These data are used to draw rainfall intensity curves from which rainfall intensity for a given return period and duration can be obtained. Figure 17.2 is an example of a set of rainfall intensity curves, and Figure 17.3 shows rainfall intensities for different parts of the United States for a storm of 1-hr duration and 1-yr frequency. The use of the rainfall intensity curves is illustrated in Figure 17.2. An intensity of 1.57 in./hr is obtained for a 10-year storm having a duration of 1.30 hours.

Note that any estimate of rainfall intensity, duration, or frequency made from these data is based on probability laws. For example, if a culvert is designed to carry a "100 year" flood, then the probability is 1 in 100 that the culvert will flow full in any one year. This does not mean that a precipitation of the designed intensity and duration will occur exactly once every 100 years. In fact, it is likely that precipitations of higher intensities could occur one or more times before a time lapse of 100 years, although the probability of this happening is low. This suggests that drainage facilities should be designed for very rare storms to reduce the chance of overflowing to a minimum. Designing for this condition, however, results in very large facilities that cause the cost of the drainage facility to be very high. The decision on what frequency should be selected for design purposes must therefore be based on a comparison of the capital cost for the drainage facility and the cost to the public in case the highway is severely damaged by storm runoff. Factors usually considered in making this decision include the importance of the highway, the volume of traffic on the highway and the population density of the area. Recommended storm

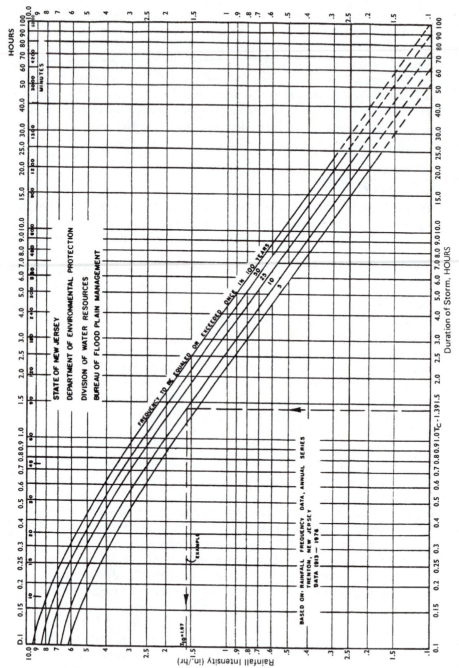

Figure 17.2   Rainfall Intensity Curves

SOURCE: G. S. Koslov, *Road Surface Drainage Design, Construction and Maintenance Guide for Pavements*, Report No. FHWA/NJ-82-004, State of New Jersey, Department of Transportation, Trenton, N.J., June 1981.

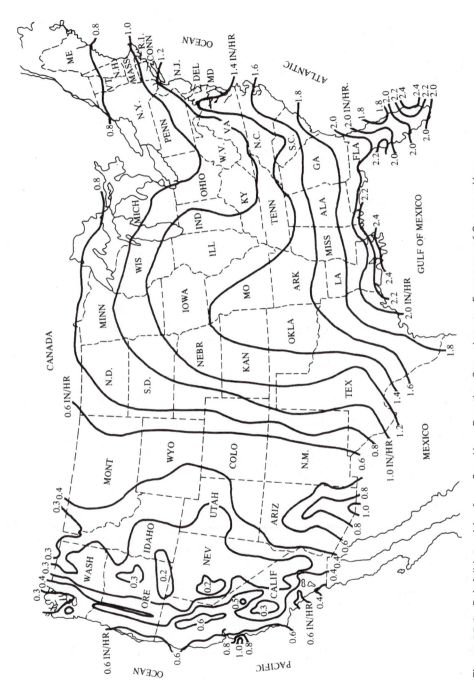

Figure 17.3   Rainfall Intensities for a One-Hour Duration Occurring at a Frequency of Once per Year

SOURCE: Redrawn from *Rainfall Frequency Atlas of the United States*, U.S. Department of Commerce, Washington, D.C., 1981.

frequencies, referred to as *return periods,* for various roadway classifications, are shown in Table 17.1.

Other hydrologic variables that the engineer uses to determine surface runoff rates are the drainage area, the runoff coefficient, and the time of concentration.

The *drainage area* is that area of land that contributes to the runoff at the point where the channel capacity is to be determined. This area is normally determined from a topographic map.

The *runoff coefficient, C,* is the ratio of runoff to rainfall for the drainage area. The runoff coefficient depends on the type of ground cover, the slope of the drainage area, storm duration, prior wetting, and the slope of the ground. Several suggestions have been made to adjust C for storm duration, prior wetting, and other factors. However, these adjustments are probably not necessary for small drainage areas. Average values assumed to be constant for any given storm are therefore used. Representative values for C for different runoff surfaces are given in Table 17.2.

In cases where the drainage area consists of different ground characteristics with different runoff coefficients, a representative value C_w is computed by determining the weighted coefficient using Eq. 17.1.

$$C_w = \frac{\sum\limits_{i=1}^{n} C_i A_i}{\sum\limits_{i=1}^{n} A_i} \tag{17.1}$$

where

C_w = weighted runoff coefficient for the whole drainage area
C_i = runoff coefficient for watershed i
A_i = area of watershed i (acres)
n = number of different watersheds in the drainage area

Table 17.1 Design Storm Selection Guidelines

Roadway Classification	Exceedence Probability	Return Period
Rural principal arterial system	2%	50-year
Rural minor arterial system	4%–2%	25–50-year
Rural collector system, major	4%	25-year
Rural collector system, minor	10%	10-year
Rural local road system	20%–10%	5–10-year
Urban principal arterial system	4%–2%	25–50-year
Urban minor arterial street system	4%	25-year
Urban collector street system	10%	10-year
Urban local street system	20%–10%	5–10-year

Note: Federal law requires interstate highways to be provided with protection from the 2% flood event, and facilities such as underpasses, depressed roadways, etc. where no overflow relief is available should be designed for the 2% event.
SOURCE: *Model Drainage Manual,* American Association of State Highway and Transportation Officials, Washington, D.C., 1991.

Table 17.2 Values of Runoff Coefficients, C

Type of Surface	Coefficient, C*
Rural Areas	
Concrete sheet asphalt pavement	0.8–0.9
Asphalt macadam pavement	0.6–0.8
Gravel roadways or shoulders	0.4–0.6
Bare earth	0.2–0.9
Steep grassed areas (2 : 1)	0.5–0.7
Turf meadows	0.1–0.4
Forested areas	0.1–0.3
Cultivated fields	0.2–0.4
Urban Areas	
Flat residential, with about 30 percent of area impervious	0.40
Flat residential, with about 60 percent of area impervious	0.55
Moderately steep residential, with about 50 percent of area impervious	0.65
Moderately steep built-up area, with about 70 percent of area impervious	0.80
Flat commercial, with about 90 percent of area impervious	0.80

*For flat slopes or permeable soil, use the lower values. For steep slopes or impermeable soil, use the higher values.
SOURCE: Adapted from George S. Koslov, *Road Surface Drainage Design, Construction and Maintenance Guide for Pavements,* Report No. FHWA/NJ-82-004, State of New Jersey, Department of Transportation, Trenton, N.J., June 1981.

The *time of concentration*, T_c, is the time required for the runoff to flow from the hydraulically most distant point of the watershed to the point of interest within the watershed. The time of concentration for a drainage area must be determined in order to select the average rainfall intensity for a selected frequency of occurrence. Time of concentration depends on several factors, including the size and shape of the drainage area, the type of surface, the slope of the drainage area, the rainfall intensity, and whether the flow is entirely over land or partly channelized. The time of concentration generally consists of one or more of three components of travel times, depending on the location of the drainage systems. These are the times for overland flow, gutter or storm sewer flow (mainly in urban areas), and channel flow.

The procedures and equations used for calculating travel time and time of concentration are as follows.

Water travels through a watershed as sheet flow, shallow concentrated flow, open channel flow, or as a combination of each separate flow. The type of flow that occurs depends upon the physical condition of the surroundings, which should be verified by field inspection. Travel time is the ratio of flow length to average flow velocity.

$$T_i = L/3600\,V \tag{17.2}$$

where

T_i = travel time for section i, in watershed (hr)
L = flow length (ft)
V = average velocity (ft/sec)

The time of concentration, T_c, is the sum of T_i for the various elements within the watershed. Thus,

$$T_c = \sum_{i}^{m} T_i \tag{17.3}$$

where

T_c = time of concentration (hr)
T_i = travel time for segment i (hr)
m = number of segments

Sheet flow velocity depends on the slope and the type of surface of the watershed area. Figure 17.4 provides average velocities for various surface types as a function of ground slope. For example, if a watercourse classified as "nearly bare ground" has an average slope of 5 percent, the average velocity is approximately 2.2 ft/sec.

The travel time for flow in a gutter or storm sewer is the sum of the travel times in each component of the gutter and/or storm sewer system between the farthest inlet and the outlet. Although velocities in the different components may be different, the use of the average velocity for the whole system does not usually result in large errors. When gutters are shallow, the curve for overland flow in paved areas in Figure 17.4 can be used to determine the average velocity.

The travel time in the open channel is determined in a similar way as that for the flow in gutter or storm sewer. In this case, however, the velocity of flow in the open channel must be determined first by using an appropriate equation such as Manning's formula. This is discussed in more detail later in the chapter under "Channel Design."

Determination of Runoffs

The amount of runoff for any combination of intensity and duration depends on the type of surface. For example, runoff will be much higher on rocky or bare impervious slopes, roofs, and pavements than on plowed land or heavy forest. The highway engineer is therefore interested in determining the proportion of rainfall that remains as runoff. This determination is not easy since the runoff rate for any given area during a single rainfall is not usually constant. Several methods for estimating runoff are available, but only two commonly used methods are presented.

Rational Method

The rational method is based on the premise that the rate of runoff for any storm depends on the average intensity of the storm, the size of the drainage area, and the type of drainage area surface. Note that for any given storm, the rainfall intensity is not usually constant over a large area, nor during the entire duration of the storm. The rational formula therefore uses the theory that for a rainfall of average intensity, I, falling over an impervious area of size A, the maximum rate of runoff at the outlet to the drainage area Q, occurs when the whole drainage area is contributing to the runoff and this runoff rate is constant. This requires that the storm duration be at least equal to the time of concentration, which is the time required for the runoff to flow from the far-

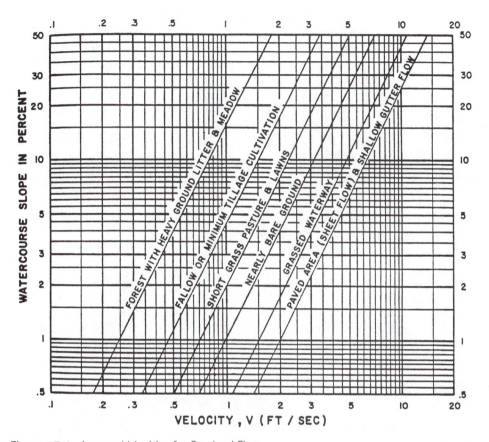

Figure 17.4 Average Velocities for Overland Flow

SOURCE: U.S. Department of Transportation, Federal Highway Administration, Hydrology, Hydraulic Engineering Circular, No. 19, 1984.

thest point of the drainage area to the outlet. This condition is not always satisfied in practice, particularly in large drainage areas. It is therefore customary for the rational formula to be used for relatively small drainage areas not greater than 200 acres. The rational formula is given as

$$Q = CIA \qquad (17.4)$$

where
$\quad Q$ = peak rate of runoff (ft³/sec)
$\quad A$ = drainage area (acres)
$\quad I$ = average intensity for a selected frequency and duration equal to at least the time of concentration (in/hr)
$\quad C$ = a coefficient representing the fraction of rainfall that remains on the surface of the ground (runoff coefficient)

Although the formula is not dimensionally correct, results of studies indicate that a rainfall of 1 in/hr falling over an area of 1 acre produces a runoff of 1.008 ft^3/sec if there are no losses. The runoff value Q is therefore almost exactly equal to the product of I and A. The losses due to infiltration and evaporation are accounted for by C. Values for C can be obtained from Table 17.2.

Example 17.1   Computing Rate of Runoff Using the Rational Formula

A 175-acre rural drainage area consists of three different watershed areas as follows:

Steep grassed areas = 50 percent
Forested areas = 30 percent
Cultivated fields = 20 percent

If the time of concentration for the drainage area is 1.5 hr, determine the runoff rate for a storm of 100 yr frequency. Assume that the rainfall intensity curves in Figure 17.2 are applicable to this drainage area.

Solution: The weighted runoff coefficient should first be determined for the whole drainage area. From Table 17.2, midpoint values for the different surface types are

Steep grassed areas = 0.6
Forested areas = 0.2
Cultivated fields = 0.3

$$C_w = \frac{175(0.5 \times 0.6 + 0.3 \times 0.2 + 0.2 \times 0.3)}{175} = 0.42$$

The storm intensity for a duration of at least 1.5 hr (time of concentration) and frequency of 50 yr is obtained from Figure 17.2 as approximately 2.2 in/hr. From Eq. 17.4,

$$Q = 0.42 \times 2.2 \times 175 = 162 \text{ (ft}^3\text{/sec)}$$

U.S. Soil Conservation Service (SCS) Method, TR-55

A method developed by the U.S. Soil Conservation Service (SCS) is commonly used in determining surface runoffs in highway engineering. This method, referred to as TR-55, can be used to estimate runoff volumes and peak rate of discharge.

The TR-55 method consists of two parts. The first part determines the runoff, h, in inches. The second part estimates the peak discharge using the value of h obtained from the first part and a graph that relates the time of concentration (hours) with the unit peak discharge (ft^3/sec/mi^2/in).

The fundamental premise used in developing this method is that the depth of runoff h in inches depends on the rainfall P in inches. Some of the precipitation occurring at the early stage of the storm, known as initial abstraction (I_a), will not be part of the run-off. The potential maximum retention, S, of the surface (similar in concept to C in the rational method) is a measure of the imperviousness of the watershed area. The SCS equation is

$$h = \frac{(P - I_a)^2}{(P - I_a) + S} = \frac{(P - 0.2S)^2}{P + 0.8S} \qquad (17.5)$$

where

$\quad h$ = runoff (in)
$\quad P$ = rainfall (in 24 hr)
$\quad S$ = potential maximum retention after runoff begins (in.)
$\quad I_a$ = initial abstraction, including surface storage, interception, and infiltration prior to runoff. The relationship between I_a and S, developed empirically from watershed data, is $I_a = 0.2S$.

To calculate the peak discharge we use the following equation:

$$q_p = q'_p A h \qquad (17.6)$$

where

$\quad q_p$ = peak discharge (ft^3/sec)
$\quad A$ = area of drainage basin (mi^2)
$\quad h$ = runoff (in.)
$\quad q'_p$ = peak discharge, for 24-hr storm at time of concentration (ft^3/sec/mi^2/in); see Figure 17.5

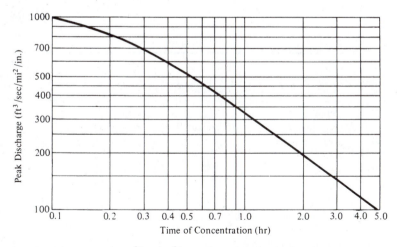

Figure 17.5 Peak Discharge in ft^3/sec/mi^2/In of Run-Off Versus Time of Concentration, T_c, for 24 hr, Type II Storm Distribution

NOTE: This type of storm distribution curve does not apply everywhere. It is not valid for most coastal regions, including regions bordering the Gulf of Mexico.

To determine h, we must first compute S, which is determined as

$$S = (1000/CN) - 10 \tag{17.7}$$

where

CN is the runoff curve number, which varies from 0 (pervious) to 100 (impervious). Note that high values of S indicate low retention of storm water, whereas low values of S indicate that most water is retained by the ground.

The runoff curve number accounts for the watershed characteristics, which include the soil type, land use, hydrologic condition of the cover, and soil moisture just prior to the storm (antecedent soil moisture). In order to determine values for CN, SCS has developed classification systems for soil type, cover, and antecedent soil moisture.

Soils are divided into four groups as follows:

- Group A: deep sand, deep loess, aggregated silts
- Group B: shallow loess, sandy loam
- Group C: clay loams, shallow sandy loam, soils low in organic content, and soils usually high in clay
- Group D: soils that swell significantly when wet, heavy plastic clays, and certain saline soils

Land use and hydrologic conditions of the cover are incorporated as shown in Table 17.3, which shows 14 different classifications for land use. In addition, agricultural land uses are usually subdivided into specific treatment or practices to account for the differences in hydrologic runoff potential among these treatments or practices. The

Table 17.3 Runoff Curve Numbers for AMC II

Land-Use Description/Treatment/Hydrologic Condition		Hydrologic Soil Group			
		A	B	C	D
Residential[1]					
Average lot size:	Average Percent Impervious:[2]				
$\frac{1}{8}$ acre or less	65	77	85	90	92
$\frac{1}{4}$ acre	38	61	75	83	87
$\frac{1}{3}$ acre	30	57	72	81	86
$\frac{1}{2}$ acre	25	54	70	80	85
1 acre	20	51	68	79	84
Paved parking lots, roofs, driveways, etc.[3]		98	98	98	98
Streets and roads					
Paved with curbs and storm sewers[3]		98	98	98	98
Gravel		76	85	89	91
Dirt		72	82	87	89
Commercial and business areas (85 percent impervious)		89	92	94	95
Industrial districts (72 percent impervious)		81	88	91	93

Table 17.3 Runoff Curve Numbers for AMC II (*continued*)

Land-Use Description/Treatment/Hydrologic Condition		Hydrologic Soil Group			
		A	*B*	*C*	*D*
Open spaces, lawns, parks, golf courses, cemeteries, etc.					
Good condition: grass cover on 75 percent or more of the area		39	61	74	80
Fair condition: grass cover on 50 percent to 75 percent of the area		49	69	79	84
Fallow					
Straight row	—	77	86	91	94
Row crops					
Straight row	Poor	72	81	88	91
Straight row	Good	67	78	85	89
Contoured	Poor	70	79	84	88
Contoured	Good	65	75	82	86
Contoured and terraced	Poor	66	74	80	82
Contoured and terraced	Good	62	71	78	81
Small grain					
Straight row	Poor	65	76	84	88
Straight row	Good	63	75	83	87
Contoured	Poor	63	74	82	85
Contoured	Good	61	73	81	84
Contoured and terraced	Poor	61	72	79	82
Contoured and terraced	Good	59	70	78	81
Close-seeded legumes[4] or rotation meadow					
Straight row	Poor	66	77	85	89
Straight row	Good	58	72	81	85
Contoured	Poor	64	75	83	85
Contoured	Good	55	69	78	83
Contoured and terraced	Poor	63	73	80	83
Contoured and terraced	Good	51	67	76	80
Pasture or range	Poor	68	79	86	89
	Fair	49	69	79	84
	Good	39	61	74	80
Contoured	Poor	47	67	81	88
Contoured	Fair	25	59	75	83
Contoured	Good	6	35	70	79
Meadow	Good	30	58	71	78
Woods or forest land	Poor	45	66	77	83
	Fair	36	60	73	79
	Good	25	55	70	77
Farmsteads	—	59	74	82	86

[1]Curve numbers are computed assuming the runoff from the house and driveway is directed toward the street with a minimum of roof water directed to lawns where additional infiltration could occur.

[2]The remaining pervious areas (lawn) are considered to be in good pasture condition for these curve numbers.

[3]In some warmer climates of the country, a curve number of 95 may be used.

[4]Close-drilled or broadcast.

SOURCE: Adapted from Richard H. McCuen, *A Guide to Hydrologic Analysis Using SCS Methods,* copyright © 1982. Reprinted by permission of Prentice Hall, Inc., Englewood Cliffs, N.J.

hydrologic condition is considered in terms of the level of land management, which is given in three classes: poor, fair, and good.

The effect of antecedent moisture condition (AMC) on rate of runoff is taken into consideration by classifying soil conditions into three categories:

- Condition I: soils are dry but not to wilting point; satisfactory cultivation has taken place
- Condition II: average condition
- Condition III: heavy rainfall, or light rainfall and low temperatures, have occurred within the last five days; saturated soils.

Table 17.4 gives seasonal rainfall limits as guidelines for determining the antecedent moisture condition (AMC). Note that the *CN* numbers in Table 17.3 are for AMC II. When antecedent moisture condition I or III exists, Table 17.3 is used to determine the condition II *CN* value, and then Table 17.5 is used to obtain the appropriate value for either condition I or III.

The first task in the second part of the procedure is to determine the unit peak discharge (q'_p) (ft^3/sec/mi^2/in) from Figure 17.5, which relates the time of concentration (hours) and the unit peak discharge for a 24-hr type II storm, which can be applied to most of the United States, excluding most of the coastal regions.

Example 17.2   Computing Peak Discharge Using the TR-55 Method

A 0.5 mi^2 drainage area consists of 20 percent residential (½-acre lots), 30 percent row crops with straight row treatment and good hydrologic condition, and 50 percent wooded area with good hydrologic condition. If the soil is classified as group C, with an AMC III, determine the peak discharge if the 24-hr precipitation is 6 in. and the time of concentration is 2 hr.

Solution: A weighted *CN* value should first be determined as follows:

$$\text{weighted } CN = 0.2 \times 94 + 0.3 \times 97 + 0.5 \times 87 = 91.4$$

$$S = \frac{1000}{CN} - 10 = \frac{1000}{91.4} - 10 = 0.94$$

From Eq. 17.5,

$$h = \frac{(P - 0.2S)^2}{P + 0.8S} = \frac{(6 - 0.2 \times 0.94)^2}{6 + 0.8 \times 0.94} = \frac{33.78}{6.75} = 5.0 \text{ in.}$$

The peak unit discharge is determined from Figure 17.5 for a time of concentration of 2 hr, $q'_p = 190$ ft^3/sec/mi^2/in. From Eq. 17.6

$$q_p = q'_p Ah$$
$$= 190 \times 0.5 \times 5.0$$
$$= 475 \text{ ft}^3/\text{sec}$$

Land Use/Treatment/Hydrologic Condition	CN for AMC II	CN for AMC III
Residential ($\frac{1}{2}$ acre lots) (20 percent)	80	94
Row crops/straight row/good (30 percent)	85	97
Wooded/good (50 percent)	70	87

Table 17.4 Total Five-Day Antecedent Rainfall (in.)

AMC	Dormant Season	Growing Season
I	≤ 0.5	<1.4
II	0.5 to 1.1	1.4 to 2.1
III	>1.1	>2.1

Table 17.5 Corresponding Runoff Curve Numbers for Conditions I and III

CN for Condition II	Corresponding CN for Condition	
	I	III
100	100	100
95	87	99
90	78	98
85	70	97
80	63	94
75	57	91
70	51	87
65	45	83
60	40	79
55	35	75
50	31	70
45	27	65
40	23	60
35	19	55
30	15	50
25	12	45
20	9	39
15	7	33
10	4	26
5	2	17
0	0	0

SOURCE: Adapted from Richard H. McCuen, *A Guide to Hydrologic Analysis Using SCS Methods,* copyright © 1982. Reprinted by permission of Prentice-Hall, Inc., Englewood Cliffs, N.J.

UNIT HYDROGRAPHS

A hydrograph is a plot that shows the relationship between stream flow (ordinate scale) and time (abscissa scale). A unit hydrograph is the hydrograph obtained for a runoff due to water of 1-in depth, uniformly distributed over the whole drainage area, for a given storm with a specified duration. The unit hydrograph represents all the effects of the different characteristics of the drainage area, such as ground cover, slope, and soil characteristics, acting concurrently. The unit hydrograph therefore gives the runoff characteristics of the drainage area.

It has been shown that similar storms having the same duration, but not necessarily the same intensity, will produce unit hydrographs that are similar in shape. This means that for a given type of rainfall, the time scale of the unit hydrograph is constant for a given drainage area and its ordinates are approximately proportional to the runoff volumes. Thus, the unit hydrograph can be used to estimate the expected runoffs from similar storms having different intensities.

A unit hydrograph can be used to determine peak discharges from drainage areas as large as 2000 mi^2.

Computer Models for Highway Drainage

The use of computer models to generate flood hydrographs is becoming increasingly common. Some of these models merely solve existing empirical formulas, whereas others use simulation techniques. Most simulation models provide for the drainage area to be divided into smaller subareas with similar characteristics. A design storm is then synthesized for each subarea and the volume due to losses such as infiltration and interception deducted. The flow of the remaining water is then simulated using an overland flow routine. The overland flows from the subareas are collected by adjacent channels, which are eventually linked together to obtain the total response of the drainage area to the design storm.

The validity of any simulation model is increased by using measured historical data to calibrate the parameters of the model. One major disadvantage of simulation models is that they usually require a large amount of input data and extensive user experience to obtain reliable results.

Increasingly, highway drainage design and analysis are completed using software packages, many of which are available on microcomputers. Some of the computer packages are as follows:

The FHWA Urban Storm Drainage Model, written for mainframe computers, consists of four modules: Precipitation, Hydraulic/Quality, Analysis, and Cost Estimation.

FHWA and AASHTO have developed a microcomputer package called HYDRAIN, which consists of several programs. The HYDRO program generates design flows, or hydrographs, from rainfall and offers several hydrological analysis options, including the rational method. The PFP/HYDRA program is a storm and sanitary sewer system analysis. The CDS (Culvert Design System) program can assist in the design or analysis of an existing or proposed culvert, and the Water Surface Profile (WSPRO) program is used for bridge waterway analysis to compute water surface profiles.

Other computer programs are SWMM, the EPA storm water management model for planning and design, and STORM, the U.S. Army Corps of Engineers model that simulates the quality and quantity of overflows for combined sewer systems.

HYDRAULIC DESIGN OF HIGHWAY DRAINAGE STRUCTURES

The ultimate objective in determining the hydraulic requirements for any highway drainage structure is to provide a suitable structure size that will economically and efficiently dispose of the expected runoff. Certain hydraulic requirements should also be met to avoid erosion and/or sedimentation in the system.

Design of Open Channels

An important design consideration is that the flow velocity in the channel should not be so low as to cause deposits of transported material, nor so high as to cause erosion of the channel. The velocity that will satisfy this condition usually depends on the shape and size of the channel, the type of lining in the channel, the quantity of water being transported, and the type of material suspended in the water. The most appropriate channel gradient range to produce the required velocity is between 1 percent and 5 percent. For most types of linings, sedimentation is usually a problem when slopes are less than 1 percent, and excessive erosion of the lining will occur when slopes are higher than 5 percent. Tables 17.6, 17.7, and 17.8 give recommended maximum velocities for preventing erosion.

Table 17.6 Maximum Permissible Velocities in Erodible Channels, Based on Uniform Flow in Continuously Wet, Aged Channels

	Velocities		
Material	*Clear Water (ft/sec)*	*Water Carrying Fine Silts (ft/sec)*	*Water Carrying Sand and Gravel (ft/sec)*
Fine sand (noncolloidal)	1.5	2.5	1.5
Sandy loam (noncolloidal)	1.7	2.5	2.0
Silt loam (noncolloidal)	2.0	3.0	2.0
Ordinary firm loam	2.5	3.5	2.2
Volcanic ash	2.5	3.5	2.0
Fine gravel	2.5	5.0	3.7
Stiff clay (very colloidal)	3.7	5.0	3.0
Graded, loam to cobbles (noncolloidal)	3.7	5.0	5.0
Graded, silt to cobbles (colloidal)	4.0	5.5	5.0
Alluvial silts (noncolloidal)	2.0	3.5	2.0
Alluvial silts (colloidal)	3.7	5.0	3.0
Coarse gravel (noncolloidal)	4.0	6.0	6.5
Cobbles and shingles	5.0	5.5	6.5
Shales and hard pans	6.0	6.0	5.0

Note: For sinuous (winding) channels, multiply allowable velocity by 0.95 for slightly sinuous channels, by 0.9 for moderately sinuous channels, and by 0.8 for highly sinuous channels.
SOURCE: Adapted from George S. Koslov, *Road Surface Drainage Design, Construction and Maintenance Guide for Pavements,* Report No. FHWA/NJ-82-004, State of New Jersey, Department of Transportation, Trenton, N.J., June 1981.

Table 17.7 Maximum Allowable Water Velocities for Different Types of Ditch Linings

	Maximum Velocity (ft/sec)
Natural Soil Linings	
Bedrock or rip-rap sides and bottoms	15–18
Gravel bottom, rip-rap sides	8–10
Clean gravel	6–7
Silty gravel	2–5
Clayey gravel	5–7
Clean sand	1–2
Silty sand	2–3
Clayey sand	3–4
Silt	3–4
Light clay	2–3
Heavy clay	2–3
Vegetative Linings	
Average turf, erosion-resistant soil	4–5
Average turf, easily eroded soil	3–4
Dense turf, erosion-resistant soil	6–8
Gravel bottom, brushy sides	4–5
Dense weeds	5–6
Dense brush	4–5
Dense willows	8–9
Paved Linings	
Gravel bottom, concrete sides	10
Mortared rip-rap	8–10
Concrete or asphalt	18–20

SOURCE: Adapted from *Drainage Design for New York State,* U.S. Department of Commerce, Washington, D.C., November 1974.

Attention should also be paid to the point at which the channel discharges into the natural waterway. For example, if the drainage channel at the point of discharge is at a much higher elevation than the natural waterway, then the water should be discharged through a spillway or chute to prevent erosion.

Design Principles

The hydraulic design of a drainage ditch for a given storm entails the determination of the minimum cross-sectional area of the ditch that will accommodate the flow due to that storm and prevent water from overflowing the sides of the ditch. The most commonly used formula for this purpose is Manning's formula, which assumes uniform steady flow in the channel and gives the mean velocity in the channel as

$$v = \frac{1.486}{n} R^{2/3} S^{1/2} \tag{17.8}$$

Table 17.8 Maximum Permissible Velocities in Channels Lined with Uniform Strands of Various Well-Maintained Grass Covers

Cover	Slope Range (%)	Maximum Permissible Velocity on[a]	
		Erosion-Resistant Soils (ft/sec)	Easily Eroded Soils (ft/sec)
Bermuda grass	0–5	8	6
	5–10	7	5
	Over 10	6	4
Buffalo grass Kentucky bluegrass Smooth brome Blue grama	0–5	7	5
	5–10	6	4
	Over 10	5	3
Grass mixture	0–5[b]	5	4
	5–10[b]	4	3
Lespedeza sericea Weeping lovegrass Yellow bluestem Kudzu Alfalfa Crabgrass	0–5[c]	3.5	2.5
Common lespedeza[d] Sudangrass[d]	0–5[c]	3.5	2.5

[a]Use velocities over 5 ft/sec only where good covers and proper maintenance can be obtained.
[b]Do not use on slopes steeper than 10 percent.
[c]Not recommended for use on slopes steeper than 5 percent.
[d]Annuals, used on mild slopes or as temporary protection until permanent covers are established.
SOURCE: Adapted from *Drainage Design for New York State*, U.S. Department of Commerce, Washington, D.C., November 1974.

where
v = average discharge velocity (ft/sec)
R = mean hydraulic radius of flow in the channel (ft)
   = $\dfrac{a}{P}$
a = channel cross-sectional area (ft^2)
P = wetted perimeter (ft)
S = longitudinal slope in channel (ft/ft)
n = Manning's roughness coefficient

Manning's roughness depends on the type of material used to line the surface of the ditch. Table 17.9 gives recommended values for the roughness coefficients for different lining materials. The flow in the channel is then given as

Table 17.9 Manning's Roughness Coefficients

	Manning's n range [2]
I. Closed conduits:	
A. Concrete pipe	0.011–0.013
B. Corrugated-metal pipe or pipe-arch:	
1. 2⅔ by ½-in. corrugation (riveted pipe): [3]	
a. Plain or fully coated	0.024
b. Paved invert (range values are for 25 and 50 percent of circumference paved):	
(1) Flow full depth	0.021–0.018
(2) Flow 0.8 depth	0.021–0.016
(3) Flow 0.6 depth	0.019–0.013
2. 6 by 2-in. corrugation (field bolted)	0.03
C. Vitrified clay pipe	0.012–0.014
D. Cast-iron pipe, uncoated	0.013
E. Steel pipe	0.009–0.011
F. Brick	0.014–0.017
G. Monolithic concrete:	
1. Wood forms, rough	0.015–0.017
2. Wood forms, smooth	0.012–0.014
3. Steel forms	0.012–0.013
H. Cemented rubble masonry walls:	
1. Concrete floor and top	0.017–0.022
2. Natural floor	0.019–0.025
I. Laminated treated wood	0.015–0.017
J. Vitrified clay liner plates	0.015
II. Open channels, lined [4] (straight alinement): [5]	
A. Concrete, with surfaces as indicated:	
1. Formed, no finish	0.013–0.017
2. Trowel finish	0.012–0.014
3. Float finish	0.013–0.015
4. Float finish, some gravel on bottom	0.015–0.017
5. Gunite, good section	0.016–0.019
6. Gunite, wavy section	0.018–0.022
B. Concrete, bottom float finished, sides as indicated:	
1. Dressed stone in mortar	0.015–0.017
2. Random stone in mortar	0.017–0.020
3. Cement rubble masonry	0.020–0.025
4. Cement rubble masonry, plastered	0.016–0.020
5. Dry rubble (riprap)	0.020–0.030
C. Gravel bottom, sides as indicated:	
1. Formed concrete	0.017–0.020
2. Random stone in mortar	0.020–0.023
3. Dry rubble (riprap)	0.023–0.033
D. Brick	0.014–0.017
E. Asphalt:	
1. Smooth	0.013
2. Rough	0.016
F. Wood, planed, clean	0.011–0.013
G. Concrete-lined excavated rock:	
1. Good section	0.017–0.020
2. Irregular section	0.022–0.027
III. Open channels, excavated [4] (straight alinement,[5] natural lining):	
A. Earth, uniform section:	
1. Clean, recently completed	0.016–0.018
2. Clean, after weathering	0.018–0.020
3. With short grass, few weeds	0.022–0.027
4. In gravelly soil, uniform section, clean	0.022–0.025
B. Earth, fairly uniform section:	
1. No vegetation	0.022–0.025
2. Grass, some weeds	0.025–0.030
3. Dense weeds or aquatic plants in deep channels	0.030–0.035
4. Sides clean, gravel bottom	0.025–0.030
5. Sides clean, cobble bottom	0.030–0.040
C. Dragline excavated or dredged:	
1. No vegetation	0.028–0.033
2. Light brush on banks	0.035–0.050
D. Rock:	
1. Based on design section	0.035
2. Based on actual mean section:	
a. Smooth and uniform	0.035–0.040
b. Jagged and irregular	0.040–0.045
E. Channels not maintained, weeds and brush uncut:	
1. Dense weeds, high as flow depth	0.08–0.12
2. Clean bottom, brush on sides	0.05–0.08
3. Clean bottom, brush on sides, highest stage of flow	0.07–0.11
4. Dense brush, high stage	0.10–0.14

	Manning's n range [2]
IV. Highway channels and swales with maintained vegetation [6][7] (values shown are for velocities of 2 and 6 f.p.s.):	
A. Depth of flow up to 0.7 foot:	
1. Bermudagrass, Kentucky bluegrass, buffalograss:	
a. Mowed to 2 inches	0.07–0.045
b. Length 4–6 inches	0.09–0.05
2. Good stand, any grass:	
a. Length about 12 inches	0.18–0.09
b. Length about 24 inches	0.30–0.15
3. Fair stand, any grass:	
a. Length about 12 inches	0.14–0.08
b. Length about 24 inches	0.25–0.13
B. Depth of flow 0.7–1.5 feet:	
1. Bermudagrass, Kentucky bluegrass, buffalograss:	
a. Mowed to 2 inches	0.05–0.035
b. Length 4 to 6 inches	0.06–0.04
2. Good stand, any grass:	
a. Length about 12 inches	0.12–0.07
b. Length about 24 inches	0.20–0.10
3. Fair stand, any grass:	
a. Length about 12 inches	0.10–0.06
b. Length about 24 inches	0.17–0.09
V. Street and expressway gutters:	
A. Concrete gutter, troweled finish	0.012
B. Asphalt pavement:	
1. Smooth texture	0.013
2. Rough texture	0.016
C. Concrete gutter with asphalt pavement:	
1. Smooth	0.013
2. Rough	0.015
D. Concrete pavement:	
1. Float finish	0.014
2. Broom finish	0.016
E. For gutters with small slope, where sediment may accumulate, increase above values of n by	0.002
VI. Natural stream channels: [8]	
A. Minor streams [9] (surface width at flood stage less than 100 ft.):	
1. Fairly regular section:	
a. Some grass and weeds, little or no brush	0.030–0.035
b. Dense growth of weeds, depth of flow materially greater than weed height	0.035–0.05
c. Some weeds, light brush on banks	0.035–0.05
d. Some weeds, heavy brush on banks	0.05–0.07
e. Some weeds, dense willows on banks	0.06–0.08
f. For trees within channel, with branches submerged at high stage, increase all above values by	0.01–0.02
2. Irregular sections, with pools, slight channel meander; increase values given in 1a–e about	0.01–0.02
3. Mountain streams, no vegetation in channel, banks usually steep, trees and brush along banks submerged at high stage:	
a. Bottom of gravel, cobbles, and few boulders	0.04–0.05
b. Bottom of cobbles, with large boulders	0.05–0.07
B. Flood plains (adjacent to natural streams):	
1. Pasture, no brush:	
a. Short grass	0.030–0.035
b. High grass	0.035–0.05
2. Cultivated areas:	
a. No crop	0.03–0.04
b. Mature row crops	0.035–0.045
c. Mature field crops	0.04–0.05
3. Heavy weeds, scattered brush	0.05–0.07
4. Light brush and trees: [10]	
a. Winter	0.05–0.06
b. Summer	0.06–0.08
5. Medium to dense brush: [10]	
a. Winter	0.07–0.11
b. Summer	0.10–0.16
6. Dense willows, summer, not bent over by current	0.15–0.20
7. Cleared land with tree stumps, 100–150 per acre:	
a. No sprouts	0.04–0.05
b. With heavy growth of sprouts	0.06–0.08
8. Heavy stand of timber, a few down trees, little undergrowth:	
a. Flood depth below branches	0.10–0.12
b. Flood depth reaches branches	0.12–0.16
C. Major streams (surface width at flood stage more than 100 ft.): Roughness coefficient is usually less than for minor streams of similar description on account of less effective resistance offered by irregular banks or vegetation on banks. Values of n may be somewhat reduced. Follow recommendation in publication cited [8] if possible. The value of n for larger streams of most regular section, with no boulders or brush, may be in the range of	0.028–0.033

Footnotes to Table 17.9 appear at the top of page 763

Table 17.9   Manning's Roughness Coefficients (*continued*)

1 Estimates are by Bureau of Public Roads unless otherwise noted.

2 Ranges indicated for closed conduits and for open channels, lined or excavated, are for good to fair construction (unless otherwise stated). For poor quality construction, use larger values of *n*.

3 *Friction Factors in Corrugated Metal Pipe*, by M. J. Webster and L. R. Metcalf, Corps of Engineers, Department of the Army; published in Journal of the Hydraulics Division, Proceedings of the American Society of Civil Engineers, vol. 85, No. HY9, Sept. 1959, Paper No. 2148, pp. 35–67.

4 For important work and where accurate determination of water profiles is necessary, the designer is urged to consult the following references and to select *n* by comparison of the specific conditions with the channels tested:

Flow of Water in Irrigation and Similar Channels, by F. C. Scobey, Division of Irrigation, Soil Conservation Service, U.S. Department of Agriculture, Tech. Bull. No. 652, Feb. 1939; and

Flow of Water in Drainage Channels, by C. E. Ramser, Division of Agricultural Engineering, Bureau of Public Roads, U.S. Department of Agriculture, Tech. Bull. No. 129, Nov. 1929.

5 With channel of an alinement other than straight, loss of head by resistance forces will be increased. A small increase in value of *n* may be made, to allow for the additional loss of energy.

6 *Handbook of Channel Design for Soil and Water Conservation*, prepared by the Stillwater Outdoor Hydraulic Laboratory in cooperation with the Oklahoma Agricultural Experiment Station; published by the Soil Conservation Service, U.S. Department of Agriculture, Publ. No. SCS-TP-61, Mar. 1947, rev. June 1954.

7 *Flow of Water in Channels Protected by Vegetative Linings*, by W. O. Ree and V. J. Palmer, Division of Drainage and Water Control, Research, Soil Conservation Service, U.S. Department of Agriculture, Tech. Bull. No. 967, Feb. 1949.

8 For calculation of stage or discharge in natural stream channels, it is recommended that the designer consult the local District Office of the Surface Water Branch of the U.S. Geological Survey, to obtain data regarding values of *n* applicable to streams of any specific locality. Where this procedure is not followed, the table may be used as a guide. The values of *n* tabulated have been derived from data reported by C. E. Ramser (see footnote 4) and from other incomplete data.

9 The tentative values of *n* cited are principally derived from measurements made on fairly short but straight reaches of natural streams. Where slopes calculated from flood elevations along a considerable length of channel, involving meanders and bends, are to be used in velocity calculations by the Manning formula, the value of *n* must be increased to provide for the additional loss of energy caused by bends. The increase may be in the range of perhaps 3 to 15 percent.

10 The presence of foliage on trees and brush under flood stage will materially increase the value of *n*. Therefore, roughness coefficients for vegetation in leaf will be larger than for bare branches. For trees in channel or on banks, and for brush on banks where submergence of branches increases with depth of flow, *n* will increase with rising stage.

SOURCE: Reproduced from *Design Charts for Open Channels*, U.S. Department of Transportation, Washington, D.C., 1980.

$$Q = va = \frac{1.486}{n} aR^{2/3} S^{1/2} \qquad (17.9)$$

where Q is the discharge (ft³/sec).

The Federal Highway Administration (FHWA) has published a series of charts for channels of different cross sections that can be used to solve Eq. 17.9. Figures 17.6 and 17.7 are two examples of these charts.

Since Manning's formula assumes uniform steady flow in the channel, it is now necessary to discuss the concepts of steady, unsteady, uniform, and nonuniform flows.

Open channel flows can be grouped into two general categories: steady and unsteady. When the rate of discharge does not vary with time, the flow is steady; conversely, the flow is unsteady when the rate of discharge varies with time. Steady flow is further grouped into uniform and nonuniform, depending on the channel characteristics. Uniform flows are obtained when the channel properties, such as slope, roughness, and cross section, are constant along the length of the channel, whereas nonuniform flow is obtained when these properties vary along the length of the channel. When uniform flow is achieved in the channel, the depth d and velocity v_n are taken as normal and the slope of the water surface is parallel to the slope of the channel. Since it is extremely difficult to obtain the exact same channel properties along the length of the channel, it is very difficult to obtain uniform flow conditions in practice. Nevertheless, Manning's equation can be used to obtain practical solutions to stream flow problems in highway engineering since, in most cases, the error involved is small.

Flows in channels can also be tranquil or rapid. Tranquil flow is similar to the flow of water in an open channel with a relatively flat longitudinal slope, whereas rapid flow is similar to water tumbling down a steep slope. The depth at which the flow in any channel changes from tranquil to rapid is known as the critical depth. When the flow depth is greater than the critical depth, the flow is known as subcritical. This type of flow often occurs in streams in the plains and broad valley regions. When the flow depth is less than the critical depth, the flow is known as supercritical. This type of flow is prevalent in steep

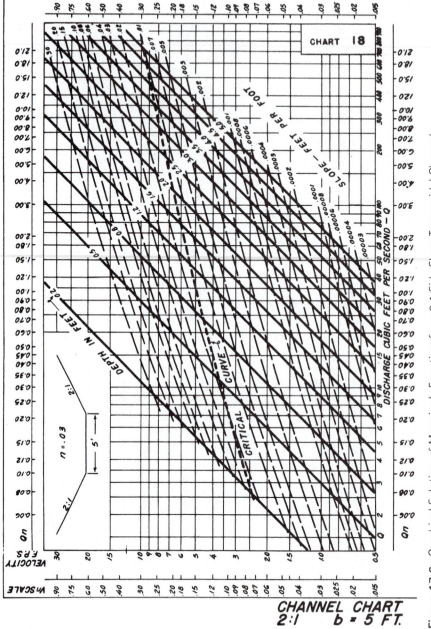

Figure 17.6   Graphical Solution of Manning's Equation for a 2:1 Side Slope Trapezoidal Channel

SOURCE: *Design Charts for Open Channels*, U.S. Department of Transportation, Washington, D.C., 1979.

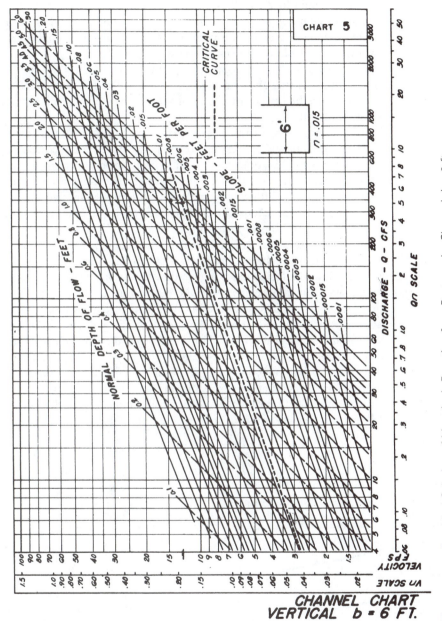

Figure 17.7   Graphical Solution of Manning's Equation for a Rectangular Channel, b = 6 ft

SOURCE: *Design Charts for Open Channels*, U.S. Department of Transportation, Washington, D.C., 1979.

flumes and mountain streams. The critical depth may also be defined as the depth of flow at which the specific energy is minimum. The critical depth depends only on the shape of the channel and the discharge. This implies that for any given channel cross section, there is only one critical depth for a given discharge.

The velocity and channel slope corresponding to uniform flow at critical depth are known as critical velocity and critical slope, respectively. The flow velocity and channel slope are therefore higher than the critical values when flow is supercritical, and lower when flow is subcritical.

Consider a channel consisting of four sections, each with a different grade, as shown in Figure 17.8. The slopes of the first two sections are less than the critical slope (although the slope of the second section is less than that of the first section), resulting in a sub-critical flow in both sections. The slope of the third section is higher than the critical slope, resulting in a supercritical flow. Flow in the fourth section is subcritical, since the slope is less than the critical slope. Let us now consider how the depth of water changes as the water flows along all four sections of the channel.

As the water flows through section A of the channel, the depth of flow is greater than the critical depth, d_c, because the flow is subcritical. The slope of section B reduces, resulting in a lower velocity and a higher flow depth ($h_B > h_A$). This increase in depth takes place gradually and begins somewhere upstream in section A. The change from subcritical to supercritical flow in section C with the steep grade also takes place smoothly over some distance. The reduction in flow depth occurs gradually and begins somewhere upstream in the subcritical section. However, when the slope changes again to a value less than the critical slope, the flow changes abruptly from supercritical to subcritical, resulting in a *hydraulic jump,* in which some of the energy is absorbed by the resulting turbulence. This shows that downstream conditions may change the depth upstream of a subcritical flow, which means that *control is downstream.* Thus, when flow is subcritical, any downstream changes in slope, cross section, or intersection with another stream will

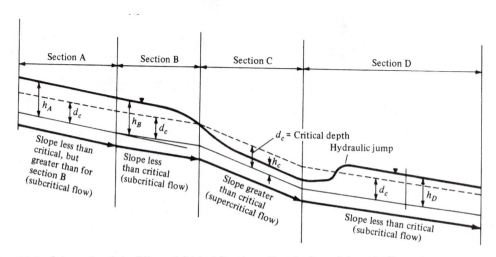

Figure 17.8 Schematic of the Effect of Critical Depth on Flow in Open Prismatic Channels

result in the gradual change of depth upstream, known as *backwater curve*. Conversely, supercritical flow is normally not affected by downstream conditions and *control is upstream*.

It should be noted that the FHWA curves for the solution of Manning's formula are only suitable for flows not affected by backwater.

Design Procedure

The design of a highway drainage channel is accomplished in two steps. The first step is to determine the channel cross section that will economically and effectively transport the expected runoff to a natural waterway. The second step is to determine whether the channel requires any erosion protection and if so, what type of lining is to be used.

Determination of Cross Section.   The Manning formula is solved for an assumed channel cross section to determine whether the channel is large enough for the runoff of the design storm. The solution may be obtained manually or by using the appropriate FHWA chart. Both methods are demonstrated in the following example.

Example 17.3   *Design of an Open Channel*

Determine a suitable cross section for a channel to carry an estimated runoff of 340 ft^3/sec if the slope of the channel is 1 percent and Manning's, *n,* is 0.015.

Solution: Select a channel section and then use Manning's formula to determine the flow depth required for the estimated runoff. Assume a rectangular channel 6-ft wide.

  Flow depth = d
  Cross sectional area = $6d$
  Wetted perimeter = $6 + 2d$
  Hydraulic radius $R = 6d/(6 + 2d)$

Using Equation 17.9,

$$Q = \frac{1.486}{n} a R^{2/3} S^{1/2}$$

$$340 = \frac{1.486}{0.015} (6d) \left(\frac{6d}{6 + 2d} \right)^{2/3} (0.01)^{1/2}$$

This equation is solved by trial and error to obtain $d \approx 4$ ft.

Alternatively, the FHWA chart shown in Figure 17.7 can be used since the width is 6 ft. Enter the chart at $Q = 340$ ft^3/sec, move vertically to intersect the channel slope of 1 percent (0.01), and then read the normal depth of 4 ft from the normal depth lines.

The critical depth can also be obtained by entering the chart at 340 ft^3/sec and moving vertically to intersect the critical curve. This gives a critical depth of about 4.6 ft, which means that the flow is supercritical.

The critical and flow velocities can be obtained similarly. The chart is entered at 340 ft^3/sec. Move vertically to the 0.01 slope and then horizontally to the velocity scale to read the flow velocity as approximately 14 ft/sec. The critical velocity is obtained in a similar manner, by moving vertically to the critical curve. The critical velocity is about 13 ft/sec. The critical slope is about 0.007.

The solution indicates that if a rectangular channel 6 ft wide is to be used to carry the runoff of 340 ft^3, the channel must be at least 4 ft deep. However, it is necessary to provide a freeboard of at least 1 ft, which makes the depth for this channel 5 ft. Note that the formula for determining critical depth for a rectangular channel is

$$y_c = \left(\frac{q^2}{g} \right)^{1/3}$$

where q is the flow per foot of width, in cfs/ft and g is 32.2 ft/sec^2. In this problem

$$y_c = \left[\frac{(340/6)^2}{32.2} \right]^{1/3} = 4.63 \text{ ft}$$

For nonrectangular sections, the critical depth occurs when

$$\frac{Q^2}{g} = \frac{A^3}{B}$$

where

   B = width of the water surface (ft)
   A = area of cross section (ft^2)

This is a trial-and-error solution, and so charts and tables are typically utilized.

The same chart can be used for different values of n when all other conditions are the same. For example, if n is 0.02 in Example 17.3, Qn is first determined as $340 \times 0.02 = 6.8$. Then enter the chart at $Qn = 6.8$ on the Qn scale, move vertically to the 0.01 slope, and read the flow depth as approximately 5 ft. The velocity is also read from the Vn scale as $Vn = 0.225$, which gives V as $0.225/0.02 = 11.25$ ft/sec.

Example 17.4   Computing Discharge Flow from a Trapezoidal Channel

A trapezoidal channel has 2 : 1 side slopes, a 5 ft bottom width, and a depth of 4 ft. If the channel is on a slope of 2 percent and $n = 0.030$, determine the discharge flow, velocity, and type of flow. The chart shown in Figure 17.6 is used to solve this problem.

Solution: The intersection of the 2 percent slope (0.02) and the 4-ft depth line is located. Move vertically to the horizontal discharge scale to determine the discharge, $Q = 600$ ft³/sec. Similarly, the velocity is 12.5 ft/sec. The intersection of the 2 percent slope and 4 ft depth lies above the critical curve, and thus the flow is supercritical.

Determination of Suitable Lining. The traditional method for determining suitable lining material for a given channel is to select a lining material such that the flow velocity is less than the permissible velocity to prevent erosion of the lining. However, research results have shown that the maximum permissible depth of flow (d_{max}) for flexible linings should be the main criterion for selecting channel lining. Rigid channels, such as those made of concrete or soil cement, do not usually erode in normal highway work and therefore have no maximum permissible depth of flow to prevent erosion. Thus, maximum flow depth for rigid channels depends only on the freeboard required over the water surface.

The maximum permissible depth of flow for different flexible lining materials can be obtained from charts given in *Design of Stable Channels with Flexible Linings.* Figures 17.9 through 17.11 show some of these charts.

The design procedure for flexible linings is described in the following steps:

Step 1. Determine channel cross section as in Example 17.3 or Example 17.4.

Step 2. Determine the maximum top width T of the selected cross section and select a suitable lining.

Step 3. Using the appropriate d_{max} for the lining selected (obtained from the appropriate chart), determine the hydraulic radius R and the cross-sectional area a for the selected cross section. This can be done by calculation or by using the chart shown in Figure 17.12.

Step 4. Determine the flow velocity V using the channel slope S and the hydraulic radius R determined in step 3.

Step 5. Determine the allowable flow in the channel using the velocity determined in step 4 and the cross-sectional area a in step 3—that is, $Q = aV$.

Step 6. Compare the flow determined in step 5 with the design flow. If there is a difference between the two flows, then the selected lining material is not suitable. For example, if the flow obtained in step 5 is less than the design flow, then the lining is inadequate; if it is much greater, the channel is overdesigned. When the flows are much different, then another lining material should be selected and steps 3 through 6 repeated.

Note that Manning's formula can be used to determine the allowable velocity in step 4, if the Manning coefficient of the material selected is known. Since it is usually not easy to determine these coefficients, curves obtained from test results and provided in *Design of Stable Channels with Flexible Linings* can be used. Figures 17.13 and 17.14 are flow velocity curves for jute mesh and fiberglass mat, for which maximum permissible depth charts are shown in Figures 17.9 and 17.10.

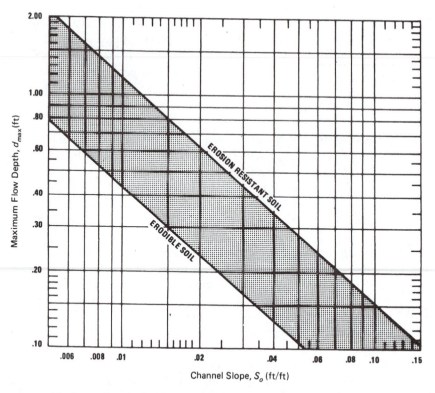

Figure 17.9   Maximum Permissible Depth of Flow, d_{max}, for Channels Lined with Jute Mesh

SOURCE: *Design of Stable Channels with Flexible Linings,* Hydraulic Engineering Circular No. 15, U.S. Department of Transportation, Washington, D.C., October 1975.

Example 17.5   Checking for Lining Suitability

Determine whether jute mesh is suitable for the flow and channel section in Example 17.3. Assume erosion-resistant soil.

Solution:

$$d_{max} = 1.20 \qquad \text{(Figure 17.9)}$$

$$\text{wetted perimeter} = 2 \times 1.20 + 6 = 8.40 \text{ (ft)}$$

$$\text{cross-sectional area} = 1.20 \times 6 = 7.20 \text{ (ft}^2)$$

$$R = \frac{7.20}{8.40} = 0.86$$

Selecting jute mesh as the lining and a channel slope of 0.01, the flow velocity for the lining is obtained from Figure 17.13 as

$$V = 61.53(0.86)^{1.028}(0.01)^{0.431} = 7.24 \text{ ft/sec}$$

and the maximum allowable flow is

$$Q = 7.24 \times 7.20 = 52.13 \text{ (ft}^3)$$

Jute mesh is not suitable. A rigid channel is probably appropriate for this flow.

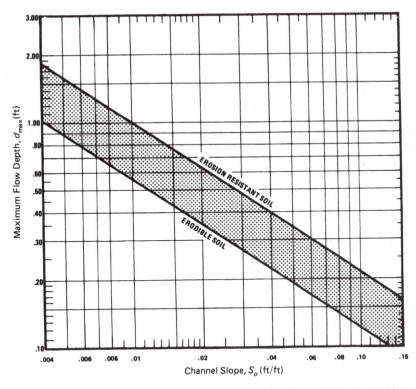

Figure 17.10 Maximum Permissible Depth of Flow, d_{max}, for Channels Lined with $3/8$ in. Fiberglass Mat

SOURCE: *Design of Stable Channels with Flexible Linings,* Hydraulic Engineering Circular No. 15, U.S. Department of Transportation, Washington, D.C., October 1975.

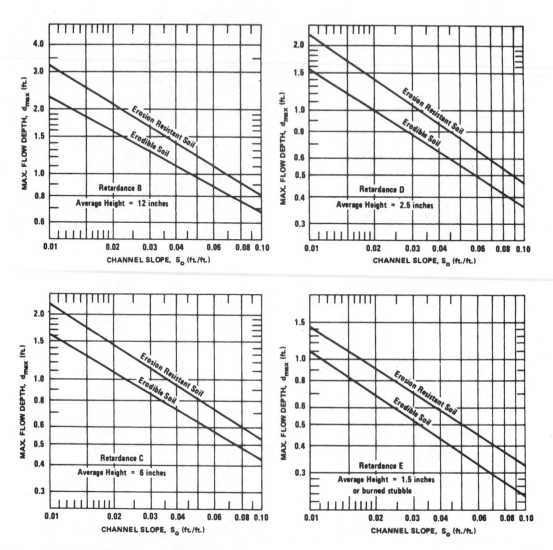

Figure 17.11   Maximum Permissible Depth of Flow, d_{max}, for Channels Lined with Bermuda Grass, Good Stand, Cut to Various Lengths

Note: Use on slopes steeper than 10 percent is not recommended.

SOURCE: *Design of Stable Channels with Flexible Linings,* Hydraulic Engineering Circular No. 15, U.S. Department of Transportation, Washington, D.C., October 1975.

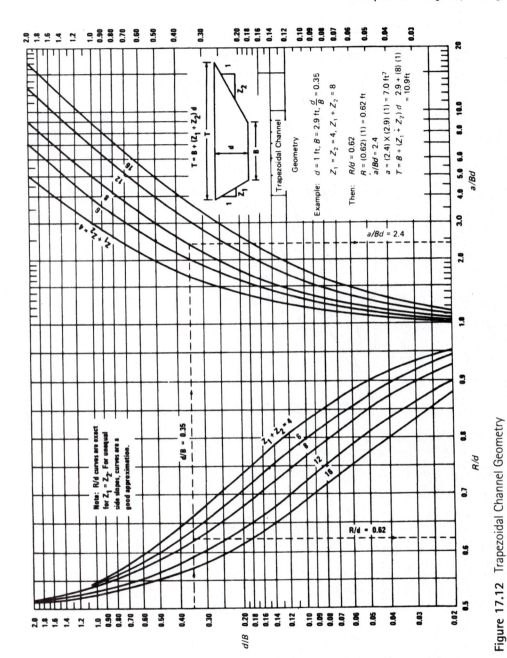

Figure 17.12 Trapezoidal Channel Geometry

SOURCE: *Design of Stable Channels with Flexible Linings*, Hydraulic Engineering Circular No. 15, U.S. Department of Transportation, Washington, D.C., October 1975.

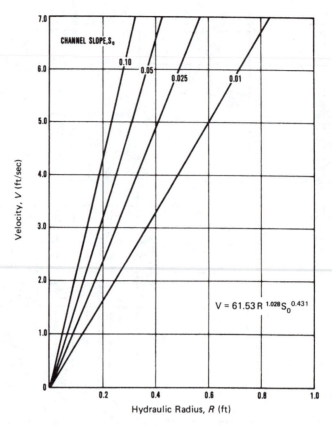

Figure 17.13 Flow Velocity for Channels Lined with Jute Mesh

SOURCE: *Design of Stable Channels with Flexible Linings,* Hydraulic Engineering Circular No. 15, U.S. Department of Transportation, Washington, D.C., October 1975.

Design of Culverts

Several complex hydraulic phenomena occur when water flows through highway culverts. The hydraulic design of culverts is therefore more complex than open channel design. The main factors considered in culvert design are the location of the culvert, the hydrologic characteristics of the watershed being served by the culvert, economy, and type of flow control.

Culvert Location

The most appropriate location of a culvert is in the existing channel bed, with the center line and slope of the culvert coinciding with that of the channel. At this location, the minimum cost associated with earth and channel work is achieved and stream flow disturbance is minimized. However, other locations may have to be selected in some cases; for example, relocation of a stream channel may be necessary to avoid an extremely long

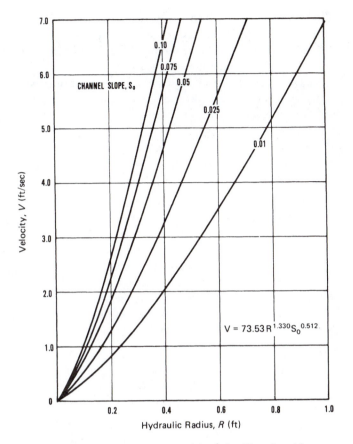

Figure 17.14 Flow Velocity for Channels Lined with 3/8 in. Fiberglass Mat

SOURCE: *Design of Stable Channels with Flexible Linings,* Hydraulic Engineering Circular No. 15, U.S. Department of Transportation, Washington, D.C., October 1975.

culvert. The basic principle used in locating culverts is that abrupt stream changes at the inlet and outlet of the culvert should be avoided.

Special consideration should be given to culverts located in mountainous areas. When culverts are located in the natural channel in these areas, high fills and long channels are usually required, which result in high construction costs. Culverts are therefore sometimes located on the side of the steep valley. When this is done, adequate measures must be taken to prevent erosion.

Hydrologic and Economic Considerations
The hydrologic and economic considerations are similar to those for open channel design, in that the design flow rate is based on the storm with an acceptable return period (frequency). This return period is selected such that construction and maintenance costs balance the probable cost of damage to adjacent properties if the storm should occur. Since

the occurrence of any given storm within a given period is a stochastic phenomenon, risk analysis is used to determine the appropriate return period. Risk analysis directly relates the culvert design to economic theory and identifies the probable financial consequences of both underdesign and overdesign.

Culverts are designed for the peak flow rate of the design storm. The peak flow rate is obtained from a unit hydrograph at the culvert site, developed from stream flow and rainfall records for a number of storm events.

Other Factors

Other factors that should be considered in culvert design are tailwater and upstream storage conditions.

Tailwater.   *Tailwater* is defined as the depth of water above the culvert outlet invert, as the water flows out of the culvert. Design of the culvert capacity must take tailwater into consideration, particularly when the design is based on the outlet conditions. High tailwater elevations may occur due to the hydraulic resistance of the channel or during flood events if the flow downstream is obstructed. Field observation and maps should be used to identify conditions that facilitate high tailwater elevations. These conditions include channel constrictions, intersections with other water courses, downstream impoundments, channel obstructions, and tidal effects. If these conditions do not exist, the tailwater elevation is based on the elevation of the water surface in the natural channel.

Upstream Storage.   The ability of the channel to store large quantities of water upstream from the culvert may have some effect on the design of the culvert capacity. The storage capacity upstream should therefore be checked using large-scale contour maps, from which topographic information is obtained.

Hydraulic Design of Culverts

As stated earlier, it is extremely difficult to carry out an exact theoretical analysis of culvert flow because of the many complex hydraulic phenomena that occur. For example, different flow types may exist at different times in the same culvert, depending on the tailwater elevation.

The design procedure presented here is that developed by FHWA and published in *Hydraulic Design of Highway Culverts.* The control section of the culvert is used to classify different culvert flows, which are then analyzed. The location at which a unique relationship exists between the flow rate and the depth of flow upstream from the culvert is the control section. When the flow is dictated by the inlet geometry, then the control section is the culvert inlet, that is, the upstream end of the culvert, and the flow is *inlet controlled.* When the flow is governed by a combination of the tailwater, the culvert inlet, and the characteristics of the culvert barrel, the flow is *outlet controlled.* Although it is possible for the flow in a culvert to change from one control to the other and back, the design is based on the *minimum performance* concept, which provides for the culvert to perform at a level that is never lower than the designed level. This means that the culvert may perform at a more efficient level; that is, a higher flow may be obtained for a given headwater level. The design procedure uses several design charts and nomographs, developed from a combination of theory and numerous hydraulic test results.

Inlet Control.   The flow in culverts operating under inlet control conditions is supercritical with high velocities and low depths. Four different flows under inlet control are shown in Figure 17.15. The flow type depends on whether the inlet and/or outlet of the culvert are submerged. In Figure 17.15(a), both the inlet and outlet are above the water surface. In this case, the flow within the culvert is supercritical, the culvert is partly full throughout its length, and the flow depth approaches normal at the outlet end. In Figure 17.15(b), only the downstream end (outlet) of the culvert is submerged, but this submergence does not result in outlet control. The flow in the culvert just beyond the culvert entrance (inlet) is supercritical, and a hydraulic jump occurs within the culvert. Figure 17.15(c) shows the inlet end of the culvert submerged, with the water flowing freely at the outlet. The culvert is partly full along its length, and the flow is supercritical within the culvert, since the critical depth is located just past the culvert inlet. Also, the flow depth at the culvert outlet approaches normal. This example of inlet control is more typical of design conditions. Figure 17.15(d) shows both the inlet and outlet of the culvert submerged, but the culvert is only partly full over part of its length. A hydraulic jump occurs within the cul-

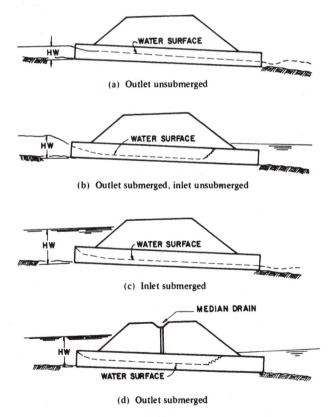

(a) Outlet unsubmerged

(b) Outlet submerged, inlet unsubmerged

(c) Inlet submerged

(d) Outlet submerged

Figure 17.15   Types of Inlet Control

SOURCE: J. M. Normann, R. J. Houghtalen, W. J. Johnston, *Hydraulic Design of Highway Culverts*, Report No. FHWA-IP-85-15, U.S. Department of Transportation, Office of Implementation, McLean, Va., September 1985.

vert, resulting in a full culvert along its remaining length. Under these conditions, pressures less than atmospheric may develop, creating an unstable situation, with the culvert alternating between partly full flow and full flow. This is prevented by providing the median inlet shown.

Several factors affect the performance of a culvert under inlet control conditions. These include the inlet area, the inlet shape, the inlet configuration, and the headwater depth. Several methods are available to increase the performance of culverts under inlet control. These include the use of special configurations for inlet edges and beveled edges at the culvert inlet. Detailed descriptions of these methods are given in *Hydraulic Design of Highway Culverts* and are briefly discussed later.

Model tests have been used to determine flow relationships between headwater (depth of water above the culvert inlet invert) and flow for culverts operating under inlet control conditions. The basic condition used to develop these equations is whether or not the inlet is submerged. The inlet performs as an orifice when it is submerged and as a weir when it is not submerged. Two equations were developed for the unsubmerged condition. The first equation (Eq. 17.10) is based on the specific head at critical depth, and the second equation (Eq. 17.11) is exponential and similar to a weir equation. Equation 17.10 has more theoretical support, but Eq. 17.11 is simpler to use. Both equations will give adequate results. Equation 17.12 gives the relationship for the submerged condition.

For the *unsubmerged condition*,

$$\frac{HW_i}{D} = \frac{H_c}{D} + K\left[\frac{Q}{(A)(D)^{0.5}}\right]^M - 0.5S \qquad (17.10)$$

$$\frac{HW_i}{D} = K\left[\frac{Q}{(A)(D)^{0.5}}\right]^M \qquad (17.11)$$

For the *submerged condition*,

$$\frac{HW_i}{D} = c\left[\frac{Q}{(A)(D)^{0.5}}\right]^2 + Y - 0.5S \qquad (17.12)$$

where
$\quad HW_i$ = required headwater depth above inlet control section invert (ft)
$\quad\quad D$ = interior height of culvert barrel (ft)
$\quad\quad V$ = flow velocity (ft/sec)
$\quad\quad V_c$ = critical velocity (ft/sec)
$\quad\quad g$ = 32.2 ft/sec^2
$\quad\quad H_c$ = specific head at critical depth—that is, $d_c + (V_c^2/2g)$(ft)
$\quad\quad d_c$ = critical depth (ft)
$\quad\quad Q$ = discharge (ft^3/sec)
$\quad\quad A$ = full cross-sectional area of culvert barrel (ft^2)
$\quad\quad S$ = culvert barrel slope (ft/ft)
K, M, c, Y = constants from Table 17.10

Table 17.10 Coefficients for Inlet Control Design Equations

Shape and Material	Inlet Edge Description	Form	Unsubmerged		Submerged	
			K	M	C	Y
Circular	Square edge w/headwall	1	0.0098	2.0	0.0398	0.67
Concrete	Groove end w/headwall		.0078	2.0	.292	.74
	Groove end projecting		.0045	2.0	.0317	.69
Circular	Headwall	1	.0078	2.0	.0379	.69
CMP	Mitered to slope		.0210	1.33	.0463	.75
	Projecting		.0340	1.50	.0553	.54
Circular	Beveled ring, 45° bevels	1	.0018	2.50	.0300	.74
	Beveled ring, 33.7° bevels		.0018	2.50	.0243	.83
Rectangular	30° to 75° wingwall flares	1	.026	1.0	.0385	.81
Box	90° and 15° wingwall flares		.061	0.75	.0400	.80
	0° wingwall flares		.061	0.75	.0423	.82
Rectangular	45° wingwall flare $d = .0430$	2	.510	.667	.0309	.80
Box	18° to 33.7° wingwall flare $d = .0830$		.486	.667	.0249	.83
Rectangular	90° headwall w/¾″ chamfers	2	.515	.667	.0375	.79
Box	90° headwall w/45° bevels		.495	.667	.0314	.82
	90° headwall w/33.7° bevels		.486	.667	.0252	.865
Rectangular	¾″ chamfers; 45° skewed headwall	2	.522	.667	.0402	.73
Box	¾″ chamfers; 30° skewed headwall		.533	.667	.0425	.705
	¾″ chamfers; 15° skewed headwall		.545	.667	.04505	.68
	45° bevels; 10°–45° skewed headwall		.498	.667	.0327	.75
Rectangular	45° non-offset wingwall flares	2	.497	.667	.0339	.803
Box	18.4° non-offset wingwall flares		.493	.667	.0361	.806
¾″ Chamfers	18.4° non-offset wingwall flares 30° skewed barrel		.495	.667	.0386	.71
Rectangular	45° wingwall flares—offset	2	.497	.667	.0302	.835
Box	33.7° wingwall flares—offset		.495	.667	.0252	.881
Top Bevels	18.4° wingwall flares—offset		.493	.667	.0227	.887
C M Boxes	90° headwall	1	.0083	2.0	.0379	.69
	Thick wall projecting		.0145	1.75	.0419	.64
	Thin wall projecting		.0340	1.5	.0496	.57
Horizontal	Square edge w/headwall	1	0.0100	2.0	.0398	.67
Ellipse	Groove end w/headwall		.0018	2.5	.0292	.74
Concrete	Groove end projecting		.0045	2.0	.0317	.69
Vertical	Square edge w/headwall	1	.0100	2.0	.0398	.67
Ellipse	Groove end w/headwall		.0018	2.5	.0292	.74
Concrete	Groove end projecting		.0095	2.0	.0317	.69
Pipe Arch	90° headwall	1	.0083	2.0	.0379	.69
18° Corner	Mitered to slope		.0300	1.0	.0463	.75
Radius CM	Projecting		.0340	1.5	.0496	.57
Pipe Arch	Projecting	1	.0296	1.5	.0487	.55
18° Corner	No Bevels		.0087	2.0	.0361	.66
Radius CM	33.7° Bevels		.0030	2.0	.0264	.75
Pipe Arch	Projecting	1	.0296	1.5	.0487	.55

Continued

Table 17.10 Coefficients for Inlet Control Design Equations (*continued*)

Shape and Material	Inlet Edge Description	Form	Unsubmerged		Submerged	
			K	*M*	*C*	*Y*
31° Corner	No Bevels		.0087	2.0	.0361	.66
Radius CM	33.7° Bevels		.0030	2.0	.0264	.75
Arch CM	90° headwall	1	.0083	2.0	.0379	.69
	Mitered to slope		.0300	1.0	.0463	.75
	Thin wall projecting		.0340	1.5	.0496	.57
Circular	Smooth tapered inlet throat	2	.534	.555	.0196	.89
	Rough tapered inlet throat		.519	.64	.0289	.90
Elliptical	Tapered inlet-beveled edges	2	.536	.622	.0368	.83
Inlet Face	Tapered inlet-square edges		.5035	.719	.0478	.80
	Tapered inlet-thin edge projecting		.547	.80	.0598	.75
Rectangular	Tapered inlet throat	2	.475	.667	.0179	.97
Rectangualr	Side tapered-less favorable edges	2	.56	.667	.0466	.85
Concrete	Side tapered-more favorable edges		.56	.667	.0378	.87
Rectangular	Slope tapered-less favorable edges	2	.50	.667	.0466	.65
Concrete	Slope tapered-more favorable edges		.50	.667	.0378	.71

SOURCE: Haestad Methods, Inc., *Computer Applications in Hydraulic Engineering,* Haestad Press, Waterbury, CT., 1997.

Note that the last term ($-0.5S$) in Eqs. 17.10 and 17.12 should be replaced by ($+0.7S$) when mitered corners are used. Equations 17.10 and 17.11 apply up to about $Q/(A)(D)^{0.5} = 3.5$. Equation 17.12 applies above about $Q/(A)(D)^{0.5} = 4.0$.

Several charts for different culvert shapes have been developed based on these equations and can be found in *Hydraulic Design for Highway Culverts.* Figure 17.16 is the chart for rectangular box culverts under inlet control, with flared wingwalls and beveled edge at top of inlet, and Figure 17.17 shows the chart for a circular pipe culvert under inlet control.

These charts are used to determine the depth of headwater required to accommodate the design flow through the selected culvert configuration under inlet control conditions. Alternatively, the iteration required to solve any of the equations may be carried out by using a computer. The use of the chart is demonstrated in Example 17.6.

Example 17.6 Computing Inlet Invert for a Box Culvert

Determine the required inlet invert for a 5 ft × 5 ft box culvert under inlet control with 45° flared wingwalls and beveled edge for the following flow conditions.

- Peak flow = 250 ft³/sec
- Design headwater elevation (EL_{hd}) = 230.5 ft (based on adjacent structures)
- Stream bed elevation at face of inlet = 224.0 ft

Solution: The chart shown in Figure 17.16 is applicable, and the solution is carried out to demonstrate the consecutive steps required.

Step 1. Select the size of the culvert and locate the design flow rate on the appropriate scales (points A and B, respectively). Note that for rectangular box culverts, the flow rate per width of barrel width is used.

$$\frac{Q}{NB} = \frac{250}{5} = 50 \text{ ft}^3/\text{sec/ft for point B}$$

Step 2. Draw a straight line through points A and B and extend this line to the first headwater/culvert height (HW/D) scale. Read the value on this scale. (Note that the first line is a turning line and that alternate values of (HW/D) can be obtained by drawing a horizontal line from this point to the other scales as shown.) Using the first line in this example, (HW/D) = 1.41.

Step 3. The required headwater is determined by multiplying the reading obtained in step 2, that is, the value for (HW/D) by the culvert depth, $HW = 1.41 \times 5 \sim 7.1$. This value is used for HW_i (required headwater depth above inlet control invert, ft) if the approach velocity head is neglected. When the approach velocity head is not neglected, then

$$HW_i = HW - \frac{V^2}{2g}$$

Neglecting the approach velocity head in this problem, HW_i = 7.1 ft.

Step 4. The required depression (fall)—that is, the depth below the stream bed at which the invert should be located—is obtained as follows:

$$HW_d = EL_{hd} - EL_{sf} \qquad (17.13)$$

and

$$\text{fal} = HW_i - HW_d \qquad (17.14)$$

where
  HW_d = design headwater depth (ft)
  EL_{hd} = design headwater elevation (ft)
  EL_{sf} = elevation at the stream bed at the face (ft)

In this case,

$$HW_d = 230.5 - 224.0 = 6.5 \text{ ft}$$

but required depth is 7.1 ft.

$$\text{fall} = 7.1 - 6.5 = 0.6 \text{ ft} \approx 7 \text{ in.}$$

The invert elevation is therefore $224.0 - 0.6 = 223.4$ ft. Note that the value obtained for the fall may be either negative, zero, or positive. When a negative or zero value is obtained, use zero. When a positive value that is regarded as being too large is obtained, another culvert configuration must be selected and the procedure repeated. In this case, a fall of 7 in. is acceptable, and the culvert is located with its inlet invert at 223.4 ft.

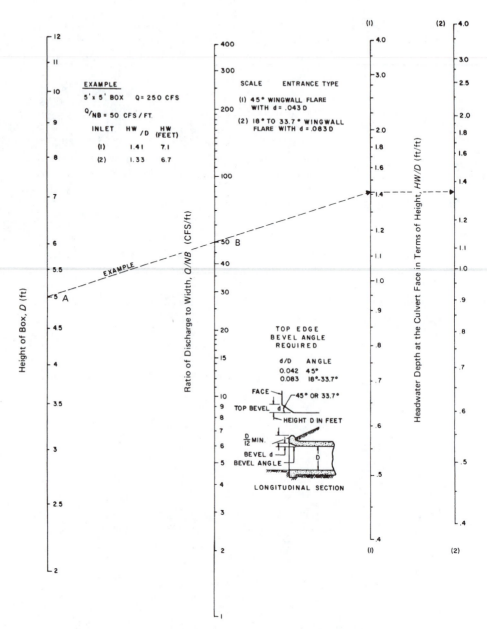

Figure 17.16 Headwater Depth for Inlet Control, Rectangular Box Culverts, Flared Wingwalls 18° to 33.7° and 45° with Beveled Edge at Top of Inlet

SOURCE: J. M. Normann, R. J. Houghtalen, W. J. Johnston, *Hydraulic Design of Highway Culverts,* Report No. FHWA-IP-85-15, U.S. Department of Transportation, Office of Implementation, McLean, Va., September 1985.

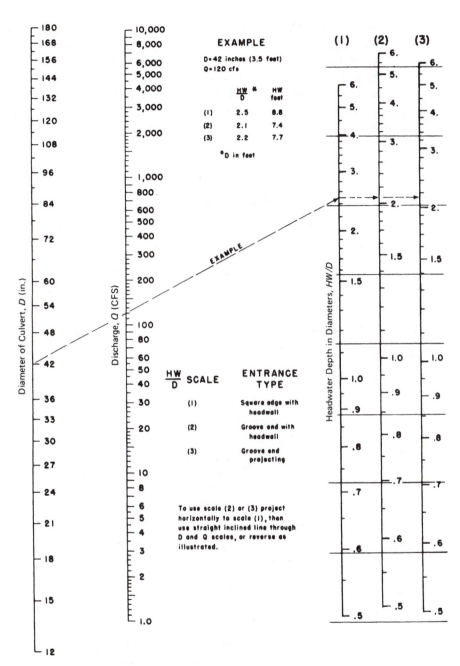

Figure 17.17   Headwater Depth for Concrete Pipe Culverts with Inlet Control

SOURCE: J. M. Normann, R. J. Houghtalen, W. J. Johnston, *Hydraulic Design of Highway Culverts,* Report No. FHWA-IP-85-15, U.S. Department of Transportation, Office of Implementation, McLean, Va., September 1985.

Outlet Control.   A culvert flows under outlet control when the barrel is incapable of transporting as much flow as the inlet opening will receive. Figure 17.18 shows different types of flows under outlet control conditions, where the control section is located at the downstream end of the culvert or beyond. In Figure 17.18(a), both the inlet and outlet of the culvert are submerged, and the water flows under pressure along the whole length of the culvert with the culvert completely full. This is a common design assumption, although it does not often occur in practice. Figure 17.18(b) shows the inlet unsubmerged

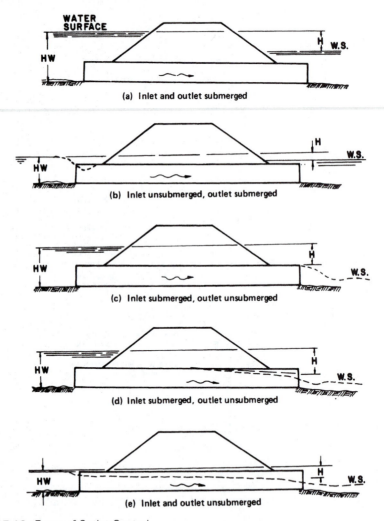

Figure 17.18 Types of Outlet Control

SOURCE: J. M. Normann, R. J. Houghtalen, W. J. Johnston, *Hydraulic Design of Highway Culverts,* Report No. FHWA-IP-85-15, U.S. Department of Transportation, Office of Implementation, McLean, Va., September 1985.

and the outlet submerged. This usually occurs when the headwater depth is low, resulting in the top of the culvert being above the surface of the water as the water contracts into the culvert. In Figure 17.18(c), the outlet is unsubmerged and the culvert flows full along its whole length because of the high depth of the headwater. This condition does not occur often, since it requires very high inlet heads. High outlet velocities are obtained under this condition. In Figure 17.18(d), the culvert inlet is submerged and the outlet unsubmerged, with a low tailwater depth. The flow in the culvert is therefore partly full along part of its length. The flow is also subcritical along part of the culvert's length, but the critical depth is passed just upstream of the outlet. Figure 17.18(e) shows both the culvert's inlet and outlet unsubmerged with the culvert partly full along its entire length, with the flow being subcritical.

In addition to the factors that affect the performance of culverts under inlet control, the performance of culverts under outlet control is also affected by the tailwater depth and certain culvert characteristics, which include the roughness, area, shape, slope, and length.

The hydraulic analysis of culverts flowing under outlet control is based on energy balance. The total energy loss through the culvert is given as

$$H_L = H_e + H_f + H_o + H_b + H_j + H_g \tag{17.15}$$

where
H_L = total energy required
H_e = energy loss at entrance
H_f = friction loss
H_o = energy loss at exit
H_b = bend loss
H_j = energy loss at junction
H_g = energy loss at safety grates

Losses due to bend, junction, and grates occur only when these features are incorporated in the culvert. For culverts without these features, the total head loss is given as

$$H_L = \left(1 + k_e + \frac{29n^2 L}{R^{1.33}}\right)\frac{V^2}{2g} \tag{17.16}$$

where
k_e = factor based on various inlet configurations (see Table 17.11)
n = Manning's coefficients for culverts (see Table 17.12)
R = hydraulic radius of the full culvert barrel = $\frac{a}{p}$(ft)
L = length of culvert barrel (ft)
V = velocity in the barrel (ft/s)

Table 17.11   Entrance Loss Coefficients

Type of Structure and Design of Entrance	Coefficient, k_e
Pipe, Concrete	
Projecting from fill, socket end (groove-end)	0.2
Projecting from fill, sq. cut end	0.5
Headwall or headwall and wingwalls	
Socket end of pipe (groove-end)	0.2
Square-edge	0.5
Rounded (radius = $\frac{1}{12}D$)	0.2
Mitered to conform to fill slope	0.7
End-section conforming to fill slope	0.5
Beveled edges, 33.7° or 45° bevels	0.2
Side- or slope-tapered inlet	0.2
Pipe or Pipe-Arch, Corrugated Metal	
Projecting from fill (no headwall)	0.9
Headwall or headwall and wingwalls square-edge	0.5
Mitered to conform to fill slope, paved or unpaved slope	0.7
End-section conforming to fill slope	0.5
Beveled edges, 33.7° or 45° bevels	0.2
Side- or slope-tapered inlet	0.2
Box, Reinforced Concrete	
Headwall parallel to embankment (no wingwalls)	
Square-edged on 3 edges	0.5
Rounded on 3 edges to radius of $\frac{1}{12}$ barrel	0.2
dimension, or beveled edges on 3 sides	
Wingwalls at 30° to 75° to barrel	
Square-edged at crown	0.4
Crown edge rounded to radius of $\frac{1}{12}$ barrel	0.2
dimension, or beveled top edge	
Wingwall at 10° to 25° to barrel	
Square-edged at crown	0.5
Wingwalls parallel (extension of sides)	
Square-edged at crown	0.7
Side- or slope-tapered inlet	0.2

SOURCE: Adapted from J. M. Normann, R. J. Houghtalen, and W. J. Johnston, *Hydraulic Design of Highway Culverts*, Report No. FHWA-IP-85-15, U.S. Department of Transportation, Office of Implementation, McLean, Va., September 1985.

Table 17.12   Manning's Coefficients for Culverts

Type of Conduit	Wall and Joint Description	Manning n
Concrete pipe	Good joints, smooth walls	0.011–0.013
	Good joints, rough walls	0.014–0.016
	Poor joints, rough walls	0.016–0.017
Concrete box	Good joints, smooth finished walls	0.012–0.015
	Poor joints, rough, unfinished walls	0.014–0.018
Corrugated metal pipes and boxes, annular corrugations	$2\frac{2}{3}$ by $\frac{1}{2}$ in. corrugations	0.027–0.022
(Manning n varies with barrel size)	6 by 1 in. corrugations	0.025–0.022
	5 by 1 in. corrugations	0.026–0.025
	3 by 1 in. corrugations	0.028–0.027
	6 by 2 in. structural plate corrugations	0.035–0.033
	9 by $2\frac{1}{2}$ in. structural plate corrugations	0.037–0.033
Corrugated metal pipes, helical corrugations, full circular flow	$2\frac{2}{3}$ by $\frac{1}{2}$ in. corrugations 24 in. plate width	0.012–0.024
Spiral rib metal pipe	$\frac{3}{4}$ by $\frac{3}{4}$ in. recesses at 12 in. spacing, good joints	0.012–0.013

SOURCE: Adapted from J. M. Normann, R. J. Houghtalen, and W. J. Johnston, *Hydraulic Design of Highway Culverts,* Report No. FHWA-IP-85-15, U.S. Department of Transportation, Office of Implementation, McLean, Va., September 1985.

When special features such as grates, bends, and junctions are incorporated in the culvert, the appropriate additional losses may be determined from one or more of the following equations.

The bend loss H_b is given as

$$H_b = k_b \frac{v^2}{2g} \tag{17.17}$$

where

k_b = bend loss coefficient (see Table 17.13)
v = flow velocity in the culvert barrel (ft/sec)
g = 32.2 ft/sec^2

The junction loss H_j is given as

$$H_j = y' + H_{v1} - H_{v2} \tag{17.18}$$

where

y' = change in hydraulic grade line through the junction
$= (Q_2 v_2 - Q_1 v_1 - Q_3 v_3 \cos \theta_j)/[0.5(a_1 + a_2)g]$

Table 17.13 Loss Coefficients for Bends

Radius of Bend	Angle of Bend		
Equivalent Diameter	90°	45°	22.5°
1	0.50	0.37	0.25
2	0.30	0.22	0.15
4	0.25	0.19	0.12
6	0.15	0.11	0.08
8	0.15	0.11	0.08

SOURCE: Adapted from Ray F. Linsley and Joseph B. Franzini, *Water Resources Engineering*, McGraw Hill Book Company, copyright © 1979.

Q_i = flow rate in barrel i (see Figure 17.19)
v_i = velocity in barrel i (ft/sec)
a_i = cross-sectional area of barrel i (ft^2)
θ_j = angle of the lateral with respect to the outlet conduit (degrees)
H_{v1} = velocity head in the upstream conduit (ft)
H_{v2} = velocity head in the downstream conduit (ft)

The head loss due to bar grate (H_g) is given as

$$H_g = k_g \frac{W}{x} \frac{v_u^2}{2g} \sin \theta_g \qquad (17.19)$$

where
x = minimum clear spacing between bars (ft)
W = maximum cross-sectional width of the bars facing the flow (ft)
θ_g = angle of grates with respect to the horizontal (degrees)
v_u = approach velocity (ft/sec)
k_g = dimensionless bar shape factor
  = 2.42 for sharp-edged rectangular bars
  = 1.83 for rectangular bars with semicircular upstream face
k_g = 1.79 for circular bars
  = 1.67 for rectangular bars with semicircular upstream and downstream faces

Note that Eqs. 17.18 and 17.19 are both empirical, and caution must be exercised in using them.

Figure 17.20 is a schematic of the energy grade lines for a culvert flowing full. If the total energies at the inlet and outlet are equated, then

$$HW_o + \frac{v_u^2}{2g} = TW + \frac{v_d^2}{2g} + H_L \qquad (17.20)$$

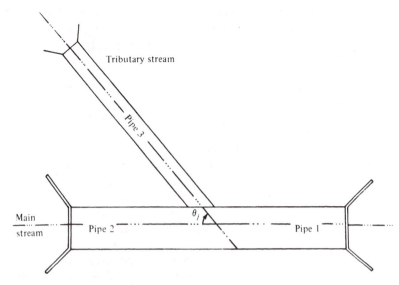

Figure 17.19 Culvert Junction

SOURCE: J. M. Normann, R. J. Houghtalen, W. J. Johnston, *Hydraulic Design of Highway Culverts,* Report No. FHWA-IP-85-15, U.S. Department of Transportation, Office of Implementation, McLean, Va., September 1985.

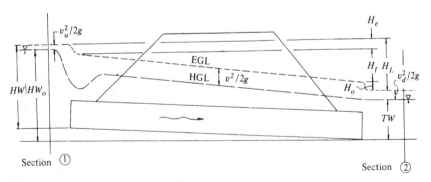

Figure 17.20 Full Flow Energy Grade Line (EGL) and Hydraulic Grade Line (HGL)

SOURCE: J. M. Normann, R. J. Houghtalen, W. J. Johnston, *Hydraulic Design of Highway Culverts,* Report No. FHWA-IP-85-15, U.S. Department of Transportation, Office of Implementation, McLean, Va., September 1985.

where
$\quad HW_o$ = headwater depth above the outlet invert (ft)
$\quad\quad v_u$ = approach velocity
$\quad\; TW$ = tailwater depth above the outlet invert (ft)
$\quad\quad v_d$ = downstream velocity (ft/sec)
$\quad\; H_L$ = sum of all losses
$\quad\quad g$ = 32.2 ft/sec^2

When the approach and downstream velocity heads are both neglected, we obtain

$$HW_o = TW + H_L \qquad (17.21)$$

Note that Eqs. 17.15, 17.16, 17.20, and 17.21 were developed for the culvert flowing full and therefore apply to the conditions shown in Figure 17.18(a), (b), and (c). Additional calculations may be required for the conditions shown in Figure 17.18(d) and (e). These additional calculations are beyond the scope of this book but are discussed in detail in *Hydraulic Design of Highway Culverts*.

Nomographs have also been developed for solving Eq. 17.20, for different configurations of culverts flowing full and performing under outlet control. Only entrance, friction, and exit losses are considered in the nomographs. Figures 17.21 and 17.22 show examples of these nomographs for a concrete box culvert and a circular concrete pipe culvert. Figures 17.23 and 17.24 show the critical depth charts for these culverts, which we also use in the design.

The nomographs for outlet control conditions are also used to determine the depth of the headwater required to accommodate the design flow through the selected culvert configuration under outlet control. The procedure is demonstrated in Example 17.7.

Example 17.7   Computing Required Headwater Elevation for a Culvert Flowing Full Under Outlet Control

Determine the headwater elevation for Example 17.6 if the culvert is flowing full under outlet control, the tailwater depth above the outlet invert at the design flow rate is 6.5 ft, the length of culvert is 200 ft, and the natural stream slopes at 2 percent. The tailwater depth is determined using normal depth or backwater depth calculations or from on-site inspection. Assume $n = 0.012$.

Solution: The charts shown in Figures 17.21 and 17.23 are used in the following steps:

Step 1.   From Figure 17.23, determine the critical depth

$$Q/B = 50 \qquad d_c = 4.3 \text{ ft}$$

Alternatively, d_c may be obtained from the equation

$$d_c = 0.315 \sqrt[3]{(Q/B)^2}$$

Step 2.   The depth (h_o) from the outlet invert to the hydraulic grade line is then determined. This is taken as $(d_c + D)/2$ or the tailwater depth TW, whichever is greater. In this case,

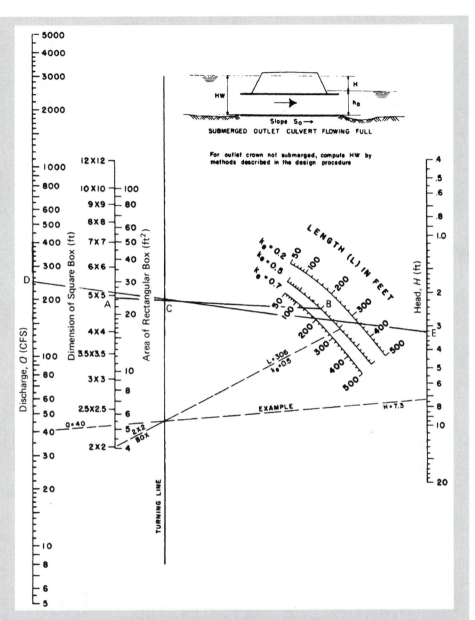

Figure 17.21   Headwater Depth for Concrete Box Culverts Flowing Full (*n* = 0.012)

SOURCE: J. M. Normann, R. J. Houghtalen, W. J. Johnston, *Hydraulic Design of Highway Culverts,* Report No. FHWA-IP-85-15, U.S. Department of Transportation, Office of Implementation, McLean, Va., September 1985.

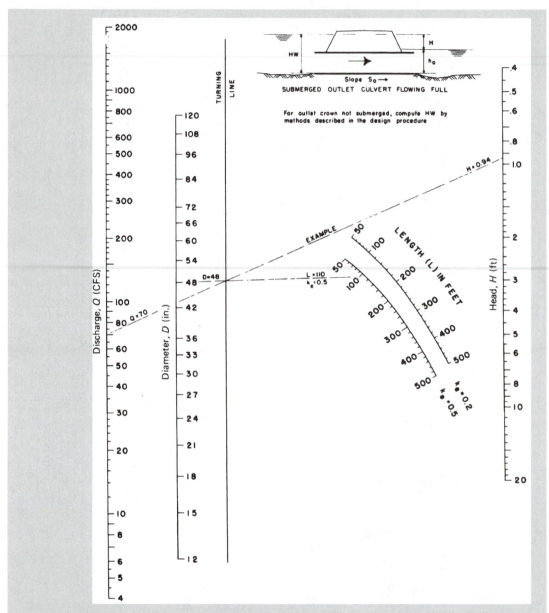

Figure 17.22   Headwater Depth for Concrete Pipe Flowing Full ($n = 0.012$)

SOURCE: J. M. Normann, R. J. Houghtalen, W. J. Johnston, *Hydraulic Design of Highway Culverts,* Report No. FHWA-IP-85-15, U.S. Department of Transportation, Office of Implementation, McLean, Va., September 1985.

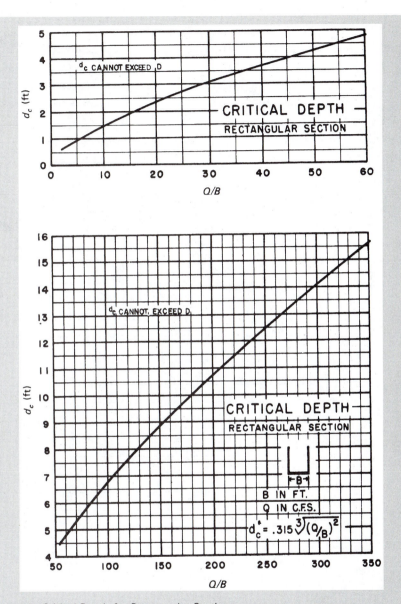

Figure 17.23 Critical Depth for Rectangular Sections

SOURCE: J. M. Normann, R. J. Houghtalen, W. J. Johnston, *Hydraulic Design of Highway Culverts,* Report No. FHWA-IP-85-15, U.S. Department of Transportation, Office of Implementation, McLean, Va., September 1985.

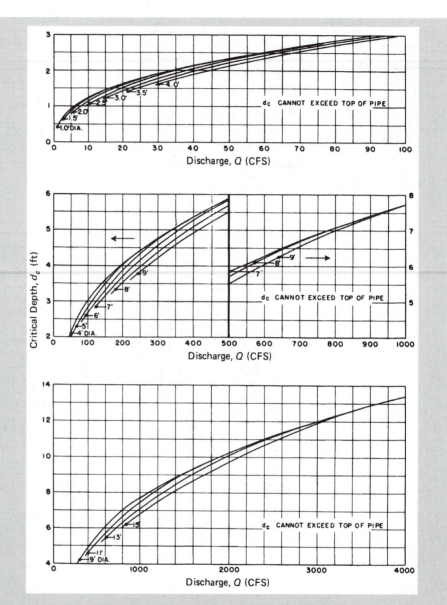

Figure 17.24 Critical Depth for Circular Pipes

SOURCE: J. M. Normann, R. J. Houghtalen, W. J. Johnston, *Hydraulic Design of Highway Culverts,* Report No. FHWA-IP-85-15, U.S. Department of Transportation, Office of Implementation, McLean, Va., September 1985.

$$TW = 6.5 \text{ ft}$$

$$(d_c + D)/2 = \frac{43+5}{2} \simeq 4.7 \text{ ft}$$

$$h_o = 6.5 \text{ ft}$$

Step 3.   The inlet coefficient k_e is obtained from Table 17.11 as 0.5.

Step 4.   Locate the size, length, and k_e of the culvert as shown at A and B in Figure 17.21. Draw a straight line through A and B and locate the intersection C of this line with the turning line. (Note that when the culvert material has a different n value than that given in the nomograph, an adjusted length L_1 of the culvert is determined as

$$L_1 = L\left(\frac{n_1}{n}\right)^2 \tag{17.22}$$

where

L = length of culvert (ft)

n_1 = desired Manning coefficient

n = Manning n value for outlet control chart

Step 5.   Locate the discharge D on the discharge scale and draw a straight line joining C and D. Extend this line to the head loss scale at E and determine the energy loss through the culvert. The total head loss (H) is obtained as 3.3 ft.

Step 6.   The required outlet headwater elevation EL_{ho} is computed as

$$EL_{ho} = EL_o + H + h_o \tag{17.23}$$

where EL_o is the invert elevation at outlet. In this case,

$$EL_o = 223.4 \text{ (inlet invert from Example 17.6)} - 0.017 \times 200 = 220.0 \text{ ft}$$

and

$$EL_{ho} = 219.4 + 3.3 + 6.5 = 229.8$$

If this computed outlet control headwater elevation is higher than the design headwater elevation, another culvert should be selected and the procedure repeated.

The design headwater elevation is 230.5 ft (from Example 17.6). The 5 ft × 5 ft culvert is therefore acceptable.

In the design of a culvert, the headwater elevations are computed for inlet and outlet controls and the control with the higher headwater elevation is selected as the controlling condition. In the above case, for example, the headwater elevation for inlet control = 223.4 + 7.1 = 230.5 ft, and the headwater elevation for outlet control is 229.8 ft, which means that the inlet control governs.

The outlet velocity is then determined for the governing control. When the inlet control governs, the normal depth velocity is taken as the outlet velocity. When the outlet control governs, the outlet velocity is determined by using the area of flow at the outlet, which is based on the culvert's geometry and the following conditions:

1. If the tailwater depth is less than critical depth, use the critical depth.
2. If the level of the tailwater is between the critical depth and the top of the culvert barrel, use the tailwater depth.
3. If the tailwater is above the top of the barrel, use the height of the culvert barrel.

In this example, since the inlet control governs, the outlet velocity should be determined based on the normal depth.

The whole design procedure can be carried out using a table similar to the one in Figure 17.25, which facilitates the trial of different pipe configurations.

Computer Programs for Culvert Design and Analysis

Examples 17.6 and 17.7 indicate how tedious the analysis or design of a culvert can be, even with the use of available charts. Several computer and hand calculator programs are presently available that can be used to increase the accuracy of the results and to significantly reduce the time it takes for design or analysis.

Programs available for hand calculators include the Calculator Design Series (CDS) 1,2,3 and the Calculator Design Series (CDS) 4, developed by FHWA. The programs in CDS 1,2,3 are suitable for the Compucorp 325 Scientist, the HP-65, and the TI-59 programmable calculators, whereas the CDS 4 is suitable only for the TX-54. The CDS 1,2,3 consists of a set of subroutines that are run sequentially, some of which provide inputs for subsequent ones. The program can be used to analyze culvert sizes with different inlet configurations. The outputs include the dimensions of the barrel, the performance data, and the outlet velocities.

The CDS 4 program can be used to analyze corrugated metal and concrete culverts. This program also consists of a series of subroutines.

FHWA has also produced various software packages, some of which are suitable for use on mainframe computers and others on personal computers. The program H4-2 can be used to design pipe-arch culverts on a mainframe computer, whereas H4-6, which is also run on a mainframe, can be used to obtain a list of optional circular and box culverts for the specified site and hydrologic conditions.

A program written in basic for use on an IBM/PC has also been developed by FHWA. The program analyzes a culvert specified by the user for a given set of hydrologic data and site conditions. It is suitable for circular, rectangular, elliptical, arch, and other geometrics defined by the user and will compute the hydraulic characteristics of the culvert.

Note, however, that both computer hardware and software in all fields of engineering are rapidly being improved, which makes it imperative that designers keep abreast of the development of new software packages and hardware equipment.

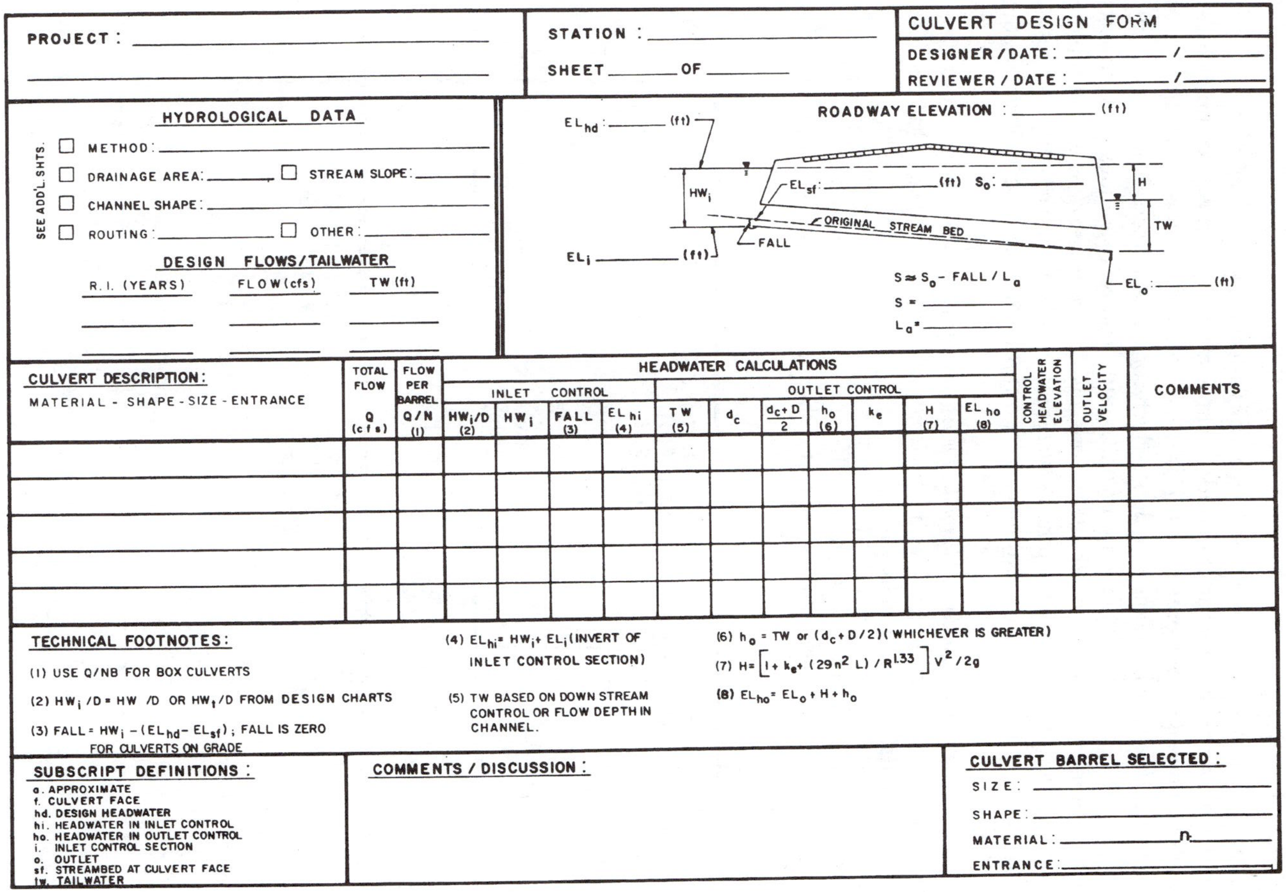

PROJECT :

STATION :

SHEET ____ OF ____

CULVERT DESIGN FORM

DESIGNER / DATE : ____ / ____

REVIEWER / DATE : ____ / ____

HYDROLOGICAL DATA

SEE ADD'L SHTS.

- ☐ METHOD :
- ☐ DRAINAGE AREA : ____ ☐ STREAM SLOPE :
- ☐ CHANNEL SHAPE :
- ☐ ROUTING : ____ ☐ OTHER :

DESIGN FLOWS/TAILWATER

R.I. (YEARS)	FLOW (cfs)	TW (ft)

ROADWAY ELEVATION : ____ (ft)

EL_{hd} : ____ (ft)

HW_i

EL_{sf} : ____ (ft) S_o :

ORIGINAL STREAM BED

FALL

H

TW

EL_i ____ (ft)

$S \approx S_o - FALL / L_a$

S =

L_a =

EL_o : ____ (ft)

CULVERT DESCRIPTION: MATERIAL - SHAPE - SIZE - ENTRANCE	TOTAL FLOW Q (cfs)	FLOW PER BARREL Q/N (1)	HEADWATER CALCULATIONS: INLET CONTROL HW_i/D (2)	HW_i	FALL (3)	EL_{hi} (4)	OUTLET CONTROL TW (5)	d_c	$\frac{d_c + D}{2}$	h_o (6)	k_e	H (7)	EL_{ho} (8)	CONTROL HEADWATER ELEVATION	OUTLET VELOCITY	COMMENTS

TECHNICAL FOOTNOTES:

(1) USE Q/NB FOR BOX CULVERTS

(2) $HW_i/D = HW/D$ OR HW_i/D FROM DESIGN CHARTS

(3) $FALL = HW_i - (EL_{hd} - EL_{sf})$; FALL IS ZERO FOR CULVERTS ON GRADE

(4) $EL_{hi} = HW_i + EL_i$ (INVERT OF INLET CONTROL SECTION)

(5) TW BASED ON DOWN STREAM CONTROL OR FLOW DEPTH IN CHANNEL.

(6) $h_o = TW$ or $(d_c + D/2)$ (WHICHEVER IS GREATER)

(7) $H = \left[1 + k_e + (29n^2 L)/R^{1.33}\right] V^2/2g$

(8) $EL_{ho} = EL_o + H + h_o$

SUBSCRIPT DEFINITIONS :

a. APPROXIMATE
f. CULVERT FACE
hd. DESIGN HEADWATER
hi. HEADWATER IN INLET CONTROL
ho. HEADWATER IN OUTLET CONTROL
i. INLET CONTROL SECTION
o. OUTLET
sf. STREAMBED AT CULVERT FACE
tw. TAILWATER

COMMENTS / DISCUSSION :

CULVERT BARREL SELECTED :

SIZE :

SHAPE :

MATERIAL : ____ n: ____

ENTRANCE :

Figure 17.25 Work Sheet for Culvert Design

SOURCE: J. M. Normann, R. J. Houghtalen, W. J. Johnston, *Hydraulic Design of Highway Culverts,* Report No. FHWA-IP-85-15, U.S. Department of Transportation, Office of Implementation, McLean, Va., September 1985.

Inlet Configuration

It was stated earlier that a culvert's performance is affected by the inlet configurations. However, since the design discharge for outlet control results in full flow, the entrance loss for a culvert flowing under outlet control is usually a small fraction of the headwater requirements. Extensive inlet configurations, which considerably increase the cost of culverts, are therefore unnecessary for culverts under outlet control conditions. With culverts flowing under inlet control, however, the culverts' hydraulic capacity depends only on the inlet configuration and headwater depth. A suitable inlet configuration can therefore result in full or nearly full flow of a culvert under inlet control. This significantly increases the capacity of the culvert.

The design charts provided in *Hydraulic Design of Highway Culverts* include charts for improved inlet configurations, such as bevel-edged, side-tapered, and slope-tapered, that help to increase the culvert capacity.

Bevel-Edged Inlets.   Figure 17.26 shows different beveled-edged configurations. The bevel edge is similar to a chamfer, except that a chamfer is usually much smaller. It has

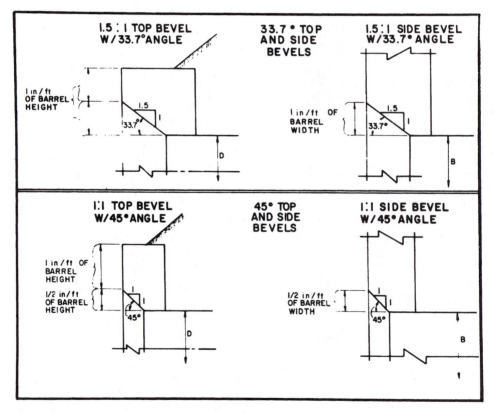

Figure 17.26 Beveled Edges

SOURCE: J. M. Normann, R. J. Houghtalen, W. J. Johnston, *Hydraulic Design of Highway Culverts,* Report No. FHWA-IP-85-15, U.S. Department of Transportation, Office of Implementation, McLean, Va., September 1985.

been estimated that the addition of bevels to a culvert having a square-edge inlet will increase the culvert capacity by 15 to 20 percent. As a minimum, therefore, all culverts operating under inlet control should be fitted with bevels. Use of bevels on culverts under outlet control is also recommended, since the entrance loss coefficient may be reduced by up to 40 percent.

Tapered Inlets.   Tapered inlets increase the capacity of culverts under inlet control, mainly by reducing the contraction at the inlet control section. Tapered inlets are more effective than bevel-edged inlets on culverts flowing under inlet control but have similar results as bevels when used on culverts under outlet control. Since tapered inlets are more expensive, they are not recommended for culverts under outlet control. Design charts are available for rectangular box and circular pipes for two types of tapered inlets: side-tapered and slope-tapered.

Figure 17.27 shows different designs of a *side-tapered inlet.* It consists of an enlarged faced section that uniformly reduces to the culvert barrel size by tapering the side walls. The inlet flow of the side taper is formed by extending the flow of the culvert barrel outward, with the face section being approximately the same height as the barrel height. The throat section is the intersection of the culvert barrel and the tapered side walls. Either the throat or face section may act as the inlet control section, depending on the inlet design. When the throat acts on the inlet control section, the headwater depth is measured from the throat section invert HW_t, and when the face acts as the inlet control section, the headwater depth is measured from the face section invert HW_f. See Figure 17.27(a). It is advantageous for the throat section to be the primary control section since the throat is usually lower than the face, resulting in a higher head on the throat for a specified headwater elevation. Figures 17.27(b) and (c) show two ways of increasing the effectiveness of the side-tapered inlet. In Figure 17.27(b), the throat section is depressed below the stream bed, and a depression constructed between the two wing walls. When this type of construction is used, it is recommended that the culvert barrel floor be extended upstream a minimum distance of $D/2$, before the steep upward slope begins. In Figure 17.27(c) a sump is constructed upstream from the face section, with the dimensional requirements given. When the side-tapered inlet is constructed as in Figures 17.27(b) or (c), a crest is formed upstream at the intersection of the stream bed and the depression slope. This crest may act as a weir if its length is too short. It should therefore be ascertained that the crest does not control the flow at the design flow and headwater.

The *slope-tapered inlet* is similar to the side-tapered inlet, in that it also has an enlarged face section, which is gradually reduced to the barrel size at the throat section by sloping the sidewalls. The slope-tapered inlet, however, also has a uniform vertical drop (fall) between the face and throat sections. (See Figure 17.28.) A third section known as the bend is also placed at the intersection of the inlet slope and barrel slope as shown in Figure 17.28.

Any one of three sections may act as the primary control section of the slope-tapered section. These are the face, the bend, and the throat. Design procedures, which are beyond the scope of this book, for the dimensions of the throat and face sections are given in *Hydraulic Design of Highway Culverts.* The only criterion given for the size of the bend section is that it should be located a minimum distance from the throat. Again, the slope-tapered inlet is most efficient when the throat acts as the primary control section.

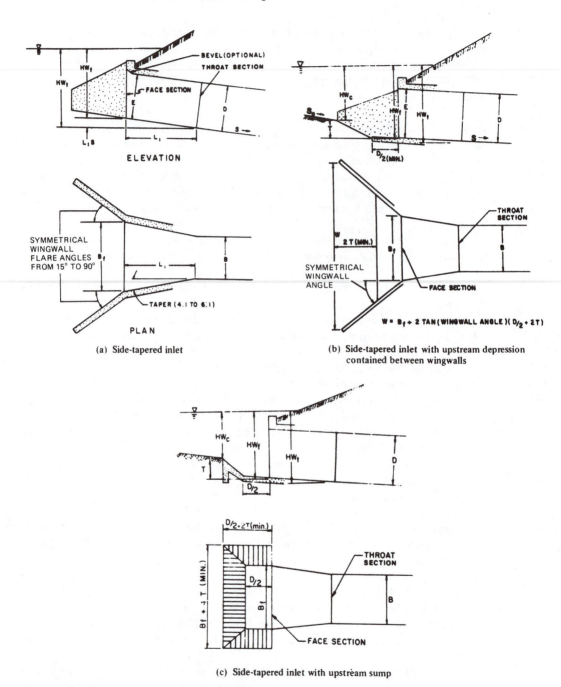

(a) Side-tapered inlet

(b) Side-tapered inlet with upstream depression contained between wingwalls

(c) Side-tapered inlet with upstream sump

Figure 17.27 Side-Tapered Inlets

SOURCE: J. M. Normann, R. J. Houghtalen, W. J. Johnston, *Hydraulic Design of Highway Culverts,* Report No. FHWA-IP-85-15, U.S. Department of Transportation, Office of Implementation, McLean, Va., September 1985.

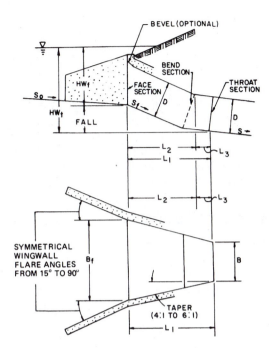

Figure 17.28 Slope-Tapered Inlet with Vertical Face

SOURCE: J. M. Normann, R. J. Houghtalen, W. J. Johnston, *Hydraulic Design of Highway Culverts,* Report No. FHWA-IP-85-15, U.S. Department of Transportation, Office of Implementation, McLean, Va., September 1985.

SUBSURFACE DRAINAGE

Subsurface drainage systems are provided within the pavement structure to drain water in one or more of the following forms:

* Water that has permeated through cracks and joints in the pavement to the underlying strata
* Water that has moved upward through the underlying soil strata as a result of capillary action
* Water that exists in the natural ground below the water table, usually referred to as ground water

The subsurface drainage system must be an integral part of the total drainage system, since the subsurface drains must operate in consonant with the surface drainage system to obtain an efficient overall drainage system.

The design of subsurface drainage should be carried out as an integral part of the complete design of the highway, since inadequate subsurface drainage may also have detrimental effects on the stability of slopes and pavement performance. However, certain design elements of the highway such as geometry and material properties are required for the design of the subdrainage system. Thus, the procedure usually adopted for subdrainage design is first to determine the geometric and structural requirements of the highway based on standard design practice, and then to subject these to a subsurface drainage analysis to determine the subdrainage requirements. In some cases, the subdrainage requirements determined from this analysis will require some changes in the original design.

It is extremely difficult, if not impossible, to develop standard solutions for solving subdrainage problems because of the many different situations that engineers come across in practice. Therefore, basic methods of analysis are given that can be used as tools to identify solutions for subdrainage problems. The experience gained from field and laboratory observations for a particular location, coupled with good engineering judgment, should always be used in conjunction with the design tools provided. Before presenting the design tools, discussions of the effects on the highway of an inadequate subdrainage system and the different subdrainage systems are first presented.

Effect of Inadequate Subdrainage

Inadequate subdrainage on a highway will result in the accumulation of uncontrolled subsurface water within the pavement structure and/or right of way, which can result in poor performance of the highway or outright failure of sections of the highway. The effects of inadequate subdrainage fall into two classes: poor pavement performance and instability of slopes.

Pavement Performance

If the pavement structure and subgrade are saturated with underground water, the pavement's ability to resist traffic load is considerably reduced, resulting in one or more of several problems, which can lead to a premature destruction of the pavement if remedial actions are not taken in time. In Portland cement concrete pavement, for example, inadequate subdrainage can result in excessive repeated deflections of the pavement (see "Pumping of Rigid Pavements" in Chapter 21), which will eventually lead to cracking.

When asphaltic concrete pavements are subjected to excessive uncontrolled subsurface water due to inadequate subdrainage, very high pore pressures are developed within the untreated base and subbase layers (see Chapter 20 for base and subbase definitions), resulting in a reduction of the pavement strength and thereby its ability to resist traffic load.

Another common effect of poor pavement performance due to inadequate subdrainage is frost action. As described later, this phenomenon requires that the base and/or subbase material be a frost-susceptible soil and that an adequate amount of subsurface water is present in the pavement structure. Under these conditions, during the active freezing period, subsurface water will move upward by capillary action toward the freezing zone and subsequently freeze to form lenses of ice. Continuous growth of the ice lenses due to the capillary action of the subsurface water can result in considerable heaving of the overlying pavement. This eventually leads to serious pavement damage, particularly if differential frost heaving occurs. Frost action also has a detrimental effect on pavement performance during the spring thaw period. During this period, the ice lenses formed during the active freeze period gradually thaw from the top down, resulting in the saturation of the subgrade soil, which results in a substantial reduction of pavement strength.

Slope Stability

The presence of subsurface water in an embankment or cut can cause an increase of the stress to be resisted and a reduction of the shear strength of the soil forming the embankment or cut. This can lead to the condition where the stress to be resisted is greater than the strength of the soil, resulting in sections of the slope crumbling down or a complete failure of the slope.

Highway Subdrainage Systems

Subsurface drainage systems are usually classified into five general categories:

- Longitudinal drains
- Transverse drains
- Horizontal drains
- Drainage blankets
- Well systems

Longitudinal Drains

Subsurface longitudinal drains usually consist of pipes laid in trenches, within the pavement structure and parallel to the center line of the highway. These drains can be used to lower the water table below the pavement structure, as shown in Figure 17.29, or to remove any water that is seeping into the pavement structure, as shown in Figure 17.30. In some cases, when the water table is very high and the highway is very wide, it may be necessary to use more than two rows of longitudinal drains to achieve the required reduction of the water table below the pavement structure (see Figure 17.31).

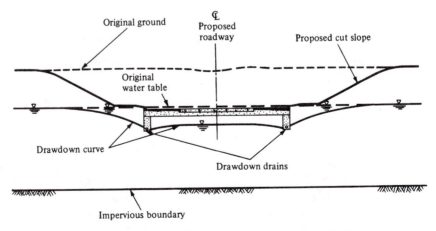

Figure 17.29 Symmetrical Longitudinal Drains Used to Lower Water Table

SOURCE: Redrawn from *Highway Subdrainage Design,* Report No. FHWA-TS-80-224, U.S. Department of Transportation, Washington, D.C., August 1980.

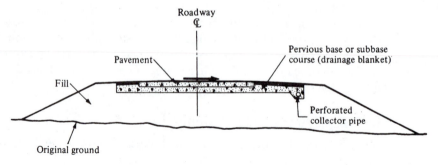

Figure 17.30 Longitudinal Collector Drain Used to Remove Water Seeping into Pavement Structural Section

SOURCE: Redrawn from *Highway Subdrainage Design,* Report No. FHWA-TS-80-224, U.S. Department of Transportation, Washington, D.C., August 1980.

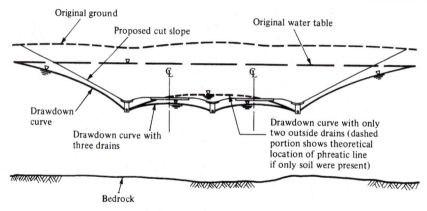

Figure 17.31 Multiple Longitudinal Drawdown Drain Installation

SOURCE: Redrawn from *Highway Subdrainage Design,* Report No. FHWA-TS-80-224, U.S. Department of Transportation, Washington, D.C., August 1980.

Transverse Drains

Transverse drains are placed transversely below the pavement, usually in a direction perpendicular to the center line, although they may be skewed to form a herringbone configuration. An example of the use of transverse drains is shown in Figure 17.32, where they are used to drain ground water that has infiltrated through the joints of the pavement. One disadvantage of transverse drains is that they can cause unevenness of the pavement when used in areas susceptible to frost action, where general frost heaving occurs. The unevenness is due to the general heaving of the whole pavement, except at the transverse drains.

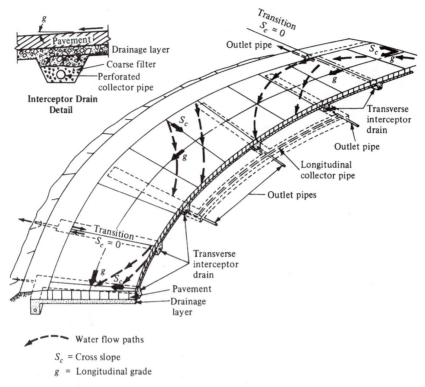

Figure 17.32 Transverse Drains on Superelevated Curves

SOURCE: Redrawn from *Highway Subdrainage Design,* Report No. FHWA-TS-80-224, U.S. Department of Transportation, Washington, D.C., August 1980.

Horizontal Drains

Horizontal drains are used to relieve pore pressures at slopes of cuts and embankments on the highway. They usually consist of small-diameter, perforated pipes inserted into the slopes of the cut or fill. The subsurface water is collected by the pipes and is then discharged at the face of the slope through paved spillways to longitudinal ditches.

Drainage Blankets

A drainage blanket is a layer of material that has a very high coefficient of permeability, usually greater than 30 ft/day, and is laid beneath or within the pavement structure such that its width and length in the flow direction are much greater than its thickness. The coefficient of permeability is the constant of proportionality of the relationship between the flow velocity and the hydraulic gradient between two points in the material (see Chapter 18). Drainage blankets can be used to facilitate the flow of subsurface water away from the pavement, as well as to facilitate the flow of ground water that has seeped through cracks into the pavement structure or subsurface water from artesian sources. A drainage blanket can also be used in conjunction with longitudinal drains to improve the stability of cut slopes by controlling the flow of water on the slopes, thereby preventing the formation of a slip surface. However, drainage blankets must be properly designed to be effective. Figure 17.33 shows two drainage blanket systems.

Well Systems

A well system consists of a series of vertical wells, drilled into the ground, into which ground water flows, thereby reducing the water table and releasing the pore pressure. When used as a temporary measure for construction, the water collected in the wells is continuously pumped out, or else it may be left to overflow. A more common construction, however, includes a drainage layer either at the top or bottom of the wells to facilitate the flow of the water collected.

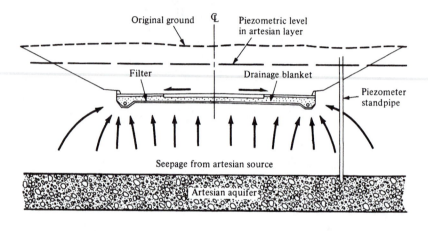

(a)

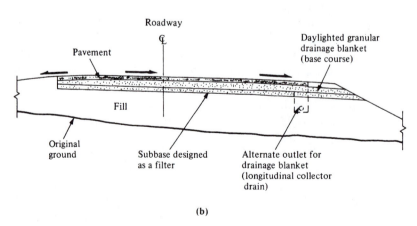

(b)

Figure 17.33 Applications of Horizontal Drainage Blankets

SOURCE: Redrawn from *Highway Subdrainage Design,* Report No. FHWA-TS-80-224, U.S. Department of Transportation, Washington, D.C., August 1980.

Design of Subsurface Drainage

The design procedure for subsurface drainage involves the following:

1. Summarize the available data.
2. Determine the quantity of water for which the subdrainage system is being designed.
3. Determine the drainage system required.
4. Determine the capacity and spacing of longitudinal and transverse drains and select filter material, if necessary.
5. Evaluate the design with respect to economic feasibility and long-term performance.

Summarize Available Data

The data that should be identified and summarized can be divided into the following four classes:

- Flow geometry
- Materials' properties
- Hydrologic and climatic characteristics
- Miscellaneous information

The flow geometry is given by the existing subsurface characteristics of the area in which the highway is located and by the geometric characteristics of the highway. These are used to determine whether any special subdrainage problems exist and what conditions must be considered in developing solutions for these problems.

The main material property required is the permeability, since this is the property that indicates the extent to which water will flow through the material.

Hydrologic and climatic characteristics will indicate precipitation rates, the sources of subsurface water, and the possibility of frost action.

Miscellaneous information includes all other information that will aid in the design of an effective and economic subdrainage system, including information of any impact the subdrainage system may have on future construction and any difficulties identified that may preclude the construction of a subdrainage system.

Determination of Discharge Quantity

The net amount of water to be discharged consists of the following components:

- Water due to infiltration
- Ground water
- Water due to thawing of ice lenses
- Water flowing vertically from the pavement structure

Water Due to Infiltration, q_i.   This is the amount of surface water that infiltrates into the pavement structure through cracks in the pavement surface. It is extremely difficult to calculate this amount of water exactly, since the rate of infiltration depends on the intensity of the design storm, the frequency and size of the cracks and/or joints in the pavement, the moisture conditions of the atmosphere, and the permeability characteristics of the materials below the pavement surface.

FHWA recommends the use of the following empirical relationship to estimate the infiltration rate:

$$q_i = I_c \left(\frac{N_c}{W} + \frac{W_c}{WC_s} \right) + K_p \tag{17.24}$$

where

q_i = design infiltration rate (ft³/day/ft² of drainage layer)
I_c = crack infiltration rate (ft³/day/ft of crack)
    2.4 ft³/day/ft is recommended for most designs)
N_c = number of contributing longitudinal cracks or joints
W_c = length of contributing transverse cracks (ft)
W = width of granular base or subbase subjected to infiltration (ft)
C_s = spacing of the transverse cracks or joints (ft)
K_p = rate of infiltration (ft³/day/ft²)
    = coefficient of permeability through the uncracked pavement surface

The suggested value of 2.4 for I_c normally should be used, but local experience should also be relied on to increase or decrease this value as necessary.

The value of N_c is usually taken as $N + 1$ for new pavements, where N is the number of traffic lanes. Local experience should be used to determine a value for C_s, although a value of 40 has been suggested for new bituminous concrete pavements. The rate of infiltration for Portland cement concrete and well-compacted, dense, graded asphaltic concrete pavements is usually very low and can therefore be taken as zero. However, when there is evidence of high infiltration rates, these should be determined from laboratory tests.

Example 17.8   Computing Infiltration Rate of a Flexible Pavement

Determine the infiltration rate for a new two-lane flexible pavement with the following characteristics:

Lane width = 11 ft
Shoulder width = 8 ft
Number of contributing longitudinal cracks $(N_c) = (N + 1) = 3$
Length of contributing transverse cracks $(W_c) = 20$ ft
K_p (from laboratory tests) = 0.03
$W = 38$ ft
Spacing of transverse cracks $(C_s) = 35$ ft

Solution: Assuming $I_c = 2.4$ ft³/day/ft², then from Eq. 17.24,

$$q_i = 2.4 \left[\frac{3}{38} + \frac{20}{38(35)} \right] + 0.03$$

$$= 0.23 + 0.03$$

$$= 0.26 \ \text{ft}^3/\text{day/ft}^2$$

Ground Water. When it is not possible to intercept the flow of ground water or lower the water table sufficiently before the water reaches the pavement, it is necessary to determine the amount of ground water seepage that will occur. Figures 17.29 and 17.33(a) demonstrate the two possible sources of ground water of interest in this case. Figure 17.29 shows a case of gravity drainage, whereas Figure 17.33(a) shows a case of artesian flow. A simple procedure to estimate the ground water flow rate due to gravity drainage is to use the chart shown in Figure 17.34. In this case the *radius of influence* L_i is first determined as

$$L_i = 3.8(H - H_o) \qquad (17.25)$$

where

$$\begin{aligned}
H_o &= \text{thickness of subgrade below the drainage pipe (ft)} \\
H &= \text{thickness of subgrade below the natural water table (ft)} \\
H - H_o &= \text{amount of draw down (ft)}
\end{aligned}$$

The chart shown in Figure 17.34 is then used to determine the total quantity of upward flow (q_2) from which the average inflow rate q_g is determined,

$$q_g = \frac{q_2}{0.5W} \qquad (17.26)$$

where

$$\begin{aligned}
q_g &= \text{design inflow rate for gravity drainage (ft}^3\text{/day/ft}^2 \text{ of drainage layer)} \\
q_2 &= \text{total upward flow into one-half of the drainage blanket (ft}^3\text{/day/linear ft} \\
&\quad\ \text{of roadway)} \\
W &= \text{width of drainage layer (ft)}
\end{aligned}$$

For the case of the artesian flow, the average inflow rate is estimated using Darcy's law:

$$q_a = K \frac{\Delta H}{H_o} \qquad (17.27)$$

where

$$\begin{aligned}
q_a &= \text{design inflow rate from artesian flow (ft}^3\text{/day/ft}^2 \text{ of drainage area)} \\
\Delta H &= \text{excess hydraulic head (ft)} \\
H_o &= \text{thickness of the subgrade soil between the drainage layer and the} \\
&\quad\ \text{artesian aquifer (ft)} \\
K &= \text{coefficient of permeability (ft/day)}
\end{aligned}$$

Example 17.9   Computing Average Inflow Rate Due to Gravity Drainage

Using the chart shown in Figure 17.34, determine the average inflow rate (q_g) due to gravity drainage, as shown in Figure 17.35, for the following data:

Thickness of subgrade below drainage pipe (H_o) = 15 ft
Coefficient of permeability of native soil (K) = 0.4 ft/day
Width of drainage layer = 40 ft
Drawdown $(H - H_o)$ = 8 ft
Radius of influence (L_i) = 3.8 × 8 (from Eq. 17.25) = 30.4 ft

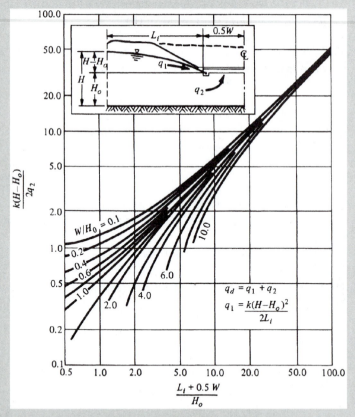

Figure 17.34   Chart for Determining Flow Rate in Horizontal Drainage Blanket

SOURCE: Redrawn from *Highway Subdrainage Design*, Report No. FHWA-TS-80-224, U.S. Department of Transportation, Washington, D.C., August 1980.

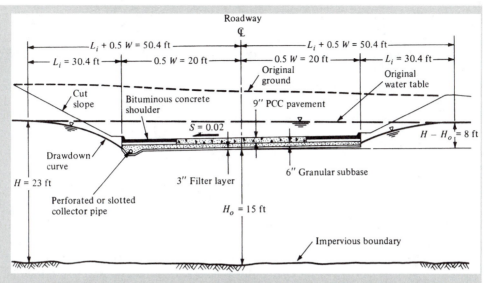

Figure 17.35 Rigid Pavement Section in Cut Dimensions and Details for Example 17.9

Solution:

$$\frac{L_i + 0.5W}{H_o} = \frac{30.4 + 0.5 \times 40}{15} = 3.36$$

$$\frac{W}{H_o} = \frac{40}{15} = 2.67$$

Entering the chart at

$$\frac{L_i + 0.5W}{H_o} = 3.36$$

and

$$W/H_o = 2.67$$

we obtain

$$\frac{K(H - H_o)}{2q_2} \approx 1.3$$

$$q_2 = \frac{0.4 \times 8}{2 \times 1.3} = 1.23 \text{ ft}^3/\text{day/ft}$$

$$q_g = \frac{1.23}{0.5 \times 40} = 0.062 \text{ ft}^3/\text{day/ft}^2$$

Example 17.10   Computing Average Inflow Rate Due to Artesian Flow

Determine the average flow rate of ground water into a pavement drainage layer due to artesian flow constructed on a subgrade soil having a coefficient of permeability of 0.05 ft/day. A piezometer installed at the site indicates an excess hydraulic head of 10 ft. The thickness of the subgrade soil between the drainage layer and the artesian aquifer is 20 ft (see Figure 17.36).

Solution: In this case,

$$q_a = \frac{0.05(10)}{20} = 0.025 \text{ ft}^3/\text{day/ft}^2$$

Note that flow net analysis may be used to estimate the flow rates due to both gravity flow and artesian flow. Flow net analysis is beyond the scope of this book, but good estimates of these flows can be obtained by the methods presented. The amount of ground water flow due to any one source is usually small, but ground water flow should not be automatically neglected, as the sum of ground water flow from all sources may be significant. It is therefore essential that estimates be made of the inflow from all sources of ground water that influence the pavement structure.

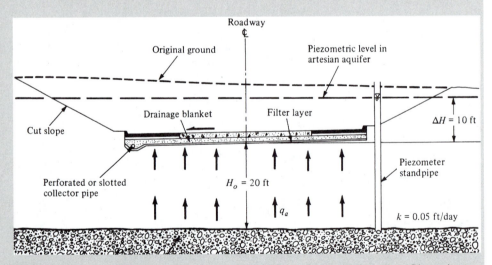

Figure 17.36 Artesian Flow of Groundwater into a Pavement Drainage Layer— Dimensions and Details for Example 17.10

Water from Ice Lenses.   As explained earlier, frost action results in ice lenses forming within the pavement subgrade. The extent to which this occurs during the active freeze period is highly dependent on the frost susceptibility of the subgrade material. These ice lenses thaw during the spring thaw period, and it is necessary to properly drain this water from the pavement environment. The rate of seepage of this water through the soil depends on several factors, which include the permeability of the soil, the thawing rate, and the stresses imposed on the soil. It is very difficult to determine the extent to which each of these factors affects the flow rate, which makes it extremely difficult to develop an exact method for determining the flow rate. However, an empirical method for determining the design flow rate q_m has been developed, using the chart shown in Figure 17.37. This requires the determination from laboratory tests of the average rate of heave of the soil due to frost action, or the classification of the soil with respect to its susceptibility to frost action and the stress σ_p imposed on the subgrade soil. The stress imposed on the subgrade in pounds per square foot due to the weight of a 1-ft square column of the pavement structure is usually taken as the value of σ_p. The design flow rate q_m obtained from the chart is the average flow during the first full day of thawing. This rate is higher than those for subsequent days because the rate of seepage decreases with time. Although the use of such a high value of flow is rather conservative, such a value may also cause the soil to be saturated for a period of up to 6 hr after thawing. In cases where saturation for this period of time is unacceptable, measures to increase the rate of drainage of the thawed water from the soil must be used. For example, a thick enough drainage layer that consists of a material with a suitable permeability that will never allow saturation to occur can be provided.

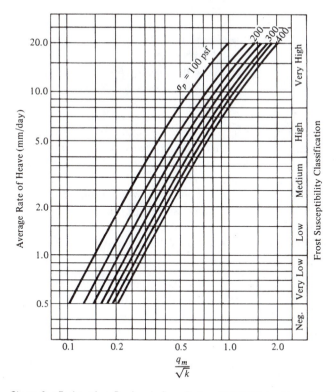

Figure 17.37 Chart for Estimating Design Inflow Rate of Melt Water from Ice Lenses

SOURCE: Redrawn from *Highway Subdrainage Design,* Report No. FHWA-TS-80-224, U.S. Department of Transportation, Washington, D.C., August 1980.

Example 17.11   Computing Flow Rate Due to Thawing of Ice Lenses

Determine the flow rate due to water thawing from ice lenses below a pavement structure that consists of a 6-in. concrete surface and a 9-in. granular base. The subgrade material is silty sand with a high frost susceptibility classification and a coefficient of permeability of 0.075 ft/day.

Solution: Assuming that concrete weighs 150 lb/ft^3 and the granular material weighs 130 lb/ft^3, then

$$\sigma_p = 0.5 \times 150 + 0.75 \times 130 = 172.5 \text{ lb/ft}^2$$

Entering the chart in Figure 17.37 at the midlevel of the high range of frost susceptibility and projecting to a stress of 172.5 on the σ_p charts and reading off the value of $q_m/\sqrt{k}$, we obtain

$$q_m/\sqrt{k} \approx 0.55$$
$$q_m = (\sqrt{0.075})0.55 = 0.15 \text{ ft}^3/\text{day/ft}^2$$

Note that if the average rate of heave of the subgrade soil is determined from either laboratory tests or local experience based on observations of frost action, then this value may be used instead of the frost susceptibility classification.

Vertical Outflow, q_v.   In some cases, the total amount of water accumulated within the pavement structure can be reduced because of the vertical seepage of some of the accumulated water through the subgrade. When this occurs, it is necessary to estimate the amount of this outflow in order to determine the net inflow for which the subdrainage system is to be provided. The procedure for estimating this flow involves the use of flow net diagrams, which is beyond the scope of this book. This procedure is discussed in detail in *Highway Subdrainage Design.* As will be seen later, however, the use of the vertical outflow to reduce the inflow is applicable only when there is neither ground water inflow nor frost action.

Net Inflow.   The net inflow is the sum of inflow rates from all sources less any amount attributed to vertical outflow through the underlying soil. However, note that all of the different flows discussed earlier do not necessarily occur at the same time. For example, it is unlikely that flows from thawed water and ground water will occur at the same time, since soils susceptible to frost action will have very low permeability when frozen. Similarly, downward vertical outflows will never occur at the same time as upward inflow from any other source. Downward vertical outflow will therefore only occur when there is no inflow due to ground water. A set of relationships for estimating the net inflow rate (q_n) have been developed, taking into consideration the different flows that occur concurrently, and is given in Eqs. 17.28 through 17.32.

$$q_n = q_i \tag{17.28}$$

$$q_n = q_i + q_g \tag{17.29}$$

$$q_n = q_i + q_a \tag{17.30}$$

$$q_n = q_i + q_m \tag{17.31}$$

$$q_n = q_i - q_v \tag{17.32}$$

Guidelines for using these equations are given in Table 17.14.

Design of Drainage Layer

The design of the drainage layer involves either the determination of the maximum depth of flow H_m when the permeability of the material k_d is known, or the determination of the required permeability of the drainage material when the maximum flow depth is stipulated. In each case, however, both the slope S of the drainage layer along the flow path and the length L of the flow path must be known. The flow through a drainage layer at full depth is directly related to the *coefficient of transmissibility*, which is the product of k_d and the depth H_d of the drainage layer. This relationship may be used to determine the characteristics of the drainage layer required, or, alternatively, the graphical solution presented in Figure 17.38 may be used. The chart shown in Figure 17.38 is based on steady inflow, uniformly distributed across the surface of the pavement section. This condition does not

Table 17.14 Guidelines for Using Eqs. 17.28 Through 17.32 to Compute Net Inflow, q_n, for Design of Pavement Drainage

Highway Cross Section	Ground Water Inflow	Frost Action	Net Inflow Rate, q_n, Recommended for Design
Cut	Gravity	Yes	Max. of Eqs. 17.29 and 17.31
		No	Eq. 17.29
Cut	Artesian	Yes	Max. of Eqs. 17.30 and 17.31
		No	Eq. 17.30
Cut	None	Yes	Eq. 17.31
		No	Eq. 17.28
Cut	None	Yes	Eq. 17.31
		No	Eq. 17.32
Fill	None	Yes	Eq. 17.31
		No	Eq. 17.32

SOURCE: Adapted from *Highway Subdrainage Design*, Report No. FHWA-TS-80-224, U.S. Department of Transportation, Washington, D.C., August 1980.

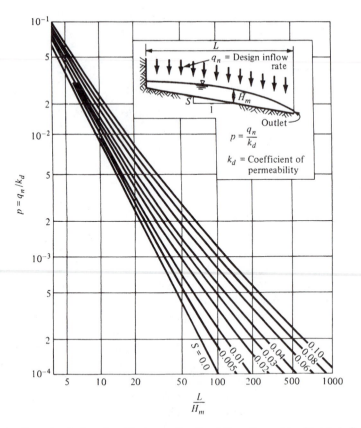

Figure 17.38 Chart for Estimating Maximum Depth of Flow Caused by Steady Inflow

SOURCE: Redrawn from *Highway Subdrainage Design,* Report No. FHWA-TS-80-224, U.S. Department of Transportation, Washington, D.C., August 1980.

normally occur in practice, but a conservative result is usually obtained when the chart is used in combination with the procedure presented herein for determining the net inflow rate q_n.

Example 17.12   Computing Required Depth for a Drainage Layer

Determine the depth required for a drainage layer to carry a net inflow of 0.50 ft³/day/ft² if the permeability of the drainage material is 2000 ft/day. The drainage layer will be laid at a slope of 2 percent, and the length of the flow path is 40 ft.

$$p = q_n / k_d \text{ (from Figure 17.38)}$$
$$= \frac{0.50}{2000} = 2.50 \times 10^{-4}$$

Solution: Entering the chart at $p = 2.50 \times 10^{-4}$ and projecting horizontally to the slope of 0.02, we determine L/H_m as 130.

$$H_m = \frac{40}{130} \quad \text{(required depth of drainage layer)}$$

$$\approx 0.31 \text{ ft}$$

$$= 3.7 \text{ in.} \quad \text{(say, 4 in.—i.e., } H_d)$$

Note that $H_d > H_m$.

Filter Requirements.   The provision of a drainage layer consisting of coarse material allows for the flow of water from the fine-grained material of the subgrade soil to the coarse drainage layer. This may result in the fine-grained soil particles being transmitted to the coarse soil and eventually clogging the voids of the coarse-grained soil. When this occurs, the permeability of the coarse-grained soil is significantly reduced, thereby making the drainage layer less effective. This intrusion of fine particles into the voids of the coarse material can be minimized if the coarse material has certain filter criteria. In cases where these criteria are not satisfied by the drainage material, a protective filter must be provided between the subgrade and the drainage layer to prevent clogging of the drainage layer.

The following criteria have been developed for soil materials used as filters.

$(D_{15})_{\text{filter}} \leq 5(D_{85})_{\text{protected soil}}$

$(D_{15})_{\text{filter}} \geq 5(D_{15})_{\text{protected soil}}$

$(D_{50})_{\text{filter}} \leq 25(D_{50})_{\text{protected soil}}$

$(D_5)_{\text{filter}} \geq 0.074 \text{ mm}$

where D_i is the grain diameter that is larger than the ith percent of the soil grains—that is, the ith percent size on the grain-size distribution curve. (See Chapter 18.)

Design of Longitudinal Collectors

Circular pipes are generally used for longitudinal collectors and are usually constructed of either porous concrete, perforated corrugated metal, or vitrified clay. The pipes are laid in trenches located at depths that will allow the drainage of the subsurface water from the pavement structure. The trenches are then backfilled with porous granular material to facilitate free flow of the subsurface water into the drains.

Design of the longitudinal collectors involves the determination of the pipe location and the pipe diameter and the identification of a suitable backfill material.

Pipe Location.   Shallow trenches within the subbase layer may be used in locations where the depth of frost penetration is insignificant and where the drawdown of the water table is low, as shown in Figure 17.39. In cases where the depth of frost penetration is high or the water table is high, thereby requiring a high drawdown, it is necessary to locate the pipe in a deeper trench below the subbase layer, as shown in Figure 17.40. Note, however, that the deeper the trench, the higher the construction cost of the system. The lateral location of the pipe depends on whether the shoulder is also to be drained. If the shoulder is to be drained, the pipe should be located close to the edge of the shoulder, as shown in Figures 17.39(b) and 17.40(b), but if shoulder drainage is not required, the pipe is located just outside the pavement surface, as shown in Figures 17.39(a) and 17.40(a).

Pipe Diameter.   The diameter D_p of the collector pipe depends on the gradient g, the amount of water per running foot (q_d) that should be transmitted through the pipe, Manning's roughness coefficient of the pipe material, and the distance between the outlets L_o. The chart shown in Figure 17.41 can be used either to determine the minimum pipe diameter when the flow depth, distance between the outlets, and the gradient are specified, or to determine the maximum spacing between outlets for different combinations of gradient and pipe diameters.

In using the chart, it is first necessary to determine the amount of flow q_d from q_n as

$$q_d = q_n L \qquad (17.33)$$

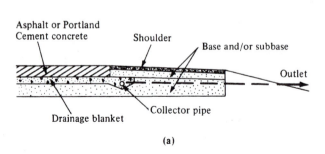

(a)

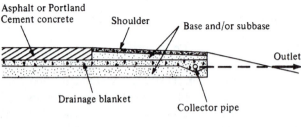

(b)

Figure 17.39 Typical Location of Shallow Longitudinal Collector Pipes

SOURCE: Redrawn from *Highway Subdrainage Design,* Report No. FHWA-TS-80-224, U.S. Department of Transportation, Washington, D.C., August 1980.

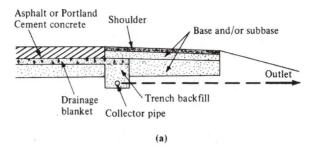

(a)

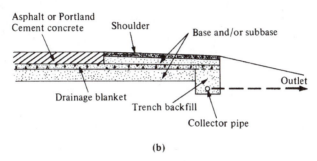

(b)

Figure 17.40 Typical Location of Deep Longitudinal Collector Pipes

SOURCE: Redrawn from *Highway Subdrainage System,* Report No. FHWA-TS-80-224, U.S. Department of Transportation, Washington, D.C., August 1980.

where

q_d = flow rate in drain (ft^3/day/ft)
q_n = net inflow (ft^3/day/ft^2)
L = the length of the flow path (ft)

Note that some variation of L may occur along the highway. The average of all values of L associated with a given pipe may therefore be used for that pipe. The use of the chart is demonstrated in the example given in Figure 17.41.

Backfill Material.   The material selected to backfill the pipe trench should be coarse enough to permit the flow of water into the pipe and also fine enough to prevent the infiltration of the drainage aggregates into the pipe. The following criteria can be used to select suitable filter material.

For slotted pipes $(D_{85})_{\text{filter}} > {}^1\!/_2$ slot width
For circular holes $(D_{85})_{\text{filter}} >$ hole diameter

ECONOMIC ANALYSIS

The economic analysis normally carried out is similar to those carried out for open channel and culvert designs. In this case, however, the cost of the subdrainage system is highly

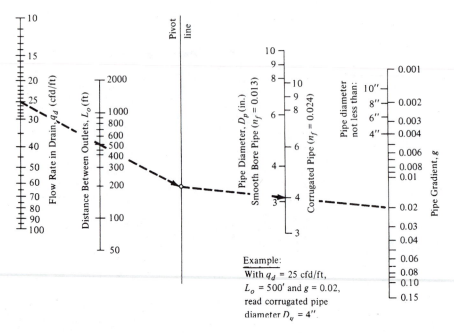

Figure 17.41 Nomogram Relating Collector Pipe Size with Flow Rate, Outlet Spacing, and Pipe Gradient

SOURCE: Redrawn from *Highway Subdrainage System,* Report No. FHWA-TS-80-224, U.S. Department of Transportation, Washington, D.C., August 1980.

dependent on the cost and availability of suitable drainage materials and the cost of the pipes used as longitudinal and transverse drains.

SUMMARY

The provision of adequate drainage facilities on a highway is fundamental and essential to the effective performance of the highway. Unfortunately, the operation of any drainage system consists of very complex hydraulic phenomena, which make it very difficult to develop exact mathematical equations for design or analysis. The analysis and design of drainage facilities are therefore mainly based on empirical relations that have been developed from extensive test results. In addition, the stochastic nature of rainfall occurrence makes it infeasible for drainage facilities to be designed for the worst-case situation. The material presented in this chapter, however, gives the reader the basic principles of analysis and design currently in use. However, the use of any methods presented should go hand in hand with experience that has been gained from local conditions.

The use of these procedures coupled with sound judgment will result in drainage facilities that effectively serve the highway.

PROBLEMS

17-1 What are the two sources of water that primarily concern a highway engineer? Briefly describe each.

17-2 Briefly describe the main differences between surface drainage and subsurface drainage.

17-3 What are the two main disadvantages of using turf cover on unpaved shoulders?

17-4 What is the difference between supercritical and subcritical flow? Under what conditions will either of these occur?

17-5 Briefly describe the three properties of rainfall that primarily concern highway engineers.

17-6 What is meant by (a) a 10-year storm, (b) a 50-year storm, (c) a 100-year storm, and (d) a 500-year storm?

17-7 Define (a) drainage area, (b) runoff coefficient (C), (c) travel time (T_i), and (d) time of concentration (T_c).

17-8 A 170-acre rural drainage area consists of four different watershed areas as follows:

Steep grass-covered area = 40 percent
Cultivated area = 25 percent
Forested area = 30 percent
Turf meadows = 5 percent

Using the rational formula, determine the runoff rate for a storm of 100-year frequency. Assume that the rainfall intensity curves in Figure 17.2 are applicable to this drainage area and that the following land characteristics apply. Use Figure 17.4 to calculate average velocity using "fallow or minimum tillage cultivation" ground cover.

Overland flow length = 0.5 miles
Average slope of overland area = 3 percent

17-9 Compute rate of runoff using the rational formula for a 225-acre rural drainage area consisting of two different watershed areas as follows:

Steep grass area = 45 percent
Cultivated fields = 55 percent

If the time of concentration for this area is 2.4 hours, determine the runoff rate for a storm of 50-year frequency. Use the rainfall intensity curves in Figure 17.2.

17-10 Using the TR-55 method, determine the depth of runoff for a 24-hour, 100-year precipitation of 9 inches if the soil can be classified as group B and the watershed is contoured pasture with good hydrologic condition and an antecedent moisture condition III.

17-11 Determine the depth of runoff by the TR-55 method for a 24-hour, 100-year precipitation of 9 inches for an antecedent moisture condition III, if the following land uses and soil conditions exist.

Area Fraction	Land/Use Condition	Soil Group
0.30	Wooded/fair condition	D
0.25	Small grain/straight row/good condition	D
0.20	Pasture/contoured/fair condition	D
0.15	Meadow/good condition	D
0.10	Farmstead	D

17-12   Determine the peak discharge that will occur for the conditions indicated in Problem 17-11 if the drainage area is 0.5 mi^2, and the time of concentration is 1.6 hours.

17-13   A trapezoidal channel of 2 : 1 side slope and 5-ft bottom width discharges a flow of 275 ft^3/sec. If the channel slope is 2.5 percent and the Manning coefficient is 0.03, determine (a) flow velocity, (b) flow depth, and (c) type of flow.

17-14   A 6-ft wide rectangular channel lined with rubble masonry is required to carry a flow of 300 ft^3/sec. If the slope of the channel is 2 percent and $n = 0.015$, determine (a) flow depth, (b) flow velocity, and (c) type of flow.

17-15   Determine a suitable rectangular flexible lined channel to resist erosion for a maximum flow of 20 ft^3/sec if the channel slope is 2 percent. Use channel dimensions given for Problem 17-14.

17-16   A trapezoidal channel of 2 : 1 side slope and 5 ft bottom width is to be used to discharge a flow of 200 ft^3/sec. If the channel slope is 2 percent and the Manning coefficient is 0.015, determine the minimum depth required for the channel. Is the flow supercritical or subcritical?

17-17   Determine whether a 5 ft × 5 ft reinforced concrete box culvert with 45° flared wingwalls and beveled edge at top of inlet carrying a 50-year flow rate of 200 ft^3/sec will operate under inlet or outlet control for the following conditions. Assume $k_e = 0.5$.

Design headwater elevation (EL_{hd}) = 105 feet
Elevation of stream bed at face of invert = 99.55 feet
Tailwater depth = 4.75 feet
Approximate length of culvert = 200 feet
Slope of stream = 1.5 percent
$n = 0.012$

17-18   Repeat Problem 17-17 using a 6 ft, 6 in. diameter circular pipe culvert with $k_e = 0.5$.

17-19   Determine the ground water infiltration rate for a new two-lane pavement with the following characteristics:

Lane width = 12 feet
Shoulder width = 10 feet
Length of contributing transverse cracks (W_c) = 20 feet
Rate of infiltration (K_p) = 0.05 ft^3/day/ft^2
Spacing of transverse cracks = 30 feet

17-20   In addition to the infiltration determined in Problem 17-19, ground water seepage due to gravity also occurs. Determine the thickness of a suitable drainage layer required to transmit the net inflow to a suitable outlet.

Thickness of subgrade below drainage pipe = 12 feet
Coefficient of permeability of native soil = 0.35 ft/day
Height of water table above impervious layer = 21 feet
Slope of drainage layer = 2 percent
Permeability of drainage area = 2,000 ft/day
Length of flow path = 44 feet

REFERENCES

Highway Drainage Guidelines, American Association of State Highway and Transportation Officials, Washington, D.C., 1992.

Model Drainage Manual, American Association of State Highway and Transportation Officials, Washington, D.C., 1991.

Normann, J. M., R. J. Houghtalen, and W. J. Johnston, *Hydraulic Design of Highway Culverts,* Report No. FHWA-IP-85-15, U.S. Department of Transportation, Office of Implementation, McLean, Va. September 1985.

Shaw L. Yu and Robert J. Kaighn, Jr., *VDOT Manual of Practice for Planning Stormwater Management,* Virginia Transportation Research Council, Charlottesville, Va., January 1992.

Shaw L. Yu, *Stormwater Management for Transportation Facilities,* Synthesis of Practice 174, Transportation Research Board, National Research Council, Washington, D.C., 1993.

Urban Hydrology for Small Watersheds, Technical Release No. 55, U.S. Department of Agriculture, Soil Conservation Service, Washington, D.C., 1975.

ADDITIONAL READINGS

Christopher, Barry R., and Verne G. McGuffey, *Pavement Subsurface Drainage Systems,* Synthesis of Highway Practice 239, Transportation Research Board, National Research Council, Washington, D.C., 1997.

Design Charts for Open Channel Flow, Hydraulic Design Series No. 3, U.S. Department of Transportation, Federal Highway Administration, Washington, D.C., December 1980.

Design and Construction of Urban Stormwater Management Systems, ASCE Manuals and Reports of Engineering Practices No. 77, American Society of Civil Engineers, Alexandria, Va., 1992.

Gray, Donald H., and Andrew T. Leiser, *Biotechnical Slope Protection and Erosion Control,* Krieger Publishing, Malabar, Fla., 1989.

Haestad Methods, Inc., *Computer Applications in Hydraulic Engineering,* Haestad Press, Waterbury, Ct., 1997.

Highway Drainage Guidance, Volume X, Evaluating Highway Effects on Surface Water Environment, Task Force on Hydrology and Hydraulics, American Association of State Highway and Transportation Officials, Washington, D.C., 1992.

Highway Subdrainage Design, Report No. FHWA-TS-80-224, U.S. Department of Transportation, Federal Highway Administration, Washington, D.C., August 1980.

Wanielista, Martin, Robert Kersten, and Ron Eaglin, *Hydrology—Water Quantity and Quality Control,* 2nd ed., John Wiley & Sons, New York, 1997.

Materials and Pavements

ighway pavements are constructed of either asphalt or concrete and ultimately rest on native soil. The engineer must be familiar with the properties and structural characteristics of materials that will be used in constructing or rehabilitating a roadway segment. The engineer must also be familiar with methods and theories for the design of heavy-duty asphaltic and concrete pavements, as well as various treatment strategies for low-volume roads.

CHAPTER 18

Soil Engineering for Highway Design

Highway engineers are interested in the basic engineering properties of soils because soils are used extensively in highway construction. Soil properties are of significant importance when a highway is to carry high traffic volumes with a large percentage of trucks. They are also of importance when high embankments are to be constructed and when the soil is to be strengthened and used as intermediate support for the highway pavement. Thus, several transportation agencies have developed detailed procedures for investigating soil materials used in highway construction.

This chapter presents a summary of current knowledge of the characteristics and engineering properties of soils that are important to highway engineers, including the origin and formation of soils, soil identification, and soil testing methods.

Procedures for improving the engineering properties of soils will be discussed in Chapter 20, Design of Flexible Pavements.

SOIL CHARACTERISTICS

The basic characteristics of a soil may be described in terms of its origin, formation, grain size, and shape. It will be seen later in this chapter that the principal engineering properties of any soil are mainly related to the basic characteristics of that soil.

Origin and Formation of Soils

Soil can be defined from the civil engineering point of view as the loose mass of mineral and organic materials that cover the solid crust of granitic and basaltic rocks of the earth. Soil is mainly formed by weathering and other geologic processes that occur on the surface of the solid rock at or near the surface of the earth. Weathering is the result of physical and chemical actions, mainly due to atmospheric factors that change the structure and composition of the rocks. Weathering occurs through either physical or chemical means. Physical weathering, sometimes referred to as *mechanical weathering,* causes the disintegration of the rocks into smaller particle sizes by the action of forces exerted on the rock. These forces may be due to running water, wind, freezing and thawing, and the activity of plants and animals. *Chemical weathering* occurs as a result of oxidation, carbonation, and other chemical actions that decompose the minerals of the rocks.

Soils may be described as residual or transported. *Residual soils* are weathered in place and are located directly above the original material from which they were formed. *Transported soils* are those that have been moved by water, wind, glaciers, and so forth, and are located away from their parent materials.

The geological history of any soil deposit has a significant effect on the engineering properties of the soils. For example, sedimentary soils, which are formed by the action of water, usually are particles that have settled from suspension in a lake, river, or ocean. These soils range from beach or river sands to marine clays. Soils that are formed by the action of wind are known as aeolian soils and are typically loess. Their voids are usually partially filled with water, and when submerged in water, the soil structure collapses.

Soils may also be described as organic when the particles are mainly composed of organic matter, or as inorganic when the particles are mainly composed of mineral materials.

Surface Texture

The texture of a soil can be described in terms of its appearance, which depends mainly on the shapes and sizes of the soil particles and their distribution in the soil mass. For example, soils consisting mainly of silts and clays with very small particle sizes are known as *fine-textured soils,* whereas soils consisting mainly of sands and gravel with much larger particles are known as *coarse-textured soils.* The individual particles of fine-textured soils are usually invisible to the naked eye, whereas those of coarse-textured soils are visible to the naked eye.

It will be seen later in this chapter that the engineering properties of a soil are related to its texture. For example, the presence of water in fine-textured soils results in significant reduction in their strength, whereas this does not happen with coarse-textured soils. Soils can therefore be divided into two main categories based on their texture. Coarse-grained soils are sometimes defined as those with particle sizes greater than 0.05 mm, such as sands and gravel, and fine-grained soils are those with particle sizes less than 0.05 mm, such as silts and clays. The dividing line of 0.05 mm (0.075 mm has also been used) is selected because that is normally the smallest grain size that can be seen by the naked eye. Since there is a wide range of particle sizes in soils, both the coarse-grained soils and fine-grained soils may be further subdivided, as will be shown later under soil classification.

The distribution of particle size in soils can be determined by conducting a sieve analysis (sometimes known as mechanical analysis) on a soil sample if the particles are sufficiently large. This is done by shaking a sample of air-dried soil through a set of sieves with progressively smaller openings. The smallest practical opening of these sieves is 0.075 mm; this sieve is designated No. 200. Other sieves include No. 140 (0.106 mm), No. 100 (0.15 mm), No. 60 (0.25 mm), No. 40 (0.425 mm), No. 20 (0.85 mm), No. 10 (2.0 mm), No. 4 (4.75 mm), and several others with openings increasing up to 125 mm or 5 in.

For soils containing particle sizes smaller than the lower limit, the hydrometer analysis is used. A representative sample of the air-dried soil is sieved through the No. 10 sieve, and a sieve analysis is carried out on the portion of soil retained. This will give a distribution of the coarse material. A portion of the material that passes through the No. 10 sieve is suspended in water, usually in the presence of a deflocculating agent, and is then left standing until the particles gradually settle to the bottom. A hydrometer is used to determine the specific gravity of the suspension at different times. The specific gravity of the suspension after any time t from the start of the test is used to determine the maximum particle sizes in the suspension as

$$D = \sqrt{\frac{18\eta}{\gamma_s - \gamma_w}\left(\frac{\gamma}{t}\right)} \qquad (18.1)$$

where

D = maximum diameter of particles in suspension at depth y—that is, all particles in suspension at depth y have diameters less than D

η = coefficient of viscosity of the suspending medium (in this case water) in poises

γ_s = unit weight of soil particles

γ_w = unit weight of water

The above expression is based on Stoke's law.

At the completion of this test, which lasts up to 24 hours, the sample of soil used for the hydrometer test is then washed over a No. 200 sieve. The portion retained in the No. 200 sieve is then oven-dried and sieved through Nos. 20, 40, 60, and 140 sieves. The combination of the results of the sieve analysis and the hydrometer test is then used to obtain the particle size distribution of the soil. This is usually plotted as the cumulative percentage by weight of the total sample less than a given sieve size or computed grain diameter, versus the logarithm of the sieve size or grain diameter. Figure 18.1 shows examples of particle size distributions of three different soil samples taken from different locations.

The natural shape of a soil particle is either round, angular, or flat. This natural shape is usually an indication of the strength of the soil, particularly for larger soil particles. Round particles are found in deposits of streams and rivers and have been subjected to extensive wear and therefore are generally strong. Flat and flaky particles have not been subjected to similar action and are usually weak. Fine-grained soils generally have flat and flaky-shaped particles, whereas coarse-grained soils generally have round or angular-shaped particles. Soils with angular-shaped particles have more resistance to deformation than those with round particles, since the individual angular-shaped particles tend to lock together, whereas the rounded particles tend to roll over each other.

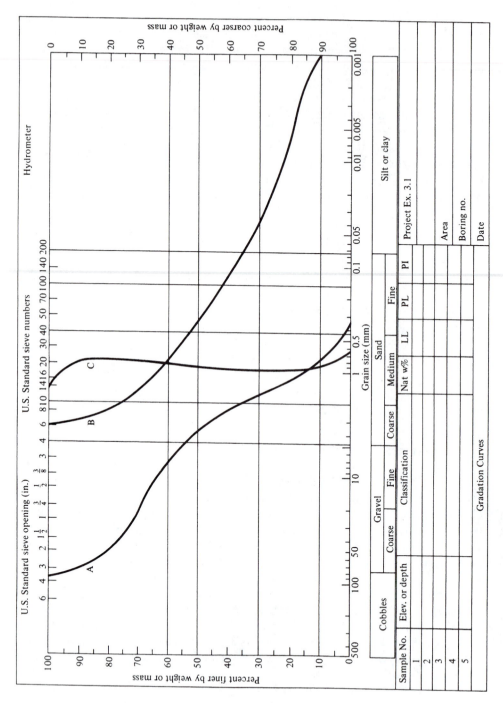

Figure 18.1 Distribution of Particle Size in Different Soils

BASIC ENGINEERING PROPERTIES OF SOILS

Highway engineers must be familiar with those basic engineering properties of soils that influence their behavior when subjected to external loads. The determination of how a specific soil deposit will behave when subjected to an external load is rather complicated because soil deposits may have heterogeneous properties. Highway engineers must always keep in mind that the behavior of any soil depends on the conditions of that soil at the time it is being tested.

Phase Relations

A soil mass generally consists of solid particles of different minerals with spaces between them. The spaces can be filled with air and/or water. Soils are therefore considered as three-phase systems that consist of air, water, and solids. Figure 18.2 schematically illustrates the three components of a soil mass of total volume V. The volumes of air, water, and solids are V_a, V_w, and V_s, respectively, and their weights are W_a, W_w, and W_s, respectively. The volume V_v is the total volume of the space occupied by air and water, generally referred to as *void*.

Porosity

The relative amount of voids in any soil is an important quantity that influences some aspects of soil behavior. This amount can be measured in terms of the *porosity* of the soil, which is defined as the ratio of the volume of voids to the total volume of the soil and is designated as n as shown in Eq. 18.2.

$$n = \frac{V_v}{V} \tag{18.2}$$

Void Ratio

The amount of voids can also be measured in terms of the *void ratio,* which is defined as the ratio of the volume of voids to the volume of solids and is designated as e as shown in Eq. 18.3.

$$e = \frac{V_v}{V_s} \tag{18.3}$$

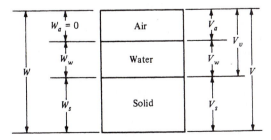

Figure 18.2 Schematic of the Three Phases of a Soil Mass

Combining Eqs. 18.2 and 18.3 we obtain

$$e = \frac{nV}{V_s} = \frac{n}{V_s}(V_s + V_v) = n(1 + e)$$

and

$$n = \frac{e}{1 + e} \qquad (18.4)$$

Similarly,

$$e = \frac{n}{1 - n} \qquad (18.5)$$

Moisture Content

The quantity of water in a soil mass is expressed in terms of the *moisture content,* which is defined as the ratio of the weight of water W_w in the soil mass to the oven-dried weight of solids W_s expressed as a percentage. It is given as

$$w = \frac{W_w}{W_s} 100 \qquad (18.6)$$

where w is the moisture content.

The moisture content of a soil mass can be determined in the laboratory by obtaining the weight of water in the soil and the weight of the dry solids, after completely drying the soil in an oven at a temperature between 100° and 110°C. The weight of the water W_w is obtained by first obtaining the weight W_1 of the soil and container. The soil and container are then placed in the oven. The weight of the dry soil and container is again obtained after repeated drying until there is no further reduction in weight. This weight W_2 is that of the dry soil and container. The weight of water W_w is therefore given as

$$W_w = W_1 - W_2$$

The weight of the empty container W_c is then obtained, and the weight of the dry soil is given as

$$W_s = W_2 - W_c$$

and the moisture content is obtained as

$$w\,(\%) = \frac{W_w}{W_s} 100 = \frac{W_1 - W_2}{W_2 - W_c} 100 \qquad (18.7)$$

Degree of Saturation

The *degree of saturation* is the percentage of void space occupied by water and is given as

$$S = \frac{V_w}{V_v} \, 100 \qquad (18.8)$$

where S is the degree of saturation.

The soil is saturated when the void is fully occupied with water, that is, when $S = 100$ percent, and partially saturated when the voids are only partially occupied with water.

Density of Soil

A very useful soil property for highway engineers is the *density* of the soil. The density is the ratio that relates the mass side of the phase diagram to the volumetric side. Three densities are commonly used in soil engineering. These are total or bulk density γ, dry density γ_d, and submerged or buoyant density γ'.

Total Density. The *total* (or *bulk*) *density* is the ratio of the weight of a given sample of soil to the volume or

$$\gamma = \frac{W}{V} = \frac{W_s + W_w}{V_s + V_w + V_a} \qquad \text{(weight of air is negligible)} \qquad (18.9)$$

The total density for saturated soils is the saturated density and is given as

$$\gamma_{\text{sat}} = \frac{W}{V} = \frac{W_s + W_w}{V_s + V_w} \qquad (18.10)$$

Dry Density. The *dry density* is the density of the soil with the water removed. It is given as

$$\gamma_d = \frac{W_s}{V} = \frac{W_s}{V_s + V_w + V_a} = \frac{\gamma}{1 + w} \qquad (18.11)$$

The dry density is often used to evaluate how well earth embankments have been compacted and is therefore an important quantity in highway engineering.

Submerged Density. The *submerged density* is the density of the soil when submerged in water, and it is the difference between the saturated density and the density of water, or

$$\gamma' = \gamma_{\text{sat}} - \gamma_w \qquad (18.12)$$

where γ_w is the density of water.

Specific Gravity of Soil Particles

The specific gravity of soil particles is the ratio of density of the soil particles to the density of distilled water.

Other Useful Relationships

The basic definitions presented above can be used to derive other useful relationships. For example, the bulk density can be given as

$$\gamma = \frac{G_s + Se}{1 + e} \, \gamma_w \qquad (18.13)$$

where

γ = total or bulk density
S = degree of saturation
γ_w = density of water
G_s = specific gravity of the soil particles
e = void ratio

Example 18.1   Determining Soil Characteristics Using the Three-Phase Principle

The wet weight of a specimen of soil is 340 g and the dried weight is 230 g. The volume of the soil before drying is 210 cc. If the specific gravity of the soil particles is 2.75, determine the void ratio, porosity, degree of saturation, and dry density. Figure 18.3 is a schematic of the soil mass.

Solution: Since the weight of the air can be taken as zero, the weight of the water is

$$W_w = 340 - 230 = 110 \text{ g}$$

Therefore, the volume of water is

$$V_w = (110/1.00) = 110 \text{ cc} \qquad \text{(since density of water = 1 gm/cc)}$$

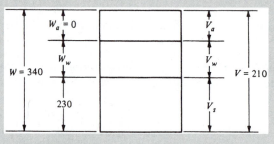

Figure 18.3 Schematic of Soil Mass for Example 18.1

For the volume of solids,

$$V_s = \frac{W_s}{\gamma_s} = \frac{230}{2.75} \qquad (\gamma_s = \text{specific gravity} \times \text{density of water} = 2.75 \times 1)$$
$$= 83.64 \text{ cc}$$

For the volume of air,

$$V_a = 210 - 110 - 83.64 = 16.36 \text{ cc}$$

For the void ratio,

$$e = \frac{V_w + V_a}{V_s} = \frac{110 + 16.36}{83.64} = 1.51$$

For the porosity,

$$n = \frac{V_v}{V} = \frac{110 + 16.36}{210} = 0.60$$

For the degree of saturation,

$$S = \frac{V_w}{V_v} = \frac{110}{110 + 16.36} = 0.87 \quad \text{or } 87\%$$

For the dry density,

$$\gamma_d = \frac{W_s}{V} = \frac{230}{210} = 1.095 \text{ g/cc}$$

Example 18.2   Determining Soil Characteristics Using the Three-Phase Principle

The moisture content of a specimen of soil is 26 percent, and the bulk density is 116 lb/ft³. If the specific gravity of the soil particles is 2.76, determine the void ratio and the degree of saturation.

Solution: The weight of 1 ft³ of the soil is 116 lb—that is,

$$W = 116 \text{ lb} = W_s + W_w = W_s + wW_s = W_s(1 + 0.26)$$

Therefore,

$$W_s = \frac{116}{1.26} = 92.1 \text{ lb}$$

$$W_w = 0.26 \times 92.1 = 23.95 \text{ lb}$$

$$V_s = \frac{W_s}{\gamma_s} \quad (\gamma_s = \text{specific gravity} \times \text{density of water} = 2.76 \times 62.4)$$

$$= \frac{92.1}{2.76 \times 62.4} = 0.53 \text{ ft}^3$$

$$V_w = \frac{W_w}{\gamma_w} = \frac{23.95}{62.4} = 0.38 \text{ ft}^3$$

$$V_a = V - V_s - V_w = 1 - 0.53 - 0.38 = 0.09 \text{ ft}^3$$

Therefore, the void ratio is

$$e = \frac{V_w + V_a}{V_s} = \frac{0.38 + 0.09}{0.53} = 0.89$$

and the degree of saturation is

$$S = \frac{Vw}{V_v} 100 = \frac{0.38}{0.38 + 0.09} 100 = 80.9\%$$

Atterberg Limits

Clay soils with very low moisture content will be in the form of solids. As the water content increases, however, the solid soil gradually becomes plastic—that is, the soil can easily be molded into different shapes without breaking up. Continuous increase of the water content will eventually bring the soil to a state where it can flow as a viscous liquid. The stiffness or consistency of the soil at any time therefore depends on the state at which the soil is, which in turn depends on the amount of water present in the soil. The water content levels at which the soil changes from one state to the other are the *Atterberg limits.* They are the shrinkage limit (SL), plastic limit (PL), and liquid limit (LL), as illustrated in Figure 18.4. These are important limits of engineering behavior because they facilitate the comparison of the water content of the soil with those at which the soil changes from one state to another. They are used in the classification of fine-grained soils and are extremely useful, since they correlate with the engineering behaviors of such soils.

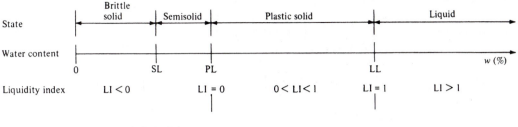

Figure 18.4 Consistency Limits

Shrinkage Limit (SL)

When a saturated soil is slowly dried, the volume shrinks but the soil continues to contain moisture. Continuous drying of the soil, however, will lead to a moisture content at which further drying will not result in additional shrinkage. The volume of the soil will stay constant, and further drying will be accompanied by air entering the voids. The moisture content at which this occurs is the *shrinkage limit,* or SL, of the soil.

Plastic Limit (PL)

The *plastic limit,* or PL, is defined as the moisture content at which the soil crumbles when it is rolled down to a diameter of $\frac{1}{8}$ in. The moisture content is higher than the PL if the soil can be rolled down to diameters less than $\frac{1}{8}$ in., and the moisture content is lower than the PL if the soil crumbles before it can be rolled to $\frac{1}{8}$ in. diameter.

Liquid Limit (LL)

The *liquid limit,* or LL, is defined as the moisture content at which the soil will flow and close a groove of $\frac{1}{2}$ in. within it, after the standard LL equipment has been dropped 25 times. The equipment used for LL determination is shown in Figure 18.5. This device was developed by Casagrande, who worked to standardize the Atterberg limits tests. It is difficult in practice to obtain the exact moisture content at which the groove will close at exactly 25 blows. The test is therefore conducted for different moisture contents and the number of blows required to close the groove for each moisture content recorded. A graph of moisture content versus the logarithm of the number of blows (usually a straight line known as the flow curve) is then drawn. The moisture content at which the flow curve crosses 25 blows is the LL.

The range of moisture content over which the soil is in the plastic state is the difference between the LL and the PL and is known as the *plasticity index* (PI).

$$PI = LL - PL \qquad (18.14)$$

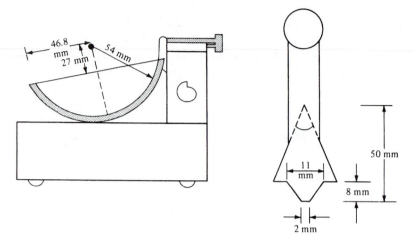

Figure 18.5   Schematic of the Casagrande Liquid Limit Apparatus

where

  PI = plasticity index
  LL = liquid limit
  PL = plastic limit

Liquidity Index (LI)

Since both the PL and LL can be determined only on remolded soils, it is quite possible that the limits determined may not apply to the undisturbed soil, since the structure of the soil particles in the undisturbed state may be different from that in the disturbed state. It is therefore possible that an undisturbed soil will not be in the liquid state if its moisture content is 35 percent, even though the LL was found to be 35 percent. The *liquidity index,* or LI, is used to reflect the properties of the natural soil and is defined as

$$LI = \frac{w_n - PL}{PI} \tag{18.15}$$

where

  LI = liquidity index
  w_n = natural moisture content of the soil

A soil with an LI less than zero will have a brittle fracture when sheared, and a soil with an LI between zero and one will be in a plastic state. When LI is greater than one, the soil will be in a state of viscous liquid if sheared.

Soils with LIs greater than one are known as *quick clays.* They stay relatively strong if undisturbed but become very unstable and can even flow like liquid if they are sheared.

Permeability

The *permeability* of a soil is the property that describes how water flows through the soil. It is usually given in terms of the coefficient of permeability (K), which is the constant of proportionality of the relationship between the flow velocity and the hydraulic gradient between two points in the soil. This relationship was first determined by a French engineer named D'Arcy, and is given as

$$u = Ki \qquad (18.16)$$

where
u = velocity of water in the soil
i = hydraulic gradient
  $= \dfrac{h}{l}$ (head loss h per unit length l)
K = coefficient of permeability

The coefficient of permeability of a soil can be determined in the laboratory by conducting either a constant head or falling head test, or in the field by pumping tests.

Clays and fine-grained soils have very low permeability; thus hardly any flow of water occurs in these soils. Coarse-grained soils, such as gravel and sands, have high permeability, which allows for water to flow easily in them. Soils with high permeability are therefore generally stable, both in the dry and saturated states. Thus, coarse-grained soils make excellent subgrade materials. Note, however, that capillary action may occur in some permeable soils such as "dirty" gravel, which may cause serious stability problems. Capillary action is the movement of free moisture by capillary forces through small diameter openings in the soil mass into pores that are not full of water. Although the moisture can move in any direction, the upward movement usually causes the most serious problems, since this may cause weakness or lead to frost heave. This is discussed further in the section "Frost Action in Soils."

Shear Strength

The *shear strength* of soils is of particular importance to the highway engineer because soil masses will usually fail in shear under highway loads. The shear strength of a soil depends on the cohesion and the angle of internal friction and is expressed as

$$S = C + \sigma \tan \phi \qquad (18.17)$$

where
S = shear strength (lb/ft^2)
C = cohesion (lb/ft^2)
ϕ = angle of internal fracture
σ = normal stress on the shear plane (lb/ft^2)

The degree of importance of either the cohesion or the angle of internal friction depends on the type of soil. In fine-grained soils such as clays, the cohesion component is the major contributor to the shear strength. In fact, it is usually assumed that the angle of internal friction of saturated clays is zero, which makes the shearing resistance on any plane of these soils equal to the cohesion C. Factors that affect the shear strength of cohesive soils include the geologic deposit, moisture content, drainage conditions, and density.

In coarse-grained soils such as sands, the shear strength is achieved mainly through the internal resistance to sliding as the particles roll over each other. The angle of internal friction is therefore important. The value of the angle of internal friction depends on the density of the soil mass, the shape of the individual soil particles, and the surface texture. In general, the angle of internal friction is high when the density is high. Similarly, soils with rough particles such as angular sand grains will have a high angle of internal friction.

The shearing strength of a soil deposit may be obtained in the laboratory by conducting either the triaxial test, the unconfined compression test, or the direct shear test. These tests may be conducted either on the undisturbed soil or on remolding soils. Note, however, that in using remolded samples, the remolding should represent conditions similar to those in the field. Details of each of these tests can be found in *Soil Mechanics Laboratory Manual*.

The in-situ shearing strengths of soils can also be obtained directly by conducting either the plate bearing test or a cone penetration test. These tests are described in *Foundation Engineering Handbook*.

CLASSIFICATION OF SOILS FOR HIGHWAY USE

Soil classification is a method by which soils are systematically categorized according to their probable engineering characteristics. It therefore serves as a means of identifying suitable subbase materials and predicting the probable behavior of a soil when used as subgrade material. (See definitions of subgrade and subbase in Chapter 20.) The classification of a given soil is determined by conducting relatively simple tests on disturbed samples of the soil; the results are then correlated with field experience.

Note, however, that although the engineering properties of a given soil to be used in highway construction can be predicted reliably from its classification, this should not be regarded as a substitute for the detailed investigation of the soil properties. Classifying the soil should be considered as a means of obtaining a general idea of how the soil will behave if used as a subgrade or subbase material.

The most commonly used classification system for highway purposes is the American Association of State Highway and Transportation Officials (AASHTO) Classification System. The Unified Soil Classification System (USCS) is also used to a lesser extent in this country. A slightly modified version of the USCS is used fairly extensively in the United Kingdom.

AASHTO Soil Classification System

The AASHTO Classification System is based on the Public Roads Classification System that was developed from the results of extensive research conducted by the Bureau of

Public Roads, now known as the Federal Highway Administration. Several revisions have been made to the system since it was first published. The system has been described by AASHTO as a means for determining the relative quality of soils for use in embankments, subgrades, subbases, and bases.

In the current publication, soils are classified into seven groups, A-1 through A-7, with several subgroups, as shown in Table 18.1. The classification of a given soil is based on its particle size distribution, LL, and PI. Soils are evaluated within each group by using an empirical formula to determine the group index (GI) of the soils, given as

$$GI = (F - 35)[0.2 + 0.005(LL - 40)] + 0.01(F - 15)(PI - 10) \qquad (18.18)$$

where
   GI = group index
    F = percent of soil particles passing 0.075 mm (No. 200) sieve in whole number based on material passing 75 mm (3 in.) sieve
   LL = liquid limit expressed in whole number
    PI = plasticity index expressed in whole number

The GI is determined to the nearest whole number. A value of zero should be recorded when a negative value is obtained for the GI. Also, in determining the GI for A-2-6 and A-2-7 subgroups, the LL part of Eq. 18.18 is not used, that is, only the second term of the equation is used.

Under the AASHTO system, granular soils fall into classes A-1 to A-3. A-1 soils consist of well-graded granular materials, A-2 soils contain significant amounts of silts and clays, and A-3 soils are clean but poorly graded sands.

Classifying soils under the AASHTO system will consist of first determining the particle size distribution and Atterberg limits of the soil and then reading Table 18.1 from left to right to find the correct group. The correct group is the first one from the left that fits the particle size distribution and Atterberg limits and should be expressed in terms of group designation and the GI. Examples are A-2-6(4) and A-6(10).

In general, the suitability of a soil deposit for use in highway construction can be summarized as follows:

1. Soils classified as A-1-a, A-1-b, A-2-4, A-2-5, and A-3 can be used satisfactorily as subgrade or subbase material if properly drained. (See definitions of subgrade and subbase in Chapter 20.) In addition, such soils must be properly compacted and covered with an adequate thickness of pavement (base and/or surface cover) for the surface load to be carried.
2. Materials classified as A-2-6, A-2-7, A-4, A-5, A-6, A-7-5, and A-7-6 will require a layer of subbase material if used as subgrade. If these are to be used as embankment materials, special attention must be given to the design of the embankment.
3. When soils are properly drained and compacted, their value as subgrade material decreases as the GI increases. For example, a soil with a GI of 0 (an indication of a good subgrade material) will be better as a subgrade material than one with GI of 20 (an indication of a poor subgrade material).

Table 18.1 AASHTO Classification of Soils and Soil Aggregate Mixtures

General Classification	Granular Materials (35% or Less Passing No. 200)									Silt-Clay Materials (More than 35% Passing No. 200)			
	A-1		A-3	A-2					A-4	A-5	A-6	A-7
Group Classification	A-1-a	A-1-b		A-2-4	A-2-5	A-2-6	A-2-7					A-7-5, A-7-6
Sieve analysis, percent passing												
No. 10	50 max.	—	—						—	—	—	—
No. 40	30 max.	50 max.	51 min.	—	—	—	—		—	—	—	—
No. 200	15 max.	25 max.	10 max.	35 max.	35 max.	35 max.	35 max.		36 min.	36 min.	36 min.	36 min.
Characteristics of fraction passing No. 40:												
Liquid limit			—	40 max.	41 min.	40 max.	41 min.		40 max.	41 min.	40 max.	41 min.
Plasticity index	6 max.		N.P.	10 max.	10 max.	11 min.	11 min.		10 max.	10 max.	11 min.	11 min.⋆
Usual types of significant constituent materials	Stone fragments, gravel and sand		Fine sand	Silty or clayey gravel and sand					Silty soils		Clayey soils	
General rating as subgrade	Excellent to good								Fair to poor			

⋆Plasticity index of A-7-5 subgroup is equal to or less than LL minus 30. Plasticity index of A-7-6 subgroup is greater than LL minus 30.

SOURCE: Adapted from *Standard Specifications for Transportation Materials and Methods of Sampling and Testing,* 20th ed., Washington, D.C.: The American Association of State Highway and Transportation Officials, copyright 2000. Used by permission.

Example 18.3 Classifying a Soil Sample Using the AASHTO Method

The following data were obtained for a soil sample.

Mechanical Analysis		Plasticity Tests:
Sieve No.	Percent Finer	
4	97	LL = 48%
10	93	PL = 26%
40	88	
100	78	
200	70	

Using the AASHTO method for classifying soils, determine the classification of the soil and state whether this material is suitable in its natural state for use as a subbase material.

Solution:

- Since more than 35 percent of the material passes the No. 200 sieve, the soil is either A-4, A-5, A-6, or A-7.
- The LL is greater than 40 percent, and therefore the soil cannot be in group A-4 or A-6. Thus, it is either A-5 or A-7.
- The PI is 22 percent (48 − 26), which is greater than 10 percent, thus eliminating group A-5. The soil is A-7-5 or A-7-6.
- (LL − 30) = 18 < PI (22%). Therefore the soil is A-7-6, since the plasticity index of A-7-5 soil subgroup is less than (LL − 30). The GI is given as

$$(70 - 35)\big[0.2 + 0.005(48 - 40)\big] + 0.01(70 - 15)(22 - 10) = 8.4 + 6.6 = 15$$

The soil is A-7-6 (15) and is therefore unsuitable as a subbase material in its natural state.

Unified Soil Classification System (USCS)

The original USCS system was developed during World War II for use in airfield construction. That system has been modified several times to obtain the current version, which can also be applied to other types of construction such as dams and foundations. The fundamental premise used in the USCS system is that the engineering properties of any coarse-grained soil depend on its particle size distribution, whereas those for a fine-grained soil depend on its plasticity. Thus, the system classifies coarse-grained soils on the basis of grain size characteristics and fine-grained soils according to plasticity characteristics.

Table 18.2 lists the USCS definitions for the four major groups of materials, consisting of coarse-grained soils, fine-grained soils, organic soils, and peat. Material that is

Table 18.2 USCS Definition of Particle Sizes

Soil Fraction or Component	Symbol	Size Range
1. Coarse-grained soils		
Gravel	G	75 mm to No. 4 sieve (4.75 mm)
Coarse		75 mm to 19 mm
Fine		19 mm to No. 4 sieve (4.75 mm)
Sand	S	No. 4 (4.75 mm) to No. 200 (0.075 mm)
Coarse		No. 4 (4.75 mm) to No. 10 (2.0 mm)
Medium		No. 10 (2.0 mm) to No. 40 (0.425 mm)
Fine		No. 40 (0.425 mm) to No. 200 (0.075 mm)
2. Fine-grained soils		
Fine		Less than No. 200 sieve (0.075 mm)
Silt	M	(No specific grain size— use Atterberg limits)
Clay	C	(No specific grain size— use Atterberg limits)
3. Organic soils	O	(No specific grain size)
4. Peat	Pt	(No specific grain size)

Gradation Symbols	Liquid Limit Symbols
Well graded, W	High LL, H
Poorly graded, P	Low LL, L

SOURCE: Adapted from *The Unified Soil Classification System,* Annual Book of ASTM Standards, vol. 04.08, American Society for Testing and Materials, West Conshohocken, Pa., 1996.

retained in the 75 mm (3 in.) sieve is recorded, but only that which passes is used for the classification of the sample. Soils with more than 50 percent of their particles being retained on the No. 200 sieve are coarse-grained, and those with less than 50 percent of their particles retained are fine-grained soils (see Table 18.3). The coarse-grained soils are subdivided into gravels (G) and sands (S). Soils having more than 50 percent of their particles larger than 75 mm—that is, retained on No. 4 sieve—are gravels and those with more than 50 percent of their particles smaller than 75 mm—that is, passed through No. 4 sieve—are sands. The gravels and sands are further divided into four subgroups, each based on grain size distribution and the nature of the fine particles in them. They can therefore be classified as either well graded (W), poorly graded (P), silty (M), or clayey (C). Gravels can be described as either well-graded gravel (GW), poorly graded gravel (GP), silty gravel (GM), or clayey gravels (GC), and sands can be described as well-graded sand (SW), poorly graded sand (SP), silty sand (SM), or clayey sand (SC). A gravel or sandy soil is described as well graded or poorly graded, depending on the values of two shape parameters known as the coefficient of uniformity, C_u, and the coefficient of curvature, C_c, given as

$$C_u = \frac{D_{60}}{D_{10}}$$

(18.19)

and

$$C_c = \frac{(D_{30})^2}{D_{10} \times D_{60}}$$

(18.20)

where

D_{60} = grain diameter at 60% passing
D_{30} = grain diameter at 30% passing
D_{10} = grain diameter at 10% passing

Gravels are described as well graded if C_u is greater than 4 and C_c is between 1 and 3. Sands are described as well graded if C_u is greater than 6 and C_c is between 1 and 3.

The fine-grained soils, which are defined as those having more than 50 percent of their particles passing the No. 200 sieve, are subdivided into clays (C) or silt (M), depending on the PI and LL of the soil. A plasticity chart, shown in Table 18.3, is used to determine whether a soil is silty or clayey. The chart is a plot of PI versus LL, from which a dividing line known as the "A" line, which generally separates the more clayey materials from the silty materials, was developed. Soils with plots of LLs and PIs below the "A" line are silty soils, whereas those with plots above the "A" line are clayey soils. Organic clays are an exception to this general rule since they plot below the "A" line. Organic clays, however, generally behave similarly to soils of lower plasticity.

Classification of coarse-grained soils as silty or clayey also depends on their LL plots. Only coarse-grained soils with more than 12 percent fines (that is, passes No. 200 sieve) are so classified (see Table 18.3). Those soils with plots below the "A" line or with a PI less than 4 are silty gravel (CM) or silty sand (SM), and those with plots above the "A" line with a PI greater than 7 are classified as clayey gravels (GC) or clayey sands (SC).

The organic, silty, and clayey soils are further divided into two groups, one having a relatively low LL (L) and the other having a relatively high LL (H). The dividing line between high LL soils and low LL soils is arbitrarily set at 50 percent.

Fine-grained soils can be classified as either silt with low plasticity (ML), silt with high plasticity (MH), clays with high plasticity (CH), clays with low plasticity (CL), or organic silt with high plasticity (OH).

Table 18.3 gives the complete layout of the USCS, and Table 18.4 shows an approximate correlation between the AASHTO system and USCS.

Table 18.3   Unified Soil Classification System

Major Divisions			Group Symbols	Typical Names	Laboratory Classification Criteria		
Coarse-grained soils (More than half of material is larger than No. 200 sieve size)	Gravels (More than half of coarse fraction is larger than No. 4 sieve size)	Clean gravels (Little or no fines)	GW	Well-graded gravels, gravel-sand mixtures, little or no fines	Determine percentages of sand and gravel from grain-size curve. Depending on percentage of fines (fraction smaller than No. 200 sieve size), coarse-grained soils are classified as follows: Less than 5 per cent — GW, GP, SW, SP; More than 12 per cent — GM, GC, SM, SC; 5 to 12 per cent — Borderline cases requiring dual symbols[b]	$C_u = \dfrac{D_{60}}{D_{10}}$ greater than 4; $C_c = \dfrac{(D_{30})^2}{D_{10} \times D_{60}}$ between 1 and 3	
			GP	Poorly graded gravels, gravel-sand mixtures, little or no fines		Not meeting all gradation requirements for GW	
		Gravels with fines (Appreciable amount of fines)	GM[a] d / u	Silty gravels, gravel-sand-silt mixtures		Atterberg limits below "A" line or P.I. less than 4	Above "A" line with P.I. between 4 and 7 are borderline cases requiring use of dual symbols
			GC	Clayey gravels, gravel-sand-clay mixtures		Atterberg limits below "A" line with P.I. greater than 7	
	Sands (More than half of coarse fraction is smaller than No. 4 sieve size)	Clean sands (Little or no fines)	SW	Well-graded sands, gravelly sands, little or no fines		$C_u = \dfrac{D_{60}}{D_{10}}$ greater than 6; $C_c = \dfrac{(D_{30})^2}{D_{10} \times D_{60}}$ between 1 and 3	
			SP	Poorly graded sands, gravelly sands, little or no fines		Not meeting all gradation requirements for SW	
		Sands with fines (Appreciable amount of fines)	SM[a] d / u	Silty sands, sand-silt mixtures		Atterberg limits above "A" line or P.I. less than 4	Limits plotting in hatched zone with P.I. between 4 and 7 are borderline cases requiring use of dual symbols
			SC	Clayey sands, sand-clay mixtures		Atterberg limits above "A" line with P.I. greater than 7	
Fine-grained soils (More than half material is smaller than No. 200 sieve)	Silts and clays (Liquid limit less than 50)		ML	Inorganic silts and very fine sands, rock flour, silty or clayey fine sands, or clayey silts with slight plasticity			
			CL	Inorganic clays of low to medium plasticity, gravelly clays, sandy clays, silty clays, lean clays			
			OL	Organic silts and organic silty clays of low plasticity			
	Silts and clays (Liquid limit greater than 50)		MH	Inorganic silts, micaceous or diatomaceous fine sandy or silty soils, elastic silts			
			CH	Inorganic clays of high plasticity, fat clays			
			OH	Organic clays of medium to high plasticity, organic silts			
	Highly organic soils		Pt	Peat and other highly organic soils			

Plasticity Chart

[a]Division of GM and SM groups into subdivisions of d and u are for roads and airfields only. Subdivision is based on Atterberg limits; suffix d used when L.L. is 28 or less and the P.I. is 6 or less; the suffix u used when L.L. is greater than 28.
[b]Borderline classifications, used for soils possessing characteristics of two groups, are designated by combinations of group symbols. For example: GW-GC, well-graded gravel-sand mixture with clay binder.

SOURCE: Joseph E. Bowles, *Foundation Analysis and Design,* McGraw-Hill, New York, 1988.

Table 18.4   Comparable Soil Groups in the AASHTO and USCS Systems

Soil Group in Unified System	Comparable Soil Groups in AASHTO System		
	Most Probable	Possible	Possible but Improbable
GW	A-1-a	—	A-2-4, A-2-5, A-2-6, A-2-7
GP	A-1-a	A-1-b	A-3, A-2-4 A-2-5, A-2-6, A-2-7
GM	A-1-b, A-2-4, A-2-5, A-2-7	A-2-6	A-4, A-5, A-6, A-7-5, A-7-6, A-1-a
GC	A-2-6, A-2-7	A-2-4, A-6	A-4, A-7-6, A-7-5
SW	A-1-b	A-1-a	A-3, A-2-4, A-2-5, A-2-6, A-2-7
SP	A-3, A-1-b	A-1-a	A-2-4, A-2-5, A-2-6, A-2-7
SM	A-1-b, A-2-4, A-2-5, A-2-7	A-2-6, A-4, A-5	A-6, A-7-5, A-7-6, A-1-a
SC	A-2-6, A-2-7	A-2-4, A-6 A-4, A-7-6	A-7-5
ML	A-4, A-5	A-6, A-7-5	—
CL	A-6, A-7-6	A-4	—
OL	A-4, A-5	A-6, A-7-5, A-7-6	—
MH	A-7-5, A-5	—	A-7-6
CH	A-7-6	A-7-5	—
OH	A-7-5, A-5	—	A-7-6
Pt	—	—	—

SOURCE: Adapted from T.K. Liu, *A Review of Engineering Soil Classification Systems—Special Procedures for Testing Soil and Rock for Engineering Purposes,* 5th ed., ASTM Special Technical Publication 479, American Society for Testing and Materials, Easton, Md., 1970.

Example 18.4   Classifying a Soil Sample Using the Unified Soil Classification System

The results obtained from a mechanical analysis and a plasticity test on a soil sample are shown below. Classify the soil using the USCS and state whether or not it can be used in the natural state as a subbase material.

Mechanical Analysis

Sieve No.	Percent Passing (by weight)	Plasticity Tests:
4	98	LL = 40%
10	93	PL = 30%
40	85	
100	73	
200	62	

Solution: The grain size distribution curve is first plotted as shown in Figure 18.6. Since more than 50 percent (62 percent) of the soil passes No. 200 sieve, the soil is fine grained. The plot of the limits on the plasticity chart is below the "A" line (PI = 40 − 30 = 10); therefore, it is either silt or organic clay. However, the LL is less than 50 percent (40 percent); therefore, it is low LL. This soil can be classified as ML or OL, which is probably equivalent to an A-4 or A-5 soil in the AASHTO classification system. It is therefore not useful as a subbase material.

Example 18.5   Classifying a Soil Sample Using the Unified Soil Classification System

Repeat Example 18.4 for the data shown below.

Mechanical Analysis

Sieve No.	Percent Passing (by weight)	Plasticity Tests:
4	95	LL = nonplastic
10	30	PL = nonplastic
40	15	
100	8	
200	3	

Solution: The grain size distribution curve is also plotted in Figure 18.6. Since only 3 percent of the particles pass through No. 200 sieve, the soil is coarse grained. Since more than 50 percent pass through the No. 4 sieve, the soil is classified as sand. Because the soil is nonplastic, it is necessary to determine its coefficient of uniformity C_u and coefficient of curvature C_c. From the particle size distribution curve,

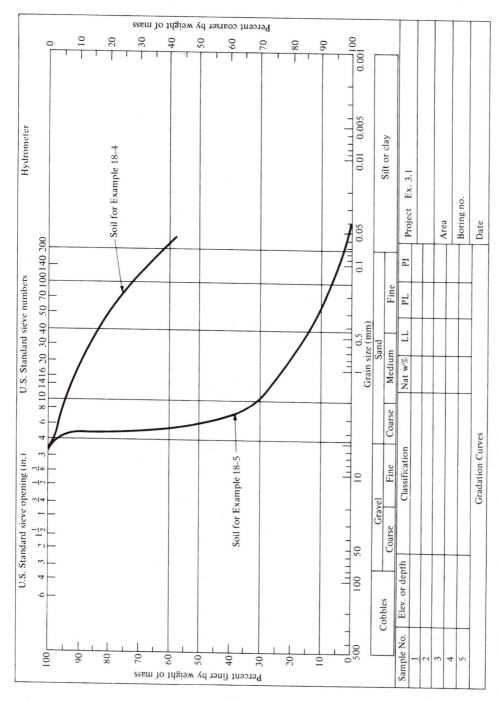

Figure 18.6 Grain Size Distribution Curve for Examples 18.4 and 18.5

$$C_u = \frac{D_{60}}{D_{10}} = \frac{3.8}{0.25} = 15.2 > 6$$

$$C_c = \frac{(D_{30})^2}{D_{10} \times D_{60}} = \frac{2^2}{0.25 \times 3.8} = 4.2$$

This sand is not well graded and is classified as SP and can therefore be used as a subbase material, if properly drained and compacted.

SOIL SURVEYS FOR HIGHWAY CONSTRUCTION

Soil surveys for highway construction entail the investigation of the soil characteristics on the highway route and the identification of suitable soils for use as subbase and fill materials. Soil surveys are therefore normally an integral part of preliminary location surveys, since the soil conditions may significantly affect the location of the highway. A detailed soil survey is always carried out on the final highway location.

The first step in any soil survey is in the collection of existing information on the soil characteristics of the area in which the highway is to be located. Such information can be obtained from geological and agricultural soil maps, existing aerial photographs, and an examination of excavations and existing roadway cuts. It is also usually helpful to review the design and construction of other roads in the area. The information obtained from these sources can be used to develop a general understanding of the soil conditions in the area and to identify any unique problems that may exist. The extent of additional investigation usually depends on the amount of existing information that can be obtained.

The next step is to obtain and investigate enough soil samples along the highway route to identify the boundaries of the different types of soils so that a soil profile can be drawn. Samples of each type of soil along the route location are obtained by auger boring or from test pits for laboratory testing. Samples are usually taken at different depths down to about 5 ft. In cases where rock locations are required, depths may be increased. The engineering properties of the samples are then determined and used to classify the soils. It is important that the characteristics of the soils in each hole be systematically recorded, including the depth, location, thickness, texture, and so forth. It is also important that the location of the water level be noted. These data are then used to plot a detailed soil profile along the highway (see Figure 18.7).

Geophysical Methods of Soil Exploration

Soil profiles can also be obtained from one of two geophysical methods of soil exploration, known as the resistivity and seismic methods.

Resistivity Method

The *resistivity method* is based on the difference in electrical conductivity or resistivity of different types of soils. An electrical field is produced in the ground by means of two current electrodes, as shown in Figure 18.8, and the potential drop between the two inter-

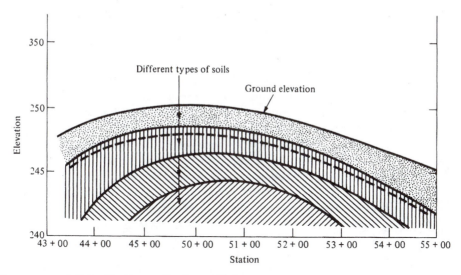

Figure 18.7 Soil Profile Along a Section of Highway

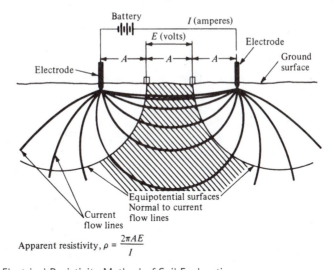

Figure 18.8 Electrical Resistivity Method of Soil Exploration

SOURCE: Redrawn from Hsai-Yang Fang, ed., *Foundation Engineering Handbook,* Routledte, Chapman & Hall, N.Y., 1990.

mediate or potential electrodes is then recorded. The apparent resistivity of the soil to a depth approximately equal to the spacing "*A*" is then computed. The resistivity equipment used is usually designed such that the apparent resistivity can be directly read on the potentiometer. Data for the soil profile are obtained by moving the electrode along the center line of the proposed highway without changing the spacing. The apparent resistivity is then determined along the highway within a depth equal to the spacing "*A.*" The resistivities obtained are then compared with known values of different soils by calibrating the instrument using locally exposed materials.

Seismic Method

The *seismic method* is used to identify the location of rock profiles or dense strata underlying softer materials (Figure 18.9 shows the layout for the seismic method). It is conducted by inducing impact or shock waves into the soil by either striking a plate located on the surface with a hammer or exploding small charges in the soil. Listening devices known as geophones then pick up the shock waves. The time lapse of the wave traveling to the geophone is then used to calculate the velocity of the wave in the surface soil. Some of the shock waves can be made to pass from the surface stratum into underlying layers and then back into the surface stratum by moving the shock point away from the geophone. This permits the computation of the wave velocity in the underlying material.

The seismic test is conducted by moving the shock-producing device along the proposed center line of the highway and the shock is produced at known distances from the geophone. A graph of the time it takes the shock waves to arrive at the geophone versus the distance of the geophone from the shock point is then drawn. These points will be on a straight line as long as the shock wave travels through the same soil material. When the distance of shock point from the geophone becomes large enough, the waves travel through

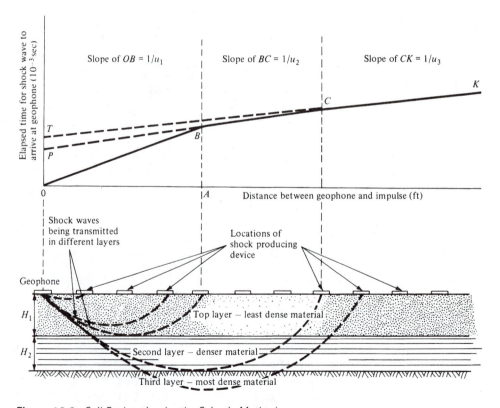

Figure 18.9 Soil Exploration by the Seismic Method

the denser material and a break is observed as shown in Figure 18.9. The inverse of the slope of each straight line will give the velocity of the wave within each layer. It can be shown that for three layers, as shown in Figure 18.9, the depths of the first and second soil layers, H_1 and H_2, are

$$H_1 = \frac{\overline{OP}u_1}{2 \cos \alpha} \tag{18.21}$$

and

$$H_2 = \frac{\overline{PT}\,u_2}{2 \cos \beta} \tag{18.22}$$

where

$\overline{OP}$ = time obtained from plot (see Figure 18.9)

u_1 = velocity of wave in upper stratum

   = 1/slope of first straight line

u_2 = velocity of wave in underlying stratum

   = 1/slope of second straight line

u_3 = velocity of wave in third stratum

   = 1/slope of the third straight line

$\overline{PT}$ = time obtained from plot (see Figure 18.9)

α = first refraction angle

$\sin \alpha = u_1/u_2$

β = second refraction angle

$\sin \beta = u_2/u_3$

The type of material within each stratum can be identified by comparing the wave velocity within each stratum, with known values of wave velocity for different types of soils. Representative values of wave velocities for different types of soils are given in Table 18.5. Note that the seismic method can be used only for cases where the underlying soil is denser than the overlying soil—that is, when u_2 is greater than u_1.

Table 18.5 Representative Values of Wave Velocities for Different Types of Soils

Material	Velocity (ft/sec)
Soil	
Sand, dry silt, and fine-grained top soil	650–3,300
Alluvium	1,650–6,600
Compacted clays, clayey gravel, and dense clayey sand	3,300–8,200
Loess	800–2,450
Rock	
Slate and shale	8,200–16,400
Sandstone	4,900–16,400
Granite	13,100–19,700
Sound limestone	16,400–32,800

SOURCE: Adapted from Braja M. Das, *Principles of Foundation Engineering,* 4th ed., PWS Publishing, 1998.

Example 18.6   Estimating Depth and Soil Type of Each Soil Stratum Using the Seismic Method

The seismic method of exploration was used to establish the soil profile along the proposed center line of a highway. The table below shows part of the results obtained. Estimate the depth of each stratum of soil at this section and suggest the type of soil in each.

Distance of Impulse to Geophone (ft)	Time for Wave Arrival (10^{-3} sec)
20	32
40	60
60	88
80	94
100	100
120	106
140	112
160	116
180	117
200	118.5
220	120
250	122

Solution: The plot of the data is shown in Figure 18.10. From Figure 18.10,

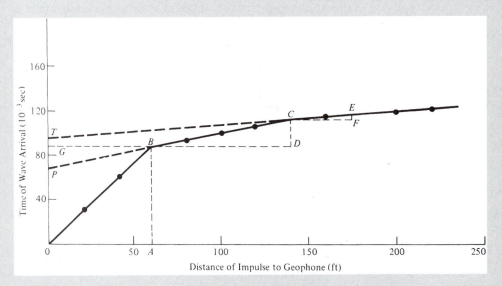

Figure 18.10   Solution for Example 18.6

$$u_1 = \frac{OA}{AB} = \frac{60}{88 \times 10^{-3}} = 681.8 \text{ ft/sec}$$

From Table 18.5, the soil in the first stratum is possibly sand.

$$u_2 = \frac{BD}{CD} = \frac{140 - 60}{(112 - 88) \times 10^{-3}} = 3333 \text{ ft/sec}$$

From Table 18.5, the soil in the second stratum can be either alluvium, compacted clay, clayey gravel, or dense clayey sand.

$$u_3 = \frac{CF}{EF} = \frac{175 - 140}{(116 - 112) \times 10^{-3}} = 8750 \text{ ft/sec}$$

The wave velocity in this material is high, which indicates some type of rock, such as sandstone, or slate and shale.

$$H_1 = \frac{\overline{OP}u_1}{2 \cos \alpha}$$

$$\sin \alpha = \frac{u_1}{u_2} = \frac{681.8}{3333}$$

$$= 0.205, \ \alpha = 11.8°$$

$$\cos \alpha = 0.979$$

$$\overline{OP} = 68 \times 10^{-3} \text{ sec} \quad \text{(from Figure 18.10)}$$

$$H_1 = \frac{(68 \times 10^{-3})681.8}{2 \times 0.979} = 23.68 \text{ ft}$$

$$H_2 = \frac{\overline{PT}u_2}{2 \cos \beta}$$

$$\overline{PT} = 26 \times 10^{-3} \text{ sec} \quad \text{(from Figure 18.10)}$$

$$\sin \beta = \frac{u_2}{u_3} = \frac{3333}{8750} = 0.381$$

$$\beta = 22.39°$$

$$\cos \beta = 0.925$$

$$H_2 = \frac{26 \times 10^{-3} \times 3333}{2 \times 0.925} = 46.84 \text{ ft}$$

SOIL COMPACTION

When soil is to be used as embankment or subbase material in highway construction, it is essential that the material be placed in uniform layers and compacted to a high density. Proper compaction of the soil will reduce to a minimum subsequent settlement and volume change, thereby enhancing the strength of the embankment or subbase. Compaction is achieved in the field by using hand-operated tampers, sheepsfoot rollers, rubber-tired rollers, or other types of rollers.

The strength of the compacted soil is directly related to the maximum dry density achieved through compaction. The relationship between dry density and moisture content for practically all soils takes the form shown in Figure 18.11. It can be seen from this relationship that for a given compactive effort, the dry density attained is low at low moisture contents. The dry density increases with increase in moisture content to a maximum value, when an optimum moisture content is reached. Further increase in moisture content results in a decrease in the dry density attained. This phenomenon is due to the effect of moisture on the soil particles. At low moisture content, the soil particles are not lubricated, and friction between adjacent particles prevents the densification of the particles. As the moisture content is increased, larger films of water develop on the particles, making the soil more plastic and easier for the particles to be moved and densified. When the optimum moisture content is reached, however, the maximum practical degree of saturation (where $S < 100\%$) is attained. The degree of saturation at the optimum moisture content cannot be increased by further compaction because of the presence of entrapped air in the void spaces and around the particles. Further addition of moisture therefore results in the voids being overfilled with water, with no accompanying reduction in the air. The soil particles are separated, resulting in a reduction in the dry density. The zero-air void curve shown in Figure 18.11 is the theoretical moisture- density curve for a saturated soil and zero-air voids, where the

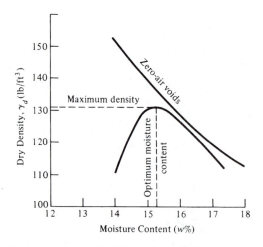

Figure 18.11 Typical Moisture-Density Relationship for Soils

degree of saturation is 100 percent. This curve is usually not attained in the field, since zero-air void cannot be attained as explained earlier. Points on the curve may be calculated from Eq. 18.23 as

$$\gamma_d \ = \ \frac{\gamma_w G_s}{1 + w G_s} \tag{18.23}$$

where

γ_w = density of water (lb/ft^3)
G_s = specific gravity of soil particles
w = moisture content of soil
γ_d = dry density of soil (lb/ft^3)

Although this curve is theoretical, the distance between it and the test moisture-density curve is of importance, since this distance is an indication of the amount of air voids remaining in the soil at different moisture contents. The farther away a point on the moisture-density curve is from the zero-air void curve, the more air voids remain in the soil and the higher is the likelihood of expansion or swelling if the soil is subjected to flooding. It is therefore better to compact at the higher moisture content—that is, the wet side of optimum moisture content—if a given dry density other than the optimum is required.

Optimum Moisture Content

The determination of the optimum moisture content of any soil to be used as embankment or subgrade material is necessary before any field work is commenced. Most highway agencies now use dynamic or impact tests to determine the optimum moisture content and maximum dry density. In each of these tests, samples of the soil to be tested are compacted in layers to fill a specified size mold. Compacting effort is obtained by dropping a hammer of known weight and dimensions from a specified height a specified number of times for each layer. The moisture content of the compacted material is then obtained and the dry density determined from the measured weight of the compacted soil and the known volume of the mold. The soil is then broken down or another sample of the same soil is obtained. The moisture content is then increased and the test repeated. The process is repeated until a reduction in the density is observed. Usually a minimum of four or five individual compaction tests are required. A plot of dry density versus moisture content is then drawn from which the optimum moisture content is obtained. The two types of tests commonly used are the standard AASHTO or the modified AASHTO.

Table 18.6 shows details for the standard AASHTO, designated T99, and the modified AASHTO, designated T180. Most transportation agencies use the standard AASHTO test.

Effect of Compacting Effort

Compacting effort is a measure of the mechanical energy imposed on the soil mass during compaction. In the laboratory it is given in units of ft-lb/in.^{3} or ft-lb/ft^3, whereas in

Table 18.6 Details of the Standard AASHTO and Modified AASHTO Tests

Test Details	Standard AASHTO (T99)	Modified AASHTO (T180)
Diameter of mold (in.)	4 or 6	4 or 6
Height of sample (in.)	5 cut to 4.58	5 cut to 4.58
Number of lifts	3	5
Blows per lift	25 or 56	25 or 56
Weight of hammer (lb)	5.5	10
Diameter of compacting surface (in.)	2	2
Free fall distance (in.)	12	18
Net volume (ft³)	1/30 or 1/13.33	1/30 or 1/13.33

the field it is given in terms of the number of passes of a roller of known weight and type. The compactive effort in the standard AASHTO test, for example, is approximately calculated as

$$\frac{(5.5 \text{ lb})(1 \text{ ft})(3)(25)}{1/30 \text{ ft}^3} = 12{,}375 \text{ ft-lb/ft}^3 \qquad \text{or } 7.16 \text{ ft-lb/in.}^3$$

Note that the optimum moisture content and maximum dry density attained depend on the compactive effort used. Figure 18.12 shows that as the compactive effort increases, so does the maximum dry density. Also, the compactive effort required to obtain a given density increases as the moisture content of the soil decreases.

Example 18.7   Determining Maximum Dry Density and Optimum Moisture Content

The table below shows results obtained from a standard AASHTO compaction test on six samples, 4 in. diameter, of a soil to be used as fill for a highway. Determine the maximum dry density and the optimum moisture content of the soil.

Sample No.	Weight Compacted Soil, W (lb)	Moisture Content, w (%)
1	4.16	4.0
2	4.39	6.1
3	4.60	7.8
4	4.68	10.1
5	4.57	12.1
6	4.47	14.0

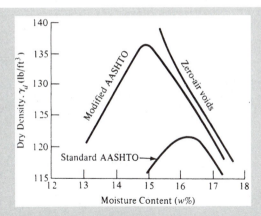

Figure 18.12 Effect of Compactive Effort in Dry Density

Solution: Since we are using the standard AASHTO test, 4 in. diameter, the volume of each sample is $1/30$ ft^3. The dry densities are calculated as shown below.

Sample No.	Bulk Density, γ $30W$ (lb/ft^3)	Moisture Content, w (%)	Dry Density, γ_d lb/ft^3 $\left(\dfrac{\gamma}{1+w}\right)$
1	124.80	4.0	120.0
2	131.70	6.1	124.1
3	138.00	7.8	128.0
4	140.40	10.1	127.5
5	137.10	12.0	122.4
6	134.10	14.0	117.6

Figure 18.13 shows the plot of dry density versus moisture content, from which it is determined that maximum dry density is 129 lb/ft^3 and the optimum moisture content is 9 percent.

Field Compaction Procedures and Equipment

A brief description of the field compaction procedures and the equipment used for embankment compaction is useful at this point because it enables the reader to see how the theory of compaction is applied in the field.

Field Compaction Procedures

The first step in the construction of a highway embankment is the identification and selection of a suitable material. This is done by obtaining samples from economically feasible

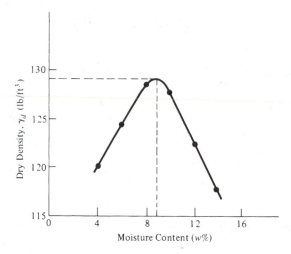

Figure 18.13 Moisture-Density Relationship for Example 18.7

borrow pits or borrow areas and testing them in the laboratory to determine the group of each. It has been shown earlier that, based on the AASHTO system of classification, materials classified as A-1, A-2-4, A-2-5, and A-3 are usually suitable embankment materials. In cases where it is necessary to use materials in other groups, special consideration should be given to the design and construction. For example, soils in groups A-4 and A-6 can be used for embankment construction if the embankment height is low, the field compaction process is carefully controlled, and the embankment is located where the moisture content is not expected to exceed that at which the construction was undertaken. A factor that also significantly influences the selection of any material is whether that material can be economically transported to the construction site. Having identified suitable materials, their optimum moisture contents and maximum dry densities are determined.

Embankment Formation.   Highway embankments are formed by spreading thin layers of uniform thickness of the material and compacting each layer at or near the optimum moisture content. End dumping of the material from trucks is not recommended. The process of constructing one layer at a time facilitates obtaining uniform strength and moisture content in the embankment. End dumping or compaction of thick layers, on the other hand, may result in variable strengths within the embankment, which could lead to differential settlement between adjacent areas.

Most states stipulate a thickness of 6 to 12 in. for each layer, although the thickness may be increased to 24 in. when the lower portion of an embankment consists mainly of large boulders.

All transportation agencies have their own requirements for the minimum density in the field. Some of these are based on the AASHTO specifications for transportation materials. Table 18.7 gives commonly used relative density values for different embankment heights. The relative density is given as a percentage of the maximum dry density

Table 18.7 Commonly Used Minimum Requirements for Compaction of Embankments and Subgrades

	Minimum Relative Density		
	Embankments		
AASHTO Class of Soil	Height Less Than 50 ft	Height Greater Than 50 ft	Subgrade
A-1, A-3	≥ 95	≥ 95	100
A-2-4, A-2-5	≥ 95	≥ 95	100
A-2-6, A-2-7	> 95	—[a]	≥ 95[b]
A-4, A-5, A-6, A-7	> 95	—[a]	≥ 95[b]

[a]Use of these materials requires special attention to design and construction.
[b]Compaction at 95 percent of T99 moisture content.

obtained from the standard AASHTO (T99) test. Some agencies base their requirements on the maximum dry density obtained in the laboratory. For example, when the maximum dry density obtained in the laboratory is less than 100 lb/ft³, the required field density is 100 percent of the laboratory density. When the maximum dry density obtained in the laboratory is 100 lb/ft³ or greater but less than 120 lb/ft³, 95 percent is required in the field, and so forth. The former practice of specifying the number of passes for different types of compacting equipment is not widely used at the present time.

Some transportation agencies also have specifications for the moisture content to be used during compaction. These specifications are usually given in general terms, although limits above and below the optimum moisture content have been given.

Control of Embankment Construction

The construction control of an embankment entails frequent and regular checks of the dry density and the moisture content of materials being compacted. The bulk density is obtained directly from measurements obtained in the field, and the dry density is then calculated from the bulk density and the moisture content. The laboratory moisture-density curve is then used to determine whether the dry density obtained in the field is in accordance with the laboratory results for the compactive effort used. These tests are conducted by using either a destructive method or a nondestructive method.

Destructive Methods. In determining the bulk density by the destructive method, a cylindrical hole of about 4 in. diameter and a depth equal to that of the layer is excavated. The material obtained from the hole is immediately sealed in a container. Care should be taken not to lose any of the excavated material. The total weight of the excavated material is obtained, usually in the field laboratory, and the moisture content determined. The compacted volume of the excavated material is then measured by determining the volume of the excavated hole.

The moisture content is determined by either rapidly drying the soil in a field oven or by facilitating evaporation of the moisture by adding some volatile solvent material such as alcohol and igniting it. The volume of the excavated hole may be obtained by one of three methods: sand replacement, oil, or balloon. In the sand replacement method, the excavated hole is carefully filled with standard sand from a jar originally filled with the standard sand. The jar can be opened and closed by a valve. When the hole is completely filled with sand, the valve is closed and the weight of the remaining sand in the jar is determined. The weight of the quantity of sand used to fill the excavated hole is then obtained by subtracting the weight of the sand remaining in the jar from the weight of the sand required to fill the jar. The volume of the quantity of standard sand used to fill the excavated hole (that is, volume of hole) is then obtained from a previously established relationship between the weight and volume of the standard sand.

In the oil method, the volume of the hole is obtained by filling the excavated hole with a heavy oil of known specific gravity.

In the balloon method, a balloon is placed in the excavated hole and then filled with water. The volume of water required to fill the hole is the volume of the excavated hole.

The destructive methods are all subject to errors. For example, in the sand replacement method, adjacent vibration will increase the density of the sand in the excavated hole, thereby indicating a larger volume hole. Large errors in the volume of the hole will be obtained if the balloon method is used in holes having uneven walls, and large errors may be obtained if the heavy oil method is used in coarse sand or gravel material.

Nondestructive Method.   The nondestructive method involves the direct measurement of the in-situ density and moisture content of the compacted soil, using nuclear equipment. The density is obtained by measuring the scatter of gamma radiation by the soil particles since the amount of scatter is proportional to the bulk density of the soil. A calibration curve for the particular equipment is then used to determine the bulk density of the soil. A plot of the amount of scatter in materials of known density measured by the equipment versus the density of the materials is generally used in the calibration curve. The moisture content is also obtained by measuring the scatter of neutrons emitted in the soil. This scatter is due mainly to the presence of hydrogen atoms. The assumption is made that most of the hydrogen is in the form of water, which allows the amount of scatter to be related to the amount of water in the soil. The moisture content is then obtained directly from a calibrated gauge.

One advantage of the nondestructive method is that results are obtained speedily, which is essential if corrective actions are necessary. Another advantage is that more tests can be carried out, which facilitates the use of statistical methods in the control process. The main disadvantages are that a relatively high capital expenditure is required to obtain the equipment, and that field personnel are exposed to dangerous radioactive material, making it imperative that strict safety standards be enforced when nuclear equipment is used.

Field Compaction Equipment

Compaction equipment used in the field can be divided into two main categories. The first category includes the equipment used for spreading the material to the desired layer or lift thickness, and the second category includes the equipment used to compact each layer of material.

Spreading Equipment.   Spreading of the material to the required thickness is done by bulldozers and motor graders. Several types and sizes of graders and dozers are now available on the market. The equipment used for any specific project will depend on the size of the project. A typical motor grader is shown in Figure 18.14.

Compacting Equipment.   Rollers are used for field compaction and apply either a vibrating force or an impact force on the soil. The type of roller used for any particular job depends on the type of soil to be compacted.

A smooth wheel or drum roller applies contact pressure of up to 55 lb/in.^{2} over 100 percent of the soil area in contact with the wheel. This type of roller is generally used for finish rolling of subgrade material and can be used for all types of soil material except rocky soils. Figure 18.15(a) shows a typical smooth wheel roller. The rubber-tired roller is another type of contact roller, consisting of a heavily loaded wagon with rows of four to six tires placed close to each other. The pressure in the tires may be up to 100 lb/in.^{2}. They are used for both granular and cohesive materials. Figure 18.15(b) shows a typical rubber-tired roller.

One of the most frequently used rollers is the sheepsfoot. This roller has a drum wheel that can be filled with water. The drum wheel has several protrusions, which may be

Figure 8.14 Typical Motor Grader

SOURCE: Photograph courtesy of Machinery Distribution, Inc., Mitsubishi Construction Equipment, Houston, Tex.

(a) Smooth wheel roller

Figure 18.15 Typical Smooth Wheel and Rubber-Tired Rollers

(b) Rubber-tired roller

SOURCE: Photograph courtesy of Caterpillar Inc., Peoria, Ill. Used with permission.

round or rectangular in shape, ranging from 5 to 12 in.² in area. The protrusions penetrate the loose soil and compact from the bottom to the top of each layer of soil, as the number of passes increases. Contact pressures ranging from 200 to 1000 lb/in.² can be obtained from sheepsfoot rollers, depending on the size of the drum and whether or not it is filled with water. The sheepsfoot roller is used mainly for cohesive soils. Figure 18.16 shows a typical sheepsfoot roller.

Tamping foot rollers are similar to sheepsfoot rollers in that they also have protrusions that are used to obtain high contact pressures, ranging from 200 to 1200 lb/in.². The feet of the tamping foot rollers are specially hinged to obtain a kneading action while compacting the soil. As with sheepsfoot rollers, tamping foot rollers compact from the bottom of the soil layer. Tamping foot rollers are used mainly for compacting fine-grained soils.

The smooth wheel and tamping foot rollers can be altered to vibrating rollers by attaching a vertical vibrator to the drum, producing a vibrating effect that makes the smooth wheel and tamping foot rollers more effective on granular soils. Vibrating plates and hammers are also available for use in areas where the larger drum rollers cannot be used.

SPECIAL SOIL TESTS FOR PAVEMENT DESIGN

Apart from the tests discussed so far, there are a few special soil tests that are sometimes undertaken to determine the strength or supporting value of a given soil if used as a subgrade or subbase material. The results obtained from these tests are used individually in the design of some pavements, depending on the pavement design method used (see

Figure 18.16 Typical Sheepsfoot Roller

SOURCE: Photograph courtesy of Caterpillar Inc., Peoria, Ill. Used with permission.

Chapter 20). The two most commonly used tests under this category are the California Bearing Ratio Test and Hveem Stabilometer Test.

California Bearing Ratio (CBR) Test

This test is commonly known as the CBR test and involves the determination of the load-deformation curve of the soil in the laboratory using the standard CBR testing equipment shown in Figure 18.17. It was originally developed by the California Division of Highways prior to World War II and was used in the design of some highway pavements. The test has now been modified and is standardized under the AASHTO designation of T193. The test is conducted as follows:

Step 1.   Disturbed samples of the selected soil material are compacted at different moisture contents in molds of 6 in. diameter and about 6 in. height using the standard AASHTO compacting method. The compacted densities should range from 95 percent or lower to 100 percent or higher. The curve of dry density versus moisture content is then plotted. The sample having the highest dry density is selected for the CBR test.

Step 2.   The selected compacted sample, still in the mold, is immersed in water for four days to obtain a saturation condition similar to what may occur in the field. During this period, the sample is loaded with a surcharge, usually 10 lb or greater, that simulates the estimated weight of pavement material the soil will support. Any expansion of the soil sample due to soaking is measured.

Step 3.   The sample is then removed from the water and allowed to drain for about 15 min.

Step 4.   The sample, still carrying the surcharge weight, is then subjected to penetration by the piston of the standard CBR equipment. The loads that cause different penetrations are recorded (in $lb/in.^2$), and a load-penetration curve is drawn. The CBR is then determined as

$$CBR = \frac{\text{(unit load for 0.1 piston penetration in test specimen) } (lb/in.^2)}{\text{(unit load for 0.1 piston penetration in standard crushed rock) } (lb/in.^2)} \quad (18.24)$$

The unit load for 0.1 piston in standard crushed rock is usually taken as 1000 $lb/in.^2$, which gives the CBR as

$$CBR = \frac{\text{(unit load for 0.1 piston penetration in test sample)}}{1000} \times 100 \quad (18.25)$$

Equations 18.24 and 18.25 show that the CBR gives the relative strength of a soil with respect to crushed rock, which is considered an excellent coarse base material.

Figure 18.17 *CBR Testing Equipment*

SOURCE: Photograph courtesy of Wykeham Farrancè International, England, 1983. Used with permission.

The main criticism of the CBR test is that it does not correctly simulate the shearing forces imposed on subbase and subgrade materials as they support highway pavements. For example, it is possible to obtain a relatively high CBR value for a soil containing rough or angular coarse material and some amount of troublesome clay if the coarse material resists penetration of the piston by keeping together in the mold. When such a material is used in highway construction, however, the performance of the soil may be poor, due to the lubrication of the soil mass by the clay, which reduces the shearing strength of the soil mass.

Hveem Stabilometer Test

This test is used to determine the resistance value R of the soil to the horizontal pressure obtained by imposing a vertical stress of 160 lb/in. on a sample of the soil. The value of

R may then be used to determine the pavement thickness above the soil to carry the estimated traffic load. (See Chapter 20.) The test was first conceived in the 1930s and was used to obtain the stability of laboratory and field samples of bituminous pavements. (See Chapter 19.) It has now been modified and made suitable for subgrade materials and is designated as T190 by AASHTO. The procedure consists of three phases: determination of the exudation pressure, determination of the expansion pressure, and determination of the resistance value R (stabilometer test). In order to carry out the test, four cylindrical specimens of 4 in. diameter and 2.5 in. height are prepared at different moisture contents by compacting samples of the soils in steel molds. The compaction is achieved by tamping or kneading the soil.

Exudation Pressure

Exudation pressure is the compressive stress that will exude water from the compacted specimen. Each specimen is pressed in the steel mold by applying a vertical load until water exudes from the soil. The base of the exudation equipment has several electrical circuits wired into it in parallel, and the exuded water completes these circuits. The pressure that exudes enough water to activate five or six of these circuits is the exudation pressure. Several tests in California have indicated that soils supporting highway pavements will exude moisture under pressure of about 300 lb/in.^{2}. The moisture content of the stabilometer test (R value) samples is therefore set to that moisture content that exudes water at a pressure of 300 lb/in.^{2}.

Expansion Pressure

At the completion of the exudation test, a perforated brass plate is placed on each sample in the steel mold and covered with water. The samples are then left to stand for a period of 16–24 hr, during which time the expansion of the soil is measured. The expansion pressure is obtained as the product of the spring constant of the steel bar of the expansion apparatus and expansion shown by the deflection gauge. This pressure indicates the load and therefore the thickness of material required above this soil to prevent any swelling if the soil is inundated with water when used as a subgrade material.

Resistance Value, R

A schematic of the Hveem stabilometer used in this phase of the test is shown in Figure 18.18. At the completion of the expansion test, the specimen is put into a flexible sleeve and placed in the stabilometer as shown in the figure. Vertical pressure is applied gradually on the specimen at a speed of 0.05 in./min until a pressure of 160 lb/in.^{2} is attained. The corresponding horizontal pressure is immediately recorded. To correct for any distortion of the results due to the surface roughness of the sample, the penetration of the flexible diaphragm into the sample is measured. This is done by reducing the vertical load on the specimen by half and also reducing the horizontal pressure to 5 lb/in.^{2}, using the screw-type pump. The number of turns of the pump required to increase the horizontal pressure to 100 lb/in.^{2} is then recorded. The soil's resistance value is given as

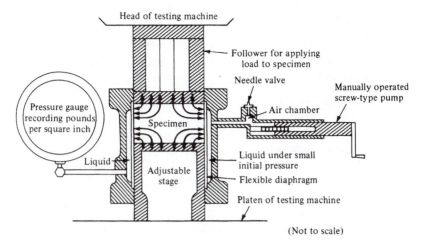

Figure 18.18 Hveem Stabilometer

$$R = 100 - \cfrac{100}{\cfrac{2.5}{D}\left(\cfrac{P_v}{P_h} - 1\right) + 1}$$ (18.26)

where

R = resistance value

P_v = vertical pressure (160 lb/in.²)

P_h = horizontal pressure at P_v of 160 lb/in.² (lb/in.²)

D = number of turns of displacement pump

FROST ACTION IN SOILS

When the ambient temperature falls below freezing for several days, it is quite likely that the water in soil pores will freeze. Since the volume of water increases by about 10 percent when it freezes, the first problem is the increase in volume of the soil. The second problem is that the freezing can cause ice crystals and lenses that are several centimeters thick to form in the soil. These two problems can result in heaving of the subgrade (frost heave), which may result in significant structural damage to the pavement.

In addition, the ice lenses melt during the spring (spring thaw), resulting in a considerable increase in the water content of the soil. This increase in water significantly reduces the strength of the soil, causing structural damage of the highway pavement known as "spring break-up."

In general three conditions must exist for severe frost action to occur:

1. Ambient temperature must be lower than freezing for several days.
2. The shallow water table that provides capillary water to the frost line must be available.
3. The soil must be susceptible to frost action.

The first condition is a natural phenomenon and cannot be controlled by humans. Frost action will therefore be more common in cold areas than in warm areas if all other conditions are the same. The second condition requires that the ground water table be within the height of the capillary rise, so that water will be continuously fed to the growing ice lenses.

The third condition requires that the soil material be of such quality that relatively high capillary pressures can be developed, but at the same time that the flow of water through its pores is restricted. Granular soils are therefore not susceptible to frost action because they have a relatively high coefficient of permeability. Clay soils are also not highly susceptible to frost action because they have very low permeability, so not enough water can flow during a freezing period to allow the formulation of ice lenses. Sandy or silty clays or cracked clay soils near the surface, however, may be susceptible to frost action. Silty soils are most susceptible to frost action. It has been determined that 0.02 mm is the critical grain size for frost susceptibility. For example, gravels with 5 percent of 0.02 mm particles are in general susceptible to frost action, whereas well-graded soils with only 3 percent by weight of their material finer than 0.02 mm are susceptible, and fairly uniform soils must contain at least 10 percent of 0.02 mm particles to be frost susceptible. Soils with less than 1 percent of their material finer than the critical size are rarely affected by frost action.

Current measures taken to prevent frost action, as discussed in Chapter 17, include removing frost-susceptible soils to the depth of the frost line and replacing them with gravel material, lowering the water table by installing adequate drainage facilities, using impervious membranes or chemical additives, and restricting truck traffic on some roads during the spring thaw.

SUMMARY

Selection of suitable soils to be used as the foundation for the highway pavement surface is of primary importance in the design and construction of any highway. Use of unsuitable material will often result in premature failure of the pavement surface and reduction of the ability of the pavement to carry the design traffic load. Chapters 20 and 21 will show that the types of material used for the base, subbase, and/or subgrade of a highway significantly influence the depth of these materials used and also the thickness of the pavement surface.

It is therefore important that transportation engineers involved in the design and/or maintenance of highway pavements be familiar with the engineering properties of soils and the procedures through which the suitability of any soil for highway construction can be determined. A summary of some of the current procedures and techniques used for soil testing and identification is presented in this chapter. The techniques presented are those currently used in highway pavement design.

One of the most important tasks in highway pavement construction is the control of embankment compaction, and discussion of the different methods used in the control of embankment compaction is presented. A brief discussion of the different types of equipment used in the compaction of embankments is also presented to cover all aspects of embankment compaction.

PROBLEMS

18-1   Determine the void ratio of a soil if its bulk density is 120 lb/ft^3 and has a moisture content of 25 percent. The specific gravity of the soil particles is 2.7. Also, determine its dry density and degree of saturation.

18-2   A soil has a bulk density of 135 lb/ft^3 and a dry density of 120 lb/ft^3, and the specific gravity of the soil particles is 2.75. Determine (a) moisture content, (b) degree of saturation, (c) void ratio, and (d) porosity.

18-3   The weight of a sample of saturated soil before drying is 3 lb and after drying is 2.2 lb. If the specific gravity of the soil particles is 2.7, determine (a) moisture content, (b) void ratio, (c) porosity, (d) bulk density, and (e) dry density.

18-4   A moist soil has a moisture content of 10.2 percent, weighs 40.66 lb, and occupies a volume of 0.33 ft^3. The specific gravity of the soil particles is 2.7. Find (a) bulk density, (b) dry density, (c) void ratio, (d) porosity, (e) degree of saturation, and (f) volume occupied by water (ft^3).

18-5   The moist weight of 0.15 ft^3 of soil is 18.6 lb. If the moisture content is 17 percent and the specific gravity of soil solids is 2.62, find the following: (a) bulk density, (b) dry density, (c) void ratio, (d) porosity, (e) degree of saturation, and (f) the volume occupied by water (ft^3).

18-6   A liquid limit test conducted in the laboratory on a sample of soil gave the following results. Determine the liquid limit of this soil from a plot of the flow curve.

Number of Blows (N)	Moisture Content (%)
20	45.0
28	43.6
30	43.2
35	42.8
40	42.0

18-7   A plastic limit test for a soil showed that moisture content was 19.2 percent. Data from a liquid limit test were as follows:

Number of Blows (N)	Moisture Content (%)
14	42.0
19	40.8
27	39.1

(a)   Draw the flow curve and obtain the liquid limit.
(b)   Determine the plasticity index of the soil.

18-8   The following results were obtained by a mechanical analysis. Classify the soil according to the AASHTO classification system and give the group index.

Sieve Analysis, % Finer				
No. 10	No. 40	No. 200	Liquid Limit	Plastic Limit
98	81	38	42	23

18-9   The following results were obtained by a mechanical analysis. Classify the soil according to the AASHTO classification system and give the group index.

Sieve Analysis, % Finer				
No. 10	No. 40	No. 200	Liquid Limit	Plastic Limit
84	58	8	—	N.P.

18-10   The following results were obtained by a mechanical analysis. Classify the soil according to the AASHTO classification system and give the group index.

Sieve Analysis, % Finer				
No. 10	No. 40	No. 200	Liquid Limit	Plastic Limit
99	85	71	55	21

18-11   The following results were obtained by a mechanical sieve analysis. Classify the soil according to the Unified Soil Classification (USCS).

Sieve Analysis, % Passing by Weight				
No. 4	No. 40	No. 200	Liquid Limit	Plastic Limit
30	40	30	33	12

18-12   The following results were obtained by a mechanical sieve analysis. Classify the soil according to the Unified Soil Classification (USCS).

Sieve Analysis, % Passing by Weight				
No. 4	No. 40	No. 200	Liquid Limit	Plastic Limit
4	44	52	29	11

18-13   The following results were obtained by a mechanical sieve analysis. Classify the soil according to the Unified Soil Classification (USCS).

Sieve Analysis, % Passing by Weight

No. 4	No. 40	No. 200	Liquid Limit	Plastic Limit
11	24	65	44	23

18-14   The following data are the results of a sieve analysis.

Sieve No.	Percent Finer
4	100
10	92
20	82
40	67
60	58
80	38
100	22
200	6
Pan	—

(a) Plot the grain-size distribution curve for this example.
(b) Determine D_{10}, D_{30}, and D_{60}.
(c) Calculate the uniformity coefficient, C_u.
(d) Calculate the coefficient of gradation, C_c.

18-15   The following data are the results of a sieve analysis:

Sieve No.	Percent Finer
4	100
10	90
20	80
40	70
60	40
80	29
100	19
200	10
Pan	—

(a) Plot the grain-size distribution curve for this example.
(b) Determine D_{10}, D_{30}, and D_{60}.
(c) Calculate the uniformity coefficient, C_u.
(d) Calculate the coefficient of gradation, C_c.

18-16   The following results are from a compaction test on samples of soil that are to be used for an embankment on a highway project. Determine the maximum dry unit weight of compaction and the optimum moisture content.

Sample No.	Moisture Content	Bulk Density (lb/ft^3)
1	4.8	135.1
2	7.5	145.0
3	7.8	146.8
4	8.9	146.4
5	9.7	145.3

18-17   The following results are from a compaction test on samples of soil that are to be used for an embankment on a highway project. Determine the maximum dry unit weight of compaction and the optimum moisture content.

Sample No.	Moisture Content	Bulk Density (lb/ft^3)
1	12	114.7
2	14	118.6
3	16	120.4
4	18	120.1
5	20	118.3
6	21	117.1

18-18   The results obtained from a seismic study along a section of the center line of a highway are shown below. Estimate the depths of the different strata of soil and suggest the type of soil in each stratum.

Distance of Impulse to Geophone (ft)	Time for Wave Arrival (10^{-3} sec)
25	20
50	40
75	60
100	68
125	74
150	82
175	84
200	86
225	88
250	90

REFERENCES

Bowles, J. E., *Foundation Analysis and Design,* McGraw-Hill Publishing Company, New York, 1988.

Das, B., *Principles of Foundation Engineering,* 4th ed., PWS Publishing, 1998.

Das, B., *Soil Mechanics Laboratory Manual,* Engineering Press, 1997.

Derucher, K. N. and Korflatis, G. P., *Materials for Civil and Highway Engineers,* Prentice-Hall, Englewood Cliffs, N.J., 1994.

Fang, H. Y., *Foundation Engineering Handbook,* Routledge, Chapman & Hall, New York, 1990.

Standard Specifications for Transportation Materials and Methods of Sampling and Testing, 20th ed., American Association of State Highway and Transportation Officials, Washington, D.C., 2000.

CHAPTER 19

Bituminous Materials

B ituminous materials are widely used all over the world in highway construction. These hydrocarbons are found in natural deposits or are obtained as a product of the distillation of crude petroleum. The bituminous materials used in highway construction are either asphalts or tars.

All bituminous materials consist primarily of bitumen, have strong adhesive properties, and have colors ranging from dark brown to black. They vary in consistency from liquid to solid; thus, they are divided into liquids, semisolids, and solids. The solid form is usually hard and brittle at normal temperatures but will flow when subjected to long, continuous loading.

The liquid form is obtained from the semisolid or solid forms by heating, dissolving in solvents, or breaking the material into minute particles and dispersing them in water with an emulsifier to form asphalt emulsion.

This chapter presents a description of the different types of bituminous materials used in highway construction, the processes by which they are obtained, and the tests required to determine those properties that are pertinent to highway engineering. The chapter also includes a description of methods of mix design to obtain a paving material known as asphaltic concrete.

SOURCES OF ASPHALTIC MATERIALS

Asphaltic materials are obtained from seeps or pools of natural deposits in different parts of the world or as a product of the distillation of crude petroleum.

Natural Deposits

Natural deposits of asphaltic materials occur as native asphalts or as rock asphalts. The largest deposit of native asphalt is known to have existed in Iraq several thousand years ago. Native asphalts have also been found in Trinidad, Bermuda, and the La Brea asphalt pits in Los Angeles, California.

Native asphalts, after being softened with petroleum fluxes, were at one time used extensively as binders in highway construction. The properties of native asphalts vary from one deposit to another, particularly with respect to the amount of insoluble material the asphalt contains. The Trinidad deposit, for example, contains about 40 percent insoluble organic and inorganic materials, whereas the Bermuda material contains about 6 percent of such material.

Rock asphalts are natural deposits of sandstone or limestone rocks filled with asphalt. Deposits have been found in California, Texas, Oklahoma, and Alabama. The amount of asphaltic material varies from one deposit to another and can be as low as 4.5 percent and as high as 18 percent. Rock asphalt can be used to surface roads, after the mined or quarried material has been suitably processed. This process includes adding suitable mineral aggregates, asphaltic binder, and oil, which facilitates the flowing of the material. Rock asphalt is not widely used because of its high transportation costs.

Petroleum Asphaltic Materials

The asphaltic materials obtained from the distillation of petroleum are in the form of different types of asphalts. These include asphalt cements, slow-curing liquid asphalts, medium-curing liquid asphalts, rapid-curing liquid asphalts, and asphalt emulsions. The quantity of asphalt obtained from crude petroleum is dependent on the American Petroleum Institute (API) gravity of the petroleum. In general, large quantities of asphalt are obtained from crude petroleums with low API gravity. Before discussing the properties and uses of the different types of petroleum asphalts, we first describe the refining processes used to obtain them.

Refining Processes

The refining processes used to obtain petroleum asphalts can be divided into two main groups, namely, *fractional distillation* and *destructive distillation* (cracking). The fractional distillation processes involve the separation of the different materials in the crude petroleum without significant changes in the chemical composition of each material. The destructive distillation processes involve the application of high temperature and pressure, resulting in chemical changes.

Fractional Distillation. The fractional distillation process removes the different volatile materials in the crude oil at successively higher temperatures until the petroleum asphalt is obtained as residue. Steam or vacuum is used to gradually increase the temperature.

Steam distillation is a continuous flow process in which the crude petroleum is pumped through tube stills or stored in batches, and the temperature is increased gradually to facilitate the evaporation of the different materials at different temperatures. Tube stills are more efficient than batches and are therefore preferred in modern refineries.

Figure 19.1 is a flow chart that shows the interrelationships of the different materials that can be obtained from the fractional distillation of crude petroleum. Immediately

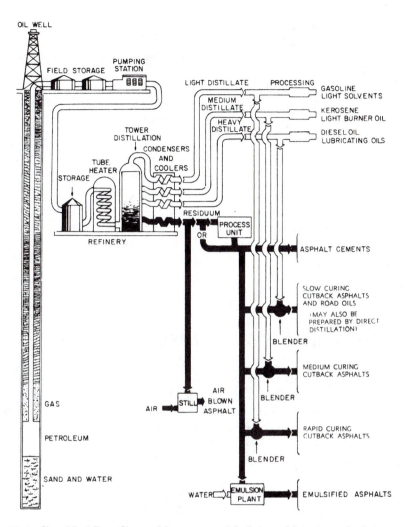

Figure 19.1 Simplified Flow Chart of Recovery and Refining of Petroleum Asphalt

SOURCE: *Introduction to Asphalt,* Manual Series No. 5, 8th ed., Asphalt Institute, Lexington, Ky., 1993.

after increasing the temperature of the crude in the tube still, it is injected into a bubble tower, which consists of a vertical cylinder into which are built several trays or platforms one above the other. The first separation of materials occurs in this tower. The lighter fractions of the evaporated materials collect on the top tray, and the heavier fractions collect in successive trays, with the heaviest residue containing asphalt remaining at the bottom of the distillation tower. The products obtained during this first phase of separation are gasoline, kerosene distillate, diesel fuel, lubricating oils, and the heavy residual material that contains the asphalt (see Figure 19.1). The various fractions collected are stored and further refined into specific grades of petroleum products. Note that a desired consistency of the residue can be obtained by continuing the distillation process. Attainment of the desired consistency is checked by measuring the temperature of the residue or by observing the character of the distillate. The residue becomes harder the longer the distillation process is continued.

It can be seen from Figure 19.1 that further processing of the heavy residue obtained after the first separation will give asphalt cement of different penetration grades—slow-curing and rapid-curing asphalts—depending on the additional processing carried out. Emulsified asphalts can also be obtained. A description of each of these different types of asphalt materials will be given later.

Destructive Distillation. *Cracking processes* are used when larger amounts of the light fractions of the materials such as motor fuels are required. Intense heat and high pressures are applied to produce chemical changes in the material. Although several specific methods of cracking exist, the process generally involves the application of temperatures as high as 1100°F and pressure higher than 735 lb/in.^{2} to obtain the desired effect. The asphaltic material obtained from cracking is not widely used in paving because it is more susceptible to weather changes than is that produced from fractional distillation.

DESCRIPTION AND USES OF BITUMINOUS BINDERS

It is necessary at this point to describe the different bituminous binders and identify the type of construction for which each is used. The commonly used bituminous binders are asphalt cement, slow-curing asphalts, medium-curing cutback asphalts, rapid-curing cutback asphalts, and asphalt emulsion. Road tars are also a type of bituminous material but are not now widely used in highway construction.

Asphalt Cements

Asphalt cements are obtained after separation of the lubricating oils, as shown in Figure 19.1. They are semisolid hydrocarbons with certain physiochemical characteristics that make them good cementing agents. They are also very viscous, and when used as a binder for aggregates in pavement construction, it is necessary to heat both the aggregates and the asphalt cement prior to mixing the two materials. For several decades the particular grade of asphalt cement has been designated by its penetration and viscosity, both of which give an indication of the consistency of the material at a given temperature. The penetration is the distance in 0.1 mm that a standard needle will penetrate a given sample under specific conditions of loading, time, and temperature. The softest grade used for highway pavement construction has a penetration value of 200–300, and the hardest has a penetration value of 60–70. For some time now, however, viscosity has been used more often than penetration to grade asphalt cements.

Asphalt cements are used for a variety of purposes, as shown in Table 19.1. The use of a given sample depends on its grade. The procedures for determining the grades of asphalt cements through standard penetration and viscosity tests will be described in subsequent sections of this chapter.

Slow-Curing Asphalts

Slow-curing (SC) asphalts can be obtained directly as *slow-curing straight run asphalts* through the distillation of crude petroleum, as shown in Figure 19.1, or as *slow-curing cutback asphalts* by "cutting back" asphalt cement with a heavy distillate such as diesel oil. They have lower viscosities than asphalt cement and are very slow to harden. Slow-curing asphalts are usually designated as SC-70, SC-250, SC-800, or SC-3000, where the numbers are related to the approximate kinematic viscosity in centistokes at 60°C (140°F). Their uses are also shown in Table 19.1. Specifications for the use of these asphalts are not now included in American Association of State Highway and Transportation Official's (AASHTO) *Standard Specifications for Transportation Materials.*

Medium–Curing Cutback Asphalts

Medium-curing (MC) asphalts are produced by *fluxing,* or cutting back, the residual asphalt (usually 120–150 penetration) with light fuel oil or kerosene. The term *medium* refers to the medium volatility of the kerosene-type dilutent used. Medium-curing cutback asphalts harden faster than slow-curing liquid asphalts, although the consistencies of the different grades are similar to those of the slow-curing asphalts. However, the MC-30 is a unique grade in this series as it is very fluid and has no counterpart in the SC and RC series.

The fluidity of medium-curing asphalts depends on the amount of solvent in the material. MC-3000, for example, may have only 20 percent of the solvent by volume, whereas MC-70 may have up to 45 percent. These medium-curing asphalts can be used for the construction of pavement bases, surfaces, and surface treatments, as shown in Table 19.1.

Rapid–Curing Cutback Asphalts

Rapid-curing (RC) cutback asphalts are produced by blending asphalt cement with a petroleum distillate that will easily evaporate, thereby facilitating a quick change from the liquid form at time of application to the consistency of the original asphalt cement. Gasoline or naphtha generally is used as the solvent for this series of asphalts.

The grade of rapid-curing asphalt required dictates the amount of solvent to be added to the residual asphalt cement. For example, RC-3000 requires about 15 percent of distillate, whereas RC-70 requires about 40 percent. These grades of asphalt can be used for jobs similar to those for which the MC series is used. Specifications for the use of these asphalts are given in AASHTO's *Standard Specifications for Transportation Materials.*

Blown Asphalts

Blown asphalt is obtained by blowing air through the semisolid residue obtained during the latter stages of the distillation process. The process involves stopping the regular distillation while the residue is in the liquid form, and then transferring it into a tank known as a *converter.* The material is maintained at a high temperature while air is blown through it. This is continued until the required properties are achieved. Blown asphalts are relatively stiff compared to other types of asphalts and can maintain a firm consistency at the maximum temperature normally experienced when exposed to the environment.

Table 19.1 Typical Uses of Asphalt

Type of Construction	Asphalt Cements														
	Viscosity Graded-Original					Viscosity Graded-Residue					Penetration Graded				
	AC-40	AC-20	AC-10	AC-5	AC-2.5	AR-16000	AR-8000	AR-4000	AR-2000	AR-1000	40-50	60-70	85-100	120-150	200-300
Asphalt-Aggregate Mixtures															
Asphalt Concrete and Hot Laid Plant Mix															
Pavement Base and Surfaces															
Highways		x	x	x	x[1]	x	x	x	x[A]	x[1]		x	x	x	x[1]
Airports		x	x	x			x	x				x	x		
Parking Areas		x	x			x	x	x				x	x		
Driveways		x	x				x	x				x	x		
Curbs	x	x	x			x	x				x	x	x		
Industrial Floors	x	x				x	x				x	x			
Blocks	x										x				
Groins	x	x									x	x			
Dam Facings	x	x									x	x			
Canal and Reservoir Linings	x	x									x	x			
Cold-Laid Plant Mix															
Pavement Base and Surfaces															
Open-Graded Aggregate★															
Well-Graded Aggregate★															
Patching, Immediate Use															
Patching, Stockpile															
Mixed-in-Place (Road Mix)															
Pavement Base and Surfaces															
Open-Graded Aggregate★															
Well-Graded Aggregate★															
Sand				x	x									x	x
Sandy Soil				x	x									x	x
Patching, Immediate Use															
Patching, Stockpile															
Asphalt-Aggregate Applications															
Surface Treatments															
Single Surface Treatment				x	x									x	x
Multiple Surface Treatment				x	x									x	x
Aggregate Seal				x	x				x					x	x
Sand Seal															
Slurry Seal															
Penetration Macadam															
Pavement Bases															
Large Voids			x											x	
Small Voids				x										x	
Asphalt Applications															
Surface Treatment															
Fog Seal															
Prime Coat, Open Surfaces															
Prime Coat, Tight Surfaces															
Tack Coat															
Dust Laying															
Mulch															
Membrane															
Canal and Reservoir Linings	x										x				
Embankment Envelopes	x	x									x	x			
Crack Filling															
Asphalt Pavements															
Portland Cement Concrete Pavements	x[4]										x[4]				

★ Evaluation of emulsified asphalt-aggregate system required to determine the proper grade of emulsified asphalt to use.
[1] For use in cold climates.
[A] For use in bases only in cold climates.
[2] Diluted with water.

Table 19.1 Typical Uses of Asphalt (*continued*)

| | Cutback Asphalts | | | | | | | | | | | | | Emulsified Asphalts[#] | | | | | | | | | | | | |
| | Rapid Curing (RC) | | | | Medium Curing (MC) | | | | | Slow Curing (SC) | | | | Anionic | | | | | | | Cationic | | | | | |
	70	250	800	3000	30	70	250	800	3000	70	250	800	3000	RS-1	RS-2	MS-1	MS-2	MS-2h	SS-1	SS-1h	CRS-1	CRS-2	CMS-2	CMS-2h	CSS-1	CSS-1h
								x						x	x	x					x	x				
		x				x	x	x		x	x	x							x	x	x				x	x
		x	x			x	x	x			x	x							x	x	x				x	x
						x	x			x	x															
		x	x	x			x	x		x	x			x	x	x					x	x				
		x				x	x			x	x								x	x	x				x	x
		x	x		x	x	x												x	x					x	x
	x	x	x			x	x												x	x	x				x	x
		x	x			x	x	x			x	x							x	x					x	x
						x	x			x	x															
		x	x	x			x	x						x	x						x	x				
		x	x	x				x						x	x						x	x				
		x	x	x			x	x						x	x						x	x				
		x				x	x							x	x						x	x				
																			x	x					x	x
		x	x											x	x						x	x				
		x												x	x						x	x				
																			x^2	x^2					x^2	x^2
	x	x				x	x																			
	x				x	x													x							
	x													x					x^2	x^2			x		x^2	x^2
							x												x^2	x^2					x^2	x^2
																			x^2	x^2						
																			x^3	x^3					x^3	x^3

[3] Slurry mix.
[4] Rubber asphalt compounds.
[#] Emulsified asphalts shown are AASHTO and ASTM grades and may not include all grades produced in all geographical areas.

SOURCE: *Introduction to Asphalt,* Manual Series No. 5, 8th ed., The Asphalt Institute, Lexington, Ky., 1993.

Blown asphalt generally is not used as a paving material. However, it is very useful as a roofing material, for automobile undercoating, and as a joint filler for concrete pavements. If a catalyst is added during the air-blowing process, the material obtained will usually maintain its plastic characteristics, even at temperatures much lower than that at which ordinary asphalt cement will become brittle. The elasticity of catalytically blown asphalt is similar to that of rubber, and it is used for canal lining.

Asphalt Emulsion

Emulsified asphalts are produced by breaking asphalt cement, usually of 100–250 penetration range, into minute particles and dispersing them in water with an emulsifier. These minute particles have like electrical charges and therefore do not coalesce. They remain in suspension in the liquid phase as long as the water does not evaporate or the emulsifier does not break. Asphalt emulsions therefore consist of asphalt, which makes up about 55 to 70 percent by weight, water, and an emulsifying agent, which in some cases may contain a stabilizer.

Asphalt emulsions are generally classified as anionic, cationic, or nonionic. The first two types have electrical charges surrounding the particles, whereas the third type is neutral. Classification as anionic or cationic is based on the electrical charges that surround the asphalt particles. Emulsions containing negatively charged particles of asphalt are classified as anionic, and those having positively charged particles of asphalts are classified as cationic. The anionic and cationic asphalts generally are used in highway maintenance and construction, although it is likely that the nonionics may be used in the future as emulsion technology advances.

Each of the above categories is further divided into three subgroups, based on how rapidly the asphalt emulsion will return to the state of the original asphalt cement. These subgroups are rapid setting (RS), medium-setting (MS), and slow setting (SS). A cationic emulsion is identified by placing the letter "C" in front of the emulsion type; no letter is placed in front of anionic and nonionic emulsions. For example, CRS-2 denotes a cationic emulsion, and RS-2 denotes either anionic or nonionic emulsion.

Asphalt emulsions are used for several purposes, as shown in Table 19.1. Note, however, that since anionic emulsions contain negative charges, they are more effective in treating aggregates containing electropositive charges such as limestone, whereas cationic emulsions are more effective with electronegative aggregates such as those containing a high percentage of siliceous material. Also note that ordinary emulsions must be protected during very cold spells because they will break down if frozen. Three grades of high-float, medium-setting anionic emulsions designated as HFMS have been developed and are used mainly in cold and hot plant mixes, and coarse aggregate seal coats. These high-float emulsions have one significant property, which is that they can be laid at relatively thicker films without a high probability of runoff.

Specifications for the use of emulsified asphalts are given in AASHTO M140 and ASTM D977.

Road Tars

Tars are obtained from the destructive distillation of such organic materials as coal. Their properties are significantly different from petroleum asphalts. In general, they are more susceptible to weather conditions than are similar grades of asphalts, and they set more quickly when exposed to the atmosphere. Because tars are rarely used now for highway pavements, this text includes only a brief discussion of the subject.

The American Society for Testing Materials (ASTM) has classified road tars into three general categories based on the method of production.

1. Gashouse coal tars are produced as a by-product in gashouse retorts in the manufacture of illuminating gas from bituminous coals.
2. Coke-oven tars are produced as a by-product in coke ovens in the manufacture of coke from bituminous coal.
3. Water-gas tars are produced by cracking oil vapors at high temperatures in the manufacture of carburated water gas.

Road tars have also been classified by AASHTO into 14 grades: RT-1 through RT-12, RTCB-5, and RTCB-6. RT-1 has the lightest consistency and can be used effectively at normal temperatures for prime or tack coat (described later in this chapter). The viscosity of each grade increases as the number designation increases to RT-12, which is the most viscous. RTCB-5 and RTCB-6 are suitable for application during cold weather since they are produced by cutting back the specific grade of tar with easily evaporating solvent. Detailed specifications for the use of tars are given by AASHTO Designation M52-78.

PROPERTIES OF ASPHALTIC MATERIALS

The properties of asphaltic materials pertinent to pavement construction can be classified into four main categories:

- Consistency
- Durability
- Rate of curing
- Resistance to water action

Consistency

The consistency properties of an asphaltic material are usually considered under two conditions: (1) variation of consistency with temperature and (2) consistency at a specified temperature.

Variation of Consistency with Temperature

The consistency of any asphaltic material changes as the temperature changes. The change in consistency of different asphaltic materials may differ considerably even for the same amount of temperature change. For example, if a sample of blown semisolid asphalt and a sample of semisolid regular paving-grade asphalt with the same consistency at a given temperature are heated to a high enough temperature, the consistencies of the two materials will be different at the high temperatures, with the regular paving-grade asphalt being much softer than the blown asphalt. Further increase in temperature will eventually result in the liquefaction of the paving asphalt at a temperature much lower than that at which the blown asphalt liquefies. If these two asphalts are then gradually cooled down to about the freezing temperature of water, the blown asphalt will be much softer than the paving-grade asphalt. Thus, the consistency of the blown asphalt is less affected by temperature changes than is the consistency of regular paving-grade asphalt. This property of asphaltic materials is known as *temperature susceptibility*. The temperature susceptibility of a given asphalt depends on the crude oil from which the asphalt is obtained, although the variation in temperature susceptibility of paving-grade asphalts from different crudes is not as high as that between regular paving-grade asphalt and blown asphalt.

Consistency at a Specified Temperature

As stated earlier, the consistency of an asphaltic material will vary from solid to liquid depending on the temperature of the material. It is therefore essential that when the consistency of an asphalt material is given, the associated temperature also be given. Different ways for measuring consistency are presented later in this chapter.

Durability

When asphaltic materials are exposed to environmental elements, natural deterioration gradually takes place, and eventually the materials lose their plasticity and become brittle. This change is caused primarily by chemical and physical reactions that take place in the material. This natural deterioration of the asphaltic material is known as *weathering*. For paving asphalt to act successfully as a binder, the weathering must be minimized as much as possible. The ability of an asphaltic material to resist weathering is described as the *durability* of the material. Some of the factors that influence weathering are oxidation, volatilization, temperature, exposed surface area, and age hardening. These factors are discussed briefly in the following sections.

Oxidation

Oxidation is the chemical reaction that takes place when the asphaltic material is attacked by the oxygen in the air. This chemical reaction causes gradual hardening and eventually permanent hardening and considerable loss of the plastic characteristics of the material.

Volatilization

Volatilization is the evaporation of the lighter hydrocarbons from the asphaltic material. The loss of these lighter hydrocarbons also causes loss of the plastic characteristics of the asphaltic material.

Temperature

It has been shown that temperature has a significant effect on the rate of oxidation and volatilization. The higher the temperature, the higher the rates of oxidation and volatilization. The relationship between temperature increase and increases in rates of oxidation and volatilization is not linear, however; the percentage increase in rate of oxidation and volatilization is usually much greater than the percentage increase in temperature that causes the increase in oxidation and volatilization. It has been postulated that the rate of organic and physical reactions in the asphaltic material approximately doubles for each 10°C (50°F) increase in temperature.

Surface Area

The exposed surface of the material also influences its rate of oxidation and volatilization. There is a direct relationship between surface area and rate of oxygen absorption and loss due to evaporation in grams per cubic centimeter per minute. An inverse relationship, however, exists between volume and rate of oxidation and volatilization. This means that the rate of hardening is directly proportional to the ratio of the surface area to the volume.

This fact is taken into consideration when asphalt concrete mixes are designed for pavement construction in that the air voids are kept to the practicable minimum required for stability to reduce the area exposed to oxidation.

Age Hardening

If a sample of asphalt is heated and then allowed to cool, its molecules will be rearranged to form a gel-like structure. This will cause continuous hardening of the asphalt over time, even though it is protected from other factors such as oxidation and volatilization that cause hardening. This process of hardening with time is known as *age hardening*. The rate at which age hardening occurs is relatively high during the first few hours after cooling, but it gradually decreases and becomes almost negligible after about a year. Age hardening does not seem to have a significant effect on pavements, except when they are laid as a thin mat.

Rate of Curing

Curing is defined as the process through which an asphaltic material increases its consistency as it loses solvent by evaporation.

Rate of Curing of Cutbacks

As discussed earlier, the rate of curing of any cutback asphaltic material depends on the distillate used in the cutting-back process. This is an important characteristic of cutback materials, since the rate of curing indicates the time that should elapse before a cutback will attain a consistency that is thick enough for the binder to perform satisfactorily. The rate of curing is affected by both inherent and external factors. The important inherent factors are

- Volatility of the solvent
- Quantity of solvent in the cutback
- Consistency of the base material

The more volatile the solvent is, the faster it can evaporate from the asphaltic material, and therefore the higher the curing rate of the material. This is why gasoline and naphtha are used for rapid-curing cutbacks, whereas light fuel oil and kerosene are used for medium-curing cutbacks.

For any given type of solvent, the smaller the quantity used, the less time is required for it to evaporate, and therefore the faster the asphalt material will cure. Also, the higher the penetration of the base asphalt, the longer it takes for the asphalt cutback to cure.

The important external factors that affect curing rate are

- Temperature
- Ratio of surface area to volume
- Wind velocity across exposed surface

These three external forces are directly related to the rate of curing in that the higher any of these factors is, the higher the rate of curing. Unfortunately these factors cannot be controlled or predicted in the field, which makes it extremely difficult to predict the expected curing time. The curing rates of different asphaltic materials are therefore usually compared with the assumption that the external factors are held constant.

Rate of Curing for Asphalt Emulsions

The curing and adhesion characteristics of emulsions (anionic and cationic) used for pavement construction depend on the rate at which the water evaporates from the mixture. When weather conditions are favorable, the water is relatively rapidly displaced and so curing progresses rapidly. When weather conditions include high humidity, low temperature, or rainfall immediately following the application of the emulsion, its ability to properly cure is adversely affected. Although the effect of surface and weather conditions on proper curing is more critical for anionic emulsions, favorable weather conditions are required to obtain optimum results for cationic emulsions also. A major advantage of cationic emulsions is that they release their water more readily.

Resistance to Water Action

When asphaltic materials are used in pavement construction, it is important that the asphalt continues to adhere to the aggregates even with the presence of water. If this bond between the asphalt and the aggregates is lost, the asphalt will strip from the aggregates, resulting in the deterioration of the pavement. The asphalt therefore must sustain its ability to adhere to the aggregates even in the presence of water. In hot-mix, hot-laid asphaltic concrete, where the aggregates are thoroughly dried before mixing, stripping does not normally occur and so no preventive action is usually taken. However, when water is added to a hot-mix, cold-laid asphaltic concrete, commercial antistrip additives are usually added to improve the asphalt's ability to adhere to the aggregates.

Temperature Effect on Volume of Asphaltic Materials

The volume of asphalt is significantly affected by changes in temperature. The volume increases with an increase in temperature and decreases with a decrease in temperature. The rate of change in volume is given as the coefficient of expansion, which is the volume change in a unit volume of the material for a unit change in temperature. Because of this variation of volume with temperature, the volumes of asphaltic materials are usually given for a temperature of 60°F (15.6°C). Volumes measured at other temperatures are converted to the equivalent volumes at 60°F by using appropriate multiplication factors published by ASTM in Petroleum Measurement tables (ASTM D-1250).

TESTS FOR ASPHALTIC MATERIALS

Several tests are conducted on asphaltic materials to determine both their consistency and their quality to ascertain whether materials used in highway construction meet the prescribed specifications. Some of these specifications given by AASHTO and ASTM have been referred to earlier, and some are listed in Tables 19.2 and 19.3. Standard specifications have also been published by the Asphalt Institute for the types of asphalts used in pavement construction. Procedures for selecting representative samples of asphalt for testing have also been standardized and are given in MS-18 by the Asphalt Institute and in D140 by ASTM.

Some of the tests used to identify the quality of a given sample of asphalt on the basis of the properties discussed will now be described, followed by the description of some general tests.

Consistency Tests

The consistency of asphaltic materials is important in pavement construction because the consistency at a specified temperature will indicate the grade of the material. It is important that the temperature at which the consistency is determined be specified, since temperature significantly affects the consistency of asphaltic materials. As stated earlier, asphaltic materials can exist in either liquid, semisolid, or solid states. This wide range dictates the necessity for more than one method for determining the consistency of asphaltic materials. The property generally used to describe the consistency of asphaltic materials in the liquid state is the viscosity, which can be determined by conducting either the Saybolt Furol viscosity test or the kinematic viscosity test. Tests used for asphaltic materials in the semisolid and solid states include the penetration test and the float test. The ring-and-ball softening point test, which is not often used in highway specifications, may also be used for blown asphalt.

Saybolt Furol Viscosity Test

Figure 19.2 shows the Saybolt Furol Viscometer. The principal part of the equipment is the standard viscometer tube, which is 5 in. long and about 1 in. in diameter. An orifice of specified shape and dimensions is provided at the bottom of the tube. The orifice is closed with a stopper, and the tube is filled with a quantity of the material to be tested. The standard tube is then placed in a larger oil or water bath, which is fitted with an electric heater and a stirring device. The material in the tube is brought to the specified temperature by heating the bath. Immediately upon reaching the prescribed temperature, the stopper is removed, and the time in seconds for exactly 60 milliliters of the asphaltic material to flow through the orifice is recorded. This time is the Saybolt Furol viscosity in units of seconds at the specified temperature. Temperatures at which asphaltic materials for highway construction are tested include 25°C (77°F), 50°C (122°F), and 60°C (140°F). It is apparent that the higher the viscosity of the material, the longer it takes for a

Table 19.2 Specifications for Rapid Curing of Cutback Asphalts

	RC-70		RC-250		RC-800		RC-3000	
	Min.	*Max.*	*Min.*	*Max.*	*Min.*	*Max.*	*Min.*	*Max.*
Kinematic viscosity at 60 C (140 F) (See Note 1) centistokes	70	140	250	500	800	1600	3000	6000
Flash point (Tag, open-cup), degrees C (F)	...	...	27 (80)	...	27 (80)	...	27 (80)	...
Water, percent	...	0.2	...	0.2	...	0.2	...	0.2
Distillation test:								
Distillate, percentage by volume of total distillate to 360 C (680 F)								
to 190 C (374 F)	10	...	...	...	...	...	...	...
to 225 C (437 F)	50	...	35	...	15	...	...	...
to 260 C (500 F)	70	...	60	...	45	...	25	...
to 315 C (600 F)	85	...	80	...	75	...	70	...
Residue from distillation to 360 C (680 F) volume percentage of sample by difference	55	...	65	...	75	...	80	...
Tests on residue from distillation:								
Absolute viscosity at 60 C (140 F) (See Note 3) poises	600	2400	600	2400	600	2400	600	2400
Ductility, 5 cm/min. at 25 C (77 F) cm	100	...	100	...	100	...	100	...
Solubility in trichloroethylene, percent	99.0	...	99.0	...	99.0	...	99.0	...
Spot test (See Note 2) with:								
Standard naphtha			Negative for all grades					
Naphtha - xylene solvent, -percent xylene			Negative for all grades					
Heptane - xylene solvent, -percent xylene			Negative for all grades					

NOTE 1—As an alternate, Saybolt Furol viscosities may be specified as follows:
  Grade RC-70—Furol viscosity at 50 C (122 F)—60 to 120 sec.
  Grade RC-250—Furol viscosity at 60 C (140 F)—125 to 250 sec.
  Grade RC-800—Furol viscosity at 82.2 C (180 F)—100 to 200 sec.
  Grade RC-3000—Furol viscosity at 82.2 C (180 F)—300 to 600 sec.
NOTE 2—The use of the spot test is optional. When specified, the Engineer shall indicate whether the standard naphtha solvent, the naphtha xylene solvent or the heptane xylene solvent will be used in determining compliance with the requirement, and also, in the case of the xylene solvents, the percentage of xylene to be used.
NOTE 3—In lieu of viscosity of the residue, the specifying agency, at its option, can specify penetration at 100 g; 5s at 25 C (77 F) of 80-120 for Grades RC-70, RC-250, RC-800, and RC-3000. However, in no case will both be required.
SOURCE: Used with permission from *Standard Specifications for Transportation Materials and Methods of Sampling and Testing*, 20th ed., Washington, D.C.: The American Association of State Highway and Transportation Officials, copyright 2000.

Table 19.3 Specifications for Medium Curing of Cutback Asphalts

	MC-30		MC-70		MC-250		MC-800		MC-3000	
	Min.	Max.	Min.	Max.	Min.	Max.	Min.	Max.	Min.	Max.
Kinematic Viscosity at 60°C (140°F) (See Note 1) mm²/s	30	60	70	140	250	500	800	1600	3000	6000
Flash point (Tag, open-cup), degrees C (F)	38 (100)	—	38 (100)	—	66 (150)	—	66 (150)	—	66 (150)	—
Water percent	—	0.2	—	0.2	—	0.2	—	0.2	—	0.2
Distillation test:										
Distillate, percentage by volume of total distillate to 360°C (680°F)										
to 225°C (437°F)	—	25	0	20	0	10	—	—	—	—
to 260°C (500°F)	40	70	20	60	15	55	0	35	0	15
to 315°C (600°F)	75	93	65	90	60	87	45	80	15	75
Residue from distillation to 360°C (680°F) volume percentage of sample by difference	50	—	55	—	67	—	75	—	80	—
Tests on residue from distillation:										
Absolute viscosity at 60°C (140°F) (See Note 4) Pa·s (Poises)	30 (300)	120 (1200)	30 (300)	120 (1200)	30 (300)	120 (1200)	30 (300)	120 (1200)	30 (300)	120 (1200)
Ductility, 5 cm/min, cm (See Note 2)	100	—	100	—	100	—	100	—	100	—
Solubility in Trichloroethylene, percent	99.0	—	99.0	—	99.0	—	99.0	—	99.0	—
Spot test (See Note 3) with:										
Standard naphtha				Negative for all grades						
Naphtha-xylene solvent, percent xylene				Negative for all grades						
Heptane-xylene solvent, percent xylene				Negative for all grades						

NOTE 1—As an alternate, Saybolt-Furol viscosities may be specified as follows:
Grade MC-70—Furol viscosity at 50°C (122°F)—60 to 120 s
Grade MC-30—Furol viscosity at 25°C (77°F)—75 to 150 s
Grade MC-250—Furol viscosity at 60°C (140°F)—125 to 250 s
Grade MC-800—Furol viscosity at 82.2°C (180°F)—100 to 200 s
Grade MC-3000—Furol viscosity at 82.2°C (180°F)—300 to 600 s
NOTE 2—If the ductility at 25°C (77°F) is less than 100, the material will be acceptable if its ductility at 15.5°C (60°F) is more than 100.
NOTE 3—The use of the spot test is optional. When specified, the engineer shall indicate whether the standard naphtha solvent, the naphtha-xylene solvent, or the heptane-xylene solvent will be used in determining compliance with the requirement, and also, in the case of the xylene solvents, the percentage of xylene to be used.
NOTE 4—In lieu of viscosity of the residue, the specifying agency, at its option, can specify penetration at 100 g; 5s at 25°C (77°F) of 120 to 250 for Grades MC-30, MC-70, MC-250, MC-800, and MC-3000. However, in no case will both be required.
SOURCE: Used with permission from *Standard Specifications for Transportation Materials and Methods of Sampling and Testing*, 20th ed., Washington, D.C.: The American Association of State Highway and Transportation Officials, copyright 2000.

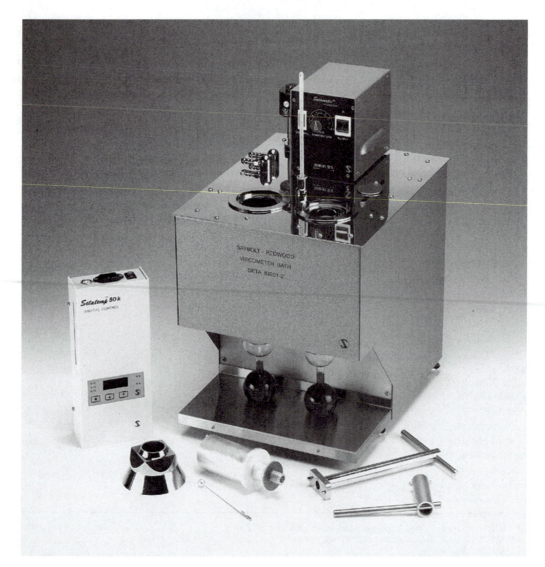

Figure 19.2 Saybolt Furol Viscometer

SOURCE: Photograph courtesy of STANHOPE-SETA, Ltd., Surrey, England. Used with permission.

given quantity to flow through the orifice. Details of the equipment and procedures for conducting the Saybolt Furol test are given in AASHTO T72-83.

Kinematic Viscosity Test

Figure 19.3 shows the equipment used to determine the *kinematic viscosity*, which is defined as the absolute viscosity divided by the density. The test uses a capillary viscometer tube to measure the time it takes the asphalt sample to flow at a specified temperature

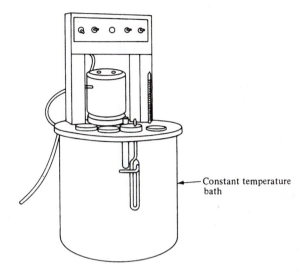

Constant temperature
bath

Figure 19.3 Kinematic Viscosity Apparatus

between timing marks on the tube. Three types of viscometer tubes are shown in Figure 19.4. Flow between the timing marks in the Zeitfuch's cross-arm viscometer is induced by gravitational forces, whereas flow in the Asphalt Institute vacuum viscometer and Cannon-Manning vacuum viscometer is induced by creating a partial vacuum.

When the cross-arm viscometer is used, the test is started by placing the viscometer tube in a thermostatically controlled constant temperature bath, as shown in Figure 19.3. A sample of the material to be tested is then preheated and poured into the large side of the viscometer tube [Figure 19.4(a)] until the filling line level is reached. The temperature of the bath is then brought to 135°C (275°F), and some time is allowed for the viscometer and the asphalt to reach a temperature of 135°C (275°F). Flow is then induced by applying a slight pressure to the large opening or a partial vacuum to the efflux (small) opening of the viscometer tube. This causes an initial flow of the asphalt over the siphon section just above the filling line. Continuous flow is induced by the action of gravitational forces. The time it takes for the material to flow between two timing marks is recorded. The kinematic viscosity of the material in units of centistokes is obtained by multiplying the time in seconds by a calibration factor for the viscometer used. The calibration of each viscometer is carried out by using standard calibrating oils with known viscosity characteristics. The factor for each viscometer is usually furnished by the manufacturer. This test is described in detail in AASHTO Designation T201.

The test may also be conducted at a temperature of 60°C (140°F) as described in AASHTO Designation T202, using either the Asphalt Institute vacuum viscometer (Figure 19.4b) or the Cannon-Manning vacuum viscometer (Figure 19.4c). In this case, flow is induced by applying a prescribed vacuum through a vacuum control device attached to a vacuum pump. The product of the time interval and the calibration factor in this test gives the absolute viscosity of the material in poises.

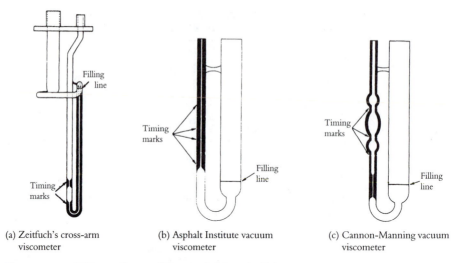

(a) Zeitfuch's cross-arm viscometer   (b) Asphalt Institute vacuum viscometer   (c) Cannon-Manning vacuum viscometer

Figure 19.4 Different Types of Kinematic Viscosity Tubes

SOURCE: *Introduction to Asphalt,* Manual Series No. 5, 8th ed., Asphalt Institute, Lexington, Ky., 1993.

Penetration Test

The penetration test gives an empirical measurement of the consistency of a material in terms of the distance a standard needle sinks into that material under a prescribed loading and time. Although more fundamental tests are now being substituted for this test, it may still be included in specifications for viscosity of asphalt cements to ensure the exclusion of materials with very low penetration values at 25°C (77°F).

Figure 19.5 shows a typical penetrometer and a schematic of the standard penetration test. A sample of the asphalt cement to be tested is placed in a container, which in turn is placed in a temperature-controlled water bath. The sample is then brought to the prescribed temperature of 25°C (77°F), and the standard needle, loaded to a total weight of 100 g, is left to penetrate the sample of asphalt for the prescribed time of exactly 5 sec. The penetration is given as the distance in units of 0.1 mm that the needle penetrates the sample. For example, if the needle penetrates a distance of exactly 20 mm, the penetration is 200.

The penetration test can also be conducted at 0°C (32°F) or at 4°C (39.2°F) with the needle loaded to a total weight of 200 g and with penetrations allowed for 60 sec. Details of the penetration test are given in AASHTO Designation T49 and ASTM Test D5.

Float Test

The float test is used to determine the consistency of semisolid asphalt materials that are more viscous than grade 3000 or have penetration higher than 300, since these materials cannot be tested conveniently using either the Saybolt Furol viscosity test or the penetration test.

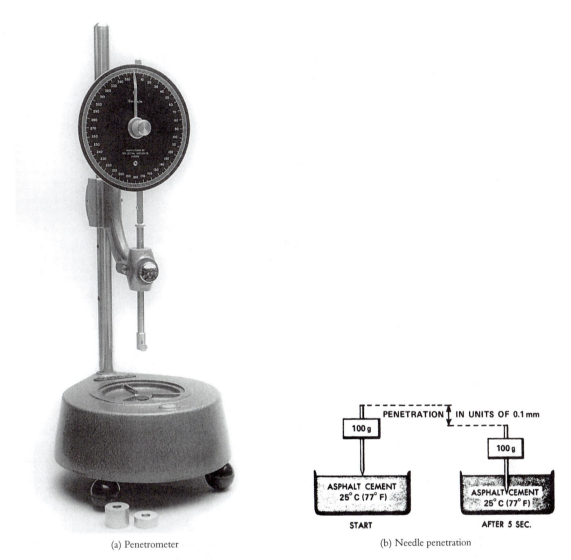

(a) Penetrometer   (b) Needle penetration

Figure 19.5 Standard Penetration Test and Equipment

SOURCE: (a) Photograph courtesy of Wykeham Farrancè International, England, 1983. Used with permission. (b) *Introduction to Asphalt,*
Manual Series No. 5, 8th ed., Asphalt Institute, Lexington, Ky., 1993.

The float test is conducted with the apparatus schematically shown in Figure 19.6.
It consists of an aluminum saucer (float), a brass collar that is open at both ends, and a
water bath. The brass collar is filled with a sample of the material to be tested and then
is attached to the bottom of the float and chilled to a temperature of 5°C (41°F) by immersing it in ice water. The temperature of the water bath is brought to 50°C (122°F), and the
collar (still attached to the float) is placed in the water bath, which is kept at 50°C (122°F).

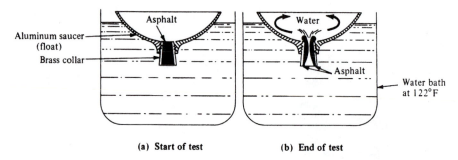

(a) Start of test   (b) End of test

Figure 19.6 Float Test

The head gradually softens the sample of asphaltic material in the collar until the water eventually forces its way through the plug into the aluminum float. The time in seconds that expires between the instant the collar is placed in the water bath and that at which the water forces its way through the bituminous plug is the *float-test value,* and it is a measure of the consistency. It is readily apparent that the higher the float-test value, the stiffer the material. Details of the float test are given in ASTM D139.

Ring-and-Ball Softening Point Test

The ring-and-ball softening point test is used to measure the susceptibility of blown asphalt to temperature changes by determining the temperature at which the material will be adequately softened to allow a standard ball to sink through it. Figure 19.7 shows the apparatus commonly used for this test. It consists principally of a small brass ring of $\frac{5}{8}$ in. inside diameter and $\frac{1}{4}$ in. high, a steel ball $\frac{3}{8}$ in. in diameter, and a water or glycerin bath. The test is conducted by first placing a sample of the material to be tested in the brass ring, which is cooled and immersed in the water or glycerin bath that is maintained at a temperature of 5°C (41°F). The ring is immersed to a depth such that its bottom is exactly 1 in. above the bottom of the bath. The temperature of the bath is then gradually increased, causing the asphalt to soften and permitting the ball to sink eventually to the bottom of the bath. The temperature in degrees Fahrenheit at which the asphaltic material touches the bottom of the bath is recorded as the softening point. The resistance characteristic to temperature shear can also be evaluated using this test.

Durability Tests

When asphaltic materials are used in the construction of roadway pavements, they are subjected to changes in temperature (freezing and thawing) and other weather conditions over a period of time. These changes cause natural weathering of the material, which may lead to loss of plasticity, cracking, abnormal surface abrasion, and eventual failure of the pavement. One test used to evaluate the susceptibility characteristics of asphaltic materials to changes in temperature and other atmospheric factors is the thin-film oven test.

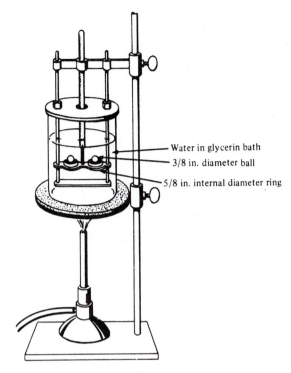

Water in glycerin bath
3/8 in. diameter ball
5/8 in. internal diameter ring

Figure 19.7 Ring-and-Ball Softening Point Test

SOURCE: *Introduction to Asphalt,* Manual Series No. 5, 8th ed., Asphalt Institute, Lexington, Ky., 1993.

Thin-Film Oven Test (TFO)

This is actually not a test but a procedure that measures the changes that take place in an asphalt during the hot-mix process by subjecting the asphaltic material to hardening conditions similar to those in a normal hot-mix plant operation. The consistency of the material is determined before and after the TFO procedure, using either the penetration test or a viscosity test, to estimate the amount of hardening that will take place in the material when used to produce plant hot-mix.

The procedure is performed by pouring 50 cc of the material into a cylindrical flat-bottom pan, 5.5 in. (14 cm) inside diameter and $\frac{3}{8}$ in. (1 cm) high. The pan containing the sample is then placed on a rotating shelf in an oven and rotated for five hours while the temperature is kept at 163°C (325°F). The amount of penetration after the TFO test is then expressed as a percentage of that before the test to determine percent of penetration retained. The minimum allowable percent of penetration retained is usually specified for different grades of asphalt cement.

Rate of Curing

Tests for curing rates of cutbacks are based on inherent factors, which can be controlled. These tests compare different asphaltic materials on the assumption that the external factors are held constant. Volatility and quantity of solvent are commonly used to indicate the rate of curing. The volatility and quantity of solvent may be determined from the distillation test; tests for consistency were described earlier.

Distillation Test for Cutbacks

Figure 19.8 is a schematic of the apparatus used in the distillation test. The apparatus consists principally of a flask in which the material is heated, a condenser, and a graduated cylinder for collecting the condensed material. A sample of 200 cc of the material to be tested is measured and poured into the flask and the apparatus is set up as shown in Figure 19.8. The material is then brought to boiling point by heating it with the burner. The evaporated solvent is condensed and collected in the graduated cylinder. The temperature in the flask is continuously monitored and the amount of solvent collected in the graduated cylinder is recorded when the temperature in the flask reaches 190°C (374°F), 225°C (437°F), 260°C (500°F), and 316°C (600°F). The amount of condensate collected at the different specified temperatures gives an indication of the volatility characteristics of the solvent. The residual in the flask is the base asphalt used in preparing the cutback. Details of this test are given in AASHTO Designation T78.

Distillation Test for Emulsions

The distillation test for emulsions is similar to that described for cutbacks. A major difference, however, is that the glass flask and Bunsen burner are replaced with an aluminum alloy still and a ring burner. This equipment prevents potential problems that may arise from the foaming of the emulsified asphalt as it is being heated to a maximum of 260°C (500°F).

Note, however, that the results obtained from the use of this method to recover the asphaltic residue and to determine the properties of the asphalt base stock used in the

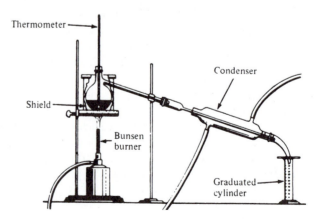

Figure 19.8 Distillation Test for Cutbacks

SOURCE: *Introduction to Asphalt,* Manual Series No. 5, 8th ed., Asphalt Institute, Lexington, Ky., 1993.

emulsion may not always be accurate because of significant changes in the properties of the asphalt. These changes are due to

- The inorganic salts from an aqueous phase concentrating in the asphaltic residue
- The emulsifying agent and stabilizer concentrating in the asphaltic residue

These changes, which are due mainly to the increase in temperature, do not occur in field application of the emulsion since the temperature in the field is usually much less than that used in the distillation test. The emulsion in the field, therefore, breaks either electrochemically or by evaporation of the water. An alternative method to determine the properties of the asphalt after it is cured on the pavement surface is to evaporate the water at subatmospheric pressure and lower temperatures. Such a test is designated AASHTO T59 or ASTM D244.

Other General Tests

Several other tests are routinely conducted on asphaltic materials used for pavement construction either to obtain specific characteristics for design purposes (for example, specific gravity) or to obtain additional information that aids in determining the quality of the material. Some of the more common routine tests are now described briefly.

Specific Gravity Test

The specific gravity of asphaltic materials is used mainly to determine the weight of a given volume of material, or vice versa, to determine the amount of voids in compacted mixes and to correct volumes measured at high temperatures. Specific gravity is defined as the ratio of the weight of a given volume of the material to the weight of the same volume of water. The specific gravity of bituminous materials, however, changes with temperature, which dictates that the temperature at which the test is conducted should be indicated. For example, if the test is conducted at 20°C (68°F) and the specific gravity is determined to be 1.41, this should be recorded as 1.41/20°C. Note that both the asphaltic material and the water should be at the same temperature. The usual temperature at which the specific gravities of asphaltic materials are determined is 25°C (77°F).

The test is normally conducted with a pycnometer, examples of which are shown in Figure 19.9. The dry weight (W_1) of the pycnometer and stopper is obtained, and then the pycnometer is filled with distilled water at the prescribed temperature. The weight (W_2) of the water and pycnometer together is determined. If the material to be tested can flow easily into the pycnometer, then the pycnometer must be completely filled with the material at the specified temperature after pouring out the water. The weight W_3 is then obtained. The specific gravity of the asphaltic material is then given as

$$G_b = \frac{W_3 - W_1}{W_2 - W_1} \tag{19.1}$$

where G_b is the specific gravity of the asphaltic material and W_1, W_2, and W_3 are in grams.

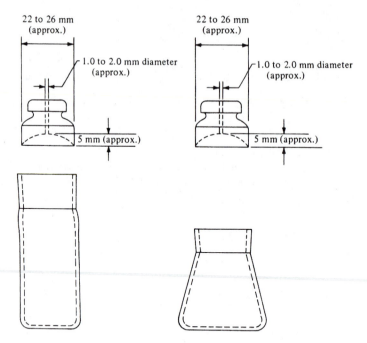

Figure 19.9 Pycnometers for Determining Specific Gravity of Asphaltic Materials

If the asphaltic material cannot easily flow, a small sample of the material is heated gradually to facilitate flow and then poured into the pycnometer and left to cool to the specified temperature. The weight W_4 of pycnometer and material is then obtained. Water is then poured into the pycnometer to completely fill the remaining space not occupied by the material. The weight W_5 of the filled pycnometer is obtained. The specific gravity is then given as

$$G_b = \frac{W_4 - W_1}{(W_2 - W_1) - (W_5 - W_4)} \tag{19.2}$$

Ductility Test

Ductility is the distance in centimeters a standard sample of asphaltic material will stretch before breaking when tested on standard ductility test equipment at 25°C (77°F). The result of this test indicates the extent to which the material can be deformed without breaking. This is an important characteristic for asphaltic materials, although the exact value of ductility is not as important as the existence or nonexistence of the property in the material.

Figure 19.10 shows the ductility test apparatus. The test is used mainly for semisolid or solid materials, which first are gently heated to facilitate flow and then are poured into

Figure 19.10 Apparatus for Ductility Test

SOURCE: Photograph courtesy of Wykeham Farrancè International, England. Used with permission.

a standard mold to form a briquette of at least 1 cm^2 in cross section. The material is then allowed to cool to 25°C (77°F) in a water bath. The prepared sample is then placed in the ductility machine shown in Figure 19.10 and extended at a specified rate of speed until the thread of material joining the two ends breaks. The distance (in centimeters) moved by the machine is the ductility of the material. The test is fully described in AASHTO Designation T51 and ASTM D113.

Solubility Test

The solubility test is used to measure the amount of impurities in the asphaltic material. Since asphalt is nearly 100 percent soluble in certain solvents, the portion of any asphaltic material that will be effective in cementing aggregates together can be determined from the solubility test. Insoluble materials include free carbon, salts, and other inorganic impurities.

The test is conducted by dissolving a known quantity of the material in a solvent, such as trichloroethylene, and then filtering it through a Gooch Crucible. The material retained in the filter is dried and weighed. The test results are given in terms of the percent of the asphaltic material that dissolved in the solvent.

Flash-Point Test

The flash point of an asphaltic material is the temperature at which its vapors will ignite instantaneously in the presence of an open flame. Note that the flash point is normally lower than the temperature at which the material will burn.

   The test can be conducted by using either the Tagliabue open-cup apparatus or the Cleveland open-cup apparatus, which is shown in Figure 19.11. The Cleveland open-cup test is more suitable for materials with higher flash points, whereas the Tagliabue open-cup is more suitable for materials with relatively low flash points, such as cutback asphalts. The test is conducted by partly filling the cup with the asphaltic material and gradually increasing its temperature at a specified rate. A small open flame is passed over the surface of the sample at regular intervals as the temperature increases. The increase in temperature will cause evaporation of volatile materials from the material being tested, until a sufficient quantity of volatile materials is present to cause an instantaneous flash when the open flame is passed over the surface. The minimum temperature at which this occurs is the flash point. It can be seen that this temperature gives an indication of the temperature limit at which

Figure 19.11 Apparatus for Cleveland Open-Cup Test

SOURCE: Photograph courtesy of Wykeham Farrancè International, England. Used with permission.

extreme care should be taken, particularly when heating is done over open flames in open containers.

Loss-on-Heating Test

The loss-on-heating test is used to determine the amount of material that evaporates from a sample of asphalt under a specified temperature and time. The result indicates whether an asphaltic material has been contaminated with lighter materials. The test is conducted by pouring 50 g of the material to be tested into a standard cylindrical tin and leaving it in an oven for 5 hr at a temperature of 163°C (325°F). The weight of the material remaining in the tin is determined, and the loss in weight is expressed as a percentage of the original weight. The penetration of the sample may also be determined before and after the test to determine the loss of penetration due to the evaporation of the volatile material. This loss in penetration may be used as an indication of the weathering characteristics of the asphalt. Details of the test procedures are given in AASHTO Designation T47.

Water Content Test

The presence of large amounts of water in asphaltic materials used in pavement construction is undesirable, and to ensure that only a limited quantity of water is present, specifications for these materials usually include the maximum percentage of water by volume that is allowable. A quantity of the sample to be tested is mixed with an equal quantity of a suitable distillate in a distillation flask that is connected with a condenser and a trap for collecting the water. The sample is then gradually heated in the flask, eventually causing all the water to evaporate and be collected. The quantity of water in the sample is then expressed as a percentage of the total sample volume.

Demulsibility Test for Emulsion

The demulsibility test is used to indicate the relative susceptibility of asphalt emulsions to breaking down (coalescing) when in contact with aggregates. Asphalt emulsions are expected to break immediately when they come in contact with the aggregate so that the material is prevented from washing away with rain that may occur soon after application. A high degree of demulsibility is therefore required for emulsions used for surface treatments such as RS1 and RS2. A relatively low degree of demulsibility is required for emulsions used for mixing coarse aggregates to avoid having the materials peel off before placing, and a very low degree is required for materials produced for mixing fine aggregates. Since calcium chloride will coagulate minute particles of asphalt, it is used in the test for anionic rapid-setting (RS) emulsions; however, dioctyl sodium sulfosuccinate is used instead of calcium chloride for testing cationic rapid-setting (CRS) emulsions. The test is conducted by thoroughly mixing the required standard solution with the asphalt emulsion and then passing the mixture through a No. 14 wire cloth sieve, which will retain the asphalt particles that have coalesced. The quantity of asphalt retained in the sieve is a measure of the breakdown that has occurred. Demulsibility is expressed in percentage as $(A/B)100$, where A is the average weight of demulsibility residue from three tests of each sample of emulsified asphalt, and B is the weight of residue by distillation in 100 g of the emulsified asphalt.

The strength of the test solution used and the minimum value of demulsibility required are prescribed in the relevant specifications. Details of this test are described in AASHTO T59 and ASTM D244.

Sieve Test for Emulsions

The sieve test is conducted on asphalt emulsions to determine to what extent the material has emulsified and the suitability of the material for application through pressure distributors. The test is conducted by passing a sample of the material through a No. 20 sieve and determining what percentage by weight of the material is retained in the sieve. A maximum value of 0.10 percent is usually specified.

Particle Charge Test for Emulsions

The particle charge test is used to identify CRS and CMS grades of emulsions. The test is conducted by immersing an anode electrode and a cathode electrode in a sample of the material to be tested and then passing an electric current through the system, as shown in Figure 19.12. The electrodes are then examined after some time to identify which one contains an asphalt deposit. If a deposit occurs on the cathode electrode, the emulsion is cationic.

ASPHALTIC CONCRETE

Asphaltic concrete is a uniformly mixed combination of asphalt cement, coarse aggregate, fine aggregate, and other materials, depending on the type of asphalt concrete. The different types of asphaltic concretes commonly used in pavement construction are hot-mix, hot-laid and cold-mix, cold-laid. Asphaltic concrete is the most popular paving material used in the United States. When used in the construction of highway pavements,

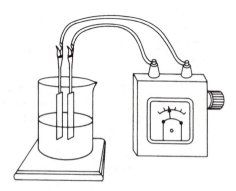

Figure 19.12 Particle Charge Test for Emulsions

it must resist deformation from imposed traffic loads, be skid resistant even when wet, and not be easily affected by weathering forces. The degree to which an asphaltic concrete achieves these characteristics is mainly dependent on the design of the mix used in producing the concrete. Three categories of asphaltic concrete will now be described, together with an appropriate mix design procedure.

Hot-Mix, Hot-Laid Asphaltic Concrete

Hot-mix, hot-laid asphaltic concrete is produced by properly blending asphalt cement, coarse aggregate, fine aggregate, and filler (dust) at temperatures ranging from about 175°F to 325°F, depending on the type of asphalt cement used. Suitable types of asphaltic materials include AC-20, AC-10, and AR-8000 with penetration grades of 60–70, 85–100, 120–150, and 200–300, as shown in Table 19.1. Hot-mix, hot-laid asphaltic concrete is normally used for high-type pavement construction, and the mixture can be described as open-graded, coarse-graded, dense-graded, or fine-graded. When produced for high-type surfacing, maximum sizes of aggregates normally range from $\frac{3}{8}$ in. to $\frac{3}{4}$ in. for open-graded mixtures, $\frac{1}{2}$ in. to $\frac{3}{4}$ in. for coarse-graded mixtures, $\frac{1}{2}$ in. to 1 in. for dense-graded mixtures, and $\frac{1}{2}$ in. to $\frac{3}{4}$ in. for fine-graded mixtures. When used as base, maximum sizes of aggregates are usually $\frac{3}{4}$ in. to $1\frac{1}{2}$ in. for open- and coarse-graded, 1 in. to $1\frac{1}{2}$ in. for dense-grades, and $\frac{3}{4}$ in. for fine-graded mixtures. As stated earlier, the extent to which an asphalt concrete meets the desired characteristics for highway pavement construction is dependent mainly on the mix design, which involves the selection and proportioning of the different material components. However, note that when designing hot-mix asphaltic concrete, a favorable balance must be found between a highly stable product and a durable one. Therefore, the overall objective of the mix design is to determine an optimum blend of the different components that will satisfy the requirements of the given specifications.

Aggregate Gradation

Aggregates are usually categorized as crushed rock, sand, and filler. The rock material is predominantly coarse aggregate retained in a No. 8 sieve, sand is predominantly fine aggregate passing the No. 8 sieve, and filler is predominantly mineral dust that passes the No. 200 sieve. It is customary for gradations of the combined aggregate and the individual fractions to be specified. Table 19.4 gives suggested grading requirements of aggregate material based on ASTM Designation 3515. The first phase in any mix design is the selection and combination of aggregates to obtain a gradation within the limits prescribed. This is sometimes referred to as *mechanical stabilization*.

The procedure used to select and combine aggregates will be illustrated in the following example.

Table 19.4 Examples of Composition of Asphalt Paving Mixtures

Dense Mixtures

Grading of Total Aggregate (Coarse Plus Fine, Plus Filler if Required) — Amounts Finer Than Each Laboratory Sieve (Square Opening), Weight %

Sieve Size	Mix Designation and Nominal Size of Aggregate								
	2 in. (50 mm)	1½ in. (37.5 mm)	1 in. (25.0 mm)	¾ in. (19.0 mm)	½ in. (12.5 mm)	⅜ in. (9.5 mm)	No. 4 (4.75 mm) (Sand Asphalt)	No. 8 (2.36 mm)	No. 16 (1.18 mm) (Sheet Asphalt)
2½ in. (63-mm)	100	...	...	...	...	...	...	...	...
2 in. (50-mm)	90 to 100	100	...	...	...	...	...	...	...
1½ in. (37.5-mm)	...	90 to 100	100	...	...	...	...	...	...
1 in. (25.0-mm)	60 to 80	...	90 to 100	100	...	...	...	...	...
¾ in. (19.0-mm)	...	56 to 80	...	90 to 100	100	...	...	...	...
½ in. (12.5-mm)	35 to 65	...	56 to 80	...	90 to 100	100	...	...	...
⅜ in. (9.5-mm)	...	...	...	56 to 80	...	90 to 100	100	...	...
No. 4 (4.75-mm)[A]	17 to 47	23 to 53	29 to 59	35 to 65	44 to 74	55 to 85	80 to 100	...	100
No. 8 (2.36-mm)[A]	10 to 36	15 to 41	19 to 45	23 to 49	28 to 58	32 to 67	65 to 100	...	95 to 100
No. 16 (1.18-mm)	...	...	...	...	...	...	40 to 80	...	85 to 100
No. 30 (600-μm)	...	...	...	...	...	...	25 to 65	...	70 to 95
No. 50 (300-μm)	3 to 15	4 to 16	5 to 17	5 to 19	5 to 21	7 to 23	7 to 40	...	45 to 75
No. 100 (150-μm)	...	...	...	...	...	...	3 to 20	...	20 to 40
No. 200 (75-μm)[B]	0 to 5	0 to 6	1 to 7	2 to 8	2 to 10	2 to 10	2 to 10	...	9 to 20

Open Mixtures

Sieve Size	Mix Designation and Nominal Maximum Size of Aggregate								
	Base and Binder Courses						Surface and Leveling Courses		
	2 in. (50 mm)	1½ in. (37.5 mm)	1 in. (25.0 mm)	¾ in. (19.0 mm)	½ in. (12.5 mm)	⅜ in. (9.5 mm)	No. 4 (4.75 mm) (Sand Asphalt)	No. 8 (2.36 mm)	No. 16 (1.18 mm) (Sheet Asphalt)
2½ in. (63-mm)	100	...	...	...	...	...	...	...	...
2 in. (50-mm)	90 to 100	100	...	...	...	...	...	...	...
1½ in. (37.5-mm)	...	90 to 100	100	...	...	...	...	...	...
1 in. (25.0-mm)	40 to 70	...	90 to 100	100	...	...	...	...	...

Sieve designation	2 to 7	3 to 8	3 to 9	4 to 10	4 to 11	5 to 12	6 to 12	7 to 12	8 to 12
¾ in. (19.0-mm)	⋯	18 to 48	⋯	40 to 70	⋯	90 to 100	100	⋯	⋯
½ in. (12.5-mm)	18 to 48	⋯	40 to 70	⋯	85 to 100	85 to 100	100	⋯	⋯
⅜ in. (9.5-mm)	18 to 48	18 to 48	⋯	40 to 70	60 to 90	40 to 70	85 to 100	100	100
No. 4 (4.75-mm)[A]	5 to 25	6 to 29	10 to 34	15 to 39	20 to 50	10 to 35	40 to 70	85 to 100	75 to 100
No. 8 (2.36-mm)	0 to 12	0 to 14	1 to 17	2 to 18	5 to 25	5 to 25	10 to 35	40 to 70	50 to 75
No. 16 (1.18-mm)	⋯	⋯	⋯	⋯	3 to 19	⋯	5 to 25	10 to 35	28 to 53
No. 30 (600-μm)	0 to 8	0 to 8	0 to 10	0 to 10	⋯	0 to 10	⋯	5 to 25	8 to 30
No. 50 (300-μm)	⋯	⋯	⋯	⋯	⋯	0 to 12	0 to 12	0 to 12	0 to 12
No. 100 (150-μm)[B]	⋯	⋯	⋯	⋯	⋯	⋯	⋯	⋯	⋯
No. 200 (75-μm)[B]	⋯	⋯	⋯	⋯	⋯	⋯	⋯	0 to 5	0 to 5
Bitumen, Weight % of Total Mixture[C]	2 to 7	3 to 8	3 to 9	4 to 10	4 to 11	5 to 12	6 to 12	7 to 12	8 to 12
Suggested Coarse Aggregate Sizes	3 and 57	4 and 67 or 4 and 68	5 and 7 or 57	67 or 68 or 6 and 8	7 or 78	8			

[A] In considering the total grading characteristics of a bituminous paving mixture, the amount passing the No. 8 (2.36-mm) sieve is a significant and convenient field control point between fine and coarse aggregate. Gradings approaching the maximum amount permitted to pass the No. 8 sieve will result in pavement surfaces having comparatively fine texture, while coarse gradings approaching the minimum amount passing the No. 8 sieve will result in surfaces with comparatively coarse texture.

[B] The material passing the No. 200 (75-μm) sieve may consist of fine particles of the aggregates or mineral filler, or both, but shall be free of organic matter and clay particles. The blend of aggregates and filler, when tested in accordance with Test Method D 4318, shall have a plasticity index of not greater than 4, except that this plasticity requirement shall not apply when the filler material is hydrated lime or hydraulic cement.

[C] The quantity of bitumen is given in terms of weight % of the total mixture. The wide difference in the specific gravity of various aggregates, as well as a considerable difference in absorption, results in a comparatively wide range in the limiting amount of bitumen specified. The amount of bitumen required for a given mixture should be determined by appropriate laboratory testing or on the basis of past experience with similar mixtures, or by a combination of both.

SOURCE: *Annual Book of ASTM Standards, Section 4, Construction*, Vol. 04.03, Road and Paving Materials; Pavement Management Technologies, American Society for Testing and Materials, Philadelphia, Pa., 1996.

Example 19.1   Determining Proportions of Different Aggregates to Obtain a Required Gradation

Table 19.5 gives the specifications for the aggregates and mix composition for highway pavement asphaltic concrete, and Table 19.6 shows the results of a sieve analysis of samples from the materials available. We must determine the proportions of the separate aggregates that will give a gradation within the specified limits.

Table 19.5   Required Limits for Mineral Aggregates Gradation and Mix Composition for an Asphaltic Concrete for Example 19.1

Passive Sieve Designation	Retained on Sieve Designation	Percent by Weight
$\frac{3}{4}$ in. (19.0 mm)	$\frac{1}{2}$ in.	0–5
$\frac{1}{2}$ in. (12.5 mm)	$\frac{3}{8}$ in.	8–42
$\frac{3}{8}$ in. (9.5 mm)	No. 4	8–48
No. 4 (4.75 mm)	No. 10	6–28
Total coarse aggregates	No. 10	48–65
No. 10 (2.0 mm)	No. 40	5–20
No. 40 (0.425 mm)	No. 80	9–30
No. 80 (0.180 mm)	No. 200	3–20
No. 200 (0.075 mm)	—	2–6
Total fine aggregate and filler	Passing No. 10	35–50
Total mineral aggregate in asphalt concrete		90–95
Asphalt cement in asphalt concrete		5–7
Total mix		100

Table 19.6   Sieve Analysis of Available Materials for Example 19.1

Passing Sieve Designation	Retained on Sieve Designation	Percent by Weight		
		Coarse Aggregate	Fine Aggregate	Mineral Filler
$\frac{3}{4}$ in. (19.0 mm)	$\frac{1}{2}$ in.	5	—	—
$\frac{1}{2}$ in. (12.5 mm)	$\frac{3}{4}$ in.	35	—	—
$\frac{3}{8}$ in. (9.5 mm)	No. 4	38	—	—
No. 4 (4.75 mm)	No. 10	17	8	—
No. 10 (2.0 mm)	No. 40	5	30	—
No. 40 (0.425 mm)	No. 80	—	35	5
No. 80 (0.180 mm)	No. 200	—	26	35
No. 200 (0.075 mm)	—	—	1	60
Total		100	100	100

Solution: It can be seen that the amount of the different sizes selected should not only give a mix that meets the prescribed limits, but should also be such that allowance is made for some variation during actual production of the mix. It can also be seen from

Table 19.5 that to obtain the required specified gradation, some combination of all three materials is required, since the coarse and fine aggregates do not together meet the requirement of 2 percent to 6 percent by weight of filler material. A trial mix is therefore selected arbitrarily within the prescribed limits. Let this mix be

Coarse aggregates = 55% (48–65% specified)

Fine aggregates = 39% (35–50% specified)

Filler = 6% (5–8% specified)

The selected proportions are then used to determine the combination of the different sizes, as shown in Table 19.7. The calculation is based on the fundamental equation for the percentage of material P passing a given sieve for the aggregates 1, 2, 3 and is given as

$$P = aA_1 + bA_2 + cA_3 + \cdots \qquad (19.3)$$

where A_1, A_2, A_3 are equal to the percentages of material passing a given sieve for aggregates 1, 2, 3, where a, b, c are equal to the proportions of aggregates 1, 2, 3 used in the combination, and where $a + b + c + \cdots = 100$. Note that this is true for any number of aggregates combined.

It can be seen that the combination obtained, as shown in the last column of Table 19.7, meets the specified limits as shown in the last column of Table 19.5. The trial combination is therefore acceptable. Note, however, that the first trial may not always meet the specified limits. In such cases, other combinations must be tried until a satisfactory one is obtained.

Table 19.7 Computation of Percentages of Different Aggregate Sizes for Example 19.1

Passing Sieve Size	Retained on Sieve Size	Percent by Weight			
		Coarse Aggregate	Fine Aggregate	Mineral Filler	Total Aggregate
$\frac{3}{4}$ in.	$\frac{1}{2}$ in.	0.55 × 5 = 2.75	—	—	2.75
$\frac{1}{2}$ in.	$\frac{3}{4}$ in.	0.55 × 35 = 19.25	—	—	19.25
$\frac{3}{8}$ in.	No. 4	0.55 × 38 = 20.90	—	—	20.90
No. 4	No. 10	0.55 × 17 = 9.35	0.39 × 8 = 3.12	—	12.47
No. 10	No. 40	0.55 × 5 = 2.75	0.39 × 30 = 11.70	—	14.45
No. 40	No. 80	—	0.39 × 35 = 13.65	0.06 × 5 = 0.3	13.95
No. 80	No. 200	—	0.39 × 26 = 10.14	0.06 × 35 = 2.10	12.24
No. 200	—	—	0.39 × 1 = 0.39	0.06 × 60 = 3.60	3.99
Total		55.0	39.0	6.0	100.00

Several graphical methods have been developed for obtaining a suitable mixture of different aggregates to obtain a desired gradation. These methods tend to be rather complicated when the number of batches of aggregates is high. They generally can be of advantage over the trial-and-error method described here when the number of aggregates is not more than two. When the number of aggregates is three or greater, the trial-and-error method is preferable, although the graphical methods can also be used.

Asphalt Content

Having determined a suitable mix of aggregates, the next step is to determine the optimum percentage of asphalt that should be used in the asphalt concrete mixture. This percentage should, of course, be within the prescribed limits. The gradation of the aggregates determined earlier and the optimum amount of asphalt cement determined combine to give the proportions of the different materials to be used in producing the hot-mix, hot-laid concrete for the project under consideration. These determined proportions are usually referred to as the *job-mix formula*.

Two commonly used methods to determine the optimum asphalt content are the *Marshall method* and the *Hveem method*. The Marshall method is described here since it is more widely used.

Marshall Method Procedure.   The original concepts of this method were developed by Bruce Marshall, who was then a bituminous engineer with the Mississippi State Highway Department. The original features have been improved by the U.S. Army Corps of Engineers, and the test is now standardized and described in detail in ASTM Designation D1559. Test specimens of 4 in. diameter and $2\frac{1}{2}$ in. height are used in this method. They are prepared by a specified procedure of heating, mixing, and compacting the mixture of asphalt and aggregates, which is then subjected to a stability-flow test and a density-voids analysis. The *stability* is defined as the maximum load resistance N in pounds that the specimen will achieve at 140°F under specified conditions. The *flow* is the total movement of the specimen, in units of 0.01 in. during the stability test, as the load is increased from zero to the maximum.

Test specimens for the Marshall method are prepared for a range of asphalt contents within the prescribed limits. Usually the asphalt content is measured by 0.5 percent increments from the minimum prescribed, ensuring that at least two are below the optimum and two above the optimum so that the curves obtained from the result will indicate a well-defined optimum. For example, for a specified amount of 5–7 percent, mixtures of 5, 5.5, 6, 6.5, and 7 are prepared. At least three specimens are provided for each asphalt content to facilitate the provision of adequate data. For this example of five different asphalt contents, therefore, a total minimum number of 15 specimens is required. The amount of aggregates required for each specimen is about 1.2 kg.

A quantity of the aggregates having the designed gradation is dried at a temperature between 105°C (221°F) and 110°C (230°F) until a constant weight is obtained. The mixing temperature for this procedure is set as the temperature that will produce a kinematic viscosity of 170 ± 20 centistokes, or a Saybolt Furol viscosity of 85 ± 10 sec, in the asphalt. The compacting temperature is that which will produce a kinematic viscosity of 280 ± 30 centistokes, or a Saybolt Furol viscosity of 160 ± 15 sec. These temperatures are determined and recorded.

The specimens containing the appropriate amounts of aggregates and asphalt are then prepared by thoroughly mixing and compacting each mixture. The compactive effort used is either 35, 50, or 75 blows of the hammer falling a distance of 18 in., depending on the design traffic category. After the application on one face, the sample mold is reversed and the same number of blows is applied to the other face of the sample. The specimen is then cooled and tested for stability and flow after determining its bulk density.

The bulk density of the sample is determined usually by weighing the sample in air and in water. It may be necessary to coat samples made from open-graded mixtures with paraffin before determining the density. The bulk specific gravity G_{bcm} of the sample—that is, the compacted mixture—is given as

$$G_{bcm} = \frac{W_a}{W_a - W_w} \qquad (19.4)$$

where

W_a = weight of sample in air (g)
W_w = weight of sample in water (g)

Stability Test

In conducting the stability test, the specimen is immersed in a bath of water at a temperature of $60°C \pm 1°C$ ($140°F \pm 1.8°F$) for a period of 30 to 40 min. It is then placed in the Marshall stability testing machine, as shown in Figure 19.13, and loaded at a constant

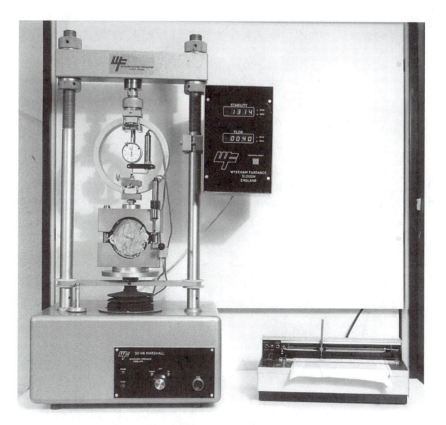

Figure 19.13 Marshall Stability Equipment

SOURCE: Photograph courtesy of Wykeham Farrancè International, England. Used with permission.

rate of deformation of 2 in. (5 mm) per minute until failure occurs. The total load N in pounds that causes failure of the specimen at 60°C (140°F) is noted as the Marshall stability value of the specimen. The total amount of deformation in units of 0.01 in. that occurs up to the point the load starts decreasing is recorded as the flow value. The total time between removing the specimen from the bath and completion of the test should not exceed 30 sec.

Analysis of Results from Marshall Test.   The first step in the analysis of the results is the determination of the average bulk specific gravity for all test specimens having the same asphalt content. The average unit weight of each mixture is then obtained by multiplying its average specific gravity by the density of water γ_w. A smooth curve that represents the best fit of plots of unit weight versus percentage of asphalt is determined, as shown in Figure 19.14(a). This curve is used to obtain the bulk specific gravity values that are used in further computations, for Example 19.2.

The percent air voids, percent voids in the mineral aggregate, and the absorbed asphalt in pounds of the dry aggregate are then calculated as shown below.

In order to compute the percent air voids, the percent voids in the mineral aggregate, and the absorbed asphalt, it is first necessary to compute the bulk specific gravity of the aggregate mixture, the apparent specific gravity of the aggregate mixture, the effective specific gravity of the aggregate mixture, and the maximum specific gravity of the paving mixtures for different asphalt contents. These different measures of the specific gravity of the aggregates take into consideration the variation with which mineral aggregates can absorb water and asphalt. See Figure 19.15.

Bulk Specific Gravity of Aggregate.   The *bulk specific gravity* is defined as the weight in air of a unit volume (including all normal voids) of a permeable material at a selected temperature, divided by the weight in air of the same density of gas-free distilled water at the same selected temperature.

Since the aggregate mixture consists of different fractions of coarse aggregate, fine aggregate, and mineral fillers with different specific gravities, the bulk specific gravity of the total aggregate in the paving mixture is given as

$$G_{bam} = \frac{P_{ca} + P_{fa} + P_{mf}}{\dfrac{P_{ca}}{G_{bca}} + \dfrac{P_{fa}}{G_{bfa}} + \dfrac{P_{mf}}{G_{bmf}}} \tag{19.5}$$

where

$\quad G_{bam}$ = bulk specific gravity of aggregates in the paving mixture

P_{ca}, P_{fa}, P_{mf} = percent by weight of coarse aggregate, fine aggregate, and mineral filler, respectively, in the paving mixture. (Note that P_{ca}, P_{fa}, and P_{mf} could be found either as a percentage of the paving mixture or as a percentage of only the total aggregates. The same results will be obtained for G_{bam}.)

$G_{bca}, G_{bfa}, G_{bmf}$ = bulk specific gravities of coarse aggregate, fine aggregate, and mineral filler, respectively

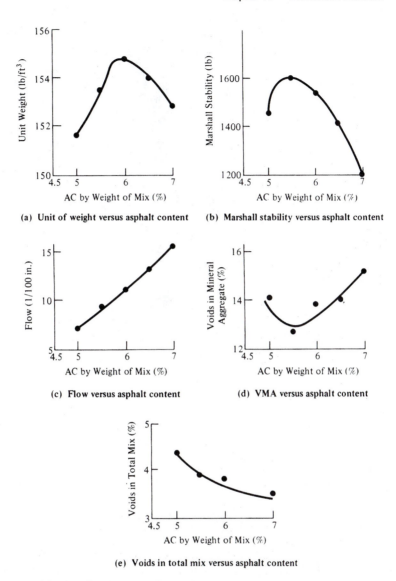

(a) **Unit of weight versus asphalt content**

(b) **Marshall stability versus asphalt content**

(c) **Flow versus asphalt content**

(d) **VMA versus asphalt content**

(e) **Voids in total mix versus asphalt content**

Figure 19.14 Marshall Test Property Curves for Example 19.2

It is not easy to determine accurately the bulk specific gravity of the mineral filler. The apparent specific gravity may therefore be used with very little error.

Apparent Specific Gravity of Aggregates.   The apparent specific gravity is defined as the ratio of the weight in air of an impermeable material to the weight of an equal volume of distilled water at a specified temperature. The apparent specific gravity of the aggregate mix is therefore obtained as

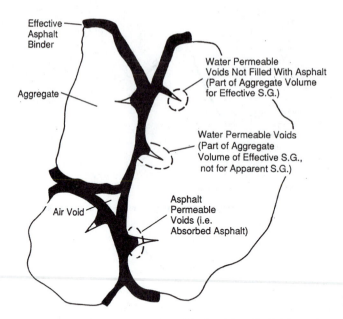

Effective
Asphalt
Binder

Water Permeable
Voids Not Filled With Asphalt
(Part of Aggregate Volume
for Effective S.G.)

Aggregate

Water Permeable Voids
(Part of Aggregate
Volume of Effective S.G.,
not for Apparent S.G.)

Air Void

Asphalt
Permeable
Voids (i.e.
Absorbed Asphalt)

Figure 19.15   Bulk, Effective, and Apparent Specific Gravities; Air Voids; and Effective Asphalt Content in Compacted Asphalt Paving Mixture

$$G_{ama} = \frac{P_{ca} + P_{fa} + P_{mf}}{\dfrac{P_{ca}}{G_{aca}} + \dfrac{P_{fa}}{G_{afa}} + \dfrac{P_{mf}}{G_{amf}}} \tag{19.6}$$

where

$\quad G_{ama}$ = apparent specific gravity of the aggregate mixture

P_{ca}, P_{fa}, P_{mf} = percent by weight of coarse aggregate, fine aggregate, and mineral filler, respectively, in the mixture

$G_{aca}, G_{afa}, G_{amf}$ = apparent specific gravities of coarse aggregate, fine aggregate, and mineral filler, respectively

Effective Specific Gravity of Aggregate.   The effective specific gravity of the aggregates is normally based on the maximum specific gravity of the paving mixture. It is therefore the specific gravity of the aggregates when all void spaces in the aggregate particles are included, with the exception of those that are filled with asphalt. (See Figure 19.15.) It is given as

$$G_{ea} = \frac{100 - P_{ac}}{(100/G_{mp}) - (P_{ac}/G_{ac})} \tag{19.7}$$

where

$\quad G_{ea}$ = effective specific gravity of the aggregates

$\quad G_{mp}$ = maximum specific gravity of paving mixture (no air voids)

$\quad P_{ac}$ = asphalt percent by total weight of paving mixture (thus $100 - P_{ac}$ is the percent by weight of the base mixture that is not asphalt)

$\quad G_{ac}$ = specific gravity of the asphalt

Maximum Specific Gravity of the Paving Mixture.   The maximum specific gravity of the paving mixture G_{mp} assumes that there are no air voids in the asphalt concrete. Although the G_{mp} can be determined in the laboratory by conducting the standard test (ASTM Designation D2041), the best accuracy is attained at mixtures near the optimum asphalt content. Since it is necessary to determine the G_{mp} for all samples, some of which contain much lower or much higher quantities than the optimum asphalt content, the following procedure can be used to determine the G_{mp} for each sample.

The ASTM Designation D2041 test is conducted on all specimens containing a selected asphalt cement content, and the mean of these determined. This value is then used to determine the effective specific gravity of the aggregates using Eq. 19.7. The effective specific gravity of the aggregates can be considered constant, since varying the asphalt content in the paving mixture does not significantly vary the asphalt absorption. The effective specific gravity obtained is then used to determine the maximum specific gravity of the paving mixtures with different asphalt cement contents using Eq. 19.8.

$$G_{mp} = \frac{100}{\left(P_{ta}/G_{ea}\right) + \left(P_{ac}/G_{ac}\right)} \qquad (19.8)$$

where

$\quad G_{mp}$ = maximum specific gravity of paving mixture (asphalt concrete)

$\quad P_{ta}$ = percent by weight of aggregates in paving mixture (asphalt concrete)

$\quad P_{ac}$ = percent by weight of asphalt in paving mixture (asphalt concrete)

$\quad G_{ea}$ = effective specific gravity of the aggregates (assumed to be constant for different asphalt cement contents)

$\quad G_{ac}$ = specific gravity of asphalt

Once these different specific gravities have been determined, the asphalt absorption, the effective asphalt content, the percent voids in mineral aggregates (*VMA*), and the percent air voids in the compacted mixture can all be determined.

Asphalt absorption is the percent by weight of the asphalt that is absorbed by the aggregates based on the total weight of the aggregates. This is given as

$$P_{aa} = 100\frac{G_{ea} - G_{bam}}{G_{bam}G_{ea}}G_{ac} \qquad (19.9)$$

where

$\quad P_{aa}$ = amount of asphalt absorbed as a percentage of the total weight of aggregates

G_{ea} = effective specific gravity of the aggregates
G_{bam} = bulk specific gravity of the aggregates
G_{ac} = specific gravity of asphalt

Effective Asphalt Content.   The effective asphalt content is the difference between the total amount of asphalt in the mixture and that absorbed into the aggregate particles. The effective asphalt content is therefore that which coats the outside of the aggregate particles and influences the pavement performance. It is given as

$$P_{eac} = P_{ac} - \frac{P_{aa}}{100} P_{ta} \qquad (19.10)$$

where

P_{eac} = effective asphalt content in paving mixture (percent by weight)
P_{ac} = percent by weight of asphalt in paving mixture
P_{ta} = aggregate percent by weight of paving mixture
P_{aa} = amount of asphalt absorbed as a percentage of the total weight of aggregates

Percent Voids in Compacted Mineral Aggregates.   The percent voids in compacted mineral aggregates, or *VMA*, is the percentage of void spaces between the granular particles in the compacted paving mixture, including the air voids and the volume occupied by the effective asphalt content. It is usually calculated as a percentage of the bulk volume of the compacted mixture, based on the bulk specific gravity of the aggregates. It is given as

$$VMA = 100 - \frac{G_{bcm} P_{ta}}{G_{bam}} \qquad (19.11)$$

where

VMA = percent voids in compacted mineral aggregates
G_{bcm} = bulk specific gravity of compacted mixture (asphalt concrete)
G_{bam} = bulk specific gravity of aggregate
P_{ta} = aggregate percent by weight of total paving mixture (asphalt concrete)

Percent Air Voids in Compacted Mixture.   This is the ratio, expressed as a percentage, between the volume of the small air voids between the coated particles and the total volume of the mixture. It can be obtained from

$$P_{av} = 100 \frac{G_{mp} - G_{bcm}}{G_{mp}} \qquad (19.12)$$

where

P_{av} = percent air voids in compacted paving mixture
G_{mp} = maximum specific gravity of the compacted paving mixture
G_{bcm} = bulk specific gravity of the compacted paving mixture

Four additional separate smooth curves are drawn: percent voids in total mix versus percent of asphalt, percent voids in mineral aggregate versus percent of asphalt, Marshall stability versus percent of asphalt, and flow versus percent of asphalt. These graphs are used to select the asphalt contents for maximum stability, maximum unit weight, and percent voids in total mix within the limits specified (usually the median of the limits). The average of the asphalt contents is the optimum asphalt content. The stability and flow for this optimum content can then be obtained from the appropriate graphs to determine whether the required criteria are met. Suggested criteria for these test limits are given in Table 19.8.

The above analysis is illustrated in Example 19.2.

Evaluation and Adjustment of Mix Design

As stated earlier, the overall objective of the mix design is to determine an optimum blend of different components that will satisfy the requirements of the given specifications, as shown in Table 19.8. This mixture should have

- An adequate amount of asphalt to ensure a durable pavement
- An adequate mix stability to prevent unacceptable distortion and displacement when traffic load is applied
- Adequate voids in the total compacted mixture to permit a small amount of compaction when traffic load is applied without loss of stability, blushing, and bleeding, but at the same time insufficient voids to prevent harmful penetration of air and moisture into the compacted mixture
- Adequate workability to facilitate placement of the mix without segregation

When the mix design for the optimum asphalt content does not satisfy all the requirements given in Table 19.8, it is necessary to adjust the original blend of aggregates. Trial mixes can be adjusted by using the following general guidelines.

Low Voids and Low Stability.   In this situation, the voids in the mineral aggregates can be increased by adding more coarse aggregates. Alternatively, the asphalt content can be reduced, but only if the asphalt content is higher than that normally used and if the excess is not required as a replacement for the amount absorbed by the aggregates. Care should be taken when the asphalt content is reduced because this can lead to both a decrease in durability and an increase in permeability of the pavement.

Low Voids and Satisfactory Stability.   This mix can cause reorientation of particles and additional compaction of the pavement with time as continued traffic load is imposed on the pavement. This in turn may lead to instability or flushing of the pavement. Mixes with low voids should be altered by adding more aggregates.

High Voids and Satisfactory Stability.   When voids are high, it is likely that the permeability of the pavement will also be high, which will allow water and air to circulate through the pavement, resulting in premature hardening of the asphalt. High voids should be reduced to acceptable limits, even though the stability is satisfactory. This can be achieved by increasing the amount of mineral dust filler in the mix.

Satisfactory Voids and Low Stability.   This condition suggests low quality aggregates; the quality should be improved.

Table 19.8 Suggested Criteria for Test Limits

(a) Maximum and Minimum Values

Marshall Method Mix Criteria[1]	Light Traffic Surface & Base[2]		Medium Traffic Surface & Base[2]		Heavy Traffic Surface & Base[2]	
	Min	Max	Min	Max	Min	Max
Compaction, number of blows each end of specimen[3]	35		50		75	
Stability, N	3336		5338		8006	
(lb.)	(750)	—	(1200)	—	(1800)	—
Flow[4], 0.25 mm (0.01 in).	8	18	8	16	8	14
Percent Air Voids[5]	3	5	3	5	3	5
Percent Voids Filled With[6] Asphalt (VFA)	70	80	65	78	65	75

NOTES

[1]All criteria, not just stability value alone, must be considered in designing an asphalt paving mix. Hot mix asphalt bases that do not meet these criteria when tested at 60°C (140°F) are satisfactory if they meet the criteria when tested at 38°C (100°F) and are placed 100 mm (4 inches) or more below the surface. This recommendation applies only to regions having a range of climatic conditions similar to those prevailing throughout most of the United States. A different lower test temperature may be considered in regions having more extreme climatic conditions.

[2]Traffic classifications

    Light   Traffic conditions resulting in a Design EAL $<10^4$

    Medium Traffic conditions resulting in a Design EAL between 10^4 and 10^6 (See Chapter 20.)

    Heavy   Traffic conditions resulting in a Design EAL $>10^6$

[3]Laboratory compaction efforts should closely approach the maximum density obtained in the pavement under traffic.

[4]The flow value refers to the point where the load begins to decrease

[5]The portion of asphalt cement lost by absorption into the aggregate particles must be allowed for when calculating percent air voids.

[6]Percent voids in the mineral aggregate is to be calculated on the basis of the ASTM bulk specific gravity for the aggregate.

(b) Minimum Percent Voids in Mineral Aggregate, VMA

Nominal Maximum Particle Size[1,2]		Minimum VMA, percent		
		Design Air Voids, Percent[3]		
mm	in.	3.0	4.0	5.0
1.18	No. 16	21.5	22.5	23.5
2.36	No. 8	19.0	20.0	21.0
4.75	No. 4	16.0	17.0	18.0
9.5	3/8	14.0	15.0	16.0
12.5	1/2	13.0	14.0	15.0
19.0	3/4	12.0	13.0	14.0
25.0	1.0	11.0	12.0	13.0
37.5	1.5	10.0	11.0	12.0
50	2.0	9.5	10.5	11.5
63	2.5	9.0	10.0	11.0

[1]Standard Specification for Wire Cloth Sieves for Testing Purposes, ASTM E11 (AASHTO M92)

[2]The nominal maximum particle size is one size larger than the first sieve to retain more than 10 percent.

[3]Interpolate minimum voids in the mineral aggregate (VMA) for design air void values between those listed.

SOURCE: Adapted from *Mix Design Methods for Asphalt Concrete and Other Hot-Mix Types,* Manual Series No. 2, 6th ed., The Asphalt Institute, Lexington, Ky., 1993.

High Voids and Low Stability.   It may be necessary to carry out two steps in this case. The first step is to adjust the voids as discussed earlier. If this adjustment does not simultaneously improve the stability, the second step is to consider the improvement of the aggregate quality.

Example 19.2   Designing an Asphalt Concrete Mixture

In designing an asphalt concrete mixture for a highway pavement to support very heavy traffic, data in Table 19.9 showing the aggregate characteristics and Table 19.10 showing data obtained using the Marshall method were used. Determine the optimum asphalt content for this mix for the specified limits given in Table 19.8.

Solution: The bulk specific gravity of the mix for each asphalt cement content is determined by calculating the average value for the specimens with the same asphalt cement content using Eq. 19.4.

For 5% asphalt content, the average bulk specific gravity is given as

$$G_{bcm} = \frac{1}{3} \left(\frac{1325.6}{1325.6 - 780.1} + \frac{1325.4}{1325.4 - 780.3} + \frac{1325.0}{1325.0 - 779.8} \right)$$

$$= \frac{1}{3} (2.43 + 2.43 + 2.43)$$

$$= 2.43$$

Therefore, the bulk density is $2.43 \times 62.4 = 151.6$ lb/ft³.

For 5.5% asphalt content,

$$G_{bcm} = \frac{1}{3} \left(\frac{1331.3}{1331.3 - 789.6} + \frac{1330.9}{1330.9 - 789.3} + \frac{1331.8}{1331.8 - 790.0} \right)$$

$$= \frac{1}{3} (2.46 + 2.46 + 2.46)$$

$$= 2.46$$

Table 19.9 Aggregate Characteristics for Example 19.2

Aggregate Type	Percent by Weight of Total Paving Mixture	Bulk Specific Gravity
Coarse	52.3	2.65
Fine	39.6	2.75
Filler	8.1	2.70

NOTE: The nominal maximum particle size in the aggregate mixture is 1 in.

Table 19.10 Marshall Test Data for Example 19.2

Asphalt % by Weight of Total Mix	Weight of Specimen (g) in Air			in Water			Stability (lb)			Flow (0.01 in.)			Maximum Specific Gravity of Paving Mixture
	1	2	3	1	2	3	1	2	3	1	2	3	
5.0	1325.6	1325.4	1325.0	780.1	780.3	779.8	1460	1450	1465	7	7.5	7	2.54
5.5	1331.3	1330.9	1331.8	789.6	789.3	790.0	1600	1610	1595	10	9	9.5	2.56
6.0	1338.2	1338.5	1338.1	798.6	798.3	797.3	1560	1540	1550	11	11.5	11	2.58
6.5	1343.8	1344.0	1343.9	799.8	797.3	799.9	1400	1420	1415	13	13	13.5	2.56
7.0	1349.0	1349.3	1349.8	798.4	799.0	800.1	1200	1190	1210	16	15	16	2.54

Therefore, the bulk density is 153.5 lb/ft³.

For 6.0% asphalt content,

$$G_{bcm} = \frac{1}{3}\left(\frac{1338.2}{1338.2 - 798.6} + \frac{1338.5}{1338.5 - 798.3} + \frac{1338.1}{1338.1 - 797.3}\right)$$

$$= \frac{1}{3}(2.48 + 2.48 + 2.47)$$

$$\cong 2.48$$

Therefore, the bulk density is 154.8 lb/ft³.

For 6.5% asphalt content,

$$G_{bcm} = \frac{1}{3}\left(\frac{1343.8}{1343.8 - 799.8} + \frac{1344.0}{1344.0 - 797.3} + \frac{1343.9}{1343.9 - 799.9}\right)$$

$$= \frac{1}{3}(2.47 + 2.46 + 2.47)$$

$$\cong 2.47$$

Therefore, the bulk density is 154.1 lb/ft³.

For 7.0% asphalt content,

$$G_{bcm} = \frac{1}{3}\left(\frac{1349.0}{1349.0 - 798.4} + \frac{1349.3}{1349.3 - 799} + \frac{1349.8}{1349.8 - 800.1}\right)$$

$$= \frac{1}{3}(2.45 + 2.45 + 2.46)$$

$$\cong 2.45$$

Therefore, the bulk density is 152.9 lb/ft³.

Average bulk density is then plotted against asphalt content, as shown in Figure 19.14(a). Similarly, the average stability and flow for each asphalt cement content are as follows:

%	Stability	Flow
5.0	1458	7.2
5.5	1602	9.5
6.0	1550	11.2
6.5	1412	13.2
7.0	1200	15.7

These values are plotted in Figure 19.14(b) and (c).

We now have to compute percent voids in the mineral aggregate VMA and the percent voids in the compacted mixture for each asphalt cement mixture. First use Eq. 19.11,

$$VMA = 100 - \frac{G_{bcm}P_{ta}}{G_{bam}}$$

For 5% asphalt content,

$$G_{bcm} = 2.43$$
$$P_{ta} = 95.0$$

Use Eq. 19.5 to calculate G_{bam}.

$$G_{bam} = \frac{P_{ca} + P_{fa} + P_{mf}}{(P_{ca}/G_{bca}) + (P_{fa}/G_{bfa}) + (P_{mf}/G_{bmf})}$$

Determine P_{ca}, P_{fa}, and P_{mf} in terms of total aggregates:

$$P_{ca} = 0.523 \times 95.0 = 49.7$$
$$P_{fa} = 0.396 \times 95.0 = 37.6$$
$$P_{mf} = 0.081 \times 95.0 = 7.7$$

Therefore,

$$G_{bam} = \frac{49.7 + 37.6 + 7.7}{(49.7/2.65) + (37.6/2.75) + (7.7/2.70)} = 2.69$$

$$P_{ta} = (100 - 5) = 95$$

and therefore,

$$VMA = 100 - \frac{2.43 \times 95}{2.69} = 14.18$$

For 5.5% asphalt content,

$$P_{ca} = 0.523 \times 94.5 = 49.4$$
$$P_{fa} = 0.396 \times 94.5 = 37.4$$
$$P_{mf} = 0.081 \times 94.5 = 7.7$$

$$G_{bam} = \frac{49.4 + 37.4 + 7.7}{(49.4/2.65) + (37.4/2.75) + (7.7/2.70)} = 2.69$$

and

$$VMA = 100 - \frac{2.46 \times 94.5}{2.69} = 13.58$$

For 6% asphalt cement,

$P_{ca} = 0.523 \times 94 = 49.2$
$P_{fa} = 0.396 \times 94 = 37.2$
$P_{mf} = 0.081 \times 94 = 7.6$

$$G_{bam} = \frac{49.2 + 37.2 + 7.7}{(49.2/2.65) + (37.2/2.75) + (7.6/2.70)} = 2.69$$

and

$$VMA = 100 - \frac{2.48 \times 94}{2.69} = 13.34$$

For 6.5% asphalt content,

$P_{ca} = 0.523 \times 93.5 = 48.9$
$P_{fa} = 0.396 \times 93.5 = 37.0$
$P_{mf} = 0.081 \times 93.5 = 7.6$

$$G_{bam} = \frac{48.9 + 37.0 + 7.6}{\dfrac{48.9}{2.65} + \dfrac{37.0}{2.75} + \dfrac{7.6}{2.7}} = 2.69$$

and

$$VMA = 100 - \frac{2.47 \times 93.5}{2.69} = 14.15$$

For 7.0% asphalt content,

$P_{ca} = 0.523 \times 93.0 = 48.6$
$P_{fa} = 0.396 \times 93.0 = 36.8$
$P_{mf} = 0.081 \times 93.0 = 7.5$

$$G_{bam} = \frac{48.6 + 36.8 + 7.5}{\dfrac{48.6}{2.65} + \dfrac{36.8}{2.75} + \dfrac{7.5}{2.7}} = 2.69$$

and

$$VMA = 100 - \frac{2.45 \times 93}{2.69} = 15.30$$

A plot of *VMA* versus asphalt content is shown in Figure 19.14(d).

We now have to determine the percentage of air voids in each of the paving mixtures using Eq. 19.12,

$$P_{av} = 100 \frac{G_{mp} - G_{bcm}}{G_{mp}}$$

For 5% asphalt content,

$$P_{av} = 100 \frac{2.54 - 2.43}{2.54} = 4.33$$

For 5.5% asphalt content,

$$P_{av} = 100 \frac{2.56 - 2.46}{2.56} = 3.91$$

For 6.0% asphalt content,

$$P_{av} = 100 \frac{2.58 - 2.48}{2.58} = 3.88$$

For 6.5% asphalt content,

$$P_{av} = 100 \frac{2.56 - 2.47}{2.57} = 3.50$$

For 7.0% asphalt content,

$$P_{av} = 100 \frac{2.54 - 2.45}{2.54} = 3.54$$

A plot of P_{av} versus asphalt content is shown in Figure 19.14(e).

The asphalt content that meets the design requirements for unit weight, stability, and percent air voids is then selected from the appropriate plot in Figure 19.14. The asphalt content having the maximum value of unit weight and stability is selected from each of the respective plots.

1. Maximum unit weight = 6.0% [Figure 19.14(a)]
2. Maximum stability = 5.5% [Figure 19.14(b)]
3. Percent air voids in compacted mixture using mean of limits [that is, (3 + 5)/2 = 4] = 5.4% [Figure 19.14(e)]. (Note the limits of 3 percent and 5 percent given in Table 19.8.)

The optimum asphalt content is determined as the average.

Therefore, the optimum asphalt cement content is

$$\frac{6.0 + 5.5 + 5.4}{3} = 5.6\%$$

The properties of the paving mixture containing the optimum asphalt content should now be determined from Figure 19.14 and compared with the suggested criteria given in Table 19.8. The values for this mixture are

Unit weight = 153.8 lb/ft^3
Stability = 1600 lb
Flow = 9.5 units of 0.01 in.
Percent void total mix = 3.9
Percent voids in mineral aggregates = 13

This mixture meets all the requirements given in Table 19.8 for stability, flow, percent voids in total mix, and percent voids in mineral aggregates.

Example 19.3   Computing the Percent of Asphalt Absorbed

Using the information given in Example 19.2, determine the asphalt absorbed for the optimum mix. The maximum specific gravity for this mixture is 2.57, and the specific gravity of the asphalt cement is 1.02. From Eq. 19.9, the absorbed asphalt is given as

$$P_{aa} = 100 \frac{G_{ea} - G_{bam}}{G_{bam} \times G_{ea}} G_{ac}$$

Solution: It is first necessary to determine the effective specific gravity of aggregates using Eq. 19.7:

$$G_{ea} = \frac{100 - P_{ac}}{(100/G_{mp}) - (P_{ac}/G_{ac})}$$

$$G_{ea} = \frac{100 - 5.6}{(100/2.57) - (5.6/1.02)} = 2.82$$

The bulk specific gravity of the aggregates in this mixture is

$$G_{bam} = \frac{0.523 \times 94.4 + 0.396 \times 94.4 + 0.081 \times 94.4}{\dfrac{0.523 \times 94.4}{2.65} + \dfrac{0.396 \times 94.4}{2.75} + \dfrac{0.081 \times 94.4}{2.7}} = 2.69$$

The asphalt absorbed is

$$P_{aa} = 100 \, \frac{2.82 - 2.69}{2.60 \times 2.82} \, 1.02 \cong 1.75\%$$

Hot–Mix, Cold–Laid Asphaltic Concrete

Asphaltic concretes in this category are manufactured hot and then shipped and immediately laid, or they can be stockpiled for use at a future date. Thus, they are suitable for small jobs for which it may be uneconomical to set up a plant. They are also a suitable material for patching high-type pavements. The Marshall method of mix design can be used for this type of asphalt concrete, but high-penetration asphalt is normally used. The most suitable asphalt cements have been found to have penetrations within the lower limits of the 200–300 penetration grade.

Hot-mix, cold-laid asphaltic concretes are produced by first thoroughly drying the different aggregates in a central hot-mix plant and then separating them into several bins containing different specified sizes. One important factor in the production of this type of asphalt concrete mix is that the manufactured product should be discharged at a temperature of 170°F ± 10°. To achieve this, the aggregates are cooled to approximately 180°F after they are dried but before they are placed into the mixer. Based on the job-mix formula, the exact amount of the aggregates from each bin is weighed and placed in the mixer. The different sizes of the aggregates are thoroughly mixed together and dried. About 0.75 percent by weight of a medium-curing cutback asphalt (MC-30), to which a wetting agent has been added, is mixed with the aggregates for another 10 sec. The high-penetration asphalt cement and water then are added simultaneously to the mixture. The addition of the water is necessary to ensure that the material remains workable after it has cooled down to normal temperatures. The amount of asphalt cement added is the optimum amount obtained from the mix design, but the amount of water depends on whether the material is to be used within a few days or to be stockpiled for periods up to several months. When the material is to be used within a few days, 2 percent of water by weight is used; if it is to be stockpiled for a long period, 3 percent of water is used. The mixture is then thoroughly mixed for about 45 sec to produce a uniform mix.

Cold–Mix, Cold–Laid Asphaltic Concrete

Emulsified asphalts and low viscosity cutback asphalts are used to produce cold mix asphaltic concretes. They also can be used immediately after production or stockpiled for use at a later date. The production process is similar to that of the hot mix, except that the mixing is done at normal temperatures, and it is not always necessary to dry the aggregates. However, saturated aggregates and aggregates with surface moisture should be dried before mixing. The type and grade of asphaltic material used depends on whether the material is to be stockpiled for a long time, the use of the material, and the gradation of the aggregates. Table 19.1 shows a suitable type of asphaltic material for different types of cold mixes.

Seal Coats

Seal coats are usually single applications of asphaltic material that may or may not contain aggregates. The three types of seal coats commonly used in pavement maintenance are fog seals, slurry seals, and aggregate seals.

Fog Seal

Fog seal is a thin application of emulsified asphalt, usually with no aggregates added. Slow-setting emulsions, such as SS-1, SS-1H, CSS-1, and CSS-1H, are normally used for fog seals. The emulsion is sprayed at a rate of 0.1 to 0.2 gal/yd^2 after it has been diluted with clean water. Fog seals are used mainly to

- Reduce the infiltration of air and water into the pavement
- Prevent the progressive separation of aggregate particles from the surface downward or from the edges inward (raveling) in a pavement. (Raveling is mainly caused by insufficient compaction during construction, which was carried out in wet or cold weather conditions.)
- Bring the surface of the pavement to its original state

Slurry Seal

Slurry seal is a uniformly mixed combination of a slow-setting asphalt emulsion (usually SS-1), fine aggregate, mineral filler, and water. The mixing can be carried out in a conventional plastic mixer or in a wheelbarrow, if the quantity required is small. It is usually applied with an average thickness of $\frac{1}{16}$ to $\frac{1}{8}$ in.

Slurry seal is used as a low-cost maintenance material for pavements carrying light traffic. Note, however, that although the application of a properly manufactured slurry seal coat will fill cracks of about $\frac{1}{4}$ in. or more and provide a fine-textured surface, the existing cracks will appear through the slurry seal in a short time.

Aggregate Seals

Aggregate seals are obtained by spraying asphalt, immediately covering it with aggregates, and then rolling the aggregates into the asphalt. Asphalts used for aggregate seals are usually the softer grades of paving asphalt and the heavier grades of liquid asphalts. Aggregate seals can be used to restore the surface of old pavements.

Prime Coats

Prime coats are obtained by spraying asphaltic binder materials onto nonasphalt base courses. Prime coats are mainly used to

- Provide a waterproof surface on the base
- Fill capillary voids in the base
- Facilitate the bonding of loose mineral particles
- Facilitate the adhesion of the surface treatment to the base

Medium-curing cutbacks are used normally for prime coating, with MC-30 recommended for priming a dense flexible base and MC-70 for more granular-type base materials. The rate of spray is usually between 0.2 and 0.35 gal/yd^2 for the MC-30, and between 0.3 and 0.6 gal/yd^2 for the MC-70. The amount of asphaltic binder used, however, should be the maximum that can be completely absorbed by the base within 24 hr of application under favorable weather conditions. The base course must contain a nominal amount of water to facilitate the penetration of the asphaltic material into the base. It is therefore necessary to lightly spray the surface of the base course with water just before the application of the prime coat if its surface has become dry and dusty.

Tack Coats

A *tack coat* is a thin layer of asphaltic material sprayed over an old pavement to facilitate the bonding of the old pavement and a new course, which is to be placed over the old pavement. In this case, the rate of application of the asphaltic material should be limited, since none of this material is expected to penetrate the old pavement. Asphalt emulsions such as SS-1, SS-1H, CSS-1, and CSS-1H are normally used for tack coats after they have been thinned with an equal amount of water. Rate of application varies from 0.05 to 0.15 gal/yd^2 of the thinned material. Rapid-curing cutback asphalts such as RC-70 may also be used as tack coats.

Sufficient time must elapse between the application of the tack coat and the application of the new course to allow for adequate curing of the material through the evaporation of most of the dilutent in the asphaltic emulsion. This curing process usually takes several hours in hot weather but can take more than 24 hr in cooler weather. When the material is satisfactorily cured, it becomes a highly viscous, tacky film.

Surface Treatments

Asphalt surface treatments are obtained by applying a quantity of asphaltic material and suitable aggregates on a properly constructed flexible base course to provide a suitable wearing surface for traffic. Surface treatments are used to protect the base course and to eliminate the problem of dust on the wearing surface. They can be applied as a single course with thicknesses varying from $\frac{1}{2}$ to $\frac{3}{4}$ in. or a multiple course with thicknesses varying from $\frac{7}{8}$ to 2 in.

A single-course asphalt treatment is obtained by applying a single course of asphaltic material and a single course of aggregates. The rate of application of the asphaltic material for a single course varies from 0.13 to 0.42 gal/yd^2, depending on the gradation of the aggregates used; the rate of application of the aggregates varies from 0.11 ft^3/yd^2 to 0.50 ft^3/yd^2. Multiple-course asphalt surface treatments can be obtained either as a double asphalt surface treatment consisting of two courses of asphaltic material and aggregates or as a triple asphalt treatment consisting of three layers. The multiple-course surface treatments are constructed by first placing a uniform layer of coarse aggregates over an initial application of the bituminous materials and then applying one or more layers of bituminous materials and smaller aggregates, with each layer having a thickness that is approximately equal to the nominal maximum size of the aggregates used for that layer. The maximum aggregate size of each layer subsequent to the initial layer is usually taken as one-half that of the aggregates used in the preceding layer. The recommended rates of application of the asphaltic material and the aggregates are shown in Table 19.11.

Table 19.11 Quantities of Materials for Bituminous Surface Treatments

Surface Treatment		Aggregate			Bituminous Material[A]
Type	Application	Size No.[B]	Nominal Size (Square Openings)	Typical Rate of Application, ft^3/yd^2	Typical Rate of Application, gal/yd^2
Single	initial	5	1 in to $\frac{1}{2}$ in.	0.50	0.42
		6	$\frac{3}{4}$ in. to $\frac{3}{8}$ in.	0.36	0.37
		7	$\frac{1}{2}$ in. to No. 4	0.23	0.23
		8	$\frac{3}{8}$ in. to No. 8	0.17	0.19
		9	No. 4 to No. 16	0.11	0.13
Double	initial	5	1 in. to in.	0.50	0.42
	second	7	$\frac{1}{2}$ in. to No. 4	0.25	0.26
Double	initial	6	$\frac{3}{4}$ in. to $\frac{3}{8}$ in.	0.36	0.37
	second	8	$\frac{3}{8}$ in. to No. 8	0.18	0.20
Triple	initial	5	1 in. to $\frac{1}{2}$ in.	0.50	0.42
	second	7	$\frac{1}{2}$ in. to No. 4	0.25	0.26
	third	9	No. 4 to No. 16	0.13	0.14
Triple	initial	6	$\frac{3}{4}$ in. to $\frac{3}{8}$ in.	0.36	0.37
	second	8	$\frac{3}{8}$ in. to No. 8	0.18	0.20
	third	9	No. 4 to No. 16	0.13	0.14

NOTE: The values are typical design or target values and are not necessarily obtainable to the precision indicated.

[A]Experience has shown that these quantities should be increased slightly (5 to 10%) when the bituminous material to be used was manufactured for application with little or no heating.

[B]According to Specification D448.

SOURCE: *Annual Book of ASTM Standards, Section 4, Construction,* Vol. 04.03, *Road and Paving Materials; Pavement Management Technologies,* American Society for Testing and Materials, Philadelphia, Pa., 1996.

Superpave Systems

As part of the Strategic Highway Research Program (SHRP), a new system for specifying the asphalt materials in asphalt concrete was developed. This system is known as *superpave,* which is a shortened form for *su*perior *per*forming asphalt *pave*ments. The research leading to the development of this new system was initiated because prior to superpave, it was difficult to relate the results obtained from laboratory analysis in the existing systems to the performance of the pavement without field experience. For example, the old systems used the results of tests performed at standard test temperatures to determine whether the materials satisfied the specifications. However, these tests are mainly empirical, and field experience is required to determine whether the results obtained have meaningful information. The superpave system includes a method for specifying asphalt binders and mineral aggregates, an asphalt mixing design, and a procedure for analyzing and predicting pavement performance. This section describes the superpave mix design procedure as given by the Asphalt Institute in Superpave Mix Design (SP-2).

The superpave system is unique in that it is performance based and engineering principles can be used to relate the results obtained from its tests and analyses to field performance.

The system consists of the following four parts:

- Selection of materials (aggregates and asphalt binder)
- Selection of design aggregate structure
- Selection of design asphalt binder content and
- Evaluation of moisture sensitivity of the design asphalt mixture

Selection of Materials

Selection of materials includes the selection of suitable mineral aggregates and asphalt binder.

Selection of Asphalt Binder.   The binder selected for a given project is based on the range of temperatures through which the pavement will be exposed and the traffic to be carried during its lifetime. The binders are therefore classified with respect to the range of temperatures at which their physical property requirements must be met. For example, a binder that is classified as PG 52-28 must satisfy all the high-temperature physical property requirements at temperatures up to at least 52°C, and all the low-temperature physical requirements down to at least 28°C. The high pavement design temperature is defined at a depth of 20 mm below the pavement surface and the low pavement design temperature at the surface of the pavement. Table 19.12 gives a listing of the more commonly used asphalt binder grades with their associated physical properties. The selection can be made in one of three ways:

1. The designer may select a binder based on the geographic location of the pavement.
2. The designer may determine the design pavement temperatures.
3. The designer may determine the design air temperatures, which are then converted to design pavement temperatures.

In the first method, the designer selects a binder grade from a map that has been prepared by her agency indicating the binder that should be used for different locations. These maps are usually based on weather characteristics and or policy decisions. In the second method, the designer determines the design pavement temperatures that should be used. In the third method, the designer uses the information given in the superpave system on the maximum and minimum air temperatures of the project location and then converts these temperatures to the design pavement temperatures.

The superpave system makes use of a temperature data base consisting of data from 6092 reporting weather stations in the United States and Canada that have been in operation for 20 years or more. In determining the maximum temperature, the hottest seven-day period for each year was identified, and the average of the maximum air temperature during each of these periods was selected as the maximum temperature. The average of the minimum one-day air temperature for each year was selected as the minimum temperature. In addition, the standard deviations of the seven-day average maximum temperature and that for the one-day minimum air temperature for each year were determined. These standard deviations are determined to facilitate the use of reliability measurements in selecting the design pavement temperatures. For example, consider a location where the mean seven-day maximum temperature is 36°C with a standard deviation of ±1°C. There is a 50 percent chance that during an average year, the seven-day maximum air temperature will exceed 36°C, but only a 2 percent chance for it to exceed 38°C (mean plus two standard deviations), assuming a normal temperature distribution. Selecting a maximum air temperature of 38°C at this location will provide a reliability of 98 percent that the maximum air temperature will not be exceeded. Table 19.13 gives asphalt binder grades and reliability for selected cities.

The superpave system also considers the fact that the pavement temperature and not the air temperature should be used as the design temperature. The system therefore uses the expression given in Eq. 19.13 to convert the maximum air temperature to the maximum design pavement temperature.

$$T_{20mm} = (T_{air} - 0.00618Lat^2 + 0.2289Lat = 42.2)(0.9545) - 17.78 \qquad (19.13)$$

where
T_{20mm} = high pavement design temperature at a depth of 20 mm
T_{air} = seven-day average high air temperature, °C
Lat = the geographical latitude of the project location in degrees

The low pavement design temperature can be selected as either the low air temperature, which is rather conservative, or it can be determined from the low air temperature using the expression:

$$T_{pav} = 1.56 + 0.72T_{air} - 0.004Lat^2 + 6.26\log_{10}(H+25) - Z(4.4 + 0.52\sigma_{air}^2)^{1/2} \qquad (19.14)$$

where
T_{pav} = Low AC pavement temperature below surface, °C
T_{air} = Low air temperature, °C
Lat = latitude of the project location, degrees
H = depth of pavement surface mm
σ_{air} = standard deviation of the mean low air temperature, °C
Z = from the standard normal distribution table, Z = 2.055 for 98% reliability

Adjusting Binder Grade for Traffic Speed and Loading. The procedure described above for selecting the asphalt binder is based on an assumed traffic condition consisting of a designed number of fast transient loads. The selected binder should therefore be adjusted for traffic conditions that are different from that assumed in the procedure, as the speed of loading has an additional effect on the ability of the pavement to resist permanent deformation at the high temperature condition. When the design loads are moving slowly, the selected asphalt binder based on the procedure described earlier should be shifted higher one high temperature grade. For example, when the standard procedure gives a PG 52-28, a PG 58-28 should be used for a slow-moving load. Also when the design load is stationary, the binder selected from the procedure should be shifted higher two high temperature grades. In addition to shifting the selected binder grade for speed, the designer should adjust the binder for the accumulated traffic load. For equivalent single axle loads (ESAL)

Table 19.12 Performance-Graded Asphalt Binder Specification

Performance Grade	PG 46			PG 52							PG 58					PG 64					
	-34	-40	-46	-10	-16	-22	-28	-34	-40	-46	-16	-22	-28	-34	-40	-10	-16	-22	-28	-34	-40
Average 7-day Maximum Pavement Design Temperature, °C[a]	<46			<52							<58					<64					
Minimum Pavement Design Temperature, °C[a]	>-34	>-40	>-46	>-10	>-16	>-22	>-28	>-34	>-40	>-46	>-16	>-22	>-28	>-34	>-40	>-10	>-16	>-22	>-28	>-34	>-40
Original Binder																					
Flash Point Temp, T48: Minimum °C						230															
Viscosity, ASTM D 4402:[b] Maximum, 3 Pas (300 cP), Test Temp, °C						135															
Dynamic Shear, TP5:[c] G★/sin δ, Minimum, 1.00 kPa Test Temperature @ 10 rad/s, °C	46					52							58					64			
Rolling Thin Film Oven (T 240) or Thin Film Oven (T 179) Residue																					
Mass Loss, Maximum, %						1.00															
Dynamic Shear, TP5: G★/sin δ, Minimum, 2.20 kPa Test Temp @ 10 rad/sec. °C	46					52							58					64			
Pressure Aging Vessel Residue (PP1)																					
PAV Aging Temperature, °C[d]	90					90							100					100			
Dynamic Shear, TP5: G★sin d, Maximum, 500 kPa Test Temp, °C	10	7	4	25	22	19	16	13	10	7	25	22	19	16	13	31	28	25	22	19	16
Physical Hardening[e]						Report															
Creep Stiffness, TP1:[f] S, Maximum, 300 MPa m-valve, Minimum, 0.300 Test Temp, @60 sec, °C	-24	-30	-36	0	-6	-12	-18	-24	-30	-36	-6	-12	-18	-24	-30	0	-6	-12	-18	-24	-30
Direct Tension, TP3:[f] Failure Strain, Minimum, 1.0% Test Temp @1.0 mm/min, °C	-24	-30	-36	0	-6	-12	-18	-24	-30	-36	-6	-12	-18	-24	-30	0	-6	-12	-18	-24	-30

932

Performance Grade	PG 70						PG 76					PG 82				
	-10	-16	-22	-28	-34	-40	-10	-16	-22	-28	-34	-10	-16	-22	-28	-34
Average 7-day Maximum Pavement Design Temperature, °C[a]			<70						<76					<82		
Minimum Pavement Design Temperature, °C[a]	>-10	>-16	>-22	>-28	>-34	>-40	>-10	>-16	>-22	>-28	>-34	>-10	>-16	>-22	>-28	>-34
Original Binder																
Flash Point Temp, T48: Minimum °C									230							
Viscosity, ASTM D 4402:[b] Maximum, 3 Pa·s (3000 cP), Test Temp, °C									135							
Dynamic Shear, TP5:[c] G★/sin δ, Minimum, 1.00 kPa Test Temperature @10 rad/s, °C			70						76					82		
Rolling Thin Film Oven (T 240) or Thin Film Oven (T 179) Residue																
Mass Loss, Maximum, %									1.00							
Dynamic Shear, TP5: G★/sin δ, Minimum, 2.20 kPa Test Temp @ 10 rad/sec, °C			70						76					82		
Pressure Aging Vessel Residue (PPI)																
PAV Aging Temperature, °C[d]			100(110)						100(110)					100(110)		
Dynamic Shear, TP5: G★sin δ, Maximum, 5000 kPa Test Temp @ 10 rad/sec, °C	34	31	28	25	22	19	37	34	31	28	25	40	37	34	31	28
Physical Hardening[e]									Report							
Creep Stiffness, TP1:[f] S, Maximum, 300 MPa m-value, Minimum, 0.300 Test Temp, @ 60 sec, °C	0	-6	-12	-18	-24	-30	0	-6	-12	-18	-24	0	-6	-12	-18	-24
Direct Tension, TP3: Failure Strain, Minimum, 1.0% Test Temp @ 1.0 mm/min, °C	0	-6	-12	-18	-24	-30	0	-6	-12	-18	-24	0	-6	-12	-18	-24

[a] Pavement temperatures can be estimated from air temperatures using an algorithm contained in the Superpave software program or may be provided by the specifying agency, or by following the procedures as outlined in PPX.

[b] This requirement may be waived at the discretion of the specifying agency if the supplier warrants that the asphalt binder can be adequately pumped and mixed at temperatures that meet all applicable safety standards.

[c] For quality control of unmodified asphalt cement production, measurement of the viscosity of the original asphalt cement may be substituted for dynamic shear measurements of G★/sin δ at test temperatures where the asphalt is a Newtonian fluid. Any suitable standard means of viscosity measurement may be used, including capillary or viscometer (AASHTO T 201 or T202).

[d] The PAV aging temperature is based on simulated climatic conditions and is one of three temperatures 90°C, 100°C, or 110°C. The PAV aging temperature is 100°C for PG 64- and above, except in desert climates, where it is 110°C.

[e] Physical Hardening–TP1 is performed on a set of asphalt beams according to Sections 13.1 of TP1, except the conditioning time is extended to 24 hrs ±10 minutes at 10°C above the minimum performance temperature. The 24-hour stiffness and m-value are reported for information purposes only.

[f] If the creep stiffness is below 300 MPa, the direct tension test is not required. If the creep stiffness is between 300 and 600 MPa the direct tension failure strain requirement can be used in lieu of the creep stiffness requirement. The m-value requirement must be satisfied in both cases.

SOURCE: *Superpave Mix Design: Superpave Series No. 2 (SP-2)*, Asphalt Institute, Lexington, Ky., 2000.

933

Table 19.13 Asphalt Binder Grades and Reliability for Selected Cities

ST	Station	Latitude	Min 50% Grade	Actual Reliability High	Low	Min 98% Grade	Actual Reliability High	Low
AL	Mobile	30.68	PG 58-10	84	99	PG 64-10	99.9	99
AK	Juneau 2	58.30	PG 40-16	91	70	PG 46-28	99.9	99
AZ	Phoenix WSFO AP	33.43	PG 70-10	99.9	99.9	PG 70-10	99.9	99.9
AR	Little Rock FAA AP	34.73	PG 58-10	69	64	PG 64-16	99.9	97
CA	Los Angeles WSO AP	33.93	PG 52-10	66	99.9	PG 58-10	99.9	99.9
	Sacramento WSO CI	38.58	PG 58-10	61	99.9	PG 64-10	99.9	99.9
	San Francisco WSO AP	37.62	PG 52-10	98	99.9	PG 52-10	98	99.9
CO	Denver WSFO AP	39.77	PG 58-22	99.9	78	PG 58-28	99.9	99
CT	Hartford WSO AP	41.93	PG 52-22	54	89	PG 58-28	99.9	99.7
DC	Wash Natl WSO AP	38.85	PG 58-10	99.9	57	PG 58-16	99.9	99
DE	Wilmington WSO AP	39.67	PG 58-16	99	84	PG 58-22	99	99.4
FL	Jacksonville WSO AP	30.50	PG 58-10	91	98.6	PG 64-10	99.9	98.6
	Miami	25.80	PG 58-10	99	99.9	PG 58-10	99	99.9
GA	Atlanta WSO AP	33.65	PG 58-10	90	64	PG 64-16	99.9	96.8
HI	Lahaina 361	20.88	PG 58-10	99.9	99.9	PG 58-10	99.9	99.9
IA	Des Moines WSFO AP	41.53	PG 58-22	98	67	PG 58-58	98	99.3
ID	Boise WSFO AP	43.57	PG 58-16	93	61	PG 64-28	99.9	99.6
IL	Chicago O'Hare WSO AP	41.98	PG 52-22	58	67	PG 58-28	99.9	99.3
	Peoria WSO AP	40.67	PG 58-22	99.9	85	PG 58-28	99.9	99.9
IN	Indianapolis SE Side	39.75	PG 58-22	99	89	PG 58-28	99	99.7
KS	Wichita WSO AP	37.65	PG 64-16	99.9	57	PG 64-22	99.9	98.5
KY	Lexington WSO AP	38.03	PG 58-16	98	50	PG 58-28	98	99.8
	Louisville WSFO	38.18	PG 58-16	96	67	PG 64-22	99.9	95
LA	New Orleans WSCMO AP	29.98	PG 58-10	97	98.5	PG 64-10	99.9	98.5
MA	Lowell	42.65	PG 52-16	58	65	PG 58-28	99.9	99.1
MD	Baltimore WSO AP	39.18	PG 58-16	99.9	91	PG 58-22	99.9	99.9
ME	Portland	43.67	PG 52-16	97	59	PG 58-28	99.9	98.7
MI	Detroit City AP	42.42	PG 52-16	56	77	PG 58-22	99.9	99.7
	Sault Ste. Marie WSO	46.47	PG 52-28	99.9	90	PG 52-34	99.9	99.9
MN	Duluth WSO AP	46.83	PG 52-28	99.9	57	PG 52-34	99.9	98.5
	Minn-St Paul WSO AP	44.88	PG 52-28	71	90	PG 58-34	99.9	99.9
MO	Kansas City FSS	39.12	PG 58-16	80	57	PG 64-22	99.9	98.5
	St. Louis WSCMO AP	38.75	PG 58-16	91	57	PG 64-22	99.9	98.5
MS	Jackson WSFO AP	32.32	PG 58-10	55	77	PG 64-16	99.9	99.7
MT	Great Falls	47.52	PG 52-28	69	67	PG 58-34	99.9	95
NC	Charlotte WSO AP	35.33	PG 58-10	92	68	PG 64-16	99.9	99.3
	Raleigh 4 SW	35.73	PG 58-10	93	68	PG 64-16	99.9	99.3
ND	Bismarck WSFO AP	46.77	PG 58-34	99.9	88	PG 58-40	99.9	99.6
NE	Omaha (North) WSFO	41.37	PG 58-22	98	68	PG 58-28	98	99.3
NH	Concord WSO AP	43.20	PG 52-22	61	50	PG 58-34	99.9	99.8
NJ	Atlantic City	39.38	PG 52-10	77	68	PG 58-16	99.9	99.3
NM	Laguna	35.03	PG 58-16	81	64	PG 64-22	99.9	96.7
NV	Reno WSFO AP	39.50	PG 58-16	97	71	PG 64-22	99.9	98
NY	Albany	42.65	PG 52-22	58	94.5	PG 58-22	99.9	94.5
	Buffalo WSFO AP	42.93	PG 52-16	96	57	PG 58-22	99.9	98.5
	New York Inter AP	40.65	PG 52-16	61	97.1	PG 58-16	99.9	97.1

Table 19.13 Asphalt Binder Grades and Reliability for Selected Cities (*continued*)

ST	Station	Latitude	Min 50% Grade	Actual Reliability High	Low	Min 98% Grade	Actual Reliability High	Low
OH	Cin Muni-Lunken Fld	39.10	PG 58-16	99.9	64	PG 58-22	99.9	96.7
	Cleveland WSO AP	41.42	PG 52-16	51	50	PG 58-28	99.9	99.8
	Columbus	39.98	PG 58-16	99.9	50	PG 58-28	99.9	99.8
OK	Oklahoma City WSFO AP	35.40	PG 64-16	99.9	91	PG 64-22	99.9	99.9
OR	Oregon City	45.35	PG 52-10	55	89	PG 58-16	99.9	99.7
PA	Philadelphia Drexel U	39.95	PG 58-10	99.9	57	PG 58-16	99.9	98.5
	Pittsburgh WSO CI	40.45	PG 58-16	99.9	77	PG 58-22	99.9	99.7
RI	Providence WSO AP	41.73	PG 52-16	71	68	PG 58-22	99.9	99.3
SC	Columbia WSFO AP	33.95	PG 58-10	61	77	PG 64-16	99.9	99.7
SD	Sioux Falls WSFO AP	43.57	PG 58-28	99.9	87	PG 58-34	99.9	99.9
TN	Knoxville U of Tenn	35.95	PG 58-16	93	92	PG 64-22	99.9	99.8
	Memphis FAA-AP	35.05	PG 58-10	65	55	PG 64-16	99.9	95
TX	Amarillo WSO AP	35.23	PG 58-16	66	68	PG 64-22	99.9	99.3
	Dallas FAA AP	32.85	PG 64-10	99	77	PG 64-16	99	99.7
	Houston FAA AP	29.65	PG 64-10	99.9	99.3	PG 64-10	99.9	99.3
UT	Salt Lake City WSO CI	40.77	PG 58-16	98	84	PG 58-22	98	99.4
VA	Norfolk WSO AP	36.90	PG 58-10	98	85	PG 58-16	98	99.9
	Richmond WSO AP	37.50	PG 58-16	95	97.1	PG 64-16	99.9	97.1
VT	Burlington WSO AP	44.47	PG 52-22	93	50	PG 58-28	99.9	97
WA	Seattle Tac WSCMO AP	47.45	PG 52-10	99.9	83	PG 52-16	99.9	98.5
	Spokane	47.67	PG 58-16	98	59	PG 58-28	98	98.7
WI	Milwaukee WSO AP	42.95	PG 52-22	77	71	PG 58-28	99.9	98
WV	Charleston WSFO AP	38.37	PG 58-16	99	71	PG 58-22	99	98
WY	Cheyenne WSFO AP	41.15	PG 52-22	68	55	PG 58-28	99.9	94.8
PR	San Juan WSFO	18.43	PG 58-10	98	99.9	PG 58-10	98	99.9

SOURCE: *Superpave Mix Design: Superpave Series No. 1 (SP-1), Asphalt Institute, Lexington, Ky., 2000.*

(see Chapter 20 for definition of ESAL) of 10,000,000 to 30,000,000, the engineer should consider shifting the binder selected based on the procedure by one high temperature binder grade, but for ESALs exceeding 30,000,000, a shift of one high temperature grade is required.

Example 19.4   Determining a Suitable Binder Grade Using High and Low Air Temperatures

The latitude at a location where a high-speed rural road is to be located is 41°. The seven-day average high air temperature is 50°C, and the low air temperature is −20°C. The standard deviation for both the high and low temperatures is ±1°C. Determine a suitable binder that could be used for the pavement of this highway, if the depth of the pavement surface is 155 mm and the expected is 9×10^6.

Solution:

- Use Eq. 19.13 to determine the high pavement temperature at a depth of 20 mm:

$$T_{20mm} = (T_{air} - 0.00618Lat^2 + 0.2289Lat + 42.2)(0.9545) - 17.78$$
$$= (50 - 0.00618(41)^2 + 0.2289(41) + 42.2)(0.9545) - 17.78$$
$$= 87.0 - 17.78$$
$$= 69.27°\,C$$

- Use Eq. 19.14 to determine low AC pavement temperature:

$$T_{pav} = 1.56 + 0.72T_{air} - 0.004Lat^2 + 6.26\log_{10}(H + 25) - Z(4.4 + 0.52\sigma_{air}^2)^{1/2}$$
$$= 1.56 + 0.72(-20) - 0.004(41)^2 + 6.26\log_{10}(155 + 25) - 2.055(4.4 + 0.52 \times 1)^{1/2}$$
$$= -10.0°\,C$$

- Select appropriate binder from Table 19.12. For minimum 50 percent reliability, binder is PG 70-10. Note: no correction is needed for ESAL < 10×10^6.
- Determine seven-day maximum temperature for 98 percent reliability: Standard deviation = ±1.
  There is a 2 percent chance that the seven-day maximum will exceed $(50 + 2 \times 1)°C$ (i.e., maximum air temperature for 98 percent reliability is 52°C).
- Use Eq. 19.13 to determine the high pavement temperature at a depth of 20 mm:

$$T_{20mm} = (T_{air} - 0.00618Lat^2 + 0.2289Lat + 42.2)(0.9545) - 17.78$$
$$= (52 - 00618 \times 41^2 + 0.2289 \times 41 + 42.2) - 17.78$$
$$= 71.18°\,C$$

- Select binder for 98 percent reliability from Table 19.12; use PG 76-10.

Example 19.5   Determining a Suitable Binder Grade

An urban interstate highway is being designed to carry an equivalent single-axle load (ESAL) of 31×10^6. It is anticipated that this road will be congested most of the time, resulting in slow moving traffic. If the road is located in Richmond, Virginia, determine an appropriate asphalt binder grade for this project.

Solution:

- Use Table 19.13 to select binder based on latitude.
  For minimum 50 percent reliability, binder is PG 58-16
  For minimum 98 percent reliability, binder is PG 64-16

- Correct for slow moving traffic: Move binder one grade higher.
  For minimum 50 percent reliability, binder is PG 64-16
  For minimum 98 percent reliability, binder is PG 70-16
- Correct for ESAL > 30×10^6: Move binder one grade higher.
  For minimum 50 percent reliability, binder is PG 70-16
  For minimum 98 percent reliability, binder is PG 76-16.

Selection of Mineral Aggregate.   Based on the results obtained from surveying pavement experts, two categories of aggregate properties were identified by the developers of superpave for use in the system. These are referred to as consensus property and source property. In addition to the consensus and source properties, superpave uses a gradation system that is based on the 0.45 power gradation chart to determine the design aggregate structure. The aggregate characteristics that were generally accepted by the experts as critical for good performance of the hot mix asphalt (HMA) are classified as *consensus properties.* These properties include the angularity of the coarse aggregates, the angularity of the fine aggregates, the amount of flat and elongated particles in the coarse aggregates, and the clay content. The angularity of the coarse aggregate is defined as the percent by weight of coarse aggregates larger than 4.75 mm with one or more fractured faces. The angularity of coarse aggregates can be determined by conducting Pennsylvania Department of Transportation test No. 621, *"Determining the Percentage of Crushed Pavements in Gravel."* The criteria for angularity of coarse aggregates depend on the traffic level and are given in Table 19.14.

The *angularity of fine aggregates* is defined as the percent of air voids in loosely compacted aggregates smaller than 2.36 mm. The AASHTO designated test TP33, "Test Method for Uncompacted Void Content of Fine Aggregate (as Influenced by Particle, Shape, Surface Texture, and Grading) (Method A)" can be used to determine the angularity of fine aggregates. In this test, a standard cylinder of known volume (V) is filled with a washed sample of the fine aggregates by pouring the aggregates through a standard funnel. The mass (W) of the fine aggregates filling the standard cylinder of volume V is then determined. The

Table 19.14   Coarse Aggregate Angularity Criteria

Traffic, Million ESALs	Depth from Surface	
	< 100 mm	> 100 mm
< 0.3	55/–	–/–
< 1	65/–	–/–
< 3	75/–	50/–
< 10	85/80	60/–
< 30	95/90	80/75
< 100	100/100	95/90
≥ 100	100/100	100/100

SOURCE: *Superpave Mix Design: Superpave Series No. 2 (SP-2),* Asphalt Institute, Lexington, Ky., 2000.

volume of the fine aggregates in the standard cylinder is then determined as (W/G_{bfa}) and the void content determined as a percentage of the cylinder volume as shown in Eq. 19.15.

$$\text{uncompacted void} = \frac{V - W/G_{bfa}}{V} \times 100\% \tag{19.15}$$

where

G_{bfa} = bulk specific gravity of the fine aggregate.

Table 19.15 gives the criteria for fine aggregate angularity.

A *flat and elongated particle* is defined as one that has its maximum dimension five times greater than its minimum dimension. The amount of fine and elongated particles in the coarse aggregate is obtained by conducting ASTM D4791 designated test *"Flat or Elongated Particles in Coarse Aggregates"* on coarse aggregates larger than 4.75 mm. This test involves the use of a proportional caliper device that automatically tells whether a particle is flat or elongated. The test is performed on a sample of the coarse aggregate and the percentage by mass determined. The maximum percentage allowed is given in Table 19.16 for different traffic loads.

The *clay content* is defined as the percentage of clayey material in the portion of aggregate passing through the 4.75 mm sieve. It is obtained by conducting the AASHTO T176 designated test *"Plastic Fines in Graded Aggregates and Soils by Use of Sand Equivalent Test."* This test is conducted by first mixing a sample of the fine aggregate in a graduated cylinder with a flocculating solution. The clayey fines coating the aggregates are then loosened from the aggregates by shaking the cylinder. The mixture is then allowed to stand for a period during which time the clayey material is suspended above the granular aggregates (sedimented sand). The heights of the suspended clay and the sedimented sand are measured. The ratio, expressed in percentage of the height of the sedimented sand to that of the suspended clay is the sand equivalent value. Table 19.17 gives minimum allowable values for the sand equivalent for different traffic loads.

Table 19.15 Fine Aggregate Angularity Criteria

	Depth from Surface	
Traffic, Million ESALs	< 100 mm	> 100 mm
< 0.3	–	–
< 1	40	–
< 3	40	40
< 10	45	40
< 30	45	40
< 100	45	45
≥ 100	45	45

SOURCE: *Superpave Mix Design: Superpave Series No. 2 (SP-2)*, Asphalt Institute, Lexington, Ky., 2000.

Table 19.16   Flat and Elongated Particles Criteria

Traffic, Million ESALs	*Maximum, Percent*
< 0.3	–
< 1	–
< 3	10
< 10	10
< 30	10
< 100	10
≥ 100	10

SOURCE: *Superpave Mix Design: Superpave Series No. 2 (SP-2),* Asphalt Institute, Lexington, Ky., 2000.

Table 19.17   Clay Content Criteria

Traffic, Million ESALs	*Sand Equivalent Minimum, Percent*
< 0.3	40
< 1	40
< 3	40
< 10	45
< 30	45
< 100	50
≥ 100	50

SOURCE: *Superpave Mix Design: Superpave Series No. 2 (SP-2),* Asphalt Institute, Lexington, Ky., 2000.

The other aggregate properties that were considered critical by the developers of the superpave system, but for which critical values were not determined by consensus, were classified as source aggregate properties. These properties are toughness, soundness, and maximum allowable percentage of deleterious materials. These properties are discussed in Chapter 21.

Gradation.   The distribution of aggregate particle sizes for a given blend of soil mixture is known as the design aggregate structure. The gradation system used for superpave is based on the 0.45 gradation plot. This is a plot of the percent passing a given sieve against the sieve size in mm raised to the 0.45 power. That is, the vertical axis of the graph is percent passing and the horizontal axis is the size of the sieve in millimeters raised to the 0.45 power. In order to understand the gradation system used, it is first necessary to define certain gradation terms that the superpave system uses. These are *maximum size, nominal maximum size,* and *maximum density gradation. Maximum size* is defined as one sieve larger than the nominal maximum size, and the *nominal maximum size* is one sieve larger than the first sieve that retains more than 10 percent of the soil. Five mixture gradations are specified in the superpave system as shown in Table 19.18. *Maximum density gradation* is obtained when the aggregate particles fit together in their densest form. An important characteristic of the 0.45 power plot is that the maximum density gradation for a sample of soil is given by a straight line joining the maximum size and the origin as shown in Figure 19.16. An

Table 19.18 Superpave Mixture Gradations

Superpave Designation	Nominal Maximum Size, mm	Maximum Size, mm
37.5 mm	37.5	50
25.0 mm	25.0	37.5
19.0 mm	19.0	25.0
12.5 mm	12.5	19.0
9.5 mm	9.5	12.5

SOURCE: *Superpave Mix Design: Superpave Series No. 2 (SP-2)*, Asphalt Institute, Lexington, Ky., 2000.

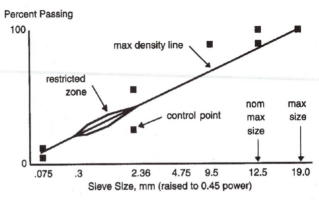

Figure 19.16 Maximum Density Gradation for 19 mm Maximum Size

SOURCE: *Superpave Mix Design: Superpave Series No. 2 (SP-2)*, Asphalt Institute, Lexington, Ky., 2000.

acceptable aggregate gradation is defined by specifying control points on the maximum density gradation chart for the smallest sieve size (0.075 mm), the nominal maximum size, and an intermediate size sieve (2.36 mm) as shown in Figure 19.16. An acceptable soil blend should therefore have a maximum density gradation line that lies within these control points.

In addition, a restricted zone, also shown in Figure 19.16, is established along the maximum density curve between the 0.3 mm sieve and the intermediate sieve (2.36 mm) as shown in Figure 19.16. For a soil blend to be acceptable, its gradation must not pass within the restricted zone. Soils that have gradations that go through the restricted zone have been found to create compaction problems during construction and tend to have inadequate VMA. Also, the superpave system recommends that the gradation pass below the restricted zone, although this is not a requirement. Table 19.19 gives control points and restricted zone boundaries or different nominal sieve sizes. Aggregates that have maximum density gradation lying between the control points and outside the restricted zones are considered as having an acceptable design aggregate structure.

Selection of Design Aggregate Structure.   In selecting the design aggregate structure, the designer must first ascertain that the individual aggregate and asphalt materials meet the

Table 19.19   Superpave Asphalt Mixture Gradation Requirements

a. 37.5 mm Nominal Size				
	Control Points		Restricted Zone Boundary	
Sieve, mm	Minimum	Maximum	Minimum	Maximum
50		100		
37.5	90	100		
25		90		
19				
12.5				
9.5				
4.75			34.7	34.7
2.36	15	41	23.3	27.3
1.18			15.5	21.5
0.6			11.7	15.7
0.3			10	10
0.15				
0.075	0	6		

b. 25 mm Nominal Size				
	Control Points		Restricted Zone Boundary	
Sieve, mm	Minimum	Maximum	Minimum	Maximum
37.5		100		
25	90	100		
19		90		
12.5				
9.5				
4.75			39.5	39.5
2.36	19	45	26.8	30.8
1.18			18.1	24.1
0.60			13.6	17.6
0.30			11.4	11.4
0.15				
0.075	1	7		

c. 19 mm Nominal Size				
	Control Points		Restricted Zone Boundary	
Sieve, mm	Minimum	Maximum	Minimum	Maximum
25		100		
19	90	100		
12.5		90		

continued

Table 19.19 Superpave Asphalt Mixture Gradation Requirements (*continued*)

	Control Points		Restricted Zone Boundary	
Sieve, mm	*Minimum*	*Maximum*	*Minimum*	*Maximum*
9.5				
4.75				
2.36	23	49	34.6	34.6
1.18			22.3	28.3
0.6			16.7	20.7
0.3			13.7	13.7
0.15				
0.075	2	8		

d. 12.5 mm Nominal Size

	Control Points		Restricted Zone Boundary	
Sieve, mm	*Minimum*	*Maximum*	*Minimum*	*Maximum*
19		100		
12.5	90	100		
9.5		90		
4.75				
2.36	28	58	39.1	39.1
1.18			25.6	31.6
0.6			19.1	23.1
0.3			15.5	15.5
0.15				
0.075	2	10		

e. 9.5 mm Nominal Size

	Control Points		Restricted Zone Boundary	
Sieve, mm	*Minimum*	*Maximum*	*Minimum*	*Maximum*
12.5		100		
9.5	90	100		
4.75		90		
2.36	32	67	47.2	47.2
1.18			31.6	37.6
0.6			23.5	27.5
0.3			18.7	18.7
0.15				
0.075	2	10		

SOURCE: *Superpave Mix Design: Superpave Series No. 2 (SP-2),* Asphalt Institute, Lexington, Ky., 2000.

criteria discussed above. Trial blends should then be prepared by varying the percentages from the different available stockpiles. The mathematical procedure described in a previous section of this chapter for determining the different proportions of different aggregates to obtain a required gradation may be used to determine trial blends that meet the control requirements. Although no specific number of trial blends is recommended, it is generally accepted that three trial blends that have a range of gradation will be adequate. The four consensus properties, the bulk and apparent specific gravities of the aggregates, and any source aggregate properties should be determined. At this stage these properties can be determined by using mathematical expressions, but actual tests should be carried out on the aggregate blend finally selected. The bulk specific gravity may be obtained using Eq. 19.5, the apparent specific gravity using Eq. 19.6.

Equation 19.16 can be used to determine the effective specific gravity (G_{ea}).

$$G_{ea} = G_{bam} + 0.8(G_{ama} - G_{bam})$$ (19.16)

where

G_{ea} = effective specific gravity of the aggregate blend
G_{bam} = bulk specific gravity of the aggregate blend
G_{ama} = apparent specific gravity of the aggregate blend

The designer may decide to change the implicit multiplication factor of 0.8, particularly when absorptive aggregates are used, as values close to 0.6 or 0.5 are more appropriate for these aggregates.

The amount of asphalt binder absorbed by the aggregates is estimated from Eq. 19.17 as:

$$V_{ba} = \frac{P_s(1-V_a)}{\left(\dfrac{P_b}{G_{ac}} + \dfrac{P_s}{G_{ea}}\right)}\left[\frac{1}{G_{bam}} - \frac{1}{G_{ea}}\right]$$ (19.17)

where

V_{ba} = volume of absorbed binder, cm³/cm³ of mix
P_b = percent of binder (assumed 0.05)
P_s = percent of aggregate (assumed 0.95)
G_{ac} = specific gravity of binder (assumed 1.02)
V_a = volume of air voids (assumed 0.04 cm³/cm³ of mix)

A trial percentage of asphalt binder is then determined for each aggregate blend using the following equation:

$$P_{bi} = \frac{G_{ac}(V_{be} + V_{ba})}{(G_{ac}(V_{be} + V_{ba})) + W_s} \times 100$$ (19.18)

where

P_{bi} = percent of binder by mass of mix i
V_{be} = volume of the effective binder and obtained from Eq. 19.19

$$V_{be} = 0.176 - 0.067\log(S_n) \tag{19.19}$$

S_n = the nominal maximum sieve size of the aggregate blend, mm
V_{ba} = volume of absorbed binder, cm³/cm³ of mix
G_{ac} = specific gravity of binder assumed to be 1.02
W_s = mass of aggregate, grams, from Eq. 19.20

$$W_s = \frac{P_s(1-V_a)}{\left(\dfrac{P_b}{G_{ac}} + \dfrac{P_s}{G_{se}}\right)} \tag{19.20}$$

V_a = volume of air voids (assumed 0.04 cm³/cm³ of mix)
P_s = percent of aggregate (assumed 0.95)
P_b = percent of binder (assumed 0.05)
G_{ac} = specific gravity of binder (assumed 1.02)
G_{ea} = effective specific gravity of the aggregate blend

At least two compacted specimens are then prepared for each trial aggregate blend using the percent of binder by mass of mix computed for that blend. The superpave gyratory compactor is used to prepare the compacted specimens. Figure 19.17 shows a schematic representation of the superpave gyratory compactor. The main components of the compactor are:

- A reaction frame, rotating base and motor
- A loading system consisting of loading ram and pressure gauge
- A control and data acquisition system and a mold and base plate

The level of compaction in the superpave system is given with respect to a design number of gyrations (N_{des}). The N_{des} depends on the average design high air temperature and the design ESAL. Two other levels of gyrations (maximum and initial) are important. The maximum number of gyrations, N_{max} is used to compact the test specimens, and the initial number of gyrations N_{ini} is used to estimate the compactibility of the mixture. N_{max} and N_{ini} are obtained from N_{des} as shown below:

$$\text{Log}N_{max} = 1.10\text{Log}N_{des} \tag{19.21}$$

$$\text{Log}N_{ini} = 0.45\text{Log}N_{des} \tag{19.22}$$

Table 19.20 shows the values of N_{des} for different ESALs and average design air temperatures.

Detailed description of the procedures for aggregates and mixture preparation and the compaction of the volumetric specimens is beyond the scope of this book. Interested readers will find this description in *Superpave Mix Design*.

Data Analysis and Preparation.   The following physical properties of the compacted specimens are first determined:

- Estimated bulk specific gravity (G_{bcm}(estimated))
- Corrected bulk specific gravity (G_{bcm}(corrected))
- Corrected percentage of maximum theoretical specific gravity

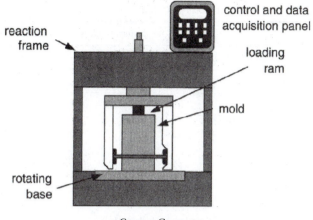

a. Gyratory Compactor

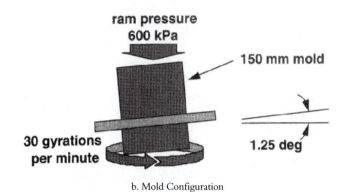

b. Mold Configuration

Figure 19.17 Superpave Gyratory Compactors

SOURCE: *Superpave Mix Design: Superpave Series No. 2 (SP-2),* Asphalt Institute, Lexington, Ky., 2000.

The following equations are used to determine these properties:

$$G_{bam} = \frac{W_m \big/ V_{mx}}{\gamma_w} \tag{19.23}$$

where
W_m = mass of specimen, grams
γ_w = density of water = 1g/cm
V_{mx} = volume of compaction mold (cm^3)

$$V_{mx} = \frac{\pi d^2 h_x}{4} x 0.001 cm^3 / mm^3 \tag{19.24}$$

Table 19.20 Superpave Design Gyratory Compactive Effort

| Design ESALs (millions) | Average Design High Air Temperature | | | | | | | | | | | |
| | < 39°C | | | 39–40°C | | | 41–42°C | | | 43–44°C | | |
	N_{ini}	N_{des}	N_{max}	N_{ini}	N_{des}	N_{max}	N_{ini}	N_{des}	N_{max}	N_{ini}	N_{des}	N_{max}
< 0.3	7	68	104	7	74	114	7	78	121	7	82	127
0.3-1.0	7	76	117	7	83	129	7	88	138	8	93	146
1.0-3.0	7	86	134	8	95	150	8	100	158	8	105	167
3.0-10.0	8	96	152	8	106	169	8	113	181	9	119	192
10.0-30.0	8	109	174	9	121	195	9	128	208	9	135	220
30.0-100.0	9	126	204	9	139	228	9	146	240	10	153	253
> 100	9	143	235	10	158	262	10	165	275	10	172	288

SOURCE: *Superpave Mix Design: Superpave Series No. 2 (SP-2),* Asphalt Institute, Lexington, Ky., 2000.

where

d = diameter of mold (150 mm)

h_x = height of specimen in mold during compaction (mm)

Since the cylinder in which the mold is compacted is not smooth and Eq. 19.24 is based on this assumption, the actual volume of the compacted mixture will be less than that obtained from the equation. The estimated bulk specific gravity is therefore corrected using the corrected factor obtained as the ratio of the measured bulk specific gravity to the estimated bulk specific gravity.

$$C = \frac{G_{bcm}(measured)}{G_{bcm}(estimated)} \tag{19.25}$$

where

C = correction factor

$G_{bcm}(measured)$ = measured bulk specific after after N_{max}

$G_{bcm}(estimated)$ = estimated bulk specific after after N_{max}

The corrected bulk specific gravity ($G_{bcm}(corrected)$) at any other gyration level can then be obtained using Eq. 19.26

$$G_{bcm}(corrected) = C \times G_{bcm}(estimated) \tag{19.26}$$

where

$G_{bcm}(corrected)$ = corrected bulk specific gravity for the specimen for a given gyration level

C = correction factor

$G_{bcm}(estimated)$ = estimated bulk specific after N_{max}

The percentage maximum theoretical specific gravity G_{mp} for each desired gyration is then computed as the ratio of $G_{bam}(corrected)$ to $G_{mp}(measured)$. The average G_{mp} value for each gyration level is obtained using the G_{mp} values for the samples compacted at that level.

The values for the properties (voids in mineral aggregate (VMA), percent variation of VMA from 4 percent (VFA) and dust ratio) are then determined. This is done at 4 percent air void content and at N_{des} to meet with the design criteria for the superpave mix.

$$P_{av} = 100 - \%Gmp@\,N_{des} \qquad (19.27)$$

where

$$P_{av} = \text{air voids @ } N_{des} \text{ percent of total volume}$$
$$\%Gmp@N_{des} = \text{maximum theoretical specific gravity @} N_{des}, \text{ percent}$$

$$\%VMA = 100 - \left(\frac{\%G_{mp}@N_{des}\,xG_{mp}\,xP_{ta}}{G_{bam}} \right) \qquad (19.28)$$

where

$$\%VMA = \text{voids in mineral aggregate, percent in bulk volume}$$
$$\%Gmp@N_{des} = \text{maximum theoretical specific gravity @} N_{des}, \text{ percent}$$
$$G_{bam} = \text{bulk specific gravity of total aggregate}$$
$$G_{mp} = \text{maximum theoretical specific gravity of the paving mixture}$$
$$P_{ta} = \text{aggregate content cm}^3/\text{cm}^3, \text{ by total mass of mixture}$$

If the percent of air voids is 4 percent, then the values obtained for the volumetric criteria are compared with the corresponding criteria values. For example, the values for VMA are shown in Table 19.21. If these criteria are met, then the blend is acceptable. When the %VMA is not 4 percent, it is necessary to determine an estimated design asphalt content with a %VMA of 4 percent using Eq. 19.29.

$$P_{ac,estimated} = P_{aci} - \left(0.4 \times \left(4 - P_{av}\right)\right) \qquad (19.29)$$

where

$$P_{ac,estimated} = \text{estimated asphalt content, percent by mass of mixture}$$
$$P_{aci} = \text{initial (trial) asphalt content, percent by mass of mixture}$$
$$P_{av} = \text{percent air voids at } N_{des} \text{ (trial)}$$

Table 19.21   VMA Criteria

Nominal Maximum Aggregate Size	Minimum VMA percent
9.5 mm	15
12.5 mm	14
19 mm	13
25 mm	12
37.5 mm	11

The VMA and VFA at N_{des}, $\%G_{mp}$ at N_{max}, and N_{ini} are then estimated for the estimated design asphalt content obtained in Eq. 19.27 using the following equations:

$$\%VMA_{estimated} = \%VMA_{initial} + C \times (4 - P_{av}) \tag{19.30}$$

$\% VMA_{initial}$ = % VMA from trial asphalt binder content
C = constant = 0.1 if P_{av} less than 4.0 percent
= 0.2 if P_{av} is greater than 4.0 percent

This value is then compared with the corresponding value given in Table 19.21.

$$\%VFA_{estimated} = 100\frac{\%VMA_{estimated} - 4.0}{\%VMA_{estimated}} \tag{19.31}$$

This value is then compared with the range of acceptable values given in Table 19.22.
For $\%G_{mp}$ (maximum specific gravity of paving mixture) at N_{max}

$$\%G_{mpestimated}@N_{max} = \%G_{mptrial}@N_{max} - (4 - P_{av}) \tag{19.32}$$

$$G_{mptrial} = G_{mp} \text{ for the trial min}$$

The maximum allowable mixture density at N_{max} is 98 percent.
For $\%G_{mp}$ at N_{ini}

$$\%G_{mpestimated}@N_{ini} = \%G_{mptrial}@N_{ini} - (4 - P_{av}) \tag{19.33}$$

The maximum allowable mixture density at N_{max} is also 98 percent.
The dust percentage is determined as the proportion of the percentage by mass of the material passing the 0.075 mm sieve to the effective binder content by mass of the mix in percent as shown in Eq. 19.34.

$$DP = \frac{P_{0.75}}{P_{eac}} \tag{19.34}$$

Table 19.22 VFA Criteria

Traffic, Million ESALs	Design VFA, percent
< 0.3	70–80
< 1	65–78
< 3	65–78
< 10	65–75
< 30	65–75
< 100	65–75
≥100	65–75

where

DP = dust percentage

$P_{0.75}$ = aggregate content passing the 0.075 mm sieve, percent by mass of aggregate

P_{eac} = effective asphalt content, percent by total mass of mixture

P_{be} is given as:

$$P_{eac} = -(P_{ac} x G_{ac}) x \left(\frac{G_{ea} - G_{bam}}{G_{ea} - G_{bam}} \right) + P_{ac,estimated} \tag{19.35}$$

where

G_{ac} = specific gravity of the asphalt

G_{ea} = effective specific gravity of the aggregate

G_{bam} = bulk specific gravity of the aggregate

P_{ac} = asphalt content, percent by total mass of the paving mixture

The range of acceptable dust proportion is from 0.6 to 1.2. Based on these results the designer will accept one or more of the trial blends that meet the desired criteria.

Design Asphalt Binder Content. After the selection of the design aggregate structure from the trial blends, a minimum of two specimens with an asphalt content of ±0.5% of the estimated asphalt content and at least two with an asphalt content of +1.0% are compacted. These specimens are then tested using the same procedure described in the section on *select design aggregate structure*. The densification data at N_{ini}, N_{des}, and N_{max} are then used to evaluate the properties of the selected aggregate blend for each of the asphalt binder contents. The volumetric properties (air voids, VAM, and VFA) at the N_{des} are calculated for each asphalt content and plotted against the asphalt content. The asphalt binder content at 4 percent air void is selected as the design asphalt binder content. The mixture with that asphalt binder content is then checked to ascertain that it meets all the other mixture property criteria.

Moisture Sensitivity. Finally, the moisture sensitivity of the design mix is established by conducting the AASHTO designated test T-283 on specimens consisting of the selected aggregate blend and the design asphalt content. This test determines the indirect tensile strength of the specimens. A total of six specimens are prepared, three of which are considered as controlled and the other three as conditioned. The conditioned specimens are first subjected to partial vacuum saturation followed by an optional freeze cycle and then a 24 hr thaw cycle at 60°C. The tensile tests of the specimens are then determined. The moisture sensitivity is given as the ratio of the average tensile strength of the conditioned specimens to that of the control specimens. The minimum acceptable value is 0.80.

This section has briefly described the superpave mix design procedure as given by the Asphalt Institute.

SUMMARY

Asphaltic concrete pavement is becoming the most popular type of pavement used in highway construction. It is envisaged that the use of asphalt in highway construction will continue to increase, particularly with the additional knowledge that will be obtained from research conducted over the next few years. The engineering properties of different asphaltic materials is therefore of significant importance in highway engineering.

This chapter has presented information on the different types of asphaltic materials, their physical characteristics, and some of the tests usually conducted on these materials when used in the maintenance and/or construction of highway pavements. Two mix design methods for asphalt concrete (Marshall method and superpave) have also been provided so that the reader will understand the principles involved in determining the optimum mix for asphalt concrete. The chapter contains sufficient material on the subject to enable the reader to become familiar with the fundamental engineering properties and characteristics of those asphaltic materials used in pavement engineering.

PROBLEMS

19-1   Briefly describe the process of distillation by which asphalt cement is produced from crude petroleum. Also describe in detail how you would obtain asphaltic binders that can be used to coat highly siliceous aggregates.

19-2   Describe both the factors that influence the durability of asphaltic materials and the effect each factor has on the material.

19-3   You have been told to produce a rapid-curing cutback asphaltic material. What factors will you need to consider?

19-4   Results obtained from laboratory tests on a sample of rapid-curing asphalt cement (RC-250) are shown below. Determine whether the properties of this material meet the Asphalt Institute specifications for this type of material. If not, where do the differences lie?

Kinematic viscosity at 140°F (60°C) = 260 (centistokes)
Flash point (Tag. open-cup) = 75°F
Distillation Test
    Distillate percent by volume of total distillate to 680°F (360°C)
        To 437°F (225°C) = 35
        To 500°F (200°C) = 54
        To 600°F (316°C) = 75
Residue from distillation to 680°F (360°C) by volume = 64
Tests on residue from distillation
    Ductility at 77°F (25°C) = 95 (cm)
    Absolute viscosity at 140°F (60°C) = 750 poises
    Solubility (%) = 95

19-5   The specifications for an asphaltic concrete mixture include the following:

Course aggregates = 60%
Fine aggregates = 35%
Filler = 5%

The table below shows the results of a sieve analysis of available materials. Determine the proportion of different aggregates to obtain the required gradation.

Passing Sieve Designation	Retained on Sieve Designation	Percent by Weight		
		Coarse Aggregate	Fine Aggregate	Mineral Filler
¾ in. (19.0 mm)	½ in.	4	—	—
½ in. (12.5 mm)	¾ in.	36	—	—
⅜ in. (9.5 mm)	No. 4	40	—	—
No. 4 (4.75 mm)	No. 10	15	6	—
No. 10 (2.00 mm)	No. 40	5	32	—
No. 40 (0.425 mm)	No. 80	—	33	5
No. 80 (0.180 mm)	No. 200	—	29	40
No. 200 (0.075 mm)	—	—	—	55
Total		100	100	100

19-6   The table below shows the particle size distributions of two aggregates A and B, which are to be blended to produce an acceptable aggregate for use in manufacturing an asphalt concrete for highway pavement construction. If the required limits of particle size distribution for the mix are as shown in the table below, determine a suitable ratio for blending aggregates A and B to obtain an acceptable combined aggregate.

Sieve Size	Percent Passing by Weight		
	A	B	Required Mix
¾ in. (19mm)	100	98	96–100
⅜ in. (9.5mm)	80	76	65–80
No. 4 (4.25mm)	50	45	40–55
No. 10 (2.00mm)	43	33	35–40
No. 40 (0.425mm)	20	30	15–35
No. 200 (0.075mm)	4	8	5–8

19-7   The table below shows four different types of aggregates that will be used to produce a blended aggregate for use in the manufacture of asphaltic concrete. Determine the bulk specific gravity of the aggregate mix.

Material	Percent by Weight	Bulk Specific Gravity
A	35	2.58
B	40	2.65
C	15	2.60
D	10	2.55

19-8   If the specific gravity of the asphalt cement used in a sample of asphalt concrete mix is 1.01, the maximum specific gravity of the mix is 2.50, and the mix contains 6.5 percent by weight of asphalt cement, determine the effective specific gravity of the mixture.

19-9   The table below lists data used in obtaining a mix design for an asphaltic concrete paving mixture. If the maximum specific gravity of the mixture is 2.41 and the bulk specific gravity is 2.35, determine: (a) the bulk specific gravity of aggregates in the paving mixture, (b) the asphalt absorbed, (c) the effective asphalt content of the paving mixture, and (d) the percent voids in the mineral aggregate VMA.

Material	Specific Gravity	Mix Composition by Weight of Total Mix
Asphalt cement	1.02	6.40
Coarse aggregate	2.51	52.35
Fine aggregate	2.74	33.45
Mineral filler	2.69	7.80

19-10   For hot-mix, hot-laid asphaltic concrete mixtures, if the asphaltic content is specified as 5 to 7 percent, how is the optimum percentage determined?

19-11   The aggregate mix used for the design of an asphaltic concrete mixture consists of 42 percent coarse aggregates, 51 percent fine aggregates, and 7 percent mineral fillers. If the respective bulk specific gravities of these materials are 2.60, 2.71, and 2.69, and the effective specific gravity of the aggregates is 2.82, determine the optimum asphalt content as a percentage of the total mix if results obtained using the Marshall method are shown in the following table. The specific gravity of the asphalt is 1.02. (Use Table 19.8 for required specifications.)

Percent Asphalt	Weight of Specimen (g)		Stability (lb)	Flow (0.01 in.)
	in Air	in Water		
5.5	1325.3	785.6	1796	13
6.0	1330.1	793.3	1836	14
6.5	1336.2	800.8	1861	16
7.0	1342.0	804.5	1818	20
7.5	1347.5	805.1	1701	25

19-12   Determine the asphalt absorption of the optimum mix of Problem 19-11.

19-13   The latitude at the location where a rural high-speed road is to be constructed is 35°. The expected ESAL is 32×10^6. The seven-day average high air temperature is 53°C, and the low air temperature is −18°C. If the standard deviations for the high and low temperatures are ± 2°C and ± 1°C respectively, and the depth of the pavement is 155 mm, determine an appropriate asphalt binder for this project for a reliability of 98 percent.

19-14   An urban expressway is being designed for a congested area in Washington, D.C. It is expected that most of the time traffic will be moving at a slow rate. If the anticipated ESAL is 8×10^6, determine an appropriate asphalt binder for this project.

19-15   The table below shows properties of three trial aggregate blends that are to be evaluated so as to determine their suitability for use in a superpave mix. If the nominal maximum sieve of each aggregate blend is 19 mm, determine the initial trial asphalt content for each of the blends.

Property	Trial Blend 1	Trial Blend 2	Trial Blend 3
G_{bam}	2.698	2.696	2.711
G_{ea}	2.765	2.766	2.764

REFERENCES

Annual Book of ASTM Standards, Section 4, Construction, Vol. 04.03, *Road and Paving Materials; Pavement Management Technologies,* American Society for Testing and Materials, Philadelphia, Pa., 1996.

Mix Design Methods for Asphalt Concrete and Other Hot-Mix Types, 6th ed., Asphalt Institute Manual Series No. 2 (MS-2), Asphalt Institute, Lexington, Ky.

Sampling Asphalt Products for Specifications Compliance MS-18, Asphalt Institute, Lexington, Ky., 1985.

Specifications for Paving and Industrial Asphalts (SS2), Asphalt Institute, Lexington, Ky., 1993.

Standard Specifications for Transportation Materials and Methods of Sampling and Testing, 16th ed., American Association of State Highway and Transportation Officials, Washington, D.C., 2000.

Superpave Asphalt Binder Specification, Asphalt Institute, Superpave Series No. 1 (SP-1), Asphalt Institute, Lexington, Ky., 2000.

Superpave Mix Design, Asphalt Institute, Superpave Series No. 2 (SP-2), Asphalt Institute, Lexington, Ky., 2000.

CHAPTER 20

Design of Flexible Pavements

Highway pavements are divided into two main categories: rigid and flexible. The wearing surface of a rigid pavement is usually constructed of Portland cement concrete such that it acts like a beam over any irregularities in the underlying supporting material. The wearing surface of flexible pavements, on the other hand, is usually constructed of bituminous materials such that they remain in contact with the underlying material even when minor irregularities occur. Flexible pavements usually consist of a bituminous surface underlaid with a layer of granular material and a layer of a suitable mixture of coarse and fine materials. Traffic loads are transferred by the wearing surface to the underlying supporting materials through the interlocking of aggregates, the frictional effect of the granular materials, and the cohesion of the fine materials.

Flexible pavements are further divided into three subgroups: high type, intermediate type, and low type. High-type pavements have wearing surfaces that adequately support the expected traffic load without visible distress due to fatigue and are not susceptible to weather conditions. Intermediate-type pavements have wearing surfaces that range from surface treated to those with qualities just below that of high-type pavements. Low-type pavements are used mainly for low-cost roads and have wearing surfaces that range from untreated to loose natural materials to surface-treated earth.

This chapter deals with the design of high-type pavements, although the methodologies presented can also be used for some intermediate-type pavements.

STRUCTURAL COMPONENTS OF A FLEXIBLE PAVEMENT

Figure 20.1 shows the components of a flexible pavement: the subgrade or prepared roadbed, the subbase, the base, and the wearing surface. The performance of the pavement depends on the satisfactory performance of each component, which requires proper evaluation of the properties of each component separately.

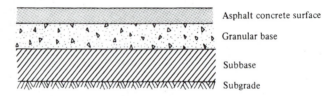

Figure 20.1 Schematic of a Flexible Pavement

Subgrade (Prepared Road Bed)

The subgrade is usually the natural material located along the horizontal alignment of the pavement and serves as the foundation of the pavement structure. The subgrade may also consist of a layer of selected borrow materials, well compacted to prescribed specifications as discussed in Chapter 18. It may be necessary to treat the subgrade material to achieve certain strength properties required for the type of pavement being constructed. This will be discussed later.

Subbase Course

Located immediately above the subgrade, the subbase component consists of material of a superior quality to that which generally is used for subgrade construction. The requirements for subbase materials are usually given in terms of the gradation, plastic characteristics, and strength, as discussed in Chapter 18. When the quality of the subgrade material meets the requirements of the subbase material, the subbase component may be omitted. In cases where suitable subbase material is not readily available, the available material can be treated with other materials to achieve the necessary properties. This process of treating soils to improve their engineering properties is known as *stabilization*.

Base Course

The base course lies immediately above the subbase. It is placed immediately above the subgrade if a subbase course is not used. This course usually consists of granular materials such as crushed stone, crushed or uncrushed slag, crushed or uncrushed gravel, and sand. The specifications for base course materials usually include stricter requirements than those for subbase materials, particularly with respect to their plasticity, gradation, and strength. Materials that do not have the required properties can be used as base materials if they are properly stabilized with Portland cement, asphalt, or lime. In some cases, high-quality base course materials may also be treated with asphalt or Portland cement to improve the stiffness characteristics of heavy-duty pavements.

Surface Course

The surface course is the upper course of the road pavement and is constructed immediately above the base course. The surface course in flexible pavements usually consists of a mixture of mineral aggregates and asphaltic materials. It should be capable of withstanding high tire pressures, resisting the abrasive forces due to traffic, providing a skid-resistant driving surface, and preventing the penetration of surface water into the underlying layers. The thickness of the wearing surface can vary from 3 in. to more than 6 in., depending on the expected traffic on the pavement. It was shown in Chapter 19 that the quality of the surface course of a flexible pavement depends on the mix design of the asphalt concrete used.

SOIL STABILIZATION

Soil stabilization is the treatment of natural soil to improve its engineering properties. Soil stabilization methods can be divided into two categories, namely, mechanical and chemical. *Mechanical stabilization* is the blending of different grades of soils to obtain a required grade. This type of stabilization was discussed in Chapter 19. *Chemical stabilization* is the blending of the natural soil with chemical agents. Several blending agents have been used to obtain different effects, as shown in Table 20.1. The most commonly used agents are Portland cement, asphalt binders, and lime. It is necessary at this point to define some of the terms commonly used in the field of soil stabilization—particularly when Portland cement, lime, or asphalt are used—to help the reader fully understand Table 20.1.

Cement-stabilized soil is a mixture of water, soil, and measured amounts of Portland cement, thoroughly mixed and compacted to a high density and then allowed to cure for a specific period, during which it is protected from loss of moisture.

Soil cement is a hardened material obtained by mechanically compacting a mixture of finely crushed soil, water, and a quantity of Portland cement that will make the mixture meet certain durability requirements.

Cement-modified soil is a semihardened or unhardened mixture of water, Portland cement, and finely crushed soil. This mixture has less cement than the soil-cement mixture.

Plastic soil cement is a hardened material obtained by mixing finely crushed soil, Portland cement, and a quantity of water, such that at the time of mixing and placing, a consistency similar to that of mortar is obtained.

Soil-lime is a mixture of lime, water, and fined-grained soil. If the soil contains silica and alumina, pozzolanic reaction occurs, resulting in the formation of a cementing-type material. Clay minerals, quartz, and feldspars are all possible sources of silica and alumina in typical fine-grained soils.

The stabilization process of each of the commonly used materials is briefly discussed. An in-depth discussion of soil stabilization is given in *Soil Stabilization in Pavement Structures*.

Cement Stabilization

Cement stabilization of soils usually involves the addition of 5 percent to 14 percent Portland cement by volume of the compacted mixture to the soil being stabilized. This type of stabilization is used mainly to obtain the required engineering properties of soils that are to be used as base course materials. Although the best results have been obtained when well-graded granular materials were stabilized with cement, the Portland Cement Association has indicated that nearly all types of soils can be stabilized with cement.

Table 20.1 Soil Types and Stabilization Methods Best Suited for Specific Applications

Purpose	Soil Type	Recommended Stabilization Methods
1. Subgrade Stabilization		
A. Improved load-carrying and stress-distributing characteristics	Coarse granular	SA, SC, MB, C
	Fine granular	SA, SC, MB, C
	Clays of low PI	C, SC, CMS, LMS, SL
	Clays of high PI	SL, LMS
B. Reduce frost susceptibility	Fine granular	CMS, SA, SC, LF
	Clays of low PI	CMS, SC, SL, CW, LMS
C. Waterproof and improve runoff	Clays of low PI	CMS, SA, CW, LMS, SL
D. Control shrinkage and swell	Clays of low PI	CMS, SC, CW, C, LMS, SL
	Clays of high PI	SL
E. Reduce resiliency	Clays of high PI	SL, LMS
	Elastic silts and clays	SC, CMS
2. Base Course Stabilization		
A. Improve substandard materials	Fine granular	SC, SA, LF, MB
	Clays of low PI	SC, SL
B. Improve load-carrying and stress-distributing characteristics	Coarse granular	SA, SC, MB, LF
	Fine granular	SC, SA, LF, MB
C. Reduce pumping	Fine granular	SC, SA, LF, MB membranes
3. Shoulders (unsurfaced)		
A. Improve load-carrying ability	All soils	See section 1A above; also MB
B. Improve durability	All soils	See section 1A above
C. Waterproof and improve runoff	Plastic soils	CMS, SL, CW, LMS
D. Control shrinkage and swell	Plastic soils	See section 1E above
4. Dust Palliative	Fine granular	CMS, CL, SA, oil, or bituminous surface spray
	Plastic soils	CL, CMS, SL, LMS
5. Ditch Lining	Fine granular	PSC, CS, SA
	Plastic soils	PSC, CS
6. Patching and Reconstruction	Granular soils	SC, SA, LF, MB

C	= Compaction	LMS =	Lime-modified soil
CMS =	Cement modified soil	MB =	Mechanical blending
CL =	Chlorides	PSC =	Plastic soil cement
CS =	Chemical solidifiers	SA =	Soil asphalt
CW =	Chemical waterproofers	SC =	Soil cement
LF =	Lime fly ash	SL =	Soil lime

SOURCE: Adapted from *Materials Testing Manual*, AFM 88-51, Department of the U.S. Air Force, Washington, D.C., February 1966.

The procedure for stabilizing soils with cement involves

- Pulverizing the soil
- Mixing the required quantity of cement with the pulverized soil
- Compacting the soil cement mixture
- Curing the compacted layer

Pulverizing the Soil

The soil to be stabilized should first be thoroughly pulverized to facilitate the mixing of the cement and soil. When the existing material on the roadway is to be used, the roadway is scarified to the required depth, using a scarifier attached to a grader. When the material to be stabilized is imported to the site, the soil is evenly spread to the required depth above the subgrade and then pulverized by using rotary mixers. Sieve analysis is conducted on the soil during pulverization, and pulverization is continued (except for gravel) until all material can pass through a 1 in. sieve and not more than 20 percent is retained on the No. 4 sieve. A suitable moisture content must be maintained during pulverization, which may require air drying of wet soils or the addition of water to dry soils.

Mixing of Soil and Cement

The first step in the mixing process is to determine the amount of cement to be added to the soil by conducting laboratory experiments to determine the minimum quantity of cement required. The required tests are described by the Portland Cement Association. These tests generally include the classification of the soils to be stabilized, using the American Association of State Highway and Transportation Officials' (AASHTO) classification system, from which the range of cement content can be deduced based on past experience. Samples of soil–cement mixtures containing different quantities of cement within the range deduced are then prepared, and the American Society for Testing and Materials moisture–density relationship test (ASTM Designation D558) is conducted on these samples. The results are used to determine the amount of cement to be used for preparing test specimens that are then used to conduct durability tests. The durability tests are described in "Standard Methods of Wetting-and-Drying Tests of Compacted Soil–Cement Mixtures," ASTM Designation D559, and "Standard Methods of Freezing-and-Thawing Tests of Compacted Soil–Cement Mixtures," ASTM Designation D560. The quantity of cement required to achieve satisfactory stabilization is obtained from the results of these tests. Table 20.2 gives suggested cement contents for different types of soils.

Having determined the quantity of cement required, the mixing can be carried out either on the road, *road mixing,* or in a central plant, *plant mixing.* Plant mixing is primarily used when the material to be stabilized is borrowed from another site.

With road mixing the cement is usually delivered in bulk in dump or hopper trucks and spread by a spreader box or some other type of equipment that will provide a uniform amount over the pulverized soil. Enough water is then added to achieve a moisture content that is 1 percent or 2 percent higher than the optimum required for compaction, and the soil, cement, and water are properly blended to obtain a uniform moisture of soil cement. Blending at a moisture content slightly higher than the compaction optimum moisture content allows for loss of water by evaporation during the mixing process.

Table 20.2 Normal Range of Cement Requirements for B- and C-Horizon Soils*

AASHTO Soil Group	Cement (percentage by weight of soil)	Cement (pounds per cubic foot of compacted soil cement)	Cement (kilograms per cubic meter of compacted soil cement)
A-1-a	3–5	5–7	80–110
A-1-b	5–8	7–8	110–130
A-2-4			
A-2-5			
A-2-6	5–9	7–9	110–140
A-2-7			
A-3	7–11	8–11	130–180
A-4	7–12	8–11	130–180
A-5	8–13	8–11	130–180
A-6	9–15	9–13	140–210
A-7	10–16	9–13	140–210

*A-horizon soils (topsoils) may contain organic or other material detrimental to cement reaction and thus require higher cement factors. For dark gray to gray A-horizon soils, increase the above cement contents 4 percentage points (4 lb/cu ft [60 kg/m^3] of compacted soil-cement); for black A-horizon soils, 6 percentage points (6 lb/cu ft [100 kg/m^3] of compacted soil-cement).

In plant mixing, the borrowed soil is properly pulverized and mixed with the cement and water in either a continuous or batch mixer. It is not necessary to add more than the quantity of water that will bring the moisture content to the optimum for compaction, since moisture loss in plant mixing is relatively small. The soil–cement mixture is then delivered to the site in trucks and uniformly spread. The main advantage of plant mixing over road mixing is that the proportioning of cement, water, and soil can be easily controlled.

Compacting the Soil–Cement Mixture

It is essential that compaction of the mixture be carried out before the mixture begins to set. Specifications given by the Federal Highway Administration (FHWA) stipulate that the length of time between the addition of water and the compaction of the mix at the site should not be greater than 2 hours for plant mixing and not more than 1 hour for road mixing. The soil–cement mixture is initially compacted with sheepsfoot rollers, or with pneumatic-tired rollers when the soil (for example, very sandy soil) cannot be effectively compacted with a sheepsfoot roller. The uppermost layer of 1 to 2 in. depth usually is compacted with a pneumatic-tired roller, and the final surface compaction is carried out with a smooth-wheel roller.

Curing the Compacted Layer

Moisture loss in the compacted layer must be prevented before setting is completed because the moisture is required for the hydration process. This is achieved by applying a thin layer of bituminous material such as RC-250 or MC-250. Tars and emulsions can also be used.

Bituminous Stabilization

Bituminous stabilization is carried out to achieve one or both of the following:

- Waterproofing of natural materials
- Binding of natural materials

Waterproofing the natural material through bituminous stabilization aids in maintaining the water content at a required level by providing a membrane that impedes the penetration of water, thereby reducing the effect of any surface water that may enter the soil when it is used as a base course. In addition, surface water is prevented from seeping into the subgrade, which protects the subgrade from failing due to increase in moisture content.

Binding improves the durability characteristics of the natural soil by providing an adhesive characteristic, whereby the soil particles adhere to each other, increasing cohesion.

Several types of soils can be stabilized with bituminous materials, although it is generally required that less than 25 percent of the material passes the No. 200 sieve. This is necessary because the smaller soil particles tend to have extremely large surface areas per unit volume and require a large amount of bituminous material for the soil surfaces to be adequately coated. It is also necessary to use soils that have a plasticity index (PI) of less than 10 because difficulty may be encountered in mixing soils with a high PI, which may result in the plastic fines swelling on contact with water and thereby losing strength.

The mixing of the soil and bituminous materials also can be done in a central or movable plant (plant mixing) or at the roadside (road mixing). In plant mixing, the desired amounts of water and bituminous material are automatically fed into the mixing hoppers, whereas in road mixing, the water and bituminous material are measured and applied separately using a pressure distributor. The materials are then thoroughly mixed in the plant when plant mixing is used, or by rotary speed mixers or suitable alternative equipment when road mixing is used.

The material is then evenly spread in layers of uniform thickness, usually not greater than 6 in. and not less than 2 in. Each layer is properly compacted until the required density is obtained using a sheepsfoot roller or a pneumatic-tired roller. The mixture must be completely aerated before compaction to ensure the removal of all volatile materials.

Lime Stabilization

Lime stabilization is one of the oldest processes of improving the engineering properties of soils and can be used for stabilizing both base and subbase materials. In general, the oxides and hydroxides of calcium and magnesium are considered as lime, but the materials most commonly used for lime stabilization are calcium hydroxide $Ca(OH)_2$ and dolomite

$Ca(OH)_2 + MgO$. The dolomite, however, should not have more than 36 percent by weight of magnesium oxide (MgO) to be acceptable as a stabilizing agent.

Clayey materials are most suitable for lime stabilization, but these materials should also have PI values less than 10 for the lime stabilization to be most effective. When lime is added to fine-grained soil, cation exchange takes place, with the calcium and magnesium in the lime replacing the sodium and potassium in the soil. The tendency to swell as a result of increase in moisture content is therefore immediately reduced. The PI value of the soil is also reduced. Pozzolanic reaction may also occur in some clays, resulting in the formation of cementing agents that increase the strength of the soil. When silica or alumina is present in the soil, a significant increase in strength may be observed over a long period of time. An additional effect is that lime causes flocculation of the fine particles, thereby increasing the effective grain size of the soil.

The percentage of lime used for any project depends on the type of soil being stabilized. The determination of the quantity of lime is usually based on an analysis of the effect that different lime percentages have on the reduction of plasticity and the increase in strength on the soil. The PI is most commonly used for testing the effect on plasticity, whereas the unconfined compression test, the Hveem Stabilometer test, or the California bearing-ratio (CBR) test can be used to test for the effect on strength. However, most fine-grained soil can be effectively stabilized with 3 percent to 10 percent of lime, based on the dry weight of the soil.

GENERAL PRINCIPLES OF FLEXIBLE PAVEMENT DESIGN

In the design of flexible pavements, the pavement structure is usually considered as a multilayered elastic system, with the material in each layer characterized by certain physical properties that may include the modulus of elasticity, the resilient modulus, and the Poisson ratio. It is usually assumed that the subgrade layer is infinite in both the horizontal and vertical directions, whereas the other layers are finite in the vertical direction and infinite in the horizontal direction. The application of a wheel load causes a stress distribution, which can be represented as shown in Figure 20.2. The maximum vertical stresses are compressive and occur directly under the wheel load. These decrease with increase in depth from the surface. The maximum horizontal stresses also occur directly under the wheel load but can be either tensile or compressive as shown in Figure 20.2(c). When the load and pavement thickness are within certain ranges, horizontal compressive stresses will occur above the neutral axis, whereas horizontal tensile stresses will occur below the neutral axis. The temperature distribution within the pavement structure, as shown in Figure 20.2(d), will also have an effect on the magnitude of the stresses. The design of the pavement is therefore generally based on strain criteria that limit both the horizontal and vertical strains below those that will cause excessive cracking and excessive permanent deformation. These criteria are considered in terms of repeated load applications, because the accumulated repetitions of traffic loads are of significant importance to the development of cracks and permanent deformation of the pavement.

The availability of highly sophisticated computerized solutions for multilayered systems, coupled with recent advances in materials evaluation, has led to the development of several design methods that are based wholly or partly on theoretical analysis. The more commonly used design methods are the Asphalt Institute method, the American

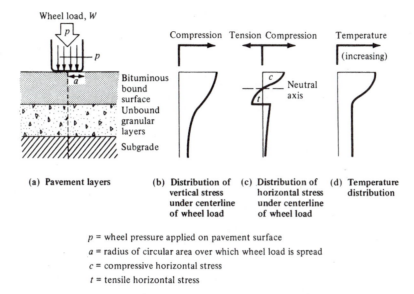

Wheel load, W

Compression   Tension   Compression

Temperature
(increasing)

p

Bituminous bound surface

Unbound granular layers

Subgrade

c
t
Neutral axis

(a) Pavement layers

(b) Distribution of vertical stress under centerline of wheel load

(c) Distribution of horizontal stress under centerline of wheel load

(d) Temperature distribution

p = wheel pressure applied on pavement surface
a = radius of circular area over which wheel load is spread
c = compressive horizontal stress
t = tensile horizontal stress

Figure 20.2 Typical Stress and Temperature Distributions in a Flexible Pavement Under a Wheel Load

Association of State Highway and Transportation Officials (AASHTO) method, and the California method. These are presented in the following sections.

Asphalt Institute Design Method

In the Asphalt Institute design method, the pavement is represented as a multilayered elastic system. The wheel load W is assumed to be applied through the tire as a uniform vertical pressure p_0, which is then spread by the different components of the pavement structure and eventually applied on the subgrade as a much lower stress p_1. This is shown in Figure 20.3. Experience, established theory, and test data are then used to evaluate two specific stress–strain conditions. The first, shown in Figure 20.3(b), is the general way in which the stress p_0 is reduced to p_1 within the depth of the pavement structure and the second, shown in Figure 20.4, is the tensile and compressive stresses and strains imposed on the asphalt due to the deflection caused by the wheel load. Thickness design charts were developed, based on criteria for maximum tensile strains at the bottom of the asphalt layer and maximum vertical compressive strains at the top of the subgrade layer. The procedure involves the use of a computer program, DAMA, but the charts presented here can be used without computers.

Design Procedure

The principle adopted in the design procedure is to determine the minimum thickness of the asphalt layer that will adequately withstand the stresses that develop for the two strain criteria discussed earlier—that is, both the vertical compressive strain at the surface of the subgrade and the horizontal tensile strain at the bottom of the asphalt layer. Design charts

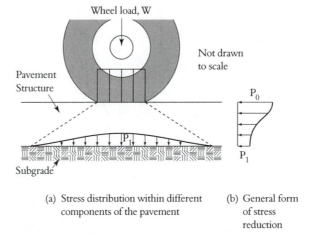

(a) Stress distribution within different   (b) General form
      components of the pavement   of stress
                                  reduction

Figure 20.3   Spread of Wheel Load Pressure Through Pavement Structure

SOURCE: *Thickness Design—Asphalt Pavements for Highways and Streets,* Manual Series No. 1, Asphalt Institute, Lexington, Ky., February 1991.

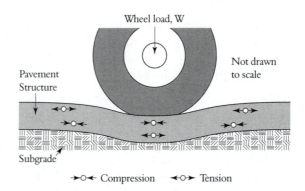

➤○◄ Compression   ◄○➤ Tension

Figure 20.4   Schematic of Tensile and Compressive Stresses in Pavement Structure

SOURCE: *Thickness Design—Asphalt Pavements for Highways and Streets,* Manual Series No. 1, Asphalt Institute, Lexington, Ky., February 1991.

have been prepared for a range of traffic loads. This range is usually adequate for normal traffic volumes encountered in practice. However, when this range is exceeded, the computer version should be used.

The procedure consists of five main steps.

1. Select or determine input data.
2. Select surface and base materials.

3. Determine minimum thickness required for input data.
4. Evaluate feasibility of staged construction and prepare stage construction plan, if necessary.
5. Carry out economic analyses of alternative designs and select the best design.

Step 1.   Determine Design Inputs. The design inputs in this method are traffic characteristics, subgrade engineering properties, and subbase and base engineering properties.

Traffic Characteristics.   The traffic characteristics are determined in terms of the number of repetitions of an 18,000-lb (80 kilonewtons (kN)) single-axle load applied to the pavement on two sets of dual tires. This is usually referred to as the *equivalent single-axle load* (ESAL). The dual tires are represented as two circular plates, each 4.51 in. radius, spaced 13.57 in. apart. This representation corresponds to a contact pressure of 70 lb/in. The use of an 18,000-lb axle load is based on the results of experiments that have shown that the effect of any load on the performance of a pavement can be represented in terms of the number of single applications of an 18,000-lb single axle. A series of equivalency factors used in this method for axle loads are given in Table 20.3.

Table 20.3 Load Equivalency Factors

Gross Axle Load		Load Equivalency Factors		
kN	*lb*	*Single Axles*	*Tandem Axles*	*Tridem Axles*
4.45	1,000	0.00002		
8.9	2,000	0.00018		
17.8	4,000	0.00209	0.0003	
26.7	6,000	0.01043	0.001	0.0003
35.6	8,000	0.0343	0.003	0.001
44.5	10,000	0.0877	0.007	0.002
53.4	12,000	0.189	0.014	0.003
62.3	14,000	0.360	0.027	0.006
71.2	16,000	0.623	0.047	0.011
80.0	18,000	1.000	0.077	0.017
89.0	20,000	1.51	0.121	0.027
97.9	22,000	2.18	0.180	0.040
106.8	24,000	3.03	0.260	0.057
115.6	26,000	4.09	0.364	0.080
124.5	28,000	5.39	0.495	0.109
133.4	30,000	6.97	0.658	0.145
142.3	32,000	8.88	0.857	0.191
151.2	34,000	11.18	1.095	0.246
160.1	36,000	13.93	1.39	0.313
169.0	38,000	17.20	1.70	0.393
178.0	40,000	21.08	2.08	0.487
187.0	42,000	25.64	2.51	0.597
195.7	44,000	31.00	3.00	0.723

Continued

Table 20.3 Load Equivalency Factors (*continued*)

| Gross Axle Load | | Load Equivalency Factors | | |
kN	lb	Single Axles	Tandem Axles	Tridem Axles
204.5	46,000	37.24	3.55	0.868
213.5	48,000	44.50	4.17	1.033
222.4	50,000	52.88	4.86	1.22
231.3	52,000		5.63	1.43
240.2	54,000		6.47	1.66
249.0	56,000		7.41	1.91
258.0	58,000		8.45	2.20
267.0	60,000		9.59	2.51
275.8	62,000		10.84	2.85
284.5	64,000		12.22	3.22
293.5	66,000		13.73	3.62
302.5	68,000		15.38	4.05
311.5	70,000		17.19	4.52
320.0	72,000		19.16	5.03
329.0	74,000		21.32	5.57
338.0	76,000		23.66	6.15
347.0	78,000		26.22	6.78
356.0	80,000		29.0	7.45
364.7	82,000		32.0	8.2
373.6	84,000		35.3	8.9
382.5	86,000		38.8	9.8
391.4	88,000		42.6	10.6
400.3	90,000		46.8	11.6

Note: kN converted to lb are within 0.1 percent of lb shown.
SOURCE: *Thickness Design—Asphalt Pavements for Highways and Streets,* Manual Series No. 1, The Asphalt Institute, Lexington, Ky., February 1991.

To determine the ESAL, the number of different types of vehicles such as cars, buses, single-unit trucks, and multiple-unit trucks expected to use the facility during its lifetime must be known. The distribution of the different types of vehicles expected to use the proposed highway can be obtained from results of classification counts that are taken by state highway agencies at regular intervals. In cases where these data are not available, estimates can be made from Table 20.4, which gives representative values for the United States. When the axle load of each vehicle type is known, these can then be converted to equivalent 18,000-lb loads using the equivalency factors given in Table 20.3. The equivalent 18,000-lb load can also be determined from the vehicle type, if the axle load is unknown, by using a truck factor for that vehicle type. The truck factor is defined as the number of 18,000-lb single-load applications caused by a single passage of a vehicle. These have been determined for each class of vehicle from the expression

$$\text{truck factor} = \frac{\sum \left(\text{number of axles} \times \text{load equivalency factor}\right)}{\text{number of vehicles}}$$

Table 20.4 Distribution of Trucks on Different Classes of Highways—United States*

| Truck Class | Rural Systems | | | | | | Urban Systems | | | | |
	Interstate	Other Principal	Minor Arterial	Collectors Major	Collectors Minor	Range	Interstate	Other Freeways	Other Principal	Minor Arterial	Collectors	Range
Single-unit trucks												
2-axle, 4-tire	43	60	71	73	80	43–80	52	66	67	84	86	52–86
2-axle, 6-tire	8	10	11	10	10	8–11	12	12	15	9	11	9–15
3-axle or more	2	3	4	4	2	2–4	2	4	3	2	<1	<1–4
All single-units	53	73	86	87	92	53–92	66	82	85	95	97	66–97
Multiple-unit trucks												
4-axle or less	5	3	3	2	2	2–5	5	5	3	2	1	1–5
5-axle**	41	23	11	10	6	6–41	28	13	12	3	2	2–28
6-axle or more**	1	1	<1	1	<1	<1–1	1	<1	<1	<1	<1	<1–1
All multiple units	47	27	14	13	8	8–47	34	18	15	5	3	3–34
All trucks	100	100	100	100	100		100	100	100	100	100	

* Compiled from data supplied by the Highway Statistics Division, Federal Highway Administration.
**Including full-trailer combinations in some states.
SOURCE: *Thickness Design—Asphalt Pavements for Highways and Streets,* Manual Series No. 1, The Asphalt Institute, Lexington, Ky., February 1991.

Table 20.5 gives values of truck factors for different classes of vehicles. However, it is advisable, when truck factors are to be used, to collect data on axle loads for the different types of vehicles expected to use the proposed highway and to determine realistic values of truck factors from that data.

The total ESAL applied on the highway during its design period can be determined only after the design period and traffic growth factors are known. The design period is the number of years the pavement will effectively continue to carry the traffic load without requiring an overlay. Flexible highway pavements are usually designed for a 20-year period. Since traffic volume does not remain constant over the design period of the pavement, it is essential that the rate of growth be determined and applied when calculating the total ESAL. Annual growth rates can be obtained from regional planning agencies or from state highway departments. These are usually based on traffic volume counts over several years. It is also advisable to determine annual growth rates separately for trucks and passenger vehicles since these may be significantly different in some cases. The overall growth rate in the United States is between 3 percent and 5 percent per year, although growth rates of up to 10 percent per year have been suggested for some interstate highways. Table 20.6 shows growth factors (G_{jt}) for different growth rates (j) and design periods (t), which can be used to determine the total ESAL over the design period.

The portion of the total ESAL acting on the design lane (f_d) is used in the determination of pavement thickness. Either lane of a two-lane highway can be considered as the design lane, whereas for multilane highways, the outside lane is considered. The identification of the design lane is important because in some cases more trucks will travel in one direction than in the other, or trucks may travel heavily loaded in one direction and empty in the other direction. Thus, it is necessary to determine the relevant proportion of trucks on the design lane. When data are not available to make this determination, percentages given in Table 20.7 can be used.

A general equation for the accumulated ESAL for each category of axle load is obtained as

$$\text{ESAL}_i = f_d \times G_{jt} \times \text{AADT}_i \times 365 \times N_i \times F_{Ei} \tag{20.1}$$

where

ESAL_i = equivalent accumulated 18,000-lb (80 kN) single-axle load for the axle category i

f_d = design lane factor

G_{jt} = growth factor for a given growth rate j and design period t

AADT_i = first year annual average daily traffic for axle category i

N_i = number of axles on each vehicle in category i

F_{Ei} = load equivalency factor for axle category i

When truck factors are used, the ESAL for each category of truck is given as:

$$\text{ESAL}_i = \text{AADT}_i \times 365 \times f_i \times G_{jt} \times f_d \tag{20.2}$$

and the accumulated ESAL for all categories of axle loads is

Table 20.5 Distribution of Truck Factors (TF) for Different Classes of Highways and Vehicles—United States[*]

Truck Factors

Vehicle Type	Rural Systems						Urban Systems					
	Interstate	Other Principal	Minor Arterial	Collectors Major	Minor	Range	Interstate	Other Freeways	Other Principal	Minor Arterial	Collectors	Range
Single-unit trucks												
2-axle, 4-tire	0.003	0.003	0.003	0.017	0.003	0.003–0.017	0.002	0.015	0.002	0.006	—	0.006–0.015
2-axle, 6-tire	0.21	0.25	0.28	0.41	0.19	0.19–0.41	0.17	0.13	0.24	0.23	0.13	0.13–0.24
3-axle or more	0.61	0.86	1.06	1.26	0.45	0.45–1.26	0.61	0.74	1.02	0.76	0.72	0.61–1.02
All single units	0.06	0.08	0.08	0.12	0.03	0.03–0.12	0.05	0.06	0.09	0.04	0.16	0.04–0.16
Tractor-semitrailers												
4-axle or less	0.62	0.92	0.62	0.37	0.91	0.37–0.91	0.98	0.48	0.71	0.46	0.40	0.40–0.98
5-axle[**]	1.09	1.25	1.05	1.67	1.11	1.05–1.67	1.07	1.17	0.97	0.77	0.63	0.63–1.17
6-axle or more[**]	1.23	1.54	1.04	2.21	1.35	1.04–2.21	1.05	1.19	0.90	0.64	—	0.64–1.19
All multiple units	1.04	1.21	0.97	1.52	1.08	0.97–1.52	1.05	0.96	0.91	0.67	0.53	0.53–1.05
All trucks	0.52	0.38	0.21	0.30	0.12	0.12–0.52	0.39	0.23	0.21	0.07	0.24	0.07–0.39

Note: Compiled from data supplied by the Highway Statistics Division, Federal Highway Administration.
*Including full-trailer combinations in some states.
**For values to be used when the number of heavy trucks is low, see original source.
SOURCE: *Thickness Design—Asphalt Pavements for Highways and Streets*, Manual Series No. 1, The Asphalt Institute, Lexington, Ky., February 1991.

Table 20.6 Growth Factors

Design Period, Years (n)	No Growth	Annual Growth Rate, Percent (r)						
		2	4	5	6	7	8	10
1	1.0	1.0	1.0	1.0	1.0	1.0	1.0	1.0
2	2.0	2.02	2.04	2.05	2.06	2.07	2.08	2.10
3	3.0	3.06	3.12	3.15	3.18	3.21	3.25	3.31
4	4.0	4.12	4.25	4.31	4.37	4.44	4.51	4.64
5	5.0	5.20	5.42	5.53	5.64	5.75	5.87	6.11
6	6.0	6.31	6.63	6.80	6.98	7.15	7.34	7.72
7	7.0	7.43	7.90	8.14	8.39	8.65	8.92	9.49
8	8.0	8.58	9.21	9.55	9.90	10.26	10.64	11.44
9	9.0	9.75	10.58	11.03	11.49	11.98	12.49	13.58
10	10.0	10.95	12.01	12.58	13.18	13.82	14.49	15.94
11	11.0	12.17	13.49	14.21	14.97	15.78	16.65	18.53
12	12.0	13.41	15.03	15.92	16.87	17.89	18.98	21.38
13	13.0	14.68	16.63	17.71	18.88	20.14	21.50	24.52
14	14.0	15.97	18.29	19.16	21.01	22.55	24.21	27.97
15	15.0	17.29	20.02	21.58	23.28	25.13	27.15	31.77
16	16.0	18.64	21.82	23.66	25.67	27.89	30.32	35.95
17	17.0	20.01	23.70	25.84	28.21	30.84	33.75	40.55
18	18.0	21.41	25.65	28.13	30.91	34.00	37.45	45.60
19	19.0	22.84	27.67	30.54	33.76	37.38	41.45	51.16
20	20.0	24.30	29.78	33.06	36.79	41.00	45.76	57.28
25	25.0	32.03	41.65	47.73	54.86	63.25	73.11	98.35
30	30.0	40.57	56.08	66.44	79.06	94.46	113.28	164.49
35	35.0	49.99	73.65	90.32	111.43	138.24	172.32	271.02

Note: Factor = $[(1 + r)^n - 1]/r$, where $r = \frac{\text{rate}}{100}$ and is not zero. If annual growth is zero, growth factor = design period.
SOURCE: *Thickness Design—Asphalt Pavements for Highways and Streets,* Manual Series No. 1, The Asphalt Institute, Lexington, Ky., February 1991.

Table 20.7 Percentage of Total Truck Traffic on Design Lane

Number of Traffic Lanes (Two Directions)	Percentage of Trucks in Design Lane
2	50
4	45 (35–48)★
6 or more	40 (25–48)★

★Probable range.
SOURCE: Adapted from *Thickness Design—Asphalt Pavements for Highways and Streets,* Manual Series No. 1, The Asphalt Institute, Lexington, Ky., February 1991.

$$\text{ESAL} = \sum_{i=1}^{n} [\text{ESAL}_i] \tag{20.3}$$

where

ESAL_i = equivalent accumulated 18,000-lb axle load for truck category i

AADT_i = first year annual average daily traffic for vehicles in truck category i

f_i = truck factor for vehicles in truck category i

G_{jt} = growth factor for a given growth rate j and design period t

f_d = design lane factor

ESAL = equivalent accumulated 18,000-lb axle loads for all vehicles

n = number of truck categories

The procedure for determining the design ESAL is demonstrated in Examples 20.1 and 20.2.

Example 20.1 Computing Accumulated Equivalent Single-Axle Load for a Proposed Eight-Lane Highway Using Load Equivalency Factors

An eight-lane divided highway is to be constructed on a new alignment. Traffic volume forecasts indicate that the average annual daily traffic (AADT) in both directions during the first year of operation will be 12,000, with the following vehicle mix and axle loads.

Passenger cars (1000 lb/axle) = 50 percent

2-axle single-unit trucks (6000 lb/axle) = 33 percent

3-axle single-unit trucks (10,000 lb/axle) = 17 percent

The vehicle mix is expected to remain the same throughout the design life of the pavement. If the expected annual traffic growth rate is 4 percent for all vehicles, determine the design ESAL, given a design period of 20 years.

The following data apply.

Growth factor = 29.78 (from Table 20.6)

Percent truck volume on design lane = 45 (assumed, from Table 20.7)

Load equivalency factors (from Table 20.3)

Passenger cars (1000 lb/axle) = 0.00002 (negligible)

2-axle single-unit trucks (6000 lb/axle) = 0.01043

3-axle single-unit trucks (10,000 lb/axle) = 0.0877

Solution: The ESAL for each class of vehicle is computed from Eq. 20.1,

$$\text{ESAL} = f_d \times G_{jt} \times \text{AADT} \times 365 \times N_i \times F_{Ei}$$

$$\text{2-axle single-unit trucks} = 0.45 \times 29.78 \times 12{,}000 \times 0.33 \times 365 \times 2 \times 0.01043$$

$$= 0.4041 \times 10^6$$

$$\text{3-axle single-unit trucks} = 0.45 \times 29.78 \times 12{,}000 \times 0.17 \times 365 \times 3 \times 0.0877$$

$$= 2.6253 \times 10^6$$

Thus,

$$\text{total ESAL} = 3.0294 \times 10^6$$

It can be seen that the contribution of passenger cars to the ESAL is negligible. Passenger cars are therefore omitted when computing ESAL values. This example illustrates the conversion of axle loads to ESAL, using axle load equivalency factors.

Example 20.2   Computing Accumulated Equivalent Single-Axle Load for a Proposed Two-Lane Highway Using Truck Factors

The projected vehicle mix for a proposed two-lane major collector rural highway during its first year of operation is given below. If the first year AADT will be 3000, the annual growth rate is 5 percent, and the design period is 20 yr, determine the accumulated ESAL, using the truck factors given in Table 20.5.

    Passenger cars = 66 percent

    Single-unit trucks

        2-axle, 4-tire = 18 percent

        2-axle, 6-tire = 8 percent

        3-axle or more = 4 percent

    Tractor-semitrailers and combination

        4 axle or less = 4 percent

Solution: Since the axle loads are not given, the truck factor is used to compute the respective ESALs. Using Eq. 20.2,

$$\text{ESAL}_i = \text{AADT}_i \times 365 \times f_i \times G_{jt} \times f_d$$

The solution of this problem is given in Table 20.8, which also demonstrates a tabular format for determining ESAL. Passenger vehicles are not considered in the calculations since their contribution to the ESAL is negligible.

Table 20.8 Tabular Solution of Example 20.2

Vehicle Type	Number of Vehicles During 1st Year 1	Truck Factor (from Table 20.5) 2	Growth Factor for 5% Annual Growth Rate (from Table 20.6) 3	ESAL $(1 \times 2 \times 3)$ 4
Single-Unit Trucks				
2-axle, 4-tire	98,550	0.017	33.06	55,387
2-axle, 6-tire	43,800	0.41	33.06	593,691
3-axle or more	21,900	1.26	33.06	912,258
Tractor-Semi-trailers and Combinations				
4-axle or less	21,900	0.37	33.06	267,885
Total				1,829,221

Note: Passenger vehicles are not considered here because their effect is negligible. Values under column 1 are obtained from the first year AADT and the design lane factor of 0.5. For example, for 2-axle, 6-tire single-unit trucks, the number of vehicles during the first year = $3000 \times 365 \times 0.08 \times 0.5 = 43,800$.

The use of load equivalency factors for facilities such as residential streets, parkways, and parking lots, where the traffic is primarily automobiles with only occasional trucks, will result in low ESALs and therefore thin pavements that may not be capable of withstanding the occasional heavy traffic or environmental effects. Thus, a reasonably accurate estimate of the occasional truck traffic for residential streets and parking lots should be made and included in the determination of the design ESAL. In cases where this is not possible, the minimum pavement thickness recommended by the Asphalt Institute should be used.

In the design of shoulders, the Asphalt Institute recommends that the principles and procedures adopted for the main travel lanes be utilized. In order to provide minimum protection against the damaging effects of occasional heavy vehicles, a minimum of 2 percent of the design ESAL for the design lane should be used as the design ESAL for the shoulders.

Subgrade Engineering Properties. As stated earlier, the subgrade consists of either the natural soil or the soil that has been imported to form an embankment. This borrowed soil may be in its natural state or may be improved by stabilization. Stabilization of the natural soil existing on the alignment is not usually done, except when it is found that the material in its natural state cannot support heavy construction equipment. Stabilization of in situ subgrade material is therefore usually carried out to provide an adequate support platform for heavy construction equipment and is not considered in the design of the pavement.

The main engineering property required for the subgrade is its resilient modulus, which gives the resilient characteristic of the soil when it is repeatedly loaded with an axial load. It is determined in the laboratory by loading specially prepared samples of the soil with a deviator stress of fixed magnitude, frequency, and load duration while the specimen is triaxially loaded in a triaxial chamber. The method of conducting this test is described in detail in the Asphalt Institute's *Soils Manual for the Design of Asphalt Pavement Structures.* To facilitate the use of the more direct CBR and Hveem Stabilometer tests described in Chapter 18, the Asphalt Institute has determined conversion factors that can be used to convert the CBR and *R* values to the resilient modulus values. These are given as

$$M_r(MP_a) = 10.342 \text{ CBR} \qquad (20.4)$$

$$M_r(\text{lb/in.}^2) = 1500 \text{ CBR}$$

or

$$M_r(MP_a) = 7.963 + 3.826 \ (R \text{ value}) \qquad (20.5)$$

$$M_r(\text{lb/in.}^2) = 1155 + 555 \ (R \text{ value})$$

where M_r is the equivalent resilient modulus.

The above conversion factors should be used only for materials that can be classified under the unified classification system as CL, CH, ML, SC, SM, or SP, or when the resilient modulus is less than 30,000 lb/in.^{2}. The Asphalt Institute recommends direct measurement for materials with higher values.

Subbase and Base Engineering Properties.   The material used for subbase or base courses must meet certain requirements, which are given in terms of the , liquid limit, PI, particle size distribution (maximum percentage passing the No. 200 sieve), and minimum sand equivalent. The sand equivalent test is a rapid field test that shows the relative proportions of fine dust or claylike material in soils or graded aggregates. A detailed description of this test is given in *Soils Manual for the Design of Asphalt Pavement Structures.* Table 20.9 gives

Table 20.9 Untreated Aggregate Base and Subbase Quality Requirements

	Test Requirements	
Test	*Subbase*	*Base*
CBR, minimum	20	80
or		
R value, minimum	55	78
Liquid limit, maximum	25	25
Plasticity index, maximum, or	6	NP
Sand equivalent, minimum	25	35
Passing No. 200 sieve, maximum	12	7

SOURCE: Adapted from *Thickness Design—Asphalt Pavements for Highways and Streets,* Manual Series No. 1, The Asphalt Institute, Lexington, Ky., February 1991.

the requirements for those engineering properties for soils that can be used as base or subbase in this method.

Step 2. Select Surface and Base Materials. The designer is free to select either an asphalt concrete surface or an emulsified asphalt surface, along with an asphalt concrete base, an emulsified asphalt base, or an untreated aggregate base and subbase for the underlying layers. This, of course, will depend on the material that is economically available.

However, the Asphalt Institute recommends certain grades of asphalt cement that should be used for different temperature conditions, as shown in Table 20.10. These temperature conditions represent the mean annual air temperature (MAAT) for the area where the pavement is to be constructed. The grade of asphalt used should be selected primarily on the basis of its ability to satisfactorily coat the aggregates at the given temperatures.

Step 3. Determine Minimum Thickness Requirements. The minimum thickness required for the design ESAL and the type of surface, base, and subbase selected is obtained either by using the computer program DAMA or by entering the appropriate table or chart with the design ESAL and M_r of the subgrade and selecting the required minimum thickness, as demonstrated for different types of pavements in the following sections.

Full-Depth Asphalt Concrete. Pavements of this type use asphalt mixtures for all courses above the subgrade. The minimum thickness is determined from charts similar to Figure 20.5 by extracting the pavement thickness for the design ESAL and subgrade M_r, both of which were determined in step 1. The procedure for determining the minimum depth required for this type of pavement is illustrated in Example 20.3.

Table 20.10 Recommended Asphalt Grades for Different Temperature Conditions

Temperature Condition	Asphalt Grades*	
Cold, mean annual air temperature 7°C (45°F)	AC-5, AR-2000 120/150 pen.	AC-10 AR-4000 85/100 pen.
Warm, mean annual air temperature between 7°C (45°F) and 24°C (75°F)	AC-10, AR-4000 85/100 pen.	AC-20 AR-8000 60/70 pen.
Hot, mean annual air temperature ≥ 24°C (75°F)	AC-20, AR-8000 60/70 pen.	AC-40 AR-16000 40/50 pen.

*Both medium-setting (MS) and slow-setting (SS) emulsified asphalts are used in emulsified asphalt base mixes. They can be either of two types: cationic (ASTM D2397 or AASHTO M208) or anionic (ASTM D977 or AASHTO M140).

The grade of emulsified asphalt is selected primarily on the basis of its ability to satisfactorily coat the aggregate. This is determined by coating and stability tests (ASTM D244, AASHTO T59). Other factors important in the selection are the water availability at the job site, anticipated weather at the time of construction, the mixing process to be used, and the curing rate.

SOURCE: Adapted from *Thickness Design—Asphalt Pavements for Highways and Streets,* Manual Series No. 1, The Asphalt Institute, Lexington, Ky., February 1991.

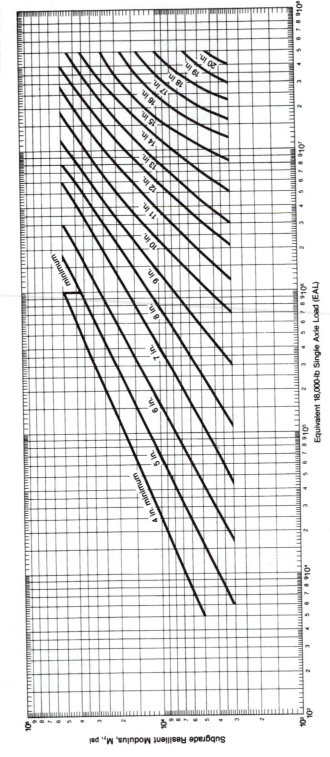

MAAT 60°F

Figure 20.5 Design Chart for Full-Depth Asphalt

SOURCE: *Thickness Design—Asphalt Pavements for Highways and Streets*, Manual Series No. 1, Asphalt Institute, Lexington, Ky., February 1991, p. 79.

Example 20.3   Computing Thickness for a Full-Depth Asphalt Pavement

A full-depth asphalt pavement is to be constructed to carry an ESAL of 2,172,042. If the CBR for the subgrade is 10 and the Mean Annual Air Temperature (MAAT) is 60°F, determine the depth required for the asphalt layer using Eq. 20.4.

$$M_r = 1500 \text{ CBR}$$
$$= 1500 \text{ lb/in.}^2 \times 10 = 15,000$$
$$\text{ESAL} = 2,172,042 = 2.172 \times 10^6$$

Solution: Using Figure 20.5, the depth required for a full-depth asphalt layer = 9 in.

Asphalt Concrete Surface and Emulsified Asphalt Base.   These pavements have asphalt concrete as surface material and emulsified asphalt as base material. Three mix types of emulsified asphalt are used in this design, and they are defined by the Asphalt Institute as:

- Type I: Emulsified asphalt mixes made with processed, dense-graded aggregates
- Type II: Emulsified asphalt mixes made with semiprocessed, crusher-run, pit-run, or bank-run aggregate
- Type III: Emulsified asphalt mixes made with sands or silty sands

Table 20.11 shows the recommended minimum thickness of asphalt concrete over types II and III emulsified asphalt bases. For pavements constructed with type I emulsified asphalt base, a surface treatment will be adequate. The depth of the emulsified asphalt base is determined as the difference between the total thickness (asphalt concrete surface and emulsified asphalt base) as obtained from the design charts and the minimum required thickness of the asphalt concrete, as obtained from Table 20.11. This depth should not be less than the recommended minimum. Figures 20.6 through 20.8 give design charts for

Table 20.11   Minimum Thickness of Asphalt Concrete over Emulsified Asphalt Bases

Traffic Level ESAL	Type II and Type III*	
	(mm)	*(in.)*
10^4	50	2
10^5	50	2
10^6	75	3
10^7	100	4
$>10^7$	130	5

*Asphalt concrete, or type I emulsified asphalt mix with a surface treatment, may be used over type II or type III emulsified asphalt base courses.
SOURCE: Adapted from *Thickness Design—Asphalt Pavements for Highways and Streets,* Manual Series No. 1, The Asphalt Institute, Lexington, Ky., February 1991.

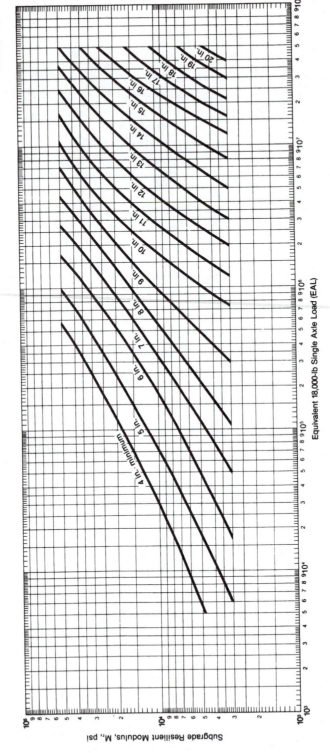

MAAT 60°F

Figure 20.6 Design Chart for Emulsified Asphalt Mix Type I

SOURCE: *Thickness Design—Asphalt Pavements for Highways and Streets*, Manual Series No. 1, Asphalt Institute, Lexington, Ky., February 1991, p. 80.

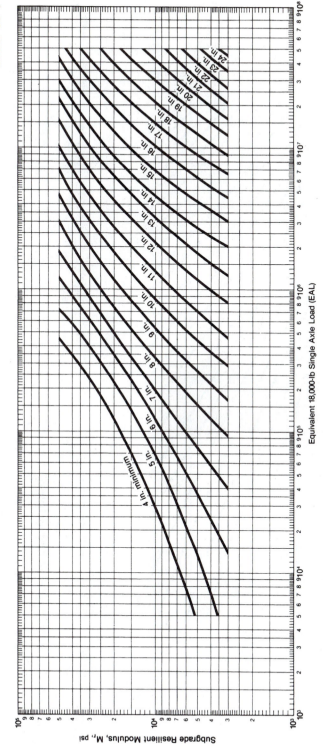

MAAT 60°F

Figure 20.7 Design Chart for Emulsified Asphalt Mix Type II

SOURCE: *Thickness Design—Asphalt Pavements for Highways and Streets*, Manual Series No. 1, Asphalt Institute, Lexington, Ky., February 1991, p. 81.

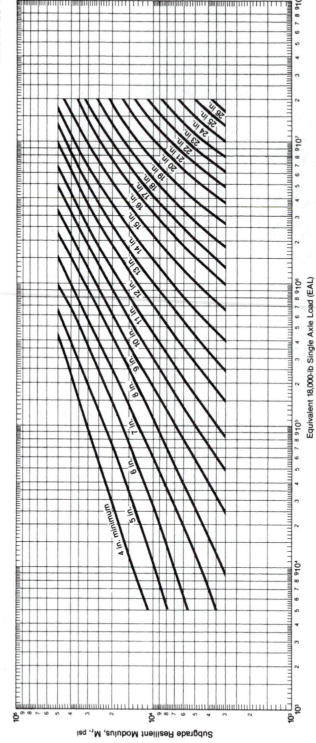

Figure 20.8 Design Chart for Emulsified Asphalt Mix Type III

SOURCE: *Thickness Design—Asphalt Pavements for Highways and Streets*, Manual Series No. 1, Asphalt Institute, Lexington, Ky., February 1991, p. 82.

minimum depths of types I, II, and III emulsified asphalt bases and MAAT of 60°F . These charts were developed based on a six-month curing period. The procedure for the design of this type of pavement is illustrated in Example 20.4.

Example 20.4   Designing a Pavement Consisting of Asphalt Concrete Surface and Emulsified Asphalt Base

Design a suitable pavement to carry a design ESAL of 1,303,225 on a subgrade having a resilient modulus of 15,000 lb/in.^{2} in an area with MAAT of 60°F.

   (a)   Determine the thicknesses for the surface and base courses for a pavement consisting of an asphaltic concrete surface and type II emulsified asphalt base.
   (b)   Repeat (a) for type III emulsified asphalt base.

Solution for **(a)**

   ESAL = 1,303,225

   M_r = 15,000 lb/in.^{2}

   Minimum asphalt concrete surface depth = 3 in. (from Table 20.11)

   Total thickness (surface and base) = 10 in. (from Figure 20.7)

   Base thickness = (10 – 3) = 7 in.

Solution for **(b)**

   Minimum asphalt concrete surface depth = 3 in. (from Table 20.11)

   Total pavement thickness = 13 in. (from Figure 20.8)

   Base thickness = 10 in.

Asphalt Concrete Surface and Untreated Aggregate Base.   These pavements consist of a layer of asphalt concrete over untreated aggregate base and subbase courses. As stated earlier, it is not always necessary to use a subbase course. Figures 20.9 through 20.14 give design charts for determining the thicknesses of asphalt concrete surfaces for different pavements constructed of asphalt concrete and untreated aggregates. These charts are given for different base thicknesses and are based on the quality requirements for base and subbase materials given in Table 20.9. The Asphalt Institute also recommends that the base course be not less than 6 in. Table 20.12 also gives the minimum recommended thicknesses for the asphalt concrete surface over the untreated aggregate base. These values depend on the design ESAL. In using the design charts, minimum thicknesses should not be extrapolated into higher traffic regions. The procedure for carrying out this design is illustrated in Example 20.5.

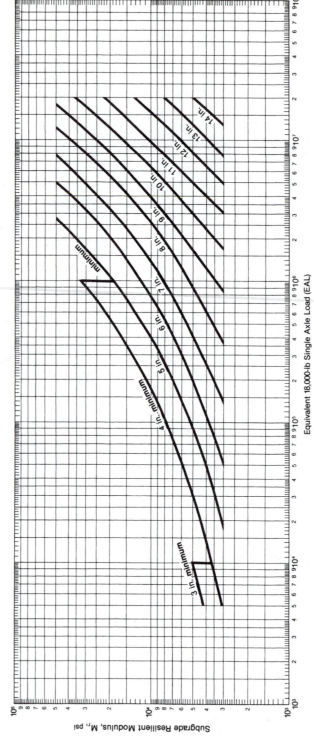

MAAT 45°F

Figure 20.9 Design Chart for Pavements with Asphalt Concrete Surface and Untreated Aggregate Base 6 in. Thick

SOURCE: *Thickness Design—Asphalt Pavements for Highways and Streets*, Manual Series No. 1, Asphalt Institute, Lexington, Ky., February 1991, p. 77.

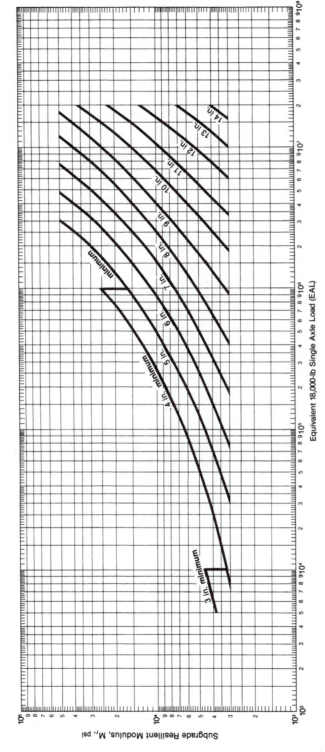

MAAT 45°F

Subgrade Resilient Modulus, M$_r$, psi

Equivalent 18,000-lb Single Axle Load (EAL)

Figure 20.10 Design Chart for Pavements with Asphalt Concrete Surface and Untreated Aggregate Base 12 in. Thick

SOURCE: *Thickness Design—Asphalt Pavements for Highways and Streets*, Manual Series No. 1, Asphalt Institute, Lexington, Ky., February 1991, p. 78.

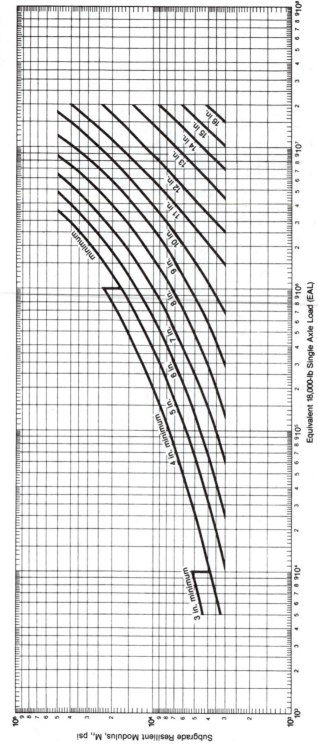

MAAT 60°F

Figure 20.11 Design Chart for Pavements with Asphalt Concrete Surface and Untreated Aggregate Base 6.0 in. Thick

SOURCE: *Thickness Design—Asphalt Pavements for Highways and Streets,* Manual Series No. 1, Asphalt Institute, Lexington, Ky., February 1991, p. 83.

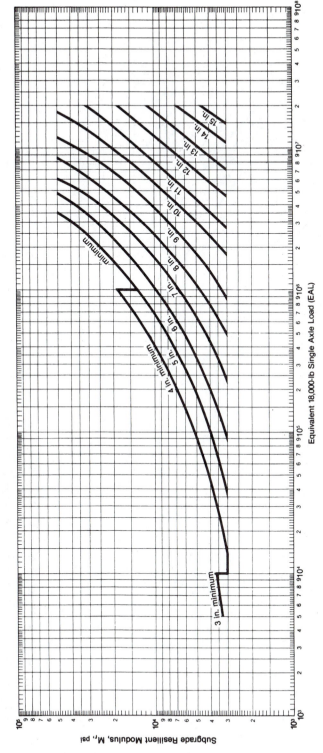

Figure 20.12 Design Chart for Pavements with Asphalt Concrete Surface and Untreated Aggregate Base 12 in. Thick

SOURCE: *Thickness Design—Asphalt Pavements for Highways and Streets*, Manual Series No. 1, Asphalt Institute, Lexington, Ky., February 1991 p. 84.

985

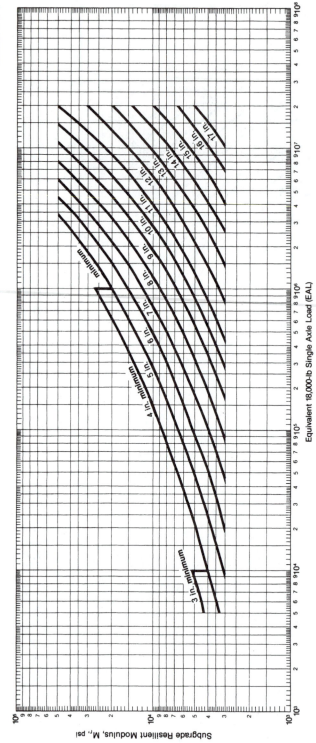

Figure 20.13 Design Chart for Pavements with Asphalt Concrete Surface and Untreated Aggregate Base 6 in. Thick

SOURCE: *Thickness Design—Asphalt Pavements for Highways and Streets*, Manual Series No. 1, Asphalt Institute, Lexington, Ky., February 1991, p. 89.

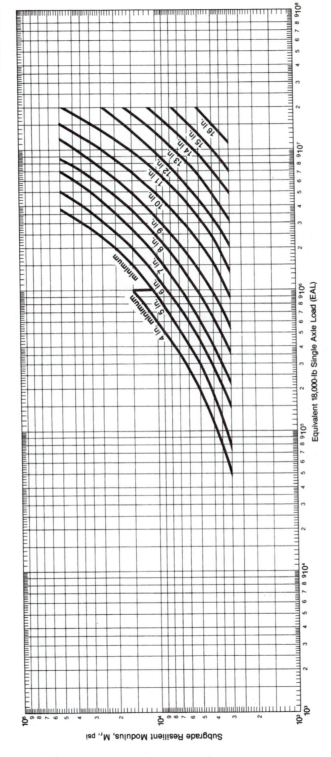

MAAT 75°F

Subgrade Resilient Modulus, M_r, psi

Equivalent 18,000-lb Single Axle Load (EAL)

Figure 20.14 Design Chart for Pavements with Asphalt Concrete Surface and Untreated Aggregate Base 12 in. Thick

SOURCE: *Thickness Design—Asphalt Pavements for Highways and Streets*, Manual Series No. 1, Asphalt Institute, Lexington, Ky., February 1991, p. 90.

Table 20.12 Minimum Thickness of Asphalt Concrete over Untreated Aggregate Base

Traffic ESAL	Traffic Condition	Minimum Thickness of Asphalt Concrete
10^4	Light traffic parking lots, driveways, and light traffic rural roads	75 mm (3.0 in.)★
10^4 but less than 10^6	Medium truck traffic	100 mm (4.0 in.)
10^6 or more	Medium to heavy truck traffic	125 mm (5.0 in.) or greater

★For full-depth asphalt concrete or emulsified asphalt pavements, a minimum thickness of 100 mm (4 in.) applies in this traffic region, as shown on the design charts.
SOURCE: Adapted from *Thickness Design—Asphalt Pavements for Highways and Streets,* Manual Series No. 1, The Asphalt Institute, Lexington, Ky., February 1991.

Example 20.5 Designing a Pavement Consisting of Asphalt Concrete Surface and Untreated Aggregate Base

Design a suitable pavement in an area with MAAT of 60°F to carry a design ESAL of 1.3×10^6 on a subgrade having a resilient modulus of 15,000 lb/in.^{2}, with the pavement constructed of an asphalt concrete surface and a 6 in. untreated granular base course.

Solution: The minimum depth of the asphalt concrete surface is obtained from Figure 20.11 as 6 in. The design will therefore consist of 6 in. of asphalt concrete surface plus 6 in. of untreated granular base course.

An alternative design can be made with a 12-in. base. The depth of the asphalt concrete surface required for this is 5 in. as obtained from Figure 20.13. Since the minimum requirement for the depth of the base material is 6 in., the pavement may consist of 5 in. of asphalt concrete surface, 6 in. of untreated aggregate base course, and 6 in. of untreated aggregate subbase course.

Step 4. Feasibility of Planning Stage Construction. Staged construction of a pavement structure involves construction in steps at specific times to meet the increase in traffic. When used as a design tool, it is presumed that a subsequent stage of the pavement will be constructed before the preceding stage shows any serious signs of distress. This method of construction usually is used when adequate funds are not available to construct the pavement to the required depth for the full design period of, say, 20 years. In such a case, the pavement may be constructed in two stages, with the first stage designed on an ESAL based on the first 10 years, and the first section overlayed sometime later to increase the depth to that required for the original design ESAL of 20 years. To ascertain that the planned overlay is placed before any serious distress occurs on the first stage, it is assumed that the cumulative damage in the first layer before the overlay will not be higher than 60 percent of its life. This ensures that at least 40 percent of the life of the first layer is still remaining.

The feasibility of staged construction should be evaluated when capital expenditure funds for the full depth of pavement are not available and when traffic growth rates are expected to be higher in the future.

The procedure for the design of a pavement using staged construction is demonstrated as follows. Let n_1 = design ESAL for the first stage and n_2 = design ESAL for the second stage. Then adjust n_1 and n_2 to take into account the hypothesis that not more than 60 percent of the cumulative damage of the first layer occurs before the second layer is applied. Thus, N_1 = adjusted design ESAL for stage 1 and N_2 = adjusted design ESAL for stage 2. It is assumed that the response of the pavement is elastic and that the damage at any stage is proportional to the ratio of the actual ESAL for that stage to the allowable ESAL for the thickness selected for that stage.

If D_1 = proportion of the life of the pavement expended (damage) at end of stage 1, then

$$D_1 = \frac{n_1}{N_1}$$ (20.6)

where
  n_1 = actual accumulated ESALs for stage 1
  N_1 = allowable number of ESALs for the initial thickness (h_1) selected for stage 1

At the end of stage 1, the proportion of the life of the pavement remaining is $(1 - D_1)$.

$$\frac{n_2}{N_2} = (1 - D_1)$$ (20.7)

where
  n_2 = design ESAL for stage 2
  N_2 = adjusted design ESAL for stage 2

If $D_1 = 0.6$, then

$$N_1 = 1.67n_1$$ (20.8)

$$N_2 = 2.50n_2$$ (20.9)

If h_1 and h_2 = design thicknesses for stage 1 and stage 2, respectively, then h_s = thickness to be added at stage 2—that is, $h_2 - h_1$. Thus, the respective design thicknesses are determined through the following steps:

1. Determine n_1 and n_2.
2. Determine N_1 and N_2.
3. Using the appropriate charts, determine h_1 and h_2.
4. Determine h_s.

Example 20.6 Designing a Full-Depth Asphalt Concrete Pavement for a Two-Stage Construction

A full-depth asphalt concrete pavement is to be constructed in two stages. The design period is 20 years, and the second stage will be constructed 10 years after the first stage. If the ESAL on the design lane during the first year is 60,000 and the growth rate for all vehicles is 5 percent, determine the asphalt thicknesses for the first and second stages of construction if the subgrade resilient modulus is 15,000 lb/in.^{2} and the MAAT of the area is 60°F.

Solution: First we must determine the ESAL for the first 10 years and the ESAL for the 20-year design period.

Growth rate = 5 percent
Growth factor for 10 yr = 12.58 (from Table 20.6)
Growth factor for 20 yr = 33.06 (from Table 20.6)
n_1 = 60,000 × 12.58 = 754,800
n_2 = 60,000(33.06 − 12.58) = 1,228,800 (Note that the ESAL is given, not the AADT)
N_1 = 754,800 × 1.67 = 1,260,516
N_2 = 1,228,800 × 2.5 = 3,072,000

From Figure 20.5, we obtain the following.

Required depth for first stage (h_1) = 7.0 in.
Required depth for first and second stages (h_2) = 9.5 in.
Depth of overlay h_s = 2.5 in.

Step 5. Economic Analysis and Design Selection. Examples 20.3 through 20.6 have demonstrated that several alternative designs can be obtained for the same design ESAL and subgrade resilient modulus, based on the type of pavement and whether or not planned stage construction is used. When alternative pavements are designed, it is necessary to carry out an economic evaluation of these alternatives, as presented in Chapter 13, to determine the best alternative.

AASHTO Design Method

The AASHTO method for design of highway pavements is based primarily on the results of the AASHTO road test that was conducted in Ottawa, Illinois. It was a cooperative effort carried out under the auspices of 49 states, the District of Columbia, Puerto Rico, the Bureau of Public Roads, and several industry groups. Tests were conducted on short-span bridges and test sections of flexible and rigid pavements constructed on A-6 subgrade material. The pavement test sections consisted of two small loops and four larger ones, each being a four-lane divided highway. The tangent sections consisted of a successive set of pavement lengths of different designs, each length being at least 100 ft. The principal flexible pavement sections were constructed of asphaltic concrete surface, a well-graded crushed limestone base, and a uniformly graded sand-gravel subbase. Three levels of surface thicknesses ranging from 1 to 6 in. were used in combination with three levels of base thicknesses ranging from 0 to 9 in. Each of these nine combinations were then combined with three levels of subbase thicknesses ranging from 0 to 6 in. In addition to the crushed limestone bases, some special sections had bases constructed with either a well-graded uncrushed gravel, a bituminous plant mixture, or cement-treated aggregate.

Test traffic consisting of both single-axle and tandem-axle vehicles were then driven over the test sections until several thousand load repetitions had been made. Ten different axle-arrangement and axle-load combinations were used, with single-axle loads ranging from 2,000 to 30,000 lb, and tandem-axle loads ranging from 24,000 to 48,000 lb. Data were then collected on the pavement condition with respect to extent of cracking and amount of patching required to maintain the section in service. The longitudinal and transverse profiles were also obtained to determine the extent of rutting, surface deflection caused by loaded vehicles moving at very slow speeds, pavement curvature at different vehicle speeds, stresses imposed on the subgrade surface, and temperature distribution in the pavement layers. These data were then thoroughly analyzed, the results of which formed the basis for the AASHTO method of pavement design.

AASHTO initially published an interim guide for the design of pavement structures in 1961, which was revised in 1972. A further revision was published in 1986, consisting of two volumes, Volume 1 being the basic design guide, and Volume 2 containing a series of appendices that provided documentation or further explanations for information contained in Volume 1. That edition incorporated new developments and specifically addressed pavement rehabilitation. Another edition was subsequently published in 1993, consisting of only one volume, which replaced Volume 1 of the 1986 guide. The major changes in the 1993 edition involved the overlay design procedure. Each edition of the guide specifically mentions that the design procedure presented cannot possibly include all conditions that relate to any one specific site. It is therefore recommended that, in using the guide, local experience be used to augment the procedures given in the guide.

Design Considerations
The factors considered in the AASHTO procedure for the design of flexible pavement as presented in the 1993 guide are

- Pavement performance
- Traffic
- Roadbed soils (subgrade material)
- Materials of construction
- Environment
- Drainage
- Reliability

Pavement Performance.   The primary factors considered under pavement performance are the structural and functional performance of the pavement. Structural performance is related to the physical condition of the pavement with respect to factors that have a negative impact on the capability of the pavement to carry the traffic load. These factors include cracking, faulting, raveling, and so forth. Functional performance is an indication of how effectively the pavement serves the user. The main factor considered under functional performance is riding comfort.

To quantify pavement performance, a concept known as the *serviceability performance* was developed. Under this concept, a procedure was developed to determine the present serviceability index (PSI) of the pavement, based on its roughness and distress, which were measured in terms of extent of cracking, patching, and rut depth for flexible pavements. The original expression developed gave the PSI as a function of the extent and type of cracking and patching and the slope variance in the two wheel paths, which is a measure of the variations in the longitudinal profile. The mean of the ratings of individual engineers with wide experience in all facets of highway engineering was used to relate the PSI with the factors considered. The scale ranges for 0 to 5, where 0 is the lowest PSI and 5 is the highest. A detailed discussion of PSI is given in Chapter 22.

Two serviceability indices are used in the design procedure: the initial serviceability index (p_i), which is the serviceability index immediately after the construction of the pavement, and the terminal serviceability index (p_t), which is the minimum acceptable value before resurfacing or reconstruction is necessary. In the AASHTO road test, a value of 4.2 was used for p_i for flexible pavements. AASHTO, however, recommends that each agency determine more reliable levels for p_i based on existing conditions. Recommended values for the terminal serviceability index are 2.5 or 3.0 for major highways and 2.0 for highways with a lower classification. In cases where economic constraints restrict capital expenditures for construction, the p_t can be taken as 1.5, or the performance period may be reduced. However, this low value should be used only in special cases on selected classes of highways.

Traffic.   The treatment of traffic load in the AASHTO design method is similar to that presented for the Asphalt Institute method, in that the traffic load application is given in terms of the number of 18,000-lb single-axle loads (ESALs). The procedure presented earlier is used to determine the design ESAL. The equivalence factors used in this case, however, are based on the terminal serviceability index to be used in the design and the structural number (SN) (see definition of SN under "Structural Design"). Table 20.13 gives traffic equivalence factors for p_t of 2.5.

Roadbed Soils (subgrade material).   The 1993 AASHTO guide also uses the resilient modulus (M_r) of the soil to define its property. However, the method allows for the conversion of the CBR or R value of the soil to an equivalent M_r value using the following conversion factors:

M_r (lb/in.^{2}) = 1500 CBR (for fine-grain soils with soaked CBR of 10 or less)
$M_r$lb/in.^{2} = 1000 + 555 × R value (for R ≤ 20)

Note that the first term in the expression for converting R value to M_r is 1000 and not 1155, as used by the Asphalt Institute. Actually, the Asphalt Institute developed the general relationship

Table 20.13a Axle Load Equivalency Factors for Flexible Pavements, Single Axles, and p_t of 2.5

Axle Load (kips)	Pavement Structural Number (SN)					
	1	2	3	4	5	6
2	.0004	.0004	.0003	.0002	.0002	.0002
4	.003	.004	.004	.003	.002	.002
6	.011	.017	.017	.013	.010	.009
8	.032	.047	.051	.041	.034	.031
10	.078	.102	.118	.102	.088	.080
12	.168	.198	.229	.213	.189	.176
14	.328	.358	.399	.388	.360	.342
16	.591	.613	.646	.645	.623	.606
18	1.00	1.00	1.00	1.00	1.00	1.00
20	1.61	1.57	1.49	1.47	1.51	1.55
22	2.48	2.38	2.17	2.09	2.18	2.30
24	3.69	3.49	3.09	2.89	3.03	3.27
26	5.33	4.99	4.31	3.91	4.09	4.48
28	7.49	6.98	5.90	5.21	5.39	5.98
30	10.3	9.5	7.9	6.8	7.0	7.8
32	13.9	12.8	10.5	8.8	8.9	10.0
34	18.4	16.9	13.7	11.3	11.2	12.5
36	24.0	22.0	17.7	14.4	13.9	15.5
38	30.9	28.3	22.6	18.1	17.2	19.0
40	39.3	35.9	28.5	22.5	21.1	23.0
42	49.3	45.0	35.6	27.8	25.6	27.7
44	61.3	55.9	44.0	34.0	31.0	33.1
46	75.5	68.8	54.0	41.4	37.2	39.3
48	92.2	83.9	65.7	50.1	44.5	46.5
50	112.0	102.0	79.0	60.0	53.0	55.0

$$M_r \ (\text{lb/in.}^2) = A + B \times (R \text{ value}) \tag{20.10}$$

where A varies from 772 to 1155 and B varies from 369 to 555. The Asphalt Institute uses the maximum values of A and B, whereas AASHTO suggests using 1000 and 555, respectively.

Materials of Construction.   The materials used for construction can be classified under three general groups: those used for subbase construction, those used for base construction, and those used for surface construction.

Subbase Construction Materials.   The quality of the material used is determined in terms of the layer coefficient, a_3, which is used to convert the actual thickness of the subbase to an equivalent SN. The sandy gravel subbase course material used in the AASHTO road test was assigned a value of 0.11. Layer coefficients are usually assigned, based on the description of the material used. Note, however, that due to the widely

Table 20.13b Axle Load Equivalency Factors for Flexible Pavements, Tandem Axles, and p_t of 2.5

Axle Load (kips)	Pavement Structural Number (SN)					
	1	2	3	4	5	6
2	.0001	.0001	.0001	.0000	.0000	.0000
4	.0005	.0005	.0004	.0003	.0003	.0002
6	.002	.002	.002	.001	.001	.001
8	.004	.006	.005	.004	.003	.003
10	.008	.013	.011	.009	.007	.006
12	.015	.024	.023	.018	.014	.013
14	.026	.041	.042	.033	.027	.024
16	.044	.065	.070	.057	.047	.043
18	.070	.097	.109	.092	.077	.070
20	.107	.141	.162	.141	.121	.110
22	.160	.198	.229	.207	.180	.166
24	.231	.273	.315	.292	.260	.242
26	.327	.370	.420	.401	.364	.342
28	.451	.493	.548	.534	.495	.470
30	.611	.648	.703	.695	.658	.633
32	.813	.843	.889	.887	.857	.834
34	1.06	1.08	1.11	1.11	1.09	1.08
36	1.38	1.38	1.38	1.38	1.38	1.38
38	1.75	1.73	1.69	1.68	1.70	1.73
40	2.21	2.16	2.06	2.03	2.08	2.14
42	2.76	2.67	2.49	2.43	2.51	2.61
44	3.41	3.27	2.99	2.88	3.00	3.16
46	4.18	3.98	3.58	3.40	3.55	3.79
48	5.08	4.80	4.25	3.98	4.17	4.49
50	6.12	5.76	5.03	4.64	4.86	5.28
52	7.33	6.87	5.93	5.38	5.63	6.17
54	8.72	8.14	6.95	6.22	6.47	7.15
56	10.3	9.6	8.1	7.2	7.4	8.2
58	12.1	11.3	9.4	8.2	8.4	9.4
60	14.2	13.1	10.9	9.4	9.6	10.7
62	16.5	15.3	12.6	10.7	10.8	12.1
64	19.1	17.6	14.5	12.2	12.2	13.7
66	22.1	20.3	16.6	13.8	13.7	15.4
68	25.3	23.3	18.9	15.6	15.4	17.2
70	29.0	26.6	21.5	17.6	17.2	19.2
72	33.0	30.3	24.4	19.8	19.2	21.3
74	37.5	34.4	27.6	22.2	21.3	23.6
76	42.5	38.9	31.1	24.8	23.7	26.1
78	48.0	43.9	35.0	27.8	26.2	28.8

Continued

Table 20.13b   Axle Load Equivalency Factors for Flexible Pavements, Tandem Axles, and p_t of 2.5 (continued)

Axle Load (kips)	Pavement Structural Number (SN)					
	1	2	3	4	5	6
80	54.0	49.4	39.2	30.9	29.0	31.7
82	60.6	55.4	43.9	34.4	32.0	34.8
84	67.8	61.9	49.0	38.2	35.3	38.1
86	75.7	69.1	54.5	42.3	38.8	41.7
88	84.3	76.9	60.6	46.8	42.6	45.6
90	93.7	85.4	67.1	51.7	46.8	49.7

SOURCE: Adapted from *AASHTO Guide for Design of Pavement Structures,* American Association of State Highway and Transportation Officials, Washington, D.C., 1993. Used by permission.

different environmental, traffic, and construction conditions, it is essential that each design agency develop layer coefficients appropriate to the conditions that exist in its own environment.

Charts correlating the layer coefficients with different soil engineering properties have been developed. Figure 20.15 shows one such chart for granular subbase materials.

Base Course Construction Materials.   Materials selected should satisfy the general requirements for base course materials given earlier in this chapter and in Chapter 18. A structural layer coefficient, a_2, for the material used should also be determined. This can be done using Figure 20.16.

Surface Course Construction Materials.   The most commonly used material is a hot plant mix of asphalt cement and dense-graded aggregates with a maximum size of 1 in. The procedure discussed in Chapter 19 for the design of asphalt mix can be used. The structural layer coefficient (a_1) for the surface course can be extracted from Figure 20.17, which relates the structural layer coefficient of a dense-grade asphalt concrete surface course with its resilient modulus at 68°F.

Environment.   Temperature and rainfall are the two main environmental factors used in evaluating pavement performance in the AASHTO method. The effects of temperature on asphalt pavements include stresses induced by thermal action, changes in the creep properties, and the effect of freezing and thawing of the subgrade soil, as discussed in Chapters 17 and 18. The effect of rainfall is due mainly to the penetration of the surface water into the underlying material. If penetration occurs, the properties of the underlying materials may be significantly altered. In Chapter 17, different ways of preventing water penetration were discussed. However, this effect is taken into consideration in the design procedure, and the methodology used is presented later under "Drainage."

The effect of temperature, particularly with regard to the weakening of the underlying material during the thaw period, is considered a major factor in determining the strength of the underlying materials used in the design. Test results have shown that the normal modulus (that is, modulus during summer and fall seasons) of materials

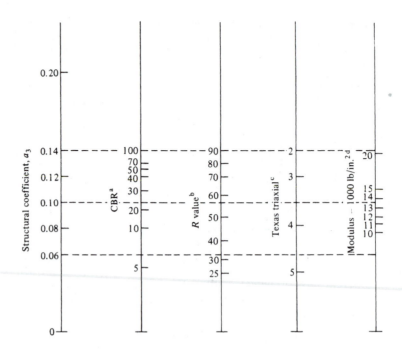

ᵃ Scale derived from correlations from Illinois.
ᵇ Scale derived from correlations obtained from The Asphalt
   Institute, California, New Mexico, and Wyoming.
ᶜ Scale derived from correlations obtained from Texas.
ᵈ Scale derived on NCHRP project 128, 1972.

Figure 20.15 Variation in Granular Subbase Layer Coefficient, a_3, with Various Subbase
Strength Parameters

SOURCE: Redrawn from *AASHTO Guide for Design of Pavement Structures*, American Association of State Highway
and Transportation Officials, Washington, D.C., 1993.

susceptible to frost action can reduce by 50 percent to 80 percent during the thaw period.
Also, resilient modulus of a subgrade material may vary during the year, even when there is
no specific thaw period. This occurs in areas subject to very heavy rains during specific
periods of the year. It is likely that the strength of the material will be affected during the
periods of heavy rains.

The procedure used to take into consideration the variation during the year of the
resilient modulus of the roadbed soil is to determine an effective annual roadbed soil
resilient modulus. The change in the PSI of the pavement during a full 12-month period
will then be the same if the effective resilient modulus is used for the full period or if
the appropriate resilient modulus for each season is used. This means that the effective
resilient modulus is equivalent to the combined effect of the different seasonal moduli
during the year.

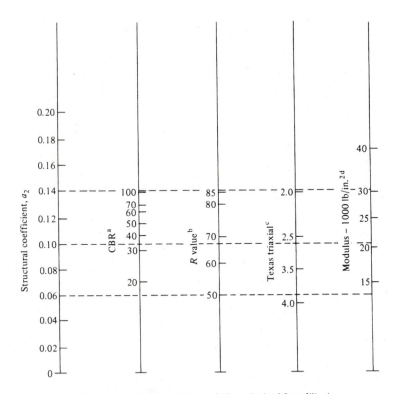

^a Scale derived by averaging correlations obtained from Illinois.
^b Scale derived by averaging correlations obtained from California,
New Mexico, and Wyoming.
^c Scale derived by averaging correlations obtained from Texas.
^d Scale derived on NCHRP project 128, 1972.

Figure 20.16 Variation in Granular Base Layer Coefficient a_2, with Various Subbase Strength Parameters

SOURCE: Redrawn from *AASHTO Guide for Design of Pavement Structures,* American Association of State Highway and Transportation Officials, Washington, D.C., 1993.

The AASHTO guide suggests two methods for determining the effective resilient modulus. Only the first method is described here. In this method, a relationship between resilient modulus of the soil material and moisture content is developed using laboratory test results. This relationship is then used to determine the resilient modulus for each season based on the estimated in situ moisture content during the season being considered. The whole year is then divided into the different time intervals that correspond with the different seasonal resilient moduli. The AASHTO guide suggests that it is not necessary to use a time interval less than one-half month. The relative damage u_f for each time period is then determined from the chart in Figure 20.18, using the vertical scale, or the equation given in the chart. The mean relative damage $\bar{u}_f$ is then computed, and the effective subgrade resilient modulus is determined using the chart and the value of $\bar{u}_f$.

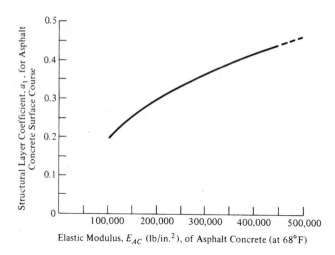

Figure 20.17 Chart for Estimating Structural Layer Coefficient of Dense-Graded/Asphalt Concrete Based on the Elastic (Resilient) Modulus

SOURCE: Redrawn from *AASHTO Guide for Design of Pavement Structures,* American Association of State Highway and Transportation Officials, Washington, D.C., 1993.

Example 20.7   Computing Effective Resilient Modulus

Figure 20.18 shows roadbed soil resilient modulus M_r for each month estimated from laboratory results correlating M_r with moisture content. Determine the effective resilient modulus of the subgrade.

Solution: Note that in this case, the moisture content does not vary within any one month. The solution of the problem is given in Figure 20.18. The value of u_f for each M_r is obtained directly from the chart. The mean relative damage $\bar{u}_f$ is 0.133, which in turn gives an effective resilient modulus of 7250 lb/in.^{2}.

Drainage.   The effect of drainage on the performance of flexible pavements is considered in the 1993 guide with respect to the effect water has on the strength of the base material and roadbed soil. The approach used is to provide for the rapid drainage of the free water (noncapillary) from the pavement structure by providing a suitable drainage layer, as shown in Figure 20.19, and by modifying the structural layer coefficient. The modification is carried out by incorporating a factor m_i for the base and subbase layer coefficients (a_2 and a_3). The m_i factors are based both on the percentage of time during which the pavement structure will be nearly saturated and on the quality of drainage, which is dependent on the time it takes to drain the base layer to 50 percent of saturation. Table 20.14 gives the general definitions of the different levels of drainage quality, and Table 20.15 gives recommended m_i values for different levels of drainage quality.

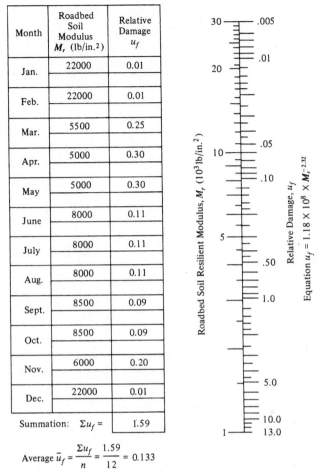

Month	Roadbed Soil Modulus M_r (lb/in.²)	Relative Damage u_f
Jan.	22000	0.01
Feb.	22000	0.01
Mar.	5500	0.25
Apr.	5000	0.30
May	5000	0.30
June	8000	0.11
July	8000	0.11
Aug.	8000	0.11
Sept.	8500	0.09
Oct.	8500	0.09
Nov.	6000	0.20
Dec.	22000	0.01
Summation: $\Sigma u_f =$		1.59

$$\text{Average } \bar{u}_f = \frac{\Sigma u_f}{n} = \frac{1.59}{12} = 0.133$$

Effective Roadbed Soil Resilient Modulus, M_r (lb/in.²) = <u>7250</u> (corresponds to $\bar{u}_f$)

Figure 20.18 Chart for Estimating Effective Roadbed Soil Resilient Modulus for Flexible Pavements Designed Using the Serviceability Criteria

SOURCE: Redrawn from *AASHTO Guide for Design of Pavement Structures,* American Association of State Highway and Transportation Officials, Washington, D.C., 1993.

Reliability.   It has been noted that the cumulative ESAL is an important input to any pavement design method. However, the determination of this input usually is based on assumed growth rates, which may not be accurate. Most design methods do not consider this uncertainty, but the 1993 AASHTO guide proposes the use of a reliability factor that considers the possible uncertainties in traffic prediction and performance prediction. A detailed discussion of the development of the approach used is beyond the scope of this book; however, a general description of the methodology is presented to allow the incorporation of reliability in the design process, if so desired by a designer. Reliability design

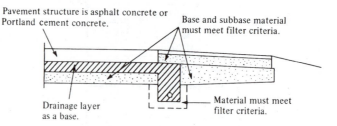

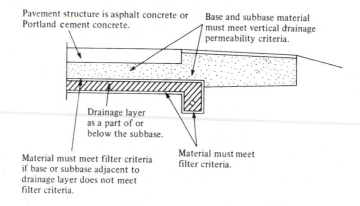

Pavement structure is asphalt concrete or Portland cement concrete.

Base and subbase material must meet filter criteria.

Drainage layer as a base.

Material must meet filter criteria.

(a) Base is used as the drainage layer.

Pavement structure is asphalt concrete or Portland cement concrete.

Base and subbase material must meet vertical drainage permeability criteria.

Drainage layer as a part of or below the subbase.

Material must meet filter criteria if base or subbase adjacent to drainage layer does not meet filter criteria.

Material must meet filter criteria.

(b) Drainage layer is part of or below the subbase.

Note: Filter fabrics may be used in lieu of filter material, soil, or aggregate, depending on economic considerations.

Figure 20.19 Example of Drainage Layer in Pavement Structure

SOURCE: Redrawn from *AASHTO Guide for Design of Pavement Structures,* American Association of State Highway and Transportation Officials, Washington, D.C., 1993.

Table 20.14 Definition of Drainage Quality

Quality of Drainage	Water Removed Within*
Excellent	2 hours
Good	1 day
Fair	1 week
Poor	1 month
Very poor	(water will not drain)

*Time required to drain the base layer to 50 percent saturation.
SOURCE: Adapted with permission from *AASHTO Guide for Design of Pavement Structures,* American Association of State Highway and Transportation Officials, Washington, D.C., 1993.

Table 20.15 Recommended m_i Values

	Percent of Time Pavement Structure Is Exposed to Moisture Levels Approaching Saturation			
Quality of Drainage	Less Than 1 Percent	1–5 Percent	5–25 Percent	Greater Than 25 Percent
Excellent	1.40–1.35	1.35–1.30	1.30–1.20	1.20
Good	1.35–1.25	1.25–1.15	1.15–1.00	1.00
Fair	1.25–1.15	1.15–1.05	1.00–0.80	0.80
Poor	1.15–1.05	1.05–0.80	0.80–0.60	0.60
Very Poor	1.05–0.95	0.95–0.75	0.75–0.40	0.40

SOURCE: Adapted from *AASHTO Guide for Design of Pavement Structures,* American Association of State Highway and Transportation Officials, Washington, D.C., 1993. Used by permission.

levels (R%), which determine assurance levels that the pavement section designed using the procedure will survive for its design period, have been developed for different types of highways. For example, a 50 percent reliability design level implies a 50 percent chance for successful pavement performance—that is, the probability of design performance success is 50 percent. Table 20.16 shows suggested reliability levels, based on a survey of the AASHTO pavement design task force. Reliability factors, $F_R \geq 1$, based on the reliability level selected and the overall variation, S_o^2 have also been developed. S_o^2 accounts for the chance variation in the traffic forecast and the chance variation in actual pavement performance for a given design period traffic, W_{18}.

The reliability factor F_R is given as

$$\log_{10} F_R = -Z_R S_o \tag{20.11}$$

where
$\quad Z_R$ = standard normal variate for a given reliability (R%)
$\quad S_o$ = estimated overall standard deviation

Table 20.17 gives values of Z_R for different reliability levels R.

Overall standard deviation ranges have been identified for flexible and rigid pavements as:

	Standard Deviation, S_o
Flexible pavements	0.40–0.50
Rigid pavements	0.30–0.40

These values were derived through a detailed analysis of existing data. However, very little data presently exist for certain design components, such as drainage. A methodology for

Table 20.16 Suggested Levels of Reliability for Various Functional Classifications

Functional Classification	Recommended Level of Reliability	
	Urban	Rural
Interstate and other freeways	85–99.9	80–99.9
Other principal arterials	80–99	75–95
Collectors	80–95	75–95
Local	50–80	50–80

Note: Results based on a survey of the AASHTO Pavement Design Task Force.
SOURCE: Adapted with permission from *AASHTO Guide for Design of Pavement Structures,* American Association of State Highway and Transportation Officials, Washington, D.C., 1993.

Table 20.17 Standard Normal Deviation (Z_R) Values Corresponding to Selected Levels of Reliability

Reliability (R%)	Standard Normal Deviation, Z_R
50	−0.000
60	−0.253
70	−0.524
75	−0.674
80	−0.841
85	−1.037
90	−1.282
91	−1.340
92	−1.405
93	−1.476
94	−1.555
95	−1.645
96	−1.751
97	−1.881
98	−2.054
99	−2.327
99.9	−3.090
99.99	−3.750

SOURCE: Adapted with permission from *AASHTO Guide for Design of Pavement Structures,* American Association of State Highway and Transportation Officials, Washington, D.C., 1993.

improving these estimates is presented in the 1993 AASHTO guide, which may be used when additional data are available.

Structural Design

The objective of the design using the AASHTO method is to determine a flexible pavement SN adequate to carry the projected design ESAL. It is left to the designer to select the type of surface used, which can be either asphalt concrete, a single surface treatment, or a double surface treatment. This design procedure is used for ESALs greater than 50,000 for the performance period. The design for ESALs less than this is usually considered under low-volume roads.

The 1993 AASHTO guide gives the expression for SN as

$$SN = a_1D_1 + a_2D_2m_2 + a_3D_3m_3 \tag{20.12}$$

where

m_i = drainage coefficient for layer i

a_1, a_2, a_3 = layer coefficients representative of surface, base, and subbase course, respectively

$D_1, D_2, D_3,$ = actual thickness in inches of surface, base, and subbase courses, respectively

The basic design equation given in the 1993 guide is

$$\log_{10} W_{18} = Z_R S_o + 9.36 \log_{10}(SN + 1) - 0.20 + \frac{\log_{10}[\Delta PSI/(4.2 - 1.5)]}{0.40 + [1094/(SN + 1)^{5.19}]}$$
$$+ 2.32 \log_{10} M_r - 8.07 \tag{20.13}$$

where

W_{18} = predicted number of 18,000-lb (80 kN) single-axle load applications

Z_R = standard normal deviation for a given reliability

S_o = overall standard deviation

SN = structural number indicative of the total pavement thickness

$\Delta PSI = p_i - p_t$

p_i = initial serviceability index

p_t = terminal serviceability index

M_r = resilient modulus in lb/in.^{2}

Equation 20.13 can be solved for SN using a computer program or the chart in Figure 20.20. The use of the chart is demonstrated by the example solved on the chart and in the solution of Example 20.8.

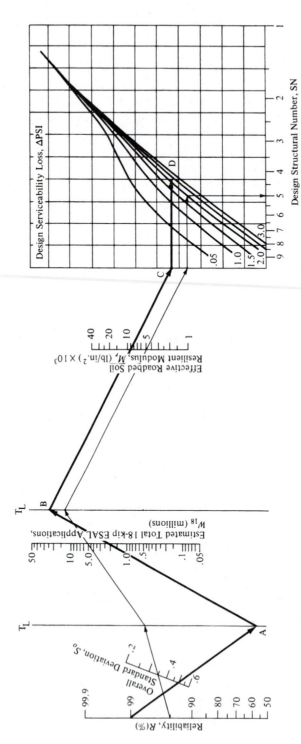

Figure 20.20 Design Chart for Flexible Pavements Based on Using Mean Values for Each Input

SOURCE: Redrawn from *AASHTO Guide for Design of Pavement Structures*, American Association of State Highway and Transportation Officials, Washington, D.C., 1993.

Example 20.8   Designing a Flexible Pavement Using the AASHTO Method

A flexible pavement for an urban interstate highway is to be designed using the 1993 AASHTO guide procedure to carry a design ESAL of 2×10^6. It is estimated that it takes about a week for water to be drained from within the pavement and the pavement structure will be exposed to moisture levels approaching saturation for 30 percent of the time. The following additional information is available:

> Resilient modulus of asphalt concrete at 68°F = 450,000 lb/in.^{2}
> CBR value of base course material = 100, M_r = 31,000 lb/in.^{2}
> CBR value of subbase course material = 22, M_r = 13,500 lb/in.^{2}
> CBR value of subgrade material = 6

Determine a suitable pavement structure, M_r of subgrade = 6×1500 lb/in.^{2} = 9000 lb/in.^{2}.

Solution: Since the pavement is to be designed for an interstate highway, the following assumptions are made:

> Reliability level (R) = 99 percent (range is 80–99.9 from Table 20.16)
> Standard deviation (S_o) = 0.49 (range is 0.4–0.5)
> Initial serviceability index p_i = 4.5
> Terminal serviceability index p_t = 2.5

The nomograph in Figure 20.20 is used to determine the design SN through the following steps:

Step 1.   Draw a line joining the reliability level of 99 percent and the overall standard deviation S_o of 0.49, and extend this line to intersect the first T_L line at point A.

Step 2.   Draw a line joining point A to the ESAL of 2×10^6, and extend this line to intersect the second T_L line at point B.

Step 3.   Draw a line joining point B and resilient modulus (M_r) of the roadbed soil, and extend this line to intersect the design serviceability loss chart at point C.

Step 4.   Draw a horizontal line from point C to intersect the design serviceability loss (ΔPSI) curve at point D. In this problem ΔPSI = 4.5 – 2.5 = 2.

Step 5.   Draw a vertical line to intersect the design SN, and read this value. SN = 4.4.

Step 6.   Determine the appropriate structure layer coefficient for each construction material.

> **(a)**   Resilient value of asphalt cement = 450,000 lb/in.^{2}. From Figure 20.17, a_1 = 0.44.
> **(b)**   CBR of base course material = 100. From Figure 20.16, a_2 = 0.14.
> **(c)**   CBR of subbase course material = 22. From Figure 20.15, a_3 = 0.10.

Step 7. Determine appropriate drainage coefficient m_i. Since only one set of conditions is given for both the base and subbase layers, the same value will be used for m_1 and m_2. The time required for water to drain from within pavement = 1 week, and from Table 20.14, drainage quality is fair. The percentage of time pavement structure will be exposed to moisture levels approaching saturation = 30, and from Table 20.15, $m_i = 0.80$.

Step 8. Determine appropriate layer thicknesses from Eq. 20.13

$$= a_1 D_1 + a_2 D_2 m_2 + a_3 D_3 m_3$$

It can be seen that several values of D_1, D_2, and D_3 can be obtained to satisfy the SN value of 4.40. Layer thicknesses, however, are usually rounded up to the nearest 0.5 in.

The selection of the different layer thicknesses should also be based on constraints associated with maintenance and construction practices, so that a practical design is obtained. For example, it is normally impractical and uneconomical to construct any layer with a thickness less than some minimum value. Table 20.18 lists minimum thicknesses suggested by AASHTO.

Taking into consideration that a flexible pavement structure is a layered system, the determination of the different thicknesses should be carried out as indicated in Figure 20.21. The required SN above the subgrade is first determined, and then the required SNs above the base and subbase layers are determined using the appropriate strength of each layer. The minimum allowable thickness of each layer can then be determined using the differences of the computed SNs as shown in Figure 20.21.

Using the appropriate values for M_r in Figure 20.20, we obtain $SN_3 = 4.4$ and $SN_2 = 3.8$. Note that when SN is assumed to compute ESAL, the assumed and computed SN values must be approximately equal. If these are significantly different, the computation must be repeated with a new assumed SN.

Table 20.18 AASHTO-Recommended Minimum Thicknesses of Highway Layers

	Minimum Thickness (in.)	
Traffic, ESALs	*Asphalt Concrete*	*Aggregate Base*
Less than 50,000	1.0 (or surface treatment)	4
50,001–150,000	2.0	4
150,001–500,000	2.5	4
500,001–2,000,000	3.0	6
2,000,001–7,000,000	3.5	6
Greater than 7,000,000	4.0	6

SOURCE: Adapted with permission from *AASTHO Guide for Design of Pavement Structures*, American Association of State Highway and Transportation Officials, Washington, D.C., 1993.

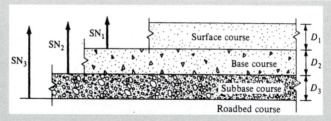

Figure 20.21 Procedure for Determining Thicknesses of Layers Using a Layered Analysis Approach

SOURCE: Redrawn from *AASHTO Guide for Design of Pavement Structures,* American Association of State Highway and Transportation Officials, Washington, D.C., 1993.

We know

$$M_r \text{ for base course} = 31{,}000 \text{ lb/in.}^2$$

Using this value in Figure 20.20, we obtain

$$SN_1 = 2.6$$

giving

$$D_1 = \frac{2.6}{0.44} = 5.9 \text{ in.}$$

Using 6 in. for the thickness of the surface course,

$$D_1^\star = 6 \text{ in.}$$
$$SN_1^\star = a_1 D_1^\star = 0.44 \times 6 = 2.64$$
$$D_2^\star \geq \frac{SN_2 - SN_1^\star}{a_2 m_2} \geq \frac{3.8 - 2.64}{0.14 \times 0.8} \geq 10.36 \text{ in.} \quad (\text{use } 12 \text{ in.})$$
$$SN_2^\star = 0.14 \times 0.8 \times 12 + 2.64 = 1.34 + 2.64$$
$$D_3^\star = \frac{SN_3 - SN_2^\star}{a_3 m_3} = 4.4 - \frac{(2.64 + 1.34)}{0.1 \times 0.8} = 5.25 \text{ in.} \quad (\text{use } 6 \text{ in.})$$
$$SN_3^\star = 2.64 + 1.34 + 6 \times 0.8 \times 0.1 = 4.46$$

The pavement will therefore consist of 6 in. asphalt concrete surface, 12 in. granular base, and 6 in. subbase.

$\star$An asterisk with D or SN indicates that it represents the value actually used, which must be equal to or greater than the required value.

California (Hveem) Design Method

This method was originally developed in the 1940s, based on a combination of information obtained from some test roads, theory, and experience. It is widely used in the western states. It has been modified several times to take into consideration changes in traffic characteristics. The objective of the original design method was to avoid plastic deformation and the distortion of the pavement surface, but a later modification includes the reduction to a minimum of early fatigue cracking due to traffic load.

The factors considered are

- Traffic load
- Strength of subgrade material
- Strength of construction materials

The traffic load is initially calculated as the ESAL—that is, the total number of 18,000-lb axle loads in one direction—as discussed earlier, and then converted to a traffic index (TI), where

$$TI = 9.0\left(\frac{ESAL}{10^6}\right)^{0.119} \tag{20.14}$$

In determining the ESAL, passenger vehicles and pickups are not considered.

The strength of the subgrade material is given in terms of the resistance value R of the subgrade soil, obtained from the Hveem Stabilometer test described in Chapter 18.

The strength characteristics of each construction material (asphalt concrete, base, and subbase materials) are given in terms of a gravel equivalent factor (G_f), which has been determined for different types of materials. The factors given for asphalt concrete depend on the TI, as shown in Table 20.19.

Table 20.19 Gravel Equivalent Factors for Different Types of Materials

Material	G_f
Cement-treated base	
Class A	1.7
B	1.2
Lime-treated base	1.2
Untreated aggregate base	1.1
Aggregate subbase	1.0
Asphalt concrete for TI of ≤5.0	2.54
5.5–6.0	2.32
6.5–7.0	2.14
7.5–8.0	2.01
8.5–9.0	1.89
9.5–10.0	1.79
10.5–11.0	1.71
13.5–14.0	1.52

SOURCE: Adapted from *California Department of Transportation Design Manual,* California Division of Highways, Sacramento, Calif., 1995.

Structural Design

The objective of the design is to determine the total thickness of material required above the subgrade to carry the projected traffic load. This thickness is determined in terms of gravel equivalent (GE) in feet, which is given as

$$GE = 0.0032(TI)(100 - R) \qquad (20.15)$$

where

GE = thickness of material required above a given layer in terms of gravel
equivalent in feet

TI = traffic index

R = resistance value of the supporting layer material, normally determined at an
exudation pressure of 300 lb/in.^{2}

The actual depth of each layer can then be determined by dividing GE for that layer by the GE factor G_f for the material used in the layer. A check is usually made to ascertain that the thickness obtained is adequate for the expansion pressure requirements.

Example 20.9   Designing a Flexible Pavement Using the Hveem Method

A flexible pavement is to be designed to carry a 1.5×10^6 ESAL during its design period. Using the California (Hveem) method, determine appropriate thicknesses for an asphalt concrete surface and a granular base course over a subgrade soil, for which stabilometer test data are shown in Table 20.20. The R value for the untreated granular base material is 70.

Solution: The design is carried out through the following steps:

Step 1.   Determine TI using Eq. 20.14

$$TI = 9.0 \left(\frac{ESAL}{10^6} \right)^{0.119}$$

$$= 9.0 \left(\frac{1.5 \times 10^6}{10^6} \right)^{0.119} = 9.44 \qquad (\text{say, } 9.5)$$

Step 2.   Determine the asphalt concrete surface thickness over the granular base. Use Eq. 20.15 to determine GE, with R = 70 and TI = 9.5.

$$GE = 0.0032(TI)(100 - R)$$
$$= 0.0032(9.5)(100 - 70) = 0.91 \text{ ft}$$

The G_f factor from Table 20.19 is 1.79.

Table 20.20 Stabilometer Test Data for Subgrade Soil of Example 20.9

Moisture Content (%)	R Value	Exudation Pressure (lb/in.)	Expansion Pressure (lb/in.)	GE (ft) (from Eq. 20.15)	Expansion Pressure Thickness (ft)
25.2	48	550	1.03	1.6	1.13
25.8	37	412	0.30	1.9	0.33
29.3	15	80	0.00	2.6	0.00

$$\text{actual depth required} = 0.91/1.79 = 0.51 \text{ ft}$$
$$= 6.1 \text{ in.} \quad (\text{say, } 6.0 \text{ in.})$$

Step 3. Determine base thickness. The total thickness required above the subgrade should first be determined for an exudation pressure of 300 lb/in.^{2}. This is obtained by plotting a graph of GE (given in Table 20.20) for all layers above the subgrade versus the exudation pressure as shown in Figure 20.22(a). The GE is obtained as 2.05 ft, and G_f for untreated aggregate base = 1.1. The base thickness is determined from the difference in total GE required over subgrade material and the GE provided by the asphalt base.

$$\text{GE of base layer} = 2.05 - \frac{6.0}{12} \times 1.79$$

Therefore,

$$\text{base thickness} = \left(2.05 - \frac{6.0}{12} \times 1.79\right)/1.1 \text{ ft}$$
$$= 12.6 \text{ in.} \quad (\text{say, } 12\frac{1}{2} \text{ in.})$$

Step 4. Check to see whether strength design satisfies expansion pressure requirements. The moisture content that corresponds with the strength design thickness (that is, at exudation pressure of 300 lb/in.^{2}) is obtained by plotting the GE values for all layers above the subgrade versus the moisture content (given in Table 20.20) as shown in curve B in Figure 20.22(b). This moisture content is determined as 27 percent. The thickness required to resist expansion at 27 percent moisture content is then determined from curve C, which is a plot of expansion pressure thickness against moisture content. Thickness required to prevent expansion at 27 percent moisture content is 0.2 ft, which shows that strength design satisfies expansion pressure requirements, since the total thickness of the materials above the subgrade is much higher than 0.2 ft.

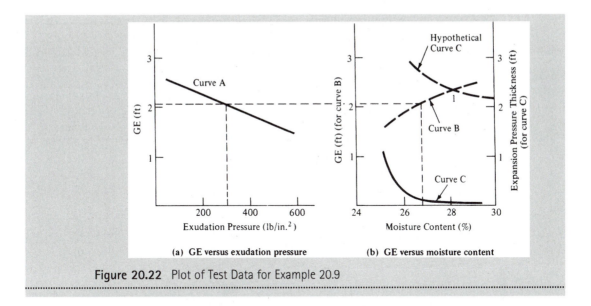

Figure 20.22 Plot of Test Data for Example 20.9

When the thickness based on strength design does not satisfy the expansion pressure requirements, a balanced design should be adopted, where the design point is the intersection of the curves of GE (for strength) versus moisture content and GE (for expansion pressure requirement) versus moisture content. A hypothetical case is shown in Figure 20.22, where the balanced point is point 1, indicating that the total thickness GE must be increased to about 2.4 ft and compaction of the embankment must take place at a moisture content of 28 percent rather than at 27 percent.

SUMMARY

The design of flexible pavements basically involves determining the strength characteristics of the materials of the pavement surface and underlying materials, and then determining the respective thicknesses of the subbase (if any), base course, and pavement surface that should be placed over the native soil. The pavement is therefore usually considered as a multilayered elastic system. The thicknesses provided should be adequate to prevent excessive cracking and permanent deformation beyond certain limits. These limits are considered in terms of required load characteristics, which can be determined as the number of repetitions of 18,000-lb single-axle loads the pavement is expected to carry during its design life.

The availability of computerized solutions for multilayered systems has led to the development of several design methods. Those commonly used in the United States are the Asphalt Institute method, the AASHTO method, and the California (Hveem) method, all of which have been presented in this chapter.

PROBLEMS

20-1   Discuss the structural components of a flexible pavement.

20-2   Describe the purpose of soil stabilization, and discuss at least three methods of achieving soil stabilization.

20-3   Discuss the process of lime stabilization, and indicate the soil characteristics for which it is most effective.

20-4   An axle weight study on a section of a highway gave the following data on axle-load distribution:

Axle Load Group: Single (1000 lb)	No. of Axles per 1000 Vehicles	Axle Load Group: Tandem (1000 lb)	No. of Axles per 1000 Vehicles
<3	678	<6	18
3–7	600	6–12	236
7–8	175	12–18	170
8–12	500	18–24	120
12–16	150	24–30	152
16–18	60	30–32	66
18–20	40	32–34	30
20–22	7	34–36	12
22–24	4	36–38	4
24–26	3	38–40	1

Determine the truck factor for this section of highway.

20-5   How does an ESAL differ from a truck factor?

20-6   A six-lane divided highway is to be designed to replace an existing highway. The present AADT (both directions) of 6000 vehicles is expected to grow at 5 percent per annum. Determine the design ESAL if the design life is 20 years and the vehicle mix is

Passenger cars (1000 lb/axle) = 60 percent
2-axle single-unit trucks (5000 lb/axle) = 30 percent
3-axle single-unit trucks (7000 lb/axle) = 10 percent

20-7   A section of a two-lane rural highway is to be realigned and replaced by a four-lane highway with a full-depth asphalt pavement. The AADT (both ways) on the existing section can be represented by 500 ESAL. It is expected that construction will be completed 5 years from now. If the traffic growth rate is 5 percent and the CBR of the subgrade on the new alignment is 10, determine the depth of the asphalt pavement, using the Asphalt Institute method. Take the design life of the pavement as 20 years. MAAT = 60°

20-8   The predicted traffic mix of a proposed four-lane urban expressway is

Passenger cars = 69 percent
Single-unit trucks
  2-axle, 4-tire = 20 percent
  2-axle, 6 tire = 5 percent
  3-axle or more = 4 percent
Tractor semitrailers and combinations
  3-axle = 2 percent

The projected AADT during the first year of operation is 3500 (both directions). If the traffic growth rate is estimated at 4 percent and the CBR of the subgrade is 9, determine the depth of a full-asphalt pavement using the Asphalt Institute method, and $n = 20$ years. MAAT $= 60°$

20-9   A roadway is under design that is expected to carry an AADT of 4500 during the first year of operation with an expected annual growth of 6 percent over the 20-year design life. The expected traffic mix is 60% passenger cars; 20% 2-axle, 4-tire single-unit trucks; 10% 2-axle, 6-tire, single-unit trucks; 5% 3-axle, single-unit trucks; 3% 3-axle tractor trailers; and 2% 4-axle tractor trailers. If the subgrade has a resilient modulus of 15,000 lb/in.^{2} and the mean ambient air temperature is 60°, design a suitable pavement consisting of an asphalt concrete surface and a type II and type III emulsified base.

20-10   Using the information given in Problem 20-6, design a suitable pavement consisting of an asphalt concrete surface and a type III emulsified asphalt base on a subgrade having a CBR of 10. MAAT $= 60°$

20-11   Repeat Problem 20-10 for a pavement consisting of an asphalt concrete surface and 6 in. untreated aggregate base. MAAT $= 60°$

20-12   Repeat Problem 20-7, using two different depths of untreated aggregate bases of 6 in. and 12 in. for MAAT of 45°F. Contact highway contractors in your area and obtain the rates for providing and properly laying asphalt concrete surface and untreated granular base. With these rates, determine the cost for constructing the different pavement designs if the highway section is 5 miles long and the lane width is 12 ft. Which design will you select for construction?

20-13   The design life of a proposed four-lane urban expressway is 20 years. The estimate ESAL (both directions) during the first year of operation is 150,000 with a growth rate of 5 percent. Design a full-depth asphalt concrete (using the Asphalt Institute method) to be constructed in two stages, with the second stage to be constructed 10 years after the first stage. The resilient modulus of the subgrade is 15,000 lb/in.^{2}.

20-14   The traffic on the design lane of a proposed four-lane rural interstate highway consists of 40 percent trucks. If classification studies have shown that the truck factor can be taken as 0.45, design a suitable flexible pavement using the 1986 AASHTO procedure if the AADT on the design lane during the first year of operation is 1000, and $p_i = 4.2$.

    Growth rate $= 4$ percent
    Design life $= 20$ years
    Reliability level $= 95$ percent
    Standard deviation $= 0.45$

The pavement structure will be exposed to moisture levels approaching saturation 20 percent of the time, and it will take about 1 week for drainage of water. CBR of the subgrade material is 7. CBR of the base and subbase are 70 and 22, respectively, and M_r for the asphalt concrete, 450,000 lb/in.^{2}

20-15   Repeat Problem 20-14, with the subgrade M_r values for each month from January through December being 20,000, 20,000, 6000, 6000, 6000, 9000, 9000, 9000, 9500, 9500, 8000, and 20,000 lb/in.^{2}, respectively. The pavement structure will be exposed to moisture levels approaching saturation for 20 percent of the time, and it will take about 4 weeks for drainage of water from the pavement. Use untreated sandy gravel with M_r of 15×10^3 lbs/in.^{2} for subbase and untreated granular material with M_r of 28×10^3 lbs/in.^{2} for the base course.

20-16   An existing two-lane rural highway is to be replaced by a four-lane divided highway on a new alignment. Construction of the new highway will commence 2 years from now and is expected to take 3 years to complete. The design life of the pavement is 20 years. The present AADT (two-way) on the two-lane highway is 3600, consisting of 60% passenger cars; 20% 2-axle, 6-tire single-unit trucks; 15% 3-axle; and 5% 4-axle tractor-semitrailer combinations. Design a flexible pavement consisting of asphalt concrete surface and lime-treated base, using the California (Hveem) method. The results of a stabilometer test on the subgrade soil are

Moisture Content (%)	R Value	Exudation Pressure (lb/in.^{2})	Expansion Pressure Thickness (ft)
19.8	55	575	1.00
22.1	45	435	0.15
24.9	16	165	0.10

Assume the R value for the lime-treated base of 80 and a traffic growth rate of 5 percent annually.

REFERENCES

AASHTO Guide for Design of Pavement Structures, American Association of State Highway and Transportation Officials, Washington, D.C., 1993.

AASHTO Guide for Design of Pavement Structures, American Association of State Highway and Transportation Officials, Washington, D.C., 1986.

AASHTO Interim Guide for Design of Pavement Structures, American Association of State Highway and Transportation Officials, Washington, D.C., 1972, Chapter III, revised 1981.

Annual Book of ASTM Standards, Section 4, American Society for Testing and Materials, Philadelphia, Pa., 1998.

California Dept. of Transportation, *Highway Design Manual: Design of the Pavement Standard Section,* 1995.

DAMA Computer Program, Asphalt Thickness Design, Asphalt Institute, Lexington, Ky.

Dorman, G.M. and C.T. Metcalf, "Design Curves for Flexible Pavements Based on Layered Systems Theory," *Highway Research Record* 71(1964).

Gerrard, C.M. and L.I. Wardle, "Rational Design of Surface Pavement Layers," *Australia Road Research* 10(1980).

Soil Cement Laboratory Handbook, Portland Cement Association, Skokie, Ill., 1971.

Soil Stabilization in Pavement Structures, A User's Manual, Vols. I and II, U.S. Department of Transportation, Federal Highway Administration, Washington, D.C., October 1979.

Soils Manual for the Design of Asphalt Pavement Structures, MS-10, 5th ed., Asphalt Institute, Lexington, Ky., January 1993.

Southgate H.F., R.C. Deen, and J.G. Moyer, "Strain Energy Analysis of Pavement Designs for Heavy Trucks," *Transportation Research Record* 949 (1983).

Thickness Design—Asphalt Pavements for Highways and Streets, Manual Series No.1, Asphalt Institute, Lexington, Ky., February 1991.

CHAPTER 21

Design of Rigid Pavements

R igid highway pavements are normally constructed of Portland cement concrete and may or may not have a base course between the subgrade and the concrete surface. When a base course is used in rigid pavement construction, it is usually referred to as a subbase course. It is common, however, for only the concrete surface to be referred to as the rigid pavement, even where there is a base course. In this text, the terms "rigid pavement" and "concrete pavement" are synonymous. Rigid pavements have some flexural strength that permits them to sustain a beamlike action across minor irregularities in the underlying material. Thus, the minor irregularities may not be reflected in the concrete pavement. Properly designed and constructed rigid pavements have long service lives and usually are less expensive to maintain than the flexible pavements.

Thickness of highway concrete pavements normally ranges from 6 in. to 13 in. Different types of rigid pavements are described later in this chapter. These pavement types usually are constructed to carry heavy traffic loads, although they have been used for residential and local roads.

The topics discussed in this chapter cover the major aspects of rigid pavement design sufficiently for an understanding of the design principles of the basic types: plain, simply reinforced, and continuous. The topics covered include a description of the basic types of rigid pavements and the materials used, a discussion of the stresses imposed by traffic wheel loads and temperature differentials on the concrete pavement, and a description of two design methods.

MATERIALS USED IN RIGID PAVEMENTS

The Portland cement concrete commonly used for rigid pavements consists of Portland cement, coarse aggregate, fine aggregate, and water. Steel reinforcing rods may or may not be used, depending on the type of pavement being constructed. A description of the quality requirements for each of the basic materials is presented in the following sections.

Portland Cement

Portland cement is manufactured by crushing and pulverizing a carefully prepared mix of limestone, marl, and clay or shale and by burning the mixture at a high temperature (about 2800°F) to form a clinker. The clinker is then allowed to cool, a small quantity of gypsum is added, and then the mixture is ground until more than 90 percent of the material passes the No. 200 sieve. The main chemical constituents of the material are tricalcium silicate (C_3S), dicalcium silicate (C_2S), and tetracalcium alumino ferrite (C_4AF).[*]

The material is usually transported in 1 ft^3 bags, each weighing 94 lb, although it can also be transported in bulk for large projects.

Most highway agencies use either the American Society for Testing Materials (ASTM) specifications (ASTM Designation C150) or the American Association of State Highway and Transportation Officials (AASHTO) specifications (AASHTO Designation M85) for specifying Portland cement quality requirements used in their projects. The AASHTO specifications list five main types of Portland cement.

Type I is suitable for general concrete construction, where no special properties are required. A manufacturer will supply this type of cement when no specific type is requested.

Type II is suitable for use in general concrete construction, where the concrete will be exposed to moderate action of sulphate or where moderate heat of hydration is required.

Type III is suitable for concrete construction that requires a high concrete strength in a relatively short time. It is sometimes referred to as *high early strength cement.*

Types IA, IIA, and IIIA are similar to types I, II, and III, respectively, but contain a small amount (4 percent to 8 percent of total mix) of entrapped air. This is achieved during production by thoroughly mixing the cement with air-entraining agents and grinding the mixture. In addition to the properties listed for types I, II, and III, types IA, IIA, and IIIA are more resistant to calcium chloride and de-icing salts and are therefore more durable.

Type IV is suitable for projects where low heat of hydration is necessary, and type V is used in concrete construction projects where the concrete will be exposed to high sulphate action. Table 21.1 shows the proportions of the different chemical constituents for each of the five types of cement.

Coarse Aggregates

The coarse aggregates used in Portland cement concrete are inert materials that do not react with cement and are usually comprised of crushed gravel, stone, or blast furnace slag. The coarse aggregates may be any one of the three materials or else a combination of any two or all three. One of the major requirements for coarse aggregates used in Portland cement concrete is the gradation of the material. The material is well graded, with the

---

[*] In expressing compounds, C=CaO, S=SiO$_2$, A=Al$_2$O$_3$, F=Fe$_2$O$_3$. For example, C$_3$A=3CaO.Al$_2$O$_3$.

Table 21.1 Proportions of Chemical Constituents and Strength Characteristics for the Different Types of Portland Cement

Cement Type[A]	I and IA	II and IIA	III and IIIA	IV	V
Silicon dioxide (SiO_2), min, percent	—	20.0[B,C]	—	—	—
Aluminum oxide (Al_2O_3), max, percent	—	6.0	—	—	—
Ferric oxide (Fe_2O_3), max, percent	—	6.0[B,C]	—	6.5	—
Magnesium oxide (MgO), max, percent	6.0	6.0	6.0	6.0	6.0
Sulfur trioxide (SO_3),[D] max, percent					
When (C_3A)[E] is 8 percent or less	3.0	3.0	3.5	2.3	2.3
When (C_3A)[E] is more than 8 percent	3.5	F	4.5	F	F
Loss on ignition, max, percent	3.0	3.0	3.0	2.5	3.0
Insoluble residue, max, percent	0.75	0.75	0.75	0.75	0.75
Tricalcium silicate (C_3S)[E] max, percent	—	55	—	35[B]	—
Dicalcium silicate (C_2S)[E] min, percent	—	—	—	40[B]	—
Tricalcium aluminate (C_3A)[E] max, percent	—	8	15	7[B]	5[C]
Tetracalcium aluminoferrite plus twice the tricalcium aluminate[E] ($C_4AF + 2 (C_3A)$), or solid solution ($C_4AF + C_2F$), as applicable, max, percent	—	—	—	—	25[C]

[A] See source below.

[B] Does not apply when the heat of hydration limit is specified (see source below).

[C] Does not apply when the sulfate resistance limit is specified (see source below).

[D] There are cases where optimum SO_3 (using ASTM C 563) for a particular cement is close to or in excess of the limit in this specification. In such cases where properties of a cement can be improved by exceeding the SO_3 limits stated in this table, it is permissible to exceed the values in the table, provided it has been demonstrated by ASTM C 1038 that the cement with the increased SO_3 will not develop expansion in water exceeding 0.020 percent at 14 days. When the manufacturer supplies cement under this provision, he shall, upon request, supply supporting data to the purchaser.

[E] The expressing of chemical limitations by means of calculated assumed compounds does not necessarily mean that the oxides are actually or entirely present as such compounds.

When expressing compounds, C = CaO, S = SiO_2, A = Al_2O_3, F = Fe_2O_3. For example, $C_3A = 3CaO \cdot Al_2O_3$.

Titanium dioxide and phosphorus pentoxide (TiO_2 and P_2O_5) shall not be included with the Al_2O_3 content.

When the ratio of percentages of aluminum oxide to ferric oxide is 0.64 or more, the percentages of tricalcium silicate, dicalcium silicate, tricalcium aluminate, and tetracalcium aluminoferrite shall be calculated from the chemical analysis as follows:

Tricalcium silicate = (4.071 × percent CaO) − (7.600 × percent SiO_2) − (6.718 × percent Al_2O_3) − (1.430 × percent Fe_2O_3) − (2.852 × percent SO_3)

Dicalcium silicate = (2.867 × percent SiO_2) − (0.7544 × percent C_3S)

Tricalcium aluminate = (2.650 × percent Al_2O_3) − (1.692 × percent Fe_2O_3)

Tetracalcium aluminoferrite = 3.043 × percent Fe_2O_3

When the alumina-ferric oxide ratio is less than 0.64, a calcium aluminoferrite solid solution (expressed as ss($C_4AF + C_2F$)) is formed. Contents of this solid solution and of tricalcium silicate shall be calculated by the following formulas:

ss($C_4AF + C_2F$) = (2.100 × percent Al_2O_3) + (1.702 × percent Fe_2O_3)

Tricalcium silicate = (4.071 × percent CaO) − (7.600 × percent SiO_2) − (4.479 × percent Al_2O_3) − (2.859 × percent Fe_2O_3) − (2.852 × percent SO_3). No tricalcium aluminate will be present in cements of this composition. Dicalcium silicate shall be caculated as previously shown.

[F] Not applicable.

SOURCE: Adapted with permission from *Standard Specifications for Transportation Materials and Methods of Sampling and Testing*, 20th ed., American Association of State Highway and Transportation Officials, Washington, D.C., 2000.

maximum size specified. Material retained in a No. 4 sieve is considered coarse aggregate. Table 21.2 shows gradation requirements for different maximum sizes as stipulated by ASTM.

Coarse aggregates must be clean. This is achieved by specifying the maximum percentage of deleterious substances allowed in the material. Other quality requirements include the ability of the aggregates to resist abrasion and the soundness of the aggregates.

A special test, known as the Los Angeles Rattler Test (AASHTO Designation T96), is used to determine the abrasive quality of the aggregates. In this test, a sample of the coarse aggregate retained in the No. 8 sieve and a specified number of standard steel spheres are placed in a hollow steel cylinder, with diameter 28 in. and length 20 in., that is closed at both ends. The cylinder, which also contains a steel shelf that projects $3\frac{1}{2}$ in. radially inward, is mounted with its axis in the horizontal position. The cylinder is then rotated 500 times at a specific speed. The sample of coarse aggregate is then removed and sieved on a No. 12 sieve. The portion of the material retained on the No. 12 sieve is weighed, and the difference between this weight and the original weight is the weight loss. This loss is expressed as a percentage of the original weight. Maximum permissible loss in weight ranges from 30 percent to 60 percent, depending on the specifications used; however, a maximum of 40 percent to 50 percent has proved to be generally acceptable.

Soundness is defined as the ability of the aggregate to resist breaking up due to freezing and thawing. This property can be determined in the laboratory by first sieving a sample of the coarse aggregate through a No. 4 sieve, and then immersing in water the portion retained in the sieve. The sample is frozen in the water for 2 hours and thawed for $\frac{1}{2}$ hr. This alternate freezing and thawing is repeated between 20 and 50 times. The sample is then dried and sieved again to determine the change in particle size. Sodium or magnesium sulphate may be used instead of water.

Table 21.2 Gradation Requirements for Aggregates in Portland Cement Concrete (ASTM Designation C33)

Sieve Designation	Percent Passing by Weight		
	Aggregate Designation		
	2 in. to No. 4 (357)	1½ in. to No. 4 (467)	1 in. to No. 4 (57)
2½ in. (63 mm)	100	—	—
2 in. (50 mm)	95–100	100	—
1½ in. (37.5 mm)	—	95–100	100
1 in (25.0 mm)	35–70	—	95–100
¾ in. (19.0 mm)	—	35–70	—
½ in. (12.5 mm)	10–30	—	25–60
⅜ in. (9.5 mm)	—	10–30	—
No. 4 (4.75 mm)	0–5	0–5	0–10
No. 8 (2.36 mm)	—	—	0–5

SOURCE: Adapted from *ASTM Standards, Concrete and Aggregates,* Vol. 04.02, American Society for Testing and Materials, Philadelphia, Pa., October 2000.

Fine Aggregates

Sand is mainly used as the fine aggregate in Portland cement concrete. Specifications for this material usually include grading requirements, soundness, and cleanliness. Standard specifications for the fine aggregates for Portland cement concrete (AASHTO Designation M6) give grading requirements normally adopted by state highway agencies (see Table 21.3).

The soundness requirement usually is given in terms of the maximum permitted loss in the material after 5 alternate cycles of wetting and drying in the soundness test. A maximum of 10 percent weight loss is usually specified.

Cleanliness often is specified in terms of the maximum amounts of different types of deleterious materials contained in the fine aggregates. For example, maximum amount of silt (material passing No. 200 sieve) is usually specified within a range of 2 percent to 5 percent of the total fine aggregates. Since the presence of large amounts of organic material in the fine aggregates may reduce the hardening properties of the cement, a standard test (AASHTO Designation T21) is also usually specified as part of the cleanliness requirements. In this test, a sample of the fine aggregates is mixed with sodium hydroxide solution and then allowed to stand for 24 hours. At the end of the 24 hours, the amount of light transmitted through the liquid floating above the test sample is compared with that transmitted through a standard color solution of reagent grade potassium dichromate ($K_2Cr_2O_2$) and concentrated sulfuric acid. The fine aggregate under test is considered to possibly contain injurious organic compounds, if less light is transmitted through the liquid floating above the test sample. In such cases, additional tests should be carried out before using the fine aggregate in the concrete mix. For example, the sand can be used only if the strength developed by 2-in. cubes made with this sand is at least 95 percent of that developed by similar cubes made with the same sand, after washing it in a 3 percent hydroxide solution.

Water

The main water requirement stipulated is that the water used should also be suitable for drinking. This requires that the quantity of organic matter, oil, acids, and alkalies should not be greater than the allowable amount in drinking water.

Table 21.3   AASHTO-Recommended Particle Size Distribution for Fine Aggregates Used in Portland Cement Concrete

Sieve (M 92)	Mass Percent Passing
3/8 in. (9.5 mm)	100
No. 4 (4.75 mm)	95 to 100
No. 8 (2.36 mm)	80 to 100
No. 16 (1.18 mm)	50 to 85
No. 30 (600 µm)	25 to 60
No. 50 (300 µm)	10 to 30
No. 100 (µm)	2 to 10

SOURCE: Adapted with permission from *Standard Specifications for Transportation Materials and Methods of Sampling and Testing,* 20th ed., American Association of State Highway and Transportation Officials, Washington, D.C., 2000.

Reinforcing Steel

Steel reinforcing may be used in concrete pavements to reduce the amount of cracking that occurs, as a load transfer mechanism at joints, or as a means of tying two slabs together. Steel reinforcement used to control cracking is usually referred to as *temperature steel,* whereas steel rods used as load transfer mechanisms are known as *dowel bars,* and those used to connect two slabs together are known as *tie bars.*

Temperature Steel

Temperature steel is provided in the form of a bar mat or wire mesh consisting of longitudinal and transverse steel wires welded at regular intervals. The mesh is usually placed about 3 in. below the slab surface. The cross-sectional area of the steel provided per foot width of the slab depends on the size and spacing of the steel wires forming the mesh. The amount of steel required depends on the length of the pavement between expansion joints, the maximum stress desired in the concrete pavement, the thickness of the pavement, and the moduli of elasticity of the concrete and steel. Equation 21.19 (developed later in this chapter) can be used to determine the area of steel required if the length of the slab is fixed. However, steel areas obtained by this equation for concrete slabs less than 45 ft may be inadequate and therefore should be compared with the following general guidelines for the minimum cross-sectional area of the temperature steel.

1. Cross-sectional area of longitudinal steel should be at least equal to 0.1 percent of the cross-sectional area of the slab.
2. Longitudinal wires should not be less than No. 2 gauge, spaced at a maximum distance of 6 in.
3. Transverse wires should not be less than No. 4 gauge, spaced at a maximum distance of 12 in.

Temperature steel does not prevent cracking of the slab, but it does control the crack widths because the steel acts as a tie holding the edges of the cracks together. This helps to maintain the shearing resistance of the pavement, thereby maintaining its capacity to carry traffic load, even though the flexural strength is not improved.

Dowel Bars

Dowel bars are used mainly as load-transfer mechanisms across joints. They provide flexural, shearing, and bearing resistance. The dowel bars must be of a much larger diameter than the wires used in temperature steel. Size selection is based mainly on experience. Diameters of 1 to $1\frac{1}{2}$ in. and lengths of 2 to 3 ft have been used, with the bars usually spaced at 1 ft centers across the width of the slab. At least one end of the bar should be smooth and lubricated to facilitate free expansion.

Tie Bars

Tie bars are used to tie two sections of the pavement together, and therefore they should be either deformed bars or should contain hooks to facilitate the bonding of the two sections of the concrete pavement with the bar. These bars are usually much smaller in diameter than the dowel bars and are spaced at larger centers. Typical diameter and spacing for these bars are ¾ in. and 3 ft, respectively.

JOINTS IN CONCRETE PAVEMENTS

Different types of joints are placed in concrete pavements to limit the stresses induced by temperature changes and to facilitate proper bonding of two adjacent sections of pavement when there is a time lapse between their construction (for example, between the end of one day's work and the beginning of the next). These joints can be divided into four basic categories:

- Expansion joints
- Contraction joints
- Hinge joints
- Construction joints

Expansion Joints

When concrete pavement is subjected to an increase in temperature, it will expand, resulting in an increase in length of the slab. When the temperature is sufficiently high, the slab may buckle or "blow up" if it is sufficiently long and if no provision is made to accommodate the increased length. Therefore, *expansion joints* are usually placed transversely, at regular intervals, to provide adequate space for the slab to expand. These joints are placed across the full width of the slab and are ¾ to 1 in. wide in the longitudinal direction. They must create a distinct break throughout the depth of the slab. The joint space is filled with a compressible filler material that permits the slab to expand. Filler materials can be cork, rubber, bituminous materials, or bituminous fabrics.

A means of transferring the load across the joint space must be provided since there are no aggregates that will develop an interlocking mechanism. The load-transfer mechanism is usually a smooth dowel bar that is lubricated on one side. An expansion cap is also usually installed, as shown in Figure 21.1, to provide a space for the dowel to occupy during expansion.

Some states no longer use expansion joints because of the inability of the load-transfer mechanism to adequately transfer the load. Other states continue to use expansion joints and may even use them in place of construction joints.

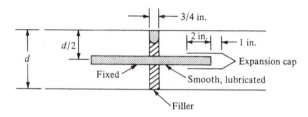

Figure 21.1 Typical Expansion Joint

Contraction Joints

When concrete pavement is subjected to a decrease in temperature, the slab will contract if it is free to move. Prevention of this contraction movement will induce tensile stresses in the concrete pavement. *Contraction joints* are therefore placed transversely at regular intervals across the width of the pavement to release some of the tensile stresses that are so induced. A typical contraction joint is shown in Figure 21.2. It may be necessary in some cases to install a load-transfer mechanism in the form of a dowel bar when there is doubt about the ability of the interlocking grains to transfer the load.

Hinge Joints

Hinge joints are used mainly to reduce cracking along the center line of highway pavements. Figure 21.3 shows a typical hinge joint (keyed joint) suitable for single-lane-at-a-time construction.

Construction Joints

Construction joints are placed transversely across the pavement width to provide suitable transition between concrete laid at different times. For example, a construction joint is usually placed at the end of a day's pour to provide suitable bonding with the start of the next day's pour. A typical butt construction joint is shown in Figure 21.4. In some cases, as shown in Figure 21.3, a keyed construction joint may also be used in the longitudinal direction when only a single lane is constructed at a time. In this case, alternate lanes of the pavement are cast, and the key is formed by using metal formwork that

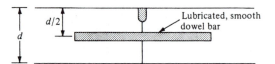

Figure 21.2 Typical Contraction Joint

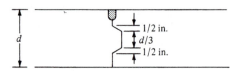

Figure 21.3 Typical Hinge Joint (keyed joint)

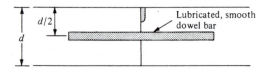

Figure 21.4 Typical Butt Joint

has been cast with the shape of the groove or by attaching a piece of metal or wood to a wooden formwork. An expansion joint can be used in lieu of a transverse construction joint in cases where the construction joint falls at or near the same position as the expansion joint.

TYPES OF RIGID HIGHWAY PAVEMENTS

Rigid highway pavements can be divided into three general types: plain concrete pavements, simply reinforced concrete pavements, and continuously reinforced concrete pavements. The definition of each pavement type is related to the amount of reinforcement used.

Plain Concrete Pavement

Plain concrete pavement has no temperature steel or dowels for load transfer. However, steel tie bars are often used to provide a hinge effect at longitudinal joints and to prevent the opening of these joints. Plain concrete pavements are used mainly on low-volume highways or when cement-stabilized soils are used as subbase. Joints are placed at relatively shorter distances (10 to 20 ft) than with the other types of concrete pavements to reduce the amount of cracking. In some cases, the transverse joints of plain concrete pavements are skewed about 4 to 5 ft in plan, such that only one wheel of a vehicle passes through the joint at a time. This helps to provide a smoother ride.

Simply Reinforced Concrete Pavement

Simply reinforced concrete pavements have dowels for the transfer of traffic loads across joints, with these joints spaced at larger distances, ranging from 30 to 100 ft. Temperature steel is used throughout the slab, with the amount dependent on the length of the slab. Tie bars are also commonly used at longitudinal joints.

Continuously Reinforced Concrete Pavement

Continuously reinforced concrete pavements have no transverse joints, except construction joints or expansion joints when they are necessary at specific positions, such as at bridges. These pavements have a relatively high percentage of steel, with the minimum usually at 0.6 percent of the cross section of the slab. They also contain tie bars across the longitudinal joints.

PUMPING OF RIGID PAVEMENTS

Pumping is an important phenomenon associated with rigid pavements. Pumping is the discharge of water and subgrade (or subbase) material through joints, cracks, and along the pavement edges. It is primarily caused by the repeated deflection of the pavement slab in the presence of accumulated water beneath it. The mechanics of pumping can be explained best by considering the sequence of events that lead to it.

The first event is the formation of void space beneath the pavement. This void forms from either the combination of the plastic deformation of the soil, due to imposed loads and the elastic rebound of the pavement after it has been deflected by the imposed load, or warping of the pavement, which occurs as a result of temperature gradient within the slab. Water then accumulates in the space after many repetitions of traffic load. The water may be infiltrated from the surface through joints and the pavement edge. To a lesser extent, ground water may settle in the void. If the subgrade or base material is granular, the water will freely drain through the soil. If the material is fine-grained, however, the water is not easily discharged, and additional load repetitions will result in the soil going into suspension with the water to form a slurry. Further load repetitions and deflections of the slab will result in the slurry being ejected to the surface (pumping). Pumping action will then continue, with the result that a relatively large void space is formed underneath the concrete slab. This results in faulting of the joints and eventually the formation of transverse cracks or the breaking of the corners of the slab. Joint faulting and cracking is therefore progressive, since formation of a crack facilitates the pumping action.

Visual manifestations of pumping include:

- Discharge of water from cracks and joints
- Spalling near the centerline of the pavement and a transverse crack or joint
- Mud boils at the edge of the pavement
- Pavement surface discoloration (caused by the subgrade soil)
- Breaking of pavement at the corners

Design Considerations for Preventing Pumping

A major design consideration for preventing pumping is the reduction or elimination of expansion joints, since pumping is usually associated with these joints. This is the main reason why current design practices limit the number of expansion joints to a minimum. Since pumping is also associated with fine-grained soils, another design consideration is either to replace soils that are susceptible to pumping with a nominal thickness of granular or sandy soils, or to improve them by stabilization. Current design practices therefore usually include the use of 3 to 6 in. layers of granular subbase material at areas along the pavement alignment where the subgrade material is susceptible to pumping or stabilizing the susceptible soil with asphaltic material or Portland cement. The Portland Cement Association method of rigid pavement design indirectly considers this phenomenon in the erosion analysis.

STRESSES IN RIGID PAVEMENTS

Stresses are developed in rigid pavements as a result of several factors, including the action of traffic wheel loads, the expansion and contraction of the concrete due to temperature changes, yielding of the subbase or subgrade supporting the concrete pavement, and volumetric changes. For example, traffic wheel loads will induce flexural stresses that are dependent on the location of the vehicle wheels relative to the edge of the pavement, whereas expansion and contraction may induce tensile and compressive stresses, which are dependent on the range of temperature changes in the concrete pavement. These different factors that can induce stress in concrete pavement have made the theoretical determination of stresses rather complex, requiring the following simplifying assumptions.

1. Concrete pavement slabs are considered as unreinforced concrete beams. Any contribution made to the flexural strength by the inclusion of reinforcing steel is neglected.
2. The combination of flexural and direct tensile stresses will inevitably result in transverse and longitudinal cracks. The provision of suitable crack control in the form of joints, however, controls the occurrence of these cracks, thereby maintaining the beam action of large sections of the pavement.
3. The supporting subbase and/or subgrade layer acts as an elastic material in that it deflects at the application of the traffic load and recovers at the removal of the load.

Stresses Induced by Bending

The ability of rigid pavement to sustain a beamlike action across irregularities in the underlying materials suggests that the theory of bending is fundamental to the analysis of stresses in such pavements. The theory of a beam supported on an elastic foundation can therefore be used to analyze the stresses in the pavement when it is externally loaded. Figure 21.5 shows the deformation sustained by a beam on an elastic foundation when it is loaded externally. The stresses developed in the beam may be analyzed by assuming that a reactive pressure (p), which is proportional to the deflection, is developed as a result of the applied load. This pressure is given as

$$p = ky \tag{21.1}$$

where

$\quad p$ = reactive pressure at any point beneath the beam (lb/in.^{2})
$\quad y$ = deflection at the point (in.)
$\quad k$ = modulus of subgrade reaction (lb/in.^{3})

The modulus of subgrade reaction is the stress (lb/in.^{2}) that will cause an inch deflection of the underlying soil. Equation 21.1 assumes that k (lb/in.^{3}) is constant, which implies that the subgrade is elastic. However, this assumption is valid for only a limited range of different factors. Research has shown that the value of k depends on certain soil characteristics such as density, moisture, soil texture, and other factors that influence the strength of the soils. The k value of a particular soil will also vary with the size of the loaded area and the amount of deflection. The modulus of subgrade reaction is directly proportional to the loaded area and inversely proportional to the deflection. In pavement design, however, minor changes in k do not have significant impact on design results, and an average value is usually assumed. The plate-bearing test is commonly used for determining the value of k in the field.

$p = ky$

Figure 21.5 Deformation of a Beam on Elastic Foundation

A general relationship between the moment and the radius of curvature of a beam is given as

$$\frac{1}{R} = \frac{M}{EI}$$

(21.2)

where
  R = radius of curvature
  M = moment in beam
  E = modulus of elasticity
  I = moment of intertia

The general differential equation relating the moment at any section of a beam with the deflection at that section is given as

$$M = EI\frac{d^2 y}{dx^2}$$

(21.3)

whereas the basic differential equation for the deflection on an elastic foundation is given as

$$EI\frac{d^4 y}{dx^4} = - ky = q$$

(21.4)

The basic differential equation for a slab is given as

$$M_x = \left[\frac{Eh^3}{12(1 - \mu^2)}\right]\frac{d^2 w}{dx^2}$$

or

$$M_y = \left[\frac{Eh^3}{12(1 - \mu^2)}\right]\frac{d^2 w}{dy^2}$$

(21.5)

where
  h = thickness of slab
  μ = Poisson ratio of concrete
  w = deflection of the slab at a given point
  M_x = bending moment at a point about the x axis
  M_y = bending moment at a point about the y axis

The EI term in Eq. 21.4 is called the *stiffness* of the beam, whereas the stiffness of the slab is given by the term within the square brackets of Eq. 21.5. This term is usually denoted as D, where

$$D = \frac{E_c h^3}{12(1 - \mu^2)} \tag{21.6}$$

In developing expressions for the stresses in a concrete pavement, Westergaard made use of the radius of relative stiffness, which depends on the stiffness of the slab and the modulus of subgrade reaction of the soil. It is given as

$$\ell = \sqrt[4]{\frac{E_c h^3}{12(1 - \mu^2)k}} \tag{21.7}$$

where

ℓ = radius of relative stiffness (in.)
E_c = modulus of elasticity of the concrete pavement (lb/in.^{2})
h = thickness of pavement (in.)
μ = Poisson ratio of the concrete pavement
k = modulus of subgrade reaction (lb/in.^{3})

It will be seen later that the radius of relative stiffness is an important parameter in the equations used to determine various stresses in the concrete pavement.

Stresses Due to Traffic Wheel Loads

The basic equations for determining flexural stresses in concrete pavements due to traffic wheel loads were first developed by Westergaard. Although several theoretical developments have been made since then, the Westergaard equations are still considered a fundamental tool for evaluating the stresses on concrete pavements. Westergaard considered three critical locations of the wheel load on the concrete pavement in developing the equations. These locations are shown in Figure 21.6 and are described as follows:

- **Case A.** Load is applied at the corner of a rectangular slab. This provides for the cases when the wheel load is applied at the intersection of the pavement edge and a transverse joint. However, this condition is not common because pavements are generally much wider. Thus no equation is presented for this case.
- **Case B.** Load is applied at the interior of the slab at a considerable distance from its edges.
- **Case C.** Load is applied at the edge of the slab at considerable distance away from any corner.

The locations shown as Cases I, II, and III in Figure 21.6 are the critical locations presently used for the relatively wide pavements now being constructed.

The equations for determining these stresses were developed taking into consideration the different day and night temperature conditions that may exist. During the day, the temperature is higher at the surface of the slab than at the bottom. This temperature gradient through the depth of the slab will create a tendency for the slab edges to curl downward. During the night, however, the temperature at the bottom of the slab is higher than at the surface, thereby reversing the temperature gradient, which results in the

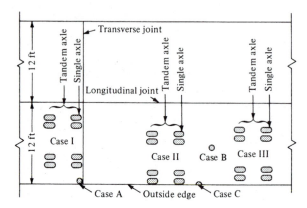

Figure 21.6 Critical Locations of Wheel Loads on Concrete Pavements

tendency for the slab edges to curl upward. The equations for stresses due to traffic load reflect this phenomenon of concrete pavements.

The original equations developed by Westergaard were modified, using the results of full-scale tests conducted by the Bureau of Public Roads. These modified equations for the different loading conditions are as follows:

1. Edge loading when the edges of the slab are warped upward at night

$$\sigma_e = \frac{0.572P}{h^2}\left[4\log_{10}\left(\frac{\ell}{b}\right) + \log_{10} b\right] \tag{21.8}$$

2. Edge loading when the slab is unwarped or when the edge of the slab is curled downward in daytime

$$\sigma_e = \frac{0.572P}{h^2}\left[4\log_{10}\left(\frac{\ell}{b}\right) + 0.359\right] \tag{21.9}$$

3. Interior loading

$$\sigma_i = \frac{0.316P}{h^2}\left[4\log_{10}\left(\frac{\ell}{b}\right) + 1.069\right] \tag{21.10}$$

where
σ_e = maximum stress (lb/in.^{2}) induced in the bottom of the slab, directly under the load P and applied at the edge and in a direction parallel to the edge
σ_i = maximum tensile stress (lb/in.^{2}) induced at the bottom of the slab directly under the load P applied at the interior of the slab

P = applied load in pounds, including allowance for impact

h = thickness of slab (in.)

ℓ = radius of relative stiffness

$$= \sqrt[4]{E_c h^3 / [12(1-\mu^2)k]}$$

E_c = modulus of elasticity of concrete (lb/in.^{2})

μ = Poisson ratio for concrete = 0.15

k = subgrade modulus (lb/in.^{3})

b = radius of equivalent distribution of pressure (in.)

$$= \sqrt{1.6a^2 + h^2} - 0.675h \ \text{(for } a < 1.724h)$$

$$= a \qquad\qquad\qquad \text{(for } a > 1.724h)$$

a = radius of contact area of load (in.) (Contact area is usually assumed as a circle for interior and corner loadings and semicircle for edge loading.)

Revised equations for the edge loadings have been developed by Ioannides et al. and are given as Eqs. 21.11 and 21.12.

For a circular loaded area

$$\sigma_e = \frac{0.803P}{h^2}\left[4\log_{10}\left(\frac{\ell}{a}\right) + 0.666\frac{a}{\ell} - 0.034\right] \tag{21.11}$$

For a semicircular loaded area,

$$\sigma_e = \frac{0.803P}{h^2}\left[4\log_{10}\left(\frac{\ell}{a}\right) + 0.282\frac{a}{\ell} + 0.650\right] \tag{21.12}$$

It should be noted that the above equations assume a value of 0.15 for the Poisson ratio of the concrete pavement.

The results obtained from Ioannides' formulas differ significantly from those obtained from Westergaard's formulas, as illustrated in Example 21.1.

Example 21.1   Computing Tensile Stress Resulting from a Wheel Load on a Rigid Pavement

Determine the tensile stress imposed by a semicircular wheel load of 900 lb imposed during the day and located at the edge of a concrete pavement with the following dimensions and properties (a) by using the Westergaard equation, and (b) by using the Ioannides equation.

Pavement thickness = 6 in.

$\mu = 0.15$

$E = 5 \times 10^6$ lb/in.^{2}

$k = 130$ lb/in.^{3}

radius of loaded area = 3 in.

Solution for **(a)** Use the Westergaard equation.

From Eq. 21.9,

$$\sigma_e = \frac{0.572P}{h^2}\left[4\log_{10}\left(\frac{\ell}{b}\right) + 0.359\right]$$

$$a = 3 \text{ in. } < 1.724h$$

Since $a < 1.724 \times 6$, we use the expressions

$$b = \sqrt{1.6a^2 + h^2} - 0.675h = \sqrt{1.6 \times 3^2 + 6^2} - 0.675 \times 6 = 7.1 - 4.05$$
$$= 3.05$$

$$\ell = \sqrt[4]{(5 \times 10^6 \times 6^3)/[12(1 - 0.15^2)130]} = 29.0$$

$$\sigma_e = [(0.572 \times 900)/6^2][4\log_{10}(29/3.05) + 0.359 = 14.3(4 \times 0.978 + 0.359)]$$
$$= 61.07 \text{ lb/in.}^2$$

Solution for **(b)** Use the Ioannides equation.

From Eq. 21.12,

$$\sigma_e = \frac{0.803P}{h^2}\left[4\log\left(\frac{\ell}{a}\right) + 0.282\left(\frac{a}{\ell}\right) + 0.650\right]$$

$$= \left[\frac{0.803 \times 900}{6^2}\right]\left[4\log_{10}\left(\frac{29}{3.05}\right) + 0.282\frac{3.05}{29} + 0.650\right]$$

$$= 92.19 \text{ lb/in.}^2$$

Stresses Due to Temperature Effects

The tendency of the slab edges to curl downward during the day and upward during the night as a result of temperature gradients is resisted by the weight of the slab itself. This resistance tends to keep the slab in its original position, resulting in stresses being induced in the pavement. Compressive and tensile stresses are therefore induced at the top and bottom of the slab, respectively, during the day, whereas tensile stresses are induced at the top and compressive stresses at the bottom during the night.

Under certain conditions these curling stresses may have values high enough to cause cracking of the pavement. They may also reduce the subgrade support beneath some sections of the pavement, which can result in a considerable increase of the stresses due to traffic loads over those pavements with uniform pavement support. Studies have shown that curling stresses can be higher than 200 lb/in.² for 10 ft slabs and much higher for wider slabs. One of the main purposes of longitudinal joints is to limit the slab width by dividing the concrete pavement into individual slabs 11 or 12 ft wide.

Tests have shown that maximum temperature differences between the top and bottom of the slab depend on the thickness of the slab, and that these differences are about 2.5° to 3°F per inch thickness for 6 to 9 in. thick slabs. The temperature differential also depends on the season, with maximum differentials occurring during the day in the spring and summer months. Another factor that affects the temperature differential is the latitude of the location of the slab. The surface temperature of the pavement tends to be high if the angle of incidence of the sun's rays is high, as in areas near the equator.

These curling stresses can be determined from Eqs. 21.13 and 21.14. Note, however, that curling stresses are not normally taken into consideration in pavement thickness design, since joints and steel reinforcement are normally used to reduce the effect of such stresses.

$$\sigma_{xe} = \frac{C_x E_c et}{2} \tag{21.13}$$

and

$$\sigma_{xi} = \frac{E_c et}{2} \left(\frac{C_x + \mu C_y}{1 - \mu^2} \right) \tag{21.14}$$

where
σ_{xe} = maximum curling stress in lb/in.² at the edge of the slab in the direction of the slab length due to temperature difference between the top and bottom of the slab

σ_{xi} = maximum curling stress in lb/in.² at the interior of the slab in the direction of the slab length due to temperature difference between the top and bottom of the slab

E_c = modulus of elasticity for concrete (lb/in.²)

μ = Poisson ratio for concrete

C_x, C_y = coefficients that are dependent on the radius of relative stiffness of the concrete and can be obtained from Figure 21.7

e = thermal coefficient of expansion and contraction of concrete per degree Fahrenheit

t = temperature difference between the top and bottom of the slab (°F)

Temperature changes in the slab will also result in expansion (for increased temperature) and contraction (for reduced temperature). The provision of suitable expansion/contraction joints in the slab reduces the magnitude of these stresses but does not entirely eliminate them, since considerable resistance to the free horizontal movement of the pavement will still be offered by the subgrade, due to friction action between the bottom of the

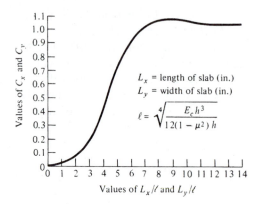

Figure 21.7 Values of C_x and C_y for Use in Formulas for Curling Stresses

SOURCE: Redrawn from R. D. Bradbury, *Reinforced Concrete Pavements,* Wire Reinforcement Institute, Washington, D.C., 1938.

slab and the top of the subgrade. The magnitude of these stresses depends on the length of the slab, the type of concrete pavement, the magnitude of the temperature changes, and the coefficient of friction between the pavement and the subgrade. Tensile stresses greater than 100 lb/in.^{2} have been reported for an average temperature change of 40°F in a slab with a length of 100 ft.

If only the effect of uniform temperature drop is being considered, the maximum spacing of contraction joints required to ensure that the maximum stress does not exceed a desired value p_c can be estimated by equating both the force in the slab and the force developed due to friction, as follows.

Considering a unit width of the pavement, the frictional force (F) developed due to a uniform drop in temperature is

$$F = f\,\frac{h}{12}\,(1)\,\frac{L}{2}\,\gamma_c \tag{21.15}$$

where
f = coefficient of friction between the bottom of concrete pavement and subgrade
h = thickness of concrete pavement (in.)
L = length of pavement between contraction joints (ft)
γ_c = density of concrete (lb/ft^3)

The force (P) developed in the concrete due to stress p_c is given as

$$P = p_c\,(12)(1)h \tag{21.16}$$

where p_c is the maximum desired stress in the concrete in lb/in.^{2}.

Equating these two forces gives

$$f \frac{h}{12} (1) \frac{L}{2} \gamma_c = p_c (12)(1)(h) \tag{21.17}$$

and

$$L = \frac{288 p_c}{f \gamma_c} \tag{21.18}$$

The effects of temperature changes can also be reduced by including reinforcing steel in the concrete pavement. The additional force developed by the steel may also be taken into consideration in determining L. If A_s is the total cross-sectional area of steel per foot width of slab, Eq. 21.15 becomes

$$f \frac{h}{12} (1) \frac{L}{2} \gamma_c = p_c (12h + n A_s) \tag{21.19}$$

or

$$L = \frac{24 p_c}{h \gamma_c f} (12h + n A_s) \tag{21.20}$$

where n = modular ratio = E_s/E_c.

THICKNESS DESIGN OF RIGID PAVEMENTS

The main objective in rigid pavement design is to determine the thickness of the concrete slab that will be adequate to carry the projected traffic load for the design period. Several design methods have been developed over the years, some of which are based on the results of full-scale road tests, others on theoretical development of stresses on layered systems, and others on the combination of the results of tests and theoretical development. However, two methods are used extensively: the AASHTO method and the Portland Cement Association (PCA) method.

AASHTO Design Method

The AASHTO method for rigid pavement design is based mainly on the results obtained from the AASHTO road test (Chapter 20). The design procedure was initially published in the early 1960s but was revised in the 1970s and 1980s. A further revision has been carried out since then, incorporating new developments. The design procedure provides for the determination of the pavement thickness and the amount of steel reinforcement when used, as well as the design of joints. It is suitable for plain concrete, simply reinforced concrete, and continuously reinforced concrete pavements. The design procedure for the longitudinal reinforcing steel in continuously reinforced concrete pavements, however, is beyond the scope of this book. Interested readers should refer to the AASHTO guide for this procedure.

Design Considerations

The factors considered in the AASHTO procedure for the design of rigid pavements as presented in the 1993 guide are:

- Pavement performance
- Subgrade strength
- Subbase strength
- Traffic
- Concrete properties
- Drainage
- Reliability

Pavement Performance. Pavement performance is considered in the same way as for flexible pavement, as presented in Chapter 20. The initial serviceability index (P_i) may be taken as 4.5, and the terminal serviceability index may also be selected by the designer.

Subbase Strength. The guide allows the use of either graded granular materials or suitably stabilized materials for the subbase layer. Table 21.4 gives recommended specifications for six types of subbase materials. AASHTO suggests that the first five types—A through E—can be used within the upper 4 in. layer of the subbase, whereas type F can be used below the uppermost 4 in. layer. Special precautions should be taken when certain conditions exist. For example, when A, B, and F materials are used in areas where the pavement may be subjected to frost action, the percentage of fines should be reduced to a minimum. Subbase thickness is usually not less than 6 in. and should be extended 1 to 3 ft outside the edge of the pavement structure.

Subgrade Strength. The strength of the subgrade is given in terms of the Westergaard modulus of subgrade reaction k, which is defined as the load in pounds per square inch on a loaded area, divided by the deformation in inches. Values of k can be obtained by conducting a plate-bearing test in accordance with the AASHTO test Designation T222 using a 30 in. diameter plate. Estimates of k values can also be made either from experience or by correlating with other tests. Figure 21.8 shows an approximate interrelationship of bearing values obtained from different types of tests.

The guide also provides for the determination of an effective modulus of subgrade reaction, which depends on (1) the seasonal effect on the resilient modulus of the subgrade, (2) the type and thickness of the subbase material used, (3) the effect of potential erosion of the subbase, and (4) whether bedrock lies within 10 ft of the subgrade surface. The seasonal effect on the resilient modulus of the subgrade was discussed in Chapter 20, and a procedure similar to that used in flexible pavement design is used here to take into consideration the variation of the resilient modulus during a 12-month period.

Since different types of subbase materials have different strengths, the type of material used is an important input in the determination of the effective modulus of subgrade reaction. In estimating the composite modulus of subgrade reaction, the subbase material is defined in terms of its elastic modulus E_{SB}. Also, it is necessary to consider the combination of material types and the required thicknesses because this serves as

Table 21.4 Recommended Particle Size Distributions for Different Types of Subbase Materials

Sieve Designation	Type A	Type B	Type C (Cement Treated)	Type D (Lime Treated)	Type E (Bituminous Treated)	Type F (Granular)
			Types of Subbase			
Sieve analysis percent passing						
2 in.	100	100	—	—	—	—
1 in.	—	75–95	100	100	100	100
$\frac{3}{8}$ in.	30–65	40–75	50–85	60–100	—	—
No. 4	25–55	30–60	35–65	50–85	55–100	70–100
No. 10	15–40	20–45	25–50	40–70	40–100	55–100
No. 40	8–20	15–30	15–30	25–45	20–50	30–70
No. 200	2–8	5–20	5–15	5–20	6–20	8–25
	(The minus No. 200 material should be held to a practical minimum.)					
Compressive strength lb/in.^{2} at 28 days			400–750	100		
Stability						
Hveem Stabilometer					20 min.	
Hubbard field					1000 min.	
Marshall stability					500 min.	
Marshall flow					20 max.	
Soil constants						
Liquid limit	25 max.	25 max.				25 max.
Plasticity index[a]	N.P.	6 max.	10 max.[b]		6 max.[b]	6 max.

[a]As performed on samples prepared in accordance with AASHTO Designation T87.
[b]These values apply to the mineral aggregate prior to mixing with the stabilizing agent.
SOURCE: Adapted with permission from *Standard Specifications for Transportation Materials and Methods of Sampling and Testing,* 20th ed., American Association of State Highway and Transportation Officials, Washington, D.C., 2000.

a basis for determining the cost-effectiveness of the pavement. The chart in Figure 21.9 is used to estimate the composite modulus of subgrade reaction for the type of subbase material, based on its elastic modulus, its resilient modulus, and the thickness of the subbase.

The effective k value also depends on the potential erosion of the subbase material. This effect is included by the use of a factor (see Table 21.5) for the loss of support (LS) in determining the effective k value. This factor is used to reduce the effective modulus of subgrade reaction, as shown in Figure 21.10.

The presence of bedrock, within a depth of 10 ft of the subgrade surface and extending over a significant length along the highway alignment, may result in an increase of the overall modulus of subgrade reaction. This effect is taken into consideration by adjusting the effective modulus subgrade using the chart in Figure 21.11. The procedure is demonstrated in the solution of Example 21.2.

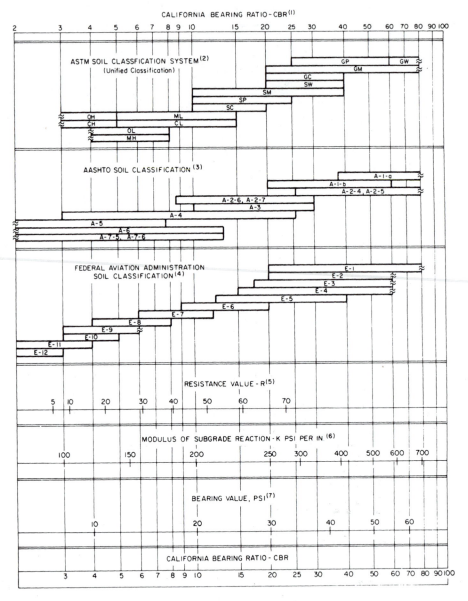

(1) For the basic idea, see O. J. Porter, "Foundations for Flexible Pavements," Highway Research Board *Proceedings of the Twenty-second Annual Meeting*, 1942, Vol. 22, pages 100-136.
(2) ASTM Designation D2487.
(3) "Classification of Highway Subgrade Materials," Highway Research Board *Proceedings of the Twenty-fifth Annual Meeting*, 1945, Vol. 25, pages 376-392.
(4) *Airport Paving*, U.S. Department of Commerce, Federal Aviation Agency, May 1948, pages 11-16. Estimated using values given in FAA *Design Manual for Airport Pavements*. (Formerly used FAA Classification; Unified Classification now used.)
(5) C. E. Warnes, "Correlation Between *R* Value and *k* Value," unpublished report, Portland Cement Association, Rocky Mountain-Northwest Region, October 1971 (best-fit correlation with correction for saturation).
(6) See T. A. Middlebrooks and G. E. Bertram, "Soil Tests for Design of Runway Pavements," Highway Research Board *Proceedings of the Twenty-second Annual Meeting*, 1942, Vol. 22, page 152.

Figure 21.8 Approximate Interrelationship of Soil Classification and Bearing Values

SOURCE: R. G. Packard, *Thickness Design for Concrete Highway and Street Pavements*, Portland Cement Association, Skokie, Ill., 1984.

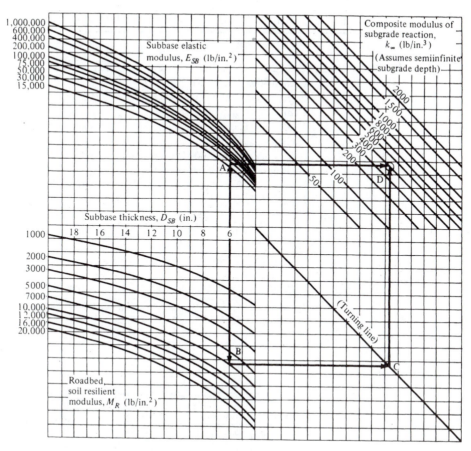

Figure 21.9 Chart for Estimating Composite Modulus of Subgrade Reaction, K$_\infty$, Assuming a Semi-Infinite Subgrade Depth*

*For practical purposes, a semi-infinite depth is considered to be greater than 10 ft below the surface of the subgrade.

SOURCE: Redrawn from *AASHTO Guide for Design of Pavement Structures,* American Association of State Highway and Transportation Officials, Washington, D.C., 1993.

Table 21.5 Typical Ranges of Loss of Support Factors for Various Types of Materials

Type of Material	Loss of Support (LS)
Cement-treated granular base $(E = 1,000,000 \text{ to } 2,000,000 \text{ lb/in.}^2)$	0.0 to 1.0
Cement aggregate mixtures $(E = 500,000 \text{ to } 1,000,000 \text{ lb/in.}^2)$	0.0 to 1.0
Asphalt-treated base $(E = 350,000 \text{ to } 1,000,000 \text{ lb/in.}^2)$	0.0 to 1.0
Bituminous stabilized mixtures $(E = 40,000 \text{ to } 300,000 \text{ lb/in.}^2)$	0.0 to 1.0
Lime-stabilized mixtures $(E = 20,000 \text{ to } 70,000 \text{ lb/in.}^2)$	1.0 to 3.0
Unbound granular materials $(E = 15,000 \text{ to } 45,000 \text{ lb/in.}^2)$	1.0 to 3.0
Fine-grained or natural subgrade materials $(E = 3,000 \text{ to } 40,000 \text{ lb/in.}^2)$	2.0 to 3.0

Note: E in this table refers to the general symbol for elastic or resilient modulus of the material.
SOURCE: Adapted from B.F. McCullough and Gary E. Elkins, *CRC Pavement Design Manual,* Austin Research Engineers, Inc., Austin, Tex., October 1979.

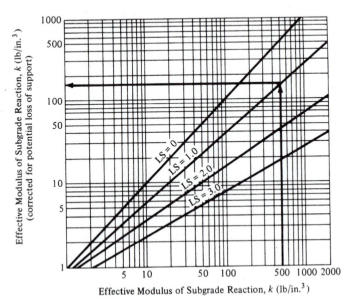

Figure 21.10 Correction of Effective Modulus of Subgrade Reaction for Potential Loss of Subbase Support

SOURCE: Redrawn from *AASHTO Guide for Design of Pavement Structures,* American Association of State Highway and Transportation Officials, Washington, D.C., 1993.

Modulus of Subgrade Reaction, k_∞ (lb/in.^{3})
Assuming Semiinfinite Subgrade Depth

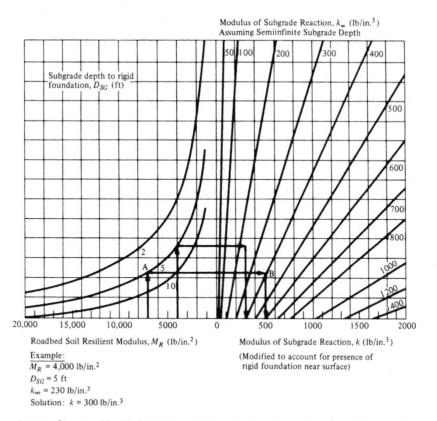

Figure 21.11   Chart to Modify Modulus of Subgrade Reaction to Consider Effects of Rigid Foundation Near Surface (within 10 ft)

SOURCE: Redrawn from *AASHTO Guide for Design of Pavement Structures,* American Association of State Highway and Transportation Officials, Washington, D.C., 1993.

Example 21.2   Computing Effective Modulus of Subgrade Reaction for a Rigid Pavement Using AASHTO Method

A 6-in. layer of cement-treated granular material is to be used as subbase for a rigid pavement. The monthly values for the roadbed soil resilient modulus and the subbase elastic (resilient) modulus are given in columns 2 and 3 of Table 21.6. If the rock depth is located 5 ft below the subgrade surface and the projected slab thickness is 9 in., estimate the effective modulus of subgrade reaction, using the AASHTO method.

Note that this is the example given in the 1993 AASHTO guide. Also note that the values for the modulus of the roadbed and subbase materials should be determined as discussed in Chapter 20, and the corresponding values shown in columns 2 and 3 of Table 21.6 should be for the same seasonal period.

Table 21.6 Data for and Solution to Example 21.2

(1) Month	(2) Roadbed Modulus M_R (lb/in.²)	(3) Subbase Modulus E_{SB} (lb/in.²)	(4) Composite k Value (lb/in.²) (Fig. 21.9)	(5) k Value (E_{SB}) on Rigid Foundation (Fig. 21.11)	(6) Relative Damage, u_r (Fig. 21.12)
January	20,000	50,000	1100	1350	0.35
February	20,000	50,000	1100	1350	0.35
March	2,500	15,000	160	230	0.86
April	4,000	15,000	230	300	0.78
May	4,000	15,000	230	300	0.78
June	7,000	20,000	400	500	0.60
July	7,000	20,000	400	500	0.60
August	7,000	20,000	400	500	0.60
September	7,000	20,000	400	500	0.60
October	7,000	20,000	400	500	0.60
November	4,000	15,000	230	300	0.78
December	20,000	50,000	1100	1350	0.35

$$\sum u_r = 7.25$$

Total Subbase
  Type: Granular
  Thickness (in.): 6
  Loss of Support, LS: 1.0
Depth to Rigid Foundation (ft): 5
Projected Slab Thickness (in.): 9
Average: $\bar{u}_r = \dfrac{\sum u_r}{n} = \dfrac{7.25}{12} = 0.60$
Effective modulus of subgrade reaction, k (lb/in.³) = 500
Corrected for loss of support: k (lb/in.³) = 170

Solution: Having determined the different moduli as given in columns 2 and 3 of Table 21.6, the next step is to estimate the composite subgrade modulus for each of the seasonal periods considered. At this stage the effect of the existence of bedrock within 10 ft from the subgrade surface is not considered, and it is assumed that the subgrade is of infinite depth. The composite subgrade k is then determined for each seasonal period using the chart in Figure 21.9. The procedure involves the following steps:

Step 1. Estimate k. For example, in September roadbed modulus $M_R = 7000$ lb/in.² and subbase modulus $E_{SB} = 20,000$ lb/in.².

  1. Enter the chart in Figure 21.9 at subbase thickness of 6 in. and draw a vertical line to intersect the subbase elastic modulus graph of 20,000 lb/in.² at A and the roadbed modulus graph of 7000 lb/in.² at B.
  2. Draw a horizontal line from B to intersect the turning line at C.
  3. Draw a vertical line upward from C to intersect the horizontal line drawn from A, as shown. This point of intersection D is the composite modulus of subgrade reaction.

$$k = 400 \text{ lb/in.}^3$$

Step 2.   Adjust k for presence of rockbed within 10 ft of subgrade surface. This involves the use of the chart in Figure 21.11. This chart takes into account the depth below the subgrade surface at which the rockbed is located, the resilient modulus of the subgrade soil, and the composite modulus of subgrade reaction (k) determined earlier. The adjustment is made as follows:

1. Enter Figure 21.11 at roadbed soil resilient modulus M_R of 7000 lb/in.^{2}. Draw a vertical line to intersect the graph corresponding to the depth of the rockbed below the subgrade surface. In this case it is 5 ft. The intersection point is A.
2. From A, draw a horizontal line to intersect the appropriate k line. In this case the intersection point is B.
3. From B, draw a vertical line downward to determine the adjusted modulus of subgrade reaction k. This is obtained as 500 lb/in.^{2}. Enter the k values in column 5.

Step 3.   Determine the effective modulus of subgrade reaction by determining the average damage ($\bar{u}_r$). The steps involved are similar to those for flexible pavements, but in this case Figure 21.12 is used.

1. Enter Figure 21.12 at the k value obtained in step 2 (that is, 500 lb/in.^{2}) and project a vertical line to intersect the graph representing the projected slab thickness of 9 in. at point A.
2. From A, project a horizontal line to determine the appropriate u_r. In this case $u_r = 60$ percent or 0.60. These values are recorded in column 6 of Table 21.6.
3. Determine the mean u_r as shown in Table 21.6, that is, $\bar{u}_r = 0.6$.
4. Using the mean u_r of 0.6, obtain the effective modulus of subgrade reaction from Figure 21.12 as 500 lb/in.^{2}.

Step 4.   Adjust the effective modulus of subgrade reaction determined in step 3 to account for the potential loss of subbase support due to erosion, using Figure 21.10. The LS factor given in Table 21.5 is used. Since the subbase consists of a cement-treated granular material, LS $= 1$ and $k = 500$ lb/in.^{2}. The corrected modulus of subgrade reaction is 170 lb/in.^{3}.

Traffic.   The treatment of traffic load is similar to that presented for flexible pavements, in that the traffic load application is given in terms of the number of 18,000-lb equivalent single-axle loads (ESALs). ESAL factors depend on the slab thickness and the terminal serviceability index of the pavement. Tables 21.7 and 21.8 give ESAL factors for rigid pavements with a terminal serviceability index of 2.5. Since the ESAL factor depends on the thickness of the slab, it is therefore necessary to assume the thickness of the slab at the start of the computation. This assumed value is used to compute the number of accumulated ESALs, which in turn is used to compute the required thickness. If the computed

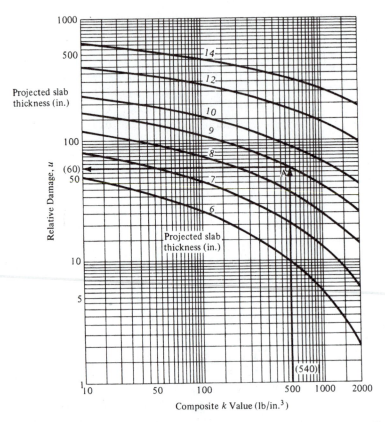

Figure 21.12 Chart for Estimating Relative Damage to Rigid Pavements Based on Slab Thickness and Underlying Support

SOURCE: Redrawn from *AASHTO Guide for Design of Pavement Structures,* American Association of State Highway and Transportation Officials, Washington, D.C., 1993.

thickness is significantly different from the assumed thickness, the accumulated ESAL should be recomputed. This procedure should be repeated until the assumed and computed thicknesses are approximately the same.

Concrete Properties.   The concrete property is given in terms of its flexural strength (modulus of rupture) at 28 days. The flexural strength at 28 days of the concrete to be used in construction should be determined by conducting a three-point loading test as specified in AASHTO Designation T97.

Drainage.   The drainage quality of the pavement is considered by introducing a factor (C_d) into the performance equation. This factor depends on the quality of the drainage as described in Chapter 20 (see Table 20.14) and the percent of time the pavement structure is exposed to moisture levels approaching saturation. Table 21.9 gives AASHTO-recommended values for C_d.

Reliability.   Reliability considerations for rigid pavement are similar to those for flexible pavement as presented in Chapter 20. Reliability levels, $R\%$, and the overall standard deviation, S_o, are incorporated directly in the design charts.

Table 21.7   ESAL Factors for Rigid Pavements, Single Axles, and P_t of 2.5

Axle Load (kip)	Slab Thickness, D (in.)								
	6	7	8	9	10	11	12	13	14
2	.0002	.0002	.0002	.0002	.0002	.0002	.0002	.0002	.0002
4	.003	.002	.002	.002	.002	.002	.002	.002	.002
6	.012	.011	.010	.010	.010	.010	.010	.010	.010
8	.039	.035	.033	.032	.032	.032	.032	.032	.032
10	.097	.089	.084	.082	.081	.080	.080	.080	.080
12	.203	.189	.181	.176	.175	.174	.174	.173	.173
14	.376	.360	.347	.341	.338	.337	.336	.336	.336
16	.634	.623	.610	.604	.601	.599	.599	.599	.598
18	1.00	1.00	1.00	1.00	1.00	1.00	1.00	1.00	1.00
20	1.51	1.52	1.55	1.57	1.58	1.58	1.59	1.59	1.59
22	2.21	2.20	2.28	2.34	2.38	2.40	2.41	2.41	2.41
24	3.16	3.10	3.22	3.36	3.45	3.50	3.53	3.54	3.55
26	4.41	4.26	4.42	4.67	4.85	4.95	5.01	5.04	5.05
28	6.05	5.76	5.92	6.29	6.61	6.81	6.92	6.98	7.01
30	8.16	7.67	7.79	8.28	8.79	9.14	9.35	9.46	9.52
32	10.8	10.1	10.1	10.7	11.4	12.0	12.3	12.6	12.7
34	14.1	13.0	12.9	13.6	14.6	15.4	16.0	16.4	16.5
36	18.2	16.7	16.4	17.1	18.3	19.5	20.4	21.0	21.3
38	23.1	21.1	20.6	21.3	22.7	24.3	25.6	26.4	27.0
40	29.1	26.5	25.7	26.3	27.9	29.9	31.6	32.9	33.7
42	36.2	32.9	31.7	32.2	34.0	36.3	38.7	40.4	41.6
44	44.6	40.4	38.8	39.2	41.0	43.8	46.7	49.1	50.8
46	54.5	49.3	47.1	47.3	49.2	52.3	55.9	59.0	61.4
48	66.1	59.7	56.9	56.8	58.7	62.1	66.3	70.3	73.4
50	79.4	71.7	68.2	67.8	69.6	73.3	78.1	83.0	87.1

SOURCE: Adapted with permission from *AASHTO Guide for Design of Pavement Structures*, American Association of State Highway and Transportation Officials, Washington, D.C., 1993.

Design Procedure

The objective of the design is to determine the thickness of the concrete pavement that is adequate to carry the projected design ESAL. The basic equation developed in the 1986 AASHTO design guide for the pavement thickness is given as:

$$\log_{10} W_{18} = Z_R S_o + 7.35 \log_{10}(D + 1) - 0.06 + \frac{\log_{10}[\Delta PSI/(4.5 - 1.5)]}{1 + [(1.624 \times 10^7)/(D + 1)^{8.46}]}$$

$$+ (4.22 - 0.32 P_t) \log_{10}\left\{ \frac{S_c' C_d}{215.63 J} \left(\frac{D^{.75} - 1.132}{D^{.75} - [18.42/(E_c/k)^{.25}]} \right) \right\} \qquad (21.21)$$

Table 21.8 ESAL Factors for Rigid Pavements, Tandem Axles, and p_t of 2.5

Axle Load (kip)	Slab Thickness, D (in.)								
	6	7	8	9	10	11	12	13	14
2	.0001	.0001	.0001	.0001	.0001	.0001	.0001	.0001	.0001
4	.0006	.0006	.0005	.0005	.0005	.0005	.0005	.0005	.0005
6	.002	.002	.002	.002	.002	.002	.002	.002	.002
8	.007	.006	.006	.005	.005	.005	.005	.005	.005
10	.015	.014	.013	.013	.012	.012	.012	.012	.012
12	.031	.028	.026	.026	.025	.025	.025	.025	.025
14	.057	.052	.049	.048	.047	.047	.047	.047	.047
16	.097	.089	.084	.082	.081	.081	.080	.080	.080
18	.155	.143	.136	.133	.132	.131	.131	.131	.131
20	.234	.220	.211	.206	.204	.203	.203	.203	.203
22	.340	.325	.313	.308	.305	.304	.303	.303	.303
24	.475	.462	.450	.444	.441	.440	.439	.439	.439
26	.644	.637	.627	.622	.620	.619	.618	.618	.618
28	.855	.854	.852	.850	.850	.850	.849	.849	.849
30	1.11	1.12	1.13	1.14	1.14	1.14	1.14	1.14	1.14
32	1.43	1.44	1.47	1.49	1.50	1.51	1.51	1.51	1.51
34	1.82	1.82	1.87	1.92	1.95	1.96	1.97	1.97	1.97
36	2.29	2.27	2.35	2.43	2.48	2.51	2.52	2.52	2.53
38	2.85	2.80	2.91	3.03	3.12	3.16	3.18	3.20	3.20
40	3.52	3.42	3.55	3.74	3.87	3.94	3.98	4.00	4.01
42	4.32	4.16	4.30	4.55	4.74	4.86	4.91	4.95	4.96
44	5.26	5.01	5.16	5.48	5.75	5.92	6.01	6.06	6.09
46	6.36	6.01	6.14	6.53	6.90	7.14	7.28	7.36	7.40
48	7.64	7.16	7.27	7.73	8.21	8.55	8.75	8.86	8.92
50	9.11	8.50	8.55	9.07	9.68	10.14	10.42	10.58	10.66
52	10.8	10.0	10.0	10.6	11.3	11.9	12.3	12.5	12.7
54	12.8	11.8	11.7	12.3	13.2	13.9	14.5	14.8	14.9
56	15.0	13.8	13.6	14.2	15.2	16.2	16.8	17.3	17.5
58	17.5	16.0	15.7	16.3	17.5	18.6	19.5	20.1	20.4
60	20.3	18.5	18.1	18.7	20.0	21.4	22.5	23.2	23.6
62	23.5	21.4	20.8	21.4	22.8	24.4	25.7	26.7	27.3
64	27.0	24.6	23.8	24.4	25.8	27.7	29.3	30.5	31.3
66	31.0	28.1	27.1	27.6	29.2	31.3	33.2	34.7	35.7
68	35.4	32.1	30.9	31.3	32.9	35.2	37.5	39.3	40.5
70	40.3	36.5	35.0	35.3	37.0	39.5	42.1	44.3	45.9
72	45.7	41.4	39.6	39.8	41.5	44.2	47.2	49.8	51.7
74	51.7	46.7	44.6	44.7	46.4	49.3	52.7	55.7	58.0
76	58.3	52.6	50.2	50.1	51.8	54.9	58.6	62.1	64.8
78	65.5	59.1	56.3	56.1	57.7	60.9	65.0	69.0	72.3
80	73.4	66.2	62.9	62.5	64.2	67.5	71.9	76.4	80.2
82	82.0	73.9	70.2	69.6	71.2	74.7	79.4	84.4	88.8

Continued

Table 21.8   ESAL Factors for Rigid Pavements, Tandem Axles, and p_t of 2.5 (continued)

Axle Load (kip)	Slab Thickness, D (in.)								
	6	7	8	9	10	11	12	13	14
84	91.4	82.4	78.1	77.3	78.9	82.4	87.4	93.0	98.1
86	102.0	92.0	87.0	86.0	87.0	91.0	96.0	102.0	108.0
88	113.0	102.0	96.0	95.0	96.0	100.0	105.0	112.0	119.0
90	125.0	112.0	106.0	105.0	106.0	110.0	115.0	123.0	130.0

SOURCE: Adapted with permission from *AASHTO Guide for Design of Pavement Structures*, American Association of State Highway and Transportation Officials, Washington, D.C., 1993.

Table 21.9   Recommended Values for Drainage Coefficient, C_d, for Rigid Pavements

Quality of Drainage	Percent of Time Pavement Structure is Exposed to Moisture Levels Approaching Saturation			
	Less Than 1 Percent	1–5 Percent	5–25 Percent	Greater Than 25 Percent
Excellent	1.25–1.20	1.20–1.15	1.15–1.10	1.10
Good	1.20–1.15	1.15–1.10	1.10–1.00	1.00
Fair	1.15–1.10	1.10–1.00	1.00–0.90	0.90
Poor	1.10–1.00	1.00–0.90	0.90–0.80	0.80
Very poor	1.00–0.90	0.90–0.80	0.80–0.70	0.70

SOURCE: Adapted from *AASHTO Guide for Design of Pavement Structures*, American Association of State Highway and Transportation Officials, Washington, D.C., 1986. Used by permission.

where

Z_R = standard normal variant corresponding to the selected level of reliability

S_o = overall standard deviation (see Chapter 20)

W_{18} = predicted number of 18 kip ESAL applications that can be carried by the pavement structure after construction

D = thickness of concrete pavement to the nearest half-inch

ΔPSI = design serviceability loss = $p_i - p_t$

p_i = initial serviceability index

p_t = terminal serviceability index

E_c = elastic modulus of the concrete to be used in construction (lb/in^2)

S'_c = modulus of rupture of the concrete to be used in construction (lb/in^2)

J = load transfer coefficient = 3.2 (assumed)

C_d = drainage coefficient

Equation 21.21 can be solved for the thickness of the pavement (D) in inches by using either a computer program or the two charts in Figures 21.13 and 21.14. The use of a computer program facilitates the iteration necessary, since D has to be assumed to determine the effective modulus of subgrade reaction and the ESAL factors used in the design.

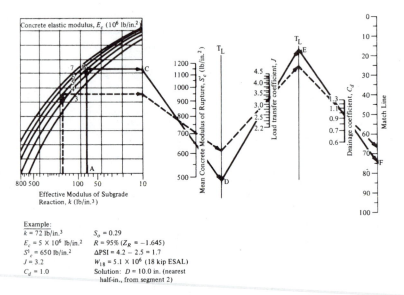

Example:
$k = 72$ lb/in.^{3} $S_o = 0.29$
$E_c = 5 \times 10^6$ lb/in.^{2} $R = 95\%$ ($Z_R = -1.645$)
$S^1_c = 650$ lb/in.^{2} $\Delta PSI = 4.2 - 2.5 = 1.7$
$J = 3.2$ $W_{18} = 5.1 \times 10^6$ (18 kip ESAL)
$C_d = 1.0$ Solution: $D = 10.0$ in. (nearest
                           half-in., from segment 2)

Figure 21.13 Design Chart for Rigid Pavements Based on Using Mean Values for Each Input Variable (segment 1)

SOURCE: Redrawn from *AASHTO Guide for Design of Pavement Structures*, American Association of State Highway and Transportation Officials, Washington, D.C., 1993.

Example 21.3   Designing a Rigid Pavement Using the AASHTO Method

The use of the charts is demonstrated with the example given in Figure 21.13. In this case, input values for segment 1 of the chart (Figure 21.13) are

- Effective modulus of subgrade reaction, $k = 72$ lb/in.^{3}
- Mean concrete modulus of rupture, $S'_c = 650$ lb/in.^{2}
- Load transfer coefficient, $J = 3.2$
- Drainage coefficient, $C_d = 1.0$

These values are used to determine a value on the match line as shown in Figure 21.13 (solid line ABCDEF). Input parameters for segment 2 (Figure 21.14) of the chart are

- Match line value determined in segment 1 (74)
- Design serviceability loss, $\Delta PSI = 4.5 - 2.5 = 2.0$
- Reliability, $R\% = 95$ percent ($Z_R = 1.645$)
- Overall standard deviation, $S_o = 0.29$
- Cumulative 18 kip ESAL = (5×10^6)

Solution: The required thickness of the concrete slab is then obtained, as shown in Figure 21.14, as 10 in. (nearest half-inch).

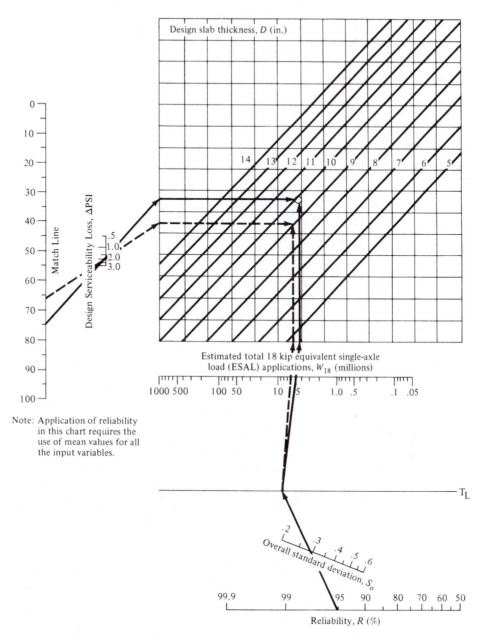

Figure 21.14 Design Chart for Rigid Pavements Based on Using Mean Values for Each Input Variable (segment 2)

SOURCE: Redrawn from *AASHTO Guide for Design of Pavement Structures,* American Association of State Highway and Transportation Officials, Washington, D.C., 1993.

Note that when the thickness obtained from solving Eq. 21.21 analytically or by use of Figures 21.13 and 21.14 is significantly different from that originally assumed to determine the effective subgrade modulus and to select the ESAL factors, the whole procedure has to be repeated until the assumed and designed values are approximately the same, emphasizing the importance of using a computer program to facilitate the necessary iteration.

Example 21.4 Evaluating the Adequacy of a Rigid Pavement Using the AASHTO Method

Using the data and effective subgrade modulus obtained in Example 21.2, determine whether the 9-in. pavement design of Example 21.2 will be adequate on a rural expressway for a 20-yr analysis period and the following design criteria.

$P_i = 4.5$

$P_t = 2.5$

ESAL on design lane during first yr of operation $= 0.2 \times 10^6$

Traffic growth rate $= 4$ percent

Concrete elastic modulus, $E_c = 5 \times 10^6$ lb/in.^{2}

Mean concrete modulus of rupture $= 700$ lb/in.^{2}

Drainage conditions are such that $C_d = 1.0$

$R = 0.95$ $(Z_R = 1.645)$

$S_o = 0.30$ (for rigid pavements $S_o = 0.3 - 0.4$)

Growth factor $= 29.78$ (from Table 20.6)

$k = 170$ (from Example 21.2)

Assume $D = 9$ in.   (from Example 21.2)

ESAL over design period $= 0.2 \times 10^6 \times 29.78 = 6 \times 10^6$

Solution: The depth of concrete required is obtained from Figures 21.13 and 21.14. The dashed lines represent the solution, and a depth of 9 in. is obtained. The pavement is therefore adequate.

PCA Design Method

The PCA method for concrete pavement design is based on a combination of theoretical studies, results of model and full-scale tests, and experience gained from the performance of concrete pavements normally constructed and carrying normal traffic loads. Tayabji and Colley have reported on some of these studies and tests. The design procedure was initially published in 1961 but was revised in 1984. The procedure provides for the determination of the pavement thickness for plain concrete, simply reinforced concrete, and continuously reinforced concrete pavements.

Design Considerations
The basic factors considered in the PCA design method are

- Flexural strength of the concrete
- Subgrade and subbase support
- Traffic load

Flexural Strength of Concrete.   The flexural strength of the concrete used in this procedure is given in terms of the modulus of rupture obtained by the third-point method (ASTM Designation C78). The average of the 28-day test results is used as input by the designer. The design charts and tables, however, incorporate the variation of the concrete strength from one point to another in the concrete slab and the gain in strength with age.

Subgrade and Subbase Support.   The Westergaard modulus of subgrade reaction (k) is used to define the subgrade and subbase support. This can be determined by performing a plate-bearing test or by correlating with other test results, using the chart in Figure 21.8. No specific correction is made for the reduced value of k during the spring thaw period, but it is suggested that normal summer or fall k values should be used. The modulus of subgrade reaction can be increased by adding a layer of untreated granular material over the subgrade. An approximate value of the increased k can be obtained from Table 21.10.

Cement-stabilized soils can also be used as subbase material when the pavement is expected to carry very heavy traffic or when the subgrade material has a low value of modulus of subgrade reaction. Suitable soil materials for cement stabilization are those classified under the AASHTO Soil Classification System as A-1, A-2-4, A-2-5, and

Table 21.10 Design k Values for Untreated and Cement-Treated Subbases

Subgrade k Value (lb/in.^{3})	Subbase k Value (lb/in.^{3})			
	4 in.	6 in.	9 in.	12 in.
(a) Untreated Granular Subbases				
50	65	75	85	110
100	130	140	160	190
200	220	230	270	320
300	320	330	370	430
(b) Cement-Treated Subbases				
50	170	230	310	390
100	280	400	520	640
200	470	640	830	—

SOURCE: Adapted from Robert G. Packard, *Thickness Design for Concrete Highway and Street Pavements,* Portland Cement Association, Skokie, Ill., 1984.

A-3. The amount of cement used should be based on standard ASTM laboratory freeze-thaw and wet-dry tests, ASTM Designation D560 and D559, respectively, and weight loss criteria. It is permissible, however, to use other methods that will produce an equivalent quality material. Suggested k values for this type of stabilized material are given in Table 21.10(b).

Traffic Load.   The traffic load is computed in terms of the cumulated number of single and tandem axles of different loads projected for the design period of the pavement. The information required to determine cumulated numbers are the average daily traffic (ADT), the average daily truck traffic (ADTT) (in both directions), and the axle load distribution of truck traffic. Only trucks with six or more tires are included in this design. It can be assumed that truck volume is the same in each direction of travel. When there is reason to believe that truck volume varies in each direction, an adjustment factor can be used as discussed in Chapter 20.

The design also incorporates a load safety factor (LSF), which is used to multiply each axle load. The recommended LSF values are:

- 1.2 for interstate and multilane projects with uninterrupted traffic flow and high truck volumes
- 1.1 for highways and arterials with moderate truck volume
- 1.0 for roads and residential streets with very low truck volume

The LSF can be increased to 1.3 if the objective is to maintain a higher-than-normal pavement serviceability level throughout the design life of the pavement. The design procedure also provides for a factor of safety of 1.1 or 1.2 in addition to the LSF to allow for unexpected truck traffic.

Design Procedure

The procedure is based on a detailed finite-element computer analysis of stresses and deflections of the pavement at edges, joints, and corners. It considers such factors as finite slab dimensions, position of axle load, load transfer at transverse joints or cracks, and load transfer at pavement and concrete shoulder joints. Load transfer characteristics are modeled using the diameter and modulus of elasticity of the dowels at dowel joints.

A spring stiffness value is used for keyway joints, aggregate interlocks, and cracks in continuously reinforced concrete pavements to represent the load deflection characteristics at such joints. Field and laboratory test results are used to determine the spring stiffness value.

Analysis of different axle load positions on the pavement reveals that the Case II position of Figure 21.6 is critical for flexural stresses, whereas Case I position is critical for deflection, but with one set of wheels at or near the corner of the free edge and the transverse joint.

The design procedure consists of two parts: fatigue analysis and erosion analysis. The objective of fatigue analysis is to determine the minimum thickness of the concrete required to control fatigue cracking. This is done by comparing the expected axle repetitions with the allowable repetitions for each axle load and ensuring that the cumulative repetitions are less than the allowable. Allowable axle repetitions depend on the stress ratio factor, which is the ratio of the equivalent stress of the pavement to the modulus of rupture of the concrete. The equivalent stress of the pavement depends on the thickness of the slab and the subbase-subgrade k. The chart in Figure 21.15 can be used to determine the allowable load repetitions based on the stress ratio factor. Tables 21.11 and 21.12 give equivalent stress values for pavements without concrete shoulders and with concrete shoulders, respectively.

The objective of the erosion analysis is to determine the minimum thickness of the pavement required to control foundation and shoulder erosion, pumping, and faulting. These pavement distresses are more closely related to deflection, as will be seen later. The erosion criterion is based mainly on the rate of work expended by an axle load in deflecting a slab, as it was determined that a useful correlation existed between pavement performance and the product of the corner deflection and the pressure at the

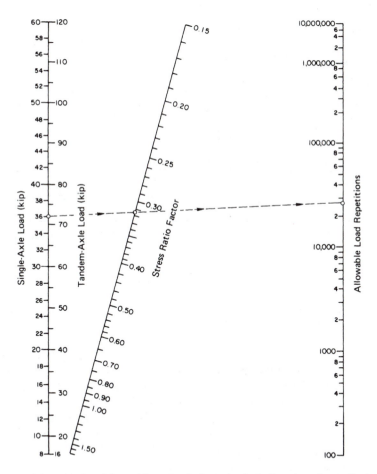

Figure 21.15 Allowable Load Repetitions for Fatigue Analysis Based on Stress Ratio

SOURCE: R. G. Packard, *Thickness Design for Concrete Highway and Street Pavements,* Portland Cement Association, Skokie, Ill., 1984.

Table 21.11   Equivalent Stress Values for Single Axles and Tandem Axles (without concrete shoulder)

Slab Thickness (in.)	k of Subgrade-Subbase (lb/in.³) (Single Axle/Tandem Axle)						
	50	100	150	200	300	500	700
4	825/679	726/585	671/542	634/516	584/486	523/457	484/443
4.5	699/586	616/500	571/460	540/435	498/406	448/378	417/363
5	602/516	531/436	493/399	467/376	432/349	390/321	363/307
5.5	526/461	464/387	431/353	409/331	379/305	343/278	320/264
6	465/416	411/348	382/316	362/296	336/271	304/246	285/232
6.5	417/380	367/317	341/286	324/267	300/244	273/220	256/207
7	375/349	331/290	307/262	292/244	271/222	246/199	231/186
7.5	340/323	300/268	279/241	265/224	246/203	224/181	210/169
8	311/300	274/249	255/223	242/208	225/188	205/167	192/155
8.5	285/281	252/232	234/208	222/193	206/174	188/154	177/143
9	264/264	232/218	216/195	205/181	190/163	174/144	163/133
9.5	245/248	215/205	200/183	190/170	176/153	161/134	151/124
10	228/235	200/193	186/173	177/160	164/144	150/126	141/117
10.5	213/222	187/183	174/164	165/151	153/136	140/119	132/110
11	200/211	175/174	163/155	154/143	144/129	131/113	123/104
11.5	188/201	165/165	153/148	145/136	135/122	123/107	116/98
12	177/192	155/158	144/141	137/130	127/116	116/102	109/93
12.5	168/183	147/151	136/135	129/124	120/111	109/97	103/89
13	159/176	139/144	129/129	122/119	113/106	103/93	97/85
13.5	152/168	132/138	122/123	116/114	107/102	98/89	92/81
14	144/162	125/133	116/118	110/109	102/98	93/85	88/78

SOURCE: Reproduced from Robert G. Packard, *Thickness Design for Concrete Highway and Street Pavements,* Portland Cement Association, Skokie, Ill., 1984.

slab-subgrade interface. It is suggested that the engineer use the erosion criterion mainly as a guideline and modify it based on experience with local conditions of climate and drainage.

The erosion analysis is similar to that of fatigue analysis, except that an erosion factor is used instead of the stress factor. The erosion factor is also dependent on the thickness of the slab and the subgrade-subbase k. Tables 21.13 through 21.16 give erosion factors for different types of pavement construction. Figures 21.16 and 21.17 are charts that can be used to determine the allowable load repetitions based on erosion.

The minimum thickness that satisfied both analyses is the design thickness. Design thicknesses for pavements carrying light traffic and pavements with doweled joints carrying medium traffic will usually be based on fatigue analysis, whereas design thicknesses for pavements with undoweled joints carrying medium or heavy traffic and pavements with doweled joints carrying heavy traffic will normally be based on erosion analysis.

Table 21.12 Equivalent Stress Values for Single Axles and Tandem Axles (with concrete shoulder)

Slab Thickness (in.)	k of Subgrade-Subbase (lb/in.³) (Single Axle/Tandem Axle)						
	50	100	150	200	300	500	700
4	640/534	559/468	517/439	489/422	452/403	409/388	383/384
4.5	547/461	479/400	444/372	421/356	390/338	355/322	333/316
5	475/404	417/349	387/323	367/308	341/290	311/274	294/267
5.5	418/360	368/309	342/285	324/271	302/254	276/238	261/231
6	372/325	327/277	304/255	289/241	270/225	247/210	234/203
6.5	334/295	294/251	274/230	260/218	243/203	223/188	212/180
7	302/270	266/230	248/210	236/198	220/184	203/170	192/162
7.5	275/250	243/211	226/193	215/182	201/168	185/155	176/148
8	252/232	222/196	207/179	197/168	185/155	170/142	162/135
8.5	232/216	205/182	191/166	182/156	170/144	157/131	150/125
9	215/202	190/171	177/155	169/146	158/134	146/122	139/116
9.5	200/190	176/160	164/146	157/137	147/126	136/114	129/108
10	186/179	164/151	153/137	146/129	137/118	127/107	121/101
10.5	174/170	154/143	144/130	137/121	128/111	119/101	113/95
11	164/161	144/135	135/123	129/115	120/105	112/95	106/90
11.5	154/153	136/128	127/117	121/109	113/100	105/90	100/85
12	145/146	128/122	120/111	114/104	107/95	99/86	95/81
12.5	137/139	121/117	113/106	108/99	101/91	94/82	90/77
13	130/133	115/112	107/101	102/95	96/86	89/78	85/73
13.5	124/127	109/107	102/97	97/91	91/83	85/74	81/70
14	118/122	104/103	97/93	93/87	87/79	81/71	77/67

SOURCE: Robert G. Packard, *Thickness Design for Concrete Highway and Street Pavements,* Portland Cement Association, Skokie, Ill., 1984.

Example 21.5   Designing a Rigid Pavement Using the PCA Method

The procedure is demonstrated by the following example, which is one of the many solutions provided by PCA. The following project and traffic data are available.

Four-lane interstate highway

Rolling terrain in rural location

Design period = 20 yr

Axle loads and expected repetitions are shown in Figure 21.18

Subbase-subgrade $k = 130$ lb/in.³

Concrete modulus of rupture = 650 lb/in²

Determine minimum thickness of a pavement with doweled joints and without concrete shoulders.

Table 12.13 Erosion Factors for Single Axles and Tandem Axles (doweled joints, without concrete shoulder)

Slab Thickness (in.)	k of Subgrade-Subbase (lb/in.³) (Single Axle/Tandem Axle)					
	50	100	200	300	500	700
4	3.74/3.83	3.73/3.79	3.72/3.75	3.71/3.73	3.70/3.70	3.68/3.67
4.5	3.59/3.70	3.57/3.65	3.56/3.61	3.55/3.58	3.54/3.55	3.52/3.53
5	3.45/3.58	3.43/3.52	3.42/3.48	3.41/3.45	3.40/3.42	3.38/3.40
5.5	3.33/3.47	3.31/3.41	3.29/3.36	3.28/3.33	3.27/3.30	3.26/3.28
6	3.22/3.38	3.19/3.31	3.18/3.26	3.17/3.23	3.15/3.20	3.14/3.17
6.5	3.11/3.29	3.09/3.22	3.07/3.16	3.06/3.13	3.05/3.10	3.03/3.07
7	3.02/3.21	2.99/3.14	2.97/3.08	2.96/3.05	2.95/3.01	2.94/2.98
7.5	2.93/3.14	2.91/3.06	2.88/3.00	2.87/2.97	2.86/2.93	2.84/2.90
8	2.85/3.07	2.82/2.99	2.80/2.93	2.79/2.89	2.77/2.85	2.76/2.82
8.5	2.77/3.01	2.74/2.93	2.72/2.86	2.71/2.82	2.69/2.78	2.68/2.75
9	2.70/2.96	2.67/2.87	2.65/2.80	2.63/2.76	2.62/2.71	2.61/2.68
9.5	2.63/2.90	2.60/2.81	2.58/2.74	2.56/2.70	2.55/2.65	2.54/2.62
10	2.56/2.85	2.54/2.76	2.51/2.68	2.50/2.64	2.48/2.59	2.47/2.56
10.5	2.50/2.81	2.47/2.71	2.45/2.63	2.44/2.59	2.42/2.54	2.41/2.51
11	2.44/2.76	2.42/2.67	2.39/2.58	2.38/2.54	2.36/2.49	2.35/2.45
11.5	2.38/2.72	2.36/2.62	2.33/2.54	2.32/2.49	2.30/2.44	2.29/2.40
12	2.33/2.68	2.30/2.58	2.28/2.49	2.26/2.44	2.25/2.39	2.23/2.36
12.5	2.28/2.64	2.25/2.54	2.23/2.45	2.21/2.40	2.19/2.35	2.18/2.31
13	2.23/2.61	2.20/2.50	2.18/2.41	2.16/2.36	2.14/2.30	2.13/2.27
13.5	2.18/2.57	2.15/2.47	2.13/2.37	2.11/2.32	2.09/2.26	2.08/2.23
14	2.13/2.54	2.11/2.43	2.08/2.34	2.07/2.29	2.05/2.23	2.03/2.19

SOURCE: Robert G. Packard, *Thickness Design for Concrete Highway and Street Pavements,* Portland Cement Association, Skokie, Ill., 1984.

Solution.   Figure 21.18 is used in working out the example through the following steps:

Step 1.   Fatigue Analysis.

1. Select a trial thickness (9.5 in.).
2. Complete all the information at the top of the form as shown.
3. Determine projected number of single-axle and tandem-axle repetitions in the different weight groups for the design period. This is done by determining the ADT, the percentage of truck traffic, the axle load distribution of the truck traffic, and the proportion of trucks on the design lane. The actual cumulative expected repetitions for each range of axle load (without converting to 18,000-lb equivalent axle load) is then computed, using the procedure presented in Chapter 20.

Table 21.14 Erosion Factors for Single Axles and Tandem Axles (aggregate interlock joints, without concrete shoulder)

Slab Thickness (in.)	k of Subgrade-Subbase (lb/in.^{3}) (Single Axle/Tandem Axle)					
	50	100	200	300	500	700
4	3.94/4.03	3.91/3.95	3.88/3.89	3.86/3.86	3.82/3.83	3.77/3.80
4.5	3.79/3.91	3.76/3.82	3.73/3.75	3.71/3.72	3.68/3.68	3.64/3.65
5	3.66/3.81	3.63/3.72	3.60/3.64	3.58/3.60	3.55/3.55	3.52/3.52
5.5	3.54/3.72	3.51/3.62	3.48/3.53	3.46/3.49	3.43/3.44	3.41/3.40
6	3.44/3.64	3.40/3.53	3.37/3.44	3.35/3.40	3.32/3.34	3.30/3.30
6.5	3.34/3.56	3.30/3.46	3.26/3.36	3.25/3.31	3.22/3.25	3.20/3.21
7	3.26/3.49	3.21/3.39	3.17/3.29	3.15/3.24	3.13/3.17	3.11/3.13
7.5	3.18/3.43	3.13/3.32	3.09/3.22	3.07/3.17	3.04/3.10	3.02/3.06
8	3.11/3.37	3.05/3.26	3.01/3.16	2.99/3.10	2.96/3.03	2.94/2.99
8.5	3.04/3.32	2.98/3.21	2.93/3.10	2.91/3.04	2.88/2.97	2.87/2.93
9	2.98/3.27	2.91/3.16	2.86/3.05	2.84/2.99	2.81/2.92	2.79/2.87
9.5	2.92/3.22	2.85/3.11	2.80/3.00	2.77/2.94	2.75/2.86	2.73/2.81
10	2.86/3.18	2.79/3.06	2.74/2.95	2.71/2.89	2.68/2.81	2.66/2.76
10.5	2.81/3.14	2.74/3.02	2.68/2.91	2.65/2.84	2.62/2.76	2.60/2.72
11	2.77/3.10	2.69/2.98	2.63/2.86	2.60/2.80	2.57/2.72	2.54/2.67
11.5	2.72/3.06	2.64/2.94	2.58/2.82	2.55/2.76	2.51/2.68	2.49/2.63
12	2.68/3.03	2.60/2.90	2.53/2.78	2.50/2.72	2.46/2.64	2.44/2.59
12.5	2.64/2.99	2.55/2.87	2.48/2.75	2.45/2.68	2.41/2.60	2.39/2.55
13	2.60/2.96	2.51/2.83	2.44/2.71	2.40/2.65	2.36/2.56	2.34/2.51
13.5	2.56/2.93	2.47/2.80	2.40/2.68	2.36/2.61	2.32/2.53	2.30/2.48
14	2.53/2.90	2.44/2.77	2.36/2.65	2.32/2.58	2.28/2.50	2.25/2.44

SOURCE: Robert G. Packard, *Thickness Design for Concrete Highway and Street Pavements*, Portland Cement Association, Skokie, Ill., 1984.

4. Complete columns 1, 2, and 3. Note that column 2 is column 1 multiplied by the LSF, which in this case is 1.2, since the design is for a four-lane interstate highway.
5. Determine the equivalent stresses for single axle and tandem axle. Table 21.11 is used in this case since there is no concrete shoulder. Interpolating for $k = 130$, for single axles and 9.5-in.-thick slab

$$\text{equivalent stress} = 215 - \frac{215 - 200}{50} \times 30$$
$$= 206 \text{ lb/in}^2$$

Similarly, for the tandem axles,

Table 21.15 Erosion Factors for Single Axles and Tandem Axles (doweled joints, concrete shoulder)

Slab Thickness (in.)	k of Subgrade-Subbase (lb/in.³) (Single Axle/Tandem Axle)					
	50	100	200	300	500	700
4	3.28/3.30	3.24/3.20	3.21/3.13	3.19/3.10	3.15/3.09	3.12/3.08
4.5	3.13/3.19	3.09/3.08	3.06/3.00	3.04/2.96	3.01/2.93	2.98/2.91
5	3.01/3.09	2.97/2.98	2.93/2.89	2.90/2.84	2.87/2.79	2.85/2.77
5.5	2.90/3.01	2.85/2.89	2.81/2.79	2.79/2.74	2.76/2.68	2.73/2.65
6	2.79/2.93	2.75/2.82	2.70/2.71	2.68/2.65	2.65/2.58	2.62/2.54
6.5	2.70/2.86	2.65/2.75	2.61/2.63	2.58/2.57	2.55/2.50	2.52/2.45
7	2.61/2.79	2.56/2.68	2.52/2.56	2.49/2.50	2.46/2.42	2.43/2.38
7.5	2.53/2.73	2.48/2.62	2.44/2.50	2.41/2.44	2.38/2.36	2.35/2.31
8	2.46/2.68	2.41/2.56	2.36/2.44	2.33/2.38	2.30/2.30	2.27/2.24
8.5	2.39/2.62	2.34/2.51	2.29/2.39	2.26/2.32	2.22/2.24	2.20/2.18
9	2.32/2.57	2.27/2.46	2.22/2.34	2.19/2.27	2.16/2.19	2.13/2.13
9.5	2.26/2.52	2.21/2.41	2.16/2.29	2.13/2.22	2.09/2.14	2.07/2.08
10	2.20/2.47	2.15/2.36	2.10/2.25	2.07/2.18	2.03/2.09	2.01/2.03
10.5	2.15/2.43	2.09/2.32	2.04/2.20	2.01/2.14	1.97/2.05	1.95/1.99
11	2.10/2.39	2.04/2.28	1.99/2.16	1.95/2.09	1.92/2.01	1.89/1.95
11.5	2.05/2.35	1.99/2.24	1.93/2.12	1.90/2.05	1.87/1.97	1.84/1.91
12	2.00/2.31	1.94/2.20	1.88/2.09	1.85/2.02	1.82/1.93	1.79/1.87
12.5	1.95/2.27	1.89/2.16	1.84/2.05	1.81/1.98	1.77/1.89	1.74/1.84
13	1.91/2.23	1.85/2.13	1.79/2.01	1.76/1.95	1.72/1.86	1.70/1.80
13.5	1.86/2.20	1.81/2.09	1.75/1.98	1.72/1.91	1.68/1.83	1.65/1.77
14	1.82/2.17	1.76/2.06	1.71/1.95	1.67/1.88	1.64/1.80	1.61/1.74

SOURCE: Robert G. Packard, *Thickness Design for Concrete Highway and Street Pavements,* Portland Cement Association, Skokie, Ill., 1984.

$$\text{equivalent stress} = 205 - \frac{205 - 183}{50} \times 30$$

$$= 192 \text{ lb} / \text{in}^2$$

Record these values in spaces provided on lines 8 and 11, respectively.

6. Determine the stress ratio, which is the equivalent stress divided by the modulus of rupture.

(a) For single axles,

$$\text{stress ratio} = \frac{206}{650} = 0.317$$

(b) For tandem axles,

Table 21.16 Erosion Factors for Single Axles and Tandem Axles (aggregate interlock joints, concrete shoulder)

Slab Thickness (in.)	k of Subgrade-Subbase (lb/in.³) (Single Axle/Tandem Axle)					
	50	100	200	300	500	700
4	3.46/3.49	3.42/3.39	3.38/3.32	3.36/3.29	3.32/3.26	3.28/3.24
4.5	3.32/3.39	3.28/3.28	3.24/3.19	3.22/3.16	3.19/3.12	3.15/3.09
5	3.20/3.30	3.16/3.18	3.12/3.09	3.10/3.05	3.07/3.00	3.04/2.97
5.5	3.10/3.22	3.05/3.10	3.01/3.00	2.99/2.95	2.96/2.90	2.93/2.86
6	3.00/3.15	2.95/3.02	2.90/2.92	2.88/2.87	2.86/2.81	2.83/2.77
6.5	2.91/3.08	2.86/2.96	2.81/2.85	2.79/2.79	2.76/2.73	2.74/2.68
7	2.83/3.02	2.77/2.90	2.73/2.78	2.70/2.72	2.68/2.66	2.65/2.61
7.5	2.76/2.97	2.70/2.84	2.65/2.72	2.62/2.66	2.60/2.59	2.57/2.54
8	2.69/2.92	2.63/2.79	2.57/2.67	2.55/2.61	2.52/2.53	2.50/2.48
8.5	2.63/2.88	2.56/2.74	2.51/2.62	2.48/2.55	2.45/2.48	2.43/2.43
9	2.57/2.83	2.50/2.70	2.44/2.57	2.42/2.51	2.39/2.43	2.36/2.38
9.5	2.51/2.79	2.44/2.65	2.38/2.53	2.36/2.46	2.33/2.38	2.30/2.33
10	2.46/2.75	2.39/2.61	2.33/2.49	2.30/2.42	2.27/2.34	2.24/2.28
10.5	2.41/2.72	2.33/2.58	2.27/2.45	2.24/2.38	2.21/2.30	2.19/2.24
11	2.36/2.68	2.28/2.54	2.22/2.41	2.19/2.34	2.16/2.26	2.14/2.20
11.5	2.32/2.65	2.24/2.51	2.17/2.38	2.14/2.31	2.11/2.22	2.09/2.16
12	2.28/2.62	2.19/2.48	2.13/2.34	2.10/2.27	2.06/2.19	2.04/2.13
12.5	2.24/2.59	2.15/2.45	2.09/2.31	2.05/2.24	2.02/2.15	1.99/2.10
13	2.20/2.56	2.11/2.42	2.04/2.28	2.01/2.21	1.98/2.12	1.95/2.06
13.5	2.16/2.53	2.08/2.39	2.00/2.25	1.97/2.18	1.93/2.09	1.91/2.03
14	2.13/2.51	2.04/2.36	1.97/2.23	1.93/2.15	1.89/2.06	1.87/2.00

SOURCE: Robert G. Packard, *Thickness Design for Concrete Highway and Street Pavements,* Portland Cement Association, Skokie, Ill., 1984.

$$\text{stress ratio} = \frac{192}{650} = 0.295$$

Record these values in spaces provided on lines 9 and 12 respectively.

7. Using Figure 21.15, determine the allowable load repetitions for each axle load based on fatigue analysis. Note that the axle loads used are those in column 2 (that is, after multiplying by the LSF). For example, for the first row of single axles, the axle load to be used is 36. Draw a line joining 36 kip on the single-axle load axis (left side) and 0.317 on the stress ratio factor line. Extend that line to the allowable load repetitions line and read allowable repetitions as 27,000. Repeat this for all axle loads, noting that for tandem axles the scale is on the right side of the axle load axis. Complete column 4 of Figure 21.18.

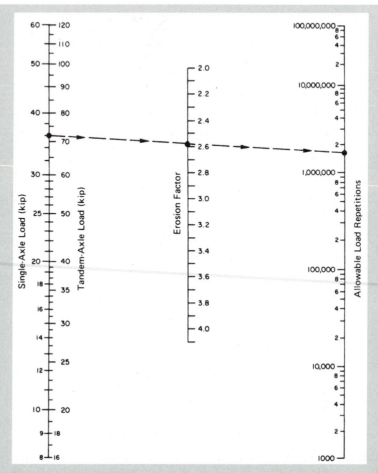

Figure 21.16 Allowable Load Repetitions for Erosion Analysis Based on Erosion Factors (without concrete shoulder)

SOURCE: R.G. Packard, *Thickness Design for Concrete Highway and Street Pavements,* Portland Cement Association, Skokie, Ill., 1984.

8. Determine the fatigue percentage for each axle load, which is an indication of the resistance consumed by the expected number of axle load repetitions.

$$\text{fatigue percentage} = \frac{\text{column 3}}{\text{column 4}} \times 100$$

Complete column 5.
9. Determine total fatigue resistance consumed by summing up column 5 (single and tandem axles). If this total does not exceed 100 percent,

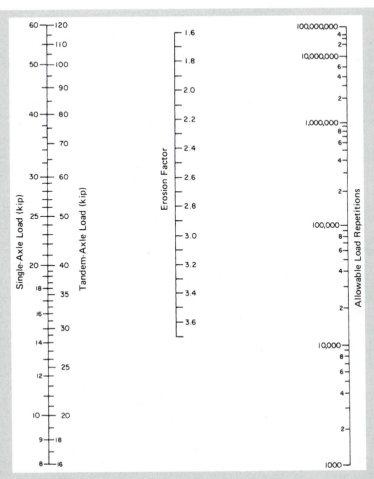

Figure 21.17 Allowable Load Repetitions for Erosion Analysis Based on Erosion Factor (with concrete shoulder)

SOURCE: R. G. Packard, *Thickness Design for Concrete Highway and Street Pavements,* Portland Cement Association, Skokie, Ill., 1984.

the assumed thickness is adequate for fatigue resistance for the design period. The total in this example is 62.8 percent, which shows that 9.5 in. is adequate for fatigue resistance.

Step 2.   **Erosion Analysis.**   Items 1 through 4 in the fatigue analysis are the same for the erosion analysis; therefore, they will not be repeated. The following additional procedures are necessary.

1. Determine the erosion factor for the single-axle loads and tandem-axle loads. Note that the erosion factors depend on the type of pavement construction—that is, doweled or undoweled joints, with or without

Project _Design 1A, four-lane Interstate, Rural_

Trial thickness _9.5_ in.

Subbase-subgrade k _130_ pci

Modulus of rupture, MR _650_ psi

Load safety factor, LSF _1.2_

Doweled joints: yes ✓ no ____

Concrete shoulder: yes ____ no ✓

Design period _20_ years

4 in. untreated subbase

Axle load, kips	Multiplied by LSF 1.2	Expected repetitions	Fatigue analysis		Erosion analysis	
			Allowable repetitions	Fatigue, percent	Allowable repetitions	Damage, percent
1	2	3	4	5	6	7

Single Axles

8. Equivalent stress _206_

9. Stress ratio factor _0.317_

10. Erosion factor _2.59_

30	36.0	6,310	27,000	23.3	1,500,000	0.4
28	33.6	14,690	77,000	19.1	2,200,000	0.7
26	31.2	30,140	230,000	13.1	3,500,000	0.9
24	28.8	64,410	1,200,000	5.4	5,900,000	1.1
22	26.4	106,900	Unlimited	0	11,000,000	1.0
20	24.0	235,800	"	0	23,000,000	1.0
18	21.6	307,200	"	0	64,000,000	0.5
16	19.2	422,500			Unlimited	0
14	16.8	586,900			"	0
12	14.4	1,837,000			"	0

Tandem Axles

11. Equivalent stress _192_

12. Stress ratio factor _0.295_

13. Erosion factor _2.79_

52	62.4	21,320	1,100,000	1.9	920,000	2.3
48	57.6	42,870	Unlimited	0	1,500,000	2.9
44	52.8	124,900	"	0	2,500,000	5.0
40	48.0	372,900	"	0	4,600,000	8.1
36	43.2	885,800			9,500,000	9.3
32	38.4	930,700			24,000,000	3.9
28	33.6	1,656,000			92,000,000	1.8
24	28.8	984,900			Unlimited	0
20	24.0	1,227,000			"	0
16	19.2	1,356,000				
			Total	62.8	Total	38.9

Figure 21.18 Pavement Thickness Calculation

SOURCE: Redrawn from R. G. Packard, _Thickness Design for Concrete Highway and Street Pavements,_ Portland Cement Association, Skokie, Ill., 1984.

concrete shoulders. The construction type in this project is doweled joints without shoulders. Table 21.13 is therefore used. For single axles, erosion factors = 2.59. For tandem axles, erosion factor = 2.79. Record these values as items 10 and 13, respectively.

2. Determine the allowable axle repetitions for each axle load based on erosion analysis using either Figure 21.16 or Figure 21.17. In this problem, Figure 21.16 will be used as the pavement has no concrete shoulder. Enter these values under column 6.

3. Determine erosion damage percent for each axle load, that is, divide column 3 by column 6. Enter these values in column 7.

4. Determine the total erosion damage by summing column 7 (single and tandem axles). In this problem total drainage is 38.9 percent.

The results indicate that 9.5 in. is adequate for both fatigue and erosion analysis. Since the total damage for each analysis is much lower than 100 percent, the question may arise whether a 9 in. thick pavement would not be adequate. If the calculation is repeated for a 9 in. thick pavement, it will show that the fatigue condition will not be satisfied since total fatigue consumption will be about 245 percent. In order to achieve the most economic section for the design period, trial runs should be made until the minimum pavement thickness that satisfies both analyses is obtained.

SUMMARY

This chapter has presented the basic design principles for rigid highway pavements and the application of these principles in the AASHTO and PCA methods. Rigid pavements have some flexural strength and can therefore sustain a beam-like action across minor irregularities in the underlying material. The flexural strength of the concrete used is therefore an important factor in the design of the pavement thickness. Although reinforcing steel may sometimes be used to reduce cracks in rigid pavement, it is not considered as contributing to the flexural strength of the pavement.

Both the AASHTO and PCA methods of design are iterative, indicating the importance of computers in carrying out the design. Design charts are included in this chapter, which can be used if computer facilities are not readily available.

A phenomenon that may have significant effect on the life of a concrete pavement is pumping. The PCA method indirectly considers this phenomenon in the erosion consideration. When rigid pavements are to be constructed over fine-grain soils that are susceptible to pumping, effective means to reduce pumping, such as limiting expansion joints or stabilization of the subgrade, should be considered.

PROBLEMS

21-1   Portland cement concrete consists of what four primary elements?

21-2   List and briefly describe the five main types of Portland cement as specified by AASHTO.

21-3   What is the main requirement for the water used in Portland cement?

21-4   Describe the basic types of highway concrete pavements and give the conditions under which each type will be constructed.

21-5   List and briefly describe the four types of joints used in concrete pavements.

21-6   Define the phenomenon of pumping and its effects on rigid pavements.

21-7   List and describe the types of stresses that develop in rigid pavements.

21-8   Determine the tensile stress imposed at night by a wheel load of 750 lb located at the edge of a concrete pavement with the following dimensions and properties.

> Pavement thickness = 8.5 in.
> $\mu = 0.15$
> $E_c = 4.2 \times 10^6$ lb/in.^{2}
> $k = 200$ lb/in.^{3}
> Radius of loaded area = 3 in.

21-9   Repeat Problem 21-8 for the load located at the interior of the slab.

21-10   Determine the maximum distance of contraction joints for a plain concrete pavement if the maximum allowable tensile stress in the concrete is 50 lb/in.^{2} and the coefficient of friction between the slab and the subgrade is 1.7. Assume uniform drop in temperature. Weight of concrete is 144 lb/ft^3.

21-11   Repeat Problem 21-10, with the slab containing temperature steel in the form of welded wire mesh consisting of 0.125 in.^{2} steel/ft width. The modulus of steel E_s is 30×10^6 lb/in.^{2}, $E_c = 5 \times 10^6$ lb/in.^{2}, and $h = 6$ in.

21-12   A concrete pavement is to be constructed for a four-lane urban expressway on a subgrade with an effective modulus of subgrade reaction k of 100 lb/in.^{3}. The accumulated equivalent axle load for the design period is 3.25×10^6. The initial and terminal serviceability indeces are 4.5 and 2.5, respectively. Using the AASHTO design method, determine a suitable thickness of the concrete pavement if the working stress of the concrete is 600 lb/in.^{2} and the modulus of elasticity is 5×10^6 lb/in.^{2}. Take the overall standard deviation S_o as 0.30, the load transfer coefficient J as 3.2, the drainage coefficient as 0.9, and $R = 95\%$.

21-13   A six-lane concrete roadway is being designed for a metropolitan area. This roadway will be constructed on a subgrade with an effective modulus of subgrade reaction k of 170 lb/in.^{3}. The ESAL used for the design period is 2.5×10^6. Using the AASHTO design method, determine a suitable thickness of the concrete pavement (to the nearest $\frac{1}{2}$ inch), provided that the working stress of the concrete to be used is 650 lb/in.^{2} and the modulus of elasticity is 5×10^6 lb/in.^{2}. Assume the initial serviceability is 4.75 and the terminal serviceability is 2.5. Assume the overall standard deviation, S_0, as 0.35, the load transfer coefficient J as 3.2, the drainage coefficient, C_d as 1.15, and R as 95%.

21-14   An existing rural four-lane highway is to be replaced by a six-lane divided expressway (3 lanes in each direction). Traffic volume data on the highway indicate that the AADT (both directions) during the first year of operation is 24,000 with the following vehicle mix and axle loads.

> Passenger cars = 50 percent
> 2-axle single-unit trucks (12,000 lb/axle) = 40 percent
> 3-axle single-unit trucks (16,000 lb/axle) = 10 percent

The vehicle mix is expected to remain the same throughout the design life of 20 years, although traffic is expected to grow at a rate of 3.5 percent annually. Using the AASHTO design procedure, determine the minimum depth of concrete pavement required for the design period of 20 years.

$$P_i = 4.5$$
$$P_t = 2.5$$
$$S'_c = 650 \text{ lb/in.}^2$$
$$E_c = 5 \times 10^6 \text{ lb/in.}^2$$
$$k = 130$$
$$J = 3.2$$
$$C_d = 1.0$$
$$S_o = 0.3$$
$$R = 95 \text{ percent}$$

21-15   Repeat Problem 21-14 for a pavement containing doweled joints, 6 in. untreated subbase, and concrete shoulders, using the design method.

21-16   Repeat Problem 21-14, using the PCA design method, for a pavement containing doweled joints and concrete shoulders. The modulus of rupture of the concrete used is 600 lb/in.^{2}.

21-17   Repeat Problem 21-14, using the PCA design method, if the subgrade k value is 50 and a 6 in. stabilized subbase is used. The modulus of rupture of the concrete is 600 lb/in.^{2} and the pavement has aggregate interlock joints (that is, no dowels) and a concrete shoulder.

21-18   Repeat Problem 21-17, using the PCA design method, assuming the pavement has doweled joints and no concrete shoulders.

REFERENCES

AASHTO Guide for Design of Pavement Structures, American Association of State Highway and Transportation Officials, Washington, D.C., 1993.

Annual Book of ASTM Standards, Concrete and Mineral Aggregates, vol. 04.02, *Road and Paving Materials; Pavement Management Technologies,* American Society for Testing and Materials, Philadelphia, Pa., October 1992.

Ioannides, A.M., M.R. Thompson, and E.J. Barenberg, "Westergaard Solutions Reconsidered," *Transportation Research Record* 1043, Transportation Research Board, Washington, D.C., 1985.

Packard, R.G., *Thickness Design for Concrete Highway and Street Pavements,* Portland Cement Association, Skokie, Ill., 1984.

Standard Specifications for Transportation Materials and Methods of Sampling and Testing, 20th ed., American Association of State Highway and Transportation Officials, Washington, D.C., 2000.

Tayabji, S.D. and B.E. Colley, "Analysis of Jointed Concrete Pavements," Report prepared by the Construction Technology Laboratories of the Portland Cement Association for the Federal Highway Administration, October 1981.

Tayabji, S.D. and B.E. Colley, "Improved Rigid Pavement Joints," Paper presented at the annual meeting of the Transportation Research Board, January 1983.

Teller, L.W., and Sutherland, E.C., "The Structural Design of Concrete Pavements," *Public Roads,* Vol. 17, Bureau of Public Roads, Washington, D.C., 1936.

Westergaard, H.M., *Theory of Concrete Pavement Design,* Proceedings, Highway Research Board, Washington, D.C., 1927.

CHAPTER 22

Pavement Management

The nation's highway system will be required to accommodate increasing numbers of motor vehicles during the coming decades. Accordingly, rehabilitation of the existing system has become a major activity for highway and transportation agencies. In this chapter we describe the various approaches for pavement management and the data required to evaluate pavement condition.

PROBLEMS OF HIGHWAY REHABILITATION

A major problem that faces highway and transportation agencies is that the funds they receive are usually insufficient to adequately repair and rehabilitate every roadway section that deteriorates. The problem is further complicated in that roads may be in poor condition but are still usable, making it easy to defer repair projects until conditions become unacceptable. Roadway deterioration usually is not the result of poor design and construction practices but is caused by the inevitable wear and tear that occurs over a period of years. The gradual deterioration of a pavement occurs due to many factors including variations in climate, drainage, soil conditions, and truck traffic. Just as a piece of cloth eventually tears asunder if a small hole is not immediately repaired, so will a roadway unravel if its surface is allowed to deteriorate. Lack of funds often limits timely repair and rehabilitation of transportation facilities, causing a greater problem with more serious pavement defects and higher costs.

Because funds and personnel are often inadequate to accommodate needs, the dilemma faced by many transportation agencies is to balance its programs between preventive maintenance activities and those projects requiring immediate corrective action. If preventive maintenance is neglected, then projects will be selected based on the extent of deterioration as seen by the user. The traveling public is unwilling to tolerate the condition of a pavement when the ride is extremely rough, vibrations cause damage to their vehicles, crashes occur, and user costs increase significantly. Usually preventive maintenance, carried out in an orderly and systematic way, will be the least expensive approach in the long run. However, when funds are extremely limited, agencies often respond to the most pressing and severe problems or the ones that generate the most vocal complaints.

Approaches to Pavement Management

The term *pavement management* is used to describe the various strategies that can be used to decide on a pavement restoration and rehabilitation policy. At one extreme there is the "squeaky wheel" approach, wherein projects are selected because they have created the greatest attention. At the other extreme is a system wherein all roads are repaired on a regular schedule, for which money is no object. In realistic terms, pavement rehabilitation or repair strategies are plans that establish minimum standards for pavement condition and seek to establish the type of treatment required and the time frame for project completion. Rehabilitation management strategies include consideration of items such as pavement condition, first cost, annual maintenance and user costs, safety, physical, environmental, and economic constraints, and life cycle costs.

The general topic of pavement management includes design, construction, maintenance, and rehabilitation. In essence, pavement management includes all activities involved in furnishing the pavement portion of a transportation system, such as planning, design, construction, maintenance, and rehabilitation. A total framework for pavement management is shown in Figure 22.1. Note that this process includes elements of network and project management and requires technical analysis in areas of planning, economics, design, and construction—topics that were described in earlier chapters of this textbook.

In this chapter we confine ourselves to the subject of rehabilitation management, which is defined as any improvement made to an existing pavement after initial construction, excluding improvements to shoulders (that is, widening or surfacing) or bridges (which is a topic more appropriate to structural design). Pavement rehabilitation can be both preventive and corrective.

Importance of Pavement Condition Data

The first step in the process of pavement management is to secure data about the condition of each pavement section in the system. Originally, condition data were obtained by visual inspections that established the type, extent, and severity of pavement condition. These inspections were subjective and relied heavily on judgment and experience for determining pavement condition and program priorities. Although such an approach can be appropriate under certain circumstances, possibilities exist for variations among inspectors, and experience is not easily transferable. In more recent years, visual ratings have been supplemented with standardized testing equipment to measure road roughness, surface condition, pavement deflection, and skid resistance.

Pavement condition data are used for the following purposes:

1. *Establishing Project Priorities.* The data on pavement condition are used to establish the relative condition of each pavement and to establish project priorities. There are several methods of data acquisition, and each state selects that combination of measures it considers most appropriate.

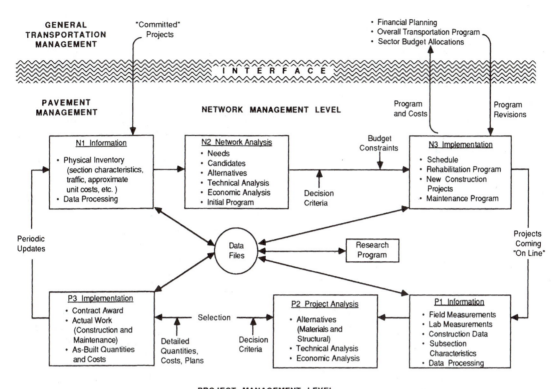

Figure 22.1   Framework for Total Pavement Management System

SOURCE: W.R. Hudson, R. Haas, and R.D. Pedigo, *Pavement Management System Development,* NCHRP Report 215, Transportation
Research Board, National Research Council, Washington, D.C., 1979. See also R. Haas, W.R. Hudson, and J. Zaniewski, *Modern Pavement*

2. *Establishing Options.* Pavement condition data can be used to develop a long-term
   rehabilitation program. Data about pavement condition, in terms of type, extent, and
   severity, are used to determine which of the available rehabilitation options should be
   selected.

3. *Forecasting Performance.* By use of correlations between pavement performance indi-
   cators and variables such as traffic loadings, it is possible to predict the likely future
   condition of any given pavement section. This information is useful for preparing
   long-range budget estimates of the cost to maintain the highway system at a minimum
   standard of performance or to determine future consequences of various funding
   levels.

METHODS FOR MEASURING ROADWAY CONDITION

Four characteristics of pavement condition used in evaluating pavement rehabilitation
needs are: (1) pavement roughness (rideability), (2) pavement distress (surface condition),

(3) pavement deflection (structural failure), and (4) skid resistance (safety). These characteristics are described in the following sections.

Pavement Roughness

Pavement roughness refers to irregularities in the pavement surface that affect the smoothness of a ride. The serviceability of a roadway was initially defined in the AASHTO road test, a national pavement research project. Two terms were defined: (1) present serviceability rating (PSR), and (2) present serviceability index (PSI).

PSR is a number grade given to a pavement section based on the ability of that pavement to serve its intended traffic. The PSR rating is established by observation and requires judgment on the part of the individual doing the rating. In the original AASHTO road test, a panel of raters drove on each test section and rated the performance of each section on the basis of how well the road section would serve if the rater were to drive his or her car over a similar road all day long. The ratings range between 0 and 5, with 5 being very good and 0 being impassable. Figure 22.2 illustrates the PSR as used in the AASHTO road test. Serviceability ratings are based on the user's perception of pavement performance and are determined from the average rating of a panel of road users.

PSI is a value for pavement condition determined as a surrogate for PSR and is based on physical measurements. PSI is not based on panel ratings. The primary measure of PSI is pavement roughness. The PSI is an objective means of estimating the PSR, which is subjective.

The performance of a pavement can be described in terms of its PSI and traffic loading over time, as is illustrated in Figure 22.3. For example, when a pavement is originally constructed, it is in very good condition with a PSI value of 4.5 (see Figure 22.4). Then, as the

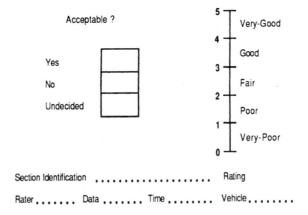

Figure 22.2 Individual Present Serviceability Rating

SOURCE: R. G. Hicks, *Collection and Use of Pavement Condition Data,* NCHRP Synthesis 76, Transportation Research Board, National Research Council, Washington, D.C., 1979. See also R. Haas, W.R. Hudson, and J. Zaniewski, *Modern Pavement Management,* Krieger Publishing, Malabar, Fla., 1994, p. 80.

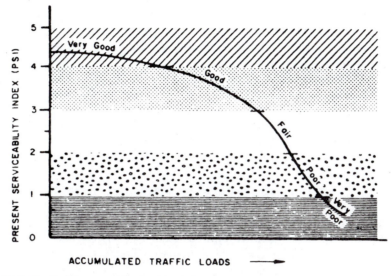

Figure 22.3 Typical Pavement Performance Curve

SOURCE: D. E. Peterson, *Pavement Management Practices,* NCHRP Synthesis 135, Transportation Research Board, National Research Council, Washington, D.C., November 1987.

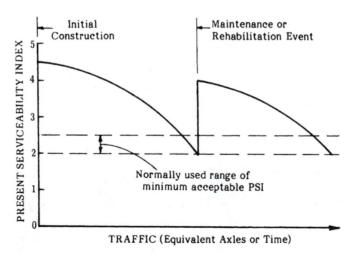

Figure 22.4 Performance History for Pavement Using PSI

SOURCE: Modified from D. E. Peterson, *Pavement Management Practices,* NCHRP Synthesis 135, Transportation Research Board, National Research Council, Washington, D.C., 1987.

number of traffic loadings increases, the PSI declines to a value of 2, which is normally the minimum acceptable range. After the pavement section is rehabilitated, the PSI value increases to 4; as traffic loads increase, the PSI declines again until it reaches 2 and rehabilitation is again required.

Equipment for measuring roughness is classified into two basic categories: response type and profilometers. Response type equipment does not measure the actual profile of the road but rather the response of the vehicle to surface roughness. Equipment used includes: (1) Mays Ride Meter (MRM), (2) Bureau of Public Roads (BPR) Roughmeter, and (3) Cox Road Meter. The Mays Ride Meter (see Figure 22.5) measures the number of inches of vertical movement per mile. The greater the vertical movement, the rougher the road.

Originally, some type of car-mounted meter was used. The advantages are low cost, simplicity, ease of operation, capability for acquisition of large amounts of data, and output correlated with PSI. However, the measure obtained is influenced by the vehicle itself, and now meters are usually towed in trailers. Response-type equipment must be calibrated frequently to ensure reliable results. Accordingly, they are not widely used.

Profilometers are devices that measure the true profile of the roadway and provide accurate and complete reproductions of the pavement profile. Profilometers also eliminate the need for the time-consuming and labor-intensive calibration required for response-type equipment performance.

There are several types of profilometers. Some equipment uses inertial profilometry, which includes the following basic components. (An example of a profilometer is illustrated in Figure 22.6.)

Figure 22.5 Mays Ride Meter Trailer Unit

SOURCE: J. A. Epps and C. L. Monismith, *Equipment for Obtaining Pavement Condition and Traffic Loading Data,* NCHRP Synthesis No. 126, Transportation Research Board, National Research Council, Washington, D.C., September 1986, p. 47.

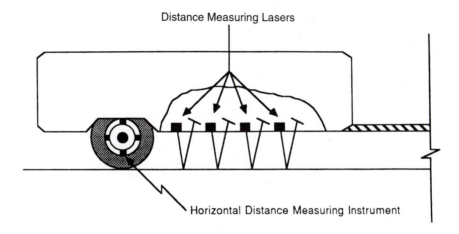

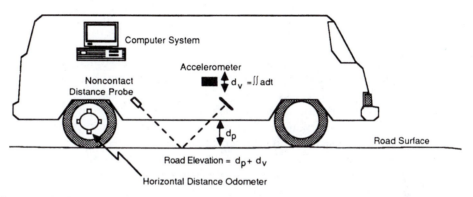

Figure 22.6 Example of Surface Dynamics Profilometer

SOURCE: R. Haas, W. R. Hudson, J. Zaniewski, *Modern Pavement Management,* Krieger Publishing, Malabar, Fla., 1994, p. 85. Used with permission.

1. A device to measure the distance between the vehicle and the road surface
2. An inertial referencing device to compensate for the vertical movement of the vehicle body
3. A distance odometer to locate the profile points along the pavement
4. An on-board processor for recording and analyzing the data

Acoustic or optic and laser devices are used for measuring the distance between the pavement surface and the vehicle, d_p. Acoustic systems are more widely used because of their lower cost and the reliability and adequacy of results.

The inertial referencing device is usually either a mechanical or electronic accelerometer mounted to represent the vertical axis of the vehicle. The accelerometer measures the vertical acceleration of the vehicle body. These accelerations are then integrated twice to determine the vertical movement, d_v. These movements are added to the distance measurements d_p to obtain elevations of the pavement profile. Some profilometers record the actual profile, whereas others process the data on board and give only a roughness summary statistic.

The two most popular models of profilometers are the K. J. Law profilometer and the South Dakota profilometer. The K. J. Law profilometer measures and records the road surface profile in each of the vehicle's two wheel paths. An optical measuring system, based on reflectivity from the road surface, and an accelerometer are used together to measure both the distance between the vehicle and the road surface and the vehicle vertical movement. Operating speeds range between 10 and 55 mi/h. The South Dakota profilometer is used by most states. It costs less than the K. J. Law profilometer, although it is not as accurate. It has two additional sensors that can be used to automatically measure rut depth. Data can be collected at normal highway speeds, and about seven hundred miles of pavement can be measured within a single week.

Pavement Distress

The term *pavement distress* refers to the condition of a pavement surface in terms of its general appearance. A perfect pavement is level and has a continuous and unbroken surface. In contrast, a distressed pavement may be fractured, distorted, or disintegrated. These three basic categories of distress can be further subdivided. For example, fractures can be seen as cracks or as spalling (chipping of the pavement surface). Cracks can be further described as generalized, transverse, longitudinal, alligator, and block. A pavement distortion may be evidenced by ruts or corrugation of the surface. Pavement disintegration can be observed as raveling (loosening of pavement structure), stripping of the pavement from the subbase, and surface polishing. The types of distress data collected for flexible and rigid pavements vary from one state to another. Figure 22.7 lists the three pavement distress groups, the measure of distress, and the probable causes.

Most agencies use some measure of cracking in evaluating the condition of flexible pavements. The most common measures are transverse, longitudinal, and alligator cracks. Distortion is usually measured by determining the extent of rutting, and disintegration is measured by the amount of raveling. Each state or federal agency has its own procedures for measuring pavement distress, consequently, there is a wide range of methods used to conduct distress surveys. Typically, each agency has a procedural manual that defines each element of distress to be observed, with instructions as to how these are to be rated on a given point scale. As an example, the pavement condition data form for the Washington Department of Transportation is illustrated in Figure 22.8. Note that the form distinguishes between bituminous and Portland cement concrete pavements. For bituminous pavements, the items observed are corrugations, alligator cracking, raveling, longitudinal cracking, transverse cracking, and patching. For Portland cement concrete, the measures are cracking, raveling, joint spalling, faulting, and patching.

Typically, distress data are obtained by trained observers who make subjective judgments about pavement condition based on predetermined factors. Often photographs

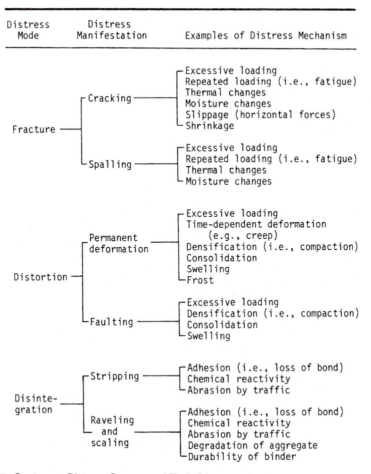

Figure 22.7 Pavement Distress Groups and Their Causes

SOURCE: R.G. Hicks, *Collection and Use of Pavement Condition Data,* NCHRP Synthesis No. 76, Transportation Research Board, National Research Council, Washington, D.C., 1981.

are used for making judgments. Figure 22.9 illustrates information to assist raters in estimating cracking. Although some agencies use full sampling, pavement sections of about 300 ft often are randomly selected to represent each mile of road. Measurements are usually made on a regular schedule about every 1–3 yr. After the data are recorded, the results are condensed into a single number, called a *distress* (or *defect*) *rating* (DR). A perfect pavement usually is given a score of 100; if distress is observed, points are subtracted. The general equation is

$$DR = 100 - \sum_{i=1}^{n} d_i w_i \qquad (22.1)$$

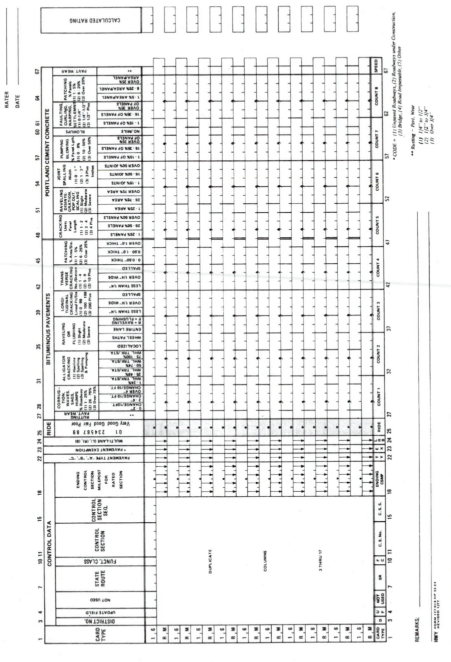

Figure 22.8 Pavement Distress Recording Form

SOURCE: *Collection and Use of Pavement Condition Data*, NCHRP Synthesis No. 126, Transportation Research Board, National Research Council, Washington, D.C., 1981, p. 10.

Name of Distress:	Transverse and Diagonal Cracks
Description:	Linear cracks are caused by one or a combination of the following: heavy load repetition, thermal and moisture gradient stresses, and drying shrinkage stresses. Medium or high severity cracks are working cracks and are considered major structural distresses. (Note: hairline cracks that are less than 6 feet (1.8 m) long are not rated).
Severity Levels:	L - Hairline (tight) crack with no spalling or faulting, a well sealed crack with no visible faulting or spalling.
	M - Working crack with low to medium severity level of spalling, and/or faulting less than 1/2 inch (13 mm). Temporary patching may be present.
	H - A crack with width of greater than 1 inch (25 mm); a crack with a high severity level of spalling; or, a crack faulted 1/2 inch (13 mm) or more.
How to Measure:	Cracks are measured in linear feet (or meters) for each level of distress. The length and average severity of each crack should be identified and recorded. Cracks in patches are recorded under patch deterioration.

(a) Medium-severity transverse crack

Figure 22.9 Guide for Rating Pavement Cracking

where

d_i = the number of points assigned to distress type i for a given severity and frequency

n = number of distress types used in rating method

w_i = relative weight of distress type i

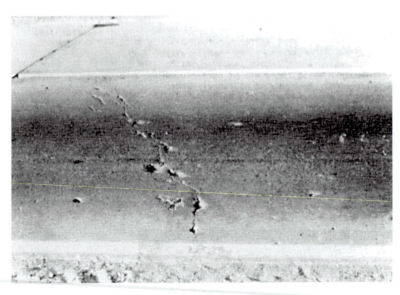

(b) Low-severity transverse crack

Figure 22.9 Guide for Rating Pavement Cracking (*continued*)

SOURCE: M. I. Darter, J. M. Becker, M. B. Snyder, and R. E. Smith, *Portland Cement Concrete Pavement Evaluation System,* NCHRP Synthesis No. 277, Transportation Research Board, National Research Council, Washington, D.C.,

One of the major problems with condition or distress surveys is the variability in results due to the subjective procedures used. Other causes of error are variations in the condition of the highway segment observed, changes in evaluation procedure, and changes in observed location from year to year. Other concerns are related to safety of pavement surveyors.

Problems with manual measurements have been minimized through development of automated techniques for evaluating pavement distress. Film and video devices are used to record continuous images of the pavement surface for later evaluation. Survey crews are not required to personally observe the pavement under hazardous traffic conditions, and a permanent record of the surface condition is created. There are many types of automated distress survey equipment available, and the technology is constantly being improved. The PASCO corporation has been developing a distress survey device since the 1960s. Figure 22.10 illustrates the PASCO ROADRECON system, which produces a continuous filmstrip recording of the pavement surface and a measure of roughness. Photographs are taken at night with a controlled amount and angle of lighting. The vehicle can operate at speeds of up to 50 mi/h. Manual interpretation of the photographs is still required for evaluating pavement distress. Eventually, however, computer vision technology will be available to analyze images without human intervention.

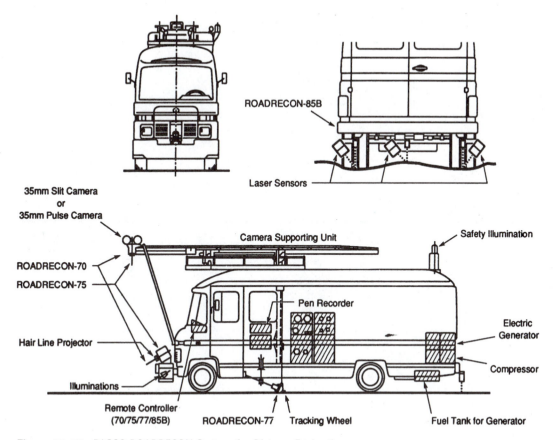

ROADRECON-85B

Laser Sensors

35mm Slit Camera
or
35mm Pulse Camera

Camera Supporting Unit

Safety Illumination

ROADRECON-70

ROADRECON-75

Hair Line Projector

Pen Recorder

Electric
Generator

Compressor

Illuminations

Remote Controller
(70/75/77/85B)   ROADRECON-77   Tracking Wheel   Fuel Tank for Generator

Figure 22.10   PASCO ROADRECON System for Distress Evaluation

SOURCE: *Report on Pavement Monitoring Methods and Equipment,* PASCO USA Inc., Lincoln Park, N.J., 1985.

Example 22.1   Computing Distress Rating of a Pavement Section

A pavement rating method for a certain state uses the following elements in its evaluation procedure: longitudinal or alligator cracking, rutting, bleeding, ravelling, and patching. The weighting factors are 2.4, 1.0, 1.0, 0.9, and 2.3, respectively. Each distress element is characterized by (1) its severity as not severe, severe, or very severe; and (2) its frequency as none, rare, occasional, or frequent. The categories for frequency are based on the percentage of area affected by a particular distress within the area of the section surveyed. For each combination of severity and distress, a rating factor is assigned, d_i, from 0 to 9, as shown in Table 22.1.

A one-mile section of roadway was observed with results shown in Table 22.2. Calculate the distress rating for the section.

Table 22.1   Rating Factor, d_i, Related to Severity and Frequency

Frequency	Severity		
	Not Severe (NS)	Severe (S)	Very Severe (VS)
None (N)	0	0	0
Rare (R)	1	2	3
Occasional (O)	2	4	6
Frequent (F)	3	6	9

Table 22.2   Observed Distress Characteristics for Road Segment

Distress Characteristic	Frequency	Severity
Cracking	R	NS
Rutting	O	S
Bleeding	F	VS
Raveling	N	NS
Patching	R	S

Solution: Using the data in Table 22.2 and the rating factors (d_i) in Table 22.1, each distress is categorized with factors as follows:

Distress Characteristic	Rating Factor, d_i	Weight, w_i
Cracking	1	2.4
Rutting	4	1.0
Bleeding	9	1.0
Raveling	0	0.9
Patching	2	2.3

Applying the weighting values for each characteristic, the distress rating for the section is determined using Eq. 22.1:

$$
\begin{aligned}
DR &= 100 - \sum_{i=1}^{n} d_i w_i \\
&= 100 - (1 \times 2.4 + 4 \times 1 + 9 \times 1 + 0 \times 0.9 + 2 \times 2.3) \\
&= 100 - 20 = 80
\end{aligned}
$$

Measuring Pavement Structural Condition

The structural adequacy of a pavement is measured either by nondestructive means, which measure deflection under static or dynamic loadings, or by destructive tests, which involve removing sections of the pavement and testing these in the laboratory. Structural condition evaluations are rarely used by state agencies for monitoring network pavement condition, due to the expense involved. However, nondestructive evaluations, which gather deflection data, are used by some agencies on a project basis for pavement design purposes and to develop rehabilitation strategies.

Nondestructive structural evaluation is based on the premise that measurements can be made on the surface of the pavement and that from these measurements in situ characteristics can be inferred about the structural adequacy of the pavement. The four basic nondestructive test methods are (1) measurements of static deflection, (2) measurements of deflections due to dynamic or repeated loads, (3) measurements of deflections from a falling load (impulse load), and (4) measurements of density of pavement layers by nuclear radiation, used primarily to evaluate individual pavement layers during construction. Deflection data are used primarily for design purposes and not for pavement management. Some states use deflection equipment solely for research and special studies.

An extensively used method for measuring static deflections is the Benkelman Beam, which is a simple hand-operated device designed to measure deflection responses of a flexible pavement to a standard wheel load. A probe point is placed between two dual tires, and the motion of the beam is observed on a dial, which records the maximum deflection. Other static devices that are used include the traveling deflectometer, the plate-bearing test, and the lacroix deflectograph. Most of these devices are based on the Benkelman Beam principle, in which pavement deflections due to a static or slowly moving load are measured manually or by automatic recording devices.

The most common dynamic loading method for measuring pavement deflections is the Dynaflect. This device, shown in Figure 22.11, basically consists of a dynamic cyclical force generator mounted on a two-wheel trailer, a control unit, a sensor assembly, and a sensor calibration unit. The system provides rapid and precise measurements of roadway deflections, which in this test are caused by forces generated by unbalanced flywheels rotating in opposite directions. A vertical force of 1000 lb is produced at the loading wheels, and deflections are measured at five points on the pavement surface, located 1 ft apart.

Most states use falling load–type equipment, referred to as "falling weight deflectometers" (FWD), because force impulses created by a falling load more closely resemble the pulse created by a moving load than that created by either the vibratory or static load devices. Figure 22.12 illustrates the basic principle of a falling weight deflectometer.

Variations in the applied load may be achieved by altering either the magnitude of the mass or the height of drop. Vertical peak deflections are measured by the FWD in the center of the loading plate and at varying distances away from the plate to draw what are known as "deflection basins."

Measuring Skid Resistance

Safety characteristics of a pavement are another measure of its condition, and highway agencies continually monitor this aspect to insure that roadway sections are operating at

Figure 22.11 Dynaflect Deflection Sensors

SOURCE: J. A. Epps and C. L. Monismith, *Equipment for Obtaining Pavement Condition and Traffic Loading Data,* NCHRP Synthesis No. 126, Transportation Research Board, National Research Council, Washington, D.C., September 1986, p. 17.

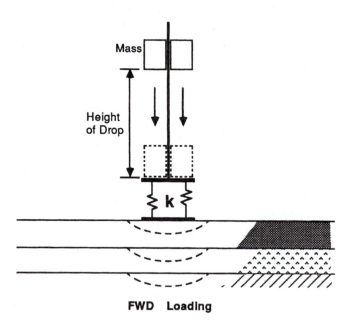

Figure 22.12 Principle of the Falling Weight Deflectometer

SOURCE: R. Haas, W. R. Hudson, and J. Zaniewski, *Modern Pavement Management,* Krieger Publishing, Malabar, Fla., 1994, p. 120. Used with permission.

the highest possible level of safety. The principal measure of pavement safety is its skid resistance. Other elements contributing to the extent in which pavements perform safely are rutting (which causes water to collect that creates hydroplaning) and adequacy of visibility of pavement markings.

Skid-resistance data are collected to monitor and evaluate the effectiveness of a pavement in preventing or reducing skid-related accidents. Skid data are used by highway agencies to identify pavement sections with low skid resistance, to develop priorities for rehabilitation, and to evaluate the effectiveness of various pavement mixtures and surface types.

The coefficient of sliding friction between a tire and pavement depends on factors such as weather conditions, pavement texture, tire condition, and speed. Since skidding characteristics are not solely dependent on the pavement condition, it is necessary to standardize testing procedures and in this way eliminate all factors but the pavement. The basic formula for friction factor f is

$$f = \frac{L}{N} \tag{22.2}$$

where
$\quad$ L = lateral or frictional force required to cause two surfaces to move tangentially to each other
$\quad$ N = force perpendicular to the two surfaces

When skid tests are performed, they must conform to specified standards set by the American Society for Testing and Materials (ASTM). The test results produce a skid number (SK)[*] where

$$SK = 100f \tag{22.3}$$

The SK is usually obtained by measuring the forces obtained with a towed trailer riding on a wet pavement, equipped with standardized tires. The principal methods of testing are (1) locked-wheel trailers, (2) Yaw mode trailers, and (3) the British Portable Tester. Table 22.3 lists the types of skid tests used by a representative group of agencies. As noted in the table, most states use locked-wheel trailers.

Locked-wheel trailers (see Figure 22.13) are the most widely used skid measuring devices. The test involves wetting the pavement surface and pulling a two-wheel trailer whose wheels have been locked in place. The test is conducted at 40 mi/h, with standard tires each with seven grooves. The locking force is measured, and from this an SK is obtained.

The Yaw mode test is done with the wheels turned at a specified angle, to simulate the effects of cornering. The most common device for this test is a Mu-Meter, which uses two

---

[*] Skid number is also referred to as SN. However, SK is used in this text to avoid confusion with the term *structural number* (SN) used in pavement design.

Table 22.3 Current Practice for Measuring Skid Resistance

Agency	Device
Arizona	Bison Mu Meter
California	Cox Towed Trailer
Florida	K. J. Law Locked-Wheel Skid Trailer
New York	ASTM Trailer
Pennsylvania	K. J. Law 1270
Utah	ASTM Trailer
Virginia	K. J. Law Locked-Wheel Skid Trailer
Washington	Cox Towed Trailer

SOURCE: Adapted from Wade L. Gramling, *Current Practices in Determining Pavement Condition,* NCHRP Synthesis 203, Transportation Research Board, National Research Council, Washington, D.C., 1994.

Figure 22.13 Locked Wheel Skid Trailer

SOURCE: J. A. Epps and C. L. Monismith, *Equipment for Obtaining Pavement Condition and Traffic Loading Data,* NCHRP Synthesis No. 126, Transportation Research Board, National Research Council, Washington, D.C., September 1986, p. 41.

wheels turned at 7.5°. The trailer is pulled in a straight line on a wetted surface with both wheels locked. Since both wheels cannot be in the wheel paths, friction values may be higher than those obtained by using a locked-wheel trailer.

As noted, the most common method for determining skid resistance is the locked-wheel trailer, and most state highway agencies own one or more of these devices. Several states use the Mu-Meter—a Yaw mode device. Figure 22.14 illustrates typical skid results for various pavement conditions. Skid resistance data typically are not used in developing rehabilitation programs. Rather they are used to monitor the safety of the highway system and to assist in reducing potential crash locations.

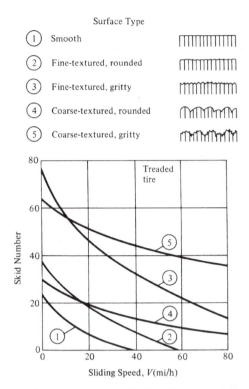

Figure 22.14 Skid Data for Various Pavement Surface Types

SOURCE: Redrawn from H. W. Kummer, W. E. Meyer, *Tentative Skid Resistance Requirements for Main Rural Highways,* NCHRP Report 37, Transportation Research Board, National Research Council, Washington, D.C., 1967.

Example 22.2   Measuring Skid Resistance

A 10,000-lb load is placed on two tires of a locked-wheel trailer. At a speed of 30 mi/h, a force of 5000 lb is required to move the device. Determine the SK and the surface type, assuming that treaded tires were used.

Solution:

$$SK = 100f = (100)\,\frac{L}{N}$$

$$= 100 \times \frac{5000}{10,000} = 50$$

From Figure 22.14, at 30 mi/h and SK = 50, the surface type is coarse-textured and gritty.

Predicting Pavement Condition

The results of pavement rating measurements can be used to predict pavement condition in future years. When data are collected over time, the results can be used to develop mathematical models that relate pavement condition to age of pavement. Two approaches have been used: (1) deterministic models, which are developed through regression analysis, and (2) probabilistic models, which are based on tables that furnish the probability of a pavement rating change from one year to the next. Each of these methods is described in the following sections.

Deterministic Models

Many states accumulate pavement distress data for specific roadway sections. As is expected, with aging of the pavement, the distress rating will decrease. Figure 22.15 illustrates the variation in pavement condition (PCR) versus age for pavements in the state of Washington. Note that there is a wide range of results, which suggests that a pavement model should be used only in circumstances similar to those which existed when the data were collected (i.e., climate, subgrade strength, axle loads, etc.).

One model, shown in Figure 22.15, states that $PCR = 100 - 0.76(Age)^{1.75}$. This model can be used to predict future pavement condition, information that is required to determine appropriate rehabilitation programs.

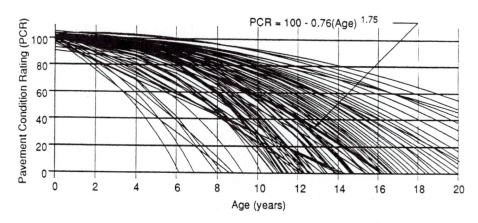

Figure 22.15 Pavement Condition Versus Pavement Age

SOURCE: An advanced Course in Pavement Management Systems, (course text), Federal Highway Administration, Washington, D.C., 1990.

Example 22.3   Developing a Pavement Prediction Model

A section of highway pavement, two miles long, was constructed in April 1992. Each year the pavement condition (PCR) was determined, with results shown in Table 22.4.

Table 22.4 PCR Ratings for a Highway Section

Year	PCR
92	100
93	100
94	99
95	98
96	97
97	95
98	93
99	90
00	87
01	84

Determine:

(a) The constants for a prediction model of the form PCR $= a + b(\text{Age})^2$
(b) The expected value of the PCR in 15 years
(c) If pavements are resurfaced when the PCR rating is less than 76, in what year should resurfacing take place?

Solution:

(a) To fit the model using linear regression, determine the square of the Age of the section, given that it was constructed in 1992.

Year	Age	Age^2	PCR
1992	0	0	100
1993	1	1	100
1994	2	4	99
1995	3	9	98
1996	4	16	97
1997	5	25	95
1998	6	36	93
1999	7	49	90
2000	8	64	87
2001	9	81	84

Use simple linear regression methods to correlate the PCR with Age^2, and determine the values of the coefficients a and b. Using a spreadsheet or any statistical software, the results of the linear regression are as follows:

$a = 100$
$b = -0.2$

with $R^2 = 0.99$. Therefore, the prediction model is given as:

$$PCR = 100 - 0.2(Age)^2$$

(b) $PCR = 100 - 0.2(Age)^2$
$= 100 - 0.2(15)^2$
$= 55$

(c)   Using the prediction model developed, determine the Age at which the section will reach a PCR of 76, as follows:

$$76 = 100 - 0.2(Age)^2$$

Therefore,

$$Age = \sqrt{(100 - 76)/0.2}$$
$$= 10.96 \text{ yr}$$

The section should be repaired 11 years from its date of construction, or in 2003.

Probabilistic Models

Probability methods are based on the assumption that future conditions can be determined from the present state if the probabilities of given outcomes are known. An example of this approach is a Markovian model, illustrated in Table 22.5. It shows the probability that the pavement condition in state i (PCR) will change to a pavement condition in state j. For example, if in the current year, the pavement condition (PCR) is within 70–79 (state 7), then the probability is 0.3 that in the next year the pavement condition will be within 60–69 (state 6). The probabilities shown in Table 22.5 can be used for successive years in a so-called Markovian chain. The assumption of Markovian models is that the future state is dependent on the current state, regardless of how the pavement reached that current state.

Example 22.4   Predicting Future Condition Using a Markovian Model

A state DOT district office is responsible for maintaining a network of 1000 mi. The results of the annual survey in a given year showed that the current network condition of these roads are:

- Group 1: 600 mi with a PCR between 90 and 100
- Group 2: 300 mi with a PCR between 80 and 89
- Group 3: 100 mi with a PCR between 70 and 79

Table 22.5 Probabilities of Pavement Condition Changes

| | | To PCR State | | | | | | | | | |
		9 *100–90*	*8* *89–80*	*7* *79–70*	*6* *69–60*	*5* *59–50*	*4* *49–40*	*3* *39–30*	*2* *29–20*	*1* *19–10*	*P*
	9 100–90	0.90	0.10								1.0
	8 89–80		0.70	0.30							1.0
	7 79–70			0.60	0.30	0.10					1.0
	6 69–60				0.50	0.30	0.15	0.05			1.0
From PCR state	5 59–50					0.30	0.40	0.30			1.0
	4 49–40						0.30	0.70			1.0
	3 39–30							0.60	0.35	0.05	1.0
	2 29–20								0.20	0.80	1.0
	1 19–10									1.0	1.0

Using the probability matrix given in Table 22.5, determine the condition of the network in year 1 and year 2.

Solution:

Step 1. Determine the current state of each of the above three groups:

Group 1: state 9
Group 2: state 8
Group 3: state 7

Step 2. Determine probabilities of outcomes for each year, in one year:

For Group 1, state 9:
  90% in state 9
  10% in state 8
For Group 2, state 8:
  70% in state 8
  30% in state 7

For Group 3, state 7:

  60% in state 7

  30% in state 6

  10% in state 5

Step 3. Determine the number of mi involved:

Group 1:

  $0.9 \times 600 = 540$ mi in state 9

  $0.1 \times 600 = 60$ mi in state 8

Group 2:

  $0.70 \times 300 = 210$ mi in state 8

  $0.30 \times 300 = 90$ mi in state 7

Group 3:

  $0.60 \times 100 = 60$ mi in state 7

  $0.30 \times 100 = 30$ mi in state 6

  $0.10 \times 100 = 10$ mi in state 5

Step 4. Summarize the mileage of highway in each PCR category:

- 540   = 540 mi will be in state 9 (PCR between 90 & 100)
- 60 + 210 = 270 mi will be in state 8 (PCR between 80 & 89)
- 90 + 60   = 150 mi will be in state 7 (PCR between 70 & 79)
- 30   = 30 mi will be in state 6 (PCR between 60 & 69)
- 10   = 10 mi will be in state 5 (PCR between 50 & 59)

For year 2, the procedure is the same. Each PCR state is determined, followed by appropriate probabilities. Then calculations are made as shown in Step 3 and summarized in Step 4. The solution is left to the reader.

PAVEMENT REHABILITATION

A variety of methods can be used to rehabilitate pavements or to correct deficiencies in a given pavement section, including using overlays, sealing cracks, using seal coats, and repairing potholes.

Classification of Techniques

Rehabilitation techniques are classified as (1) corrective, which involves the permanent or temporary repair of deficiencies on an as needed basis, or (2) preventive, which involves surface applications of either structural or nonstructural improvements intended to keep the quality of the pavement above a predetermined level. Corrective maintenance is analogous to repairing a small hole in a cloth, whereas preventive maintenance can be thought of as sewing a large patch or replacing the lining in a suit. Just as with the sewing analogy, corrective measures can serve as prevention measures as well. To illustrate, sealing a crack is done to correct an existing problem, but it also prevents further deterioration that would occur if the crack were not repaired. Similarly, a "chip seal coat," which is a layer of gravel placed on a thin coating of asphalt, often is used to correct a skid problem but also helps prevent further pavement deterioration.

Rehabilitation Strategies

Pavement rehabilitation strategies can be categorized in a variety of ways. One approach is in terms of the problem being solved, such as skid resistance, surface drainage, unevenness, roughness, or cracking. Another approach is in terms of the type of treatment used, such as surface treatment, overlay, or recycle. A third approach is in terms of the type of surface that will result from the process, such as asphalt overlay, rock seal coat, or liquid seal coat. The latter approach is the most commonly used because it enables the designer to consider each maintenance alternative in terms of a final product and then select the most appropriate one in terms of results desired and cost. Table 22.6 is a partial list of some of the pavement repair strategies used in California for both flexible and rigid pavements. Also listed are proper and improper uses of these strategies, their service life, and cost per lane mile.

For example, the table shows that asphalt concrete overlays (repair strategy 3) are used to restore the structural adequacy of a pavement, restore its surface texture, and improve ride quality. It is used where cracking is due to loads, ride score > 45, rut depth > ¾ in., and coarse aggregate ravels. (Ride score is defined in the pavement rehabilitation programming example later in this chapter.) This treatment is not used on steep grades, and it will last 10 yr and cost $12,500/lane mi/0.1 ft. In contrast, a rock seal coat (repair strategy 11) is used to waterproof pavements and decrease crack spalling. Its proper use is to seal dried-out pavements and correct skids but should not be used if the road is curved, where high volumes of traffic occur, to seal cracks, or where ride score > 45. This treatment lasts only 1–3 yr and costs $3000/lane mi.

To illustrate how this information can be used, consider a situation in which a surface treatment is desired that will serve to waterproof an existing pavement. Two repair strategies are considered: (1) a rock seal coat and (2) a liquid seal coat. (These are strategies 11 and 15 in Table 22.6.) As is noted in the table, the service life for both strategies is 1–3 yr. However, strategy 11 costs $3000/lane mi per year, whereas strategy 15 costs only $300/lane mi per year. The rock seal coat would also serve to improve the skid resistance characteristics, whereas the liquid seal coat should not be used if the skid number is low. Neither treatment should be used if the ride score is > 45. In this instance, if the road is in generally good condition and simply requires rejuvenation, then the liquid seal coat is appropriate. As noted in the final column of Table 22.6, both treatments are extensively used.

Thus, the repair strategy information provides a means by which appropriate strategies for the problem at hand can be identified and the most economical one (based on annual cost and life expectancy) can be selected.

Table 22.6 Pavement Repair Strategies for Flexible and Rigid Pavements

Repair Strategy	Function (Objective)	Proper Use	Improper Use	Service Life	Cost per Lane Mile*	California's Experience
		Flexible Pavements				
1. Lane reconstruction	Restore structural adequacy	A. Where more cost effective than alternates B. Ride score > 45 C. Vertical grade constraints		20 yr	$90,000 ①	Extensive
2. PCC overlays	Restore structural adequacy	Where more cost effective than alternates (0.55' minimum thickness)	A. Unstable terrain B. Vertical grade constraints	10 yr	$65,000 ②	Limited
3. AC overlays	A. Restore structural adequacy B. Repair cracked pavement C. Restore surface texture D. Improve ride quality	A. Load associated cracking B. Ride score > 45 C. Rut depth > 3/4″ D. Coarse aggregate ravel	A. Vertical grade constraints	10 yr	$12,500/0.10	Extensive
4. Inverted overlays	A. Restore structural adequacy B. Repair cracked pavement C. Restore surface texture D. Improve ride quality	A. Where more cost effective than conventional overlay B. Provide drainage blanket	Freeze-thaw areas	10 yr target	$35,000 ③	Limited experimental installations
5. Pavement reinforcing fabric and overlay	A. Restore structural adequacy B. Repair cracked pavement C. Restore surface texture D. Improve ride quality E. Water resistant membrane	A. Where more cost effective than conventional overlay B. Vertical grade constraints		10 yr target	$35,000 ④	Limited experimental installations
6. Rubberized asphalt interlayer and overlay	A. Restore structural adequacy B. Repair cracked pavement C. Restore surface texture D. Improve ride quality E. Water resistant membrane	A. Where more cost effective than conventional overlays B. Vertical grade constraints		10 yr target	$35,000 ⑤	Limited experimental installations
7. Hot recycling	A. Restore structural adequacy B. Repair cracked pavement C. Restore surface texture D. Conserve natural resources	A. Where more cost effective than alternates B. Vertical grade constraints	Air quality constraint at plant	10 yr	$24,000/0.10	None to date
8A. Heater remix 8B. Cutler process	A. Restore structural adequacy B. Repair cracked pavement C. Restore surface texture	A. Where more cost effective than alternates B. Vertical grade constraints	Air quality constraint at site	5–10 yr	$25,000 ⑥	Heater remix-extensive cutler process-none to date
9. Cold planing	A. Conform to elevation control B. Remove deteriorated and/or contaminated layer	A. Where more cost effective than alternates B. Prepare for overlay C. Air quality controls preclude hot planing D. Vertical grade controls		Not applicable	$15,000 ⑦	Moderate

Treatment	Purpose	Application	Conditions	Life	Cost	Usage
10. Rubberized asphalt chip seal coat	A. Waterproof pavement B. Decrease crack spalling	A. Seal dried-out pavement B. Stage construction preceding a planned overlay C. Fine aggregate ravel	A. High degree of road curvature B. High volume turning moves C. Heal cracks D. > hairline cracks unless filled E. Ride score > 45	Unknown: est 2–5 yr	$10,000	Limited experimental installations
11. Rock seal coat	A. Waterproof pavement B. Decrease crack spalling C. Texture surface	A. Seal dried-out pavement B. Fine aggregate ravel C. Skid resistance correction	A. High degree of road curvature B. High volume turning moves C. Heal cracks D. > hairline cracks unless filled E. Ride score > 45	1–3 yr	$3000	Extensive
12. Open graded seal coat	A. Decrease crack spalling B. Texture surface C. Correct bleeding	A. Seal dried-out pavement B. Coarse aggregate ravel C. Skid resistance correction D. Correct bleeding	A. Heal cracks B. > hairline cracks unless filled C. Ride score > 45 D. Frequent tire chain use required	5 yr	$5000	Extensive
13. Slurry seals	A. Stop ravel B. Waterproof pavement C. Decrease crack spalling D. Texture surface	A. Seal dried-out pavement B. Fine or coarse aggregate ravel	A. Heal cracks B. Ride score > 45 C. > hairline cracks unless filled D. Frequent tire chain use required	4 yr	$4000	Limited
14. Seal coat with (sand) cover	A. Waterproof pavement B. Decrease crack spalling C. Stop ravel D. Restore binder flexibility	A. Seal dried-out pavement B. Fine aggregate ravel	A. Heal cracks B. Coarse ravel C. Ride score > 45 D. Low to moderate skid number E. Rutting F. Highly impermeable pavement	1–3 yr	$1500	Extensive
15. Liquid seal coat	A. Waterproof pavement B. Decrease crack spalling C. Stop ravel D. Restore binder flexibility	A. Seal dried-out pavement B. Fine aggregate ravel	A. Heal cracks B. Coarse ravel C. Ride score > 45 D. Low to moderate skid number E. Rutting F. Highly impermeable pavement	1–3 yr	$300	Extensive

Continued

Table 22.6 Pavement Repair Strategies for Flexible and Rigid Pavements (*continued*)

Repair Strategy	Function (Objective)	Proper Use	Improper Use	Service Life	Cost Per Lane Mile*	California's Experience
Flexible Pavements						
16. Binder modifiers (rejuvenating agent)	A. Waterproof pavement B. Decrease crack spalling C. Stop ravel D. Restore binder flexibility	A. Seal dried-out pavement B. Fine aggregate ravel	A. Heal cracks B. Coarse ravel C. Ride score > 45 D. Low to moderate skid number E. Rutting F. Highly impermeable pavement	1–3 yr	$500	Extensive
Rigid Pavements						
17. Lane reconstruction	Restore structural adequacy	A. > 10% third-stage cracking B. Where more cost effective than alternates C. Vertical grade constraints D. Ride score > 45		20 yr	$100,000	Limited
18. PCC overlays	Restore structural adequacy	A. > 10% third stage cracking B. Where more cost effective than alternates C. No vertical grade constraints D. When traffic handling permits E. Ride score > 45		20 yr	$65,000 ①	Limited
19. AC overlays	A. Restore structural adequacy B. Repair cracked pavement C. Restore surface texture D. Improve ride quality	A. > 10% third stage cracking B. Where more cost effective than alternates C. Faulting if slabs stabilized D. No vertical grade constraints E. Ride score > 45	Rocking slabs	10 yr	$12,500/0.10″	Moderate
20. Inverted overlays	A. Restore structural adequacy B. Repair cracked pavement C. Restore surface texture D. Improve ride quality	A. > 10% third stage cracking B. Where more cost effective than alternates C. Faulting if slabs stabilized D. No vertical grade constraints E. Ride score > 45 F. Provide drainage blanket	Freeze–thaw areas	10 yr	$40,000 ②	Limited
21. Pavement reinforcing fabric and overlay	A. Repair badly cracked pavement B. Water resistant membrane	A. Where more cost effective than alternates B. Vertical grade constraints		10 yr	$35,000 ③	One experimental installation

No. Strategy	Purpose	Recommendations	Condition	Service life	Cost	Experience
22. Slab replacement	Replace random slabs which are severely distressed	A. Less than 35 slabs per mi B. Where more cost effective than overlay or reconstruction C. Severe crack spalling D. Ride score > 45		5 yr	$15,000 ④	Moderate
23. Mudjacking	A. Fill cavities under pavement B. Restore pavement gradeline	A. Improve ride score B. Faulted or vertically displaced slabs C. Where more cost-effective than alternates	Badly cracked	5–10 yr	$55,000 ⑤	Extensive
24. Subsealing	Fill cavities under pavement	Faulted pavement	Badly cracked slabs > 10% third stage cracking	5–10 yr	$55,000	No recent experience
25. Grinding	A. Relieve faulting B. Improve ride quality	A. Faulting > 1/4″ B. Ride score > 45		More than 5 yr	$20,000	Extensive
26. Pavement subdrainage	Dewater structural section	A. At lower pavement edge B. In wet climate C. Indications of faulting and/or pumping		Unknown: est 10–15 yr	$20,000/mi	Limited experimental installations
27. Crack filling	Waterproof pavement	A. Clean cracks ≥ 1/4″ wide B. Appropriate sealant	A. Dirty cracks B. < 1/4″ cracks	1–2 yr	$200	Extensive
28. Grooving	A. Reduce hydroplaning B. Improve vehicle tracking	A. Abnormal rate of wet-pavement crashes due to hydroplaning	Badly cracked pavement	10–15 yr	$5000	Extensive

Costs do not include traffic handling.

Assumptions for Flexible Pavement Cost Estimates: ① 0.35′ AC over 0.70′ Class A CTB ② 0.55′ PCC ③ 0.08′ O.G. with 0.20′ AC ④ Reinforcing fabric with 0.20′ AC ⑤ Rubberized chip seal with 0.20′ AC ⑥ Scarify, add rejuvenator and 0.08′ AC ⑦ 0.10″ depth

Assumptions for Rigid Pavement Repair Cost Estimates: ① 0.60′ PCC ② 0.08′ O.G. plus 0.25′ AC ④ Reinforcing fabric with 0.20′ AC ④ 30 slabs/mi @ $500 each (12″ depth PCC) ⑥ $8 per square yard.

SOURCE: Dale E. Peterson, *Evaluation of Pavement Maintenance Strategies,* NCHRP Synthesis 77, Transportation Research Board, National Research Council, Washington D.C., September 1981.

Alternatives for Repair and Rehabilitation

Figure 22.16 illustrates a variety of pavement repair and rehabilitation alternatives and differentiates between preventive and corrective approaches. For example, preventive strategies for pavement surfaces include fog-seal asphalt, rejuvenators, joint sealing, seal coats (with aggregate), and a thin blanket. This figure is helpful in understanding the purpose for which a given treatment is intended.

Table 22.7 lists a variety of pavement rehabilitation techniques for both flexible and rigid pavements and indicates both if the technique is commonly used in corrective or preventive situations and its degree of effectiveness. For example, patching is always considered to be corrective and can be effective if properly done. Many patching materials are available. At the other extreme, overlays are both corrective and preventive and considered to be an effective technique. Surface treatments can be either preventive or corrective and are considered an effective means of maintaining roads on a regular basis.

Another method of describing pavement rehabilitation alternatives is to tabulate the possible types of deficiencies and show the most appropriate treatment. Table 22.8 is a summary of repair methods for flexible and rigid pavements. The distress types are shown in terms of severity (moderate or severe) and whether the repair is considered temporary or permanent. For example, for flexible pavements alligator cracking is repaired by removal and replacement of the surface course, by permanent patching or scarifying, and by mixing the surface materials with asphalt.

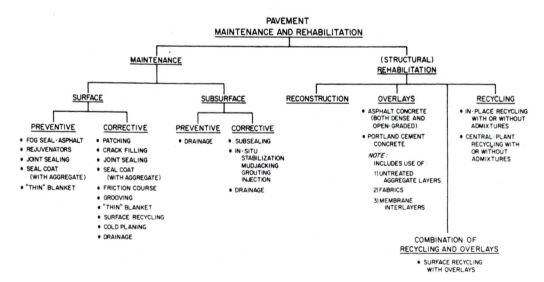

Figure 22.16 Pavement Maintenance and Rehabilitation Alternatives

SOURCE: C. L. Monismith, *Pavement Evaluation and Overlay Design Summary of Methods,* Transportation Research Record No. 700, Transportation Research Board, National Research Council, Washington, D.C., 1979.

Table 22.7   Flexible and Rigid Pavement Maintenance Techniques

Treatment Type	Techniques	Corrective or Preventive	Effectiveness	Remarks
Patching	Temporary Permanent Spot seal (spray) Cold mix Hot mix Level	All patching is considered corrective. Some agencies said patching could be preventive in some cases when preventing more serious deterioration.	When properly done, patching is considered to be effective and to serve purpose. Temporary patching is considered moderately effective to serve short period until permanent repairs can be made.	There is a wide range of materials that can be used for patching. See NCHRP Synthesis 64 (56).
Surface Treatments	Seal coating with cover aggregate, chips, etc. Sand seal coat Flush coating (black seal, liquid asphalt, etc.) Dilute sealing (fog seal) Slurry seal Rejuvenators (binder, modifiers, Reclamite, etc.) Hot-weather sanding Open-graded seal coat	Approximately 70% of replies indicated that surface treatments were preventive. Others indicated that surface treatments were corrective for skid resistance, etc. Some said surface treatments serve dual function.	Most of those providing ratings said surface treatments are effective for purpose intended. Several agencies have a schedule of every so many years for placing a surface treatment.	There is a considerable variation in materials and techniques used by various agencies for surface treatments. Local conditions vary extensively (climate, traffic, etc.) and must be considered for each treatment used.
Crack Maintenance	Crack cleaning Crack sealing Crack filling	Crack sealing is used as a preventive technique about 2/3 of the time. It is used 1/3 of the time as a corrective technique. Some consider it to serve a dual function.	Crack sealing is considered to be relatively effective. It has a fairly short life (1 to 2 yr) and so must be repeated often.	There are several different materials that can be used for crack sealing. Rubber asphalt appears to be the most cost-effective if properly used.
Surface Planing	Burn plane Cold plane	Corrective	Generally effective in correcting corrugations and in reducing effect of high asphalt content or soft mix. May excessively harden asphalt.	There are different types of equipment that can be used. See NCHRP Synthesis 54(60).
Other Localized Repairs	Blankets Base repair Remove and replace	Corrective	No evaluation given.	None
Recycling	Plant recycling In-place recycling (cold) In-place recycling (hot)	Corrective	Very few agencies identified this technique. Those that did considered it effective.	A relatively new technique that uses different softening agents, different equipment, etc.
Overlays	Thick overlays Thin overlays Pavement reinforcing Fabric and overlay Rubberized asphalt interlayer and overlay Inverted overlays	Most agencies (69%) consider overlays to be a corrective technique. Thirty-one percent considered them to be preventive. Some believe they serve a dual function.	All agencies rating overlays said they were effective.	The timing of when an overlay is placed determines whether it is preventive or corrective. If the overlay is placed after failure, then it becomes corrective. Some agencies consider overlays to be rehabilitation rather than maintenance.
Patching	Temporary patching Permanent patching PCC patching	Corrective	Temporary patching considered fair to poor. Permanent patching rated good.	There are numerous patching materials available. NCHRP Synthesis 45(57) gives data on some.
Crack Maintenance	Crack cleaning Crack sealing	Classed as preventive and/or corrective	Rated as generally effective.	There was limited input on crack maintenance in rigid pavements.

Continued

Table 22.7 Flexible and Rigid Pavement Maintenance Techniques (*continued*)

Treatment Type	Techniques	Corrective or Preventive	Effectiveness	Remarks
Joints	Pressure relief joints	Corrective and/or preventive	Rated as effective in helping maintain rigid pavement and in protecting pavement and bridges.	Limited input.
	Repair joints	Corrective		Limited input.
	Cleaning joints	Mostly classed corrective; some considered preventive	Effective	There are a number of different materials that can be used.
	Sealing joints		Generally effective depending on materials and workmanship	
Faulting Repair	Grinding (planing)	Corrective	Effective	The most common repair method identified was mudjacking.
	Mudjacking (pressure grouting)	Mostly corrective, some preventive	Moderately effective	
	Subsealing	Preventive	Not rated	
Blowup Repair	Temporary	Corrective	Temporary	There was minimal input in this area.
	Remove and replace	Corrective	Effective	
Surface Planing	Grinding	Corrective	Effective for correcting ride but may not solve basic problem creating roughness.	Limited input from agencies.
	Grooving	Corrective	Effective for correcting skid resistance.	
Other Localized Repairs	Drainage correction Underdrain installation	Preventive	Rated effective	It was felt by some agencies that water was serious problem and that proper drainage of system would extend performance.
	Remove and replace Surface/base replacement	Corrective	Effective	Limited input.
Overlays	Thick bituminous overlay Thin bituminous overlay Thick PCC overlay Thin PCC overlay Pavement reinforcing fabric and overlap Rubberized asphalt interlayer and overlay	Corrective	Generally effective; thicker overlays considered more effective.	There was limited input on overlays over Portland cement concrete pavement.

SOURCE: Dale E. Peterson, *Evaluation of Pavement Maintenance Strategies,* NCHRP Synthesis 77, Transportation Research Board, National Research Council, Washington, D.C., September 1981.

The Use of Expert Systems in the Selection of Maintenance and Rehabilitation Strategies

Expert systems (a branch of artificial intelligence) can be used in the selection of a subset of maintenance and rehabilitation strategies. Expert systems (ES) are computer models that exhibit, within a specific domain, a degree of expertise in problem solving that is comparable to that of a human being. The knowledge required to build the expert system is obtained by interviewing pavement engineers who have long experience and knowledge about pavement management. The acquired knowledge is stored in the expert system, which can then be used to recommend appropriate maintenance or rehabilitation strategies. Since the information is stored in a computer knowledge base, diagnostic advice is available to those users who may not be as experienced.

Table 22.8 Summary of Repair Methods for Flexible and Rigid Pavements

Flexible Pavement — Repair Methods

Distress	Seal Coat (1)	Spray Patching (1)	Spread Blotting Material (1)	Remove with Blade or Heater-Planer (1)	Removal and Replacement of Surface Course (2)	Asphaltic Concrete Leveling and Surfacing (3)	Temporary Hole Filling (4)	Permanent Patching (4)	Scarifying and Mixing (5)	Crack Filling (6)	Crack Filling with Slurry (7)	Asphalt Emulsion Slurry Seal (8)
Abrasion	MT					BP			BP			
Bleeding			MB		SP							
Char		MB			SP							
Indentation				MB		SP						
Loss of cover agg.	BB		MT	SP		SP						
Polished aggregates	BT			BB		BP						
Pothole		MT					BT	BP				
Raveling	BB			SP								
Streaking				BB	SP	SP						
Weathering	BB											MT
Corrugation				MB		MB			BP			
Alligator cracking					MT			SP	SP			
Contraction cracking					SP				SP	BT	BT	BT
Edge cracking						SP				MT	BT	
Edge joint cracking						SP		SP		MT		
Lane joint cracking					SP				SP	MT	MT	MT
Reflection cracking										BT		
Shrinkage cracking	MT				SP						SB	
Slippage cracking		MT			SP			SP				
Depression						SP						
Butting					SP	MB	MT	SP				
Shoving						MT		SP				
Upheaval					MT			SP				
Utility cut depression						MB		SP				

Rigid Pavement — Repair Methods

Distress	Asphalt Emulsion Slurry Seal (8)	Joint Filling (9)	Thin Asphaltic Concrete Overlay (10)	Concrete Patching: Shallow (11)	Concrete Patching: Deep (11)	Slab Jacking (12)
Crazing	MT				SP	
Joint filler extrusion/stripping		BB				
Scaling	MT		BP	SP	SP	
Buckling blow-up					BP	
Shattering blow-up					BP	
Corner cracking		MT			BP	MP
Diagonal cracking		MT			BP	BP
Longitudinal cracking		MT			SP	BP
Transverse cracking		MT			SP	BP
2nd stage cracking		MT			SP	BP
Progressive cracking		MT			SP	SP
Random cracking	MT	ST			SP	
"D" cracking	MT		BP	MP		
Faulting		BT				BP
Joint failure		BT		BP	BP	
Pumping						BP
Spalling			MT	MP	BP	

Severity of Distress: (M) moderate, (S) severe, (B) both.
Permanency of Repair: (T) temporary, (P) permanent, (B) both.
SOURCE: Adapted from Dale E. Peterson, *Evaluation of Pavement Maintenance Strategies,* NCHRP Synthesis 77, Transportation Research Board, National Research Council, Washington, D.C., September 1981.

PAVEMENT REHABILITATION PROGRAMMING

In the previous sections, we described how pavement condition is measured and discussed the alternatives and strategies available to repair and rehabilitate these surfaces. In this section we describe the process used to decide upon a specific program for rehabilitation. The program requires decisions about the type of repair or restoration technique that should be used for a given pavement section and about the timing (or programming) of the project. These decisions consider the design life of the pavement, the cost and benefits of the project, and other physical and environmental conditions.

Methods Used to Develop Pavement Improvement Programs

Transportation agencies use various methods for selecting a program of pavement rehabilitation. According to the *AASHTO Guidelines for Pavement Management Systems,* the analysis techniques are divided into three main groups: (1) condition assessment, (2) priority assessment, and (3) optimization. A study completed in 1991 determined the analysis techniques used or in the planning stage by 47 state highway agencies, and is shown in Table 22.9. Each method is used to determine the type of treatment needed and the schedule for rehabilitation. The following sections describe each method.

Condition Assessment

This method is used to develop single-year programs. The agency establishes criteria for the different measures of the pavement condition against which comparison of the actual measurements can be made. If the measurement exceeds this limit or "trigger point," then a deficiency or need exists. Figure 22.17 illustrates this concept. For example, if a limit of PSI of 2.5 has been set as the minimum acceptable roughness level for a particular class of pavements, then any section with a PSI less than 2.5 will represent a current deficiency. The fixed trigger point thus resolves the timing issue in a simple manner. Whenever the condition index falls below the given trigger point (criteria), it is assumed that rehabilitation is needed. Therefore, by using the trigger criteria, all sections are separated into two groups: "now needs" and "later needs."

In this method, we are concerned only with the "now needs." Having decided upon the sections that need an action, the next step is then to select a treatment for each of these "now needs." This could involve simple economic analysis methods, such as the net pres-

Table 22.9   Pavement Management Analysis Techniques for State Highway Agencies

Technique	Number
None	4
Condition or priority	37
Optimization	22

SOURCE: Adapted from F.C. Irrgang and T.H. Maze, "Status of Pavement Management Systems and Data Analysis Models at State Highway Agencies," *Transportation Research Record* 1397, 1993.

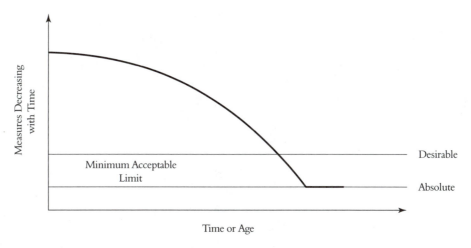

Figure 22.17 Determining Rehabilitation Needs Based on Established Criteria

ent worth (NPW) or the benefit-cost ratio (BCR), based on approximate estimates of the expected life of the different alternatives. After deciding on the treatment, three possible situations may arise: (1) the needs match the budget, (2) the needs exceed the budget, or (3) the needs are less than the budget.

Of the three situations, the most common is when needs exceed the budget, and so a ranking of these projects is often needed. The purpose of such ranking is to determine which of the needs could be deferred to the next year. There are several possible alternatives for ranking. Projects may be ranked by distress, a combination of distress and traffic, the net present worth (NPW), or the benefit-cost ratio (BCR). Sections are then selected until the budget is exhausted.

The Rational Factorial Ranking Method (FRM) is an example of a ranking method. It uses a priority index that combines climatic conditions, traffic, roughness, and distress. The priority index is expressed as

$$Y = 5.4 - (0.0263X_1) - (0.0132X_2) - [0.4 \log(X_3)] \\ + (0.749\,X_4) + (1.66X_5)$$

(22.4)

where

Y = The priority index ranging from 1 to 10, with 1 representing very poor, and 10 representing excellent. Thus a low value indicates a pavement has a high priority for treatment.

X_1 = average rainfall (in./yr)
X_2 = freeze and thaw (cycle/yr)
X_3 = traffic (AADT)
X_4 = present serviceability index
X_5 = distress (a subjective number between -1 and $+1$)

Example 22.5   Determining the Order of Priority for Rehabilitation

Three sections of highway have been measured for surface condition, with results as shown in Table 22.10. Also shown are climatic and traffic conditions for each section. Use the Rational Factorial Rating Method to determine the order of priority for rehabilitation.

Solution: Using Eq. 22.4,

$$Y = 5.4 - (0.0263 \, X_1) - (0.0132 X_2) - [0.4 \log(X_3)] + (0.749 X_4) + (1.66 X_5)$$

the priority index (Y) for each section is

- Section 1:

$$Y = 5.4 - (0.0263 \times 10) - (0.0132 \times 5) - [0.4 \log(10,000)] + (0.749 \times 2.4)$$
$$+ (1.66 \times 0.5) = 6.099$$

- Section 2:

$$Y = 5.4 - (0.0263 \times 30) - (0.0132 \times 15) - [0.4 \log(5000)] + (0.749 \times 3.2)$$
$$+ (1.66 \times (-0.2)) = 4.998$$

- Section 3:

$$Y = 5.4 - (0.0263 \times 15) - (0.0132 \times 0) - [0.4 \log(20,000)] + (0.749 \times 3.0)$$
$$+ (1.66 \times 0.8) = 6.86$$

Since low index values indicate poor condition, section 2 should receive highest priority, followed by section 1, and lastly section 3.

Table 22.10   Condition, Climatic, and Traffic Data for Highway Sections

Section	Rainfall	Freeze/Thaw	AADT	PSI	Distress
1	10 in./yr	5 cycles/yr	10,000	2.4	+0.5
2	20 in./yr	15 cycles/yr	5,000	3.2	−0.2
3	30 in./yr	0 cycles/yr	20,000	3.0	+0.8

Priority Assessment Models

Priority assessment is used to develop multiyear programs, an extension of the single-year model described. It addresses the question of when the "latter needs" are to be addressed and what action is required. To use this method, performance prediction models are required. For developing multiyear programs using ranking methods, either fixed-trigger-point or variable-trigger-point methods are used.

Trigger-Point Ranking.   Models are used to predict when each road section will reach its trigger point. That is, instead of separating the network into two groups (present needs and later needs), this method separates the pavement sections into, say, six groups, according to the year where action will be needed, as illustrated in Figure 22.18. The process from this point is essentially the same as the previous method, except that more accurate economic analysis is possible as a result of the availability of prediction models. The use of a fixed-trigger-point method was illustrated in Example 22.3.

A variable-trigger-point ranking uses economic analysis models to establish the treatment as well as the timing. The advantage of using this method is that timing and treatment selection decisions are made simultaneously, instead of being treated as two distinct stages in a sequential process, as was the case with the previous methods. Figure 22.19 illustrates the process.

Optimization

Optimization models provide the capability for simultaneous evaluation of an entire pavement network. Rehabilitation alternatives are selected to satisfy a specific objective function subject to certain constraints. For example, strategies are selected to maximize the total network benefits (or performance) or minimize the total network cost subject to network-level constraints such as budget limits and desired performance standards. Formulations used in these methods include linear programming (LP) and dynamic and integer programming.

Example of Pavement Rehabilitation Programming

To illustrate the process of pavement rehabilitation programming, the procedures that have been used by the California Department of Transportation (CALTRANS) are described. This approach illustrates a condition assessment model and typifies methods used by many states.

Two types of data are collected for use in establishing a pavement maintenance program: a pavement distress survey and a ride survey. In addition, average daily traffic (ADT) volumes are measured and used to establish priorities. (Pavement deflection data are used only for design.) Distress data are obtained by field observation, using highly trained observers. Ride survey data are obtained, using a custom-manufactured Cox ultrasound road meter to determine ride quality. A ride score is determined as the sum of $1/8$ in. vertical movements accumulated over a measured distance divided by 50 times length in mi. It has been established by observations that if the ride score is 45 or greater, the pavement section should be considered for improvement because of the poor ride. Thus a ride score of 45 is a trigger value that delineates a good ride from a bad one and is useful in establishing priorities. Ride surveys and distress surveys are conducted at the same time on a two- year rotating basis.

Analysis of the data provides guidance concerning the need for repair. For example, if longitudinal and transverse cracks are wider than $1/4$ in., repairs are deemed necessary. Also ride score values exceeding 45 establish the need for repair. The use of trigger values

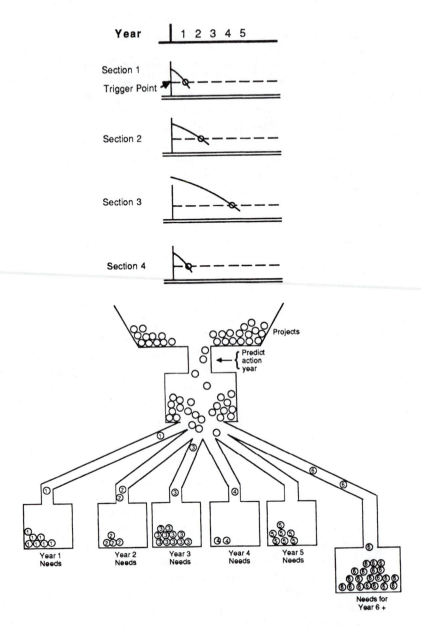

Figure 22.18 Fixed-Trigger-Point Ranking for Pavement Rehabilitation Programming

SOURCE: W. D. Cook, R. L. Lytton, "Recent Developments and Potential Future Directions in Ranking and Optimization Procedures for Pavement Management," *Second North American Conference for Managing Pavements,* vol. 2, 1987, p. 2.144.

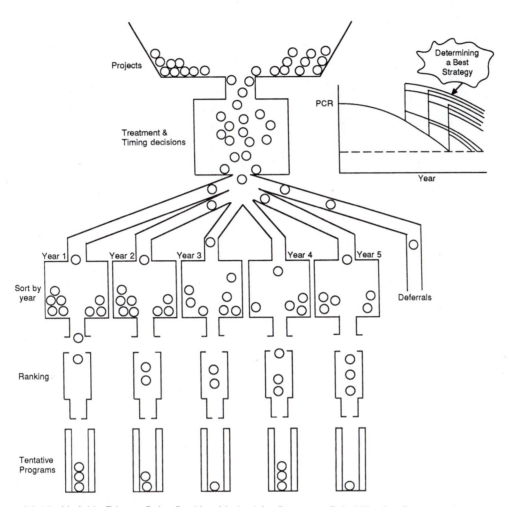

Figure 22.19 Variable-Trigger-Point Ranking Method for Pavement Rehabilitation Programming

SOURCE: W. D. Cook, R. L. Lytton, "Recent Developments and Potential Future Directions in Ranking and Optimization Procedures for Pavement Management," *Second North American Conference for Managing Pavements,* vol. 2, 1987, p. 2.145.

establishes that a pavement section should be repaired. Further analysis is now required to determine the appropriate type of repair and the order of priorities.

Determination of the type of repair required is based on a decision tree analysis. The tree path is determined by trigger values, and these lead to an appropriate repair strategy. Figure 22.20 illustrates a decision tree for a flexible pavement section in which alligator/ block cracking has been observed. For example, if a given pavement section has class B cracking (alligator cracking in wheel paths), with less than 10 percent cracked and more than 10 percent of the area patched, then the decision tree path indicates that the type of repair should be a thin asphalt cement overlay with local dig-outs.

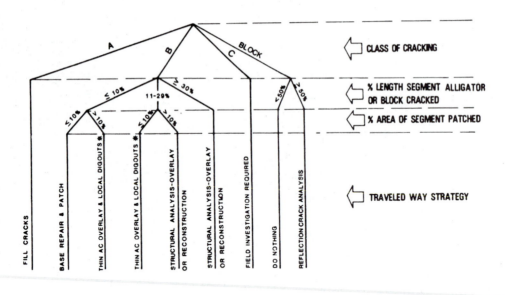

\# THIN AC OVERLAY = < 0.10' DENSE GRADED OR OPEN GRADED MIX

Figure 22.20   Decision Tree for Alligator Cracking

SOURCE: R. G. Hicks, *Collection and Use of Pavement Condition Data,* NCHRP Synthesis No. 76, Transportation Research Board, National Research Council, Washington, D.C., July 1981.

A similar analysis is carried out for sections along the traveled way and in adjoining lanes and shoulders. This may result in different pavement rehabilitation strategies for several of the sections. Figure 22.21 illustrates the results for various sections along a single traveled way. As can be seen, alligator cracking may require strategy A, longitudinal cracking, strategy B, and so forth. These results are compared and a *dominant* strategy is selected, one that will correct all the defects in that lane. A similar analysis is conducted for the other lanes of the roadway, and a dominant strategy is produced for each. These dominant strategies for each lane are then compared, and a single pavement repair strategy is chosen (called a *compatible* strategy) to correct as many defects as possible for the entire roadway segment (see Figure 22.22). The rehabilitation strategy selected is then assigned a cost and life, which is used in the next step of the procedure. The process described for selecting pavement strategies is computerized, and reports are produced that furnish data about each lane, including dominant strategy, location, ADT, road type, and cost.

The final step in the process is to establish priorities for needed rehabilitation work that has been previously identified. This step is necessary since sufficient funds are not always available to address each pavement need. California's prioritization method is based on three variables: ride score, distress rating, and ADT. These figures are combined and a priority matrix is used, as illustrated in Figure 22.23. Priority rankings increase with traffic

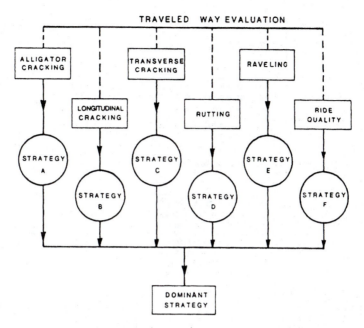

Figure 22.21 Results of Flexible Pavement Evaluation Along One Lane of a Highway

SOURCE: R. G. Hicks, *Collection and Use of Pavement Condition Data,* NCHRP Synthesis No. 76, Transportation Research Board, National Research Council, Washington, D.C., July 1981.

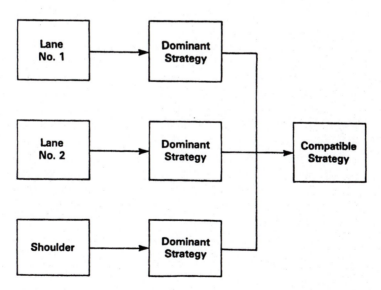

Figure 22.22 Results of a Flexible Pavement Evaluation for All Lanes of a Highway

SOURCE: *Pavement Management Rehabilitation Programming: Eight States' Experiences,* U.S. Department of Transportation, Washington, D.C., August 1983.

Pavement Problem Type		Candidate Category		
		ADT Range		
		>5,000	1,000 to 5,000	<1,000
Ride >45	Major Structural Problem and Unacceptable Ride Flex: Allig.B = 11-29% & Patch>10% or Allig.B>30% Rigid: 3rd Stg. Crk.≥10%	1	2	11
	Minor Structural Problem and Unacceptable Ride Flex: Allig.B = 11-29% & Patching<10% Allig.B<10% & Patching>10%	3	4	12
	Unacceptable Ride Only	5	6	✕
Ride <45	Major Structural Problem Only Flex: Allig.B = 11-29% & Patch>10% or Allig.B>30% Rigid: 3rd Stg. Crk.≥10%	7	8	13
	Minor Structural Problem Only Flex: Allig.B = 11-29% & Patching<10% Allig.B<10% & Patching>10%	9	10	14

Figure 22.23 Priority Matrix for Pavement Maintenance

SOURCE: *Pavement Management Rehabilitation Programming: Eight States' Experiences,* U.S. Department of Transportation, Washington, D.C., August 1983.

volume and severity of the problem. For example, if ADT on a roadway section (with no surface problems) exceeds 5000 and its ride quality exceeds 45, the section will be assigned a priority ranking of 5. However, if traffic volumes are below 1000, it is not considered. There are 14 possible rankings. If ADT exceeds 1000, a high ride score will assure a ranking of at least 6, which reflects CALTRANS' commitment to serve as many users as possible. Also, a high ADT will increase the priority for a similar category of pavement distress.

When pavement sections are given the same priority ranking, a "tie-breaker" value is computed, which is then added to the priority ranking of each of the tied sections. This value is cost per mile divided by ADT. The factor gives higher priority to projects with a lower per-mile cost and with a higher ADT. The process produces a computer-generated report that shows the priority value for each candidate segment within the state.

The rehabilitation management process that has been developed in California is used for (1) making cost-effectiveness comparisons among repair strategies, (2) determining repair costs by district, and (3) determining the appropriate statewide funding level for all categories of pavement repair programs. The results of this process are used at the district level to establish specific programs.

SUMMARY

This chapter has described the procedures used to develop a pavement rehabilitation program. The specifics may vary from one state to another, but there is agreement on the need for a rational and objective process to ensure that funds are efficiently used for pavement improvements. The benefits of using pavement management systems are (1) improved performance monitoring, (2) a rational basis for legislative support, (3) determination of various funding consequences, (4) improved administrative credibility, and (5) engineering input in policy decisions.

PROBLEMS

22-1   What is meant by the term *pavement rehabilitation management*? Describe three strategies used by public agencies to develop restoration and rehabilitation programs.

22-2   What are the three principal uses for pavement condition data?

22-3   What is the difference between PSI and PSR?

22-4   Draw a sketch showing the relationship between pavement condition (expressed as PSI) and time for a service life of 20 yr, if the PSI values range from 4.5 to 2.5 in a 6-year period, and then the pavement is resurfaced such that the PSI is increased to 4.0. After another 6 years, the PSI has reached 2.0, but with a rehabilitated pavement in place, PSI is increased to 4.5. At the end of its service life the PSI value is 3.4.

22-5   Describe the four characteristics of pavement condition used to evaluate whether a pavement should be rehabilitated, and if so, determine the appropriate treatment required.

22-6   A given pavement rating method uses six distress types to establish the DR. These are corrugation, alligator cracking, raveling, longitudinal cracking, rutting, and patching. For a stretch of highway, the number of points assigned to each category were 6, 4, 2, 4, 3, and 3. If the weighing factors are 2, 1, 0.75, 1, 1, and 1.5, determine the DR for the section.

22-7   What are some of the problems with distress surveys, and how can such problems be solved?

22-8   How are computers used in pavement management?

22-9   Describe the methods used to determine static and dynamic deflection of pavements. To what extent are these tests used in pavement rehabilitation management?

22-10   A 5000-lb load is placed on two tires, which are then locked in place. A force of 2400 lb is necessary to cause the trailer to move at a speed of 20 mi/h. Determine the value of the skid number. If treaded tires were used, how would you characterize the pavement type?

22-11   The PMS database has accumulated information regarding the performance of a pavement section before being overlaid, as well as the performance of the overlay. The following data were recorded.

Before Overlay		Overlay Performance	
PCR	Age (yr)	PCR	Age (yr)
100	0.5	98	0.5
98	1.5	96	1.5
94	3.5	92	3.5
91	4.5	86	5.5
89	5.5	83	6.5
86	6.5	80	7.5
83	8.0		

Develop a linear prediction model PCR = a + b (Age) for the two cases. If a criterion was established that maintenance or rehabilitation should occur whenever the PCR reaches a value of 81, when should such actions take place in the two cases?

22-12   Referring to the Markovian transition matrix, Table 22.5, what is the probability that a section with a PCR value of 77 in the current year will have a PCR value between 60 and 69:

(a) one year later

(b) two years later

22-13   Differentiate between corrective and preventive rehabilitation techniques. Cite three examples of surface treatments in each category. What is the best preventive technique for subsurface maintenance?

22-14   Describe the techniques used to repair flexible and rigid pavements and their effectiveness for the following treatment types: (a) patching, (b) crack maintenance, and (c) overlays.

22-15   A pavement section has lost its seal, due to drying out, and the fine aggregates have raveled. The ride score is 27. The section is part of a road where tire chains are used and there are considerable vehicle turning movements. The skid number is 83. Which repair strategy would you recommend? Explain your answer.

22-16   What is the basic difference between an expert system and a conventional computer program?

22-17   Discuss the differences between condition and priority assessment models used in developing pavement improvement programs.

22-18   Describe six methods that can be used to select a program of pavement rehabilitation. Under what circumstances would each method be used?

22-19   A pavement section has been observed to exhibit alligator cracking in the wheel paths (class B) for approximately 20 percent of its length. About 15 percent of the section has been previously patched. The ride score is 32. Use a decision tree analysis to determine the appropriate treatment required.

22-20   If the ADT on the roadway section described in Problem 22-19 is 3500 and cost per mile is $5000, what priority value would the road receive?

22-21   Another road segment has been examined and is given the same priority ranking as that determined in Problem 22-20. If the ADT is 4000 and the cost per mile is 10 percent greater than the alternative, what is the priority value of this road segment?

REFERENCES

Federal Highway Administration, *An Advanced Course in Pavement Management Systems,* course text, 1990.

Haas, R., W.R. Hudson, and J. Zaniewski, *Modern Pavement Management,* Krieger Publishing Co., Malabar, Florida, 1994.

ADDITIONAL READINGS

AASHTO Guidelines for Pavement Management Systems, American Association of State Highway and Transportation Officials, Washington, D.C., 1990.

Epps, J.A., and C.L. Monismith, *NCHRP Synthesis of Highway Practice, Report 126: Equipment for Obtaining Pavement Condition and Traffic Loading Data,* Transportation Research Board, National Research Council, Washington, D.C., 1986.

Gramling, W.L., *NCHRP Synthesis of Highway Practice, Report 203: Current Practices in Determining Pavement Condition,* Transportation Research Board, National Research Council, Washington, D.C., 1994.

Pavement Design and Management Guide, Ralph Haas, ed., Transportation Association of Canada, Ottawa, Ont., 1997.

Pavement Management Rehabilitation Programming: Eight States' Experiences, U.S. Department of Transportation, Federal Highway Administration, Washington, D.C., August 1983.

Pavement Management Systems, Transportation Research Record 1455, Transportation Research Board, National Research Council, Washington, D.C., 1995.

Pavement Management Systems for Streets, Highways and Airports, Transportation Research Record 1524, Transportation Research Board, National Research Council, Washington, D.C., 1996.

Pavement Management and Monitoring of Traffic and Pavements, Transportation Research Record 1643, Transportation Research Board, National Research Council, Washington, D.C., 1998.

Peterson, D.E., *NCHRP Synthesis of Highway Practice, Report 135: Pavement Management Practices,* Transportation Research Board, National Research Council, Washington, D.C., 1987.

Proceedings, 1st North American Conference for Managing Pavements, Vols. 1–2, 1985.

Proceedings, 2nd North American Conference for Managing Pavements, Vols. 1–3, 1987.

Proceedings, 3rd International Conference for Managing Pavements, Vols. 1–2, 1994.

Shahin, M.Y., *Pavement Management for Airports, Roads and Parking Lots,* Chapman and Hall, New York, 1996.

Zimmerman, K.A., and ERES Consultants, Inc., *NCHRP Synthesis of Highway Practice, Report 222: Pavement Management Methodologies to Select Projects and Recommend Practices,* Transportation Research Board, National Research Council, Washington, D.C., 1995.

Critical Values for the Student's t Distribution

	Level of Significance for One-Tailed Test							
	.250	.100	.050	.025	.010	.005	.0025	.0005
	Level of Significance for a Two-Tailed Test							
Degrees of Freedom	.500	.200	T .100	.050	.020	.010	.005	.001
1.	1.000	3.078	6.314	12.706	31.821	63.657	27.321	536.627
2.	.816	1.886	2.920	4.303	6.965	9.925	14.089	31.599
3.	.765	1.638	2.353	3.182	4.541	5.841	7.453	12.924
4.	.741	1.533	2.132	2.776	3.747	4.604	5.598	8.610
5.	.727	1.476	2.015	2.571	3.365	4.032	4.773	6.869
6.	.718	1.440	1.943	2.447	3.143	3.707	4.317	5.959
7.	.711	1.415	1.895	2.365	2.998	3.499	4.029	5.408
8.	.706	1.397	1.860	2.306	2.896	3.355	3.833	5.041
9.	.703	1.383	1.833	2.262	2.821	3.250	3.690	4.781
10.	.700	1.372	1.812	2.228	2.764	3.169	3.581	4.587
11.	.697	1.363	1.796	2.201	2.718	3.106	3.497	4.437
12.	.695	1.356	1.782	2.179	2.681	3.055	3.428	4.318
13.	.694	1.350	1.771	2.160	2.650	3.012	3.372	4.221
14.	.692	1.345	1.761	2.145	2.624	2.977	3.326	4.140
15.	.691	1.341	1.753	2.131	2.602	2.947	3.286	4.073
16.	.690	1.337	1.746	2.120	2.583	2.921	3.252	4.015
17.	.689	1.333	1.740	2.110	2.567	2.898	3.222	3.965

Continued

Level of Significance for One-Tailed Test

	.250	.100	.050	.025	.010	.005	.0025	.0005
				Level of Significance for a Two-Tailed Test				
Degrees of Freedom	.500	.200	T .100	.050	.020	.010	.005	.001
18.	.688	1.330	1.734	2.101	2.552	2.878	3.197	3.922
19.	.688	1.328	1.729	2.093	2.539	2.861	3.174	3.883
20.	.687	1.325	1.725	2.086	2.528	2.845	3.153	3.850
21.	.686	1.323	1.721	2.080	2.518	2.831	3.135	3.819
22.	.686	1.321	1.717	2.074	2.508	2.819	3.119	3.792
23.	.685	1.319	1.714	2.069	2.500	2.807	3.104	3.768
24.	.685	1.318	1.711	2.064	2.492	2.797	3.091	3.745
25.	.684	1.316	1.708	2.062	2.485	2.787	3.078	3.725
26.	.684	1.315	1.706	2.056	2.479	2.779	3.067	3.707
27.	.684	1.314	1.703	2.052	2.473	2.771	3.057	3.690
28.	.683	1.313	1.701	2.048	2.467	2.763	3.047	3.674
29.	.683	1.311	1.699	2.045	2.462	2.756	3.038	3.659
30.	.683	1.310	1.697	2.042	2.457	2.750	3.030	3.646
35.	.682	1.306	1.690	2.030	2.438	2.724	2.996	3.591
40.	.681	1.303	1.684	2.021	2.423	2.704	2.971	3.551
45.	.680	1.301	1.679	2.014	2.412	2.690	2.952	3.520
50.	.679	1.299	1.676	2.009	2.403	2.678	2.937	3.496
55.	.679	1.297	1.673	2.004	2.396	2.668	2.925	3.476
60.	.679	1.296	1.671	2.000	2.390	2.660	2.915	3.460
65.	.678	1.295	1.669	1.997	2.385	2.654	2.906	3.447
70.	.678	1.294	1.667	1.994	2.381	2.648	2.899	3.435
80.	.678	1.292	1.664	1.990	2.374	2.639	2.887	3.416
90.	.677	1.291	1.662	1.987	2.368	2.632	2.878	3.402
100.	.677	1.290	1.660	1.984	2.364	2.626	2.871	3.390
125.	.676	1.288	1.657	1.979	2.357	2.616	2.858	3.370
150.	.676	1.287	1.655	1.976	2.351	2.609	2.849	3.357
200.	.676	1.286	1.653	1.972	2.345	2.601	2.839	3.340
∞	.6745	1.2816	1.6448	1.9600	2.3267	2.5758	2.8070	3.2905

Source: Reproduced from Richard H. McCuen, *Statistical Methods for Engineers,* copyright © 1985. Reprinted by permission of Prentice-Hall, Inc., Englewood Cliffs, N.J.

Computing Regression Coefficients

Let a dependent variable Y and an independent variable x be related by an estimated regression function

$$Y = a + bx \qquad \text{(B.1)}$$

Let Y_i be an estimate and y_i be an observed value of Y for a corresponding value x_i for x. Estimates of a and b can be obtained by minimizing the sum of the squares of the differences (R) between Y_i and y_i for a set of observed values, where

$$R = \sum_{i=1}^{n} (y_i - Y_i)^2 \qquad \text{(B.2)}$$

Substituting $(a + bx_i)$ for Y_i in Eq. B.2, we obtain

$$R = \sum_{i=1}^{n} (y_i - a - bx_i)^2 \qquad \text{(B.3)}$$

Differentiating R partially with respect to a, and then with respect to b, and equating each to zero, we obtain

$$\frac{\partial R}{\partial a} = -2 \sum_{i=1}^{n} (y_i - a - bx_i) = 0 \qquad \text{(B.4)}$$

$$\frac{\partial R}{\partial b} = -2\sum_{i=1}^{n} x_i(y_i - a - bx_i) = 0 \tag{B.5}$$

From Eq. B.4 we obtain

$$\sum_{i=1}^{n} (y_i) = na + b \sum_{i=1}^{n} (x_i) \tag{B.6}$$

giving

$$a = \frac{1}{n}\sum_{i=1}^{n} y_i - \frac{b}{n}\sum_{i=1}^{n} (x_i) \tag{B.7}$$

From Eq. B.5 we obtain

$$\sum_{i=1}^{n} x_i y_i = a \sum_{i=1}^{n} x_i + b \sum_{i=1}^{n} x_i^2 \tag{B.8}$$

Substituting for a, we obtain

$$b = \frac{\displaystyle\sum_{i=1}^{n} x_i y_i - \frac{1}{n}\left(\sum_{i=1}^{n} x_i\right)\left(\sum_{i=1}^{n} y_i\right)}{\displaystyle\sum_{i=1}^{n} x_i^2 - \frac{1}{n}\left(\sum_{i=1}^{n} x_i\right)^2} \tag{B.9}$$

where
$\quad n$ = number of sets of observation
$\quad x_i$ = ith observation for x
$\quad y_i$ = ith observation for y

Equations B.8 and B.9 may be used to obtain estimated values for a and b in Eq. B.1. To test the suitability of the regression function obtained, the coefficient of determination, R^2, which indicates to what extent values of Y_i obtained from the regression function agree with observed values y_i, is determined from the expression

$$R^2 = \frac{\displaystyle\sum_{i=1}^{n} (Y_i - \bar{y})^2}{\displaystyle\sum_{i=1}^{n} (y_i - \bar{y})^2} \tag{B.10}$$

which is also written as

$$R^2 = \frac{\left(\sum\limits_{i=1}^{n} x_i y_i - n\bar{x}\bar{y} \right)^2}{\left(\sum\limits_{i=1}^{n} x_i^2 - n\bar{x}^2 \right)\left(\sum\limits_{i=1}^{n} y_i^2 - n\bar{y}^2 \right)} \tag{B.11}$$

The closer the value of R^2 is to 1, the more suitable the estimated regression function for the data.

Metric Conversion Factors for Highway Geometric Design

This appendix provides information regarding areas critical to basic geometric design and provides the metric version of formulas that appear in Chapters 3 and 16, as given by the AASHTO publication, *A Policy on Geometric Design of Highways and Streets,* 4th edition, 2001, and the *Guide to Metric Conversion,* 1993.

There are nine areas critical to basic geometric design, for which AASHTO has provided metric values. These are speed, lane width, shoulders, vertical clearance, clear zone, curbs, sight distance, horizontal curvature, and structures. Recommended values are listed in Table C1.0 of the following section.

Table C1.1 provides conversion factors for length, area, and volume, and formulas converted to metric are listed together with the corresponding equation in the text.

METRIC FORMULAS USED IN HIGHWAY DESIGN
Stopping Sight Distance

$$S = 0.278vt + \frac{v^2}{254\left(\left(\dfrac{a}{9.81}\right) \mp G\right)} \tag{3.24}$$

where
    t = brake reaction time, 2.5 secs
    v = design speed km/h
    a = deceleration rate m/s^2

Table C1.0 Selected Metric Values for Geometric Design

Metric		US Customary
Design Speed (km/h)		Corresponding Design Speed (mi/h)
20		15
30		20
40		25
50		30
60		40
70		45
80		50
90		55
100		60
110		70
120		75
130		80
II. Lane Width		
2.7 m	(8.86 ft)	(1.56% less than 9′ lane)
3.0 m	(9.84 ft)	(1.60% less than 10′ lane)
3.3 m	(10.83 ft)	(1.55% less than 11′ lane)
3.6 m	(11.81 ft)	(1.58% less than 12′ lane)
III. Shoulders		
0.6 m	(1.97 ft)	
1.2 m	(3.94 ft)	
1.8 m	(5.91 ft)	
2.4 m	(7.87 ft)	
3.0 m	(9.84 ft)	
IV. Vertical Clearance		
3.8 m	(12.47 ft)	
4.3 m	(14.11 ft)	
4.9 m	(16.08 ft)	

The 4.9 m value is seen to be the critical value since the federal legislation required Interstate design to have 16 feet vertical clearance. In view of the fact that the Interstate, now virtually complete, is based on this minimum clearance, the metric value should provide this clearance as a minimum.

V. Clear Zone

Urban Conditions	0.5 m (1.64 ft)
Locals/Collectors	3.0 m minimum (9.84 ft)

With two exceptions, the *Green Book* refers to the *Roadside Design Guide* for clear zone values. The two critical values are the clear zone for urban conditions and locals and collectors.

Continued

Table C1.0 Selected Metric Values for Geometric Design (*continued*)

VI. Curbs

A. Curb Heights
1. Mountable Curb, 150 mm max (5.91")
2. Barrier Curb, 225 mm max (8.86")
B. The definition of high speed/low speed has an impact on where curb is used.
      Low speed: 60 km/h or less design speed
      High speed: 80 km/h or more design speed

VII. Sight Distance

Stopping Sight Distance	
Eye Height	1070 mm (3.51 ft)
Object Height	150 mm (5.91 in.)
Headlight Height	610 mm (2 ft.)
Passing Sight Distance	
Eye Height	1070 mm (3.51 ft)
Object Height	1300 mm (4.27 ft)

VIII. Horizontal Curvature

Radius definition should be used in lieu of degree of curve. Radius should be expressed in multiples of 5 m increments.

IX. Structures

Long bridges will be those over 60 m in length.

Minimum Radius of Curve

$$R_{\min} = \frac{v^2}{127(e + f_s)} \tag{3.31}$$

v = initial speed (km/h)
f_{so} = coefficient of friction (Table 3.4)
e = rate of superelevation

Length of Spiral

$$L = \frac{0.0214\, v^3}{RC} \tag{16.41}$$

where C is the rate of increase of centripetal acceleration, m/s^2 (use 1–3 m/s^2).

L = minimum length of spiral, m
v = speed km/h
R = curve radius, m

Table C1.1   Area, Length, and Volume Conversion Factors

Quantity	From Inch-Pound Units	To Metric Units	Multiply By
Length	mile	km	1.609344
	yard	m	0.9144
	foot	m	0.3048
		mm	304.8
	inch	mm	25.4
Area	square mile	km^2	2.59000
	acre	m^2	4,046.856
		ha (10,000 m^2)	0.4046856
	square yard	m^2	0.83612736
	square foot	m^2	0.09290304
	square inch	mm^2	645.15
Volume	acre foot	m^3	1,233.49
	cubic yard	m^3	0.764555
	cubic foot	m^3	0.0283168
	cubic foot	cm^3	28,316.85
	cubic foot	L(1000 cm^3)	28.31685
	100 board feet	m^3	0.235974
	gallon	L(1000 cm^3)	3.78541
	cubic inch	cm^3	16.387064
	cubic inch	mm^3	16,387.064

Middle Ordinate

$$M = R\left[1 - \cos\frac{28.65S}{R}\right] \tag{16.25}$$

where M = middle ordinate in meters.

S = stopping sight distance, m
R = radius of curve, m

Length of Crest Vertical Curves

S = sight distance, m
L = length of vertical curve (m)
A = algebraic difference in grades (percent)
h_1 = height of eye = 1080 mm
h_2 = height of object = 600 mm

$$S < L: L = \frac{AS^2}{658} \tag{16.5}$$

$$S > L: L = 2S - \frac{658}{A} \tag{16.3}$$

Sag Vertical Curves

H_e = headlight height = 600 mm
$\alpha = 1°$

$$S < L: L = \frac{AS^2}{120 + 3.5S} \tag{16.9}$$

$$S > L: L = 2S - \left(\frac{120 + 3.5S}{A}\right) \tag{16.7}$$

S = stopping sight distance, m
L = length of vertical curve, m
A = algebraic difference in grades, percent

Comfort Criteria

$$L = \frac{Av^2}{395} \tag{16.10}$$

where

v = design speed, km/h
L = length of sag vertical curve, m
v = design speed, km/h

Decision Sight Distance

Table C1.3 shows the metric values for the components of safe passing sight distance on two-lane highways (see Table 3.7).

Table C1.3 Components of Safe Passing Distances on Two-Lane Highways

Speed Group (km/h) *Average Passing Speed (km/h)*	50–65 56.2	66–80 70.0	81–95 84.5	96–110 99.8
Initial Maneuver:				
a = average acceleration (km/h/s)	2.25	2.30	2.37	2.41
t_1 = time (s)	3.6	4.0	4.3	4.5
d_1 = distance traveled (m)	45	65	90	110
Occupation of left lane:				
t_2 = time (s)	9.3	10.0	10.7	11.3
d_2 = distance traveled (m)	145	195	250	315
Clearance length:				
d_3 = distance traveled (m)	30	55	75	90
Opposing vehicle:				
d_4 = distance traveled (m)	95	130	165	210
Total distance, $d_1 + d_2 + d_3 + d_4$ (m)	315	445	580	725

SOURCE: Adapted from *A Policy on Geometric Design of Highways and Streets,* American Association of State Highway and Transportation Officials, Washington, D.C., copyright 1995. Used with permission.

REFERENCES

A Policy on Geometric Design of Highways and Streets, 4th ed., American Association of State Highway and Transportation Officials, Washington, D.C., 2001.

Guide to Metric Conversion, American Association of State Highway and Transportation Officials, Washington, D.C., 1993.

Ibid.

Index